3rd Maritime Safety International Conference (MASTIC 2022)

IOP Conference Series: Earth and Environmental Science
Volume 1081

Online
15-17 July 2022

ISBN: 978-1-7138-6166-9
ISSN: 1755-1307

Printed from e-media with permission by:

Curran Associates, Inc.
57 Morehouse Lane
Red Hook, NY 12571

Institute of Physics
Dirac House, Temple Back
Bristol BS1 6BE UK

Phone: 44 1 17 929 7481
Fax: 44 1 17 920 0979

techtracking@iop.org

Additional copies of this publication are available from:

Curran Associates, Inc.
57 Morehouse Lane
Red Hook, NY 12571 USA
Phone: 845-758-0400
Fax: 845-758-2633
Email: curran@proceedings.com
Web: www.proceedings.com

TABLE OF CONTENTS

IOP Conf. Series: Earth and Environmental Science 1081 (2022) 011001 doi:10.1088/1755-1315/1081/1/011001

Preface of MASTIC 2022 Proceedings

Welcome to 3rd Maritime Safety International Conference (MASTIC) 2022. After successfully held first and second MASTIC, Department of Marine Engineering, Faculty of Marine Technology, Institut Teknologi Sepuluh Nopember (ITS), Surabaya collaborated with The Center of Excellence for Maritime Safety and Marine Installation (PUI KEKAL) proudly presents the 3rd MASTIC 2022. Since we are still facing the global COVID-19 pandemic, the committee decided to hold this conference virtually on July 16, 2022, while the organizer was in Surabaya, Indonesia. This virtual conference theme is **"Future Challenges for the improvement of Maritime Safety Systems, Marine Operation and Environment".**

There are three prominent keynotes speakers. First speaker is Ir. Hengki Angkasawan who will deliver keynotes about the concept of E-Navigation and E-pilotage. Second is Prof. Takeshi Nakazawa, who delivers keynotes about New initiative in Improving Human Resources for Safer Maritime Transport and Operations. The last speaker is Mr. Denzal John Hargreaves, with the topic the role of Classification Societies to achieve a balance between Safety and Technological Advances. The committee of MASTIC received 77 full paper submissions. From those submissions, there were 60 accepted papers after some rigor review process and were presented in MASTIC 2022 which consists of three parallel sessions with four rooms. The paper represents 6 different countries (Indonesia, Japan, Germany, South Korea, United Arab Emirates dan Timor Leste). Hence, on behalf of MASTIC committee, I would like to extend gratitude to all the authors who contribute to the submission of the result of their latest research. Also, I would like to express my deep appreciation to the Technical Program Committee and the reviewers for their support in the reviewing process to keep the quality of manuscripts. The virtual MASTIC 2022 was conducted via the Zoom platform and was attended by more than 220 participants, including presenters, authors, and co-authors.

MASTIC could be a prestigious forum for educational institutions, researchers, and practitioners to exchange information and share research in the field of safety in ship operation and maritime industry. This conference could be successfully held due to the supports from our partnership sponsors, which are PT. Samudera Indonesia, Indonesian Classification Bureau (BKI), Petrokimia Gresik, Semen Gresik, and Indonesia Port Corporation (Pelindo).

We would also appreciate for the IOP conference publisher so that we could make an agreement to publish the MASTIC proceeding to the IOP Conference Series: Earth and Environmental Science (EES).

Once again, thank you for your support, I wish that the aims of MASTIC 2022 still can be achieved and gives great benefits for all of authors though we could not face each other's in person. I also hope that the MASTIC 2022 proceeding can be useful for the readers in supporting their research related in marine safety or marine engineering and further contributing to develop the technology in maritime field.

General Chair of MASTIC 2022
Dr. Dhimas Widhi Handani
Institut Teknologi Sepuluh Nopember, Surabaya, Indonesia
July 16, 2022

IOP Conf. Series: Earth and Environmental Science 1081 (2022) 011001 doi:10.1088/1755-1315/1081/1/011001

List of Conference Committee
The 3rd Maritime Safety International Conference (MASTIC) 2022

ADVISORY COMMITTEE

1	Bambang Pramujati ST MSc Eng PhD	Vice Rector of Institut Teknologi Sepuluh Nopember
2	Agus M. Hatta, ST., M.Sc., Ph.D.	Director of DIKST Institut Teknologi Sepuluh Nopember
3	Fadillatul Taufany, S.T., Ph.D.	Director of DRPM Institut Teknologi Sepuluh Nopember
4	Beny Cahyono, S.T., M.T., Ph.D	Head of Department of Marine Engineering, Institut Teknologi Sepuluh Nopember
5	D.Eng.Ir.AA Masroeri, M.Eng	Department of Marine Engineering, Institut Teknologi Sepuluh Nopember
6	Prof. Himsar Ambarita	Department of Mechanical Engineering, Unversitas Sumatera Utara

EDITORS

1	Prof. Dr. Ketut Buda Artana. S.T., M.Sc	Department of Marine Engineering, Institut Teknologi Sepuluh Nopember
2	Dr. I Made Ariana,S.T.,M.Sc	Department of Marine Engineering, Institut Teknologi Sepuluh Nopember
3	AAB.Dinariyana Dwi P.,S.T.,M.ES.,Ph.D.	Department of Marine Engineering, Institut Teknologi Sepuluh Nopember
4	Dr. Eng. Trika Pitana,S.T.,M.Sc	Department of Marine Engineering, Institut Teknologi Sepuluh Nopember
5	Dr. Eng. I Putu Sindhu Asmara, S.T., M.T.	Surabaya State Shipping Polytechnic
6	Dr. Emmy Pratiwi, S.T.	Department of Marine Engineering, Institut Teknologi Sepuluh Nopember

ORGANIZING COMMITTEE

1	Dr. Dhimas Widhi Handani	Department of Marine Engineering, Institut Teknologi Sepuluh Nopember
2	Dr. Fadilla Indrayuni Prastyasari M.Sc.	Department of Marine Engineering, Institut Teknologi Sepuluh Nopember
3	Thariq A. Akbar S.T.	Department of Marine Engineering, Institut Teknologi Sepuluh Nopember
4	Dwi Suasti Antara, M.MT.	Department of Marine Engineering, Institut Teknologi Sepuluh Nopember
5	Hayy Nur Abdillah, S.T., M.T	Department of Marine Engineering, Institut Teknologi Sepuluh Nopember
6	Putu Widhi Aprilia, S.T.	Department of Marine Engineering, Institut Teknologi Sepuluh Nopember
7	Ni Luh Pujiyanti	Department of Marine Engineering, Institut Teknologi Sepuluh Nopember
8	Nur Azizah	Department of Marine Engineering, Institut Teknologi Sepuluh Nopember

IOP Conf. Series: Earth and Environmental Science 1081 (2022) 011001 doi:10.1088/1755-1315/1081/1/011001

9	Qatrinnada Tiara M.	Department of Marine Engineering, Institut Teknologi Sepuluh Nopember
10	Syifa Asillah Putri L	Department of Marine Engineering, Institut Teknologi Sepuluh Nopember
11	Muhammad Birawa Balapati	Department of Marine Engineering, Institut Teknologi Sepuluh Nopember
12	Andreas Mahendri Putra	Department of Marine Engineering, Institut Teknologi Sepuluh Nopember
13	I Gusti Ngurah Raditya P	Department of Marine Engineering, Institut Teknologi Sepuluh Nopember
14	Sucahyo Bawono	Department of Marine Engineering, Institut Teknologi Sepuluh Nopember
15	Fatahillah Syach	Department of Marine Engineering, Institut Teknologi Sepuluh Nopember

KEYNOTE SPEAKER

1	Ir. Hengki Angkasawan	Indonesia Ministry of Transportation
2	Prof. Takeshi Nakazawa	International Association of Maritime Universities
3	Dr. Denzal John Hargreaves	Regional Manager – Southeast Asia, Pacific and India DNV

CHAIRPERSON

1	A.A.B. Dinariyana DP., Ph.D.	Department of Marine Engineering, Institut Teknologi Sepuluh Nopember
2	Dr. I G. N. Sumanta Buana	Department of Naval Architecture, Institut Teknologi Sepuluh Nopember
3	Dr. Emmy Pratiwi	Department of Marine Engineering, Institut Teknologi Sepuluh Nopember
4	Dr. Fadilla Indrayuni P.	Department of Marine Engineering, Institut Teknologi Sepuluh Nopember
5	Sunarsih, S.T., M.Eng., PH.D	Department of Marine Engineering, Institut Teknologi Sepuluh Nopember
6	Handi Rahmanuri, M.Sc	Department of Marine Engineering, Institut Teknologi Sepuluh Nopember
7	Dr. I Putu Sindhu A.	Surabaya State Shipping Polytechnic
8	Dr. Adi Kurniawan	Department of Marine Engineering, Institut Teknologi Sepuluh Nopember
9	Dr. Septia Hardy S.	Department of Naval Architecture, Institut Teknologi Sepuluh Nopember
10	Dr. Nurhadi Siswantoro	Department of Marine Engineering, Institut Teknologi Sepuluh Nopember
11	Dr. Dhimas W. H.	Department of Marine Engineering, Institut Teknologi Sepuluh Nopember
12	Dr. Achmad Baidowi	Department of Marine Engineering, Institut Teknologi Sepuluh Nopember

Peer Review Statement

All papers published in this volume have been reviewed through processes administered by the Editors. Reviews were conducted by expert referees to the professional and scientific standards expected of a proceedings journal published by IOP Publishing Publishing.

- **Type of peer review:** Double Anonymous
- **Conference submission management system:** Morressier
- **Number of submissions received:** 77
- **Number of submissions sent for review:** 74
- **Number of submissions accepted:** 60
- **Acceptance Rate (Submissions Accepted / Submissions Received × 100):** 77.9
- **Average number of reviews per paper:** 2.05
- **Total number of reviewers involved:** 39
- **Contact person for queries:**
  Name: Dhimas Widhi Handani
  Email: dhimas@ne.its.ac.id
  Affiliation: Institut Teknologi Sepuluh Nopember (ITS)

IOP Conf. Series: Earth and Environmental Science 1081 (2022) 012001 doi:10.1088/1755-1315/1081/1/012001

System Dynamics Modelling Scenarios for The Import Receiving Infrastructures Capacity Plan & Development in anticipating Future Deficits of Natural Gas Supply in Indonesia.

A A Purwosaputra[1], K B Artana[2], A A B Dinariyana[2]

[1] Graduate Student Department of Management Technology, Institut Teknologi Sepuluh Nopember, Surabaya,
[2] Department of Marine Engineering, Institut Teknologi Sepuluh Nopember, Surabaya, 60111, Indonesia.

[1]E-mail: ahmada0505@yahoo.com

Abstract. Indonesia is one amongst the biggest natural gas reserves country in south East Asia, with total reserve of about 142,71 TSCF in 2018. Indonesian demand for energy is also increasing over the time in line with its economic and population growth. The improved of global consciousness toward the sustainable environmental issues, has led in to the choice over an affordable source of energy with a less carbon footprint. On the other hand the natural gas supply and domestic production is depleting. At the current rate of consumption and assuming no new reserves discoveries, sooner or later Indonesia will have to face deficit in gas supply. This study, is first aimed to forecast the Indonesia supply and demand situation of natural gas in years to come, and then performing an analysis and assessment toward the availability and adequacy of the required infrastructures especially receiving terminals facilitating for gas import in anticipation of future deficit of gas supply in Indonesia. From which, it will be further analyzed, the geographical distribution of the demand and then potential short of supply in several regions and areas where the demand are most concentrated. Thus few basis System Dynamics scenarios of public policies were simulated for 30 years running simulation. System Dynamics model simulation has been well known as a powerful tool to represent a complex industrial problems such as national energy systems. The result is unveiled the fact that additional import receiving terminals and storage facility is required at certain areas where the deficits most likely exist, and most of them in fact are in Western Zone of Indonesian.

ABBREVIATIONS

LNG	: Liquefied Natural Gas	**GHG**	: Green House Gas
TSCF	: Trillion Standrad Cubic Feet	**GWH**	: Gigawatt-hour
BAU	: Business As Usual	**SD**	: System Dynamics
FDU	: Full Domestic Utilization	**BCM**	: Billion Cubic Meter
CLD	: Causal Loop Diagram.	**MT**	: Metric Ton
BSCFD	: Billion Standard Cubic Feet per-day		
MTOE	: Million Ton Oil Equivalent	**FSRU**	: Floating Storage Regass Unit
APERC	: Asia Pacific Energy Research Centre	**IEA**	: Interntl Energy Agency
MEMR	: Ministery of Energy & Mineral Resources		
LNGC	: LNG Carrier	**SFD**	: Stock Flow Diagram

1. Introduction

Innovations in the LNG technologies, allows the trade to possible reaches places even to the area where pipeline couldn't reach. The global Liquid Natural Gas (LNG) trades development has triggered out an increasing global demand of LNG for electricity generation mainly due to natural gas is the efficient and the cleanest energy source compared to other fossil types.[1]

Natural gas plays a prominent role as one of energy source in Indonesian energy-mixed for the power generation sector. The Presidential Decree No. 22/2017 concerning to the road-map of the National Energy Plan [2], the energy-mixed composition are targeted as follows:

- Group sources of low-carbon energy and renewables – is aimed to be at least 23% by 2025 and 31% by 2050
- Oil – is expected to be less than 25% by 2025 and kept as low as 20% by 2050
- Coal – is expected to be at least 30% by 2025 and 25% by 2050
- Natural Gas – is projected to be at least 22% by 2025 and 24% by 2050

Indonesia is one of Southeast Asia's biggest gas producers, with exports level is up to 45% of its current production. The largest undeveloped gas reserves are located in the offshore East Natuna Block, which holds approximately 49.6 TSCF, there are also many high potential areas like West Papua and Maluku. At the current rate of consumption and assuming no new discoveries, existing reserves would last for around 50 years [3].

Report of the Asia Pacific Energy Research Centre (APERC) on "APEC Energy Supply–Demand Outlook 2019" [4] foresaw the developing gas demand to meet up the domestic and export commitment from 58.8 Mtoe in 2020 to 60.9 Mtoe by 2040 as outlined in the Table.1 below. Gas demand for power generation will remain dominant, about 30% in 2020 and going up to 40%. Industry sector is the second-largest off-taker. Even thought Indonesia will keep it's export commitment, yet the volume will be decreasing significantly up to only 4.4 Mtoe by 2035. And by 2040, Indonesia is expected to be one of nett importer of natural gas.

Table 1. Indonesian Gas demand outlook in MTOe

Year	2020	2025	2030	2035	2040
Total Demand of Gas	58,8	54,3	55,8	59,7	60,9
Electricity	18	17,1	19,1	24,8	33,1
Energy Industry (Owned Used)	3,4	3,8	4,2	4,6	4,9
Transport	0,1	0,3	0,7	1,1	1,6
Building	1,2	1,6	2,1	2,8	3,4
Non energy use	3,4	3,8	4,2	4,6	4,9
Industry	11,8	13,8	15,8	17,5	19
Exports	20,9	13,7	9,6	4,4	-6

Source: APERC, 2019

Government Regulation No. 79/2014 concerning to the National Energy Policy has governed the role of natural gas in Indonesia's energy-mixed. Domestic demand of natural gas until 2050, will increase from 1.84 TSCF (trillion standard cubic feet) in 2015 to 3.29 TCF in 2025 and to 9.21 TCF in 2050 as it's outlined in

Table 2. The average growth rate from 2015 to 2020 is about 6% per year; in the period of 2020–2025 is about7% per year; period 2025–2030 is 5% per year; in the period 2030–204 is, 5% per year; and in the period of 2040–2050 is about 3% per year.

The natural gas demand will grow significantly in the period of 2015 to 2025 (about 6%–7% per annum). During such period, the government will aim to maximize the use of gas for domestic use to boost up higher economic value added and set it out as a transition toward such as new and renewable kind of energy.

Table 2. Projected Gas Demand of Indonesia (Mtoe)

Year	**2015**	**2020**	**2025**	**2030**	**2040**	**2050**
Gas Demand						
Share (%)	22	22	22	23	24	24
Volume (Mtoe)	47	64	84	110	178	235
Volume (TCF)	1,84	2,51	3,29	4,31	6,98	9,21
AAGR (%)		6	7	5	5	3

Source: Government of Indonesia, DEN (2014), TCF: Trillion Cubic Feet

Indonesia's electricity consumption was 234.6 gigawatt-hour (GWh) in 2018 according to the Directorate General of Electricity (2019). Most of this consumption was from the residential (42%) and industry (33%) sectors. By region, Java electricity consumption accounted for about 71% of the total consumption, and Eastern Indonesia shares were 9%. These are the consumption of Sulawesi, Maluku, North Maluku, Papua, West Papua, Bali, West and East Nusa Tenggara (NTB and NTT).

The MEMR issued Ministerial Decree No. 13 K/13/MEM/2020, mandating PT PLN, the State Electricity Company, to convert it's 52 diesel power plants in to gas burnt. The programme's purpose, as noted in the decree, is to cut down the trade deficit and to help in the energy diversification programme. The conversion from diesel to gas would cut down diesel consumption from 2.6 million kilolitres (kL) per annum to 1.6 million kL per annum, with estimated operational cost savings at Rp 4 trillion (US$286 million).The ministry also assigning the state-owned oil company, Pertamina, to supply the gas. This includes the development of the LNG infrastructure to receive, store, and regasify LNG. PT Pertamina is obligated to set the price of the regasified gas at 'plant gate', which will result in a lower production cost of PT PLN compared to using high-speed diesel.

The gas reserves in Indonesia spread throughout the region as shown in figure.1 According to the General Directorate of Oil and Gas, Ministry of Energy and Mineral Resources of Indonesia (2012), in 2012, the country had 103.35 tcf proven reserves and 47.35 potential resources. Natuna holds the majority of the reserves. It is also the biggest gas field in Southeast Asia. Currently, the West Natuna gas field has been well developed and is used to export gas to Singapore and Malaysia via subsea pipelines (Offshore-Technology). It is estimated that the East Natuna has more gas reserves than the West Natuna. However, according to the feasibility study by ExxonMobil, the natural gas from East Natuna contains 71% CO2, which make it too difficult and expensive to be developed. In 2007, the Indonesian government decided to terminate its contract with Exxon, leaving Pertamina (the Indonesian state-owned oil and gas company) in charge (Offshore-Technology).

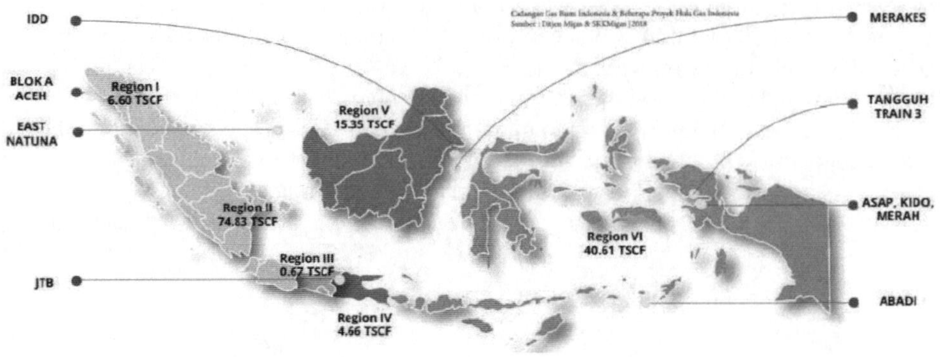

Source: Indonesian Gov't, DG Oil&Gas, 2018

Figure 1. Indonesian Gas reserves

2. Literature Review

A though-rough scientific and systematic Literature Review process has been conducted toward articles published within the search period of 2010-2020, the articles were searched from it's abstract and keywords in the field area of Natural gas supply and demand study using System Dynamics as tool. Searching was made based upon a generated search strings that forms series of key-words such as; "Natural gas supply & demand", "System dynamics model for natural gas supply & demand", and "System dynamics model for infrastructure capacity plan". Search ws done toward several database journal such as SCOPUS, Google Scholars, and Microsoft academic. And the process had eventually resulted in total of not more than 8 articles only that are really in relevant topics.

It reveals that more than half of them are in the research topics of "forecasting future natural gas supply and demand using System dynamics model based simulation", whilst the rest of them are in the research topics of "Gas Infrastructure capacity plan using System dynamics model simulation".

Amongst the few of those who has focusing their research in the topics study of forecasting future supply & demand of natural gas, some articles come of great interest, a group of researchers in China had came up with a supply & demand model on the basis of System dynamics model, and the result was a prediction that china would have an increasing demand which led in to a recommendation to increase the gas supply from various sources including import.[5]. Another researcher from China also created and established a system dynamics model to predict future natural gas supply & demand in China [6]. An Iranian researcher has used system dynamics model simulation to forecast the future supply and demand of natural gas in Iran and to assess the public policy in determining natural gas pricing in the entire Iranian Energy system.[7], a more recent study has been conducted by group of Indonesian researchers toward the natural gas supply & demand situation in one of Indonesian Province i.e East Java, they used system dynamics model simulation to predict the future shortage of supply and recommending an infrastructure development plant to anticipate the short of supply in the future [8] and [9].

The rest of the articles that focusing more on the Infrastructure capacity plan & development are, one Latvian researcher who established a system dynamics model for natural gas infrastructures in Latvia [10], a group of Brazilian researchers who developed a model for natural gas infrastructures development at Southern Brazil [11], another germany researchers who developed a model for natural gas infrastructures expansion in Germany [12], and also one Indonesian researcher who has developed a system dynamics model for gas infrastructures pathways in one Indonesian province i.e Eastern Java [9].

3. Methodology

A certain model of natural gas supply & demand is first built upon the basis of the data and information of the current supply and demand situation. The built up model has to be able to represent well what the real condition in the real life situation is, of which each elements in the system are affected one in to another. Once the model has been established then system dynamics will do the simulation. The objective of such simulation is to assess and then analyze the current and upcoming supply & demand conditions, how certain fulfillment level is met and how to get it optimized.

System Dynamics is a method and tool that is believed to be the most fits to represent a complex system and to determine feedback behaviors over such complex models that represented a real-life situation in real-world system [9]. System Dynamics model was first built under certain modeling structures, main elements and variables are set and a causal loop is further defined. Then under certain assumptions and pre-defined scenarios, stock flow diagram and model formulation are then presented [7].

In this research the system dynamics model simulation was aimed not only to predict the future natural gas supply & demand situation yet instead also to simulate and assessing the infrastructures fulfillment level and scenarios. In general natural gas infrastructure is a complex system, thus entail a complex model either to represent it. A model that works under specific policy regimes pre-determined by technical parameters set out in the system [13]

The stages of process on how the simulation and the analysis to be done are set as per illustration given in Figure 2 below, whereby an outlined process flowchart describing each and every steps taken.

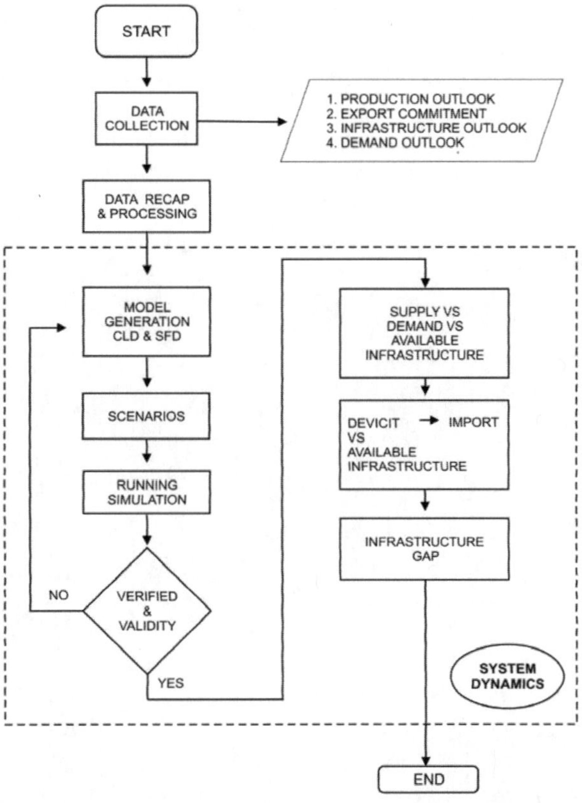

Figure 2. Flowchart of the stages methodology

IOP Conf. Series: Earth and Environmental Science 1081 (2022) 012001 doi:10.1088/1755-1315/1081/1/012001

4. Data Collection & Processing

All the data and information forming the current situation of the natural gas supply and demand is collected from various sources (mostly from government agency such as Ministry of EMR, National Energy Board, SKK MIGAS, BPH MIGAS, etc.)). Once they're obtained, then get it compiled and categorized them in to 2 main groups i.e. Group of data supply and production, as well as Group of demand data. Then we employ them as a baseline value for further analysis. A certain basis growth assumption is also applied in accordance to the guidance given by the MEMR & BPH MIGAS with regard to the gas consumption rate and it's growth rate per-annum (*NERACA GAS BUMI INDONESIA*, 2018) The demand growth rate assumption is set as outlined in the Table.3 below.

Table 3 Demand growth scenarios

Industry	Transportation	Household	Electrical Generation
1,1%/yr	5% correspondent to each region	5% correspondent to each region	in line with the information extracted from RUPTL 2019-2028 and 1,1%/annum in regard to the additional demand of 52 converted power plants, starting from 2022 onward is 1,1%/annum

On the other hand, there are also 3 main factors that are affecting the supply level in the supply data. Those are Domestic Production, International Import and International Export. For Domestic Production, the variable is connected to the production rate of all gas reserves in Indonesia both existing and newly monetized reserves and also the volume of gas in related to the LNG production from the Liquefaction plants both existing as well as those expected new plants coming onstream (only those who has secured the Government's principal approval and License is granted). Figure.2 below is giving an illustration of the spread location of the Natural Gas Production in Indonesia.

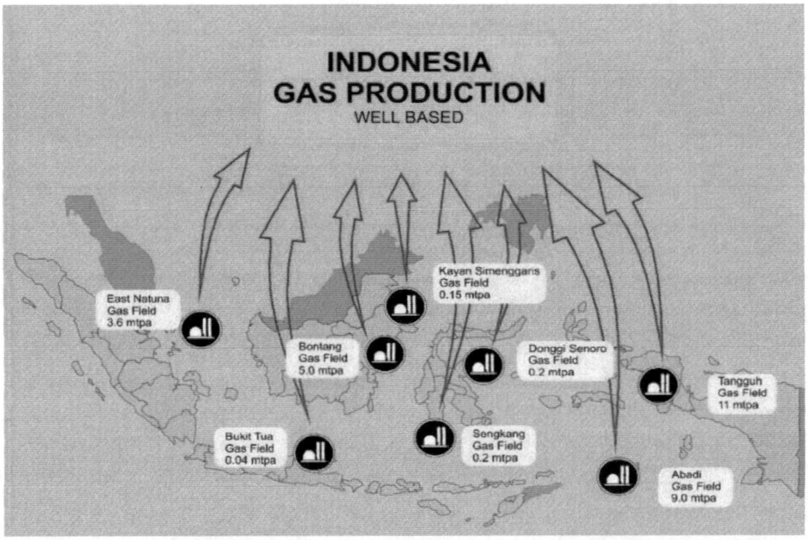

Figure 3 Indonesian Well based natural gas production - existing & future development

IOP Conf. Series: Earth and Environmental Science 1081 (2022) 012001 doi:10.1088/1755-1315/1081/1/012001

We will employ data available within "Neraca Gas Bumi" (*NERACA GAS BUMI INDONESIA*, 2018) as main data references and employ them as baseline for further analysis. It is also assumed that there is no new reserves discoveries during the simulation period. Meanwhile for the International import, certain assumption whereby there is no import made until then when it's really required to do so, is applied. Whilst for Export, it assumed there is no new export commitment is made as the Indonesia government has decided to focus on more allocating all gas production to serve domestic need only, thus as aforementioned all the figure of gas production volume as well as LNG production volume were made in such away its net off after deducted off the export volume, from the current and existing contract commitment until it's expiration. However for the sake of discussion, we deliberately maintain the option of revisit back the policy in regard to the export gas commitment as one of the alternative of gas fulfillment scenario offered in this study.

In regard to the spread of the existing gas infrastructures especially related to the liequefaction plants, storage and re-gas plants are as illustrated in Table.4 and 5 below. We will then apply this as baseline for further analyze purposes.

Table 4 Indonesian Gas Infrastructures – Liquefaction Plants

Project - Plants	Capacity		Status	On Stream
	MT p.a	Bcm p.a		
Existing Plants				
Badak LNG - Bontang (East Kal)	21.6	29.4	Operating	1978
PAG - Arun, Aceh*	4.8	6.4	Operating, converted	1978
Tangguh LNG, Papua	7.6	10.3	Operating, expanded	2009
Donggi-Senoro, Sulawesi	2.0	2.7	Operating	2014
Sengkang LNG, South Sulawesi	2.0	2.7	Final Stage construction	2020
New Developing Plants				
Masela, Abadi Field	9.0		FEED	2028
Kayan LNG (Blok Simenggaris)	0.15		FEED	2024
Paraamartha Energy (Sidoarjo)	0.62		FEED	2027
Sumber Aneka Gas LNG (Tuban)	0.10		Pre-FEED	2030
Natgas LNG (Batam)	0.18		Pre-FEED	2030
IndoJaya Energy (Bukit Tua Field)	0.41		FEED	2025

* Arun has been converted in to receiving & regasification plant

Source: [15], and other various

Table 5 Indonesian Gas Infrastructures – Receiving & Regasification Facilities

Project - Plants	Capacity		Status	On Stream
	MT p.a	Bcm p.a		
NR FSRU - Jakarta Bay	3.7	5.2	Operating	1978
PGN LNG FSRU, Lampung	2.0	2.8	Operating	1978
PAG - Arun, Aceh	1.5	2.1	Operating	2009
Java One FSRU, Cilamaya	3,7	5.2	Operating	2021
Amurang, Sulawesi	0,35		Operating	2020
PEL FSRU, Benoa - Bali	0.35		Operating	2017
New Developing Plants				
PGN LNG FSRU, Teluk Lamong	2.0	2.8	FEED	2025
FSRU Grati, Bangil	2.0	2.8	FEED	2027

Source: [15], and other various

The illustration given in the Figure.3 below is showing that most of the current existing gas receiving and regassification infrastructures are located at western part of Indonesia, hence the eastern part is very lacking and left behind in term of Infrastructures adequacy.

Figure 4 The spread location of Indonesian existing Gas Receiving and Regassification Infrastructures

5. Model Construction

Natural gas infrastructure model is a real complex system, that works under specific regimes that are determined by technical aspects of the gas infrastructure such as gas flows, transmission system capacities, the storage (stock) level, and some other social aspects, resembles demand and consumption. Each of those elements are directly influenced by a set up energy policies. It is hard to figure out a set of variety of these aspects in one tool or method. For this purpose then SD modelling is introduced. Objective of SD modelling is to analyze systems behavior and its structure together, by implementing 2 different scenarios representing 2 extreme opposite conditions, and the way it is creating a dynamic environment [13]

5.1 Causal Loop Diagram (CLD)

CLD is a representation of a conceptualization of the dynamics in the model that will be built up. Through the CLD, the built up model then is further explained by showing the causal relationship between variables and elements entered in to the model beforehand. By describing the correlation amongst elements in the system, the causal loop diagram would show how the system that being modelled works. The causal loop was crafted using the help of VENSIM software as tool.

IOP Conf. Series: Earth and Environmental Science 1081 (2022) 012001 doi:10.1088/1755-1315/1081/1/012001

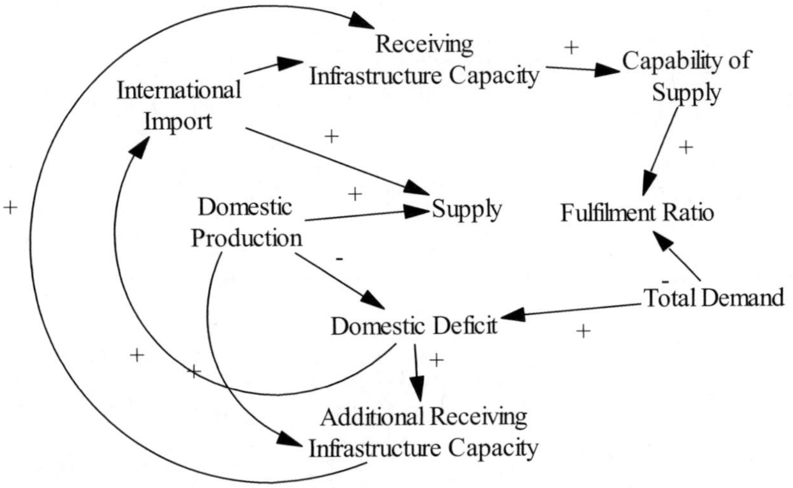

Figure 5. Causal Loop Diagram (CLD) for Gas Infrastructures Capacity Plan

5.2 Stock Flow Diagram (SFD)

Once the causal loop diagram is ready and the connections amongst all the elements are known, then next following step is to ensemble the stock and flow diagram. It is a conceptualization diagram of a system dynamics model that is representing the stock and flow of goods in a system. In this case we use stock flow diagram to assess and find out how is the supply and demand condition looks like. We employ POWERSIM as software tool to craft the model.

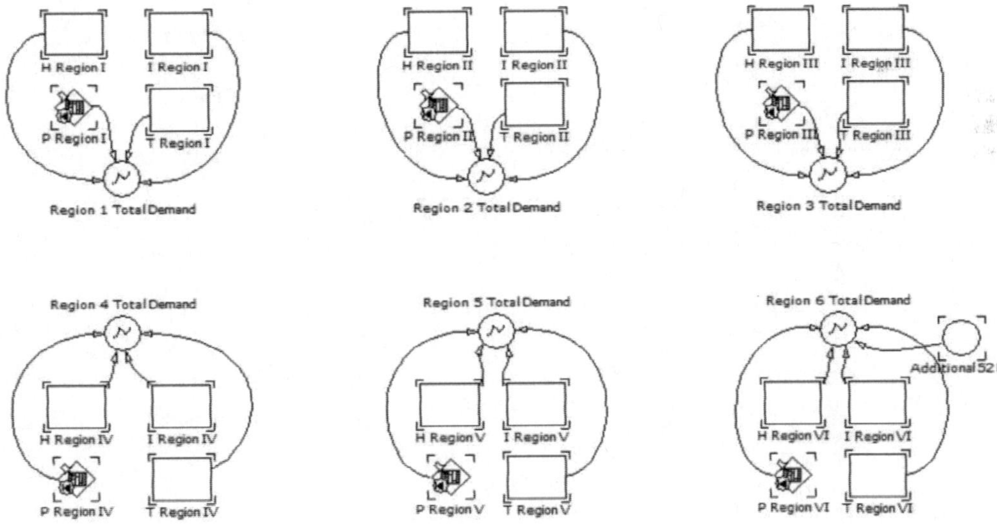

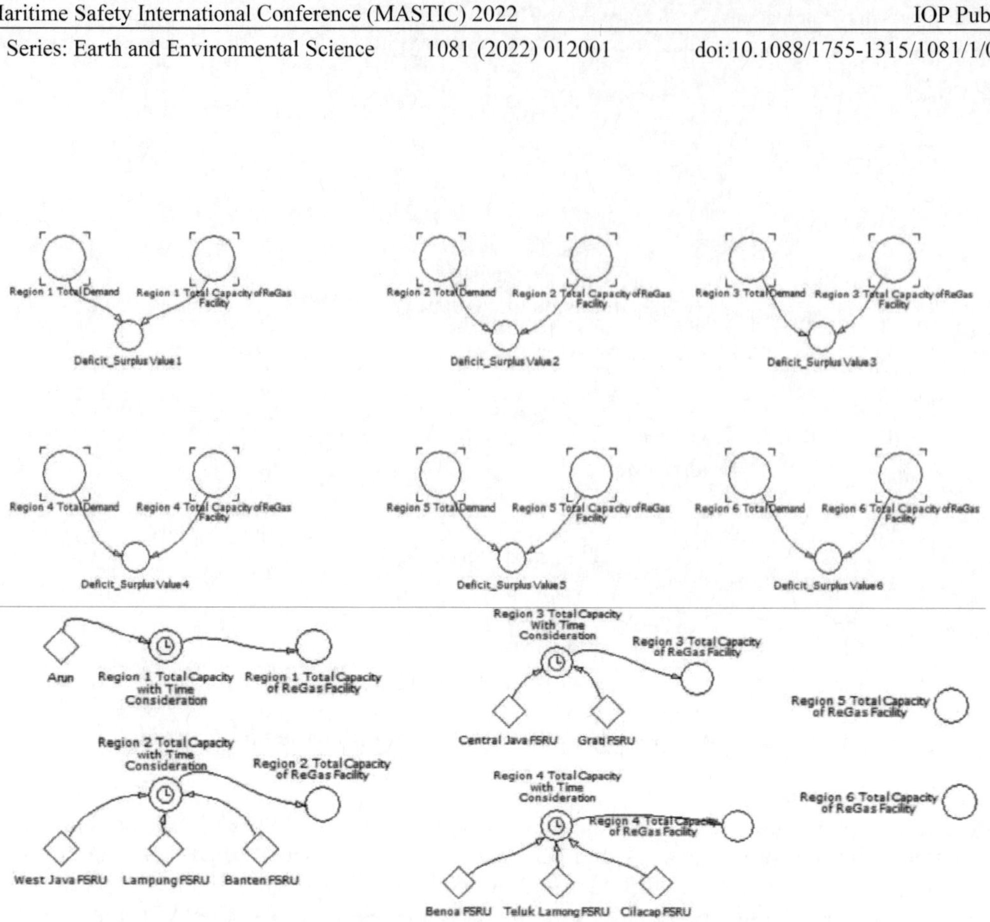

Figure 6 Stock Flow Diagram (SFD) for Gas demand Vs Infrastructures

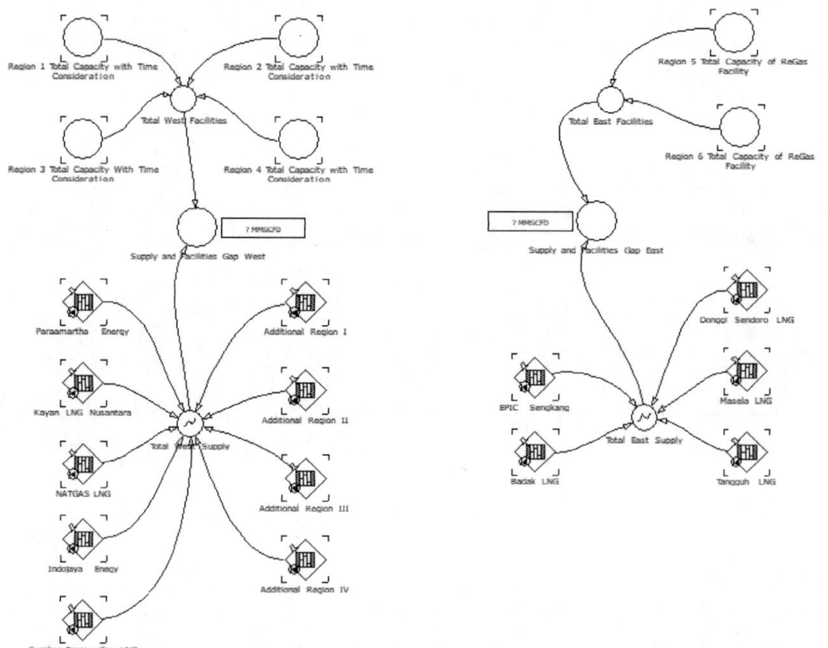

Figure 7 Stock Flow Diagram (SFD) for the Gas Supply Vs Infrastructures

In general, as the way how it's structured, the natural gas infrastructures model in this study is a complex system which consists of at least 3 main elements in the system i.e first is Gas supply (GS), Gas Demand (GD) and the availability of the infrastructures i.e storage (IA). All of them are working under certain regimes that was set out by the policy.

Within this study there are 2 main scenarios prepared and being used as the basis for the simulation, the 2 scenarios were prepared on the basis implementation of 2 different policies in regard to the intentional use of the gas production capacity in Indonesia. Those 2 main scenarios are outlined as follows;

1. Business As Usual (BAU) scenario; In this base-case scenario, the intentional use of gas production capacity were majority set for serving export commitment, and only 25% out of it, serves for domestic market obligation (DMO).
2. Full Domestic Utilization (FDU) scenario; In this scenario, the whole national gas production capacity would be set to serve wholly the domestic market need. In consequence there would be no more export of natural gas in the future, thus expanding the national gas infrastructures capacity is considered a must.

These 2 prepared scenarios were thought to be well representing 2 extreme opposite conditions that could be of choice considerations for the policy makers. Figure 5 and 6 are describing stock and flows diagram representing the fluctuation of the stock variables in the level of natural gas demand and supply against the availability of the gas infrastructures. The stock and flows in the developed model works under set of equations, as follows;

$$\text{Tot GD} = \sum_{k=1}^{n} GDn \tag{1}$$

$$\text{Tot GS} = \sum_{k=1}^{n} GSn \tag{2}$$

$$\text{Tot AI} = \sum_{k=1}^{n} AIn \tag{3}$$

Whereby;

GD: is Gas Demand
GS: is Gas Supply
AI: is Available (receiving) infrastructures
n: is referring to each Region (from I to VI)

$$\text{Tot IG} = \sum_{k=1}^{n} GDn - \sum_{k=1}^{n} GSn - \sum_{k=1}^{n} AIn \tag{4}$$

Whereby;

IG: is Infrastructure Gap
n: is referring to each Region (from I to VI)

6. Result & Discussion

A 30-year long simulation was done, the model was ran under 2 different scenarios to reveal how the system would behave and to assess further how the intactness of its structure is. Then for a better observation and for the sake of a closer look, the Indonesian territory that covered 6 different Regions, within this study is further divided geographically in to 2 main Zones i.e West Zone and East Zone.

The result of a 30 year-long simulation as it is shown in Figure 8 reveals the fact that in the Western Zone area of Indonesia, a significant deficit occurs between the natural gas supply & production which keeps on depleting in one side over the natural gas demand on the other hand that keeps on growing in this respective Zone. This situation has led in to another fact, that there will be a potential huge gap in the availability of the gas receiving infrastructures need in this area. By end period of the simulation or in the early year of 2050, it is expected that the volume gap will grow and come closer to 8 BSCFD.

Result of the 30-year long simulation also reveals another fact that in the Western Zone the 2 different scenarios ran, have in fact does not give quite significant different in impact. The only rational explanation behind is that there are not much volume of export arranged from this Zone area so far, and the whole local gas production has been absorbed by high local demand in this respective Zone. That is why the supply growth does not much influenced with the policy decision as wheatear the whole capacity of gas production is dedicated for domestic (local) market demand or not.

In the long run Government has to think of and take necessary action in anticipating the building up volume gap between the availability of the local receiving infrastructure and high local demand in this respective Zone by investing and building up more receiving infrastructures capacity in this Zone area to fill in and close up the growing gap.

The presence of this study and the result of the model simulation promoted within this study will provide the help and guidelines to the Policymaker at least in which area they should be focusing themselves to and get them to the idea on how much volume they need to incorporate within their future capacity building plan. Further on they could even come in to the right time prediction on when exactly do they need to start the planning and executing the development.

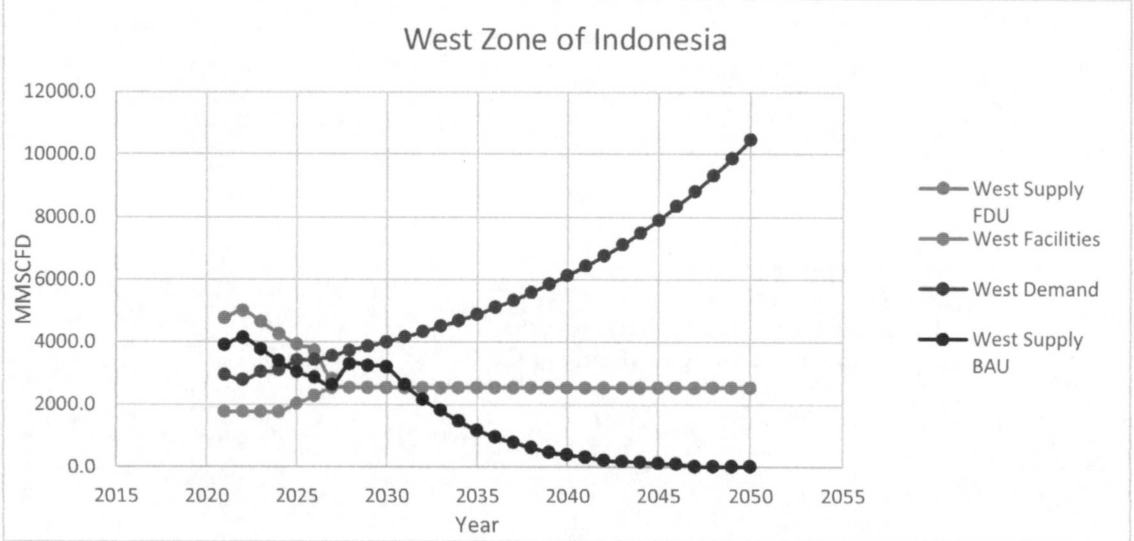

	2021	2025	2030	2035	2040	2045	2050
Supply BAU	3897,0	3046,2	3189,8	1176,5	387,7	112,3	14,9
Supply FDU	4763,7	3912,9	3189,8	1176,5	387,7	112,3	14,9

Figure 8. Profile of The Infrastructures Gap against Local Supply & Demand in Western Zone area of Indonesia

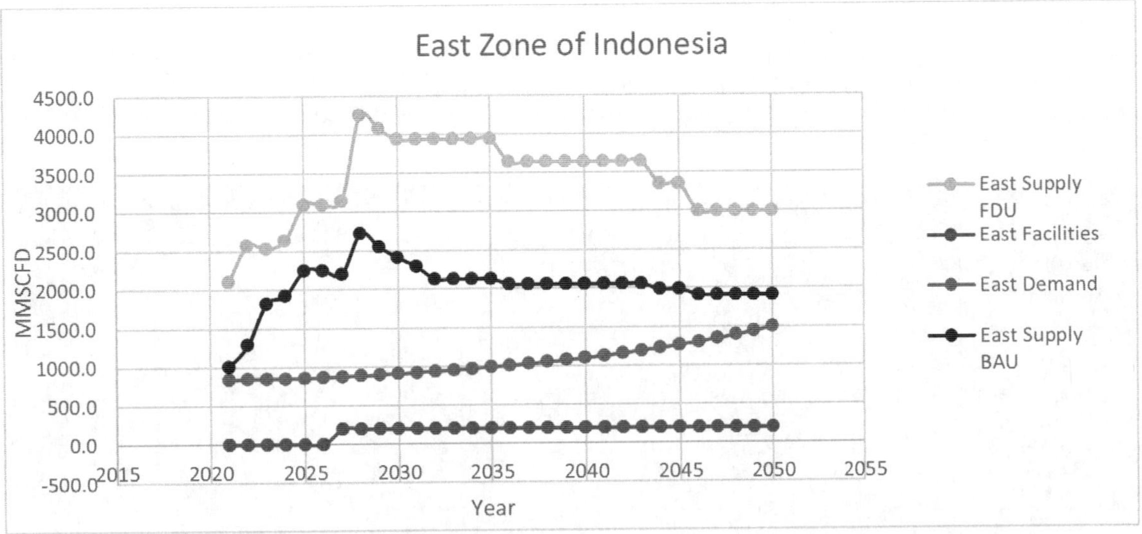

	2021	2025	2030	2035	2040	2045	2050
Supply BAU	1003,7	2244,4	2406,3	2124,4	2051,1	1977,9	1904,6
Supply FDU	2102,8	3094,4	3930,0	3930,0	3636,9	3343,9	2988,8

Figure 9 Profile of The Infrastructures Gap against Local Supply & Demand in Eastern Zone area of Indonesia

In contrary to the result given in the Western Zone, Figure.9 showing the result of 30-year long simulation done toward the Eastern Zone of Indonesia, it reveals the fact that in this respective Zone area, the local demand is not as much as what it is in the Western Zone. It is also reveals the fact, there is no such gap of that requires more receiving infrastructures in place in this respective Zone, as the gas supply & production in this Zone is growing much more than the growth rate of the local demand, apparently whatever the local demand is the local production and supply level is able to meet up such demand, that has made this respective Zone self-sustained.

Another important fact revealed is that, there is a surplus of gas production and supply in this respective Zone. The 2 different scenarios ran in the simulation giving a different impact in term of how much volume of gas surpluses each scenarios resulted. The result of both scenarios are elaborated as follows;

- The BAU scenario, gives a surplus result up to the volume of abt.1 BSCFD in the end period of simulation or in the early year of 2050.
- Whilst the FDU scenario, offers bigger surplus result, up to the volume of abt. 1.5 BSCFD in the end period of simulation or in the early year of 2050.

Eventually the surpluses could be further utilized to help in filling up the deficit gap occurs on the other Zone i.e Western Zone, now it is a matter of the policymaker's choices to consider and decide which policy scenario to run in general, a though decision to make based upon the information made up available by this study.

7. Conclusions & Future works Recommendations

In general there are 3 main conclusions could be drawn up from this study, first is that The Western Zone of Indonesia, it happens to be the area where most of the gas demand are concentrated would require

IOP Conf. Series: Earth and Environmental Science 1081 (2022) 012001 doi:10.1088/1755-1315/1081/1/012001

additional receiving infrastructures at the volume of abt. 3 BSCFD by 2040 and it grows up to abt. 8 BSCFD by 2050, as illustration given in Figure.9 below. Second is that The East Zone of Indonesia will not require an additional receiving infrastructures, at least until year 2050. And third is that there would be surplus in gas production & supply in the East at volume of about 1-1.5 BSCFD that could eventually lead the trade flows from East to West.

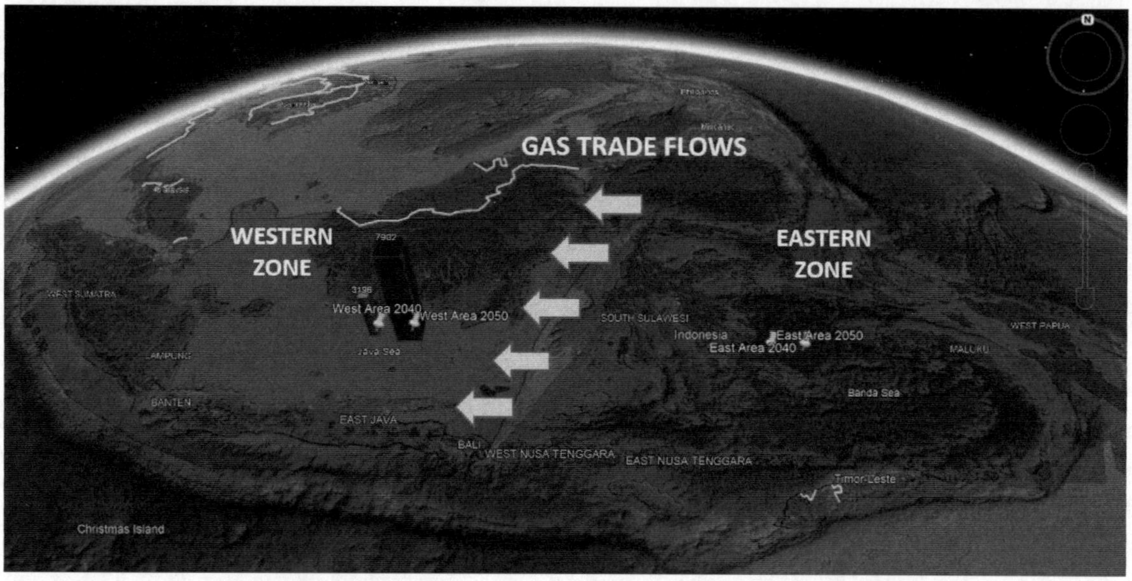

Figure 10 Additional required receiving Infrastructures and Gas Trade flows

This research work applied an assumption that any gas demand in wherever market location would be supplied and fulfilled by any gas producers or suppliers located in nearby vicinity, wherelse in actual real life situation this might not always be the case, sometime due to one and another reason such as long term commitment pricing, spot urgency, or other commercial reasons, etc., market in anyhow have to absorb gas supplied from sources that located far away. This could be one of the limitation applied to this study in general.

Future research work recommendation is given to pursue the quest for the appropriate supply chain model that could represent well the situation and condition of Indonesia archipelago.

REFFERENCE

[1] "2020 WORLD LNG REPORT," 2020.

[2] Perpres, "Peraturan Presiden No 63 Tahun2017," *Setkab*, vol. 5, no. 1. pp. 1–8, 2017, [Online]. Available: https://sipuu.setkab.go.id/PUUdoc/175301/Perpres No 63 Tahun 2017.pdf.

[3] J. Kusuma, "Market Analysis of Natural Gas Supply and Demand in Java Island To Enable a Sustainable Scenario With System," 2019.

[4] Asia Pacific Energy Research Centre, *APEC Energy Demand and Supply Outlook – 7th edition*, vol. II, no. May. 2019.

[5] S. Jingchun, L. Ding, and W. Fan, "The simulated system dynamics analysis of the natural gas supply and demand," *Kybernetes*, vol. 39, no. 8, pp. 1262–1269, 2010, doi: 10.1108/03684921011063547.

[6] X. Z. Mu, G. H. Li, and G. W. Hu, "Modeling and scenario prediction of a natural gas demand system based on a system dynamics method," *Pet. Sci.*, vol. 15, no. 4, pp. 912–924, 2018, doi: 10.1007/s12182-018-0269-3.

[7] F. Daneshzand, M. R. Amin-Naseri, A. Elkamel, and M. Fowler, "A System Dynamics Model for Analyzing Future Natural Gas Supply and Demand," *Ind. Eng. Chem. Res.*, vol. 57, no. 32, pp. 11061–11075, 2018, doi: 10.1021/acs.iecr.8b00709.

[8] J. Kusuma, K. B. Artana, and A. A. B. Dinariyana, "Market Analysis of Natural Gas Supply and Demand in East Java Province to Enable a Sustainable Scenario Using System Dynamics Simulation," *IOP Conf. Ser. Earth Environ. Sci.*, vol. 557, no. 1, 2020, doi: 10.1088/1755-1315/557/1/012041.

[9] M. Haekal, K. B. Artana, and D. W. Handani, "A Gas Infrastructure Pathways in East Java Using System Dynamic Approach," *IOP Conf. Ser. Earth Environ. Sci.*, vol. 557, no. 1, 2020, doi: 10.1088/1755-1315/557/1/012025.

[10] M. Feofilovs, F. Romagnoli, and A. Gravelsins, "ScienceDirect ScienceDirect ScienceDirect ScienceDirect temperature function for Francesco a long-term district heat Gravelsins demand forecast Maksims System dynamics model for natural gas Heating infrastructure with on District and Cooling with System d," *Energy Procedia*, vol. 147, pp. 549–557, 2018, doi: 10.1016/j.egypro.2018.07.070.

[11] I. García Kerdan, F. Jalil-Vega, J. Toole, S. Gulati, S. Giarola, and A. Hawkes, "Modelling cost-effective pathways for natural gas infrastructure: A southern Brazil case study," *Appl. Energy*, vol. 255, no. August, p. 113799, 2019, doi: 10.1016/j.apenergy.2019.113799.

[12] B. Gillessen, H. Heinrichs, J. F. Hake, and H. J. Allelein, "Natural gas as a bridge to sustainability: Infrastructure expansion regarding energy security and system transition," *Appl. Energy*, vol. 251, no. April, p. 113377, 2019, doi: 10.1016/j.apenergy.2019.113377.

[13] M. Feofilovs, F. Romagnoli, and A. Gravelsins, "System dynamics model for natural gas infrastructure with storage facility in Latvia," *Energy Procedia*, vol. 147, pp. 549–557, 2018, doi: 10.1016/j.egypro.2018.07.070.

[14] "NERACA GAS BUMI INDONESIA."

[15] Yuliana, "Southeast Asian gas outlook to 2045: Prospects of gas network expansion to increase security of supply," 2015.

IOP Conf. Series: Earth and Environmental Science 1081 (2022) 012002 doi:10.1088/1755-1315/1081/1/012002

Dynamic Safety Modelling for Ship Management Performance

Adi Mas Nizar[a], Masumi Nakamura[b], Takashi Miwa[a], Makoto Uchida[a]

[a]Graduate School of Maritime Sciences, Kobe University, Japan
[b]National Institute of Technology (KOSEN), Yuge College, Japan

email: 185w604w@stu.kobe-u.ac.jp

Abstract. Recent development and projection of ship operations as a sociotechnical system is getting more complex. In order to successfully emulate a high-reliability organization with a balance between operational efficiency and safety, the shipping companies have to grab a well understanding of the operating performance. However, because the teams are distributed spatially and temporally, a misalignment of shared situation awareness casually exists. We extended Rasmussen's dynamic safety theory and adjusted it in the context of ship management performance. A modelling study using system dynamics was done to illuminate how the feedback loops construct the interaction between safety, efficiency, and workload. The simulation result shows that the operations behave following the safety and efficiency pressures created by existing goals and boundaries. The model is also able to capture these trade-offs in different variety of operation scenarios. Application such modelling may provide the managers with a better understanding and valuable insight to implement the strategies to sustain safe operations.

1. Introduction

Balancing between efficiency and safety in ship operation is common practice, as well as other general industries. While doing it, each level of the organization's components in ship operation frequently makes a trade-off between efficiency and thoroughness to manage the available resources [1]. Naturally, an organization such as a shipping or ship management company based their decision to pursue optimum cost-effectiveness, but on the other hand, it also prepares the stage for the accident [2]. In a sociotechnical system where safety is viewed as a control problem, an accident occurs whenever this control system cannot handle component failure or external troubles [3]. Therefore, examining the behaviour in each interaction between components becomes more prominent.

Ship management operation has a characteristic of the distributed team between shore staff and onboard seafarers. Different time operations make the team distributed not only geographically but also temporally [4]. Even the accident number in maritime operations is decreasing, human error, especially the failure of situation awareness, remains dominant [5]. The failure to attain situation awareness occurs not only because the difficulty in communicating mental models between the team members onboard, but also between onboard and shore side [6]. The overall situation awareness is perceived differently by an individual based on incomplete and inaccurate information. Such condition makes the team remain locked into a false picture of the situation until accidents or incidents occur [4].

The interaction between onboard seafarers and shore management and its behavior in terms of efficiency, safety, and workload needs to be closely observed in a feedback loop environment.

However, the traditional approach is unable to cover the dynamic behavior of these interactions. To address the challenge, the developed system dynamic model in this study provides a holistic understanding of safety and efficiency pressure allocation in sustaining a safe operation. Modelling with system dynamics focuses on system behaviour made by nonlinear feedback loops [7,8]. It is well used in safety studies for complex systems, where the decision-making system analysis is involved in planning a safety policy [9].

2. Method

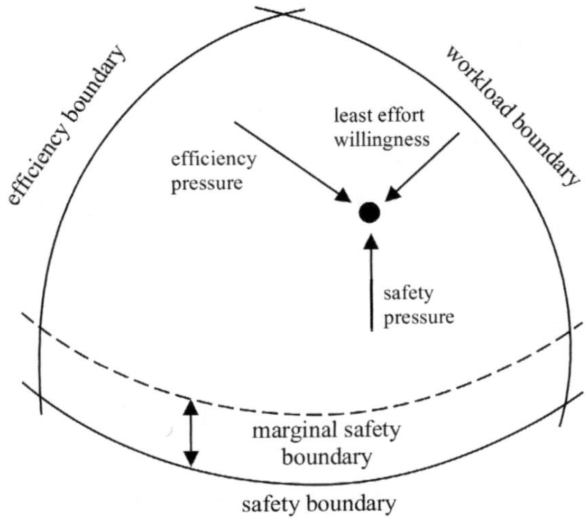

Figure 1. Operating point and three boundaries, as adopted from Rasmussen [2]

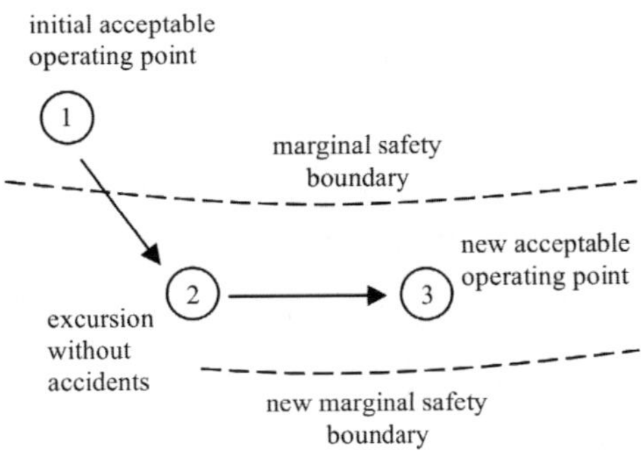

Figure 2. Operating point movement, as adopted from Cook and Rasmussen [10]

We developed a system dynamics model based on Rasmussen's dynamic safety theory [2] and what Morrison and Wears already have formulated [11]. As shown in Figure 1, Rasmussen's theory consists of three boundaries where a sociotechnical system operates: safety, efficiency, and workload. Later, this concept was formalized into the system dynamics model by Morrison and Wears [11] with simplification of the workload and efficiency as a single boundary. We see the model should be extended regarding the ship management operation where the teams are divided into the shore team who act as the management, and the seafarers onboard as the front-line worker. Thus, the constructed model within this paper is based on the previous study [11], together with the related literature in ship operation management. The developed model consists of several loops as shown in Figure 3; its variables construction is explained more extensively in each subsection below.

2.1. Operating Point

The operating point represents the state of the organization's operation from the perspective of three boundaries. The continuous movement of the operating point is made by a trade-off between the management who will apply an effective cost gradient, and the worker who will find an effort gradient. The migration is defined as the management's tendency to find a more suitable position to gain performance. Likewise, it describes the aggressivity of ship management in doing business. Rasmussen explained this practice with an analogy of Brownian Gas movement, where the operating point's variation will randomly move within the available space made by these three boundaries [2]. Further, stability is achieved when movement is small and random [10]. This practice pushes the operating point move away from the efficiency and workload boundary toward the safety boundary. As a result, the migration will likely cross the safety boundary, and an accident may occur.

In Figure 3, the operating point is described as the input function of safety pressure and efficiency pressure to decide its magnitude and direction of movement. The symbol "+" in the arrowhead from loop B1 means the operating point will move to a higher position (greater safety) as the safety pressure increase, and vice versa. Contrary, the symbol "–" from loop B2 indicates the operating point will move to a lower position (less safety) as the efficiency pressure increase. Safety and efficiency work similarly but in a different direction of consequences. The aforementioned management strategy to pursue more suitable operation performance is represented as the excursion. The negative arrow input from the excursion indicates that the operating point will move closer to the safety boundary as the excursion value increase.

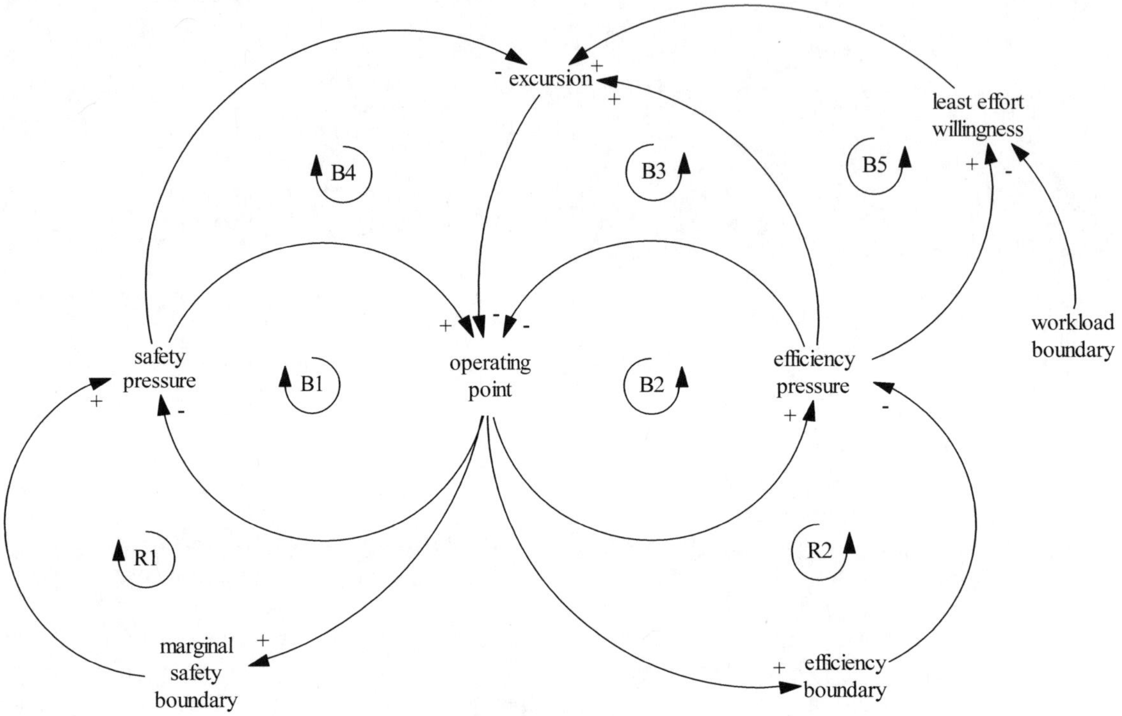

Figure 3. Causal loop diagram of the proposed model

2.2. Safety Pressure

Since the location of the safety boundary is unknown unless the organization is experiencing an accident, the marginal safety boundary is created to estimate and give the margin error from the safety boundary. It is normal if the organization, also in this context, is a ship management company, to operate near the marginal safety boundary given the advantage of higher efficiency and lower workload. Unintentional crossing the marginal safety boundary is recognized as a near-miss since no accident occurred [10]. Immediately an effort is created to bring the operating point move upwards. Crossing the marginal safety boundary also gives the organization a new understanding that the current operation state is acceptable. Hence, the practice of lowering the marginal safety boundary is unavoidable.

Loop R1 explains the safety pressure created by the marginal safety boundary. Illustrated in Figure 2, when the operating point moves beyond the marginal safety boundary, the new boundary value will exist lower than the previous one. The new lower marginal safety boundary generates lower safety pressure since the operating point is more distant from the marginal safety boundary. Operating with lower safety pressure stimulates the operating point to move even lower, close to the safety boundary. Therefore, the feedback mechanism in this marginal safety boundary is categorized as the reinforcing loop.

2.3. Efficiency Pressure

Seeking more profit is the nature of the company in doing the business. It is not uncommon for the organization to cut off safety corners or boundaries to meet operational demand [12]. The manager's job, such as a marine superintendent, is to allocate resources such as budget and time efficiently. Allocating the resources is not without obstacles. Even shore staff have had a long sea experience before, demands to meet the target and performance indicators make the actual situation onboard often remain unconsidered [13]. These are accumulated into the pressure to pursue greater efficiency, but on the other hand, also lead the condition into less safety.

In an extended theory by Morrison and Wears [11], not only the marginal safety boundary that is moveable but also the efficiency boundary will adapt to the recent organization experience and pressure. In the model, efficiency pressure is defined as the exact opposite of safety pressure. Explained by the loop R2, the efficiency boundary will migrate to the lower side, which means more efficiency but less safety because of the feedback from the lower operating point without safety breach or accident. On the other hand, having a lower efficiency boundary means the efficiency pressure will increase and push the operating point even lower. Similar to safety pressure, the mechanism in this feedback is a reinforcing loop.

2.4. Excursion

As explained in Figure 2, an excursion makes the operating point move closer to the safety boundary or to the condition of less safety. The position across the marginal boundary does not immediately trigger an accident, but instead, it states that the current operating point position is acceptable. Further, it made the organization move the marginal boundary to a lower position. Although the direction of the excursion itself was explained by Rasmussen [2], and its magnitude was explained by Morrison and Wears [11], we argue that its magnitude should also be affected by safety pressure and efficiency pressure. In Figure 3, Loop B4 describes that when safety pressure increases, it reduces the excursion. The same balancing feedback explains with loop B3: when efficiency pressure increases, it also triggers a higher excursion.

2.5. Least Effort Workload

Given the circumstances of ship operation where operator onboard and shore management are distributed spatially and temporally, some unintended resources allocation may exist. In large scope, ship companies also consider the efficiency in manning allocation by cutting the number of seafarers onboard; a common practice for some ships not to employ a third mate to save the budget. The implication directly adds the work demand to remaining seafarers [13]. It settles unsuitable conditions to maintain safe operation since the remaining seafarers tend to limit their rest hours and have to work longer. In several cases, as reported by Rajapakse et al., work tasks were assigned to the seafarers onboard sometimes without considering the seafarers [1]. It makes the allocation task not adjust to the limitation of seafarers as a human. A higher workload will make the operators tend to find the strategy to lower it to achieve a suitable workload level. Such a strategy is generally understood as a shortcut or workaround. Moreover, because of the lack of understanding between shore and onboard operators, it is common for seafarers to keep back particular events or activities from the shore-based staff [13]. This deviation in the actual work environment was not recognized by shore management.

The point we want to stress is that kind of additional pressure excursion is the actual condition on the field but beyond management estimation. We argue this matter from the original model and define workload as a separate entity besides economic as efficiency pressure. Explained in loop B5, the increasing efficiency pressure will increase the front-line operator's workload because of the human limitation that defines the workload boundary. Higher workload naturally will make the operators develop a tendency to find the strategy to lower it beyond workload boundary and achieve a suitable workload level. In the loop feedback, this practice brings the least effort willingness more significant, directly increases the excursion, and lowers the operating points (less safety).

3. Simulation Result

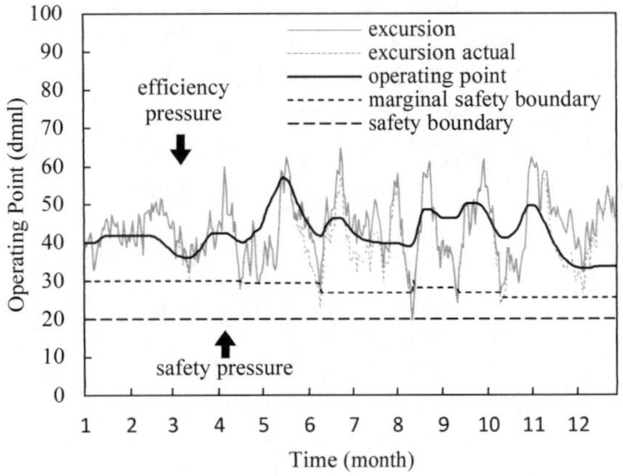

Figure 4. Single running simulation result for Scenario 1

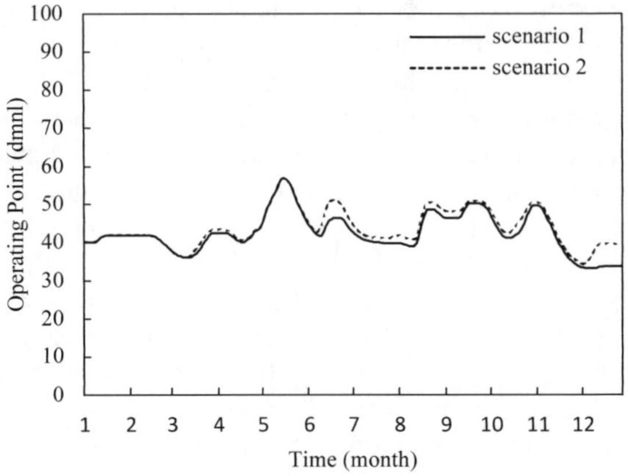

Figure 5. Operating point comparison between scenarios

Figure 4 shows the base model simulated in a one-year period. We followed the Morrison and Wears [11] representation model into one dimension vertical axis. Because this study focuses on the theoretical construction and dynamic behavior pattern, the variables are made on a theoretical basis and assumption, including its unit defined in dimensionless (dmnl). An operation point on the vertical axis, for instance, illustrates the higher value with greater safety and vice versa. The safety and marginal safety boundaries are shown in the lower part. The horizontal axis shows the operating point over the time variable; it demonstrates the objective of explaining the behavior dynamically.

The model starts from the equilibrium initial state, with the operating point being the average function of safety and efficiency pressure. The excursion combined with the safety pressure brings the operating point upward to higher safety, and efficiency pressure balances it to a downward position. When the excursion crosses the marginal safety boundary, safety pressure suddenly increases and brings the operating point upward, but we can notice the marginal safety boundary moved to a lower state. In a condition where the excursion moves further lower and crosses the safety boundary, the system experienced an accident.

In Scenario 1 (S1), the onboard excursion (excursion actual) is assumed not to be noticed by the shore management. Hence, the management only used the partial excursion to decide the operating point's direction and magnitude. While in Scenario 2 (S2), we assume the shore management has sufficient situation awareness about the excursion actual and uses it in allocating safety and efficiency pressure. The different operating point between these two scenarios is shown in Figure 5. Two types of excursion lines indicate the difference between shore side perception and actual onboard excursion affected by the least effort workload.

To test the model's sensitivity, we run the simulation 100 times for each scenario in different simulation duration. Table 1 shows the total accidents on the average and normalized ratio for each running scenario. The most noticeable difference is that Scenario 2 experienced fewer accidents compared to Scenario 1 in all simulation duration. Comparison between different simulation duration also indicates the consistency is established in long simulation duration, but not with the short one.

Table 1. Simulation result

	1 year		4 years		10 years		15 years	
	S1	S2	S1	S2	S1	S2	S1	S2
Accidents (mean)	0.40	0.26	3.12	1.84	8.56	4.99	13.10	7.79
Accidents (SD)	0.65	0.52	1.58	1.16	2.74	1.86	3.68	2.33
Ratio	0.40	0.26	0.78	0.46	0.86	0.50	0.87	0.52
Reduction		35%		41%		42%		41%

4. Discussion

Rasmussen's dynamic safety model has a significant contribution to understanding the influences of safety, workload, and efficiency in a sociotechnical system to determine the position of the observed organization over time. To adopt the theory into the ship operation management context, we extended Morrison and Wears's formulation model. Instead of combining the workload and efficiency into a single effect, we translated the workload as a separate entity that depends on efficiency pressure but still acts upon safety. In adjusting the model, we tried to stay in line with the original theory.

Viewing safety as a control system in sociotechnical, what is needed is a leading indicator; therefore, an act to intervene can be performed before this system's state leads to an accident. The developed system dynamics model in this study attempted to provide an understanding of management strategies in allocating safety and efficiency pressure to approximate that leading indicator. As a result, having sufficient situation awareness of onboard conditions allows the management to put the operating position in a state where accidents are less likely to occur. It amplified the essential of maintaining good coordination between the onshore and onboard teams. The modelling with system dynamics also allows observing this behavior in the long-horizon time span.

As an effort to analyze the system holistically, this paper focuses on behavior prediction rather than point prediction. Validation is limited to model construction, and the test is conducted only with extreme scenarios. Like other models, many simplifications and assumptions were made. Future studies should involve actual data representation as an attempt to represent each dimension into the countable unit. We agree with the previous study that safety, for instance, can define as a risk factor, but the feedback system likely will show a similar behavior pattern.

Acknowledgment
This research was supported by JSPS KAKENHI Grant. No. 21K04517.

References

[1] Rajapakse A, Emad G R, Lützhöft M and Grech M 2019 A study on time constraints and task deviations at sea leading to accidents – a cultural-historical perspective *Maritime Policy & Management* **46** 436–52
[2] Rasmussen J 1997 Risk management in a dynamic society: a modelling problem *Safety Science* **27** 183–213
[3] Leveson N G 2011 *Engineering a safer world: systems thinking applied to safety* (Cambridge, Massachusetts: The MIT Press)
[4] Endsley M R 2012 *Designing for situation awareness: an approach to user-centered design 2nd ed* (CRC Press)
[5] Grech M, Horberry T and Koester T 2008 *Human factors in maritime domain* (CRC Press)
[6] Barnett P M, Gatfield D and Pekcan C 2006 Non-technical skills: the vital ingredient in world maritime technology? *Proceedings of the International Conference on World Maritime Technology 2006* p 11
[7] Sterman J D 2000 *Business dynamics: systems thinking and modeling for a complex world* (Boston: Irwin McGraw-Hill)
[8] Forrester J W 1961 *Industrial Dynamic* (Cambridge, Massachusetts: The MIT Press)
[9] Ibrahim Shire M, Jun G T and Robinson S 2018 The application of system dynamics modelling to system safety improvement: Present use and future potential *Safety Science* **106** 104–20
[10] Cook R and Rasmussen J 2005 "Going solid": a model of system dynamics and consequences for patient safety *Quality and Safety in Health Care* **14** 130–4
[11] Morrison J B and Wears R L 2021 Modeling Rasmussen's dynamic modeling problem: drift towards a boundary of safety *Cognition, Technology & Work*
[12] Reason J 1998 *Managing the risk of organizational accidents* (Aldershot: Ashgate)
[13] Sampson H, Acejo I, Ellis N, Tang L and Turgo N 2016 *The relationships between seafarers and shore-side personnel* (Seafarers International Research Centre (SIRC), Cardiff University)

Risk Assessment of Subsea Power Cables in South Sumatera Province Waters

Zulfaidah Ariany[1, 2], Sarwoko[2], Budhi Santoso [3], and Winarno D. Rahardjo[4]

[1]Departement of Marine Engineering, Institut Teknologi Sepuluh Nopember, Surabaya, Indonesia
[2]Department of Industry, Teknologi Rekayasa Konstruksi Perkapalan, Sekolah Vokasi, Universitas Diponegoro, Semarang, Indonesia
[3] Department of Naval Architecture Engineering Polytecnic of Bengkalis, Riau, Indonesia
[4]Department of Machinery Engineering, Universitas Negeri Semarang, Semarang, Indonesia

Abstract. Risk Assessment to assess the potential risk of subsea power cable diameter 120 mm with a capacity of 70 kVA and a length of about 1.6 km from the Sungai Juaro Palembang city – Sungai Kedukan Banyuasin district in South Sumatra province. A subsea power cable risk assessment case study using Det Norske Veritas-RP-F107 offshore pipeline standard. Potential Hazards risk sinking vessels, risk of dropped anchor, risk of a dragged anchor. The power cable has a length of 2 (two) 1900m lines without connection. This condition applies to 3 (three) acceptable variations of ship speed. Due to the ship's sinking, the risk is at medium risk and is still in an acceptable area. The risk of the lower cable being dragged by the anchor with the windlass indicates that group A and B vessels are at very low risk, while groups C, D, E, and F are at medium risk. The results of the study and risk assessment of the subsea cable line in the Musi River, Palembang, the power cable must be buried as deep as 2 meters or more. The security of the submarine power cable can still be maintained safely.

1. Introduction

This Risk Assessment was prepared to assess the potential risk of an subsea power cable diameter of 120 mm^2 with a capacity of 70 kVA, the length of about 1.6 km from the Juaro River – Kedukan River area on the Musi River. Please note that the Juaro River and Kedukan River are regional names and not river names. Both sites are located on the banks of the Musi River. The Juaro River itself is located adjacent to the city of Palembang, while the Kedukan River is located in the Banyuasin district. There are 2 (two) existing cable lines at that location with almost the exact specifications as the specifications to be deployed. However, one line is no longer in use because it has been damaged, while the other line is considered unable to load the load at peak conditions. The age of the existing cable has reached 30 years so it should be replaced. The concept development, design, manufacturing, testing, storage, load-out, transport, installation commissioning, operation, maintenance, and decommissioning phases of the cable system shall be conducted in compliance with applicable international and national legislation and company policies. The selection of materials and processes will be undertaken regarding health, safety, environmental, and quality aspects [4]. This subsea power cable connects the electricity network from the Palembang city area and the electricity network in the Banyuasin area. The increasing population and economic activities in the Banyuasin area have increased electricity demand. With predictions in the next few years that the installed power cable capacity has exceeded, it is deemed necessary to increase the ability of the electricity network in the area. This electricity network connection was also a backup for the electricity needs of the Asian Games activities.

2. Methodology

Cable Burial Risk Assessments (CBRA) are becoming increasingly common for subsea telecommunications, electricity, and offshore wind installations. A further variable that requires regular cable surveys, regional fisheries analysis, sea liaison efforts, and deployment of strategic guard vessels, a comprehensive CBRA forms a vital component of a holistic risk mitigation strategy. In principle, risk assessment calculation is the multiplication between frequency and consequences [6]. The importance of the cabling position may depend on the operating location. While each project has a different frame, it can be argued that the choice of deployment method will significantly affect the available weather window [7]. CBRA focuses on optimizing the depth of drop or cable burial. It has been suggested that many previous projects, particularly for telecommunications and power cables, need not be conservative in the desired depth of cable burial. This proves to be much more expensive and can complicate installation, increasing the chance of cable damage during the installation process.
CBRA considers many elements of the proposed cable route when optimizing the route and depth of descent. These factors relate to the proposed cable route's physical, geological, and hydrological impacts. An outline of the proposed method is given in reference to the appropriate section, which contains further details. The new CBRA method works the opposite of the BPI method. Once the cable condition has a risk of developing, the depth of the drop target is determined. The number of ships carrying a sufficiently large anchor is then determined and the probability of failure on the cable calculated. The calculated probabilities are then assessed by critical stakeholders, including a review of the implications on the entire cost of cable life. It is difficult to determine an acceptable level of risk and will be an area for which significant discussion and review should be anticipated [10]. Cable routing is the first and most efficient method of reducing threats to submarine cables. During routing, avoid seabed features that could complicate burial, mooring, primary shipping lane operations, etc.

Data collection and review obtaining survey data, this subsea power cable connects the electricity network from the Palembang city area and the electricity network in the Banyuasin area. Of course, the cable line passes through the river and the urban regions and various obstacles along the pipeline. Each Burial Risk assessment of seabed conditions with confirmed cable lines and appropriate survey data obtained begins with a detailed analysis of seabed conditions along the route. Risk list/threat assessment to determine a proper depth of drop for a submarine power cable, a threat assessment should be carried out. This is best done by preparing a risk register that considers the likelihood and consequences of external threats to the cable. Probabilistic risk assessment process probabilistic of shipping activities such as ships passing over the cable line. The subsequent probabilistic failure to remove the anchor is unusual. Probabilistic collision of ships that occur around the cable line.

a. Cable routing is the first and most efficient method of reducing threats to submarine cables. During routing, avoiding seabed features that could complicate burial, mooring, or major shipping lane operations etc.
b. Data collection and review, obtaining survey data, this subsea power cable connects the electricity network from the Palembang city area and the electricity network in the Banyuasin area. The cable line passes through river and urban areas, of course, various obstacles along the pipeline.
c. Assessment of seabed conditions, with confirmed cable lines and appropriate survey data obtained, each Burial Risk Assessment begins with a detailed analysis of seabed conditions along the route.

d. Risk List / Threat Assessment, in order to determine an appropriate Depth of Drop for a submarine power cable, a threat assessment should be carried out, this is best done by preparing a risk register that considers the likelihood and consequences of external threats to the cable.
e. Probabilistic risk assessment process, probabilistic of shipping activities such as ships passing over the cable line. The next probabilistic failure to remove the anchor is unusual. Probabilistic collision of ships that occur around the cable line.
The following is the Cable Burial Risk Assessments (CBRA) risk assessment methodology:

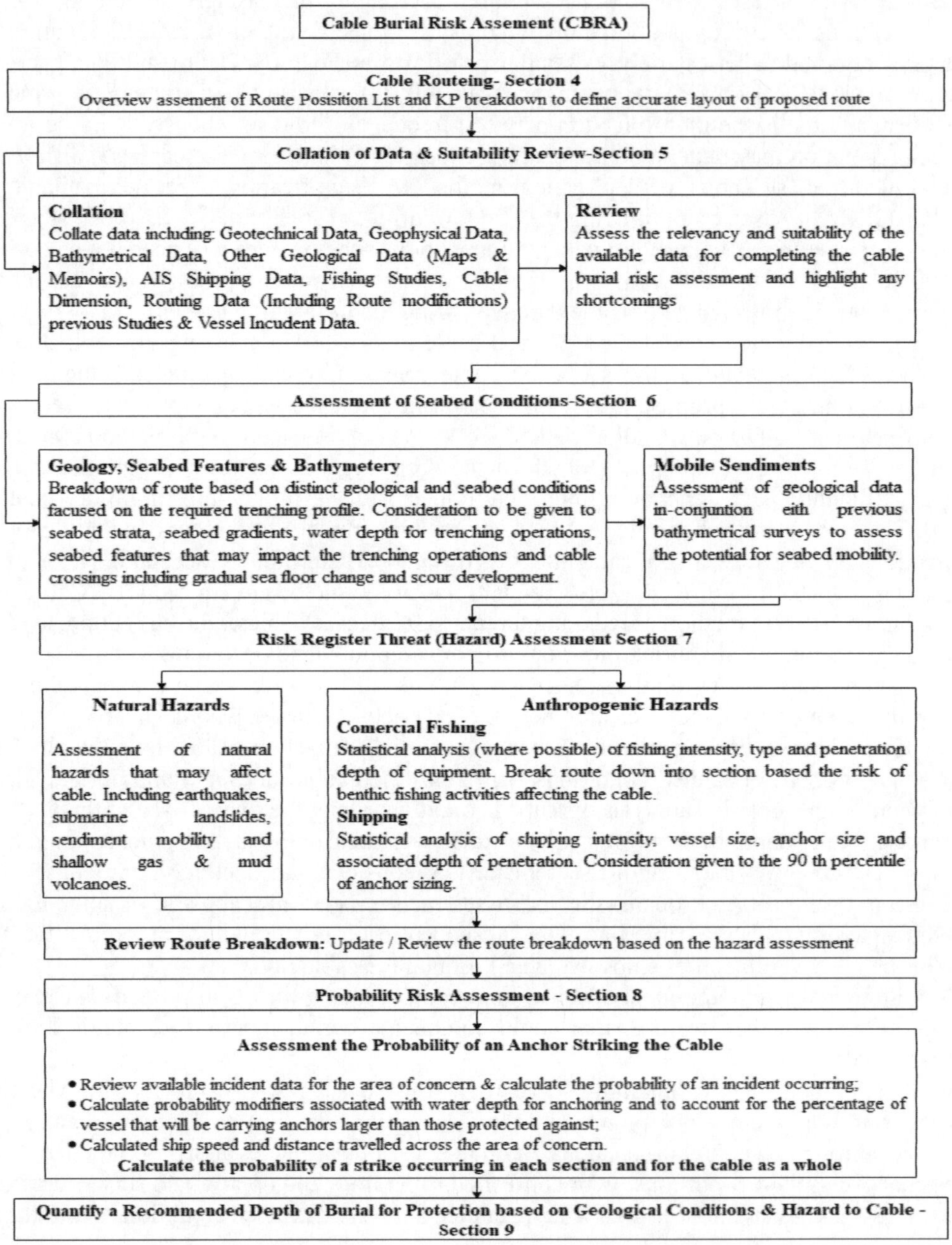

Figure 1. Burial Risk Assessment Method Flowchart [10]

2.1. Equipment Number Calculation

Ship data obtained from KSOP Class II Palembang is ship data based on the amount of Gross Tonnage (GRT) of ships that pass through subsea power lines. Using the GRT data, the size of the ship is predicted based on commonly used empirical standards and formulas. After knowing the estimated main size of the ship, the weight and size of the anchor attached to the ship are calculated. To simplify the calculation, the main size of the ship is taken from the most considerable GRT value. Analysis of anchor weight begins with calculating the equipment number of the ship. Rule BKI obtained a formula for the value of this equipment, namely [1]:

$$Z = D^{\wedge 2/3} + 2.h.B + A/10 \quad (1)$$

Where:

Z : Equipment value
D : Displacement (Tons)
H : Height from water line to deck line (m)
B : Ship width (m)
A : Cross-sectional area of the shipbuilding (m^2)

2.2. The energy that Cable is holding

The amount of energy that the cable line can hold can be calculated based on the formula in the standard DNV-RPF-107 as in the equation [2].

$$E = 16\left(\frac{2\pi}{9}\right)^{\frac{1}{2}} m_P \left(\frac{D}{t}\right)^{\frac{1}{2}} D \left(\frac{\delta}{D}\right)^{\frac{3}{2}} \quad (2)$$

Where:

M_p = plastic moment capacity of the wall (1/4 $\sigma_y r^2$)
δ = Pipe deformation, dent depth
t = Wall thickness
σ_y = Yield stress
D = Steel outer diameter

2.3. Dropped Anchor Energy Analysis

The impact area and drag coefficient have the highest sensitivity and mass, and the additional mass coefficient has the lowest sensitivity on the probability of failure [8]. Calculating the probability and depth of penetration of the anchor deployment near the submarine cable is studied. This method can reduce the cost of protection while providing lines with better protection and guiding the specification of the buried depth of subsea power cables [12]. Combining these to equations gives the following expression for the terminal energy as stated in DNV-RP-F107 is as follows [2][8]:

a. Energy Terminal

$$E_T = \frac{m.g}{C_D.A} \cdot \left(\frac{m}{\rho_{water}} - V\right) \quad (3)$$

b. Velocity Terminal

$$E_T = \frac{1}{2}.m.v_T^2 \quad (4)$$

c. Added Mass

$$E_T = \rho_w.C_a.V \quad (5)$$

d. Impact Energy

$$E_T = E_T + E_A = \frac{1}{2}(m + m_a).v_T^2 \quad (6)$$

Where: m = Anchor Massa (kg); ρ_w = Sea Water Mass (1025 kg/m3); V = Anchor Volume (m^3); C_D = Drag coefficient; A = Anchor Surface Area (m^2)[2].

2.4. Dragged Anchor Energy Analysis

The tensile energy due to dropped anchor that pulls to the pipe, as stated in DNV-RP-F107 and DNV-RP-F111, is as follows [3]:

a. Impact Energy

Steel Impact Energy Dragged Anchor as following calculation

$$E_s = R_{fs} \cdot \frac{1}{2} m_1 (C_h + V)^2 \tag{7}$$

Hydrodinamic Mass Energi

$$E_a = R_{fa} \cdot \frac{\alpha . m_p . D}{(\beta+1)} \cdot \left| \frac{F_b}{a . m_p} \right|^{\frac{(\beta+1)}{\beta}} \leq \frac{1}{2} m_a (C_h . V)^2 \tag{8}$$

Impact Energy

Eimpact = Es + Ea (9)

b. Pull-Over Energy

Downward Force

$$E_z = R_P \, (0{,}2 + 0{,}8 \, . \, e^{-2.5 \, . \, H}) \tag{10}$$

Pull Over Load Duration

$$E_P = C_T \, C_F (m_1/k_w)^{\ 1/2} + \delta_P / V \tag{11}$$

Pull Over Energy

Epo = Fz + Tp (12)

c. Hooking Energy

EH = EP + EK (13)

Where: R_{fs} = Steel mass associated; m = Anchor Weight (kg); Ch = Coefficient of Impact Velocity; Mh = Hydrodynamic Weight; V = Speed (m/s); H = Dimensionless Height; OD = Out Side Diameter of (m). The energy of total from anchor withdrawal is [3]:

$E_{total} = E_{impact} + E_{po} + E_H$ (14)

2.5. Energy Analysis of Anchor Penetration On Seabed

Therefore the energy needed by the anchor to land before reaching the pipeline needs to be calculated [3].

a. Energy Penetration

$$E_P = 0{,}5 . \gamma' D \, N_\gamma . A_P . z + \gamma' . z^2 . N_q A_P \tag{15}$$

Where: γ = Sea Water Mass kg/m^3; Ap = Surface Area Pipe; D = Diameter of pipe (m); z = Anchor penetration on seabed. Based on the above equation, we can find out the penetration energy and the penetration depth of the anchor on the seabed.

2.6. Consequence Standard

Analysis of hazards that can occur to determine the hazard impact on the cable path, the impact due to dropped anchors, dragged anchors and sinking vessels. After the required data is obtained, it is calculated by referring to the DNV-RPF-107 standard. Furthermore, after all the consequence calculation have been obtained, they are then ranked based on the consequence categories contained in the DNV-RPF-107 standard.

Table 1. Impact Capacity and Damage Clasification

Dent/Diameter (%)	Damage Description	Conditional Probability					
		D1	D2	D3	R0	R1	R2
<5	Minor damage	1,0	0	0	1,0	0	0
5 – 10	Minor damage Leakage anticipated	0,1	0,8	0,1	0,9	0,1	0
10 – 15	Major damage Leakage and rupture	0	0,75	0,25	0,75	0,2	0,05
15 – 20	Major damage Leakage and rupture	0	0,25	0,75	0,25	0,5	0,25
> 20	Rupture	0	0,1	0,9	0,1	0,2	0,7

3. Result and Discussion

3.1. Design Data dan Assumption

This subsea power cable has a cross-sectional area of 120 mm^2 cu, with a maximum design voltage of 70 kVA and a cable life that can be used for 30 years. The power cable will be used for two lines of 1900 meters without any connection. According to the results of the bathymetry survey, the maximum depth of this cable line is 12 meters.

This power cable will be buried as deep as 1 meter inland/coastal areas by PLN standards. In riverbed areas, it will be buried as deep as 2 meters by or exceeding the provisions in Ministerial Regulation no. 129 of 2016.

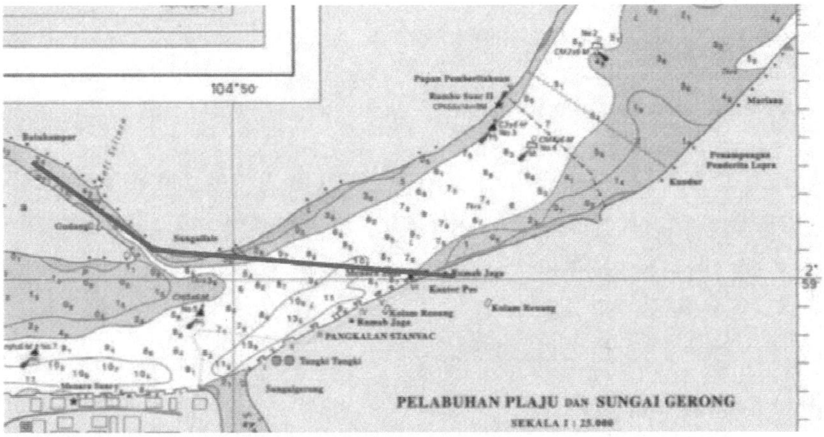

Figure 2. Map of Incoming Cable Submarine Coordinates

3.2. Cable Specification

The cable specifications data used in this research are subject to incoming cable submarines. Complete data can be seen in the following table.

Table 2. Cable Submarine Specifications

Parameter	Unit	Values
Material Jacket	-	XLPE Rigid
Outer Diameter	Mm	137
Wall Thickness	Mm	Full cover
Cable grade	-	High Voltage
Specified Minimum	psi (MPa)	6500 (43)

Parameter	Unit	Values
Yield Strength		
Specified Minimum	psi (MPa)	7500 (50)
Tensile Strength		
Steel Density	$lb/ft^3(kg/m^3)$	490 (7850)
Young Modulus	psi (MPa)	6.0 x 105 (4 x 103)
Thermal Expansion	/oK	1.17x10-5
Coefficient		
Poisson Ratio	-	0,4

3.3. Shipping Potential hazards
Based on shipping conditions and the geographical area of the Musi River waters, the following are some of the potential hazards that may occur:

Table 3. Potential Hazards DNV-RP-F107 [2]

No	Hazards
	Hazards due to the shipping lane
	1.1 Ship aground
1	1.2 Ship sink
	1.3 Dropping of anchors/other objects
	1.4 Anchor dragging
	Hazards due to fishing activities
4	4.1. Current fishing activities
	4.2. Development/improvement of fishing activities (use of Trawling)
	Hazards of force major from the activities of foreign ship and aircraft
	6.1. Airplane accident
6	6.2. Anchor drop
	6.3. Loading/unloading activities

3.4. The Result of Frequency
In calculating this frequency estimate, the ship's speed, the probability of the ship deviating from the shipping lane, and the probability of the ship dropping anchor or the ship sinking are assumed. Then the estimated frequency of each group of vessels is ranked according to the frequency ranking of the DNV RP F 107. The potential hazards to be analyzed include: dropped anchor, dragged anchor, and sinking vessel.
Ship data collection aims to determine the number and capacity of ships that pass through the subsea power line. The ship's data was taken from the official Palembang Class II Port Authority and Port Authority Office.

Table 4. Ship Exit Frequency

Vessel Class	Annual Frequency	Weigh Range (DWT)	Anchor weight (Ton)	Max Vessel's Draft (m)
A	623	<2500	1,29	3,6
B	504	5.000	2,64	05.02
C	156	7.500	04.05	06.03
D	37	10.000	5,25	07.01
E	19	20.000	6	08.05
F	35	30.000 >	7,8	09.06

Determination of the Opportunity of Ships Entering the Critical Anchorage Dangerous Zone (CADZ). The calculation of the probability of ships passing through the CADZ is by combining the average probability of each type of ship leaving the shipping lane and the CADZ area compared to the width of the shipping lane. Critical Anchor Damage Zone (CADZ) = 2 x largest anchor width + pipe diameter + layer thickness [5].

Where
Width of the largest anchor: Depends on the size of the anchor (meters)
Cable diameter : 137 mm (0.137 meters)
Coating thickness : 0.01 meters
The length of the mandatory sailing route on the Musi River = +/- 28 miles

3.5. Consequence Analysis.

At this stage, an analysis of the hazards is carried out to determine the hazard impact on the cable path, namely the effect of dropped anchors, dragged anchors, and sinking vessels. After the required data is obtained, it is calculated by referring to the DNV-RPF-107 standard. Furthermore, after all the consequence calculations have been received, they are then ranked based on the consequence categories in the standard DNV-RPF-107 equation 2.

Table 5. Risk sinking vessels

Ship group	A	B	C	D	E	F
Anchor weight (kg)	1290	2640	4050	5250	6000	7800
Dent	6.84%	7.81%	9.81%	12.23%	16.02%	22.4%
Level of	2	2	2	3	3	4
Frequency	6,16E-06	1,81E-06	2,32E-07	1,68E-07	5,34E-08	6,31E-08
Level of Frequency	1	1	1	1	1	1

Table 6. Risk of Dropped Anchor

Ship group	A	B	C	D	E	F
Anchor weight (kg)	1290	2640	4050	5250	6000	7800
Dent	Low	Low	Low	Low	Low	Low
Level of Consequence	1	1	1	1	1	1
Frequency	4,62E-06	2,71E-06	6,65E-07	3,48E-07	6,81E-08	6,65E-08
Level of Frequency	1	1	1	1	1	1

3.6. Risk Matrix

Based on the results of the estimated frequency and the estimated consequences that have been obtained, a risk matrix is then made that refers to the DNV RP F107. If it turns out that the risk is in an unacceptable zone, then a risk analysis must be carried out again until the risk obtained is acceptable. To minimize risk, it is essential for management to pay attention to the steps contained in the risk mitigation plan [11]. The way to reduce risk is to reduce the frequency of events or reduce the consequences of events and maybe reduce both. The following is a risk matrix due to falling anchors, being dragged by anchors, and sinking ships. In the risk matrix, dropped anchors with categories A, B, C, D, E & F are in the acceptable green zone so that the risk is still acceptable. Ship speed 2.5 knots. The risk matrix sinking vessel in categories A, B, C, D, and E is in the acceptable green zone, so the risk is still acceptable. Meanwhile, category F ships are in ALARP and can be considered mitigation. In the risk matrix, the dragged anchors with categories A, B, C, D, E & F are in the green zone are acceptable, so the risk is still acceptable.

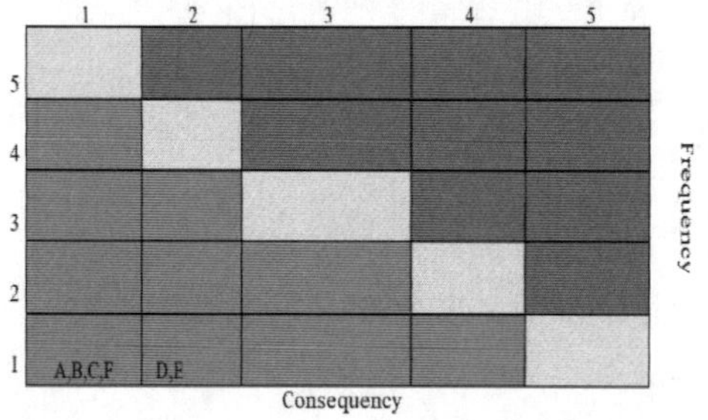

Figure 3. Risk Matrix Dropped Anchor

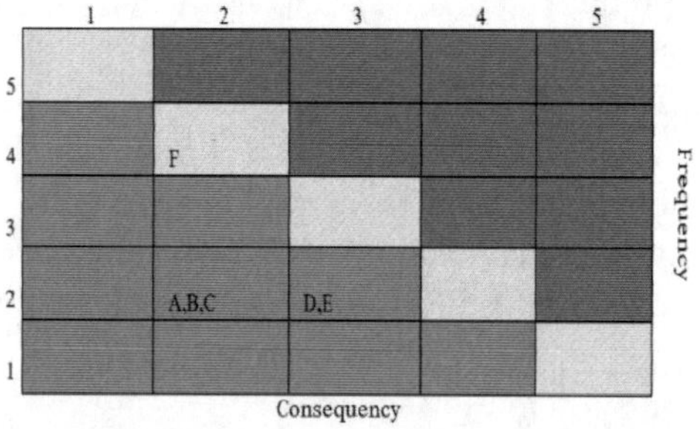

Figure 4. Risk Matrix Sinking Vessel

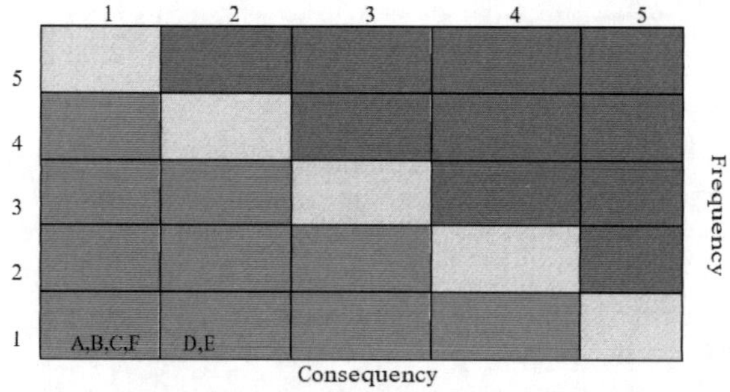

Figure 5. Risk Matrix Dragged Anchor

3.7. Risk Mitigation

After conducting a risk assessment of the subsea power line, some are in the Acceptable and ALARP areas. To further ensure the safety of the operation of the 70 kVA power line, it is necessary to make efforts to reduce the possible risks that arise so that an acceptable area can be maintained. As soon as modifications to the protection of the cableway are made, the company must ensure that the route map is updated where the protection of the pipeline is marked on the map. Even though the shipping route map has been updated, not all ships will use the map or use the latest version. PT PLN Palembang Region should increase awareness of protection on cable lines by establishing better relations with the Port Authority in the area. Pipeline markers may also be used and provide an apparent visual reference for ships traveling in areas with a seabed hazard. This can increase user awareness of the flow because the signs are permanently posted and visible. The laying of subsea power lines is marked on the map. Regular maritime patrols are also required. These patrols have the dual purpose of enabling routine visual inspection of the cableway, and patrols can identify potential anomalies.

4. Conclusions

Based on the risk assessment that has been carried out on the underwater power cable line, the following conclusions can be drawn. From the risk assessment that has been carried out on the underwater power line, a risk matrix is obtained. From the risk matrix analysis obtained. Risks due to cables by showing that the anchor with a weight of 1290 Kg, 2640 kg, 4050 kg, 5250 kg, 6000 kg and 7000 kg at very low risk, and are still in acceptable areas. This condition applies to three variations of ship speed, namely 2.5 knots, 5 knots and 7.5 knots. The risk due to the sinking of the ship is at medium risk and is still in an acceptable area. This condition is obtained for all ship sizes at variations of ship speed, namely, 2.5 knots, 5 knots and 7.5 knots. The risk due to the bottom cable being dragged by the anchor with the windlass indicates that group A and B ships are at very low risk while groups C, D, E and F are in the green zone, both of which are still in the acceptable area. This condition applies to all variations of ship speed, namely, 2.5 knots, 5 knots and 7.5 knots.

Based on the results of the study and risk assessment of the underwater cable line in the Musi River, Palembang, the power cable must be buried as deep as 2 meters or more, according to applicable regulations, the security of the underwater power cable can still be maintained safely. With the results of the study and mitigation as well as the conclusions above, it is hoped that all risks in this study can be lowered one level below so that the worst conditions in the results of this study are in low risk areas.

Acknowledgments

This research work was collaborated and supported by PT. Lautan Sinergi Perkapalan,

References

[1] BKI Volume II section 14.A.3. 2009
[2] Det Norsk Veritas Marine Risk Assessment - Risk Assessment of Pipeline Protection, (DNV-RP-F107), 2010.
[3] Det Norsk Veritas Marine Risk Assessment - Interference Between Trawl Gear & Pipelines. (DNV-RP-F111), 2010.
[4] Det Norsk Veritas Marine Subsea Power Cables in Shallow Water Renewable Energy Applications. (DNV-RP-J301), February 2014.
[5] Kinzo Inoue. Theory and Practice of Ship Handling, January 1, 2014. ISBN-10: 6054778919. ISBN-13: 978-6054778911
[6] Muhlbauer, W. Kent. Pipeline Risk Management Manual, Third Edition. 2004
[7] Okkerstrom, L. Simulation of peak tension loads in Subsea power cables during installation. EOTech & OGTech 2021. IOP Conf. Series: Materials Science and Engineering1201 (2021) 01201 DOI:10.1088/1757-899X/1201/1/012010
[8] R Waskito, at all. Risk Assessment of Subsea pipeline using standard DNVGLRP-F107. Maritime Safety International Conference. IOP Conf. Series: Earth and Environmental Science 557 (2020) 012069. DOI:10.1088/1755-1315/557/1/012069
[9] Sina Taghizadeh Edmollaii, at all. Reliability sensitivity analysis of dropped objects on submarine pipelines. Ocean Systems Engineering, Vol. 9, No. 2 (2019) 135-155. DOI: https://doi.org/10.12989/ose.2019.9.2.135
[10] The Carbon Trust, Application Guide for the specification of the Depth of Lowering using the

Cable Burial Risk Assessment (CBRA) methodology, UK, 2015.

[11] Wibowo, A, and Sumirat, E. Investment Analysis for Submarine Power Cable Project Between Belakang Padang Island and Sambu Island. European Journal of Business and Management Research. www.ejbmr.org. October 06, 2021. ISSN: 2507-1076. DOI: 10.24018/ejbmr.2021.6.5.1098

[12] ZHANG, Z, at all. The Study of Burial Depth and Risk Assessment of Submarine Power Cable. 2018 4th International Conference on Green Materials and Environmental Engineering (GMEE 2018). ISBN: 978-1-60595-592-6

A study of interceptor performance for deep-v planing hull

S. Samuel[1], Serliana Yulianti[1], Parlindungan Manik[1], Abubakar Fathuddiin[2]

[1]Department of Naval Architecture, Faculty of Engineering, Diponegoro University, Semarang, 50275, Indonesia
[2]Blue Gulf Cat, Sheikh Rashid bin Saeed St, Abu Dhabi, 41655, United Arab Emirates

Coresponding author: serliana.yuliantii@gmail.com

Abstract. The acting on the planing hull is the most complex hydrodynamics simulation. Therefore, an analysis was done to evaluate drag, lift force, and seakeeping in two degrees of freedom (2-DOF) which is heave and trim. It was fundamental aspects of the overall high-speed vessel. This article focused on the hydrodynamic performance of a complete interceptor configuration that could control the motion behavior of deep-V planing hull in calm water conditions. The benchmark study was undertaken by comparing numerical results with experimental study by Park at al. Models with and without interceptors had been analyzed by numerical simulation performed using Reynold Averaged Navier Stokes (RANS) to describe turbulence model with k epsilon based on computational fluid dynamic (CFD). In this study, the interceptor proper applies at a speed of less than Froude number 0.87. Interceptor reduce by 21% drag at Froude number 0.87 and also reduce by 16% trim and 6% heave at Froude number 0.58. Nevertheless, applied interceptor in high Froude number such as more than Froude number 1.16 caused interceptor lose effectiveness due to producing a decisive moment which made negative trim (bow-down) and increase total drag.

Keywords: CFD, drag, full interceptor, heave, lift force, planing craft

1. Introduction

The interceptor is a thin rectangle mounted on the vessel's transom to modify local flow near the stern. The interceptor generates substantial additional pressure towards the vessel’s stern. This concept is adopted from an aerodynamic device called a gurney flap on race cars to increase down-force on the car's rear wing. The interceptor mechanism aims to increase the lift to drag ratio pressure on the vessel, which is slightly different from the situation on race cars. The aerodynamic aspect of a racing car does not affect drag friction. However, drag friction is significantly affected even in steady flow on the vessel. The interceptor will modify the pressure distribution in the same way as the gurney flap. However, the interceptor can affect the balance, which impacts the generation of the stern wave system and the resistance due to the waves. In addition, the increase in lift force also reduces sinkage and wetted surface areas as well as frictional resistance to the hull.

An experimental study conducted by Day and Cooper reported that the interceptor could reduce drag on sailing yachts [1]. Furthermore, Van Oossanen conducted a 45 m motor yacht with an interceptor dimension of 50 mm using the CFD approach. The study resulted in a reduction in the trim of 1 degree and decrease in drag of 7% at Froude number 0.6 [2].

In recent years, the interceptor has become increasingly recognized as a trim controller on medium to high-speed vessels, especially for semi-planing and planing hulls. The complexity of the hydrodynamics of planing hull often causes the vessel to move coarse when it is operated. Excessive trim and drag when the vessel is at high speed is a challenge for architects to improve performance of the vessel.

Interceptors can reduce trim and drag on high-speed planing craft [3]. The interceptor system can reduce pitch motion up to 32.4% and heave motion 12.1% on irregular waves at a certain speed [4]. The hydrodynamic performance of the transom-interceptor configuration can control the movement of speed boats on waves. The transom-interceptor configuration experiment resulted in a positive and negative lifting force as a controlled force in the opposite direction so that the vessel's motion became better [5]. Another study also compared trim tabs and interceptors of the same size on a planing hull. The study reported that the interceptor provided better trim control [6].

This research used numerical simulation to get the vessel's resistance. The research of Yousefi discussed the numerical method using Computational Fluid Dynamics (CFD) related to planing hull. The research was divided into three methods, namely Finite Volume Method (FVM), Finite Element Method (FEM), and Boundary Element Method (BEM). It was reported that FVM was the most accurate method compared to other methods [7]. In 2015, research on CFD was successfully carried out to predict catamaran ship drag by ignoring ship movement [8]. The research was continued by using a numerical method to modify the addition of a center bulb to reduce the total drag [9]. Furthermore, the numerical simulation was also carried out to reduce the drag by applying a spray strip to the Fridsma hull [10]. In another analysis also compared experimental and numerical method to give a reliable and fairly accurate prediction of the flow around the ship[11]. The growing world of computing made numerical methods an efficient choice.

2. Method

In this study, the Aragon 2 vessel tested experimentally by Park [4] was used, and the interceptor size was selected based on the vessel's dimensions. Table 1 shows the main dimension of the vessel that would be used as this research topic. Table 2 as shown data of interceptor's dimension. The simulation was performed in the scaled model (1: 5.33) . Figure 1 represents the lines plan on the deep-V planing hull, such as side view 1(a), front view 1(b) and bottom view 1(c). Figure 2 represents dimension of interceptor. Figure 3 as shown position of interceptor toward of the ship.

Table 1. The main dimension of deep-V planing hull [4]

Dimension	Full-scale	Model scale 1/5.33	Unit
Length overall (LOA)	8.00	1.50	Meter
Length waterline (LWL)	7.54	1.41	Meter
Breadth overall	2.30	0.43	Meter
Draft	0.45	0.08	Meter
Weight	3000	19.7	Kg
LCG	2.64	0.49	Meter
C.G from baseline	0.80	0.15	Meter
k	21.5	27.2	degree

IOP Conf. Series: Earth and Environmental Science 1081 (2022) 012004 doi:10.1088/1755-1315/1081/1/012004

Table 2. Dimension of interceptor

Dimension	Rectangular Interceptor (full scale)	V-Shape Interceptor (full scale) for Current Study	Unit
Span	300	450	mm
Height	50	Ha: 115 Hb: 200.5	mm
Thickness	20	20	mm

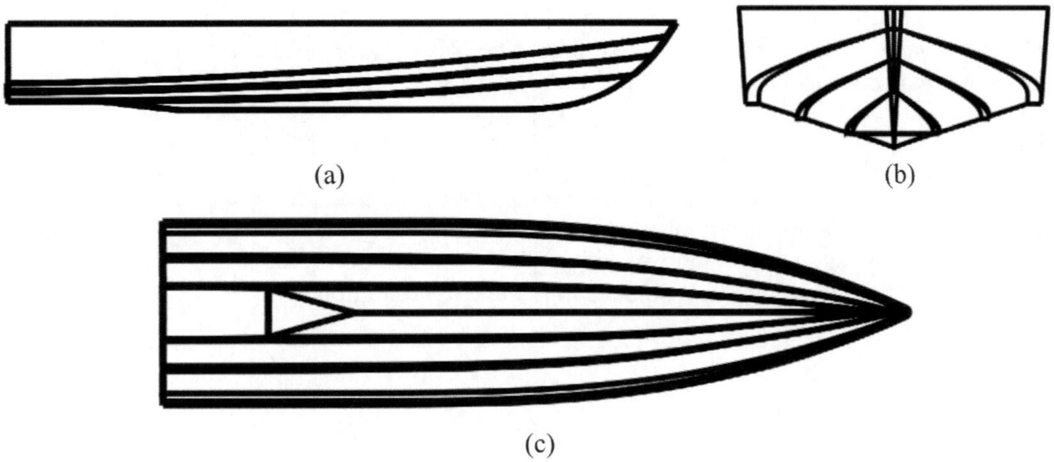

Figure 1. (a) Side view, (b) front view, and (c) bottom view of Aragon 2

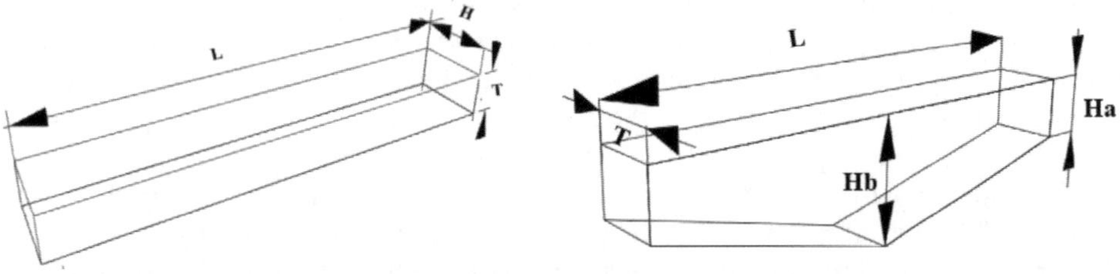

Figure 2. Dimension of (a) rectangular interceptor, (b) v-shape interceptor

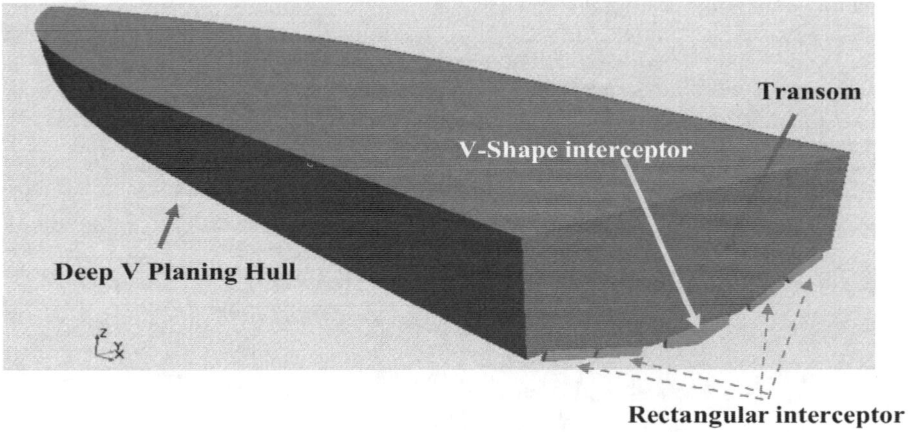

Figure 3. Position of interceptor toward of the ship

The fluid flow modeling used the RANS equation to carry out a predictor-corrector approach that connected the continuity and turbulence equations with the standard k-epsilon model according to the ITTC recommendation [12]. The RANS equation consisted of three main steps, namely:

1. Integration of equations governing fluid flow overall control volumes of the solution domain.
2. Discretization of integral equations into a system of algebraic equations.
3. Solving algebraic equations using an iterative method with the RANS equation and numerical turbulence. The discrete equations were derived using the volume method until then solved by the SIMPLE method and the Implicit Pressure-Correction method, which were used for flow and pressure equations, respectively.

Free surface effects are represents using that called the "Volume-of-Fluid" approach: the solution domain is assumed to be filled by a sufficient fluid whose properties vary according to volume fraction. In CFD, assumed that the volume fraction of fluid in each cell is represented by a function, F, and its value is unity in a cell full of water, and zero value represents no water. Hence, each fluid which has a value between 0 and 1 must contain a free surface. The free surface field can be represented by the transport equation in conservative form as follow:

$$\frac{\delta F}{\delta t}+\frac{\delta uF}{\delta x}+\frac{\delta vF}{\delta y}+\frac{\delta wF}{\delta z}=0 \quad (1)$$

This study referred to the ITTC as an international standard that provides recommendations for predicting ship hydrodynamics using numerical and experimental methods. As for the case of speed vessel, ITTC recommends several things, including [12]:

1. The size of the computational domain and boundary condition
2. Grid sizing
3. Convergence
4. Time-step
5. Grid on the ship wall (y+)

Figure 4 illustrates the size of the computational domain and boundary conditions. Domain length was calculated from -2.5 LOA to 1 LOA with the coordinates of the zero point at the stern and vessel's draft. The width was set to 1.5 LOA vessel. The next was the definition of boundary conditions. The inlet velocity condition was set in the negative x-direction, where the wave was simulated. The positive x-direction was modeled as outlet pressure to show the static pressure at the outlet. The upper limit was chosen as the inlet velocity to represent the infinite air condition, while the lower limit was selected as the inlet velocity. The simulation was carried out on half the hull. It was to reduce time consumption.

The grid size is shown in Figure 5 to see the effect of installing the interceptor. The grid size on the vessel's hull was shown with more mesh to measure the dynamics of the vessel's movement to be more accurate. Figure 5 (a) shows grid size on the side view, and figure 5 (b) shows grid size on the top view. The same thing was done on the water surface.

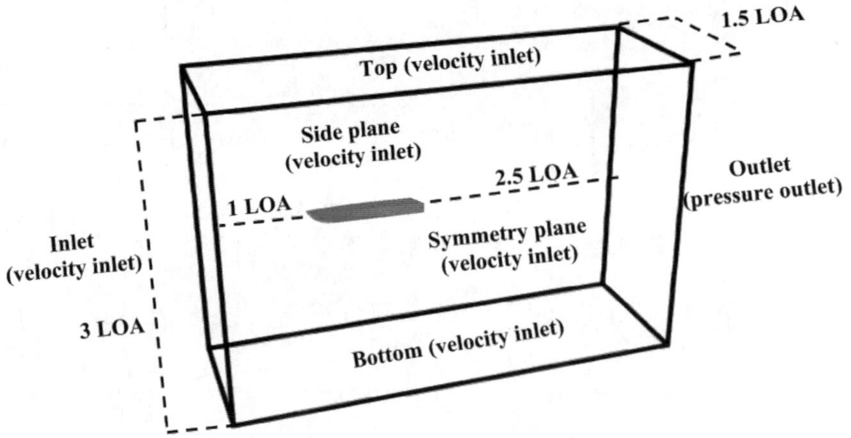

Figure 4. Computational domain and boundary condition

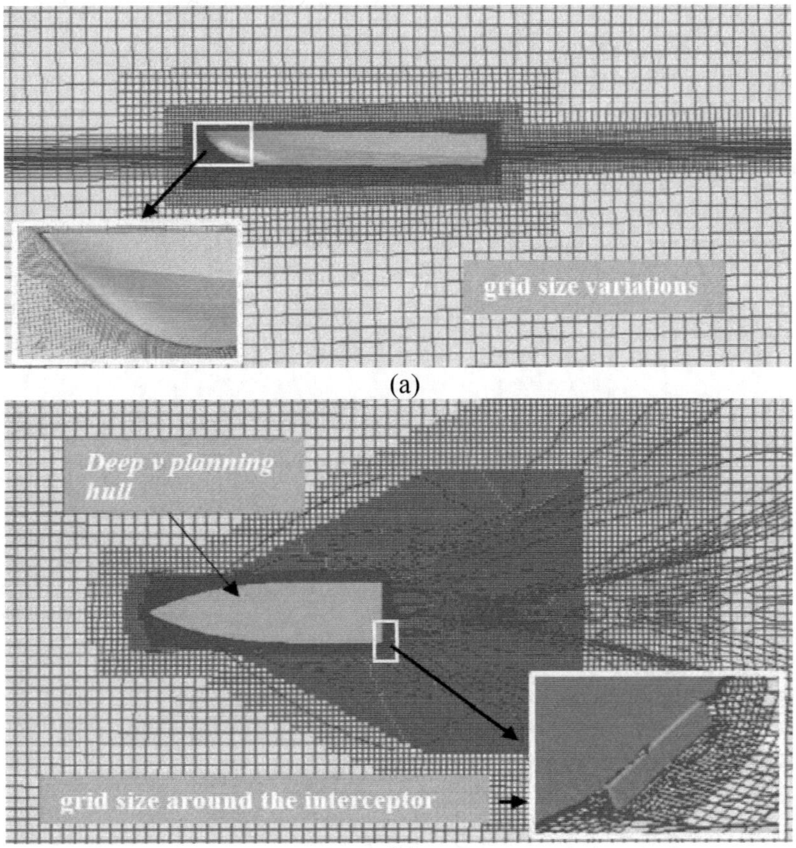

(b)

Figure 5. Variation of grid size (a) side view, (b) top view

In some cases, the dispersion error often resulted in an oscillating solution due to coarse mesh quality. The result was that the residual plot could show that the solution was not convergent. Based on this reason, it was essential to do a grid study to find the best mesh to use for each simulation. In this study, a grid independence test was conducted. The test was carried out using a five-level grid. In Table 3, it was reported that the higher the cells, the higher the time required in one test simulation. The simulation selection was determined based on the drag, heave, and trim values. On grids 1 to 3 produce fluctuating values and illustrate that the results are not good. Grids 4 and 5 produce better values with the drag, trim, and heave stable. But grid 5 takes a relatively long time to generate data. So that grid 4 is selected with good results and does not require relatively much time. It taken 2 seconds to get a convergent iteration. The drag, trim and heave, as shown in figure 6 (a), 6 (b), 6 (c) results showed stable results at 2 seconds. Simulations were carried out on Froude number 1.072. Drag and heave are non-dimensional by the weight and draft after perpendicular, respectively.

Table 3. Grid independence

No	Number of cells (Million)	Time processing (hours)	Drag	Heave	Trim (deg)
Grid 1	0.52	8	0.143	0.081	5.29
Grid 2	0.66	14	0.141	0.086	5.30
Grid 3	0.87	24	0.143	0.083	5.20
Grid 4	1.24	32	0.141	0.085	5.29
Grid 5	1.47	48	0.143	0.082	5.15

IOP Conf. Series: Earth and Environmental Science 1081 (2022) 012004 doi:10.1088/1755-1315/1081/1/012004

The time step was the interval period for each iteration. The time step was determined based on the CFL (Courant-Froderichs Lewy) number, which showed the number of points travelled by the fluid particles in the time interval. The time step value was inversely proportional to the speed of the vessel so that the higher the speed of the vessel, the smaller the time step used. In this study, the time step value used was 0.008.

Another parameter to validate the CFD analysis was the value of y+. The value of y+ (wall function) was expressed in dimensionless units to capture the boundary layer phenomenon. The presence of y+ made it easier to predict the thickness of the first layer of the mesh and reduced numerical simulation inaccuracies. Figure 7 describes the value of y+ between 1 to 6. Based on the ITTC guide [12], the value of y+ can be formulated as follows:

$$\frac{y}{L} = \frac{y^+}{Re\sqrt{\frac{C_f}{2}}} \tag{2}$$

Meanwhile, Y was the thickness of the first layer, L was the vessel's length. Then, Re was the Reynolds number, and C_f was the coefficient of friction.

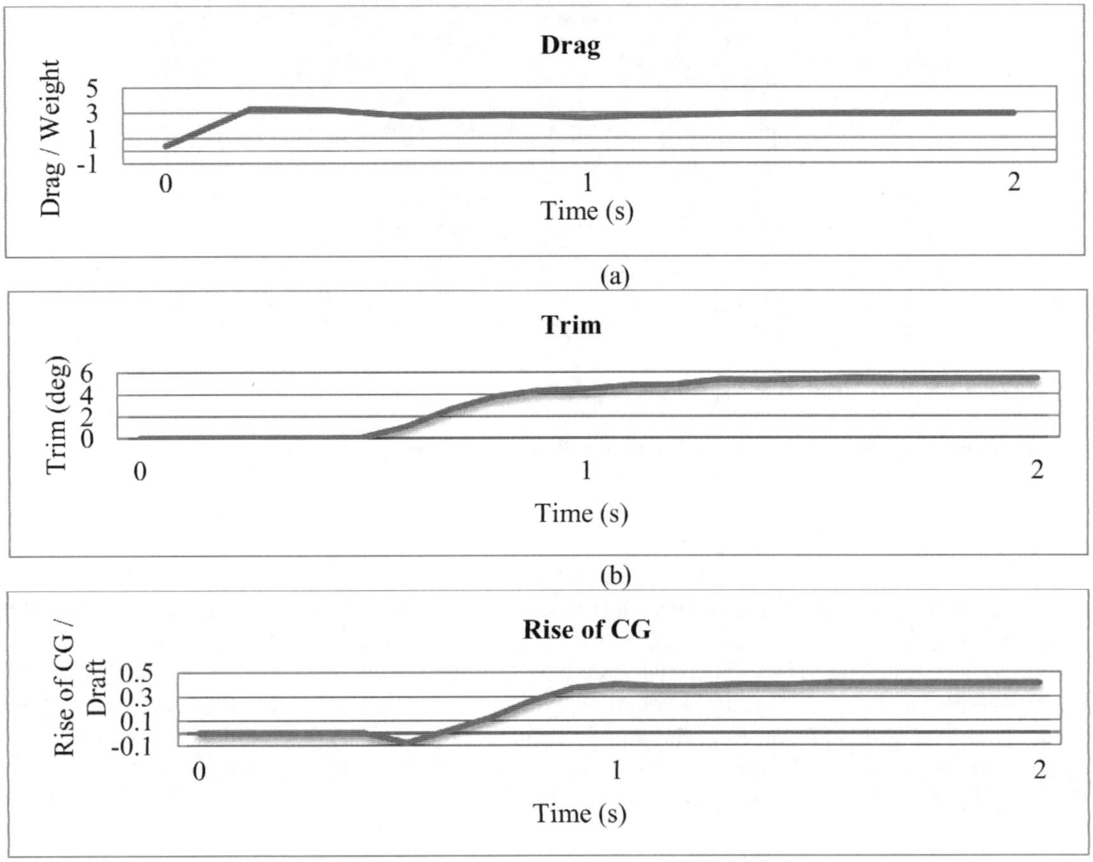

(c)

Figure 6. A sample of convergence of (a) drag, (b) trim, (c) heave parameters, showing how a CFD model calculates the equilibrium condition for the deep-V hull operating in planing mode at Froude number of 1.072

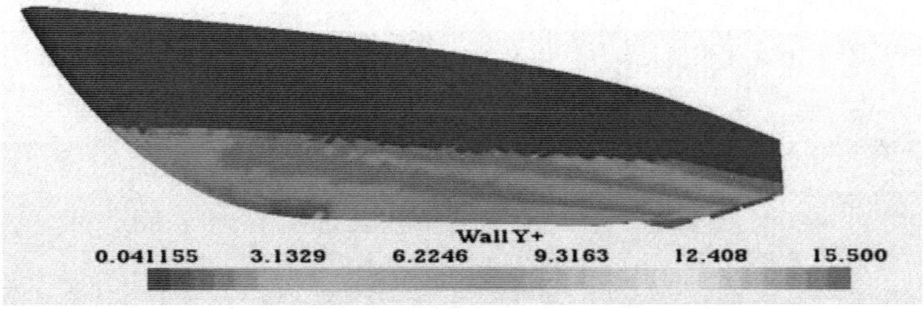

Figure 7. Wall y+

3. Results and Discussion

3.1 CFD verification

The vessel's motion was assumed to heave and trim motion as two degrees of freedom (2-DOF). The study focused on the hydrodynamic performance of a interceptor configuration that could control the motion of deep v planing hull in condition of calm water.

In this study obtained a maximum difference of 10.7%. Figure 8 shows the drag, trim, and heave patterns using the CFD and comparison with experimental approaches[4]. The same pattern had obtained between the results of numerical simulation analysis and experiments. However, there have slightly differences value of the CFD calculation and the experimental results due to the limitations of CFD in modeling according to actual conditions. This also has a similar study in Brizzola and Serra. They investigated the accuracy of numerical simulations by predicting ship planning. They obtained the variation of drag values and compared it with the experimental data of Savitsky[13] and resulting in an average error of 10% in predicting barriers[14]. The percentage difference between Park's experimental results and CFD analysis is shown in table 4.

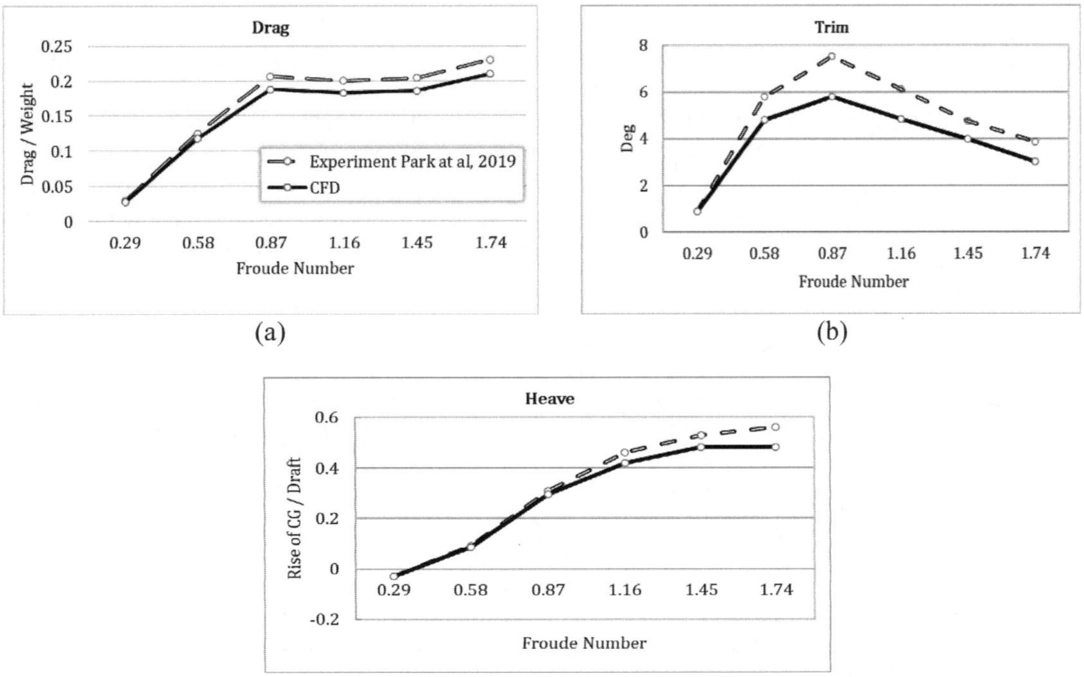

(a) (b)

(c)

Figure 8. Validation results (a) drag, (b) trim, (c) heave

IOP Conf. Series: Earth and Environmental Science 1081 (2022) 012004 doi:10.1088/1755-1315/1081/1/012004

Table 4. CFD analysis and experiment Park at al 2019

Froude Number	Drag	Heave	Trim
0.29	6.1	3.2	0.9
0.58	7.5	4.6	7.5
0.87	9.9	5.7	10.7
1.16	9.5	9.6	9.5
1.45	9.9	9.8	6.9
1.74	9.5	10.0	7.4

3.2 Numerical simulation

The line graph compares the study analyzed variation with and without an interceptor. The results of this study as shown in Figure 9. Overall figure 9 (a) had represented both with and without interceptor had increased from their initial values by the end of the Froude number. Interestingly that using of an interceptor showed improvement in drag for Froude numbers 0.29, 0.58 to 0.87. Although, the drag increased significantly at the Froude number 1.16.

Figure 9 (b) shows that with the interceptor, the trim value increased even at Froude number 0.29, but at Froude number 0.58, there was an improvement in the trim. Along with the increase in the vessel's speed from Froude number 0.87, there was a significant decrease in the trim value.

Figure 9 (c) shows that the value of lift force on the vessel without an interceptor decreased as the Froude number increased. The lowest lift value was at Froude number 0.29 and experienced the lowest lift force at Froude number 1.74. However, when the vessel was applied to the interceptor, the lift force value decreased at Froude number 0.29. Then, the vessel experienced an increase in the value of the lift force at Froude number 0.58 to Froude number 1.16.

Furthermore, from figure 9 (d) the vessel experienced an increase in the heave value at Froude number 0.29. However, the vessel experienced an improvement in the heave value at Froude numbers 0.58 and 0.87. Then, there was a negative heave on Froude number 1.16.

The vessel 's characteristics can be seen in figure 10. This figure explains the vessel's condition with and without using interceptor for each Froude number variation. On a vessel without using an interceptor, it can be seen that the trim increases as the Froude number increases. However, on the vessel using an interceptor, it is found that the trim value improvement at the Froude number was low and effective at the Froude numbers were 0.58 and 0.87. However, when the vessel was in the planing phase, namely at Froude number 1.16, there was excessive trim due to the interceptor moment being more significant than the pressure moment of the fluid while causing a change in the pressure distribution as described in figure 11.

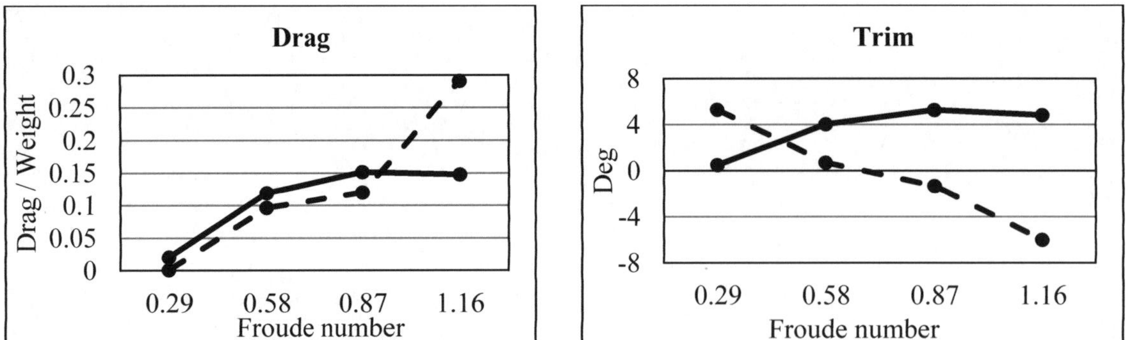

(a) (b)

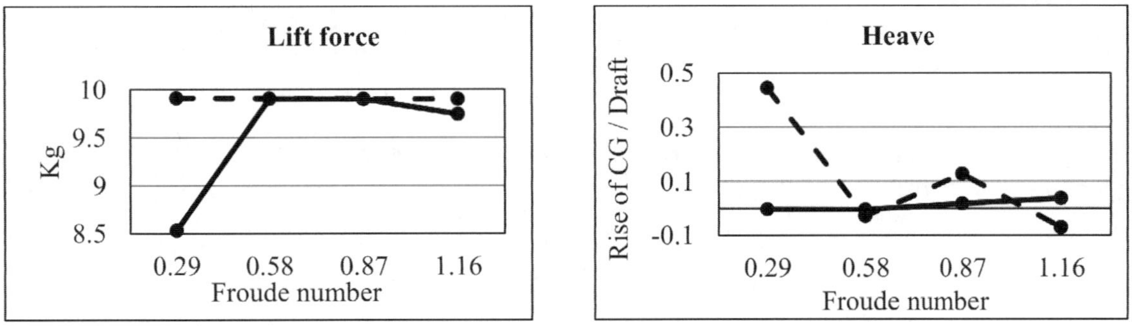

(c) (d)

without interceptor — with interceptor

Figure 9. Result analysis (a) drag, (b) trim, (c) lift force, (d) heave

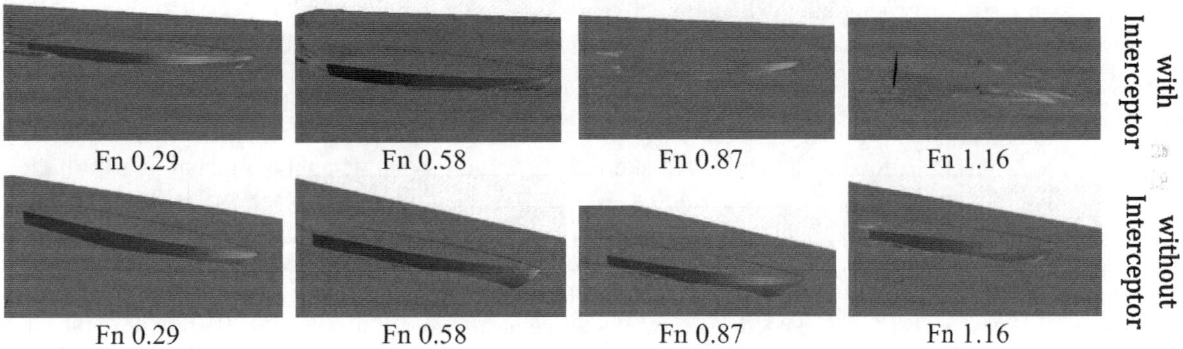

Figure 10. Snapshot interceptor performance

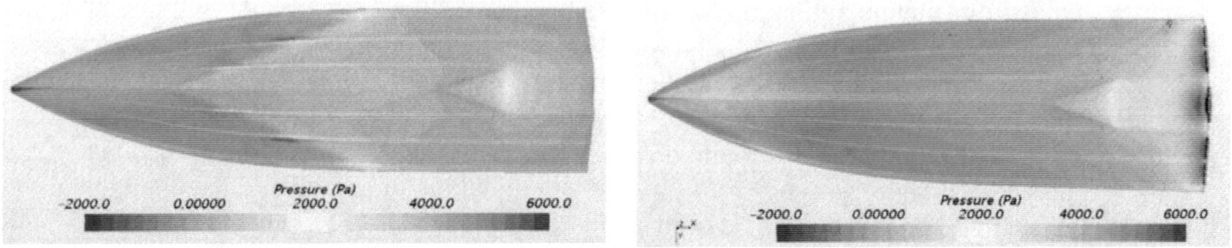

(a) (b)

Figure 11. Pressure distribution at Froude number 1.16 (a) without interceptor, (b) with interceptor

4. Conclusion

A vessel without an interceptor experienced pressure in the bow area resulting in trim. Vessel using interceptors experienced moments at the stern and bow. The interceptor became ineffective if the aft moment was more significant than the bow moment. However, the interceptor was not needed at Froude numbers more than 1.16 because it could create a big enough moment at the vessel's stern to result in negative trim. The interceptor could be used at a certain speed. In this study, the interceptor suggestion applies at a speed of less than Froude number 0.87. Interceptor can decrease 21% drag at Froude number 0.87, 16% trim and 6% heave at Froude number 0.58.

Acknowledgement
This research was financially supported by The Faculty of Engineering, University of Diponegoro, Indonesia, through Strategic Research Grant 2021 No. 195/UN7.5.3.2/HK/2021.

References
[1] A. H. Day and C. Cooper, 2011, An experimental study of interceptors for drag reduction on high-performance sailing yachts, *Ocean Eng,* **38**, pp. 983–994. doi: 10.1016/j.oceaneng.2011.03.006.
[2] P. van Oossanen, J. Heimann, J. Henrichs, and K. Hochkirch, 2019, Motor yacht hull form design for the displacement to semi-displacement speed range, *FAST 2009 10th Int. Conf. Fast Sea Transp,* pp. 629–643.
[3] M. H. Karimi, M. S. Seif, and M. Abbaspoor,2013, An experimental study of interceptor's effectiveness on hydrodynamic performance of high-speed planing crafts, *Polish Marit. Res*, **20**, pp. 21–29. doi: 10.2478/pomr-2013-0013.
[4] J. Y. Park *et al,* 2019. An experimental study on vertical motion control of a high-speed planing vessel using a controllable interceptor in waves, *Ocean Eng,* **173,** pp. 841–850. doi: 10.1016/j.oceaneng.2019.01.019.
[5] A. A. K. Rijkens, H. M. A. Cleijsen, and J. A. Keuning, 2013, On the hydrodynamic performance of an improved motion control device for fast ships, *FAST 12th Int. Conf. Fast Sea Transp.*
[6] M. Mansoori and A. C. Fernandes, 2015, Interceptor and trim tab combination to prevent interceptor's unfit effects, *Ocean Eng,* **134**, pp. 140–156, doi: 10.1016/j.oceaneng.2017.02.024.
[7] R. Yousefi, R. Shafaghat, and M. Shakeri,2013, "Hydrodynamic analysis techniques for high-speed planing hulls," *Appl. Ocean Res,* **42**, pp. 105–113, doi: 10.1016/j.apor.2013.05.004.
[8] S. Samuel, M. Iqbal, and I. K. A. P. Utama, 2015, An investigation into the resistance components of converting a traditional monohull fishing vessel into catamaran form, *Int. J. Technol,* **6**, pp. 432–441, 2015, doi: 10.14716/ijtech.v6i3.940.
[9] S. Samuel, D. J. Kim, M. Iqbal, A. Bahatmaka, and A. R. Prabowo, 2016, Modification of traditional catamaran to reduce total resistance: configuration of centerbulb, *Marit. Transp. Harvest. Sea Resour,* **1**, pp. 541–546.
[10] S. Samuel, A. Trimulyono, P. Manik, and D. Chrismianto, 2021. A numerical study of spray strips analysis on fridsma hull form, *Fluids*, **6**, no. 11, p. 420, 2021, doi: 10.3390/fluids6110420.
[11] A. Firdhaus, I. K. Suastika, K. Kiryanto, and S. Samuel,2021. "Benchmark study of finetm/marine cfd code for the calculation of ship resistance," *Kapal J. Ilmu Pengetah. dan Teknol. Kelaut,* **18**, pp. 111–118. doi: 10.14710/kapal.v18i2.39727.
[12] ITTC, "Practical Guidelines for Ship CFD Applications," *ITTC – Recomm. Proced. Guidel. ITTC*, pp. 1–8, 2011.
[13] D. Savitsky, 1964, Hydrodynamic design of planing hulls, *Mar. Technol. SNAME*, **1**, pp. 71–95.
[14] S. Brizzolara and F. Serra, 2007,Accuracy of CFD codes in the prediction of planing surfaces hydrodyamic characteristics, *2nd Int. Conf. Mar. Res. Transp,* pp. 147–158.

IOP Conf. Series: Earth and Environmental Science 1081 (2022) 012005 doi:10.1088/1755-1315/1081/1/012005

The Control of PMSM Motor to Drive Propeller in Ship Propulsion System

M.A. Jami'in, M.B. Rahmat, P.P.A. Nugroho, M. Santoso, E. Julianto

Politeknik Perkapalan Negeri Surabaya, Surabaya, Indonesia

jammy@ppns.ac.id

Abstract. Electric propulsion has advantages such as high maneuverability, good safety and reliability, being able to operate optimally with a control system. It is a future green technology which is environmentally friendly. The application of electric propulsion for ship is that the propeller is driven by using electric motor by regulating speed and/or torque. in this research, an permanent magnet synchronous motor (PMSM) is used to drive propeller of ship propulsion. The controller of proportional integral (PI) scheme is proposed to drive electric motor. The method of Ziegler Nichols stability margin tuning rule algorithm is performed to search the controller parameters. The optimal working point of propeller is determined by the cross intersection between the propeller loading curve to the coefficient of thrust curve of open water test propeller. The speed from the result of propeller working point analysis is set as control input reference. Based on the results of simulation, the proposed control of electric propulsion using PMSM motor is able to achieve the reference of speed.

1. Introduction

Climate change affected by pollution of vehicle transportation has become a worldwide concern. The International Maritime Organization has issued in order to limit the emissions gas from the combustion of marine vessels [1]. However, the emission of gas in maritime transport become more stringent in the future. Hence, the alternative energy source will have potential application in maritime transport such green energy source like solar, wind energy, and hydrogen energy to drive electric propulsion of ship. Technological engineering of electric propulsion is a green technology which is to reduce environmental pollution to zero carbon emissions [2]. Supported by the advanced technology of power electronics, the sources of electricity has a high potential to be developed and applied for propulsion system using many alternative energy sources [3].

Technological developments in electronics, propulsion systems, electric motors, control systems and information technology have a dominant role in the application of electric propulsion system technology. Several electric motors have been proposed for propulsion drives such as BLDC motors [4], induction motors [5] and synchronous motors with permanent magnets [6]. The PMSM has the advantages such as light weight, small size, , economical fuel consumption, high power density and has a good dynamic response, making it very suitable for the use of ship propulsion systems [7]. In addition, compared to induction motors, PMSM has higher efficiency with smaller rotor losses. However, the challenge of control system to use PMSM motors are non-linear model. It is a complex functional relationship between output torque and stator current and model transformations for control applications.

Several research applications have been proposed regarding the use of electric motors for ship electrical propulsion including sensor-less of PMSM applied in real time to electric propulsion prototypes as propulsors [6], PMSM motor starting techniques for ship propulsion [8], speed tracking controller based on prediction model was proposed by using PMSM for propulsion [9]. Several control methods have been developed to drive PMSM with proportional integral (PI) [10][11], Sliding mode controller on PMSM motor [12], and neural network-based control system [13]. In this research, we develop the design of a speed control system to drive propeller of ship propulsion system. The controller is designed using PI control method. The optimal working point of the propeller is calculated using ANLOVA software. Hence, the working point of the propeller is used as a reference to adjust the speed of the PMSM motor.

2. Methodology

The steps to design the speed control system in ship propulsion consists of 1) The calculation related to determine the optimum operating point of the propeller, 2) Design of PMSM motor speed control.

2.1. Ship resistance and propeller selection

The working point of the propeller is based on the characteristics of the ship resistance using the resistance data from the results of towing tank test of model. The data of model and full scales ship with the length ratio between ship and model is 8 are shown in Table 1. Meanwhile, the resistance of model by towing test and the calculation results of ship resistance in full scale are shown in Table 2.

Table 1. Dimension of ship vs model

PARAMETER	SHIP	MODEL
Loa (m)	11.4	1.425
Lwl (m)	10.73	1.34125
Lpp (m)	10.34	1.2925
B (m)	5.7	0.7125
H (m)	2	0.25
T (m)	1.8	0.225
Vs (knot)	7	
Prism Coef	0.643	0.643
Block coef	0.505	0.505

Table 2. The resistance of model from the results of towing tank test and ship resistance

RUN	Vm (m/s)	Vs (m/s)	Rtm (kg)	Rts (kg)
1	0.177	0.501	0.0634	423.67
2	0.284	0.803	0.0932	650.6
3	0.403	1.14	0.1914	1325.73
4	0.432	1.222	0.3121	2094.88
5	0.585	1.655	0.5013	3387.40
6	0.773	2.186	0.7425	5070.71

The performance of the propeller is calculated using ANLOVA software. The inputs of ANLOVA software to calculate the operating point of propeller are ship resistance, gearbox ratio, propeller properties like diameter, expanded area ratio, number of blades and pitch of propeller. The ship resistance is calculated based on model test results using towing test or is calculated based on numerical method the Holtrop, Van Oortmerssen, Savitsky, Latiharju and others. However, the type of propeller used is the B series propeller in which the propeller performance modelled under mathematic polynomial model. Polynomial model is a good model to express the open water test characteristic of B series propeller model. The Polynomial model expresses the thrust coefficient and the torque coefficient with the input variables can be pitch-diameter ratio (P/D), number of blades (Z), blade area ratio (AE/A0), and advance coefficient (J) [14]. The operating point of propeller is the point of intersection between the propeller loading curve and the thrust coefficient curve of open water test propeller curve.

To calculate the operating point of the propeller, the first step is the calculation of the ship's resistance which can be expressed as

$$Rt = kV^2 \tag{1}$$

V = Ship speed (m/s)

k = coefficient of drag
Rt = total resistance (kg)

The performance characteristics of propeller can be expressed by constants that describe the performance of the propeller which can be represented by an open water test propeller graph written by

$$Kt = \frac{T}{\rho.n^2 D^4} \tag{2}$$

$$Kq = \frac{Q}{\rho.n^2 D^5} \tag{3}$$

$$J = \frac{V_a}{nD} \tag{4}$$

$$\eta = \frac{Kt.J}{2.\pi.Kq} \tag{5}$$

$$R = T(1-t) \tag{6}$$

$$V_a = V(1-w) \tag{7}$$

Where the notations are expressed as Kt = Thrust coefficient, Kq = torque coefficient, J = Speed of advance, η = propeller efficiency. The equation of propeller loading curve is the subtitution between the equation (1) until equation (7) that can be written by,

$$Kt = \frac{c}{\rho.(1-t).(1-w)^2.D^2}\left[\frac{Va}{n.D}\right]^2 \tag{8}$$

The characteristics of the open water test propeller is a propeller performance curve presented by propeller efficiency, torque constant and thrust constant as a function of speed of advance. The propeller loading curve presented in equation (8) is plotted against the open water test propeller characteristic graph. The point of intersection between the thrust constant of the propeller loading curve and the thrust constant of the open water test propeller characteristics is the propeller working point. Based on the working point of the propeller, the performance of propeller will be known such as efficiency, torque, thrust, delivered horse power (DHP) and propeller cavitation symptoms. The requirements of cavitation are also considered to select the propeller. It is calculated using Burril method that can be expressed as follow,

$$\frac{P}{D_{\max}} \leq \frac{1.067 - \left(\frac{A_p}{A_d}\right)}{0.229} \tag{9}$$

2.2. Modelling and control system of PMSM motor

PMSM has a constant magnetic flux so that the control system is more efficient designed using vector control techniques on the magnetic rotor. The PMSM motor dynamic system modeling with d-q coordinates is as follow [15],

$$\frac{di_d}{dt} = \frac{1}{L_d}u_d - \frac{R}{L_d}i_d + \frac{L_q}{L_d}p\omega_r i_q \tag{10}$$

$$\frac{di_q}{dt} = \frac{1}{L_q}u_q - \frac{R}{L_q}i_q + \frac{L_d}{L_q}p\omega_r i_d - \frac{mp\omega_r}{L_q} \tag{11}$$

$$T_e = 1.5[\lambda i_q + (L_d - L_q)i_d i_q] \tag{12}$$

$$\frac{d\omega_r}{dt} = \frac{1}{J_L}(T_e - F\omega_r - T_m) \tag{13}$$

$$\frac{d\theta}{dt} = \omega_r \tag{14}$$

Where, L_d, L_q, u_d and u_q denote the inductance and stator voltage of d-q axes respectively. The notation of R, J_L, T_e, T_m, F, p, m, ω_r denote the resistance of stator, inertia of motor, electromagnetic torque, torque of load, number of poles, the amplitude of flux, and the rotor speed respectively.

To detect the angular position of the rotor shaft, the synthesis of a three-phase stator current vector controller on the q-axis is carried out using the d-q transformation. This technique is simpler than vector control applied to asynchronous motors. The synthesis of the current vector controlling inverter with the d-q transformation technique is expressed as

$$i_a = i_s \cos\left(\theta + \frac{\pi}{2}\right) = -i_s \sin(\theta) \tag{15}$$

$$i_b = i_s \cos\left(\theta + \frac{\pi}{2} - \frac{2\pi}{3}\right) = -i_s \sin(\theta - \frac{2\pi}{3}) \tag{16}$$

$$i_c = i_s \cos\left(\theta + \frac{\pi}{2} + \frac{2\pi}{3}\right) = -i_s \sin(\theta + \frac{2\pi}{3}) \tag{17}$$

Where, i_a , i_b and i_c are the currents of three-phase coil. The stator currents in the d-q vector has a phase angle difference of 90^0. There are two control schemes using vector control method, namely 1) if i_{sd} =0, then the speed setting for controller is set as constant torque, and 2) if i_{sd}<0, then speed setting for controller is set as constant power. Therefore, for speed regulation with constant torque, it can be indicated by the block diagram in Figure 1.

Vector control applied to PMSM motor allows for better control performance compared to that is applied to AC motor in order to reduce torque pulses for the motor. It is due to the lack of slip frequency current of the induction motor and being slightly affected by the rotor parameters. Hence, vector control is more feasible for PMSM. The control structure of PMSM is presented on Figure 2.

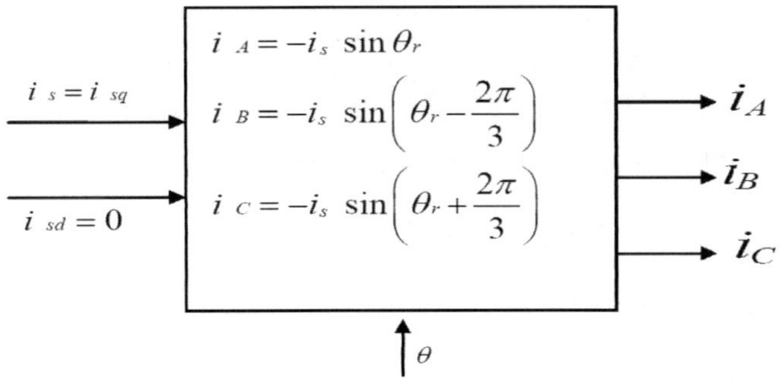

Figure 1. The rotor flux orientation of PMSM with $i_{sd} = 0$

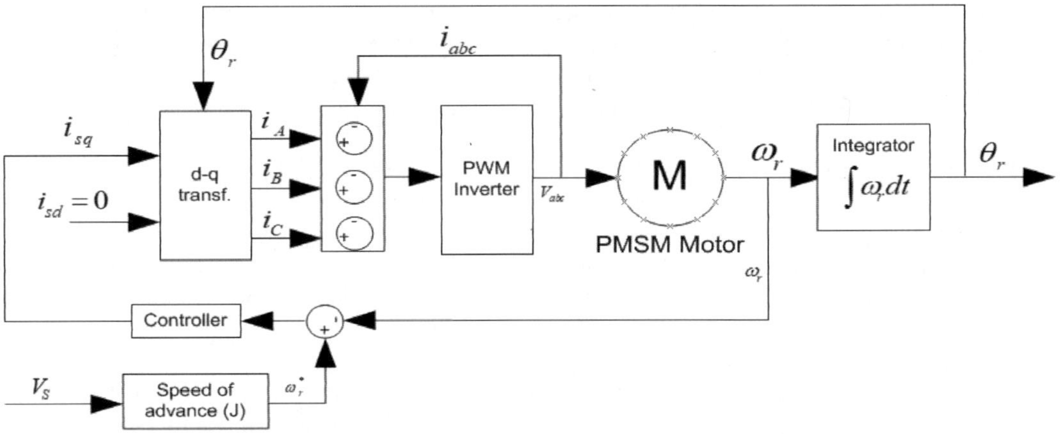

Figure 2. Block diagram of vector control system of PMSM motor

The vector control is derived using the direction coordinates of force field. This component consists of an excitation current to produce the magnetic flux and torque to produce current. The magnetic flux position of PMSM is determined by checking the actual position of the rotor. Hence, the important point of PMSM vector control is the control of phase and spatial amplitude of the vector rotor electricity.

In order to design a speed controller of PMSM motor, a PI controller scheme is proposed. The continuous PI controller can be stated as,

$$u(t) = K_p e(t) + K_i \int e(t)dt \tag{18}$$

where the controller parameters of PI consist of Kp is the proportional gain and Ki is the integral gain. The auto tuning method is used to regulate the parameter based on Ziegler Nichols stability margin tuning rules. The selection of controller parameter is based on gain and phase margin of Ziegler – Nichols Stability Margin Tuning Rule that the parameters are shown in Table 3. In order to search gain margin and frequency margin, the gain of closed system is regulated until the response of system has full oscillation then the gain is set as gain margin and the frequency of oscillation is set as frequency margin.

Table 3. The tuning rule parameters of PI controllers

Parameter	Kc	Ti
P	0.5 Kcu	.-
PI	0.45Kcu	Pu/1.2
PID	0.6Kcu	Pu/2

The parameter of Kcu is the ultimate gain margin and $Pu = \frac{2\pi}{\omega_c}$, ω_c denotes frequency margin. The transfer function form of PI controllers can be written as,

$$g_c(s) = Kc\left(1 + \frac{1}{Ti.s}\right) \tag{19}$$

3. Results and discussion

The Analysis and optimization of the ship propulsion is related to search the propeller working point with optimum efficiency. The tool used to optimize the optimum working point of propeller is ANLOVA. The input of software is ship's resistance, the propeller size and specification, and gearbox ratio. The output of software is propeller working point, the estimated propulsion power, and cavitation. Furthermore, the optimum propeller working point is set as the reference to design the controller of electric propulsion using PMSM motor. The steps to design the controller of ship electric propulsion is as follows,

1. PMSM motor dynamic modelling: In this step is the identification of PMSM motor dynamic modelling. The components of the PMSM motor modelling system are the inverse d-q coordinate transformation. These models are formed in a block diagram of the overall model for speed and/or torque regulation.
2. Design of control law: Control law is derived based on the model obtained from identifying the system with numeric simulation. The controller parameters will be searched to obtain the best transient and steady state response.
3. The last step is to simulate the controller design parameters. Controller performance is measured by several control system performance such as maximum overshoot and undershoot, time to steady state, steady state error, rooted mean square error.

3.1. Propeller working point analysis

The data used in order to estimate the propeller working point is the existing propeller of traditional shipyard that is as follows: type of B Series Propeller, number of propellers = two, number of blades = 3 blades, expanded area ratio = 0.5, pitch per diameter = 0.8, diameter = 0.6 m. The input of ship resistance is that shown in Table 1. and Table 2. Using the ANLOVA software, the results of propeller working point are shown in Figure 3. The results obtained by plotting the propeller loading curve against the propeller open water test in Fig 3. is as follows: propeller efficiency = 0.389, propeller rpm = 2800/3337 = 840 rpm, power = 40 kW, speed of advance (J)= 0.31, thrust coefficient (Kt) = 0.228, coefficient of torque (Kq) = 0.029. The ship's propulsion characteristics can be stated by the function of power to rpm written as

$$DHP = 6.96x10^{-8} N_p^3 \tag{20}$$

Where DHP denotes delivered power in kW and N_p denotes the propeller speed in rpm.

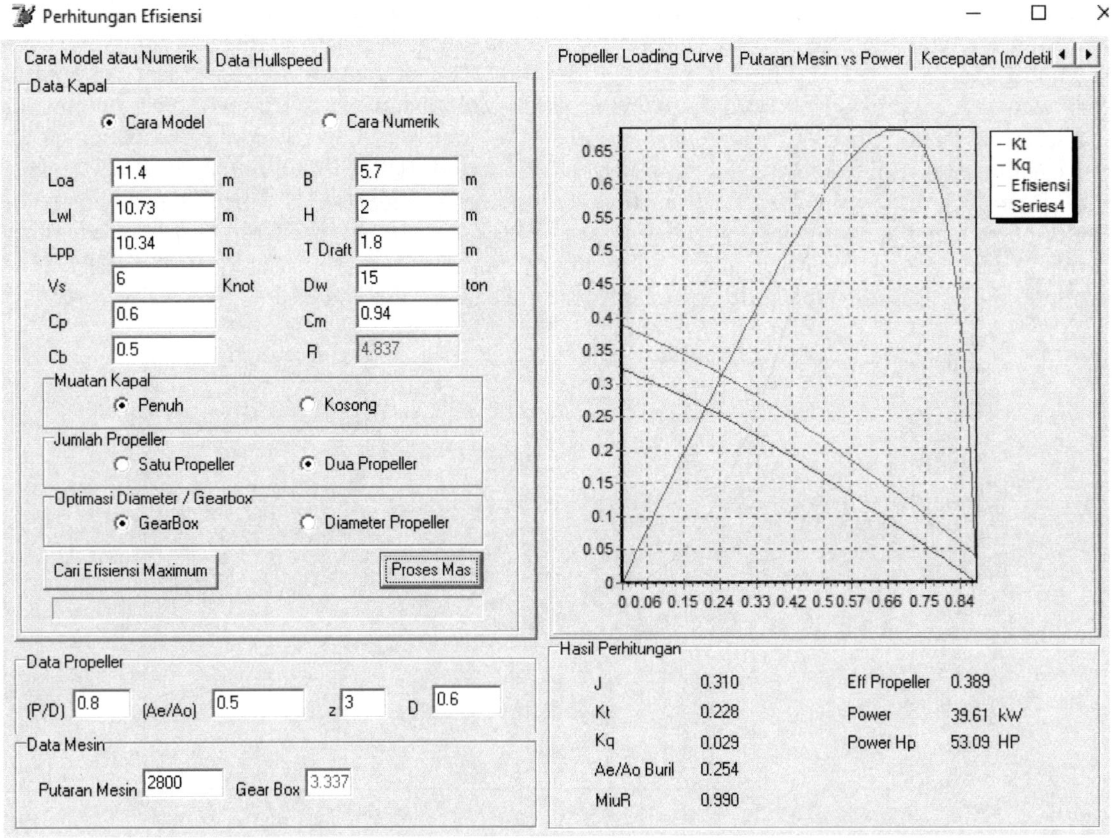

Figure 3. Tool of ANLOVA software used to estimate working point of propeller

3.2. Control system simulation

The PMSM vector control is modelled by using SIMULINK-MATLAB under PI control strategy. The transformation using Clarke and Park coordinate is performed in vector control. Vector control is derived according to the transformations between three-phase stator coordinates (a,b,c) to stationary two-phase coordinates (α,β) and two-phase rotational coordinates (d, q). The block diagram simulation of PMSM control under simulink-matlab is shown in Figure 4.

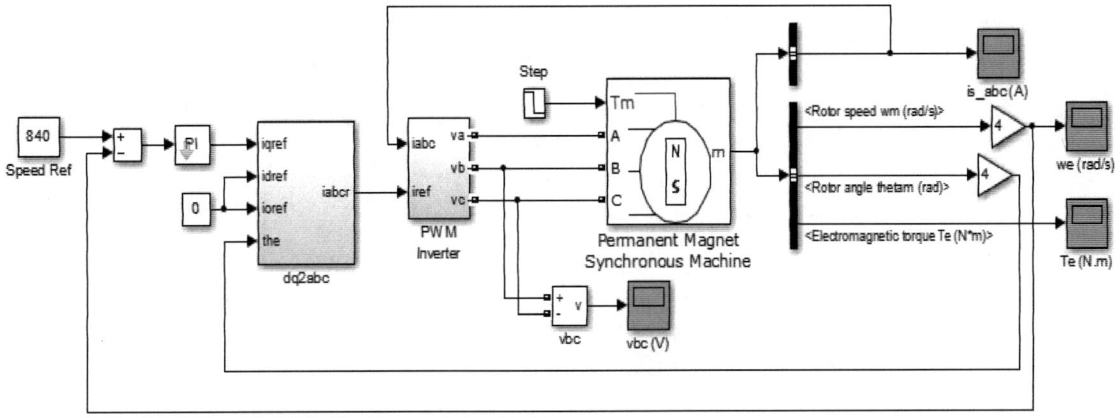

Figure 4. The simulation of PMSM control by matlab simulink

The motor parameters are selected by voltage rating of 380 V, winding resistance of stator Rs=2.875, the inductance of winding phase of d-q denoted by Ld equal to Lq = 8.5 10^{-3} H, magnetic flux of rotor f = 0.175 Wb, inertia J= 0.8 10^{-3} kg·m2, pole number P=4. The simulation is used to test the control system with a low rotor speed according to the ship's propulsion characteristics. The running of simulation is under conditions with a fixed ship speed of 6 knots related to 840 rpm of propeller speed. The simulation results obtained are shown in Figure 5, Figure 6, Figure 7 and Figure 8. The reference speed can be achieved by 0.26 second from first start shown in Figure 5. By changing the torque of propeller shown in Figure 6, the controller is also able to keep the speed reference by changing electromagnetic torque shown in Figure 7. The control performance is measured by the RMS error shown in Figure 8. Based on the RMS error curve, it can be stated that the error converges to zero.

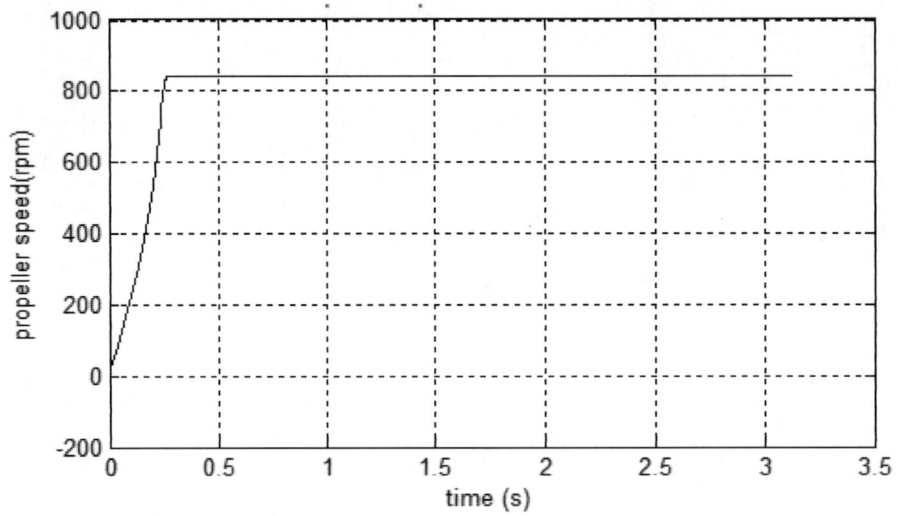

Figure 5. Response of control system of speed controller

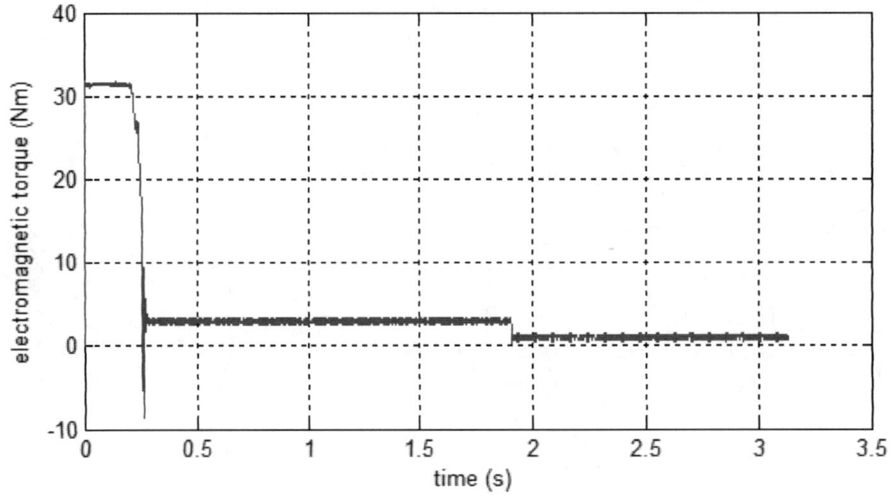

Figure 6. Response of electromagnetic torque

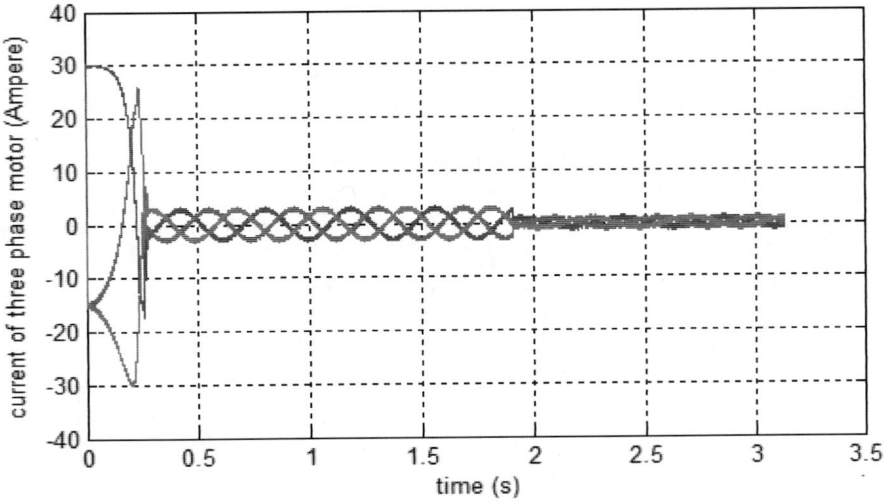

Figure 7. Response of three phase currents Iabc

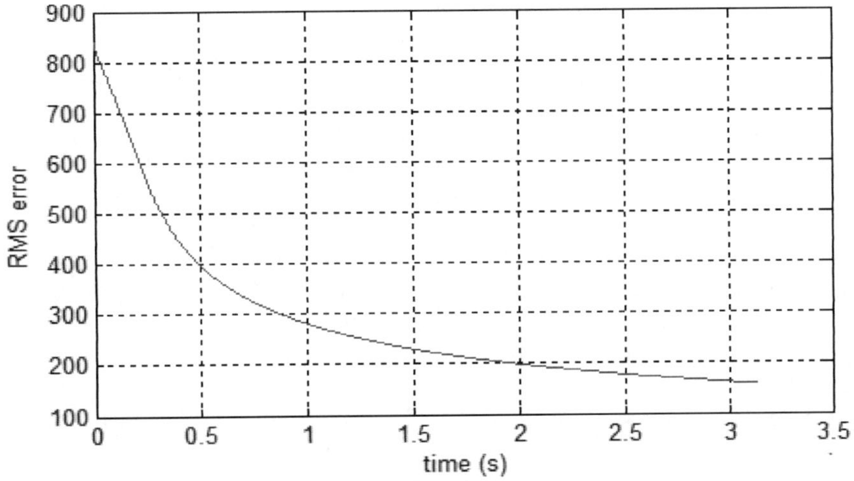

Figure 8. RMS error of speed controllers

4. Conclusions

In this research, a PMSM motor speed control system is studied for driving an electric propulsion system. Ship data is used to design the control system of electric propulsion is as follows: the overall length of the ship 11.4 meters, width 5.7 m, draft 1.8m, with a full load speed of 6 knots. The propeller data used is Type B Series with two propellers, 3 blades, expanded area ratio = 0.5, pitch per diameter = 0.8 and diameter = 0.6m. Based on the results using ANLOVA software the propeller should be operated by 840 rpm. Hence the control system is set by 840 rpm of propeller speed. From the simulation results shows that the controller is able to follow the speed reference with RMS error 158.85. The RMS error of control system converges to zero that the controller has stable performance.

IOP Conf. Series: Earth and Environmental Science 1081 (2022) 012005 doi:10.1088/1755-1315/1081/1/012005

5. Reference

[1] Nuchturee, C., Li, T., Xia, H., "Energy efficiency of integrated electric propulsion for ships–A review," *Renew. Sustain. Energy Rev.*, vol. 134, p. 110145, 2020.

[2] Nguyen, H. P. *et al.*, "The electric propulsion system as a green solution for management strategy of CO2 emission in ocean shipping: A comprehensive review," *Int. Trans. Electr. energy Syst.*, vol. 31, no. 11, p. e12580, 2021.

[3] Jain, S., Kumar, L., "Fundamentals of power electronics controlled electric propulsion," in *Power Electronics Handbook*, Elsevier, 2018, pp. 1023–1065.

[4] Vinida, K., Chacko, M., "An optimized H infinity strategy for robust control of sensorless BLDC propulsion motor in submarines for improved maneuverability," in *2016 IEEE International Conference on Power Electronics, Drives and Energy Systems (PEDES)*, 2016, pp. 1–6.

[5] Kumar, A., Fernandes, B. G., Chatterjee, K., "Direct torque control of open-end winding induction motor drive using the concept of imaginary switching times for marine propulsion systems," in *31st Annual Conference of IEEE Industrial Electronics Society, 2005. IECON 2005.*, 2005, pp. 6-pp.

[6] Batzel, T. D., Lee, K. Y., "Electric propulsion with sensorless permanent magnet synchronous motor: implementation and performance," *IEEE Trans. energy Convers.*, vol. 20, no. 3, pp. 575–583, 2005.

[7] Liu, X., Chen, H., Zhao, J., Belahcen, A., "Research on the performances and parameters of interior PMSM used for electric vehicles," *IEEE Trans. Ind. Electron.*, vol. 63, no. 6, pp. 3533–3545, 2016.

[8] Ji, Q., Liu, G., "A starting method of ship electric propulsion permanent magnet synchronous motor," *Procedia Eng.*, vol. 15, pp. 655–659, 2011.

[9] Li, H., Xie, H., Xie, Y., "Research on PMSM model predictive control for ship electric propulsion," in *2018 Chinese Control And Decision Conference (CCDC)*, 2018, pp. 1392–1398.

[10] Lee, S.-B., "Closed-loop estimation of permanent magnet synchronous motor parameters by PI controller gain tuning," *IEEE Trans. Energy Convers.*, vol. 21, no. 4, pp. 863–870, 2006.

[11] Zheng, S., Tang, X., Song, B., Lu, S., Ye, B., "Stable adaptive PI control for permanent magnet synchronous motor drive based on improved JITL technique," *Isa Trans.*, vol. 52, no. 4, pp. 539–549, 2013.

[12] Mohd Zaihidee, F., Mekhilef, S., Mubin, M., "Robust speed control of PMSM using sliding mode control (SMC)—A review," *Energies*, vol. 12, no. 9, p. 1669, 2019.

[13] El-Sousy, F. F. M., "Adaptive hybrid control system using a recurrent RBFN-based self-evolving fuzzy-neural-network for PMSM servo drives," *Appl. Soft Comput.*, vol. 21, pp. 509–532, 2014.

[14] Ariana, I. M., Cahyono, B., "Optimization of Propeller Design Through Polynomial Approach to Optimize The Ship Energy Efficiency," in *IOP Conference Series: Earth and Environmental Science*, 2020, vol. 557, no. 1, p. 12051.

[15] Kyslan, K., Šlapák, V., Petro, V., Marcinek, A., Ďurovský, F., "Speed Sensorless Control of PMSM with Unscented Kalman Filter and Initial Rotor Alignment," in *2019 International Conference on Electrical Drives & Power Electronics (EDPE)*, 2019, pp. 373–378.

Experimental study of a ship with the self-righting moment in extreme condition

A Trimulyono[1]*, A F Zakki[1], M A Fuadi[1]

[1]Department of Naval Architecture, Faculty of Engineering, Universitas Diponegoro, Semarang, Indonesia.

*Correspondence: anditrimulyono@lecturer.undip.ac.id

Abstract. Patrol vessel is an essential part of offshore security archipelago countries such as Indonesia. The main feature of patrol vessel is operating in rough waves during operation. Ship stability is one of the challenging aspects of patrol vessel because, in extreme conditions, the ship can capsize due to losing a self-righting moment. The present paper carried out a study of a patrol vessel with a self-righting moment in extreme conditions. The condition is a ship with rolling up to above 180^0. Thus, the ship can capsize because of losses of the self-righting moment in rough condition. An experimental study is carried out to model a ship with a rolling angle above 180^0. The principal dimension of ships Lpp, B, H, T, are 13.0 m, 4.2 m, 2.19 m and 1.15 m, respectively. The study was carried out with a physical model of ship 1:27.4. The model is made using 3D printing to maintain the hull integrity. In the present paper, only the full load condition was tested in the experiment condition. It was showed the ship design was proven to have a self-righting moment in the rolling angle above 180^0.

1. Introduction

Patrol vessel is an essential part of offshore security archipelago countries such as Indonesia. The main feature of patrol vessels is operating in rough waves during operation. Ship stability is one of the challenging aspects of high-speed craft because the ship has a small breadth that leads to small self-righting moments in roll motion. This situation could endanger the vessel when a vessel is rolling above 180^0; as a result, a ship can capsize due to losing a self-righting moment. Priohutomo has studied patrol vessels with control manoeuvering ships using an experimental approach in wave basins [1]. The study of the number of bilge keel in patrol vessels was done by Widyatmoko [2]. The study of ship construction and strength of patrol vessels has been conducted by Koostanto [3], moreover CFD simulation was performed by Samuel [4]. The effect of spray strip on on high-speed craft has been conducted by Samuel [5]. Self-righting moment of the patrol vessel was carried out by Putra using a numerical approach. The results showed self-righting moments could be achieved by extending the breadth of the upper deck structure [6]. Previously, the self-righting moment study was done using a numerical method using Hydromax [7]. It shows that the study of a physical model of patrol vessel with a self-righting moment is rolling above 180^0 still few. Thatcher has divided some remarks of self-righting methods into three methods, i.e., inherent stability self-righting, inflatable bag, and ballast movement [8]. Capsize is one cause of marine accidents in the world, as shown in Fig. 1 [9], although it is a very small portion based on Fig.1. Moreover, a patrol vessel is commonly used as a search and rescue (SAR) vessel; as a result, this vessel has to operate in a rough wave. Thus, the vessel could undergo rolling in extreme conditions (above 180^0).

This study aims to conduct self-righting moments in patrol vessels using an experimental approach, the ship's condition rolling above 180^0. Firstly, ship modeled in CAD model and calculation of the center of gravity (CG), Buoyancy (B), and metacenter (M) conducted in full scale. The next step is ship was scaled 1:27,4 and recalculated of CG, B, and M. Finally, an experiment was carried out in towing tank to verify self-righting moment is rolling 180^0. It was found the self-righting moment was proved to exist during rolling above 180^0.

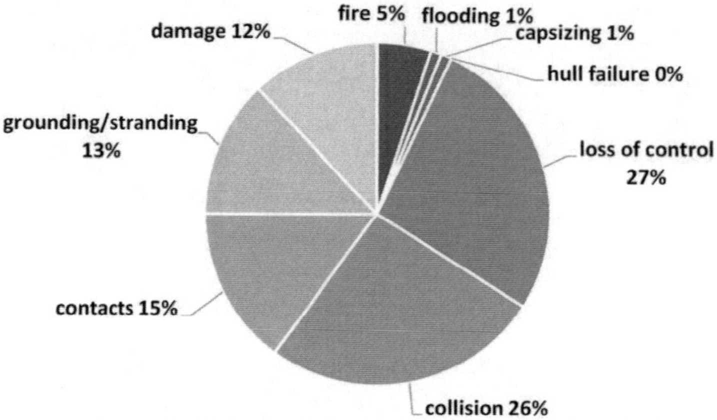

Figure 1. Marine accident of ship during 2011-2019 [10]

2. Methods

The experimental study was started scaled the full-scale model into a scale model (see Fig. 2), as shown in Table 1. Table 1 shows the dimension of the patrol vessel that will be modeled in the laboratory. The next step is to model patrol vessels using 1:27,4. In this study, the condition of the ship is the maximum load, which is a heavier weight, higher center of gravity, and a higher draft compared to other shipload conditions. LCG is a longitudinal center of gravity, and KG is the distance of keel to gravity.

Table 1. Principal dimension of patrol vessel

Principal Particular	Dimension	Scaled model
LOA	13,7 m	50,0
LPP	13,0 m	47,4
B	4,20 m	15,0
D	1,15 m	4,20
H	2,19 m	8,00
Displacement	26,023 ton	1268,606 gr
LCG	5,318 m	19,40 cm
KG	1,21 m	4,42 cm

Fig. 3 (a) shows a scale prototype model of the actual ship made using 3D printing. Fig. 3 (b) is the solid ballast for the prototype model to make the ship is in a maximum load condition and made the center of gravity according to the actual ship. Fig. 3 (b) shows the model of solid ballast placed at 14.55 cm from the AP. The position has been designed so that the prototype model has the same center of gravity as the condition of the ship with maximum load.

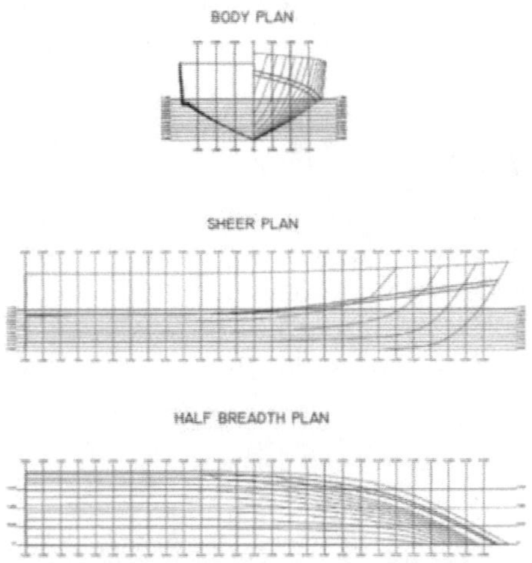

Figure 2. Lines plan of the patrol boat

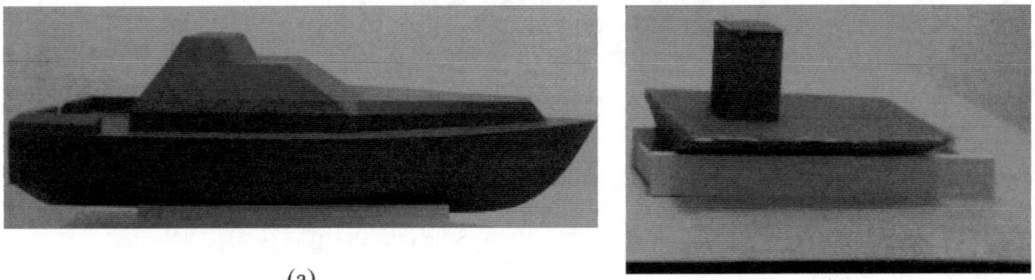

(a) (b)

Figure 3. The modeled patrol vessel with a scale 1: 27,4.

Fig. 4 shows a measurement of the weight of each part on the prototype model ship. Fig. 4 (a) is calibrating the scales. The purpose is to avoid errors in calculating the load. The calibration process uses a 500 gr load with a correction value of 0% showed in scale. Fig. 4 (b) shows the process of setting 0 gr when calculating the load using a seat booth. It shows the load of the seat booth is counted 0 gr. Fig. 4 (c) is the process of calculating the weight of the superstructure with the result is 159 gr. Fig. 4 (d) is the process of calculating the weight of the hull with the result is 377 gr. Fig. 4 (e) calculates the additional weight with the result 726 gr.

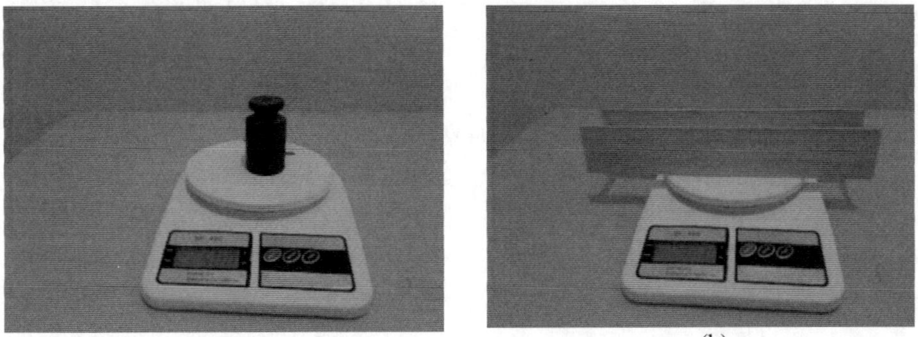

(a) (b)

(c)

(d)

(e)

Figure 4. Measurement of the physical model of the patrol boat.

The center of gravity is calculated using the moment calculation formula. Equation (1) is the equation for the moment of force measurement of the physical model of a patrol vessel.

$$CG = F \; x \; l \tag{1}$$

Where
F = Force (N)
l = arm distance (m)
F = m.a
Where: m = mass (kg); a = acceleration (m/s^2)

3. Results and Discussion

Table 2 shows the longitudinal center of gravity and transverse center of gravity calculation. Data of Weight is to obtain the center of gravity and the total weight of the prototype model. This calculation uses the formula moment of force.

Table 2. Calculation of the center of gravity and the total weight of the model

Item	m (gr)	F (N)	LCG (cm)	Moment LCG (N.cm)	KG (cm)	Moment KG (N.cm)
Hull	377.00	3694.60	19.39	71627.21	4.92	18184.82
Upper Deck	159.00	1558.20	22.13	34484.52	11.57	18022.14
Load	726	7114.8	18,77	133523.4516	2,63	18726.1536
Total	1262.00	5252.80	19.38	239635.186	4,44	54933.116

The results obtained from the calculation of the center of gravity and the total load of the prototype model are compared with the actual ship with maximum load conditions. Table 3 is a comparison of the calculation results with the actual ship in maximum condition with the aim of seeing the error value that occurs after the printing process and calculating the weight with a scale.

Table 3. Comparison calculation of center gravity

Item	Full scale ship	Scaled ship	Difference (%)
LCG	19,4	19,38	0,165%
KG	4,42	4,44	0,581%
Mass	1268,606	1262	0,363%

The result shows the comparison calculation value in Table 3. It shows an error value below 5% so that the prototype model can be used in the experiment. In this research, the test was set up in the towing tank, where the prototype of the model was placed on calm waves. In the first step, the prototype model is rolled over to an angle of 180^0 from the prototype model's initial position (0^0). When the prototype model has reached 180^0 position, the prototype model will be released freely. It can be seen that the prototype model can return to its original position, i.e., the initial position / 0^0 position. Fig. 5 shows the process of self-righting recovery experiment.

From Fig. 5, the ship can return to its original position without stopping between 0^0-180^0 after being overthrown 180^0. From these results, it can be concluded that the patrol boat has self-righting capabilities during rolling above 180^0.

Figure 5. Self-righting recovery experiment

4. Conclusion

A verification test was successfully conducted to prove the self-righting capability of the patrol vessel. This verification was carried out using the experimental approach of making a prototype 3D printing model with a scale of 1:27,4. The principal dimension of ships Lpp, B, H, T, are 13.0 m, 4.2 m, 2.19 m and 1.15 m, respectively. This prototype ship was conducted in the maximum load of vessel that had been designed with a correction below 5%. Then the prototype model was tested for rolling 180^0. It was found that the prototype ship model was able to return to its original position. It was proven the patrol boats have self-righting capabilities based on experiment study. Future works of the numerical method of ship stability will be carried out to verify the present result.

5. Acknowledgments

The authors wish to thank the ship hydrodynamics laboratory, Department of Naval Architecture for the experimental facility.

References

[1] K. Priohutomo, A.A. Masroeri and C. Permana 2017 Perancangan Kendali Manuver Untuk Menghindari Tabrakan Pada Kapal Patroli Cepat Berbasis Pengujian Model *Kapal: Jurnal Ilmu Pengetahuan dan Teknologi Kelautan*, **14**(3) p 71-78

[2] A. Widyatmoko, S. Samuel, P. Manik and A. Trimulyono 2021 Analisis Pengaruh Jumlah Bilge Keel terhadap Gerakan Rolling pada Kapal Patroli 14 m *Warta Penelitian Perhubungan,* **33**(1) p 1-10

[3] H. K. Koostanto, A. F. Zakki and S. Samuel 2021 Studi Perencanaan Konstruksi Dan Analisa Kekuatan Kapal Patroli 165 DWT Pada Perairan Indonesia *Jurnal Teknik Perkapalan*, **9** (1) p 91-104.

[4] S. Samuel, A. Trimulyono, A.W.B. Santosa 2019 Simulasi CFD pada Kapal Planing Hull *Kapal: Jurnal Ilmu Pengetahuan dan Teknologi Kelautan*, **16**(3) p 123-128

[5] S. Samuel, A. Trimulyono, P. Manik and D. Chrismianto 2021 A Numerical Study of Spray Strips Analysis on Fridsma Hull Form *Fluids*, **6**(11).

[6] A. B. C. Putra and H. Hasanudin 2019 Desain *Self-Righting Rescue Boat* Untuk Basarnas *Jurnal Teknik ITS,* **8** (2) p 99-105.

[7] A. Zuhdi, S. Sunaryo and M. Baqi 2014 Analisis Kemampuan *Self Righting* Kapal Model Tanpa Awak Dengan Pendekatan Stabilitas Menggunakan Hydromax Pro. Final Project *Naval Architecture Department University of Indonesia*

[8] K. C. Thatcher 2013 Self-Righting Craft Basic Principles And Design Requirements *The International Journal of Small Craft Technology*, **155** (B1): p B1-B11.

[9] D. Vassalos and D. Paterson 2021 Towards unsinkable ships *Ocean Engineering* **232**.

[10] Clarkson 2019 Shipping Intelligence Network. *UK Clarkson Research Service Limited.*

Risk Assessment of Delay for CEMS and WHRU Installation Project on Plant Shutdown at Central Processing Gas Gundih

Ivan Ali Sanjaya[1], Ketut Buda Artana[2], AAB Dinariyana[2]

[1] Department of Technology Management, Institut Teknologi Sepuluh Nopember
[2] Department of Marine Engineering, Institut Teknologi Sepuluh Nopember

Abstract. This paper addresses a risk assessment of delay for installation of Continuous Emission Monitoring System and Waste Heat Recovery Unit as an effort to fulfil compliance with the Regulation of the State Minister of the Environment Number 13 of 2009 concerning Quality Standards for Emissions from Immovable Sources for Oil and Gas Businesses and Activities. The installation of the unit consists of three steps, pre-shutdown, window shutdown and post shutdown. In accordance with Gas Sale Agreement, that the entire period of normal maintenance work that causes gas to not flow, should not be more than 15 (fifteen) days every year. The planning and implementation of maintenance is very important by utilizing the condition of the window shutdown to avoid material and non-material losses. Reviewing the vulnerability of this project, it is necessary to conduct research that assesses the level of risk of project delays using the ISO 31000:2018 risk management concept approach, through identification of risk variables from several literature studies, risk analysis through Group Discussion Forums, and risk management efforts to reduce risk of delays in the installation project. From the risk identification, 30 risk variables that affect project delays consist of 2 High categories, 3 Medium to High categories, 15 Medium categories, 6 Low to Medium categories and 4 Low categories. The Risk Treatment carried out in the high and Medium to high categories resulted in 9 mandatory RCOs to reduce delays and 1 RCO to speed up the implementation schedule from 19 days to 13 days and earn a profit of $615,000 because the process of selling gas to consumers can be accelerated by 2 days.

Keywords: Risk Assessment, Project Delay, Continuous Emission Monitoring System, Waste Heat Recovery Unit, ISO 31000:2018

1. Introduction

Continuous Emission Monitoring System (CEMS) and Waste Heat Recovery Unit (WHRU) are the part of Thermal Oxidizer (TOX) that planned to be replaced at window shutdown. Figure 1 shows the concept of maintenance replacing the WHRU unit which will be equipped with the installation of CEMS as an effort to fulfil compliance with the Regulation of the State Minister of the Environment Number 13 of 2009 concerning Quality Standards for Emissions from Immovable Sources for Oil and Gas Businesses and Activities. The replacement project which consists of three time-sharing steps, pre-shutdown, window shutdown and post shutdown. Referring to the Gas Sale Agreement between PT PEP and PT SPP, the entire period of normal maintenance work which causes gas to not flow is not more than 15 days per year. Referring to the agreement, each hour of delay that will be borne by the company is $307,500 per day or $12,812 per hour.

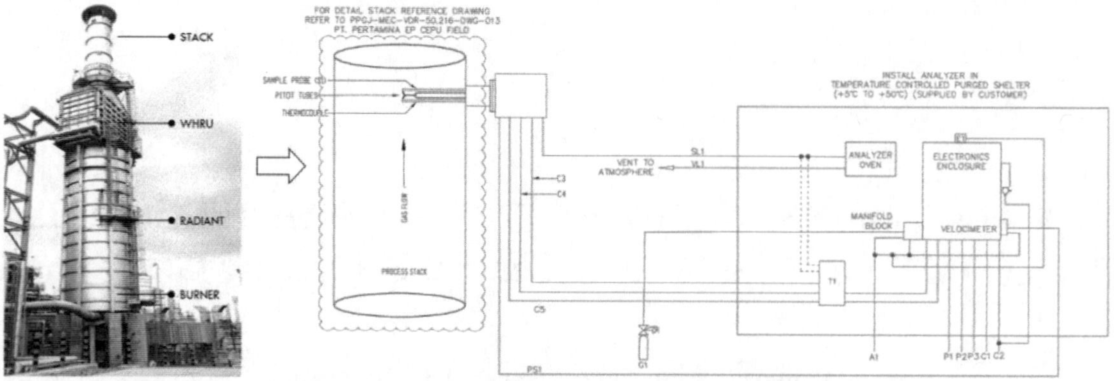

Figure 1. Replacement Concept CEMS & WHRU

In oil and gas projects, only 5 percent of them are completed according to initial budget and schedule. For completed projects, the average cost overrun is 37 percent and the average schedule overrun is 53 percent [1]. Based on the project performance, it is very urgent and important to organize a project delay risk assessment where the expected result is to be able to find out early the risks of project delays as a basis for optimizing and measuring the opportunities for losses from the installation project on shutdown period.

For implementation of a construction project, there are three important things that must be considered including time, cost and quality [2]. Technically the level of success of a project is judged by the extent to which these three limitations can be met. In addition, the delay resulted in lost opportunity for another project work. For the owner, the delay in the completion of the project work will caused losses to the operating time of the results of the project, so that the use of the results of the development project was delayed. Completing projects on time is an indicator of efficiency, but the construction process is subject to many variables and unpredictable factors, which result from many sources. These sources include the performance of parties, resources availability, environmental conditions, involvement of other parties, and contractual relations. However, it is rarely happened that a project is completed within the specified time [3]. The main objectives of this study include how to obtain a project delay risk map through a risk assessment of the factors causing delays, how to provide proposals for handling high risks that can cause delays in the project and estimating the cost of risk management.

2. Literature Review

Many articles and studies conducted on causes of delay in construction projects, in construction projects in Saudi Arabia, identified 73 factors causing delays which were grouped into nine categories, project-related factors, project owners, contractors, consultants, project design, materials, equipment, human resources and related external factors. The results of survey data conducted on 23 contractors, 19 consultants and 15 project owners, 76% contractors and 56% consultants show that the average completion time of a project is around 10% to 30% of its initial duration. The most common cause of delays is a change in demand during construction [3]. In 31 high-rise projects in Indonesia found that cost overruns occur more frequently and are more severe problem than time overruns. They pointed out that the major factors influencing cost overrun are material cost increase due to inflation, inaccurate material estimation and degree of complexity. While in time overrun, the most important factors causing delays are design changes, poor labour productivity, inadequate planning, and resource shortages [4]. The causes of delays in the construction industry are divided into several groups, including: project owners, contractors, consultants, subcontractors, contacts, project materials/materials, work equipment, labour and external factors [5]. In construction projects Jordan the causes of delays are grouped into Internal Factors (labour, materials and equipment), Internal Environmental Factors (contractors, owners, and consultants) and Exogenous Factors (weather and government regulations) [6]. Lack of engagement/commitment to contract, inefficient construction management, poor site coordination, inadequate planning, lack of clarity of project scope, lack of communication; and substandard contracts

IOP Conf. Series: Earth and Environmental Science 1081 (2022) 012007 doi:10.1088/1755-1315/1081/1/012007

are the most influential factors for construction delays in India [7]. In the study conducted there is a statistical analysis process of delay factors with several construction project stakeholders. 10 most important causes of delays: Decision making, improper site management and supervision, complexity in financing the project by contractor, shortage of construction materials, Shortage of equipment and/or equipment failure, improper communication and coordination with other parties, financing and payments by owner Inexperience of contractor (Poor qualification of contractors' staff), Lack of experience of consultants Shortage of utilities on site such as (water, electricity, etc.) [8]

3. Methodology

The research methodology refers to ISO 31000:2018 and PMI:2017 [9][10], it is contained 67 causes of delay were identified through literature review and discussion with 16 technical experts in PEP construction project. Due to the Covid-19 Pandemic, the implementation of Risk Identification held through an online Forum Group Discussion (FGD) activity. Data collected and processed to seek or obtain how much the inhibiting factors can affect the progress and delays in the completion of construction projects.

Time performance scale is defined as follows:

1. No effect: there was an insignificant delay
2. Less influential: there is a small amount of delay (<5%)
3. Influential: there was a significant delay (5%-10%)
4. Moderately influential: there is quite a serious delay (10%-20%)
5. Very influential: there was a big delay (>20%)

Frequency scale is defined as follows:

1. Very Rare: Very rarely occurs only under certain conditions
2. Rarely: Sometimes occurs in certain conditions
3. Sometimes: Occurs under certain conditions
4. Often: Often occurs in every condition
5. Always: Always happens in every condition

After risk identification is implemented, the data analysis process begins by calculating the risk value obtained from the calculation of the probability and consequences. The results of the assessment are then calculated the mean value using the Geometric Mean for each variable on probability and consequences so that the values and risk maps are obtained. Risk evaluation is carried out by comparing the risk value obtained with the company's ability to accept risk (risk acceptance). If the risk is unacceptable, then additional control recommendations are made on that risk then include a list of risks and controls provided in one risk register document file in order to obtain a comprehensive document in explaining the description of the risk of delays in the project.

4. Results

Based on the validation results of the delay variables that has been carried out with the team from the integration project department, the next step is to identify the qualitative impact of the delay variables which are summarized as shown in Table 1.

Table 1. Tabulation of delay risk variables

Risk ID	Risk Event	Probability					Consequences					Risk
		1	2	3	4	5	1	2	3	4	5	
R12	Uncertain weather conditions				x					x		16
R15	Covid-19 infected during construction			x							x	15
R1	Insufficient availability of labour				x				x			12
R2	Low labour productivity				x				x			12
R6	Late delivery of materials			x					x			12
R14	Restrictions on work in the field			x					x			9

Table 1. Tabulation of delay risk variables (continuous)

Risk ID	Risk Event	Probability					Consequences					Risk
		1	2	3	4	5	1	2	3	4	5	
R16	Lack of workforce competence			x					x			9
R17	Less labour experience			x					x			9
R18	Insufficient amount of material			x					x			9
R19	Limited number of work equipment			x					x			9
R20	Work Equipment Damage			x					x			9
R3	Adding project scope			x					x			9
R4	There is a rework due to an execution error			x					x			9
R5	Improper project planning			x					x			9
R8	Poor project management from the contractor			x					x			9
R9	Inability to cope with work progress			x					x			9
R21	Poor site supervision and management		x						x			6
R22	Safety regulations that are not met by contractors			x				x				6
R25	Lack of communication between parties		x						x			6
R26	The duration of decision making on the project		x						x			6
R10	The difficulty of obtaining a work permit		x					x				4
R11	Changes to government regulations		x					x				4
R13	Problems with local people		x					x				4
R23	There is a termination of work by the project owner		x					x				4
R29	Delay in producing design documents		x					x				4
R30	Insufficient data collection completes at survey		x					x				4
R24	Not able to provide imported materials		x				x					2
R27	Poor interaction at engineering and procurement	x						x				2
R28	Delays in processing work payments	x						x				2
R7	Poor management of project cash arrangements	x						x				2

After tabulating the risk variable data above, we plotting risk mapping all variable on using 5x5 risk matrix that has been agreed upon by the company's standard. Referring to the Risk-Based HSSE Management Guidelines No. A7-003/ S00100/2019-S0 [11], risk level of each Risk ID as shown in Table 2. The main focus of risk level is High and High to Moderate with a risk level value of 10 to 25, a series of risk mitigation plans must be carried out to reduce risk to a medium level. Based on the Risk Map at Figure 2, it seen that the majority of risks are in the moderate category which can be tolerated by conducting continuous monitoring and control.

Table 2. List of delay risk variables

Risk Category	Risk ID List	Remarks
High	12, 15	Unacceptable
Moderate to High	1, 2, 6	Not tolerable
Moderate	3, 4, 5, 8, 9, 14, 16, 17, 18, 19, 20, 21, 22, 25, 26	Tolerable with control & monitoring
Low to Moderate	10, 11, 13, 23, 29, 30	Tolerable
Low	7, 24, 27, 28	Acceptable

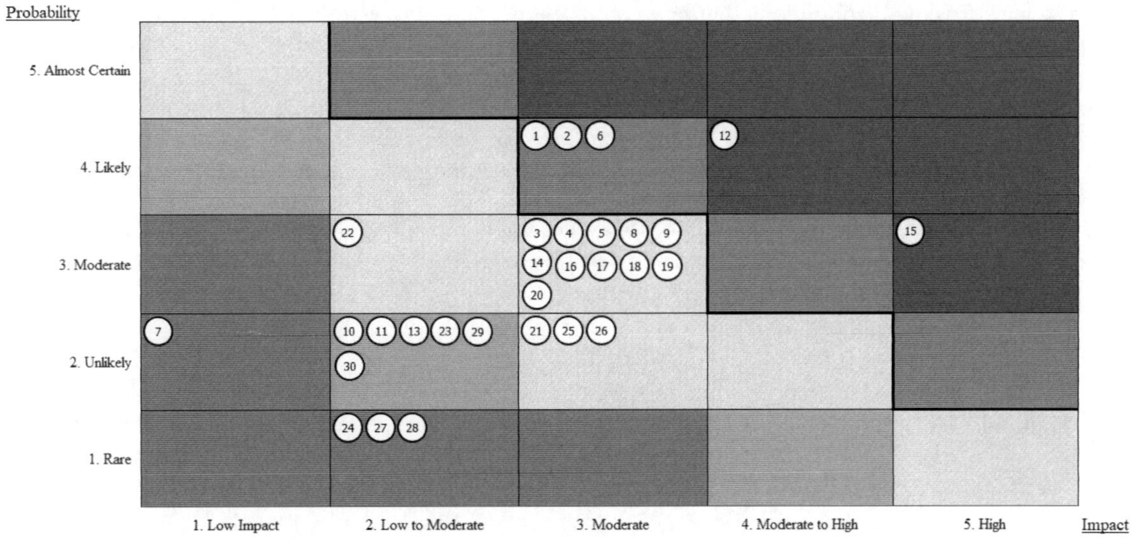

Figure 2. Risk map result

Based on Table 2 and Figure 2, the main focus in implementing risk management is at high and moderate to high levels, where based on the FGD implementation there are risk control options including:

1. R12 (Uncertain weather conditions) is mandatory to add platform, install safety net and tarpaulin canvas with the total cost required to fulfil this mitigation is $4,604.
2. R15 (Covid-19 infected during construction) is mandatory. Implementation of random check using Antigen Swab on core workers who work during the project is carried out by 30% of the total core workers. The total cost required to fulfil this mitigation is $5,794. This is more effectively implemented compared to PCR which is not yet available in health facilities in the area closest to the plant.
3. R1 (Insufficient availability of labour) and R2 (Low labour productivity) Changes in the 3-shift work method in critical activities (welding work) will certainly cause costs for the provision of labour, additional welders, fitters, helpers to supervision, and giving reward. The total cost required to fulfil this mitigation is $5,794. With the change in work methods from 2 shifts to 3 shifts, this can speed up the work schedule to 6 working days.
4. R6 (Late delivery of materials), The mitigation carried out is the appointment of a PIC from Logistics & Project Support who will monitor the shipment process through the vessel tracker or marine traffic application that is already available and can be accessed regularly.

5. Conclusions

Based on the results of data analysis and discussion that implemented there are 30 risk events that can affect the delay in the WHRU installation project consisting of 2 High categories, 3 Moderate to high categories, 15 Moderate categories, 6 Low to moderate categories and 4 Low categories. In the risk variable due to weather factors, the mitigation needed is the creation of a platform, installation of a safety net and tarpaulin canvas where an implementation fee of $ 4,604 is required so that it can reduce the risk to Moderate. Risk variables due to accidents and Covid-19 factors, the mitigation carried out is by making platforms, installing safety nets, tarpaulin canvases, tightening Covid-19 procedures, implementing random antigen checks, localizing worker mobilization, and tightening supervision where an implementation fee of $ 5,794 is required so that it can reduce the risk to Moderate. Risk variable due to labour and productivity factors that are lacking, the mitigation carried out is by adding a welder team, changing the work pattern from 2 shifts to 3 shifts and adding supervision where an implementation cost of $ 2,415 is needed so that it can reduce the risk Low to Moderate. Risk variable due to material delivery factors, the mitigation carried out is by appointing a PIC for delivery supervision and customs management with an implementation fee of $14 so that the risk can be reduced to Moderate. From the list of risk control options that have recommended adding a welder team, changing the work system to 3 shifts and adding supervision can speed up the work schedule from 19 days to 13 days.

References

[1] McKinsey Global Institute (2017). The art of project leadership: Delivering the world's largest projects. Available from: https://www.mckinsey.com/business-functions/operations/our-insights/

[2] Pinto, J. K. (2007). Project Management Best Practices: Achieving Global Excellence (Kerzner, H.; 2006). IEEE Transactions on Engineering Management, 54(2). https://doi.org/10.1109/tem.2007.893984

[3] Assaf, S. A., & Al-Hejji, S. (2006). Causes of delay in large construction projects. International Journal of Project Management, 24(4), 349–357. https://doi.org/10.1016/j.ijproman.2005.11.010

[4] Kaming, P. F., Olomolaiye, P. O., Holt, G. D., & Harris, F. C. (1997). Factors influencing construction time and cost overruns on high-rise projects in Indonesia. Construction Management and Economics, 15(1), 83–94. https://doi.org/10.1080/014461997373132

[5] Taha, G., Badawy, M., & El-Nawawy, O. (2007). A Model for Evaluation of Delays in Construction Projects. International Journal of Innovative Research in Science, Engineering and Technology (An ISO, 3297. https://doi.org/10.15680/IJIRSET.2016.0503048

[6] Sweis, G., Sweis, R., Abu Hammad, A., & Shboul, A. (2008). Delays in construction projects: The case of Jordan. International Journal of Project Management, 26(6), 665–674. https://doi.org/10.1016/j.ijproman.2007.09.009

[7] Doloi, H., Sawhney, A., Iyer, K. C., & Rentala, S. (2012). Analysing factors affecting delays in Indian construction projects. International Journal of Project Management, 30(4), 479–489. https://doi.org/10.1016/j.ijproman.2011.10.004

[8] Verma, M. (2020). Statistical Analysis of Delay Factors in Construction Projects. International Journal for Research in Applied Science and Engineering Technology, 8(9), 730–738. https://doi.org/10.22214/ijraset.2020.31531

[9] International Standard for Organization (ISO) 31000:2018 Risk Management – Principles and Guidelines. From https://www.iso.org/standard/65694.html

[10] Project Management Institute. (2017). A guide to the Project Management Body of Knowledge (PMBOK guide) (6th ed.). Project Management Institute.

[11] Pertamina (2019). Pedoman Pengelolaan Hsse Berbasis Risiko No. A7-003/ S00100/2019-S0 Revisi 0, Pertamina, Jakarta.

The Design of Ship Operation Cost Estimation Simulator Uses a Case Study of The Bung Tomo Trainer Ship

U Widyaningsih[1], Sutoyo[1], A A N P Yuda[1], A Mirianto[1], Z Zuhri[2], N V Harini[3]

[1]Deck Department, Politeknik Pelayaran Surabaya, East Java, Indonesia
[2]Department of Mathematics Education, Institut Teknologi dan Sains Nahdlatul Ulama, Pasuruan, East Java, Indonesia
[2]Department of Mathematics Education, Universitas Islam Negeri Sunan Ampel, Surabaya, East Java, Indonesia

novitavindriharini@gmail.com

Abstract. Calculation of ship operating costs is the main element in determining tariffs, since determining the high and low rates is very important in the shipping business. In other words, determining the right ship operating costs is the key to the success of the shipping business. Vessel operating costs are determined by direct and indirect operational costs. Direct operational costs include capital costs, insurance costs, insurance premiums, crew costs, fuel costs, lubricant costs, grease costs, fresh water costs, port costs and Repair, Maintenance, Supply (RMS) costs. The indirect operational costs include branch office costs and admin costs. Calculation of ship operating costs can actually be done manually, but it will take a lot of time. So, we need a way to calculate the operational costs of ships that are easy and efficient. In this study we developed an application to calculate ship operating costs. Then we call this application a simulator. From the simulations carried out, the results obtained that the results of calculations using the simulator are the same as the results of manual calculations, namely the total cost in rupiah. The advantage of this simulator is that it saves calculation time, so that the results obtained are easier and more efficient. In addition, the appearance of the simulator is very good in accordance with the demands of the digitalization of sea transportation and this simulator can be used by all groups because it is easy to operate.

1. Introduction

Calculation of ship operating costs is the main element in determining tariffs, because it affects the determination of high and low rates which are very important in the shipping business. In other words, determining the right ship operating costs is the key to the success of the shipping business. Vessel operating costs are determined by direct and indirect operational costs. Direct operational costs include capital costs, insurance costs, insurance premiums, crew costs, fuel costs, lubricant costs, grease costs, fresh water costs, port costs and RMS costs. Indirect operational costs include branch office costs and admin costs.

The problem is that the determination of the amount of operational costs still uses the manual method, namely how to calculate directly with a formula. This method is considered by researchers to

take a long time and can only be done by some groups or cannot be determined by the general public. So that researchers aim to shorten the calculation time so that it is faster.
The way to speed up the calculation of ship operating costs that will be carried out in research is to make an application for calculating ship operating costs. For the next application that we make, we will call it Simulator.

A simulator is a tool for the process of implementing a model into a computer program (software) or electronic circuit and executing the software in such a way that its behavior imitates or resembles a real system. The simulator is designed in the form of software to calculate the estimated operating costs of the ship when the ship will sail. This study aims to describe the process and results of the simulator design to calculate ship operational costs. A good simulator is that the output of calculating ship operating costs is the same as manually using a formula from the ministry of transportation but the calculation time is faster.

2. Method

There are three stages in making this simulator. The first stage is the search for ship data (ship type data, ship engine, ship engine fuel consumption data, direct cost data, and indirect cost data). In this study, the Bung Tomo training boat was used as a case study. The second step is storing data that has been inputted as a database in my SQL. The third stage is programming the simulator in PHP my Admin. At this stage, the researcher enters the formula to calculate the operational costs of the ship, which is consists of:

a. Direct Cost

[1] Fixed Cost

Based on Decree of Ministry of Transportation No. KM. 58 of 2003, fixed cost consists of:

i. Ship Depreciation Cost

$$C_{SD} = \frac{Ship\ price - Residual\ Value}{Depreciation\ period}$$

With,
Residual value is 5% out of ship price
Depreciation period is 25 years for new ship and 20 years for second hand ship

ii. Capital Interest Cost

$$C_{CI} = \frac{\frac{N+1}{2}(60\%\ ship\ price)(interest\ rate\ in\ a\ year)}{N}$$

With,
N = the term of the loan is 10 years. The loan capital is calculated at 65% of the ship price, meaning a down payment of 35% (depending on the policy of each bank).
The interest rate is based on the generally accepted price level and Insurance Fee.

[2] Variable Cost

i. Fuel Cost

$$W_{FL} = \frac{(Pbme.bme + Pae.bae)\,S}{V.10^{-6}.Add}$$

$$W_{FE} = (Pae.bme).wp.10^{-6}$$

With,
W_{FL} = Large fuel consumption at sea (Kw)
W_{FE} = Large fuel consumption at the port (Kw)
Pbme = Main Engine Power
Pae = Auxiliary Engine Power
bme = Main engine oil weight
bae= Auxiliary engine fuel weight
S = Cruise distance (Mile)
V = Ship speed (Knot)
Add = Reserve factor (1,3 – 1,5)
wp = Time at the port (Hours)

The map display used in this simulator is google map. The user can utilize the map display to determine the shipping route by clicking on the map. Besides, it can be inputted by the location coordinates for the shipping route.

ii. Lubricating Oil Cost
The amount of lubricating oil consumption can be determined using the following equation:

$$W_{LI} = \frac{Pbme\ x\ bme\ x\ S}{V\ x\ 10^{-6} + Add}$$

$$W_{LP} = Pae\ x\ bae\ x\ wp\ x\ 10^{-6} + Add$$

With,
Pbme = Main Engine Power
Pae = Auxiliary Engine Power
bme = Main engine oil weight (1,2 – 1,6 gr/Kwh)
bae = Auxiliary engine fuel weight (1,2 – 1,6 gr/Kwh)
Add = Reserve factor (10 – 20)%
Annual consumption of lubricating oil (ML) is the amount of use of lubricating oil multiplied by the frequency of sailing per year (f)

$$ML = (W_{LI} + W_{LP})\ x\ f$$

The annual cost of lubricating oil (BL) is the amount of lubricating oil used per year (ML) times the price of lubricating oil (HL)

$$BL = HL\ x\ ML$$

iii. Ship grease
In the Decree of the Minister of Transportation No. KM. 58 of 2003, the cost of grease lubricants, namely:

$$B_G = \frac{Amount\ of\ grease\ use}{month\ x} \frac{number\ of\ ship\ operations}{month\ x\ \frac{price\ of\ grease}{Kg}}$$

iv. Freshwater fee
a. Fresh water for main engine cooling

$$W_{OP} = Pbme\ x\ me\ x\ \frac{S}{V\ x\ 10^{-3}}$$

With,
me = amount of water for boiler (steam boiler) = 0,14 kg/Kwh
b. Fresh water for auxiliary engine cooling

$$W_{OP'} = pae\ x\ me\ x\ \frac{S}{V\ x\ 10^{-3}}$$

c. Fresh water for consumption and bathing

$$W_{fw} = P\ x\ Z_{fw}\ x\ \frac{t}{1000}$$

With,
Z_{fw} = Consumption of drinking water + washing and bathing water kg/person/day
P = Number of crew
t = Time for Round Trip

The cost of using fresh water is calculated by multiplying the amount of fresh water used (W_{fw}) for a year multiplied by the price of water based on the current price of fresh water. So, the formula used is:

$$BAT = \left(W_{op} + W_{op} + W_{fw}\right) BAT_{PB}$$

With:

BAT_{PB} = Price of water per ton (Rp)

v. Ship Fee at Port
The amount of this fee depends on the GRT of the ship and the length of time the ship arrives until it leaves the port.
$UL = WL\ x\ anchorage\ rate\ \ x\ frequency$
With:
UL = anchorage fee
WL = anchor time of the ship

b. Indirect Cost
The amount of management fee is 12% of crew costs, RMS, insurance with the equation:

$$B_{TM} = 0{,}12\left(B_{TAK} + RMS_{pv} + BA_{pv}\right)$$
$$B_{TAK} = (G_{AKT} + B_{KAKT} + B_{AAKT})$$

With:
B_{TM} = fixed costs of management activities (Rp/year)
B_{TAK} = crew fixed costs (Rp/year)
G_{AKT} = Salary of crew (Rp/year)
B_{KAKT} = crew consumption costs (Rp/year)
B_{AAKT} = fresh water cost for crew (Rp/year)
RMS_{pv} = average RMS cost present value (Rp/year)
BA_{pv} = average present value insurance cost (Rp/year)

3. Results and Discussion

Our discussion begins by identifying the diesel engines of ships that consume fuel. Here are the details (see Table 1):

Table 1 Data of Bung Tomo Training Boat

NO	Machine Type	Status	RPM	Fuel Consumptio n (Litre)	Water Consumptio n (Litre)
1.	Main Engine I (SS)	Full Speed	1200	25/Jam	10
		Idle Speed	700	54/Jam	10
		Manoeuvre	900	25/Jam	10
2.	Main Engine II (PS)	Full Speed	1200	25/Jam	10
		Idle Speed	700	54/Jam	10
		Manoeuvre	900	25/Jam	10
3.	Generator SS (Stbd)	Idle Speed	720	30/Jam	10
4.	Generator MS (Middle)	Idle Speed	720	30/Jam	10
5.	Generator PS (port)	Idle Speed	720	30/Jam	10
6.	Generator Harbour	Idle Speed	600	25/Jam	10
7.	Main Engine AX	Full Speed	50	15/Jam	10
8.	Main Engine AX	Full Speed	100	15/Jam	10

Direct costs used in this study are as follows (see Table 2):

Table 2 Direct Cost Data

No	Description	Cost (Rupiah)	Unit
1	Ship price	55.567	Million
2	Depreciation Period	20	Year
3	Cost of depreciation	2.639,43	Million/Year
4	Loan percentage	0	Percent
5	Loan length	0	Year
6	Interest rate	0	Percent
7	Interest fee	0	Million/Year
8	Insurance presentation	1,1	Percent
9	Insurance fee	611,24	Million/Year
10	Salary Crew	1.486	Million/Year
11	Allowance	0	Million/Year
12	Fuel price	14.000	Rupiah/Liter
13	Lubricant consumption	0,5	Liter/Hour
14	Lubricant price	26.500	Rupiah/Liter
15	Fresh water consumption	10	Liter/Day
16	Fresh water price	40.000	Rupiah/Ton
17	Grease lube cost	0	Rupiah/Month
18	Anchorages rates	0	Rupiah/Hour
19	Mooring rates	0	Rupiah/Hour
20	Guides rates	0	Rupiah/Act
21	Delay rates	0	Rupiah/Act
22	Maintenance		Rupiah/Hour

The indirect costs used in this study are as follows (see table 3):

Table 3 Indirect Cost Data

No	Description	Cost (Rupiah)	Unit
1	Ground staff salary	492	Million/Year
2	Allowance	0	Million/Year
3	Working hours	2112	Hour/Year
4	Crew consumption	412	Million/Year
5	RMS average	0	Million/Year
6	Insurance average	611,24	Million/Year

This simulator development uses PHP My Admin as a prototype/display, My SQL as a database, and Google map as a digital map [3.4.5]. The simulator design is made/designed to facilitate the user in operating it. This simulator is a software to calculate ship engine fuel that can be used easily by users [8.9.10]. Figure 1 follows the simulator display design.

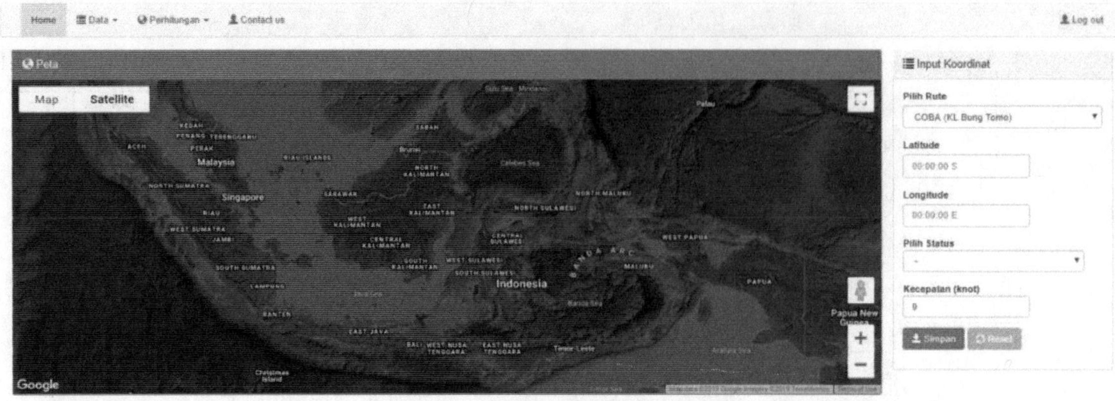

Figure 1. Main Page Map

To be able to determine Bung Tomo Training Boat's engine fuel consumption, firstly we input the ship's data as follows (see figure 2, 3,4 and 5):

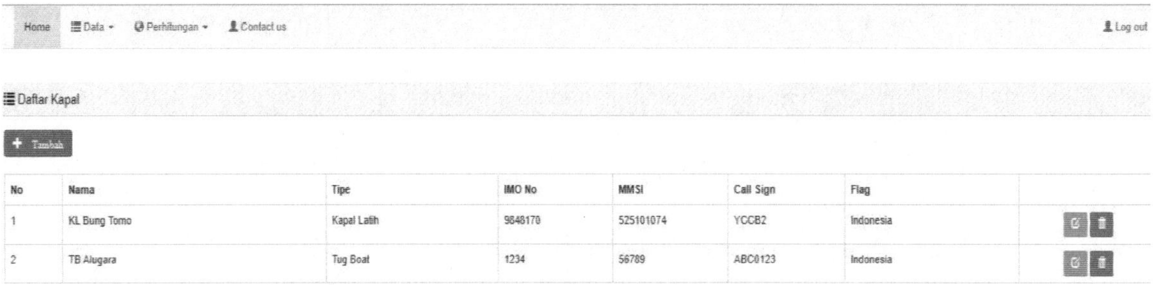

Figure 2. Input Data of Boat

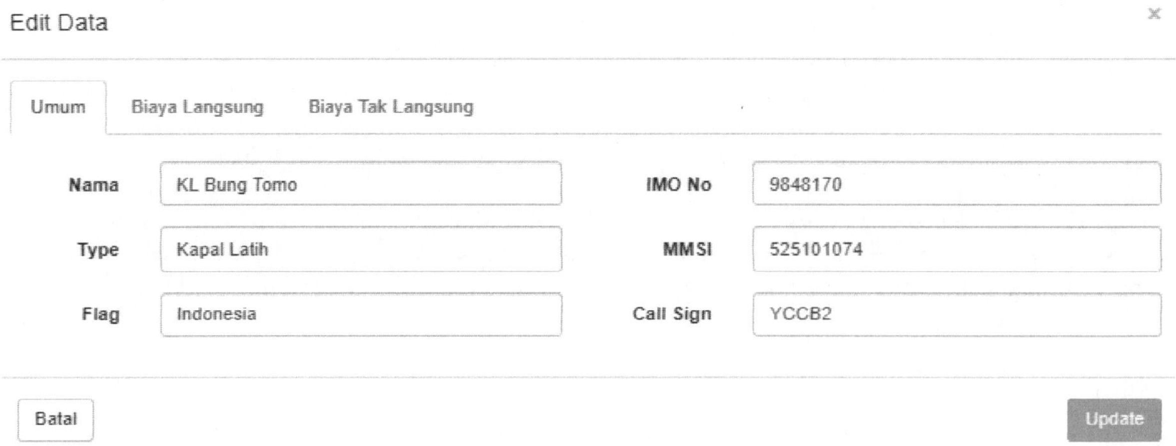

Figure 3. Input Data of Bung Tomo Training Boat's

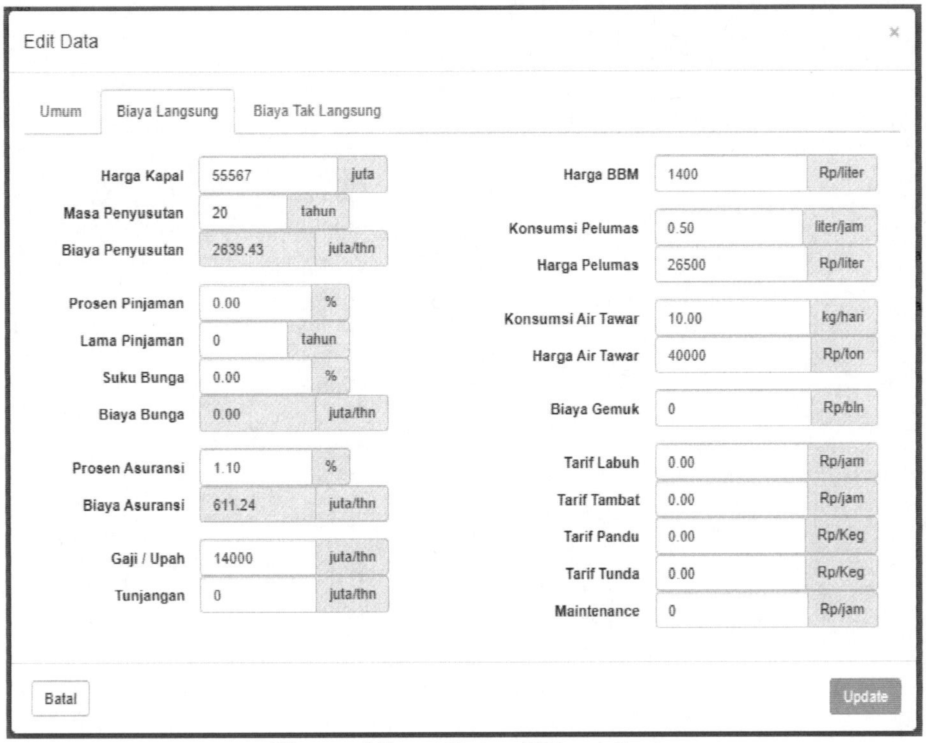

Figure 4 Input Data of Direct Cost

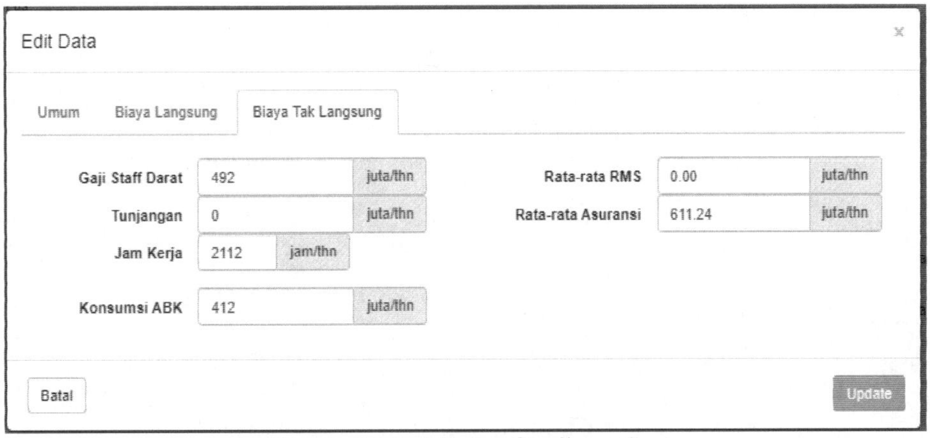

Figure 5 Input Data of Indirect Cost

Then we input Machine Rate Consumption Data (see figure 6) and Data for Ships Route Plans (see figure 7) as follows:

No	Kapal	Komponen	Status	RPM	BBM	Air Tawar	ON / OFF
1	KL Bung Tomo	Main Engine SS	full speed	1200	25	10.00	On
2	KL Bung Tomo	Main Engine PS	full speed	1200	25	10.00	On
3	KL Bung Tomo	Main Engine SS	manouver	900	25	10.00	On
4	KL Bung Tomo	Main Engine PS	manouver	900	25	10.00	On
5	KL Bung Tomo	Main Engine SS	idle speed	700	54	10.00	On
6	KL Bung Tomo	Main Engine PS	idle speed	700	54	10.00	On
7	KL Bung Tomo	Generator SS (Stbd)	idle speed	720	30	10.00	On
8	KL Bung Tomo	Generator MS (Middle)	idle speed	720	30	10.00	Off
9	KL Bung Tomo	Generator PS (port)	idle speed	720	30	10.00	Off
10	KL Bung Tomo	Generator Harbour	harbour	600	25	10.00	On
11	KL Bung Tomo	Main engine AX	full speed	50	15	10.00	On
12	KL Bung Tomo	Main engine AX	full speed	100	15	10.00	On

Figure 6. Input Data of Machine Rate Consumption

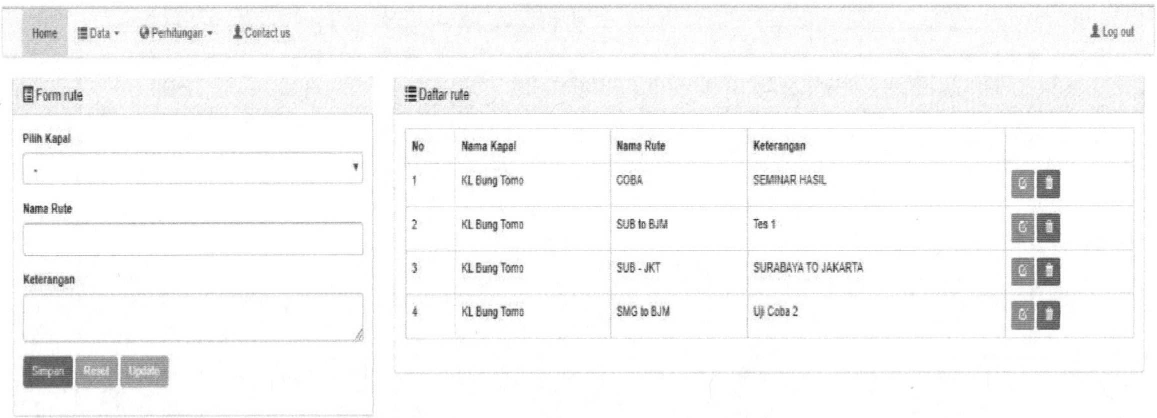

No	Nama Kapal	Nama Rute	Keterangan
1	KL Bung Tomo	COBA	SEMINAR HASIL
2	KL Bung Tomo	SUB to BJM	Tes 1
3	KL Bung Tomo	SUB - JKT	SURABAYA TO JAKARTA
4	KL Bung Tomo	SMG to BJM	Uji Coba 2

Figure 7. Input Data of Ships Route Plans

After the data input process is done, then we test the simulator that we have made.

a. First Trial

The first trial was conducted by shipping routes from the port of Surabaya to the port of Makassar. We can see in figure 8.

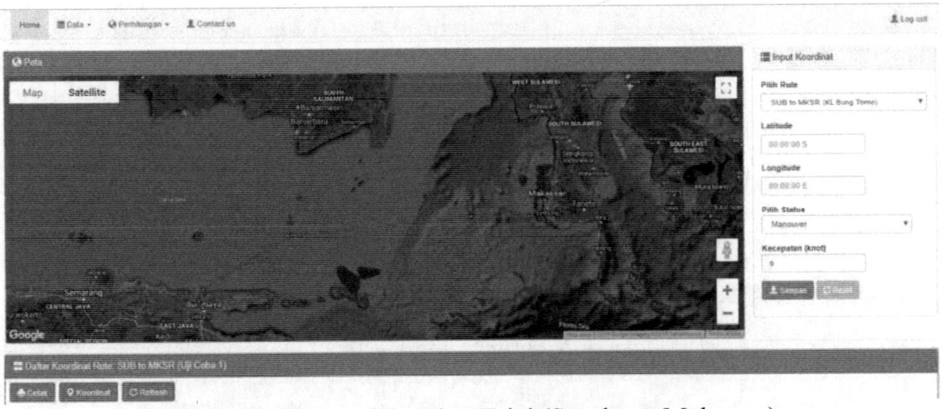

Figure 8. Route of the First Trial (Surabaya-Makassar)

Daftar Koordinat Rute: SUB to MKSR (Uji Coba 1)

WP	Latitude	Longitude	Status	Jarak	Haluan	Waktu	BBM	Air Mesin
0	7:11:52.74 S	112:43:59.83 E	harbour	0	310.4	0	0	0
1	7:8:57.38 S	112:40:32.19 E	manouver	4.51	350.57	01:30:11	75	0.045
2	7:3:2.93 S	112:39:32.86 E	manouver	5.99	19.51	01:59:46	100	0.06
3	6:49:33.3 S	112:44:26.56 E	idle speed	14.34	58.98	02:52:06	396.06	0.1292
4	5:47:38.91 S	114:27:37.49 E	full speed	119.79	84.57	13:18:37	1064.8	0.7986
5	5:27:8.43 S	117:57:33.18 E	full speed	209.92	89.98	20:59:32	1679.2	1.2594
6	5:27:2.97 S	119:4:37.61 E	full speed	66.77	45.05	06:40:37	534.4	0.4008
7	5:8:55.02 S	119:22:51.75 E	idle speed	25.66	58.07	05:07:56	707.94	0.2309
8	5:8:23.03 S	119:23:43.3 E	manouver	1.01	28.61	00:12:06	10	0.006
9	5:6:51.62 S	119:24:33.35 E	manouver	1.74	0	00:34:42	29	0.0174
		Total :		449.73 millaut		53.2592	4596.4 liter	2.9472 liter

Figure 9. Consumption List of Fuel

Based on the first trial data (see figure 9) the trip from Surabaya to Makassar obtained a distance of 449.73 miles with a travel time of 53 hours 25 minutes, thus requiring 4596.4 liters of fuel on the engine and dan consumption of fresh water as much as 2,9 liters.

Estimasi Biaya Operasi Kapal

Pilih Rute: SUB to MKSR (KL Bung Tomo)

Jarak	Waktu	BBM	Air Tawar
449.73 millaut	53.2592 jam	4596.4 liter	2.9472 liter

	SUB to MKSR
BIAYA OPERASIONAL LANGSUNG	
1. Biaya Tetap	
Biaya Penyusutan	16.047.257
Bunga Modal	0
Premi Asuransi	3,716,207
Biaya Awak Kapal	9.034.803
2. Biaya Tidak Tetap	
Biaya BBM	64,349,600
Biaya Pelumas	705.684
Biaya Gemuk	0
Biaya Air Tawar	44,972
Biaya Pelabuhan	0
Biaya RMS	2,503,873
BIAYA OPERASIONAL TIDAK LANGSUNG	
Biaya Kantor Cabang	0
Biaya Manajemen & Admin	57,133,467
TOTAL	153,535,864

Figure 10. List of Cost Estimation Tables in First Trial

IOP Conf. Series: Earth and Environmental Science 1081 (2022) 012008 doi:10.1088/1755-1315/1081/1/012008

Based on the data above (first trial) for a trip from Surabaya to Makassar, an estimated cost of Rp. 153,535.664,-.

a. Second Trial

The second trial was conducted by shipping routes from the port of Semarang to the port of Banjarmasin. We can see in figure 11.

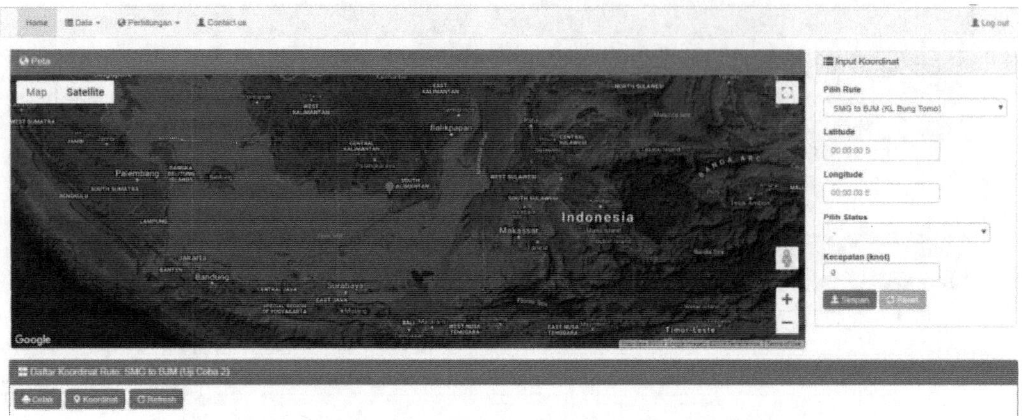

Figure 11. Route of the Second Trial (Semarang-Banjarmasin)

Daftar Koordinat Rute: SMG to BJM (Uji Coba 2)

Cetak | Koordinat | Refresh

WP	Latitude	Longitude	Status	Jarak	Haluan	Waktu	BBM	Air Mesin
0	6:56:36.32 S	110:25:24.74 E	harbour	0	359.18	0	0	0
1	6:56:14.85 S	110:25:24.43 E	manouver	0.36	2.65	00:07:09	6	0.0036
2	6:50:42.35 S	110:25:39.95 E	idle speed	5.55	18.43	01:06:34	153.18	0.05
3	6:8:43.06 S	110:39:44.11 E	full speed	44.25	73.81	04:25:31	354.4	0.2656
4	5:34:22.42 S	112:37:53.59 E	full speed	122.45	52.95	12:14:44	980	0.735
5	4:7:50.42 S	114:32:35.43 E	full speed	143.35	355.14	14:20:05	1146.4	0.8596
6	3:34:0.62 S	114:29:42.39 E	manouver	33.95	8.22	06:47:26	339.5	0.2037
7	3:27:10.43 S	114:30:41.72 E	manouver	6.91	0	02:18:09	115	0.069
		Total :		**356.82 millaut**		**41.3272**	**3094.48 liter**	**2.1869 liter**

Figure 12. Consumption List of Fuel

Based on the second trial data (see figure 12) the trip from Semarang to Banjarmasin obtained a distance of 356.82 miles with a travel time of 41 hours 32 minutes, thus requiring 3094.48 liters of fuel on the engine and consumption of fresh water as much as 218 liters.

Estimasi Biaya Operasi Kapal

Pilih Rute: SMG to BJM (KL Bung Tomo)

Jarak	Waktu	BBM	Air Tawar
356.82 millaut	41.3272 jam	3094.48 liter	2.1869 liter

	SMG to BJM
BIAYA OPERASIONAL LANGSUNG	
1. Biaya Tetap	
Biaya Penyusutan	12,452,102
Bunga Modal	0
Premi Asuransi	2,883,645
Biaya Awak Kapal	7,010,531
2. Biaya Tidak Tetap	
Biaya BBM	43,322,720
Biaya Pelumas	547,586
Biaya Gemuk	0
Biaya Air Tawar	34,877
Biaya Pelabuhan	0
Biaya RMS	1,942,917
BIAYA OPERASIONAL TIDAK LANGSUNG	
Biaya Kantor Cabang	0
Biaya Manajemen & Admin	44,333,542
TOTAL	112,527,919

Figure 13 List of Cost Estimation Tables in Second Trial

Based on the data above (second trial) for a trip Semarang to Banjarmasin, an estimated cost of Rp. 112,527,919-.

4. Conclusion

Based on the description and analysis of the results and discussion, it can be seen that the Simulator can display the ship's mileage, the ship's fuel consumption, the engine's fresh water consumption. So that the simulator can calculate the exact operating costs of the ship. The time required is also not as long as manual calculations using formulas and results can be obtained immediately after data input is carried out.

5. References

[1] Li, X., Sun, B., Zhao, Q., Li, Y., Shen, Z., Du, W., & Xu, N. (2018). Model of speed optimization of oil tanker with irregular winds and waves for given route. Ocean Engineering, 164, 628-639.

[2] Bernacki, D. (2021). Assessing the link between vessel size and maritime supply chain sustainable performance. Energies, 14(11), 2979.

[3] Alzaman, C., Zhang, Z. H., & Diabat, A. (2018). Supply chain network design with direct and indirect production costs: Hybrid gradient and local search based heuristics. International Journal of Production Economics, 203, 203-215.

[4] Zhukov, V. A., Sherban, S. A., Melnik, O. V., Sokolov, S. S., & Kolesnichenko, S. V. (2019, January). Improvement of methods and means of thermal control of ship power plants. In 2019 IEEE Conference of Russian Young Researchers in Electrical and Electronic Engineering (EIConRus) (pp. 389-392). IEEE.

[5] Wiratno, D., Mirianto, A., Cahyadi, T., Zuhri, Z., & Harini, N. V. (2021). The design of the ship's fuel estimation simulator uses a case study of the bung tomo trainer ship. In IOP Conference Series: Materials Science and Engineering (Vol. 1010, No. 1, p. 012027). IOP Publishing.

[6] Purba, D., Mirianto, A., Zuhri, Z., & Harini, N. V (2019). Applying Spherical Triangle Concept in Simulator to Determine Distance and Direction of Ship.

[7] Kamel, E., & Memari, A. M. (2019). Review of BIM's application in energy simulation: Tools, issues, and solutions. Automation in construction, 97, 164-180.

[8] Ang, J. H., Goh, C., Saldivar, A. A. F., & Li, Y. (2017). Energy-efficient through-life smart design, manufacturing and operation of ships in an industry 4.0 environment. Energies, 10(5), 610.

[9] Kolm, P. N., & Ritter, G. (2019). Dynamic replication and hedging: A reinforcement learning approach. The Journal of Financial Data Science, 1(1), 159-171.

[10] Laura, V. R., Bernardo, P., Alberto, L. J., Pablo, F. J., Juberer, M., & Esteban, C. (2019). Implementation of IT in Systems Simulation for the Registration Process. Journal of Education and Human Development, 8(4), 177-184.

Case Study of Ship Traffic Crowds in The Malacca Strait-Singapore by Using Vessel Traffic System

F Nofandi[1], U Widyaningsih[1], R A Rakhman[1], A Mirianto[1], Z Zuhri[2], and N V Harini[3]

[1]Deck Department, Politeknik Pelayaran Surabaya, East Java, Indonesia
[2]Department of Mathematics Education, Institut Teknologi dan Sains Nahdlatul Ulama, Pasuruan, East Java, Indonesia
[2]Department of Mathematics Education, Universitas Islam Negeri Sunan Ampel, Surabaya, East Java, Indonesia

novitavindriharini@gmail.com

Abstract. The Malacca Strait as one of the main routes of world trade has the potential for the emergence of transnational crime. The strategic location of the Malacca Strait makes it fragile and vulnerable to piracy practices. The occurrence of this piracy has the potential to disrupt shipping lane traffic so that shipping safety is threatened and causes substantial losses for ship owners. But it is not only piracy that causes considerable losses, there is other factor that are also very influential besides piracy, namely accidents. One of the causes of accidents is the flow of ship traffic that is too crowded and is not immediately addressed. Therefore, it is necessary to increase the safety of ship traffic, one of which can be by increasing the Vessel Traffic Service (VTS). Vessel Traffic Service (VTS) serves to help provide reports on navigational and meteorological information for ships crossing a water. If the Vessel Traffic Service (VTS) is implemented and improved in the Malacca Strait, it is hoped that accidents and piracy can be minimized. This study aims to analyze the traffic congestion of ships in the Straits of Malacca and Singapore to optimize maritime security and safety and improve the Vessel Traffic System (VTS) in these straits. From the results of the analysis obtained data in the Strait of Malacca-Singapore there is an average trend of increasing the volume of ship traffic every year by 2% and has the potential to cause a lot of ship traffic. This data is very useful in order to improve the Vessel Traffic System (VTS) in the Malacca-Singapore Strait so that this traffic jam can be overcome.

Keywords: Vessel Traffic Service (VST), Ship Traffic Density, Ship Traffic Line

1. Introduction

Indonesia is the largest archipelagic country in the world and is a maritime country because it has a wide sea and many islands. Indonesia has more than 17,000 islands, strung together by 95,181 km of coastline (the second longest after Canada), and about 70% of its territory is sea [1]. The Malacca Strait plays an important role in trade routes from the sea sector in Indonesia because the Malacca Strait is the main route connecting the east and west. Traders from various countries used the Malacca Strait as a shipping and trade route between India and South China and the nations that inhabited the plains of Southeast Asia [2].

The Singapore Strait which is located east of the Malacca Strait is also a deep water crossing to ports in Singapore. This strait separates the Riau Islands and Singapore. The Straits of Malacca and Singapore which are located in the territory of three littoral countries "Indonesia, Malaysia and Singapore" have the most strategic shipping lanes in the world with a fairly large traffic density [3]. The Straits of Malacca and Singapore have the densest shipping routes in the world and are the most important international shipping traffic lanes for trade in the world. This route is passed by thousands of ships from various countries every day [4].

The waters of the Strait of Malacca have a length of 550 nautical miles. this is a very busy narrow lane. about 90 thousand ships per year or about nearly 200 ships every day pass through this strait. This makes the Malacca strait area a very congested, narrow, and shallow international sea traffic so that it has a high risk of accidents at sea. This condition makes scouting in the Malacca Strait and Singapore Strait very important, especially in ensuring shipping safety for sailing ships [5]. To reduce the occurrence of accidents, a prevention area was built so that ships sailing with the Traffic Separation Scheme (TSS) when passing through this strait could increase awareness of the movement of other ships that were sailing. In addition to regulating the flow of ship traffic, this TSS also provides guidelines regarding ships that want to cross the existing lanes. According to the TSS rules on the wikipedia page, at least the ship that wants to pass makes an angle of 90 degrees like a zebra crossing on the road [6].

The increase in shipping traffic makes the task of supervising sea lanes even more difficult. The three littoral states of Singapore, Malaysia and Indonesia have agreed to improve the security system and shipping protection in the waters around the Straits of Malacca and Singapore [7]. Because of this, the Government of Indonesia and the Government of Japan built a Vessel Traffic Service (VTS) which was built in 2006 and completed in 2016. The VTS was built to improve the security and shipping safety system in the Straits of Malacca and Singapore. This tool will be able to accommodate 13,000 ships from Japan that pass through the Malacca and Singapore Straits [8].

Vessel Traffic Service (VTS) short based system serves to provide reports on navigational and meteorological information for ships passing through certain waters. Reports are currently made by voice over a very high frequency radio channel (VHF) [9]. The International Maritime Organization mandates an Automatic Identification System (AIS), which will provide and receive ship identity and position information to shore stations. ships and aircraft will monitor and track ships, and exchange data with the VTS Authority. As AIS reports are generated automatically. AIS-equipped vessels also have a hidden warning system that can send a signal or secret message to the VTS Authority in the event of an emergency. With the VTS, it is hoped that Indonesia can create a safe water area for ships passing through the Straits of Malacca and Singapore [10].

The Makmur Motor Ship collided in the waters of Nipah Island, Singapore Strait on December 10, 2018 because the signal for the AIS Makmur device was not caught by VTS Batam or VTS Singapore [11]. The tanker MT Antea (Hong Kong nationality) collided with the special underwater pipeline work ship MV Star Centurion (Vanuatu Nationality) which then capsized and sank on January 13, 2019. This accident was due to delayed action by officers who did not understand the traffic lane and ship speed around it (KNKT KN, 2020). The Malacca-Singapore Strait is the main route for cargo traffic, especially oil and people. Data shows that in 2006 the amount of oil transported was around 15 million tons/day. Research by the Japan International Transport Institute in 2020, ships passing through the Malacca Strait will increase by 50% (including some oil carriers). There have been several tanker accidents that resulted in oil spills in the Malacca-Singapore strait. This shows that sea accidents often occur along the Malacca Strait [12].

The Malacca Strait as one of the main routes of world trade has the potential for trans-national crime. The strategic location of the Malacca Strait makes it fragile and vulnerable to piracy practices. Large merchant ships and tankers are often attacked by robbers in the waters of the Malacca Strait. Ships and tankers passing through the Malacca Strait were forced to slow down to a speed of 10 knots. The occurrence of accidents and robberies will potentially disrupt shipping lane traffic. This results in the safety of shipping being threatened and causing considerable losses for ship owners [13].

The density of ship traffic is one of the factors that causes marine collisions and several crimes that occur in sea traffic . The density of ship traffic flows includes the flow of volume of goods, passengers, ship visits, and the speed of ship flow. Therefore, researchers want to analyse the density of ship traffic in the Malacca and Singapore straits to optimize maritime security and safety in the Malacca Strait and Singapore Strait to help improve the ship traffic system (VTS) [14].

2. Method

Based on the problems discussed, the researcher used a descriptive qualitative research approach. Descriptive method is a method used to describe or analyze a research result but is not used to make broader conclusions. This research was conducted by researchers for a correlational study (relationship) between shipping traffic density and ship traffic system (VTS) performance. This research activity includes data collection, data analysis, data interpretation, and finally a conclusion is formulated that refers to the data analysis.

3. Results and Discussion

The condition of the Malacca-Singapore Strait can be seen in the following (see Picture 1):

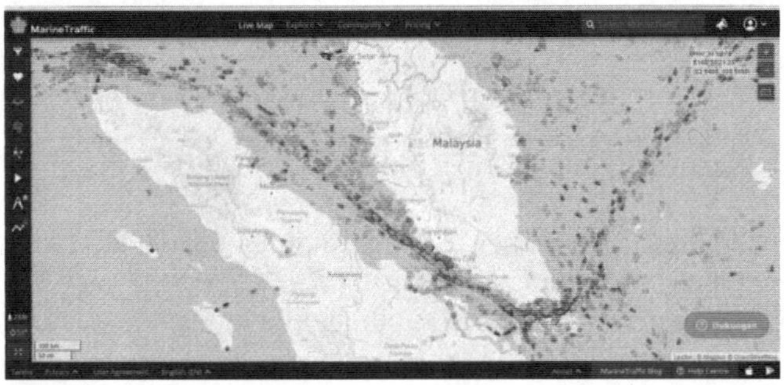

Picture 1. The condition of the Malacca-Singapore Strait

Based on data obtained from Marine Traffic, the conditions of the Straits of Malacca and Singapore are very busy. The Straits of Malacca and Singapore are among the choke points for global shipping and the busiest ship traffic. Data obtained from the Ministry of Transportation shows that a third of the volume of world trade is distributed through the Malacca-Singapore Strait (more than 200 ships per day and 80% of the oil supply for East Asia). This has resulted in increased safety and security aspects of global shipping, in 2020 there were cases of accidents and piracy in the Malacca-Singapore strait (The Maritime Executive, 2020).

Table 1. Number of Ship Visits

Length	Ship type				
	Tanker	Containe r	General Cargo	Passenger	Support Ship
0-25	0	0	0	0	59
26-50	7	0	19	15326	64
51-75	62	0	53	0	42
76-100	246	8	248	0	56
101-125	623	11	334	0	0
126-150	427	34	194	0	0
151-200	708	68	2038	0	0
201-250	907	50	1189	0	0
251-300	397	83	1088	0	0
301-400	786	56	527	0	0

Based on the table above (see Table 1), it is known that ships passing through the Malacca-Singapore Strait based on direct observation through merinetraffict from July to October 2021 were 4163 tankers, 310 container ships, 5690 general cargo ships, 15326 passenger ships and 221 support ships.

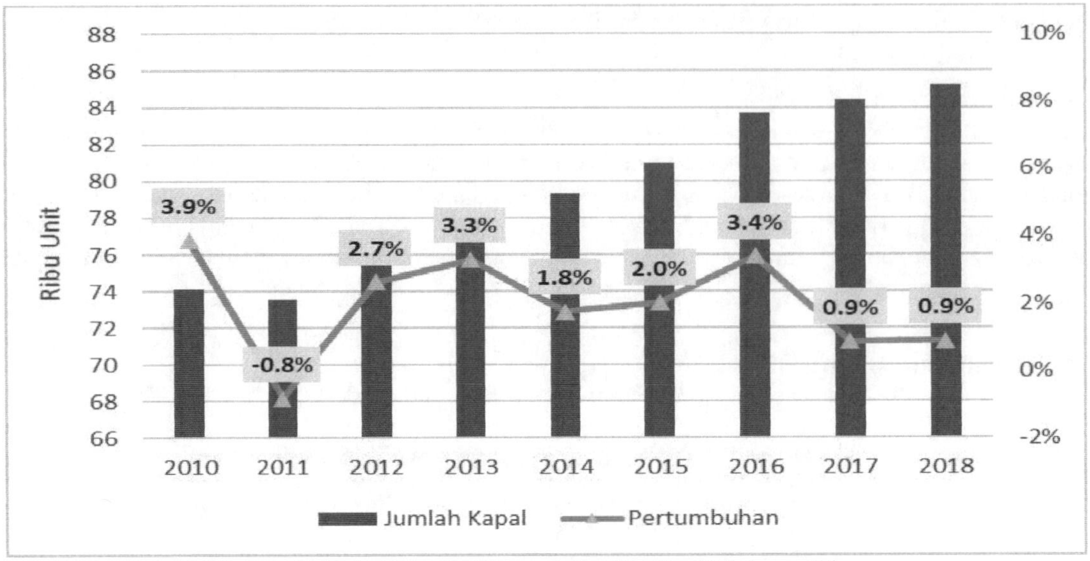

Picture 2. Ship Growth Through the Strait of Malacca-Singapore

Meanwhile, based on Picture 2 the number of ships passing through this area continues to increase with an annual growth trend that tends to fluctuate. In 2011, the trend of ship growth declined to -0.8%. Then, ship growth continued to increase until 2018, with an average growth rate in the last year of 0.9%. In 2010, the number of ships passing through this strait was recorded at 74 thousand units, and continues to increase every year until 2018 the number of ships served is 85 thousand units.

Table 2. Average speed of ships passing through the Strait of Malacca-Singapore

DWT	V (Kn)
Kurang 40.000 DWT	**NORM(10.8, 3.64)**
40.000 - 55.000 DWT	**TRIA(7, 11.6, 16)**
55.000 - 80.000 DWT	**TRIA(2, 9.92, 21)**
80.000 - 120.000 DWT	**NORM(10.9, 2.58)**
120.000 - 160.000 DWT	**5 + 12 * BETA(1.3, 0.983)**
160.000 - 320.000 DWT	**NORM(10.8, 2.52)**

Based on the table data above (see Table 2), the average speed of ships crossing the Malacca-Singapore Strait is 12 knots. The calculation of the shipping lane capacity used refers to the safe distance between ships in the shipping lane as released by PIANC (World Association for Waterborne Transport Infrastructure) in 2014. The safe distance in question is 8 times the LoA of the largest ship that passes through the shipping lane. Furthermore, by simulating the arrival of ships in certain time intervals assuming a ship speed of 12 knots, the result is that the capacity of Singapore's shipping lanes is 118,487 ships/year. If analysed further by reviewing a sample of ship arrival intervals in the Singapore Strait in 2019, with input data in the form of ship arrival intervals, and the actual speed of ships when crossing shipping lanes, the distribution of each simulation result variable is obtained as follows (see Table 3, Table 4, and Table 5):

Table 3. Ship Arrival Time

DWT	Inter Arrival Time
< 40.000	-0.001 + EXPO(0.218)
40.000 - 55.000	EXPO(0.623)
55.000 - 80.000	EXPO(0.469)
80.000 - 120.000	EXPO(0.485)
120.000 - 160.000	EXPO(0.547)
160.000 - 320.000	EXPO(0.568)

Table 4. DWT with Ship LoA

DWT	LoA
< 40.000	73 + 222 * BETA(1.56, 3.24)
40.000 - 55.000	182 + WEIB(17.2, 0.619)
55.000 - 80.000	189 + 111 * BETA(0.474, 0.575)
80.000 - 120.000	228 + LOGN(36.6, 158)
120.000 - 160.000	273 + 94 * BETA(0.0282, 0.0444)
160.000 - 320.000	277 + 123 * BETA(0.776, 1.21)

Table 5. DWT with Ship LoA

DWT	V (Kn)
< 40.000	NORM(10.7, 3.18)
40.000 - 55.000	TRIA(7, 11.6, 16)
55.000 - 80.000	TRIA(2, 9.92, 21)
80.000 - 120.000	NORM(10.7, 2.67)
120.000 - 160.000	5 + 12 * BETA(1.3, 0.983)
160.000 - 320.000	NORM(10.8, 2.52)

Based on ship traffic data in the Singapore Strait, there is an average trend of increasing ship traffic volume every year by 2%. Based on the sensitivity analysis of the decrease in speed to the capacity of the Singapore Strait shipping lanes, the following results were obtained (see Picture 3);

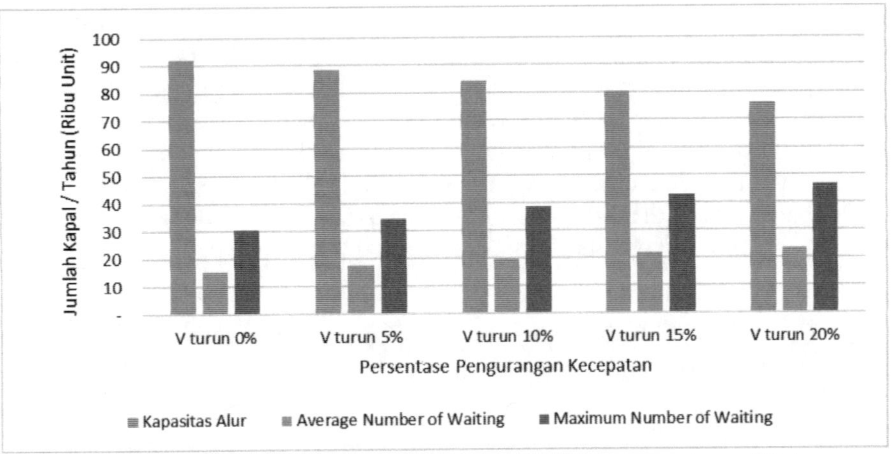

Picture 3. Presentation of Speed Reduction vs Number of Ship

Based on the sensitivity analysis table for the decrease in speed to the capacity of the Singapore Strait shipping lane, the channel capacity is 91,000 units with an average number of waiting 15,000 units and a maximum number of waiting 30,000 with a decrease of 0%. There was a 5% decrease with line capacity of 88,000 units, with an average number of waiting 17,000 units and a maximum number of waiting 33,000 units. A 10% decrease with a line capacity of 82,000 units with an average number of waiting 20,000 units and a maximum number of waiting 38,000 units. A 15% reduction with a line capacity of 80,000 units with an average number of waiting 21,000 units and a maximum number of waiting 42,000 units. 20% decrease with line capacity of 78,000 units with an average number of waiting 22,000 units and a maximum number of waiting 47,000 units.

The number of ships passing through the Malacca-Singapore Strait based on the marine traffic application from July to October was 4163 tankers, 310 container ships, 5690 general cargo ships, 15326 passenger ships, and 221 support ships. There are 5 types of ships that pass through the Straits of Malacca-Singapore Tanker, Container, General Cargo, Passenger, and Support Ship. For tankers, the average number of ships with a length of 201-250 m is 907 ships. The most container ships with a ship length of 251-300 m are 83 ships. General Cargo ships at most with a length of 151-200 m as many as 20388 ships. There are 15326 passenger ships with a length of 26-50 m and a support ship with a length of 26-50 m as many as 64 ships. The most common types of ships that pass through the Malacca-Singapore Strait are passenger ships. The growth of ships in the Malacca-Singapore Strait from 2010 to 2019 has decreased and increased. From 2012 to 2018 there was a constant increase.

The shipping traffic density in this study is calculated using an equation that requires data on the number of ships passing or passing by, the average speed of the ship, and the average width of the shipping lane. From the analysis of data obtained in the Singapore Strait, there is an average trend of increasing the volume of ship traffic every year by 2%. Along with the increase in traffic volume, it has the potential to cause congestion in the Singapore Strait.

The density of ship traffic is one of the factors for the occurrence of marine collisions and several crimes that occur in marine traffic. The density of ship traffic flows includes the flow of volume of goods, passengers, and ship visits as well as the speed of ship flow. The occurrence of traffic congestion in the Singapore Strait is indicated by a decrease in the speed of the shipping lane capacity. This will cause various kinds of accidents caused by traffic congestion. The use of VTS at ports in the Malacca-Singapore strait which has heavy traffic is very useful for providing traffic information such as incoming and outgoing ship data to assist shipping safety management. VTS in the Straits of Malacca and Singapore will optimize pilotage services to reduce the risk of accidents. By increasing the role of VTS in the Straits of Malacca and Singapore, it is able to reduce the threat of ship accidents that often occur. VTS plays a very significant role in improving shipping safety, especially on congested water routes.

4. Conclusion

From the analysis of data obtained in the Singapore Strait, there is an average trend of increasing the volume of ship traffic every year by 2%. Along with the increase in traffic volume, it has the potential to cause congestion in the Singapore Strait. So that the use of VTS at ports in the Malacca-Singapore strait is very useful for providing traffic information such as incoming and outgoing ship data to assist shipping safety management. VTS in the Straits of Malacca and Singapore will optimize pilotage services to reduce the risk of accidents. By increasing the role of VTS in the Straits of Malacca and Singapore, it is able to reduce the threat of ship accidents that often occur. VTS plays a very significant role in improving shipping safety, especially on congested water routes.

5. References

[1] Kusuma, L. T. W. N., & Tseng, F. S. (2020). IOT Framework to Support Maritime Highway Program: A Case Study in Indonesia. *Journal of Cases on Information Technology (JCIT)*, *22*(3), 35-50.

[2] Abd Rahman, N. A. (2020). China trade. *Singapore Infomedia, http://eresources. nlb. gov. sg/infopedia/articles/SIP_377_2005-01-18. html, accessed, 16.*

[3] Rusli, M. H. M. (2020). Navigational hazards in international maritime chokepoints: A study of the Straits of Malacca and Singapore. *Journal of International Studies*, *8*, 47-75.

[4] Zhang, L., Meng, Q., & Fwa, T. F. (2019). Big AIS data based spatial-temporal analyses of ship traffic in Singapore port waters. *Transportation Research Part E: Logistics and Transportation Review*, *129*, 287-304.

[5] Fatima, D. Q., & Jamshed, A. (2020). The political and economic significance of Indian Ocean: An analysis. *South Asian Studies*, *30*(2).

[6] Tichavska, M., Tovar, B., Gritsenko, D., Johansson, L., & Jalkanen, J. P. (2019). Air emissions from ships in port: Does regulation make a difference?. *Transport Policy*, *75*, 128-140.

[7] Khalid, N. (2012). Sea lines under strain: the way forward in managing sea lines of communication. *IUP Journal of International Relations*, *6*(2), 57.

[8] Song, B., Itoh, H., & Kawamura, Y. (2021). Development of training method for vessel traffic service based on cognitive process. *Cognition, Technology & Work*, 1-19.

[9] Kim, M., Takada, J. I., & Saito, K. (2017). Multi-dimensional radio channel measurement, analysis and modeling for high frequency bands. *IEICE Transactions on Communications*.

[10] Mao, S., Tu, E., Zhang, G., Rachmawati, L., Rajabally, E., & Huang, G. B. (2018). An automatic identification system (AIS) database for maritime trajectory prediction and data mining. In *Proceedings of ELM-2016* (pp. 241-257). Springer, Cham.

[11] Liu, L., Gu, Q., Li, L., & Lai, X. (2020, June). Research on Maritime Search and Rescue Recognition Based on Agent Technology. In *2020 International Conference on Artificial Intelligence and Electromechanical Automation (AIEA)* (pp. 201-205). IEEE.

[12] Ba, A. D. (2018). Governing the safety and security of the Malacca strait: the nippon foundation between states and industry. *Journal of Contemporary Asia*, *48*(2), 252-277.

[13] Banik, K. (2019). Europe and China in a Globalized World: The Geopolitical Impacts of Belt and Road. *China and the World*, *2*(01), 1950003.

[14] Van Westrenen, F., & Baldauf, M. (2020). Improving conflicts detection in maritime traffic: Case studies on the effect of traffic complexity on ship collisions. *Proceedings of the Institution of Mechanical Engineers, Part M: Journal of Engineering for the Maritime Environment*, *234*(1), 209-222.

Analysis The Effect of Outrigger (Cadik) Variations on Motion Response of Fishing Boat Using CFD Method

Akbar Bagus Darmawan[1], Yeddid Yonatan Eka Darma[1] and Jangka Rulianto[1]

[1]State Polytechnic of Banyuwangi

akbarbagusd@gmail.com

Abstract. Jukung is a type of traditional fishing boat used as transportation for fishing activities in Muncar, Banyuwangi, which has a wave as high as 2 m with a period of 4.6 s which is classified as 2nd sea state. One of the boats used has a length of 7.2 m, width 0.75 m, and height of 0.8 m. The characteristics of the hull which with an L/B ratio of 9.6 require outriggers to help balance while sailing. The outrigger itself is a cylindrical component that is placed at a certain distance on both sides of the boat. Since ancient times, the construction of outrigger boats has been carried out by local artisans by relying on experience. This is certainly very risky because it is not known for certain whether the construction, in this case, the outrigger can keep the boat stable. For this reason, we researched the boat's motion using the CFD method to provide a reference for the construction of the outrigger design. Using one of the existing boats with LOA 7.2 m, B 0.671 m, and H 0.8 m, and Vs 4.385 kt as a simulation model, we varied the outrigger-centerline distance by 2.5 m and 2.8 m; diameters of 90 mm, 120 mm, and 150 mm; which tested under conditions of full load and light load with the wave directions of headsea (180°), headquarter (135°), and beamsea (90°), resulting in Response Amplitude Operator (RAO) as follows. The model with an outrigger-CL distance of 2.8 m and diameter of 150 mm has the lowest heaving value of 0.02 m/m at a frequency of 4.1 rad/s from the headsea direction, 0.44 m/m at a frequency of 3.7 rad/s from the headquarter direction, and 0.4 m/m at a frequency of 6.2 rad/s from the beamsea direction. This model also has the lowest value in pitching and rolling responses from all directions with each value of 0.4 rad/m to 0.15 rad/m in pitching and relatively 0 rad/m in rolling responses.

1. Introduction

Based on observations on Muncar beach, Banyuwangi, the majority of local fishermen use jukung as a means of transportation used in sailing. Jukung is a term for a type of small boat that is widely used as a fishing vessel in Indonesia [1]. Jukung in Muncar has an outrigger that serves to maintain the stability of the boat while sailing.

The construction of the outrigger jukung was carried out by local carpentry fishermen relying on their experience and estimates in calculating the diameter of the outrigger and the distance between the outrigger and the centerline of the hull, so it is not known for certain whether the outrigger used is sufficient to keep the boat stable. This is quite risky considering that there are quite a several cases of jukung sinking while sailing in the waters of Muncar due to outriggers that are unable to withstand the waves.

Puspita's research [2] states that a boat with a length of 9.1m; 0.8m wide; 0.35m waterline; and an outrigger with a diameter of 0.14m, has good seakeeping characteristics in the waters of Puger, Jember. They used the Response Amplitude Operator (RAO) as a reference in determining how well the characteristics of the boat's motion in facing waves. For this reason, in this study, a boat motion simulation will be carried out on several outrigger variants using the CFD (Computational Fluid Dynamic) method to determine the size and distance of the outrigger that is suitable for use on jukung in the waters of Muncar, Banyuwangi.

2. Method
The methodology used in this research is a simulation-based method, which will be used to solve the problem. More details are described as follows.

2.1. Hull Model
Jukung Jenggirat Tangi is a boat that was used in the waters of Muncar, Banyuwangi. This boat is used as a model in the simulation.

Figure 1. Jukung Jenggirat Tangi.

The boat with a speed of 4.385kt is 7.2m long; 0.75m wide; a height of 0.8m; with an outrigger 9.2m long; 0.09m diameter; and 2.8m from the centerline of the hull. To create a 3D model of this boat, there are several steps as follows.

2.1.1. Hull point measurement. This measurement is carried out using a laser distance meter (LDM) by determining the vertical line of each station on the boat, then taking shots at 7 points on each station. The LDM results are the distance measured from the tool to the point on the boat and the angle of the shot.

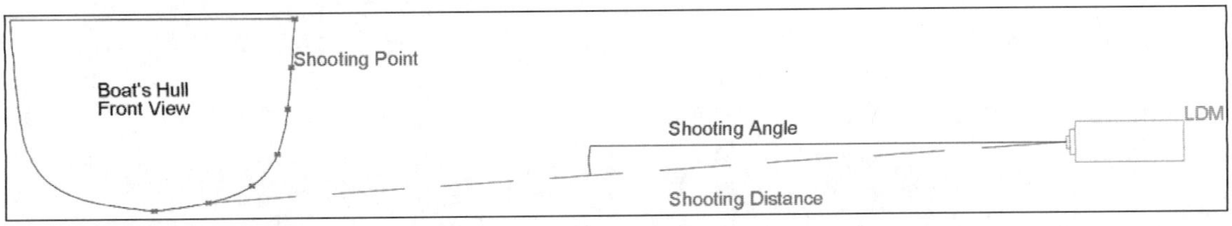

Figure 2. Boat's hull shooting scheme.

IOP Conf. Series: Earth and Environmental Science 1081 (2022) 012010 doi:10.1088/1755-1315/1081/1/012010

2.1.2. Linesplan drawing. The results are then drawn on CAD software to create a drawing of the boat's station, which is combined into a body plan. This body plan drawing is then projected up and to the side to become a linesplan.

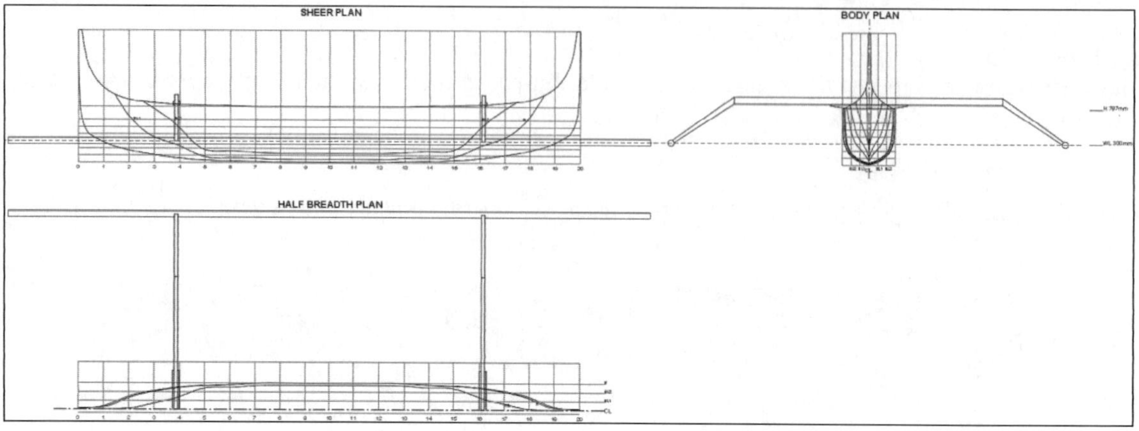

Figure 3. Linesplan drawing.

2.1.3. 3D model design. 2D drawings or linesplan and their sizes ae used as the basis for creating 3D models in the modeler software.

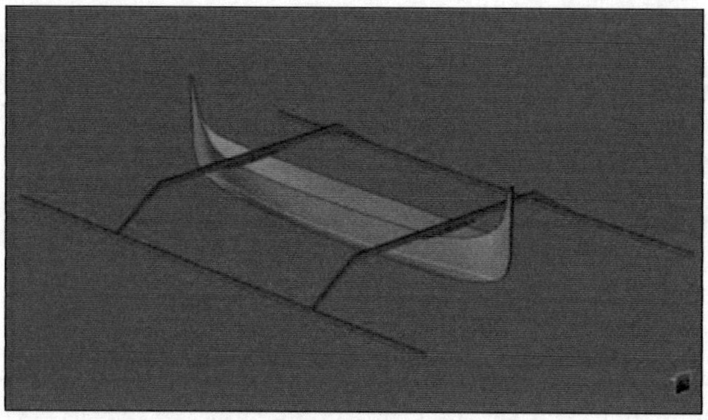

Figure 4. 3D model.

2.1.4. Model validation. Model validation was carried ou to ensure that the 3D model had the same size and parameters as Jukung Jenggirat Tangi. In practice, the tolerance value used is 0.5%.

Figure 5. Outrigger on the jukung.

The outrigger used has a brayungan and a brayungan support as a supporting component to connect the outrigger to the boat's hull. Puspita's research [2] states that jukung boats with a T/B value of 0.44 have good motion using outriggers with a diameter of 0.4T, while Santoso's research [3] said that the distance of the outrigger also affects the boat motion.

Table 1. Model validation of the jukung.

Components	Model 3D	Field Data	Gap	Gap (%)	Tolerance (%)	Status
LWL (m)	6.571	6.57124	0.00024	0.0037	0.5	*Passed*
B (m)	0.671	0.67064	0.00036	0.0537	0.5	*Passed*
T (m)	0.3	0.3	0	0	0.5	*Passed*
H (m)	0.8	0.8	0	0	0.5	*Passed*
Displacement Volume (m^3)	0.5344	0.5349	0.00050	0.0915	0.5	*Passed*
Displacement Weight (T)	0.5210	0.5218	0.00080	0.1616	0.5	*Passed*
C_b	0.394	0.395	0.00100	0.1808	0.5	*Passed*

2.2. Boat motion response
Ships/boats that are above sea level will always get external force, especially waves that cause the ship to experience free motion as shown in the figure below [4].

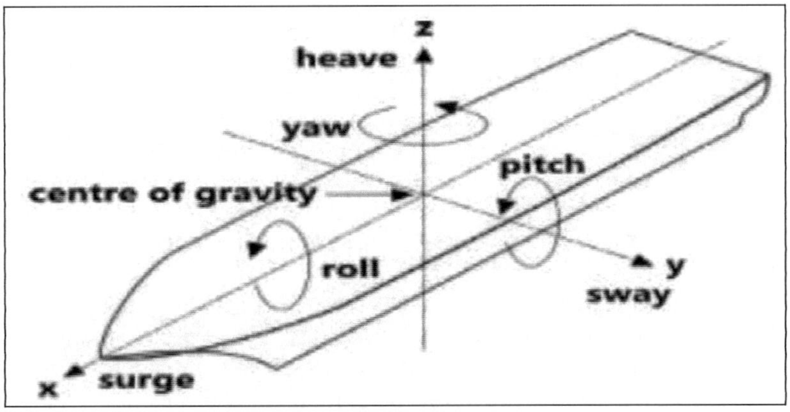

Figure 6. Six degrees of freedom of the ship.

In obtaining wave treatment, the ship experiences two types of motion, there are rotational motion and translational motion. Rotational motion is an angular motion that includes rolling (angular movement according to the X-axis in the form of swinging to the starboard and the portside), pitching (angular movement according to the Y-axis in the form of nodding by the bow and the stern), and yawing motion (angular motion according to the Z-axis of rotation). Meanwhile, translational motion is a regular motion along its axis. These motions include surging (movement about the X-axis), swaying (movement about the Y-axis), and heaving (movement about the Z-axis) [5].

2.3. Response Amplitude Operator (RAO)
Evaluation of boat motion is only carried out on oscillatory motion. Of the six motions of the ship, three motions are truly pure oscillatory motions, there are heaving, pitching, and rolling. This is because

the motions will return the ship to its original position when the ship is not in a state of equilibrium. Thus, the motion works under the influence of a force or a turning moment. This is different from the three other ship motions, which are surging, swaying, and yawing, the ship does not return to its original position when it is not in a balanced position unless there is a back force or moment that causes it to work in the opposite direction [6]. The results of the prediction of ship motion are generally displayed in the form of a comparison between the amplitude of the incident wave and the amplitude of a particular mode of motion, which serves as a function of the change in wave frequency. The amplitude comparison is called the Response Amplitude Operator (RAO) [2].

RAO is the ratio between the amplitude of the ship's motion (both translational and rotational) to the wave amplitude at a certain frequency [7]. RAO for translational motion is a direct comparison between the amplitude of the ship's motion with the wave amplitude, both of which are in units of length. For rotational motion, the RAO value is the ratio between the amplitude of the rotational motion to the slope of the wave which is the product of the wavenumber. So, to get the motion value on the CFD, it is necessary to reprocess the results of the motion value [8].

The general form of the response graph of a ship's motion which has a stiffness factor, in this context are heave, roll, and pitch, due to the excitation of harmonic wave loads is more clearly shown in the following figure.

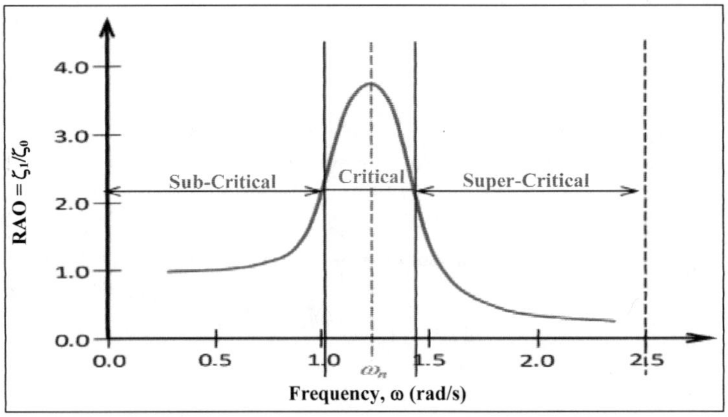

Figure 7. The general form of RAO graph.

Referring to the figure above, Djatmiko's research [9] explains that the response curve for the ship's motion can be divided into three areas, there are sub-critical, critical, and super-critical. The low-frequency portion, or wavelength (with the period), is called the sub-critical region. In this area, the marine structure will move following the pattern or contour of the long wave elevation, so that the amplitude of the motion is approximately equivalent to the wave amplitude, which is known as contouring. In the hydrodynamic correlation equation, in the low-frequency region, the motion will be dominated by the stiffness factor.

The second is the critical region, covering the mid-arm of the curve on the low-frequency side to the top of the curve and extending to the mid-arm of the curve on the high-frequency side. The peak of the curve is at the natural frequency, which is the resonance region, so the motion response is magnified, or the motion amplitude will be several times greater than the wave amplitude. Hydrodynamically in the natural frequency region, the motion will be dominated by the damping factor.

Third, is the super-critical region, namely the region of high-frequency (waves with a short period). In this area, the response to motion will be reduced. The higher the frequency or the denser the crests of the successive waves, the effect such as ocean structures moving over relatively flat water will occur. Therefore, the motion of marine structures is referred to as platforming. In terms of hydrodynamic correlation, in this high-frequency region, the motion will be dominated by the mass factor.

2.4. *Waves direction toward the ship*

The direction of the waves in Figure 8 is divided into five directions towards the ship, the five directions are head sea at an angle of 180°, bow quarter at an angle of 135° and 225°, a beam sea at an angle of 90° and 270°, a stern quarter at an angle of 45° and 315°, and stern sea at an angle of 0° [4].

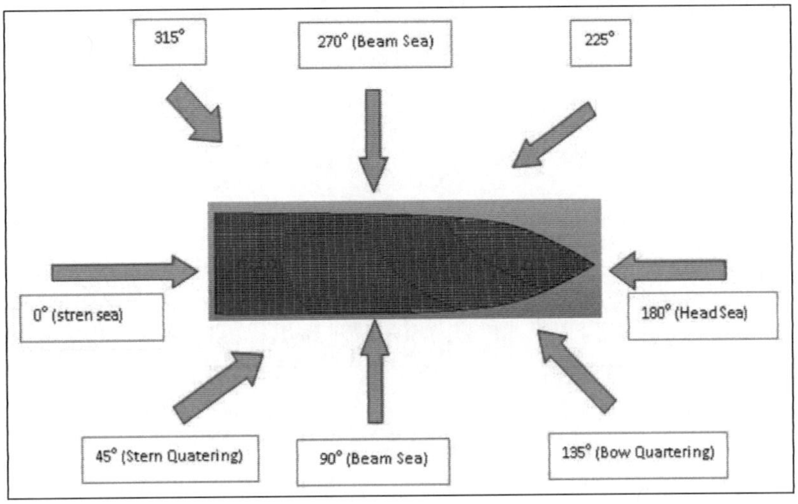

Figure 8. Waves angle orientation.

2.5. *Waves in the waters of Muncar*

Based on data from the Indonesian Meteorology, Climatology and Geophysics Agency (BMKG) on June 17th, 2021, it was stated that the wave amplitude value in the waters of Muncar, Banyuwangi ranged from 1.25 meters to 2.50 meters as shown in the following figure.

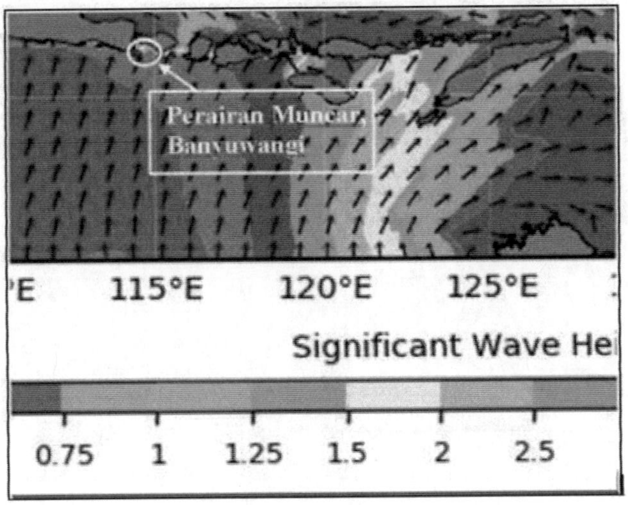

Figure 9. Wave height in the waters of Muncar.

According to Romadhoni's research [10], regarding the conditions of wave characteristics that have been determined by WMO (World Meteorological Organization) in the Sea State Code and also shown by Yim's research [11] in table 2 below, this value belongs to state 2 with a period of 4.6s.

Table 2. WMO sea state.

Sea State Code	Significant Wave Height (m)	Spectral Peak Period (s)
1	0.5	2.4
2	2.0	4.6
3	4.0	6.0
4	6.5	7.5
5	10.0	8.9
6	16.0	10.8
7	30.0	13.6
8	50.0	17.0
9	100.0	22.4

3. Result and discussion

3.1. Model variants

Following the planning of outrigger variations, six models were made with variations as shown in table 3 below and also in the following figures as the display of the variants.

Table 3. Outrigger variations for the model.

No.	Distance of Outrigger to the Centerline of Hull	Outrigger Diameter	Models
1	Original (2.8m)	Original (0.09m)	RD-2800-0090
2	Original (2.8m)	0.12m	RD-2800-0120
3	Original (2.8m)	0.15m	RD-2800-0150
4	2.5m	Original (0.09m)	RD-2500-0090
5	2.5m	0.12m	RD-2500-0120
6	2.5m	0.15m	RD-2500-0150

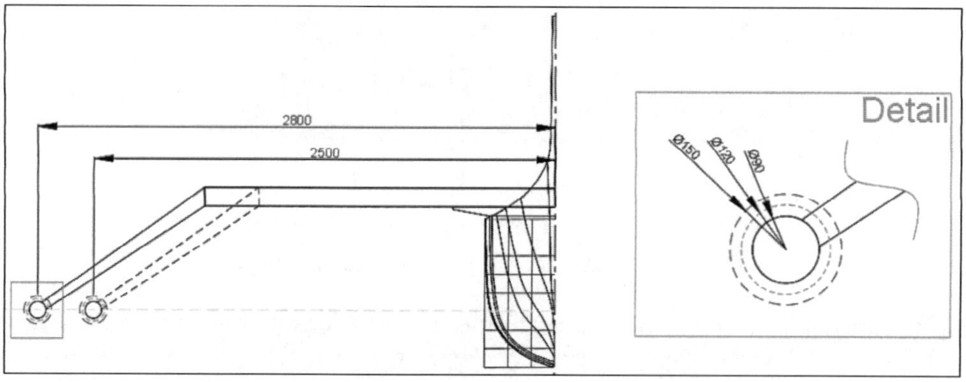

Figure 10. Outrigger variation scheme.

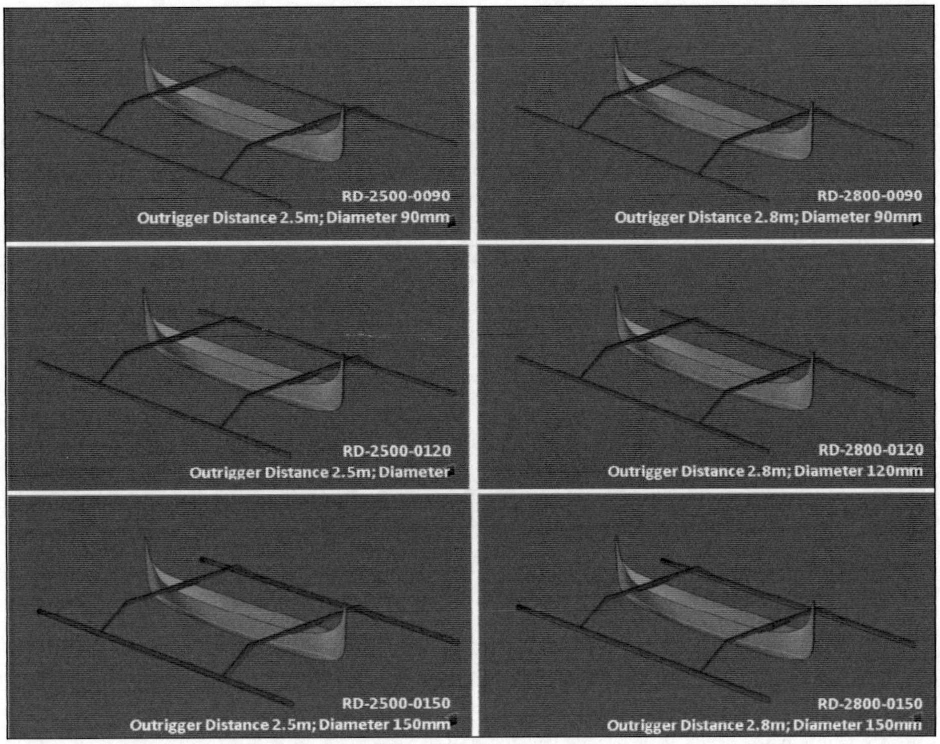

Figure 11. Display in perspective view.

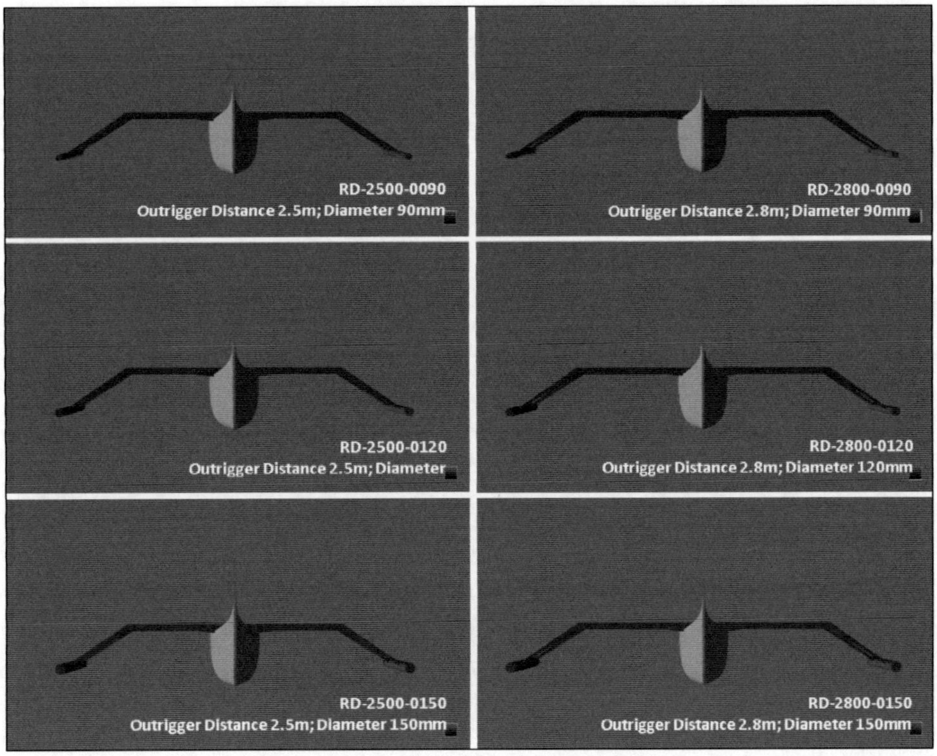

Figure 12. Display in front view.

3.2. Resistances

Before analyzing the boat motion, another simulation was carried out to determine the resistance of each model. In hydrodynamic terms, resistance is the magnitude of the fluid force acting on the ship in such a way that it opposes the motion of the boat. Resistance in the shipping world is very important to be calculated accurately [12]. By using the Holtrop method, the resistance graph is obtained as follows.

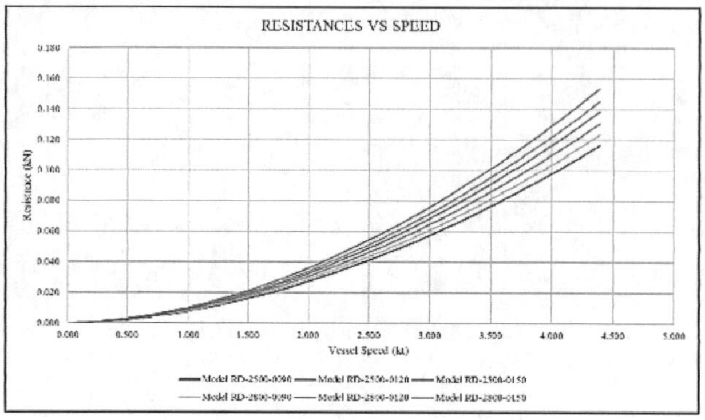

Figure 13. Resistances graph for six models.

Based on the graph above, it can be seen that the farther the outrigger distance and the larger the diameter, the resistance value will also increase. However, the difference of each model's value at a speed (Vs) of 4.385 kt is not significant, with the lowest value on the RD-2500-0090 model of 0.117 kN and the highest value on the RD-2800-0150 model of 0.154 kN. The difference between the lowest and the highest value is only 0.037 kN, so the resistance does not significantly affect the model selection.

Table 4. Resistance of each model in kN.

Model	Speed (kt)				
	0.987	**1.973**	**3.015**	**4.001**	**4.385**
RD-2500-0090	0.008	0.027	0.058	0.098	0.117
RD-2500-0120	0.008	0.030	0.065	0.110	0.131
RD-2500-0150	0.009	0.033	0.073	0.123	0.145
RD-2800-0090	0.008	0.028	0.062	0.104	0.123
RD-2800-0120	0.009	0.032	0.069	0.117	0.139
RD-2800-0150	0.010	0.035	0.077	0.130	0.154

3.3. RAO numerical simulation

Numerical computation of hydrodynamic factors which include added mass coefficient, damping, stiffness, excitation force, and wave diffraction can be done by applying strip theory, where hydrodynamic problem solving is carried out on each 2D section of the hull of the submerged floating building and then integrates it towards the longitudinal floating building [9]. The simulation this time uses the numerical method of strip theory which uses a measure hull method based on Lewis mapping theory to calculate the hydrodynamic characteristics of each cross-section [13]. The numerical parameters used include load line, frequency limit, vessel speed, wave directions, and wave characteristics, which are shown in the following table.

Table 5. Parameter used for numerical simulation.

Parameters	Value	Parameters	Value
Load line		Wave Characteristics	
Full load	0.3m	Type	Jonswap
Light load	0.2m	Char. Height	2.0m
Frequency limit	21 frequencies	Modal Period	5.493s
Vessel Speed	4.385kt	Average Period	4.6s
Wave Direction			
Head sea	180°		
Headquarter	135°		
Beam sea	90°		

The numerical simulation results include RAO graphs and boat motion displays, which are shown in the following figures.

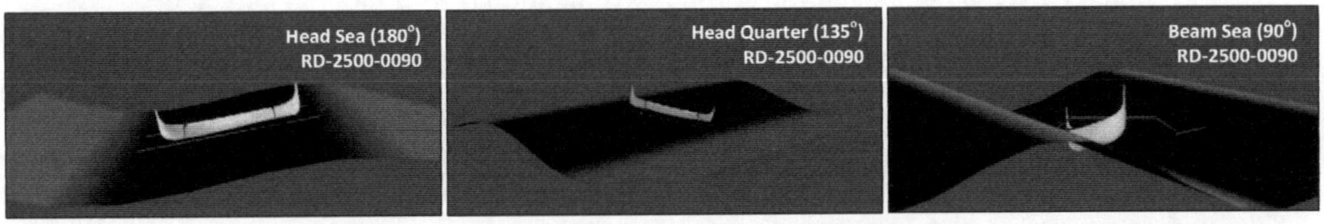

Figure 14. Display of simulation results in three directions of incoming waves using numerical methods.

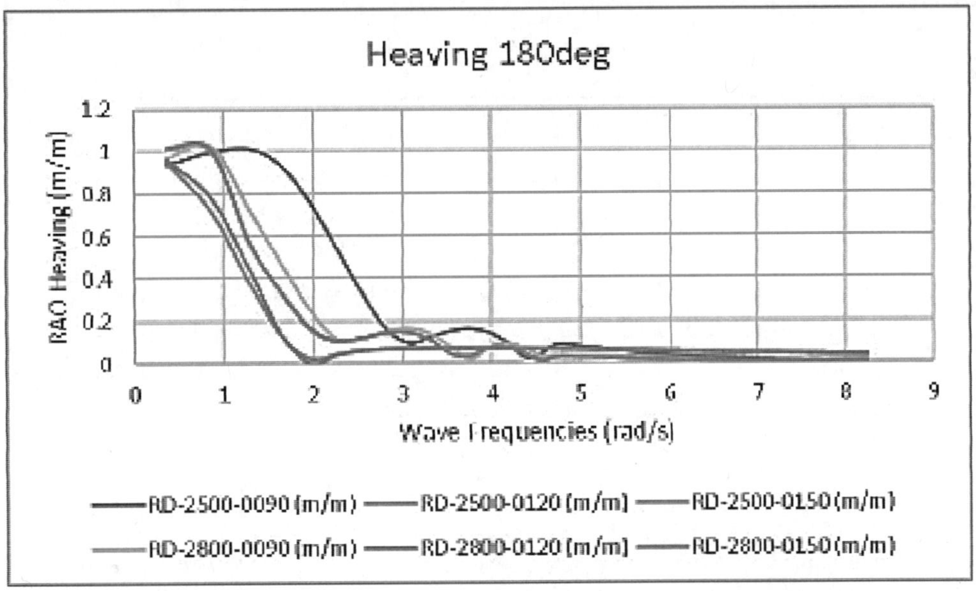

Figure 15. RAO heaving graph at full load condition using numerical method.

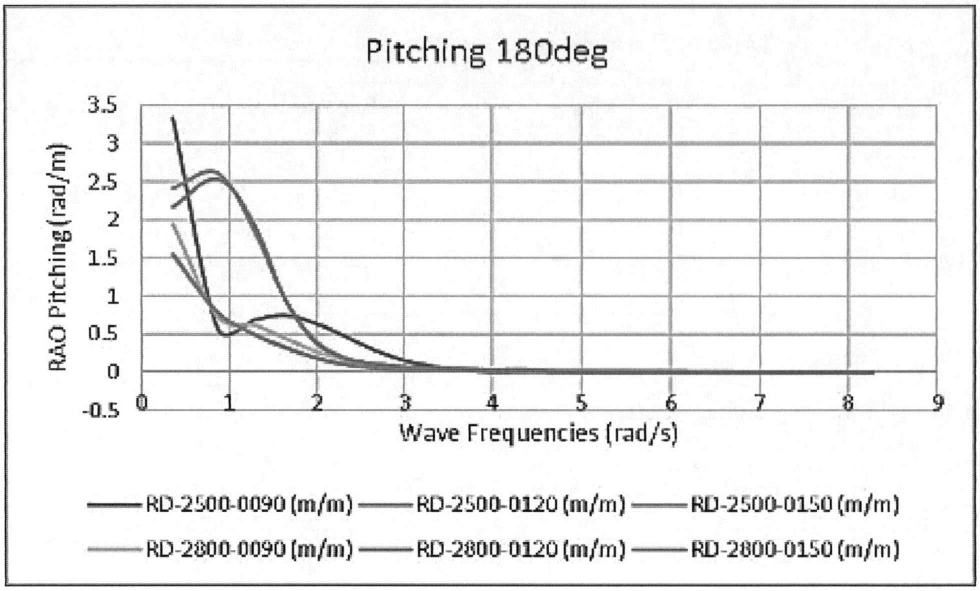

Figure 16. RAO pitching graph at full load condition using numerical method.

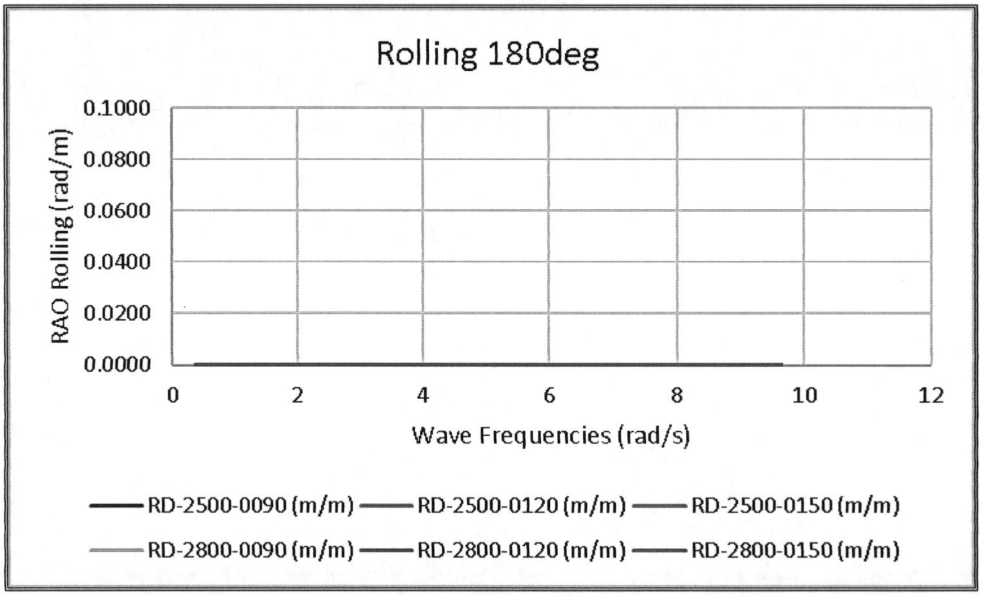

Figure 17. RAO rolling graph at full load condition using numerical method.

3.4. RAO CFD (computational fluid dynamic) simulation

CFD method is a method to determine the characteristics of fluid dynamics computationally. The stages of applying this method start from design, modeling, application of conditions, then simulation according to the model and conditions used. Parameters used in this simulation include load line, frequency range, vessel speed, wave directions, and wave characteristics. This method uses units of m/s for vessel speed and uses a range for frequencies with values of 0.369rad/s (low) and 8.269rad/s (high).

Table 6. Parameter used for CFD simulation.

Parameters	Value	Parameters	Value
Load line		Wave Direction	
Full load	0.3m	All Directions	180°; 135°; 90°; 45°; 0°; -45°; -135°; -90°; -180°
Light load	0.2m	Wave Characteristics	
Frequency range		Height	2.0m
Lowest	0.369rad/s	Period	4.6s
Highest	8.269rad/s		
Vessel Speed	2.364m/s		

From the simulation, the ship motion display and RAO graphic data are obtained with the parameters shown in table 6 above, which are shown in the following figure.

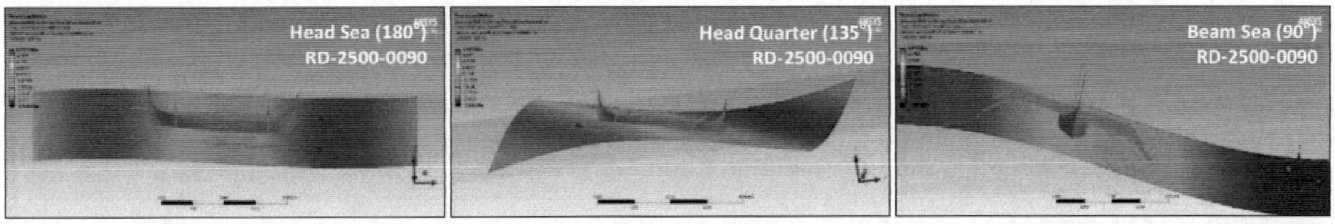

Figure 18. Display of simulation results in three directions of incoming waves using CFD methods.

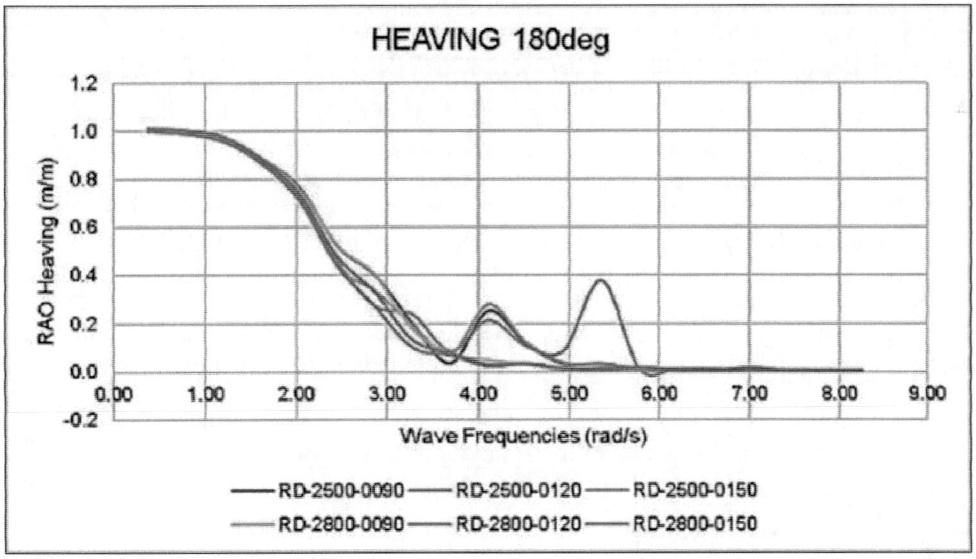

Figure 19. RAO heaving graph at full load condition using CFD method.

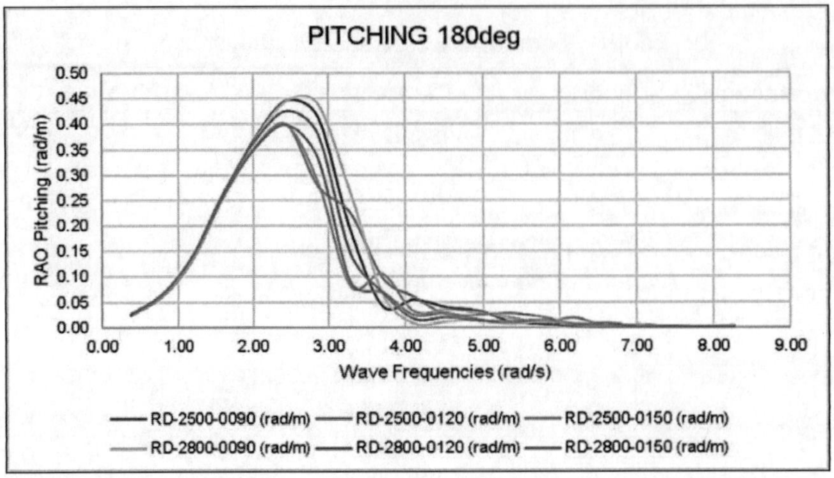

Figure 20. RAO pitching graph at full load condition using CFD method.

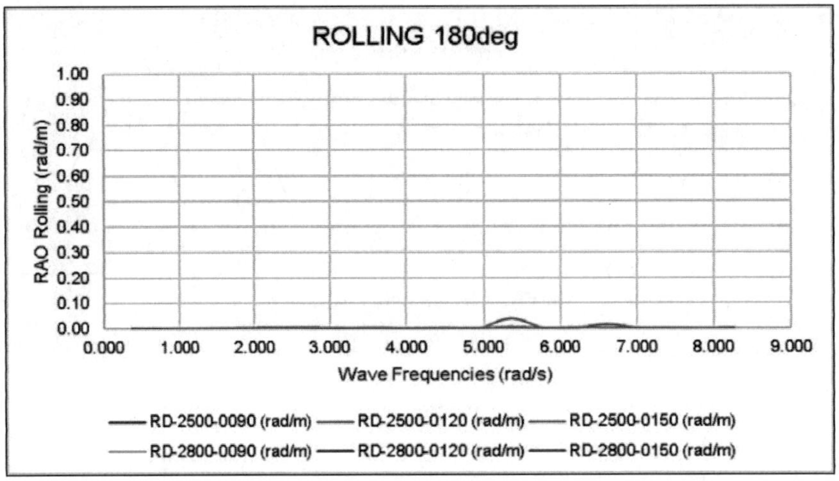

Figure 21. RAO rolling graph at full load condition using CFD method.

3.5. Simulation results validation

Ghadimi et al [14] stated in their study that a maximum error of 8.16% can indicate the fact that the simulation runs quite well, so, in this paper, we use a tolerance of 10% for simulation error.

Table 7. Data of the simulation results validation.

Wave Details			CFD	Numeric	Gap	Gap (%)	Tolerance (%)	Status
Direction	Freq	Height						
Head sea	1.3 rad/s	2 m	0.98	1.04	0.042	5.7751	10	*Passed*
Headquarter	1.3 rad/s	2 m	0.96	1.02	0.043	5.7747	10	*Passed*
Beam sea	1.3 rad/s	2 m	0.91	0.94	0.010	3.1109	10	*Passed*

The RAO heaving values of the RD-2800-0150 model at full load for the head sea, headquarter, and beam sea directions are 0.984 m/m, 0.963 m/m, and 0.910 m/m using the CFD method, respectively. This result has an average difference of 4.887% when compared to the results using the numerical method. As can be seen in the table above, this difference meets the tolerance value of 10%, so the parameters used are appropriate and the simulation results obtained can be accounted for.

3.6. Results of analysis

The results of the RAO analysis of heaving motion in the direction of the head sea (180°) show that the RD-2800-0150 model has the lowest value of 0.02m/m at a frequency of 4.1rad/s. With the headquarter direction (135°) the model RD-2500-0090 has the lowest value of 0.33m/m at a frequency of 4.5rad/s and the model RD-2800-0150 is 0.44m/m at a frequency of 3.7rad/s. With the direction of the beam sea (90°), the model RD-2800-0150 also has the lowest value of 0.4m/m at a frequency of 6.2rad/s.
The results of the RAO analysis of the pitching motion in the direction of the head sea (180°) show that the RD-2800-0150 model has the lowest value of 0.4rad/m at a frequency of 2.45rad/s, while at a frequency of 3.7rad/s, this model has a value of 0.07 rad/m. With the direction of the headquarter (135°) the model RD-2800-0150 has the lowest value of 0.15rad/m at a frequency of 4.1rad/s. With the direction of the beam sea (90°) the RD-2800-0150 model has the lowest value at each resonance and decreases at high frequencies.

The results of the RAO analysis of rolling motion with the direction of the head sea (180°) show that each model has a relatively low value close to 0 rad/m at each frequency, but is different from the RD-2500-0120 model which has a higher value than the other models of 0.039 rad/m at a frequency of 5.36rad/s. With the direction of the headquarter (135°) the model RD-2500-0090 has the highest value of 0.26rad/m at a frequency of 3.7rad/s. Then, with the direction of the beam sea (90°) the model with a variant of the outrigger-CL distance of 2.8m has a stable value at low frequencies up to 6.19rad/s.

Table 8. Result of the analysis.

Models	Descriptions	
RD-2800-0090	Jukung model with the outrigger-center distance of 2.8m and outrigger diameter of 90mm	-
RD-2800-0120	Jukung model with the outrigger-center distance of 2.8m and outrigger diameter of 120mm	-
RD-2800-0150	Jukung model with the outrigger-center distance of 2.8m and outrigger diameter of 150mm	Has the best RAO motion
RD-2500-0090	Jukung model with the outrigger-center distance of 2.5m and outrigger diameter of 90mm	-
RD-2500-0120	Jukung model with the outrigger-center distance of 2.5m and outrigger diameter of 120mm	-
RD-2500-0150	Jukung model with the outrigger-center distance of 2.5m and outrigger diameter of 150mm	-

4. Conclusion

Based on the results of the analysis and discussion of the research topics that have been carried out, the following conclusions can be drawn.

- Based on the results of the redrawing of the model and its variations, there were 12 outrigger jukung models according to the existing Jenggirat Tangi boat with LWL 6.571m, B 0.671m, T 0.3m, and H 0.8m. Along with variations of outriggers with diameters of 90mm, 120mm, and 150mm, as well as outrigger distances of 2.5m, and 2.8m CL.
- The boat model with FRP (Fiber Reinforced Plastic) material has parameters following the 1964 BKI rules regarding Rules for Fiberglass Reinforced Plastic Ships with an ivory distance of 0.36m, side hull skin thickness 20mm, bottom hull thickness 20mm, and girder thickness 8mm.

- The model with an outrigger-CL distance of 2.5m and an outrigger diameter of 150mm has a good heaving motion with the lowest values in the direction of the head sea, headquarter, and beam sea direction. This model also has the lowest pitching and rolling motion in the three directions of incoming waves, so this model is recommended for use in the waters of Muncar, Banyuwangi.

5. References

[1] Nasty A Z, Syahril S and Nabawi R A 2019 Stabilitas kapal jukung dengan jenis lambung pelat datar *Jurnal AERASI* **1** 68–75

[2] Puspita H I D 2017 Analisa gerakan kapal ikan bercadik semi-circular di perairan puger jember *Jurnal Elemen* **4** 94–101

[3] Santoso B, Helmi M and Nurhasanah N 2017 Optimasi panjang cadik kapal nelayan 3gt *Jurnal IPTEK* **21** 11–6

[4] Romadhoni R 2016 Analisa olah gerak kapal di gelombang reguler pada kapal tipe axe bow *KAPAL: Jurnal Ilmu Pengetahuan dan Teknologi Kelautan* **13** 61–8

[5] Rudiyansah H, Sulisetyono A and Ali B 2013 Analisa pengaruh seakeeping terhadap kinerja dan kenyamanan kapal ro-pax berbasis uji model di perairan laut jawa *Jurnal Teknik POMITS* **2** 1–6

[6] Hutauruk R M and Rengi P 2014 Respons gerakan kapal perikanan hasil optimisasi terhadap gelombang *Jurnal Perikanan dan Kelautan* **19** 13–22

[7] Hidayat I N, Zakki A F and Manik P 2019 Analisa performa fin stabilizer untuk memperbaiki gerakan rolling pada kapal longline 90 gt *Jurnal Teknik Perkapalan* **7** 248–54

[8] Heriyanty E, Hadi E S and Kiryanto K 2019 Analisa seakeeping ponton berbentuk prisma segi enam pada gelombang reguler dengan simulasi computational fluid dynamic (cfd) *Jurnal Teknik Perkapalan* **7** 548–56

[9] Djatmiko E B 2012 *Perilaku dan operabilitas bangunan laut di atas gelombang acak* (Surabaya: ITS-Press)

[10] Romadhoni R 2019 Analisa seakeeping ferformance kapal cepat model planing hull chine *INOVTEK POLBENG* **9** 30–7

[11] Yim S C, Nakhata T and Huang E T 2004 Coupled nonlinear barge motions, part ii: stochastic models and stability analysis *23rd International Conference on Offshore Mechanics and Arctic Engineering* (Vancouver B.C., Canada: OMAE2004-51131) pp 1–10

[12] Hadi E S 2009 Komparasi hull performance pada konsep design kapal ikan multi fungsi dengan lambung katamaran *KAPAL: Jurnal Ilmu Pengetahuan dan Teknologi Kelautan* **6** 212–7

[13] Bentley 2013 *Maxsurf motion user manual* (Bentley System, Incorporated 2013)

[14] Ghadimi P, Pourmostafa M and Najafi S 2020 Investigating the response amplitude operator of a heaving pontoon under the influence of a submerged trapezoidal breakwater *Advances in Civil Engineering* 2020

IOP Conf. Series: Earth and Environmental Science 1081 (2022) 012011 doi:10.1088/1755-1315/1081/1/012011

Current research outlook on solar-assisted new energy ships: representative applications and fuel & GHG emission benefits

Tuswan Tuswan[1,*], Saefulloh Misbahudin[1], Sony Junianto[2], Hartono Yudo[1], Ari Wibawa Budi Santosa[1], Andi Trimulyono[1], Ocid Mursid[1], Deddy Chrismianto[1]

[1] Department of Naval Architecture, Universitas Diponegoro, Semarang 50275, Indonesia
[2] Department of Mechanical and Energy Engineering, Politeknik Elektronika Negeri Surabaya, Surabaya 60111, Indonesia

Corresponding author: tuswan@lecturer.undip.ac.id

Abstract. Shipping industries, one of the efficient modes of freight transport for global trade, are now being forced to demonstrate environmental commitment to reduce greenhouse gas emissions progressively. Stakeholders in the maritime community have proposed various countermeasures and action plans, including applying new energy resources for the future development of ship powers. This paper introduces a comprehensive review to categorize the pathways and highlight the core technological concepts, technical issues, current applications, implementation barriers, and future outlooks regarding integrating new energy resources into ship power systems, including single solar-powered and hybrid new energy resources. Eco-friendly energy sources were promising energy savings and GHG emission benefits. Still, integrating new energy source generation systems would significantly depend on several critical factors. Because ships require a large surface area for installing PV panels, the utilization of solar-assisted power generation on large-scale boats is currently uncommon. The energy generated by the PV generating system is mostly utilized for lighting and electricity. Integrating new energy source generating systems with current ship power systems is a viable approach for enhancing energy efficiency will likely be the focus of future ship-based energy system research. Furthermore, due to the variety of applications, the ship type, power system structure, and navigation area should be considered while selecting a suitable new energy source.

1 Introduction

Maritime transportation has been a significant economic development promoter by facilitating trade between continents for centuries. 80–90% share global business via sea transportation [1], [2]. Maritime transport has expanded by 250 percent over the last 40 years, expanding at the same rate as worldwide GDP and faster than energy consumption (170%) and world population (90%) [3]. Shipping is recognized as a more energy-efficient mode of transportation than other transportation methods due to its vast harvest capacity and low fuel consumption per tonne moved [4]. According to the International Maritime Organization (IMO), ships use about 300 million tons of fossil fuels every year. Although other forms of transportation primarily employ fossil fuels, the shipping sector produces about 3.1% (1130 Mt CO_2) of the global total GHG emissions due to the high composition of the fossil fuel. Heavy fuel oil (HFO) accounts for 72%, marine diesel oil (MDO) for 26%, and

liquefied natural gas (LNG) for 2% [5]. Also, ships release pollutants such as sulphur oxides (SOx), nitrogen oxides (NOx), and particulate matter (PM), all of which contribute to air pollution [6].
Since the emissions produced by ships have become a severe concern, developing efficient technologies to reduce emissions in the shipping sector has attracted the attention of the world community. In response to this, IMO has implemented strict standards through the International Convention for the Prevention of Pollution from Ships (MARPOL Annex VI), such as the Energy Efficiency Design Index (EEDI) and the Ship Energy Efficiency Management Plan (SEEMP) [7]. Xing et al. [8] divided alternative strategies for reducing CO_2 emissions into four groups: technical actions, operational actions, eco-friendly fuels, alternative resources. Technological/technical measures imply CO_2 emission reductions with upgraded hardware, and active actions indicate achieving targets through operational efforts. Eco-friendly fuels and alternative resources are, to some extent, included in the technology measure area because they require hardware upgrades and new designs.
Alternative energy sources are non-polluting, clean, and infinite. LNG, nuclear, biofuel, wind energy, and solar energy are just a few of the new energy sources that have been integrated into ship power systems. Using eco-friendly fuels and alternative resources to replace fossil fuels on ships may be a potential strategy to help shipping become more eco-friendly. Research on alternative sources is conducted both as a hybrid and stand-alone system [9]. Solar energy can now be utilized as a primary power source to propel small-scale ships and as an additional power source to power lights, communication devices, and navigation systems on large-scale ships. Modern sails are more typically utilized to drive ships using wind energy as an auxiliary power source. Wind-generated electricity is a different way of using wind energy that ships are not yet routinely employed. Most experts believe mixing diverse new energy sources increases overall output power.
A summary of the utilization of new energy sources on the vessel, especially solar energy, will be discussed in this article; the progress achieved to date on solar-powered ships, whether fully solar-powered or assisted solar-powered ships until hybrid ships, will be comprehensively reviewed. Furthermore, the discussion will start from critical technology aspects to general core technology for ship power systems integrated with various new energy sources. The remainder of this paper is organized as follows: Section 2 introduces greenhouse gases on shipping activities. Section 3 discusses potential countermeasures for CO_2 emission from alternative energy resources. Section 4 presents representative applications on solar-powered ships. Section 5 summarizes typical applications on solar-powered ships integrated with new energy resources. Section 6 draws conclusions and prospects.

2 Greenhouse Gases on Shipping Activities

Maritime transportation produces 3% of CO_2 emissions of the total annual anthropogenic greenhouse gas (GHG) emissions. These emissions are assumed to increase by 150-250% by 2050. Therefore, maritime transportation needs to face the challenge of reducing GHG emissions for a more controlled climate change [10]. According to the Intergovernmental Panel on Climate Change (IPCC), a 40-70% reduction in GHG emissions could limit the increase in Earth's surface temperature to below 2 C [11]. As before, international shipping provides 3.1% of global CO_2 emissions of 1130 Mt CO_2 [12]. The largest source of GHG emissions in shipping comes from container ships, bulk carriers, and oil tankers, as shown in Figure 1. The emission contribution from these ships is due to the extended shipping routes. These ships sailed on international and intercontinental routes rather than domestic and coastline routes [13]. However, in 2007, the contribution to total emissions has decreased until 2014.
Figure 2 shows the spatial distribution of emissions. Different fuels have different emission levels, and inefficient ships use more fuel. Of the approximately 300 Mt of global maritime fuel consumption in 2015, 72% was residual fuel (e.g., HFO heavy fuel oil), 26% refined oil (e.g., marine diesel oil), and 2% was liquefied natural gas (LNG) [14]. HFO usually has a higher sulphur content when compared to other fuels [15] and accounts for about 13% of global SOx emissions per year (2012) in international shipping [16]. In 2009, The Guardian issued a report stating that the 15 largest ships produce more sulphur pollutants than all the cars in the world combined (about 760 m of cars) [17]. Sulphur and nitrogen oxide emissions have a short-lived climate cooling effect, which means the net

impact of shipping over 20 years (based on one year's emissions) is on global temperature reductions [18].

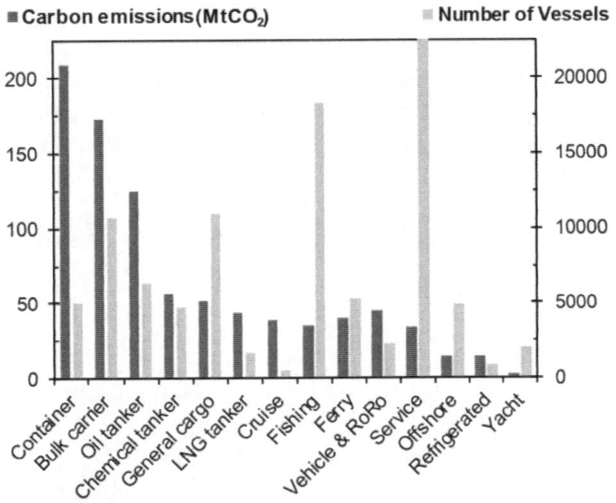

Figure 1. Several merchant ships and their carbon emissions, by category in 2017. Ferry includes passenger and passenger-RoRo (roll-on roll-off) data [13].

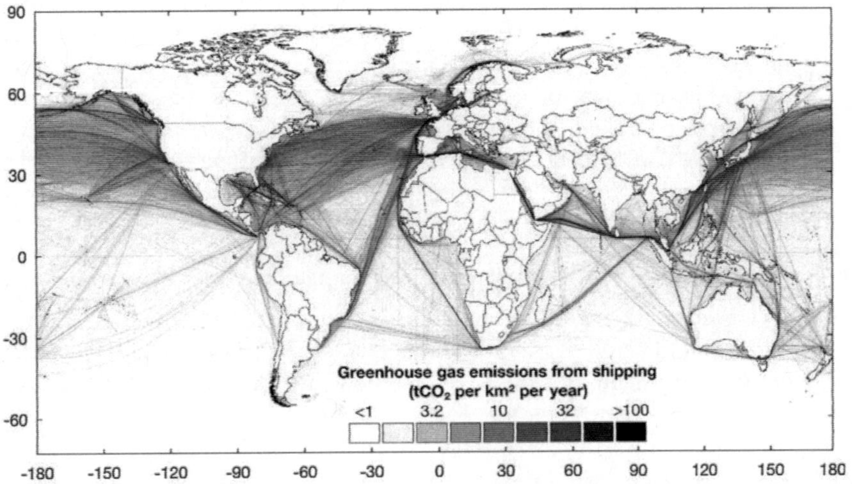

Figure 2. Map showing the global distribution of greenhouse gas emissions from shipping. Based on the intensity of shipping lane usage from [13] normalized to 2013 emissions from the sector.

To improve ship energy efficiency and control CO_2 emissions, MARPOL Annex VI introduces two mandatory mechanisms, namely the Energy Efficiency Design Index (EEDI) for new ships [19] and the Ship Energy Efficiency Management Plan (SEEMP) for all ships [20]. This strategy aims to reduce the carbon intensity of international shipping by at least 40% by 2030 and reduce total annual GHG emissions by at least 50% by 2050, compared to previous years' levels, and phase out GHGs.

CO_2 emission restriction regulations that have given effect since 2013 reduce fuel consumption, which has a high impact on emissions. Since then, all new vessels above 400 GRT must meet the minimum energy efficiency level in terms of the EEDI value [21]. On the other hand, the SEEMP ensures optimal ship operations and services. Achieving favorable EEDI scores is defined as simplifying total CO_2 emissions from fuel oil combustion in main engines, auxiliary engines, and boilers about transportation works.

These emission restrictions may be done by developing regulations for renewable energy on a shipping vessel. The use of renewable energy needs to be studied comprehensively, especially regarding the quality and quantity of its service in maritime transportation aspects. The study refers to

EEDI and SEEMP scores that must be met. To reduce GHG, renewable energy regu in the shipping aspect cannot be separated from the EEDI and SEEMP factors.

3 Potential countermeasures for CO_2 emission from alternative energy resources

Research on new energy sources for vessels to reduce CO_2 emissions is a highly developed topic among researchers. These energy sources include wind energy, solar energy, wave energy, and others. Solar energy and wind energy are more set globally than other energy sources, which can supply the world's primary energy consumption [9]. The author will be more inclined to discuss solar-powered ships in this research. Still, as a comparison, the author will also review other alternative energy sources used in the maritime world. Several new energy applications on several ship types are depicted in Figure 3.

3.1. Solar energy

Solar energy is a cheap, abundant, and low-risk energy source. Christopher J. Rhodes [22] stated that 30% (about 1015 W) of the total solar energy that illuminates the Earth is reflected into space. This value is greater than the total energy produced by humans (oil, gas, coal, nuclear, hydro, etc.). Energy from sunlight is generated from "photovoltaic power generation." The photovoltaic effect was first recognized in 1839 by French physicists. There are two main operating modes for PV (photovoltaic) generating systems: stand-alone and grid-connected modes. In addition, a hybrid mode has also been developed, which is a combination of the two operating systems [9].

A stand-alone PV generation system is the most straightforward photovoltaic generation system, not connected to the system from other energy sources. Simply put, this system can be realized with a PV module, a battery as a power store, and a load. The batteries used must have sufficient capacity to store the energy generated by the solar panels during the day for use at night and during bad weather conditions [23]. Meanwhile, grid-connected PV generation systems are more complex. This system can be applied to ships that use several energy sources, such as fuel and solar panels, to generate electricity. They are connected to the grid via an inverter and do not require batteries because the grid can receive all the electricity that a PV generator can supply [23].

There are already many small fully solar-powered boats that have been mass-produced and marketed; some examples are EcoCat EcoBoat (18m length) from Metaltec Naval Shipyard [24] and SunCRZ 22 from Navgathi [25]. These boats are 100% zero GHG emissions and 0% fuel. However, it's very inefficient when fully solar-powered for larger ships, especially for commercial usage. The overall potential for reducing CO_2 emissions for the solar-assisted ship has been estimated to be 0,2 – 12% [8].

3.2. Wind energy

Wind energy is the first energy source used to propel a sailing ship; the difference is that in previous times wind energy was used directly to move a ship, while now wind energy is converted into electrical energy with wind turbines. There are two main types of modern wind turbines, horizontal axis wind turbines (HAWTs) and vertical axis wind turbines (VAWTs) [26].

HAWTs are widely used to generate electricity on land due to their greater energy output, but for turbines used on ships, the VAWTs are more suitable. The horizontal turbine will cause more excellent resistance, reducing the ship's speed. An example of this usage is the E-Ship [27]; this vessel uses a vertical flatter rotor by relying on the Magnus effect to generate 6,3 MW of electricity. The concept of a sailing ship also developed, for example, is the ocean bird vessel [28]. The biggest challenge of sailing ships using wind energy is the efficiency of their speed and cruising range. Potential CO_2 countermeasure from the wind-powered ship is 1-50% [10].

3.3. Liquified natural gas (LNG)

Liquefied Natural Gas (LNG) is one of the "eco-friendly fuels" is because LNG still produces GHG emissions, but not as much as fuel (HFO and MDO). Theoretically, the use of LNG can reduce CO_2 emissions by about 20 - 25% compared to conventional ship fuel, or about a 12 - 20% reduction in net GHG emissions due to methane slip [8].

IOP Conf. Series: Earth and Environmental Science 1081 (2022) 012011 doi:10.1088/1755-1315/1081/1/012011

LNG is the leading choice for fleet companies to switch from fossil fuels to more eco-friendly energy; they modify the main engine and add more space as fuel gas storage. Farianto [29] stated that the modification of the ship's main engine from conventional fuel conversion to gas fuel would provide significant benefits if the modified engine is relatively new, this conversion cost will be more efficient when compared to the investment cost to install gas emission purifiers on the ship, especially for the older engine in. Several ship engines are currently being produced using dual-fuel technology with LNG fuel as energy sources. For example, is Wärtsilä 20DF marine engine [30] is used in offshore construction vessels, Orion.

3.4. Wave energy

Wave energy conversion (WEC) is the definition for ocean wave energy conversion into the desired form, usually electrical or mechanical energy [31]. The use of wave energy as a source of energy on ships is very little use, still in the experimental stage and application for small ships.

One of the concepts of wave energy conversion in the ship is a wave devouring propulsion system (WDPS) using two fins (hydrofoil) below the ship. The foil generates direct thrust from the wave power and reduces the roll and pitch motion of the ship in rough sea conditions [32]. The ship that uses this concept is the SUNTORY Mermaid II [33].

(a) Toranor Planet Solar [34]

(b) E-Ship 1 [27]

(c) Orion, offshore construction vessel [30]

(d) SUNTORY Mermaid II [33]

Figure 3. Several new energy applications on several ship types

4 Representative application on solar-powered ships

As explained by previous authors, the topic of research on solar-powered ships is fascinating; this is not limited to the research stage but has entered the production stage. Since the world's first solar-powered boat, Solar craft 1, made by Mr. Alan Freeman [35], many shipyards and shipbuilding companies are starting to develop and market solar-powered ships.

4.1 Fully Solar-Powered Ships

The Turanor planet solar has become the largest fully solar-powered vessel (displacement up to 85 t) until now and the first to circumnavigate the w. This vessel is supported by a 537m^2 solar panel and

Li-Ion battery capable of supplying a maximum motor power of up to 120 kW and three days auto world success without sunshine [34].

Fully solar-powered vessels have the capability of zero-emission and zero fuel. The author has collected several boats that use 100% solar-powered energy to represent the progress of the development of solar-powered ships, as presented in Table 1. Solar energy on ships is highly promising, but how to install more PV panels in the restricted space on the ship deck in order to increase the installed capacity of the PV generation system remains a challenge. Because of the poor conversion efficiency of PV panels, even in places with copious solar radiation, it is not possible to connect a PV generating system directly to the ship's main grid. Standalone PV generation systems coupled with an electrical propulsion unit, on the other hand, may offer the prospect of mainstream development of solar energy-based systems for small-scale ships. Solar energy might be employed as the primary energy source in small-scale solar-powered ships, giving a truly green option.

Table 1. Application on fully solar-powered ship

Vessel	**Photovoltaic System**				
Name (Type)	**Dimension (L ×B) (m)**	**PV generation System**	**Battery**	**Stage**	**Ref.**
Turanor Planet Solar (Catamaran Yacht)	35 × 16	537 m^2 of PV panels, 93 kWp	8.5 t of Li-ion battery, 1129 kWh	Already produced	[34]
Sun 21 (Catamaran Yacht)	14 × 6.6	65 m^2 of PV panel, 10 kWp	520 Ah 48 V DC in each hull, 49,92 kWh	Already produced	[36]
Solaris Solar Sal 27 (River Boat)	8.84 × 2.44	8,14 m^2 of PV panel, 6 kWp	LiPeFO4 Battery, 10 kWh	Already produced	[37]
Aquaboat 1050 (Catamaran Boat)	10.5 × 2,5	6.6 m^2 of PV panel, 1,5 kWp	48 V Battery	Already produced	[38]
Aditya (Ferry Boat)	21 × 7	140 m^2 of PV panel, 20 kWp	Li-ion battery, 50 kWh	Already produced	[39]
Ship Munster (Catamaran Ferry)	15.6 × 4.6	20.287 m^2 of PV panel, 5.18 kWp	Li-ion battery, 115,2 kWh	Already produced	[40]
Ferries Berliner VG (Catamaran Ferry)	18.5 × 5.2	58.606 m^2 of PV panel, 10.6 kWp	2 x Lead - gel battery, 69.12 kWp	Already produced	[40]
Rappbode (Dam Inspection Boat)	10.5 × 5.1	20.287 m^2 of PV panel, 3.7 kWp	Lithium polymer battery, 38.4 kWp	Already produced	[40]
Ship Sünje (Catamaran Ferry)	18.5 × 5.4	58.606 m^2 of PV panel, 10.6 kWp	2 x Lead - gel battery, 70.08 kWp	Already produced	[40]
Aquawatt 550 (Boat)	5.5 × 1.84	0.4 kWp	AGM Battery, 5,04 kWp	Already produced	[41]
Seminarchiff (Catamaran Ferry)	35 × 8	123.98 m^2 of PV panel, 23 kWp	Lithium polymer battery, 21.6 kWp	Already produced	[40]
Type ASF-25 (Inland Car Ferry)	28 × 8.9	16.905 m^2 of PV panel, 5,4 kWp	Lithium polymer battery, 21.6 kWp	Already produced	[42]
Type ASP-80 (Passenger Ferry)	19.9 × 6.6	21.5 m^2 of PV panel, 10.8 kWp	252 kWh battery	Already produced	[43]
Type ASP-20 (Passenger Ferry)	14.6 × 4.5	21.5 m^2 of PV panel, 4.3 kWp	Lithium polymer battery, 80 kWh	Already produced	[44]
Type ASP-300 (Passenger Ferry)	33 × 9	134 m^2 of PV panel, 20 kWp	Lithium polymer battery, 1100 kWh	Already produced	[45]
MobiCat (Catamaran Ferry)	33 × 11.8	180 m^2 of PV panel, 20 kWp	2 x 5000 kg battery, 480 kWh	Already produced	[46]
SoelCat 12 (Catamaran Yacht)	11.8 × 5.8	45 m^2 of PV panel, 60 kWp	Li-ion battery, 120 kWh	Already produced	[47]
iCat (Passenger Catamaran)	17 × 7	7.5 kW of PV panel	50 kWh	Design Stage	[48]
Aquanima 40 (Catamaran Yacht)	13.25 × 7	56 m^2 of PV panel, 20 kWp	Li-ion battery, 60 kWh	Already produced	[49]

Passeur La Rochelle (Catamaran Boat)	10 × 3.5	9 m^2 of PV panel, 20 kWp	Ni/Cd battery, 20 kWh	Already produced	[50]

There are general characteristics of fully solar-powered vessels from the above table. The fully solar-powered ships produced and operated are mostly passe vessels. The average dimensions of an entirely solar-powered vessel are approximately 18,7 m × 6,3 m. The effective area for a PV generation system is 84,6 m^2, with a nominal power of 16,6 kWp. And a battery capacity of 197 kWh.

4.2 Solar-Assisted Powered on Large Scale Vessels

Several large vessels have used photovoltaic technology. Installing large-scale onboard PV generator installations can be integrated into the ship's electrical propulsion system network to reduce fuel consumption; this is likely to create significant new ship designs in the future [9]. The author has collected data from various sources regarding solar-assisted vessels that have been produced or are still in the development stage, along with their advantages, as shown in Table 2.

Table 2. Application on solar-assisted powered on large scale vessels

Vessel		**Photovoltaic System**		**Solar Energy Usage**	**Benefits**	**Ref.**
Name (Type)	**Dimension (L × B) (m)**	**PV generation System**	**Battery**			
Auriga Leader (Car Carrier)	199 × 32	328 PV panels, 40 kW rated power	Nickel hydrogen battery (Gigacell)	Lighting, thruster, cargo hold fan, engine room	• 13 t/year fuel reduction • 60 t/year CO_2 emission reduction • 10% energy reduction	[51]
Emerald Ace (Car Carrier)	200 × 32	768 PV panels,160 kW rated power	Li-ion battery, 2.2 MWh	Navigation, power system	• Fuel reduction • CO_2 emission reduction • Energy efficiency	[52]
COSCO "Tengfei" (Car Carrier)	182.8 × 32.2	1050 m^2 PV panel,143.1 kW rated power	Li-ion battery, 750 kWh	Cabin's lighting	• 1.8 % fuel consumption reduction • 1,5% CO_2 emission reduction • Saving US$ 133626 fuel cost/year	[53] [54]
Blue Star Delos (Passenger Ferry)	145 × 22	14.4 m^2 PV panel, 1.32 kWp	VLRA battery, 5.4 kWh	Ship electrical	• Reduce fuel consumption • Decrease NOx, Sox, and CO_2 emissions.	[55]
AN JI 204 (Ro-Ro Car Ship)	110 × 18.8	135 PV panels, 37.12 kWp rated power	Li-ion battery, 128 kW	Hull lighting system	• Reduce 10 t/year fuel consumption • Reduce 630 kg/year SOx • Reduce 50 kg/year NOx • Reduce 28.5 t/year CO_2	[9]
Drive Green Highway (Car Carrier)	200 × 38	900 solar frontier CIS panel, 150 kWp rated power	-	The deck's all-LED lighting system	• Reduce SOx 90% /vehicle transported • Reduce NOx 50% /vehicle transported • Reduce CO_2 25% /vehicle transported	[56]

A few ships use solar-assisted power generation technology, from Table 2. The fact is that application still of solar-assisted power generation is widely applied to vessels with a large upper deck surface. The usage of solar energy is mostly for lighting and electrical purposes. Solar-assisted power generation has proven to reduce fuel consumption and GHG emissions on large-scale vessels with potential CO_2 reduction up to 0.2 – 12% [10]. The installation of the solar-Assisted Powered Large Scale Vessels can reduce 8 tons of fuel/day or 2400 tons/year assuming the ship is active for 50 from 52 weeks per year, this is not only good for the environment but can reduce fuel consumption costs, if the ship is operated for 20 years, it can reduce the cost of 100 M USD or about 5 M USD / year, this is very good as an investment [57].

4.3 Application on solar-powered ships integrated with new energy resources

Significant developments have also occurred in hybrid energy-powered ships. A hybrid ship power system combines two or more alternative energy sources, such as solar, wind, wave, and fuel cells, with a conventional ship power system. The use of energy from more than one source is intended to complement the other energy deficiencies. For example, solar and wind energy are not commonly used due to their instability and dependence on weather conditions. This problem can be partially overcome by combining solar and wind energy to create a hybrid ship power system. The summary of hybrid new energy ship is resumed in Table 3.

Solar Sailor is the world's first 100-person solar/wind-powered tourist catamaran for commercial operation on Sydney Harbour; this ship has unique solar wings that can direct the PV panels towards the sun. The vessel is silent, has a low wash, and emits no water pollution in any mode of operation. NASA wing foil incorporating solar collectors engineered to withstand 40 knots of wind. It is estimated to save about 250,000 L of diesel fuel and reduce annual GHG emissions by 670 tons [58].

Alastair Callender has developed a radical concept, the 58m rigid wing superyacht, namely Soliloquy-Super green yacht that uses energy from wind, solar, and Hybrid Marine Power (HMP) technology from Solar Sailor Holdings Ltd. the ship is equipped with 600 m^2 rigid -solar sail wings that rotate independently 360 degrees. According to the British yachting carbon offset firm, an equivalent-sized super yacht consumes between 250 and 600 gallons of marine diesel per hour, depending on speed and fuel efficiency, and releases three times lower the amount of CO_2 [59].

Moreover, the future ship design of the super eco ship is designed by Keiko Kosugi, who won the Japan Good Design Award. 800 TEU NYK excellent eco ship 2030 is powered by a 44000 kW power plant combined from LNG, wind, and solar power with 31000 sqm. It is estimated to achieve a 70% GHG emission reduction [60]. The Eoseas Concept Cruise (105000 Tons) is a concept design developed by STX Europe in collaboration with Stirling Design International (SDI). The vessel has four sets of dual-fuel LNG diesel-electric generators for propulsion, 8,300m^2 photovoltaic panels with a maximum power of 108MW, and an innovative sail concept patented by STX France. This design concept is claimed to have the potential to reduce power consumption by 50%, CO_2 emissions by 50%, SO2 by 100%, NOX by 90%, and ash by 100% [61].

The Hornblower Hybrid is the first hybrid-powered multi-hull ferry class in the United States operated by Hornblower Cruises in San Francisco and New York. Hybrid engines use solar, wind, diesel, or a combination of the three energy to generate electricity. It can also be powered by a hydrogen fuel cell, making it the first hydrogen ferry. The propulsion system enables the captain to monitor the vessel's energy requirements and choose the most appropriate power sources [62].

At last, among these proposals, one that has stood out is the E/S Orcelle, which aims to build the world's first zero-emission vessel [63]. The cruise ship designed by Wallenius Wilhelmsen Logistics will use wind, solar, and wave energy as energy sources. The conceptual design of the Orcelle began in 2004 with a pentamaran hull with a capacity of 13,000 tons. Orcelle uses three rigid sails, each measuring 1,400 m^2 made of composite material, which can be rotated and positioned to catch the wind. Furthermore, this screen will include 800 m^2 of photovoltaic panels to generate a maximum of 2,500 kW of solar electricity; when not in use for propulsion, the sails would fold back onto the ship's top deck to maximize the amount of solar energy they received. More power generation components can be found under the ship. The sponsor to the main hull is 12 horizontal fins, three on each sponge. It is designed to move up and down as the boat moves through the water, with a hydraulic motor converting the motion into power. Representative application on hybrid new energy ships is depicted in Figure 4.

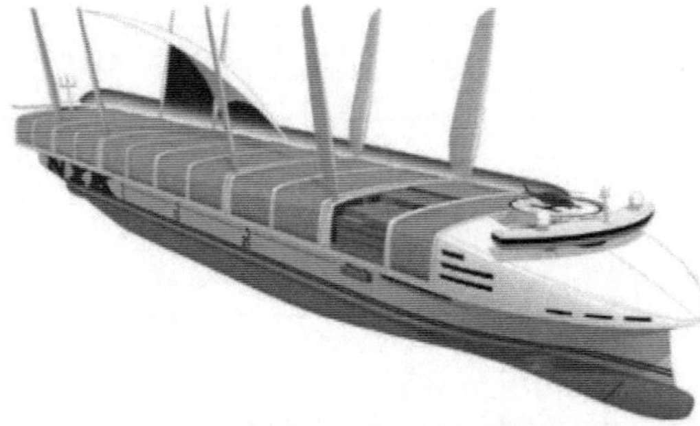

(a) NYK Super Eco [59]

(b) Eoseas Concept Cruise Ship [60]

(c) Hornblower Hybrid [61]

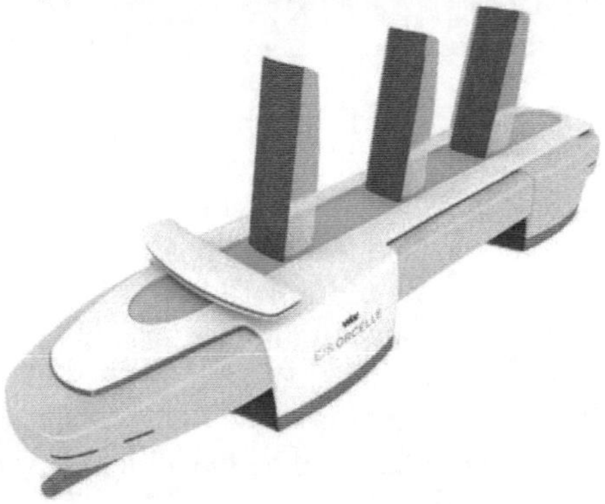

(d) E/S Orcelle [62]

Figure 4. Representative application on hybrid new energy ships

Table 3. Summary of hybrid-powered ships

Vessel Name (Type)	**Energy Source**	**Hybrid Generation System**	**Benefits**	**Ref.**
Solar Sailor (Tourism Ferry)	Solar / Wind / Battery / LPG	• Lead-acid gel battery (80 × 70 Ah) • 2 × 40 KW electrical motor	• Save 250.000 liters diesel fuel consumed / vessel annum • Decrease 670-ton GHG emission/vessel	[58]
Soliloquy- Super Green Yacht (Yacht)	Solar / Wind / Fuel Cells	• 600m² solar sail	• Three times lower emission compared to marine diesel	[59]
NYK Super Eco Ship 2030 (Cruise - Design Concept)	Solar / Wind / Fuel Cell	• LNG fuel cells 50 MW rated • 8 lightweight sails (capacity 2.5 MW power/sail) • PV panel (capacity 1.6 MW)	• A potential reduction near 70% CO_2 emission	[60]
Eoseas Cruise (Cruise – Design Concept)	Solar / Wind / Fuel Cell	• 4 dual-fuel LNG diesel/electric generator sets (each providing 8 MW power) • 300m² PV panels	• Reduce 50% fuel consumption respectively • Reduce 100% SOx, respectively • Reduce 90% NOx	[61]

		(capacity 108 MW) • 12.440 m^2 innovative sail concept mounted on five masts (patented by STX France)	respectivel • Reduce 50% CO_2 respectively • No ash	
Hornblower Hybrid (Passenger Ferry)	Solar / Wind / Diesel / Battery	• Tier II Diesel Engine • 380 VDC AGM Battery Banks • Wind turbines (capacity 5 kW) • PV array panels (capacity 20 kW)	• Reduce 70% fuel consumption • 75 – 100% potential reduction of GHG emissions • No ecological impact	[62]
E/S Orcelle (Ycht - Design Concept)	Solar / wind, wave energy / fuel cells / and LNG	• 3 rigid sail (1400 m^2) • 800 m^2 PV array (produces 2500 kW power) • 12 horizontal fins	• Zero-emission ship	[63]

Hybrid solar-powered ships are a long-term solution to overcome the dependence of ships on carbon fuel consumption. although most ships of this type are still in the design concept phase, significant developments from various companies prove that hybrid solar powered ships have very promising prospects in the future. although in reality it requires a very large initial investment cost, but this is commensurate with the benefits that can be obtained afterward, no carbon fuel, no GHG emission, and minimum environmental impact.

5 Conclusion and prospects

From this paper presented by the author, several conclusions can be drawn as follows:

1. The greenhouse gasses (GHG) produced by ships, although not as much as those produced by land vehicles, must be significantly reduced using renewable energy; this is because maritime transportation is the vital transportation in global trade.
2. The fully solar-powered vessels have average dimensions of 18.7 m × 6.3 m, with an effective area for PV panels of around 84.5 m^2 with a nominal power of 16.6 kWp and a battery capacity of 197 kWh. Challenges mainly from technical and economic aspects on solar panels such as the need for solar panels with high conversion efficiency and the large-scale storage application have some technical and market barriers that often cause higher capital costs than traditional power generation resources on ships.
3. The usage of solar-assisted power generation on large-scale vessels is still infrequent because ships require a large surface area as a medium for placing PV panels. The energy produced by the PV generation system is used more for lighting and electrical purposes. Solar-assisted power generation has proven to reduce fuel consumption up to 8 tons/day, reduce GHG emissions on large-scale vessels with potential CO2 reduction up to 0.2 – 12%, and reduce the cost of 100 M USD or about 5 M USD / year for 20 year operation.
4. Any alternative energy may be incorporated in a ship power system, allowing ships to run on more than one alternative energy source. However, because of the variety of uses, it must choose which new energy source is most suited for usage in ships. In a nutshell, integrating new energy source generation systems with existing ship power systems is a promising solution for increasing energy efficiency, improving grid reliability, and improving power quality onboard, and is thus likely to be the focus of future research on ship-based energy systems.

6 References

[1] H.-K. Ellycia, “Shipping Impacts on Climate: a Source with Solution,” 2008.
[2] J. Ling-Chin and A. P. Roskilly, “A comparative life cycle assessment of marine power systems,” *Energy Convers. Manag.*, vol. 127, pp. 477–493, Nov. 2016, doi: 10.1016/J.ENCONMAN.2016.09.012.

[3] G. S. Eskeland and H. Lindstad, "Environmental Taxation in the Transport Sector," 2015. [Online]. Available: http://www.greengrowthknowledge.org/resource/environmental-taxation-transport.

[4] S. Kumar and J. Hoffmann, "Globalisation: The Maritime Nexus," *Handb. Marit. Econ. Bus.*, no. May, pp. 35–62, 2010.

[5] N. Gray, S. McDonagh, R. O'Shea, B. Smyth, and J. D. Murphy, "Decarbonising ships, planes and trucks: An analysis of suitable low-carbon fuels for the maritime, aviation and haulage sectors," *Adv. Appl. Energy*, vol. 1, p. 100008, Feb. 2021, doi: 10.1016/J.ADAPEN.2021.100008.

[6] Ø. Buhaug *et al.*, "Second IMO GHG Study 2009," London, 2009.

[7] E. Bal Beşikçi, O. Arslan, O. Turan, and A. I. Ölçer, "An artificial neural network based decision support system for energy efficient ship operations," *Comput. Oper. Res.*, vol. 66, pp. 393–401, 2016, doi: 10.1016/j.cor.2015.04.004.

[8] H. Xing, S. Spence, and H. Chen, "A comprehensive review on countermeasures for CO2 emissions from ships," *Renew. Sustain. Energy Rev.*, vol. 134, no. August, p. 110222, 2020, doi: 10.1016/j.rser.2020.110222.

[9] P. Pan, Y. Sun, C. Yuan, X. Yan, and X. Tang, "Research progress on ship power systems integrated with new energy sources: A review," *Renew. Sustain. Energy Rev.*, vol. 144, no. April, p. 111048, 2021, doi: 10.1016/j.rser.2021.111048.

[10] E. A. Bouman, E. Lindstad, A. I. Rialland, and A. H. Strømman, "State-of-the-art technologies, measures, and potential for reducing GHG emissions from shipping – A review," *Transp. Res. Part D Transp. Environ.*, vol. 52, pp. 408–421, 2017, doi: 10.1016/j.trd.2017.03.022.

[11] S. Wang, Q. Meng, and Z. Liu, "Bunker consumption optimization methods in shipping: A critical review and extensions," *Transp. Res. Part E Logist. Transp. Rev.*, vol. 53, no. 1, pp. 49–62, 2013, doi: 10.1016/j.tre.2013.02.003.

[12] T. W. P. Smith *et al.*, "Third IMO Greenhouse Gas Study 2014," London, 2015. doi: 10.1007/s10584-013-0912-3.

[13] P. Balcombe *et al.*, "How to decarbonise international shipping: Options for fuels, technologies and policies," *Energy Convers. Manag.*, vol. 182, no. February, pp. 72–88, 2019, doi: 10.1016/j.enconman.2018.12.080.

[14] N. Olmer, B. Comer, B. Roy, X. Mao, and D. Rutherford, "Greenhouse Gas Emissions From Global Shipping, 2013-2015," Washington DC, 2017. [Online]. Available: https://www.theicct.org/sites/default/files/publications/Global-shipping-GHG-emissions-2013-2015_ICCT-Report_17102017_vF.pdf.

[15] C. Wang and J. J. Corbett, "The costs and benefits of reducing SO2 emissions from ships in the US West Coastal waters," *Transp. Res. Part D Transp. Environ.*, vol. 12, no. 8, pp. 577–588, Dec. 2007, doi: 10.1016/J.TRD.2007.08.003.

[16] C. Deniz and B. Zincir, "Environmental and economical assessment of alternative marine fuels," *J. Clean. Prod.*, vol. 113, no. X, pp. 438–449, 2016, doi: 10.1016/j.jclepro.2015.11.089.

[17] J. Vidal, "Health Risks of Shipping Pollution Have Been 'Underestimated,'" *Guard.*, 2009, Accessed: Feb. 08, 2022. [Online]. Available: https://www.scirp.org/%28S%28vtj3fa45qm1ean45vvffcz55%29%29/reference/referencespapers.aspx?referenceid=2199811.

[18] T. F. Stocker *et al.*, *Climate Change 2013 : The Physical Science Basis. Contribution of Working Group 1 to the Fifth Assessment Report of the Intergovermental Panel on Climate Change*. Cambridge, United Kingdom and New York, USA: Cambridge University Press, 2013.

[19] Internasional Maritime Organization, *Resolution MEPC.212(63) Adopted on 2 March 2012 : 2012 Guidlines on the Method of Calculation of the Attained Energy Efficiency Design Index (EEDI) for New Ships*. 2012.

[20] International Maritime Organization, *Resolution MEPC.213 (63) Adopted on 2 March 2012 : 2012 Guidlines for the Development of a Ship energy efficiency Management Plan (SEEMP)*. 2012.

[21] Z. Zapałowicz and W. Zeńczak, "The possibilities to improve ship's energy efficiency through the application of PV installation including cooled modules," *Renew. Sustain. Energy Rev.*, vol. 143, p. 110964, Jun. 2021, doi: 10.1016/J.RSER.2021.110964.
[22] C. J. Rhodes, "Solar energy: Principles and possibilities," *Sci. Prog.*, vol. 93, no. 1, pp. 37–112, 2010, doi: 10.3184/003685010X12626410325807.
[23] S. Algburi, *Compendium of Energy Science and Technology (Vol. 6 : Solar Engineering-II)*. United States: Studium Press LLC, 2015.
[24] "ECOCAT | Zero Emission | naval metaltech | Construction of aluminum boats." https://www.metaltecnaval.com/ecoboat/ecocat-zero-emission/ (accessed Jan. 24, 2022).
[25] "SunCRZ 22 – Navgathi." https://navgathi.com/suncrz-22/ (accessed Jan. 24, 2022).
[26] Y. Kumar *et al.*, "Wind energy: Trends and enabling technologies," *Renew. Sustain. Energy Rev.*, vol. 53, pp. 209–224, 2016, doi: 10.1016/j.rser.2015.07.200.
[27] "E-Ship 1 - Wikipedia." https://en.wikipedia.org/wiki/E-Ship_1 (accessed Jan. 12, 2022).
[28] "Home - The Oceanbird." https://www.theoceanbird.com/ (accessed Jan. 12, 2022).
[29] F. Fachruddin, "Conversion Cost for the Diversion from Fuel Oil to Fuel Gas in Pioneer Ships," *Int. J. Eng. Res. Technol.*, vol. 5, no. 12, pp. 286–289, 2016.
[30] "Wärtsilä 20DF - fuel flexibility in a small unit." https://www.wartsila.com/marine/products/engines-and-generating-sets/dual-fuel-engines/wartsila-20df (accessed Jan. 12, 2022).
[31] L. Klinteberg, "Wave Energy Propulsion for Pure Car and Truck Carriers (PCTCs)," KTH Royal Institute of Technology, 2009.
[32] Y. Terao and N. Sakagami, "Design and development of an autonomous wave-powered boat with a wave devouring propulsion system," *Adv. Robot.*, vol. 29, no. 1, pp. 89–102, 2015, doi: 10.1080/01691864.2014.970229.
[33] "Kenichi Horie official site." http://www.suntorymermaid.com/english/mermaid.html (accessed Jan. 12, 2022).
[34] "The boat – Fondation PlanetSolar." https://www.planetsolar.swiss/en/world-premiere/boat/ (accessed Jan. 26, 2022).
[35] K. Desmond, *Electric Boat and Ship, A history*. North Carolina: McFarland & Company, Inc., Publisher, 2017.
[36] H. Gürsu, "Solar and wind powered concept boats: The example of volitan," *Metu J. Fac. Archit.*, vol. 31, no. 2, pp. 109–123, 2014, doi: 10.4305/METU.JFA.2014.2.6.
[37] "Solar Sal 27 – the Latest Launch from our Shop | Devlin Designing Boat Builders." https://devlinboat.com/solar-sal-27-the-latest-launch-from-our-shop/ (accessed Jan. 25, 2022).
[38] "Aquabus 1050 | Marservis." https://marservis.hr/en/aquabus-1050/ (accessed Jan. 27, 2022).
[39] "Aditya (boat) - Wikipedia." https://en.wikipedia.org/wiki/Aditya_(boat) (accessed Jan. 25, 2022).
[40] "History - Ampereship EN." http://www.ampereship.com/en/history.html (accessed Jan. 25, 2022).
[41] "Solar boat aquawatt 550 | aquawatt Green Marine Technologies." https://www.aquawatt.at/elektroboote/solarboote/solarboot-aquawatt-550 (accessed Jan. 27, 2022).
[42] "ASF-25 - Ampereship EN." http://www.ampereship.com/en/asf-25.html (accessed Jan. 27, 2022).
[43] "ASP-80 Solarferry for 80 passengers - Ampereship EN." http://www.ampereship.com/en/ASP-80-passengerferry-for-80pax-allelectric.html (accessed Jan. 27, 2022).
[44] "ASP-20 Solarferry for 20 passengers - Ampereship EN." http://www.ampereship.com/en/asp-20.html (accessed Jan. 27, 2022).
[45] "ASP-300 Solar-Electric Ferry for 300 passengers - Ampereship EN." http://www.ampereship.com/en/asp-300.html (accessed Jan. 27, 2022).
[46] "MobiCat – Wikipedia." https://de.wikipedia.org/wiki/MobiCat (accessed Jan. 25, 2022).
[47] "SoelCat 12 - Naval DC." https://www.navaldc.com/portfolio/soelcat-12/ (accessed Jan. 28,

2022).
[48] "iCat - iCat." https://www.icat.hr/en/icat (accessed Jan. 28, 2022).
[49] "AQUANIMA 40 | Azura." https://www.azura-marine.com/aquanima-40 (accessed Jan. 28, 2022).
[50] "Passenger boat - Passeur zéro-émission - Alternatives Energies - catamaran / electro solar / emission-free." https://www.nauticexpo.com/prod/alternatives-energies/product-30419-558313.html (accessed Jan. 28, 2022).
[51] Nippon Yusen Kabushiki Kaisha, K. H. Industries, M. T. Institute, and N. K. Kyokai, "World's First Solar-Power-Assisted Vessel Further Developed - Car Carrier Auriga Leader to be Fitted with Hybrid Power Supply System and Ballast-Water Management System, and Adapted to Use Low-Sulfur Fuel," 2011.
[52] Japan Ship Exporters ' Association, "MHI delivers hybrid car carrier EMERALD ACE to MOL," *www.jsea.or.j*, no. 376, Tokyo, pp. 1–6, Oct. 2012.
[53] "COSCO Group built the world's largest off-grid solar photovoltaic system for ships - Shipping Industry." http://www.ship.sh/news_detail.php?nid=12332 (accessed Feb. 02, 2022).
[54] Y. Yuan, J. Wang, X. Yan, Q. Li, and T. Long, "A design and experimental investigation of a large-scale solar energy/diesel generator powered hybrid ship," *Energy*, vol. 165, pp. 965–978, 2018, doi: 10.1016/j.energy.2018.09.085.
[55] G. M. Atkinson, "Analysis of marine solar power trials on Blue Star Delos," *J. Mar. Eng. Technol.*, vol. 15, no. 3, pp. 115–123, 2016, doi: 10.1080/20464177.2016.1246907.
[56] News Hound, "Eco-Ship Drive Green Highway launched by K Line," Feb. 15, 2016. https://www.iims.org.uk/eco-ship-drive-green-highway-launched-by-k-line/ (accessed Feb. 02, 2022).
[57] T. Hammarlund, J. Sundin, and J. Kövamees, "Solar Cell Powered Boat," Uppsala University, 2017.
[58] "SOLAR SAILOR HYBRID SAIL PANELS CATAMARAN FERRY SYDNEY HARBOUR AUSTRALIA ROBERT DANE FERRIES ZERO CARBON," Jan. 05, 2005. http://www.solarnavigator.net/solar_sailor.htm (accessed Feb. 08, 2022).
[59] "SYT World Exclusive: Soliloquy - The Super-Green Superyacht," May 29, 2009. https://www.superyachttimes.com/yacht-news/syt-world-exclusive-soliloquy-the-super-green-superyacht (accessed Feb. 08, 2022).
[60] ridhika naidoo, "NYK super eco ship 2030," Nov. 17, 2009. https://www.designboom.com/design/nyk-super-eco-ship-2030/ (accessed Feb. 08, 2022).
[61] "Eoseas Concept Cruise Ship - Ship Technology," Oct. 24, 2011. https://www.ship-technology.com/projects/eoseas-concept-cruise-ship/ (accessed Feb. 08, 2022).
[62] "Hornblower Hybrid - Wikipedia." https://en.wikipedia.org/wiki/Hornblower_Hybrid (accessed Feb. 08, 2022).
[63] kate cummins, "Harnessing the sun, the wind and the sea with E/S Orcelle | The Engineer The Engineer," May 09, 2011. https://www.theengineer.co.uk/harnessing-the-sun-the-wind-and-the-sea-with-es-orcelle/ (accessed Feb. 08, 2022).

The Adaptive Engineering Asset Management Framework for Sustainable Gas Terminal Infrastructures

Oktaviani Turbaningsih[1] and Ulfa Mutaharah[2]
[1]Marine Transportation Engineering Department, Institut Teknologi Sepuluh Nopember, Indonesia
[2]Civil Engineering University of Indonesia, Depok, West Java, Indonesia
Corresponding author's email: oktav@its.ac.id

Abstract. The global initiative for decarbonisation keeps growing at a different rate for each country. The global decarbonisation rate, which presents the carbon intensity reduction or energy-related CO2 emissions per dollar GDP, reached 2.5% in 2020. Indonesia recorded contributing as the second-highest rate of emissions reduction in the global decarbonisation rate during 2019-2020 with a total of 10.6% less carbon intensity. Solid infrastructure asset management shall significantly improve to support the further decarbonisation movement. As one of the cleanest carbon energy, LNG could be the fastest solution to keeping the initiative. Unfortunately, it has a drawback: methane produces higher Green House gases (GHG) than CO2. Furthermore, LNG acts as a transition fuel while the global industry prepares for renewable energy resources. The authors propose Adaptive Port Planning (APP) and Engineering Asset Management (EAM) framework for managing the existing LNG infrastructure terminal assets in this study. The EAM combines a multidisciplinary approach to provide a framework for effectively managing all engineered physical assets within the organisation. In contrast, the APP will give the idea to count the uncertainty in infrastructure demand, allowing flexibility in planning. The author overviews the conversion or transition pathway for existing terminal gas infrastructure to support future renewable energy resources such as hydrogen and ammonia. The Adaptive EAM will guide the sustainable business mindset and provide cost-effective decarbonising solutions.

1. Introduction: Climate Change and Decarbonisation Policy

The global decarbonisation rate, which presents the carbon intensity reduction or energy-related CO2 emissions per dollar GDP, reached 2.5% in 2020. The IPCC (Intergovernmental Panel on Climate Change) Assessment Report 2021 shows that the marine heatwaves will continuously increase. The relative sea-level around Asia has increased faster than global averages, with the coastal area loss and shoreline retreat [1]. The global initiatives toward climate change started in 1988 with the establishment of IPCC, then continuously with Kyoto Protocol in 1997 and Copenhagen Accord in 2009. The Paris Agreement on Climate Change, implemented starting in December 2015, has influenced the rapid growth of the clean energy movement. It became the starting point where the action to deal with environmental issues became mandatory for every nation by their commitment to National Determined Contribution.

The decarbonisation initiatives for each country keep growing at a different rate. Indonesia contributed the second-highest rate of emissions reduction in the global decarbonisation rate during

IOP Conf. Series: Earth and Environmental Science 1081 (2022) 012012 doi:10.1088/1755-1315/1081/1/012012

2019-2020, with a total of 10.6% less carbon intensity. Figure 1 presents a regulatory framework for Indonesia's long-term strategy to achieve low carbon and climate resilience [2].

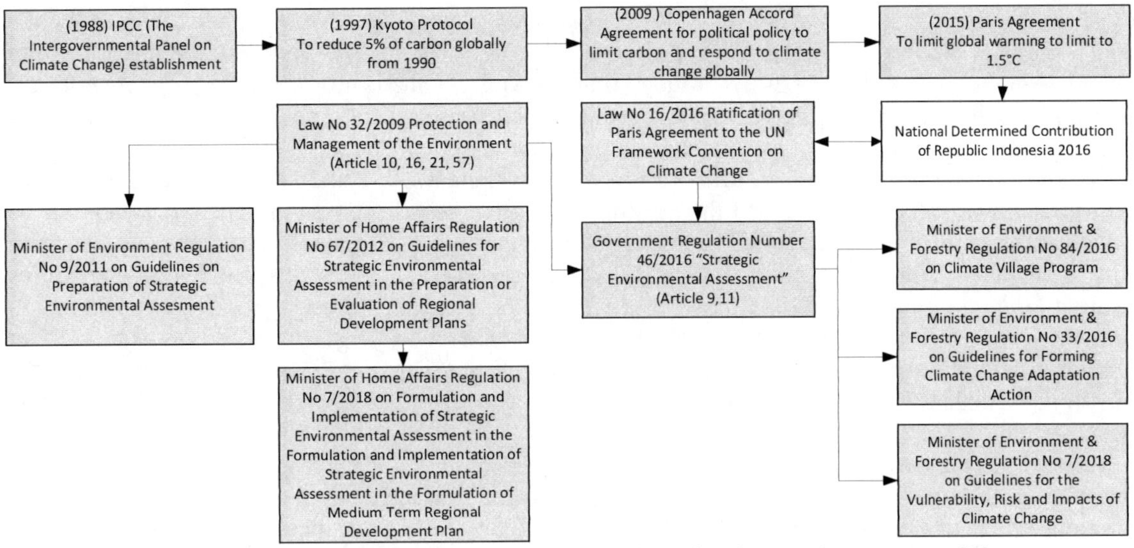

Figure 1. The regulatory framework for Indonesia's low carbon strategy [2].

Specifically for the maritime industry, in 2018, the International Maritime Organisation (IMO) started adopting a strategy to impose a decarbonisation policy on shipping organisations. The policy includes IMO Sulphur Cap, which mentioned the urge to reduce the sulphur content from 3.5 % m/m to 0.5% m/m under MARPOL Annex VI [3] became mandatory for IMO members starting from January 1st, 2020. Referring to IMO guidelines[3], IMO recommend the use of marine fuels with the following characteristics: (1) Distillate marine fuels (DM) such as DMA, DMB, DMX, and DMZ; (2) Residual marine fuels (RMD 80, RMD 380); (3) Ultra-low sulphur fuel oil (ULSFO) as specified in ISO 8217:2017; (4) Very-low sulphur oil (VLSFO) with maximum 0.5% Sulphur; and (5) High sulphur heavy fuel oil (HSHFO) exceeding 0.5% S.

Recently, researchers focused on studying the impact on the fuel and machinery system and the how-to-produce higher GHG emission reduction from the current ship operating system. The strategy includes optimisation logistics, hydrodynamics hull form modification, and ship machinery modification [4]. The latest trend is developing alternative energy towards cleaner resources such as LNG, LPG, biofuels, electrification, renewable energy, methanol, ammonia, and hydrogen [5]. Figure 2 shows Indonesia's current primary energy sources by type (excluded biomass) [6].

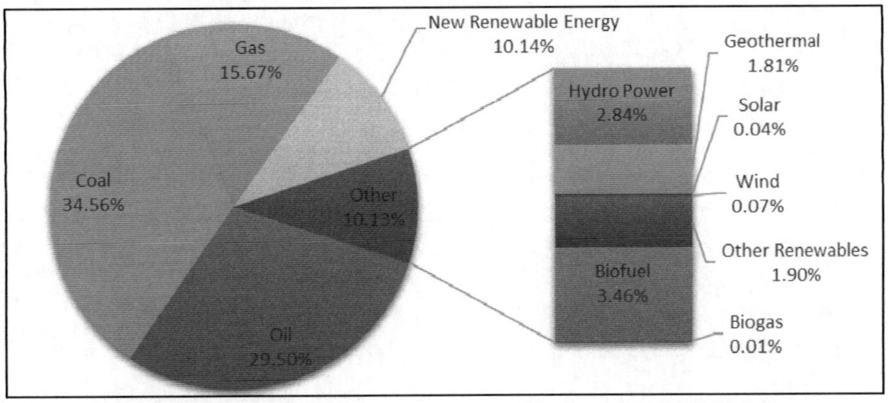

Figure 2. Primary energy resource of Indonesia by 2020 [6].

The growth of national initiatives to achieve net-zero emissions by 2050 has increased significantly over the past few years [7]. To this day, the highest potential low-carbon (or zero-carbon) energy source is hydrogen [7]. Green hydrogen ("e-Hydrogen") is hydrogen produced using renewable energy that powers the electrolysis process of water [8]. E-Hydrogen potentially be a significant source for the energy industry and is predicted to solve the EU's energy problems by 2050, as the European Commission has published its 2030 Hydrogen Strategy [8]. Benchmarking New Zealand's Government to be in line with the EU, they have set their goal to reach 100 per cent renewable electricity by 2035 with transition goals to a clean, green and carbon-neutral economy by 2050[9]. In 2019, according to IEA records, only France, Japan, and Korea had hydrogen strategies. As of 2021, the number has grown to 17 governments that have released their hydrogen strategies, with more than 20 governments have publicly announced they are developing strategies for this subject. Numerous companies seek to dive into hydrogen business opportunities [10]. Achieving the decarbonisation targets with hydrogen requires the collaboration of many parties such as the private sector, government, regulators, and consumers in the energy industry to drive the required policies and behaviours[7]. Recent studies still argue that combining centralised hydrogen production with reconfiguring existing gas infrastructure (transport grid and storage tanks) is cheaper than producing hydrogen from smaller electrolysers for on-site consumption[8].

Meanwhile, the current situation shows that hydrogen production from gas is a relatively mature technology. The CCS (Carbon Capture Storage) part of the hydrogen supply chain requires the most attention[7]. There are four energy preservation technologies: electrical, mechanical, electrochemical, and chemical. The future of long-term energy storage is more often associated with the electrochemical (batteries) and chemical (e.g. natural gas, hydrogen, and ammonia) options[7].

The existing dedicated terminals in the energy industry, such as coal, petroleum, and liquified gas, might require some transformations to handle the renewable clean energy system. The energy infrastructure facilities management is facing many challenges such as (1) the implementation of the decarbonisation policy, (2) the pursuance of sustainable development goals (SDGs) idea, (3) the application of digitalisation and automation technology as part of the Industry 4.0, and (4) ageing infrastructures condition itself and long-time issues (5) abandonment and idle assets.

By recognising these challenges, the authors adopt the Adaptive Port Planning (APP) and Engineering Asset Management (EAM) framework to manage the existing LNG infrastructure terminal assets. The EAM combines a multidisciplinary approach to provide a framework for effectively managing all engineered physical assets within the organisation. In this study, the decision-making should also focus on climate change. Addressing climate change issues has been one of the United Nations (UN) 's fundamental backgrounds in developing the Sustainability Development Goals (SDGs). The authors propose the adaptive engineering asset management framework with the impact aligned with 10 out of 17 sectors of the UN SDGs, as shown in Figure 3.

Figure 3. SDGs alignment with sustainable gas infrastructures.

2. Methods

This study uses qualitative analysis based on exploratory research reviewing the literature on sustainable infrastructure and asset management development using computerized search engines from the publication papers under Elsevier and MDPI with prioritised research from 2012 to 2022. Kusumawardhani et al. [11] performed this approach to produce a general idea to provide Asset Integrity Management framework in the petroleum industry.

This study considered applied research design to improve our understanding of asset management in the LNG industry. Using the epistemology research philosophy, the authors perform literature reviews and interviews with LNG practitioners. The research question that we used for this study such as:

1. How is the asset management of LNG infrastructures in Indonesia? Is ISO 55001 been applied in the existing practices?
2. What influences environmental policy and innovation technology on the existing company operational system?
3. What is the existing organizational strategy in business decision-making to correspond with uncertain market trends?
4. What is the organizational approach to renewable energy trends?
5. What is the company strategy to achieve sustainable infrastructure?

In the discussion session, the authors will provide the possible solution and monitor performance criteria for sustainability in the LNG supply chain. These findings will be used as guidance to provide the Engineering Asset Management framework for Sustainable Gas Infrastructure.

3. Sustainable Asset Management for LNG Supply Chain

3.1. Asset Management

The definition of asset management is a set of systematic and coordinated activities that an organisation performs continuously to manage its asset and their associated performance, risk, and expenditures over its lifecycle to achieve its organisational strategic plan[11]. The engineering assets integrate their physical assets' capability value and financial cost model. Engineering Asset Management is essential to be part of business management for the infrastructure with high capital investment. Ensuring asset productivity and process sustainability highly influenced its competitiveness in the market [12]. Srinivasan et al. (2020) recommend adapting a value-based management approach [13]. Table 1 shows the difference between the traditional or cost-based approach and the value-based approach.

Table 1. The difference between traditional asset management and value-based management [13].

	Traditional (Cost-based)	**Value-based**
Core focus	Cost	Cost, risk and performance
Stakeholder focus	A decision-maker or asset owner	All stakeholders of the assets (Meeting SDGs goals)
Decision focus	Focus on asset-specific issues	Focus on the system-level dependencies and business value
Difficulty	Well-established concept system	Relatively new concept
Impact on service	Meet minimum service level	Improve service level by approaching technology innovation
Management philosophy	Minimise expenditure Maintaining performance requirement	Maximise performance while meeting budget constraints

Crespo Marquez et al. (2020) stated that asset managers should focus on the value-based EAM considering a core focus on its assets' cost efficiency, opportunity and risk, and operational performance.

Value-based decision-making considers an organisation's resources, business process, skills, competencies, information technology, and methods for transforming existing infrastructure[14].
Yafei et al. (2021) suggested that the LNG supply chain's lifecycle sustainability is necessary [15]. It is crucial to define the whole life cycle definition of infrastructure. The whole life cycle determination depends on the type of lifecycle, planning methods, decision-making horizon, business driver, regulatory cycle and contract duration [13].

The lifecycle phase of infrastructure usually consists of Phase 1 (Design, Build, & Commissioning), Phase 2 (Operation and Maintenance), and Phase 3 (Decommissioning)[11]. Ideally, the asset management strategy starts from the beginning of the lifecycle from conceptual development, engineering design, construction, and operational phases. Maintenance and asset management in the energy industry generally have three types of maintenance: (1) Preventive maintenance, (2) Predictive maintenance and (3) Corrective maintenance. Corrective maintenance only applies to assets with low criticality and often lacks historical data [16]. The asset managers usually carry out the preventive maintenance for the higher criticality of equipment clustered into the following categories [16]: (1) Time-Based Maintenance (TBM) settled up by the asset manufacturer; (2) Risk-Based Maintenance (RBM), by implementing this type of maintenance would allow asset manager to formulate integrated multi-year and maintenance plan; (3) condition-based maintenance; (4) reliability-centred maintenance. For the gas sector asset management, the preference of the maintenance model is RBM. Hence, asset knowledge management is considered crucial in asset management, especially for critical infrastructures.

In this study, the main focus of EAM is the LNG receiving terminal. The critical infrastructure for LNG terminals includes a loading arm, cryogenic pipeline, LNG storage tank, Low-Pressure Pumps, BOG compressors and recondense, high-pressure (HP) pumps, vaporisers, and gas metering, and a command & control system [10].

3.2. LNG Infrastructure Sustainability

The primary usage of LNG is as a transportation fuel, electricity production, and chemical industry. LNG is considered the fastest solution as it provides a cleaner alternative for energy sources to replace coal and oil to reduce Green House Gas emissions. LNG is considered a 'bridging' energy from fossil energy sources to renewable energy. However, the high capital investment in the LNG infrastructure makes it challenging to manage the asset management functionality transition [17]. Table 2 shows the enabler, barriers, and drivers for sustainable gas infrastructure obtained from literature studies and interviews with practitioners.

Table 2. Barriers, enablers, and drivers for sustainability LNG supply chain

Enablers	Barriers	Drivers
• IMO Sulphur Cap 2020 for marine fuel oils implementation	• Existing machinery systems are still adapting to the fossil fuel source and require [15]	• Afraid of stranded assets (high sunk costs)
• National Determined Contribution towards net zero carbon	• Lack of standardised policy and standards for renewable resources per country region [15]	• Carbon pricing and taxation
• Investment trend toward sustainability business	• The high capital investment for renewable marine fuel infrastructure • Lack of human resources knowledge • Ageing infrastructures and a lack of asset management system • Availability of supply of fossil resources	• The need for a sustainable business market • Limited availability of supply of fossil resources

IOP Conf. Series: Earth and Environmental Science 1081 (2022) 012012 doi:10.1088/1755-1315/1081/1/012012

The Engineering Asset Management should be aligned with the portfolio business of its organisation and follow the market trend. Figure 4 shows the Ansoff Matrix [18] below shows how the LNG industry reacted to the market volatility and rapid changes on the supply and demand side.

Markets		Existing	Modified	New
New		**Market Development**	**Partial Diversification**	**Diversifications**
		New Feed Gas Supply	Reloading/Transhipment	Bunkering
		Small Scale LNG	Gasing Up/Cooldown	Renewable Energy
Expanded		**Market Expansion**	**Limited Diversification**	**Partial Diversification**
		Conversion to FLNG	Export to Import Conversion	Truck LNG Loading
		Conversion to FSRU	Import to Export Conversion	Breakbulk of LNG (Peak Shaving)
Existing		**Market Penetration**	**Product Extension**	**Product Development**
		Maximize Product Extraction	LNG Quality Adjustment	Modification of Jetty & Storage Facilities
		Redeployment of FLNG	BOG Management	ISO Tank Truck LNG
		Redeployment of FSRU	Cold Utilization	Direct Fuel Vehicles

Products or Services

Legend: Liquefaction Terminal; Regas Terminals; Peak Shaving; Applicable to all

Figure 4. Modified Ansoff matrix for LNG business market[19] [18].

Bunkering businesses and renewable energy are two alternatives for diversification in the LNG business market; thus, examining the available option for alternative transportation fuels becomes crucial. Table 3 shows the development of LNG and other alternative marine fuels based on DNVGL prediction based on its processing system from energy sources such as NG (Natural Gas)/Renewable/Fossil/Biomass) processed and converted with ICE (International Combustion Engine) as shown in Table 1.

Table 3. Marine Fuel Technology Options and Adaptation [20]

Fuel Type	Key Challenges	Pathway
LNG	• Methane slips from LNG fuelled engines • Cryogenic materials needed • Fuel storage tanks are 2 – 3 times larger than petroleum fuels	Pathway 1: NG – LNG – ICE – 4S LBSI/LPDF Pathway 2: NG – LNG – ICE – 2S LPDF Pathway 3: NG – LNG – ICE – 2S HPDF Pathway 4: NG – LNG – FC
Hydrogen	• Cryogenic Materials needed for LH2 (more expensive than LNG) • Fuel storage tanks are massive compared to petroleum fuels • Required CCS technology	Pathway 5: NG – H2 – FC Pathway 6: Renewable – H2 – FC Pathway 7: NG – H2 – ICE Pathway 8: Renewable – H2 – ICE
Ammonia	• Fuel storage tanks required are more extensive than petroleum fuels	Pathway 9: NG – NH3 – ICE Pathway 10: Renewable – NH3 – ICE Pathway 11: NG – NH3 – FC Pathway 12: Renewable – NH3 – FC

Fuel Type	Key Challenges	Pathway
Methanol	• Fuel storage tanks are 2 – 3 times larger than petroleum fuels	Pathway 13: NG – Methanol – ICE
LPG	• Fuel storage tanks are 2 – 3 times larger than petroleum fuels	Pathway 14: Fossil – LPG – ICE
Advanced Biodiesel (HVO)	• Limited production of biofuels • Fuel cost	Pathway 15: Biomass – Advanced biodiesel (HVO) – ICE
Fully electrics	• High power charging requirement • Cost • Size and weight	Pathway 16: Energy mix – Electricity – Battery-electric system

Safari et al. (2021) suggested that countries with existing gas infrastructure may consider hydrogen an alternative for their petroleum industry. Figure 5 shows the diagram for implementing hydrogen as a renewable energy resource for future applications.

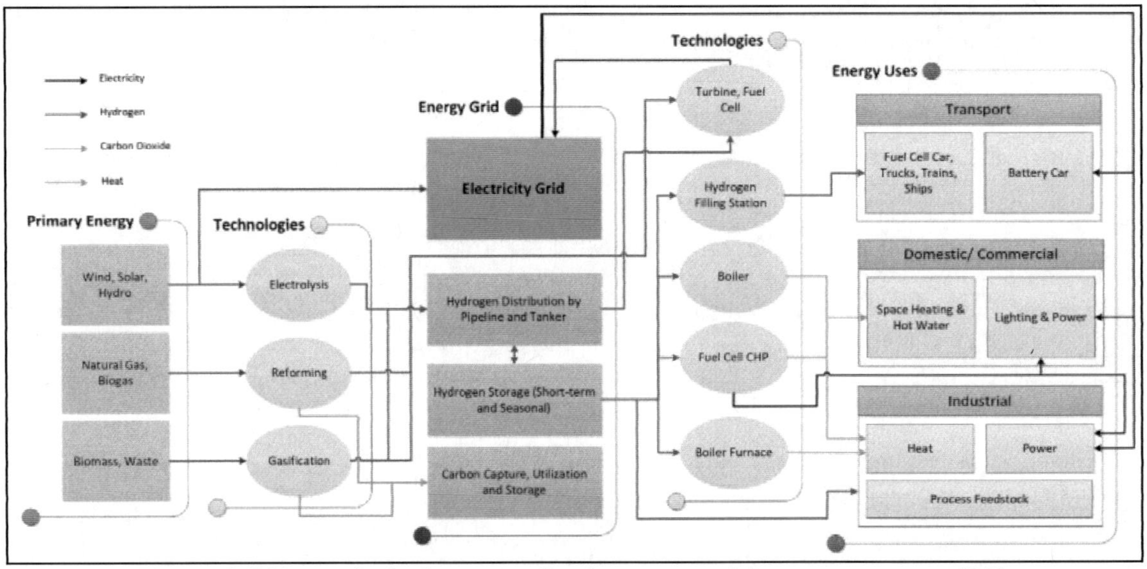

Figure 5. Diagram for hydrogen application as potential energy resources [21].

Hydrogen technology can be more competitive than lithium battery storage technologies in job adaptation, asset infrastructures, and cost competitiveness. The transition from the petroleum industry to a hydrogen-based economy has possible long-term impacts on its sustainability in terms of social (job losses), cost-effectiveness (asset infrastructures sunk costs) and environmental sustainability [21]. The possibility of transforming LNG infrastructure to hydrogen infrastructure application in Indonesia required further study. The details of hydrogen infrastructures are excluded in this study.

3.3. Sustainability Framework

The primary usage of fossil sources is for transportation fuel purposes, thus measuring the sustainability of the LNG supply chain from its view as marine fuel considered necessary. The sustainability criteria are economic, social, and environmental [15]. Ashrafi et al. (2021) conducted a study to identify the sustainability criteria for evaluating maritime fuels, while Cherepovitsyn et al. (2020) discussion on the parameters of sustainable development of the LNG Project at Arctics, combining research with the IAM Competence Framework for sustainable gas infrastructures as shown in Figure 6.

The logical sequence for assessment of project LNG infrastructures such as follows [22]: (1) analysing the current needs of stakeholders from business society perspectives; (2) identifying the risk and opportunities related to the project goals; (3) formulating the indicators of the sustainable performance; (4) indicating weightage to the groups of criteria; and (5) evaluating the project on the development of indicators using normalisation methods. The results aimed to obtain a minimum negative impact on the environment, develop an infrastructure framework, maximise the economic interests of project participants, and optimise the use of natural resources for sustainability [22].

IAM (The Institute of Asset Management) [23] provide the seven (7) critical roles for the Asset Management (AM) framework that are aligned with ISO 55000 standards to achieve the optimisation of delivery and performance of physical assets. This framework focuses on (1) analysing policy requirements and AM policy; (2) analysing the strategy requirements, forecasts the future requirement, and developing an implementation of AM strategy; (3) appraising investment evaluation with whole life costing principles, producing business case and contingency planning; (4) acquire, control, maintain, optimise and renew/repose assets (5) develop and manage stakeholders and organisational changes; (6) manage risk and quality of AM processes while review and audit the AM compliance with social and regulatory requirements; and (7) manage data and integrate information systems throughout the whole lifecycle.

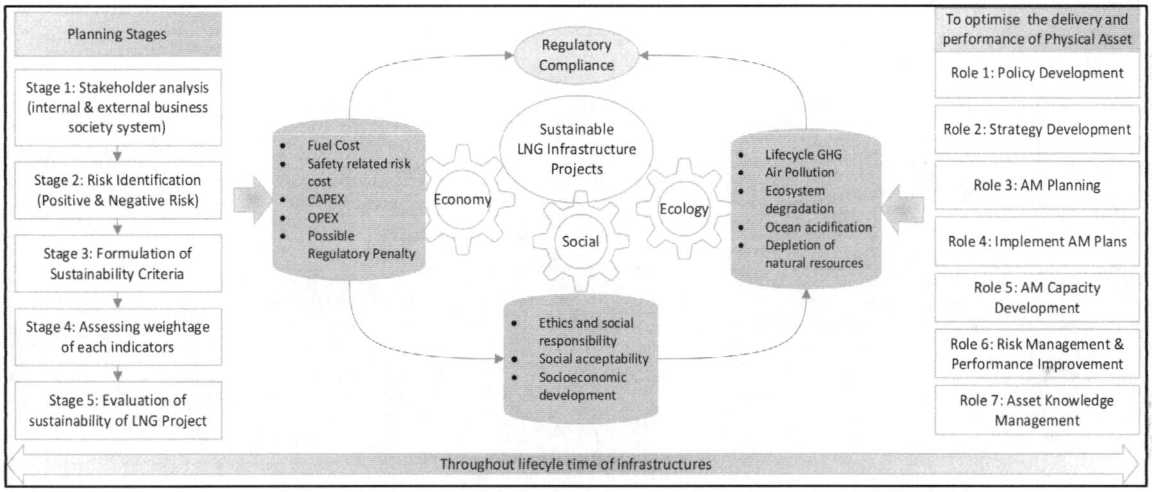

Figure 6. Sustainable LNG infrastructure projects framework [22–24].

4. Discussion: Adaptive Engineering Asset Management Framework for Sustainable Gas Terminal Infrastructures

The decision-making for strategic planning for creating sustainable gas infrastructure should be started by defining the correct definition of sustainability. The sustainability action definition is the strategy to meet the existing demand and maintain the triple bottom line sustainability balance without inhibiting the future generations from meeting their own needs [25,26]. Based to the AAPA (American Association of Port Authorities (2007), port sustainability is defined as strategies to meet the current and future needs of port stakeholders while ensuring the quality of life and sustaining natural resources [27,28]. The sustainability framework includes social, economic, and environmental issues considering integrating stakeholder engagement, environmentally friendly, cost-effective development, and efficient performance [27,28]. Those sustainability definitions become a basis for developing a decision-making framework for asset management of gas infrastructures. The decision-making process with the value driver process is considered the most suitable method for high capital infrastructures.

The necessity to transform the traditional method into a more flexible arrangement is because the current method is no longer viable to answer the rapid changes in the infrastructure's ecosystem. The ignorance of uncertainty and market volatility may cause stranded assets and sunk costs. The implementation of APP required contribution from multidisciplinary allowing the asset manager to monitor and set up the trigger for reactive, thus allowing sufficient time to be reactive on this condition [29][30].

The planning stages start from (1) context establishment based on the type of assets and their functionality; (2) mapping the target values from the stakeholder's perspectives, internal factors, and external factors and influence the decisions; and (3) value assessment where quantification modelling based on risk and sensitivity analysis between business impacts of each alternative over the time [13].

Practitioners in the LNG industry defined essential aspects to be included in the EAM as follows: (1) Production management; (2) Inventory, logistic and material management (3) Shutdown planning and scheduling (4) Maintenance management (5) Reliability analysis (6) Inspection management (7) Automation, control, equipment, monitoring and analysis (8) Major maintenance planning and scheduling. One interesting finding is the lack of interest in the asset knowledge management system. The current reporting system is still manual asset management reporting. The organisation or the asset manager should understand the necessity of an integrated database and information system to provide relevant data to monitor critical performance, forecast, and analyse maintenance records statistics, leading to asset optimisations [31].

Figure 7 shows the engineering asset management framework adapted from the value-based power and gas asset management approach.

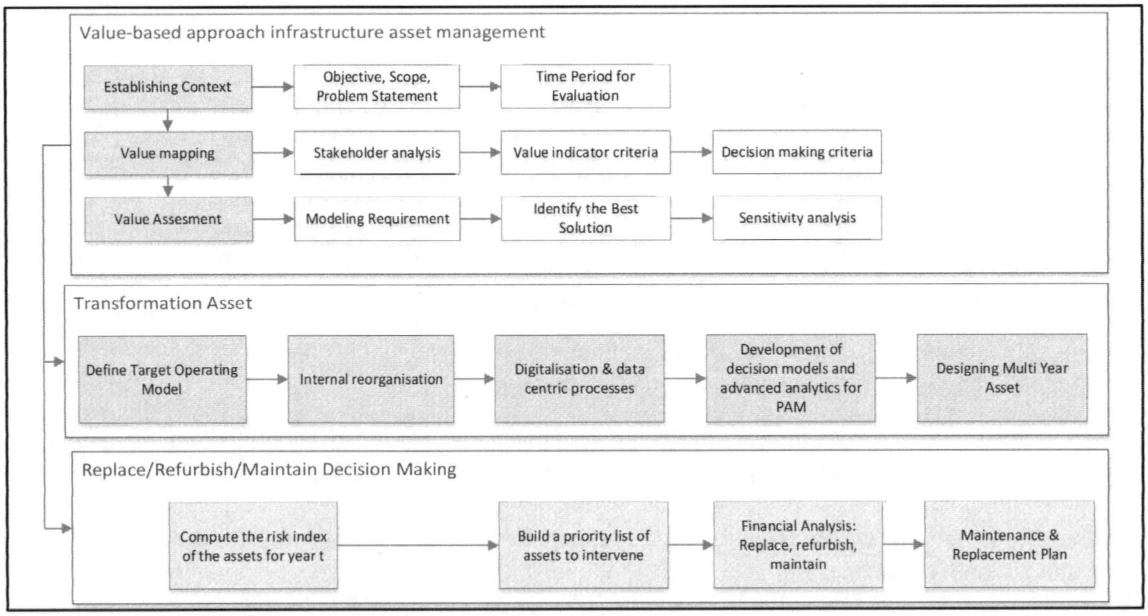

Figure 7. Engineering Asset Management (EAM) framework [13,16].

4.1. Modelling Risk and Business Impact

In the scenario analysis, the calculation of total business impact is based on the weightage of individual value drivers in the case of disruption onto the system, if caused by service obstruction, company reputation, safety and cost [13], as discussed in equation 1.

$$BI(t) = w_1 S_D + w_2 S_{REP} + w_3 S_R + w_4 Cost \quad (1)$$

Where:

$S_D = DI_D$ [Service disruption = Duration of disruption (D) and impact of disruption (I_D)] (2)

$S_R = FI_R$ [Safety risk = Fatality (F) and financial impact per fatality (I_R)] (3)

$S_{REP} = S_D/10$ [Repution risk = Service disruption/10] (4)

The suitable method to measure uncertainties, risk and cost is scenario analysis. There is an option for asset management strategy to replace, refurbish, or transform. The asset manager should only consider the asset replacement if met the following conditions: (1) a refurbishment is no longer an option (e.g. age), technology obsolesce, (2) lack of expertise for O&M (Operation & Maintenance), and (3) high operation and maintenance costs[16]. For the asset management transformation, the asset managers should perform the risk modelling within a specific band time to prioritise assets. To calculate the risk index by multiplying the criticality index and fragility index. Taleb (2012) stated that fragility relates to how the system will suffer from its environment's variability beyond a certain pre-set[32]. The calculation for the risk index such as follows[16]:

$$RI_a^y = CI_a^y FI_a^y \tag{5}$$

Where:
RI_a^y = risk index
CI_a^y = criticality index
FI_a^y = fragility index

The criticality index depends on the factors that influence the operating system, such as the impact of an asset failure, impact on the people and cargo, and environmental impact. Meanwhile, the fragility is the function of asset age, operation, performance, and product quality[16]. Based on EN 13306 Maintenance Terminology (2017), maintainability is the ability of equipment or infrastructures to be retained or kept in a particular state of functionality when performing maintenance using a specific set of procedures and resources [33]. The maintainability measured the meantime to repair (MTTR), suggesting an effective and efficient maintenance strategy[33]. In this study, the authors also propose a strategy for managing gas infrastructures that would adapt to business needs by monitoring and preparing contingency plans to adapt to an uncertain business environment.

4.2. Real Option Analysis

In the volatile market, the ability to take flexibility in the planning process becomes necessary. The Real Option Value (ROV) is a suitable tool for early financial assessment [34]. The ROA will improve business strategies in sectors emphasising opportunities, promoting strategic leverages, and maximising right to all the profits[35]. For uncertainty-free decision-making, the asset managers must calculate the Expected Net Present Value (ENPV) for each scenario. The objective function for ENPV determination is as follows [36]:

$$\text{Maximise ENPV} = \sum_{\text{i}}^{s} (P_s\, NPV_s) \tag{6}$$

where:
P is a probability of occurrence, and s is the set of uncertainty
NPV = Total of (Sales of gas – Tax – Initial Gas Cost – Fixed Cost (Initial Capital Cost and Expansion Capital Cost) – Operating Cost)

The energy industry is moving towards net-zero carbon, circular economy (waste to energy), and renewable. Built infrastructure is sensitive to multiple risks and uncertainties; thus, extending Real Option models with multiple factors of uncertainty is essential[37,38].

4.3. Recommendation Action

The contingencies plan shown in Table 4 would be the alternative plan if something happened not according to the original plan. Four categories of actions as a contingency plan once the signpost is activated. The defensive actions are actions performed to clarify the plan to meet outside challenges, while the corrective actions are adjustments to the original plan on the specific issues. The capitalising actions are actions to seize an opportunity, and the reassessments are the process initiation when the critical assumption on the original plan is no longer valid[30].

Table 4. Contingency plan for the sustainable LNG business.

Type of Action	Strategies for Sustainable Infrastructure Asset Management
Defensive Action	• Increasing connectivity between terminals to a customer by pipeline networks and LNG trucks [39] • Minimalization of total compression energy in LNG terminal by improving cold energy utilisation system and optimising BOG handling processes [39] • Optimisation of an existing system to reduce operational cost and carbon emission • Perform Risk-Based Maintenance for Gas Infrastructure Asset Management
Corrective Action	• Financial incentives for promoting the usage of LNG vessels or marine fuels [15] • Provision of bunkering infrastructures [15] at the optimum location covering a wide range of area • Diversification businesses process with a focus on the minimum investment • The integration between LNG terminal, NGL plants, and LNG plants as synergy equipment (to reduce the duplicate cost, distribution time, and transport costs) [15]
Capitalising Action	• Evaluating the LNG business hub for small-scale LNG operation • Pilot project for renewable energy for internal usage (Solar PV) • Human resources skill adaptation to Industry 4.0 or transition technology • Joint venture with a foreign company for the transfer of knowledge and transfer technology • Investing in Research & Development for innovation & technology
Reassessment	• Exploring the opportunities for transforming to renewable energy • Scenario analysis for each alternative of future marine fuels infrastructure

5. Conclusion and Future Research

5.1. Conclusion

1. The existing system is still missing the asset knowledge management system. The asset manager should understand the necessity of an integrated database and information system to provide relevant data to monitor critical performance and forecast and advance analysis with maintenance records statistics, leading to asset optimisations.
2. The asset managers should include the organisation's EAM framework: production management; material management; shutdown planning and scheduling; inspection and maintenance management; reliability analysis; control and monitoring system; and major maintenance planning and scheduling.
3. The LNG infrastructure project assessment should aim to obtain a minimum negative impact on the environment, develop an infrastructure framework, maximise the economic interests of project participants, and optimise the use of natural resources for sustainability purposes. The sequence for assessment should start from stakeholder analysis, risk identification, formulating sustainability criteria, weighing the score for each criterion and performing project assessment.
4. The transition from the fossil energy industry to a hydrogen-based economy has possible long-term impacts on its sustainability in terms of social (job losses), cost-effectiveness (asset infrastructures sunk costs) and environmental sustainability. The possibility of transforming LNG infrastructure to hydrogen infrastructure application in Indonesia required further study.

5.2. *Further Research*

The sustainability of built gas infrastructure is an immersing topic. As the energy shifting will undoubtedly happen in the near future, the best new energy alternatives are still under research; having some insights about both integration and coordination mechanism of infrastructure asset management would benefit future energy industry development.

The deepening analysis of development scenarios for transforming existing gas infrastructure facilities for energy alternatives should be a priority in future studies. Building a whole new plant system might cost more than transforming the existing facility. The authors recommend studying the stochastics model and deepening the criticality and fragility of gas assets.

References

[1] IPCC. Regional fact sheet - Asia: IPCC, 2021: Climate Change 2021: The Physical Science Basis. Contribution of Working Group I to the Sixth Assessment Report of the Intergovernmental Panel on Climate Change. 2021.

[2] Government of the Republic of Indonesia. Indonesia Long-Term Strategy for Low Carbon and Climate Resilience 2050 (Indonesia LTS-LCCR 2050) [Internet]. https://unfccc.int/sites/default/files/resource/Indonesia_LTS-LCCR_2021.pdf. 2021. Available from: https://unfccc.int/sites/default/files/resource/Indonesia_LTS-LCCR_2021.pdf

[3] IMO. MEPC 1/Circ 878. London: IMO Publication; 2018 p. 2.

[4] DNV GL. Maritime Forecast To 2050. Energy Transition Outlook 2019. 2021.

[5] IEA. Net Zero by 2050: A Roadmap for the Global Energy Sector [Internet]. International Energy Agency. 2021. Available from: https://www.iea.org/reports/net-zero-by-2050

[6] MEMR. Handbook of Energy and Economic Statistics of Indonesia [Internet]. Resources, Ministry of Energy and Mineral of Republic Indonesia. Jakarta; 2021. Available from: https://www.esdm.go.id/assets/media/content/content-handbook-of-energy-and-economic-statistics-of-indonesia-2020.pdf

[7] Forum OE. The role of hydrogen in the energy transition [Internet]. 2021. Available from: https://www.oxfordenergy.org/publications/oxford-energy-forum-the-role-of-hydrogen-in-the-energy-transition-issue-127/

[8] Gandolfo, Alberto; Patel, Ajay; Vigna, Michele Della; Pembeiro, Mafaldo; Pidoux M. Green Hydrogen: The next transformational driver of the Utilities industry [Internet]. 2020. Available from: https://www.goldmansachs.com/insights/pages/green-hydrogen.html

[9] New Zealand Government. A vision for hydrogen in New Zealand: Green Paper [Internet]. Wellington; 2019. Available from: https://www.mbie.govt.nz/dmsdocument/6798-a-vision-for-hydrogen-in-new-zealand-green-paper

[10]International IEA, Agency E. Global Hydrogen Review. 2021;

[11]Heaton, James; Parlikad AK. A Conceptual Framework for the Alignment of Infrastructure Asset to Citizen Requirements in Smart Cities. In: Parlikad AK, editor. Value-Based and Intelligent Asset Management. 1st ed. Cham: Springer Nature Switzerland; 2020. p. 39–63.

[12]Somia Alfatih MS, Leong MS, Hee LM. Definition of Engineering Asset Management: A Review. Appl Mech Mater. 2015;773–774(2008):794–8.

[13]Srinivasan RPAK. An Approach to Value-Based Infrastructure Asset Management. In: Value-Based and Intelligent Asset Management. Cambridge, UK: Springer Nature Switzerland; 2020. p. 123–38.

[14]Marquez; Adolfo Crespo; Macchi, Marco; Parlikad AK. Fundamental Concept and Framework. In: Parlikad AK, editor. Value-Based and Intelligent Asset Management. 1st ed. Springer; 2020. p. 23.

[15]Al-Yafei H, Aseel S, Kucukvar M, Onat NC, Al-Sulaiti A, Al-Hajri A. A systematic review for the sustainability of global liquified natural gas industry: A 10-year update. Energy Strategy Rev [Internet]. 2021;38(October):100768. Available from: https://doi.org/10.1016/j.esr.2021.100768

[16]Moreira da Silva M. Power and gas asset management : regulation, planning and operation of digital energy systems. 2020. 165 p.

[17]Swanson C, Center S, Levin A, Energy Program C, Stevenson A, Mall A, et al. Sailing To Nowhere: Liquefied Natural Gas Is Not an Effective Climate Strategy. 2020;(December). Available from: www.suerossi.com

[18]Turbaningsih O, Nur W, Nisa H, Mustakim A. The Study of Adaptive Planning Application for LNG Regasification Terminal Infrastructure in Indonesia. 6th Int Conf Mar Technol (SENTA 2021). 2021;1:0–14.

[19]International Gas Union (IGU). Flexible LNG Facilities Enhancing Functionality Across the LNG Value Chain. 27th World Gas Conf. 2018;1–102.

[20]System JA, Wold M, Sverud T. Comparison of Alternative Marine Fuels [Internet]. Oslo; 2019. Available from: https://sea-lng.org/wp-content/uploads/2019/09/19-09-16_Alternative-Marine-Fuels-Study_final_report.pdf

[21]Safari A, Roy J, Assadi M. Petroleum sector-driven roadmap for the future hydrogen economy. Appl Sci. 2021;11(21).

[22]Cherepovitsyn A, Evseeva O. Parameters of sustainable development: Case of arctic liquefied natural gas projects. Resources. 2021;10(1):1–27.

[23]The Institute of Asset Management. The IAM Competences Framework Part II. Guide using 2012 Asset Manag Competence Required Framew [Internet]. 2014;2(3):20. Available from: www.theIAM.org 3

[24]Ashrafi M, Lister J, Gillen D. Toward harmonising sustainability criteria for alternative marine fuels. Marit Transp Res [Internet]. 2022;3(January):100052. Available from: https://doi.org/10.1016/j.martra.2022.100052

[25]Al-Haidous S, Al-Ansari T. Sustainable liquefied natural gas supply chain management: A review of quantitative models. Sustain. 2020;12(1):1–23.

[26]Rodrigues V, Russo M, Sorte S, Reis J, Oliveira K, Dionísio AL, et al. Harmonizing sustainability assessment in seaports: A common framework for reporting environmental performance indicators. Ocean Coast Manag [Internet]. 2021;202(October 2020):105514. Available from: https://doi.org/10.1016/j.ocecoaman.2020.105514

[27]Taneja P, van der Kloot G van R, van Koningsveld M. Sustainability performance of port infrastructure—a case study of a quay wall. Sustain. 2021;13(21):1–13.

[28]Hossain T, Adams M, Walker TR. Role of sustainability in global seaports. Ocean Coast Manag [Internet]. 2021;202(October 2020):105435. Available from: https://doi.org/10.1016/j.ocecoaman.2020.105435

[29]Notteboom T. Chapter 7.3 Port Planning and Development. In: Port Economics, Management and Policy [Internet]. 1st ed. New York: Routledge; 2020. p. 133–64. Available from: https://porteconomicsmanagement.org/pemp/contents/part7/port-planning-and-development/#5_Adaptive_Port_Planning

[30]Taneja P. The Flexible Port [Internet]. 2013. Available from: http://www.narcis.nl/publication/RecordID/oai:tudelft.nl:uuid:a9f0c128-d4c3-41a2-8790-

13aec89dca63
[31]Kusumawardhani M, Markeset T. Asset Integrity Knowledge Management: A Case Study from the Petroleum Industry. Oper Supply Chain Manag An Int J. 2015;8(3):146–53.
[32]Taleb NN. Antifragile: things that gain from disorder. New York: Random House; 2012.
[33]Barabady J, Gao X, Markeset T. Criticality analysis from maintainability point of view. IEEE Int Conf Ind Eng Eng Manag. 2018;2017-Decem:2140–3.
[34]van Koningsveld, M.; Verheij, H.J.; Taneja, P.; de Vriend HJ. Por ts and Water ways Navigating the changing world. Final Publ. Delft: TU Delft Open; 2021.
[35]Leslie KJK, Michaels MMP. The real power of real options. McKinsey Q [Internet]. 1997;3(3):4–23. Available from: http://elibrary.ru/item.asp?id=3015843%5Cnwww.mckinseyquarterly.com/corpfina /repo97.asp
[36]Aseeri A, Bagajewicz MJ. New measures and procedures to manage financial risk with applications to the planning of gas commercialization in Asia. Comput Chem Eng. 2004;28(12):2791–821.
[37]Compernolle T, Huisman KJM, Kort PM, Lavrutich M, Nunes C, Thijssen JJJ. Investment Decisions with Two-Factor Uncertainty. J Risk Financ Manag. 2021;14(11):534.
[38]Castro JG, Tito EA, Brandão LE. Optimization of a Portfolio of Investment Projects: A Real Options Approach Using the Omega Measure. J Risk Financ Manag. 2021;14(11):530.
[39]Ye Z, Mo X, Zhao L. MINLP model for operational optimization of LNG terminals. Processes. 2021;9(4).

IOP Conf. Series: Earth and Environmental Science 1081 (2022) 012013 doi:10.1088/1755-1315/1081/1/012013

Illegal, Unreported and Unregulated (IUU) Fishing in Indonesia: Problems and Solutions

Adam Leonardo[1] and Nowar Deeb[1]

[1]Department of Marine Engineering, Institut Teknologi Sepuluh Nopember

Email: adamleonardo105@gmail.com

Abstract. Illegal, unreported, and unregulated (IUU) fishing is one of seven major threats to global maritime security listed by the United Nations. IUU fishing can be caused by economic factors, social factors, institutional factors, and organized IUU fishing operations. This review article discusses the topic of IUU fishing in Indonesia, as well as remedies to the problem that have been previously investigated and potentially can be applied in Indonesia. IUU fishing can result in economic losses, environmental degradation, and economic loss for coastal communities, as well as a reduction in fish stocks. Indonesia has vast marine resources, and making better use of them can boost the country's economic progress. IUU fishing is still happening in Indonesia and it caused an economic loss. According to reports, IUU fishing caused Indonesia to lose up to IDR 101 trillion annually. Therefore, appropriate strategies are needed. For this reason, the government has given IUU fishing a high priority for resolution. As it is critical to resolving the IUU fishing issue. It can be inferred that, in terms of combating IUU fishing, there are three critical aspects to consider. They are law and regulation, social and government cooperation, and technology investment, especially in monitoring, control, and surveillance.

1. Introduction

Illegal, Unreported, and Unregulated (IUU) fishing has been recognized by the united nation as one of seven major threats to world maritime security [1–3]. IUU fishing is defined by the Food and Agriculture Organization of the United Nations (FAO) as involving three categories of illicit fishing: "1. *Illegal Fishing* refers to fishing activity carried out without permission by foreign vessels in waters under the jurisdiction of another state, or in any other way that violates that state's fisheries law and regulations; 2. *Unreported Fishing* refers to fishing that is not officially recorded; and 3. *Unregulated Fishing* refers to fishing activities carried out in areas where no applicable management measures exist."[4–7].

IOP Conf. Series: Earth and Environmental Science 1081 (2022) 012013 doi:10.1088/1755-1315/1081/1/012013

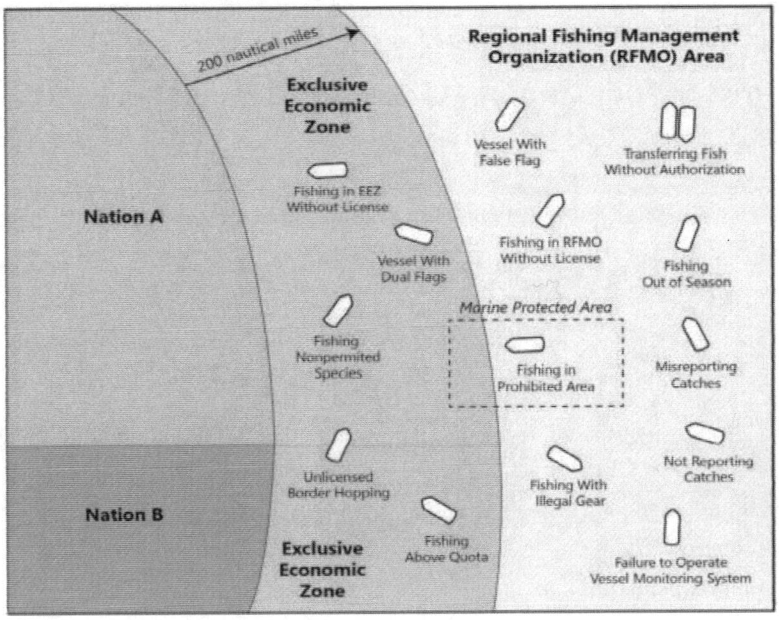

Figure 1. Common forms of IUU fishing [6].

Figure 1 shows the common forms of IUU fishing. There are many kinds of IUU fishing forms. For example, Illegal fishing occurs when another state conducts fishing activities in a state without permission. Unreported, which refers to catch that is not properly reported to authorities, for example, fisheries ministry. Unregulated refers to fishing activities conducted by fishers with vessels without nationality or vessels using a flag that is not part of RFMO (Regional Fisheries Management Organisations). IUU fishing can happen either in the exclusive economic zone or RFMO zone.

Having an Exclusive Economic Zone (ZEE) of around 7,9 million km^2 makes Indonesia one of the largest maritime domains in the world. Indonesia sea has commercial fish resources such as tuna, skipjack, etc. however, weak MCS (marine control surveillance) in ample ZEE area made Indonesia with the highest degree of IUU Fishing activities in the world [8]. Recently, there are some illegal fishing reported by the Marine and Fisheries Ministry of Indonesia (KKP). On 6th September 2021, it has been reported that KKP arrested Malaysian-flagged illegal foreign fishing boats in the Malaka strait. By 2021, KKP has arrested 135 fishing boats where 88 Indonesian boats disobey the regulation and 47 illegal foreign fishing boats. These illegal fishing boats are coming from Malaysia (16 boats), the Philippines (6), and Vietnam (25) [9]. It's been reported that IUU fishing made Indonesia lose up to IDR 101 Trillion per year in 2014. During a pandemic, 2020 until 2021 it's reported that KKP has arrested 100 fishing boats. This IUU fishing cost is IDR 30 Trillion [10].

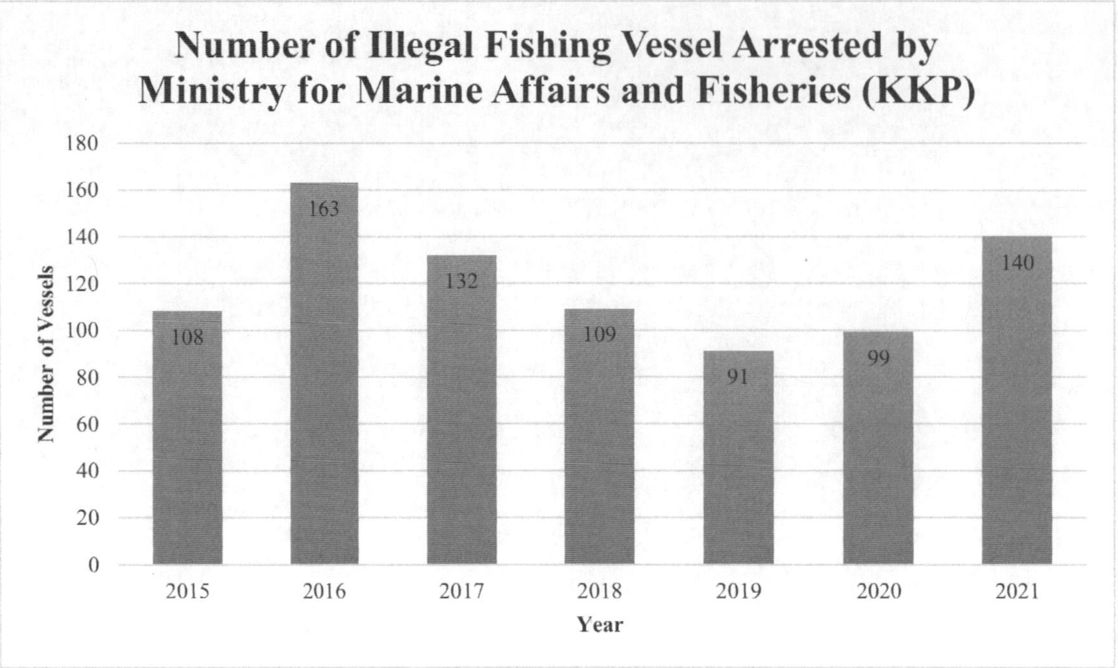

Figure 2. Number of Illegal Fishing Vessel Arrested by KKP from 2015- September 2021 [9,11].

Figure 2 depicts the number of illegal fishing vessels arrested by KKP from 2015-September 2021. It shows that the peak point happened in 2016 with a total of 163 vessels. And after that, the number of arrested vessels decreased gradually until 2019. It may be caused by the implementation of the "sinking vessel" regulation by Susi Pudjiastuti, Ministry of KKP during the 2014-2019 period. This regulation made the encouragement to engage in IUU fishing operations decrease. Thus, fewer IUU fishing cases happened.

IUU fishing can cause loss of economic revenue, environmental damage, and economic loss for coastal communities, and also it reduces fish stock levels [1,4,6,8,12]. IUU fishing also prevents governments and resource managers to gain the full economic rent from fisheries and disturbs the marine ecosystem sustainability [13]. IUU fishing is a major threat to sustainable seafood production in a country [3,14].

IUU fishing can indeed harm a country both in terms of economic, environmental, and marine reserves. From an economic perspective, for example in Indonesia, IUU fishing costs Indonesia Trillion rupiahs. From the environmental side, usage of illegal fishing equipment such as mesh nets, set lines, drum nets, explosives, etc can harm the marine ecosystem. And from the marine reserve perspective, fish exploitation can reduce the fish stock level. Unreported fishing makes it difficult for the government to track or manage marine resources. This can lead to the nescience of fish stock level in a country which can cause fish scarcity.

Joko Widodo, Indonesia's president, declared an anti-IUU fishing plan in 2014. Illegal fishing is not tolerated in Indonesia, and the captured ship will be sunk. In addition, a Task Force (Satgas 115) has been established to combat IUU fishing in Indonesian waters. The Indonesian Ministry of Marine Affairs and Fisheries (KKP) conducted audits on 1.132 fishing vessels in 2015 and discovered that all of them were breaking fishing laws and regulations [1]. The topic of IUU fishing in Indonesia is discussed in this review article, as well as potential solutions to the problem that have been previously researched and could be implemented in Indonesia.

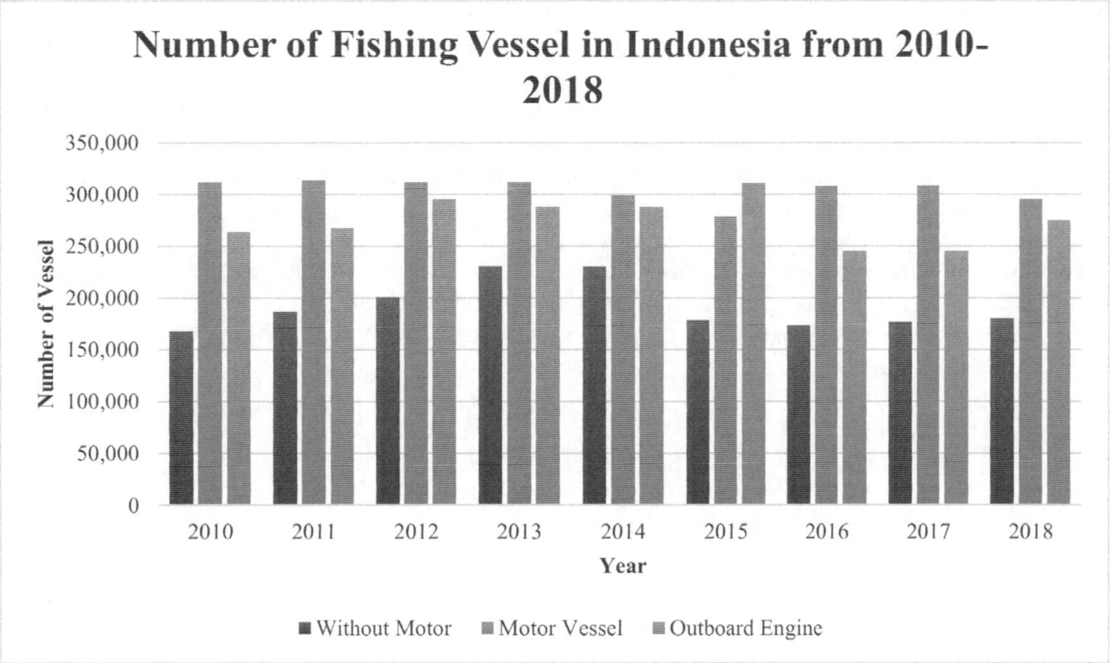

Figure 3. Number of fishing vessels in Indonesia from 2010-2018 [15].

Figure 3 depicts the total number of fishing vessels in Indonesia, it's clustered into three categories; vessels without a motor, motor vessels, and any vessel equipped with an outboard engine. Usually, the motor vessels are divided into some classes based on their GT. Table 1 below shows the number of fishing vessels based on their GT.

Table 1. Number of fishing vessels in Indonesia based on the GT [16,17].

Type (GT)	Year				
	2010	2011	2012	2013	2014
<5	110.163	123.748	137.587	151.939	152.493
5-10	31.460	35.877	37.694	46.358	41.374
10-20	10.988	13.201	11.583	15.208	14.301
20-30	7.264	8.022	7.611	8.782	9.578
30-50	857	914	917	1.074	1.029
50-100	1.747	1.801	1.641	1.727	1.766
100-200	1.290	1.204	1.176	1.127	840
>200	381	354	338	358	176

Table 1 shows the number of fishing vessels in Indonesia based on its GT. It shows that small vessel, less than 5 GT is still dominant in Indonesia.

2. What drives IUU Fishing?

IUU fishing is categorized as economic activity. It means that encouragement to engage in IUU fishing operations will exist when the expected profit of IUU fishing is in a good state. The tendency of engaging in IUU fishing will be higher if the marine species cached are in good price in the market and vice versa. There are some causes of IUU fishing according to [2]. They are economic factors, institutional factors, organized IUU fishing operations, and also social factors.

2.1. Economic factors

There are three major economic drivers of IUU fishing: overcapacity, inefficient management, and subsidies. Overcapacity in the domestic fleet can result from a general imbalance between fishing capacities and fishing opportunities, or from an incorrect allocation of fishing rights, both of which are the result of poor management systems [2].

In general, the economy can be the main driver of crime. The chance of IUU fishing happening might be closely related to the fish trade or fish market globally or locally. For example, IUU fishing tends to happen in an area where there live a bunch of tuna fish since this kind of fish is at a high price at the market. And about the capacity, as mentioned earlier, since fishing vessel in Indonesia is dominated by small fishing vessels, overcapacity in the domestic fleet must be invalid as a factor, especially in illegal fishing in Indonesia. These small fishing vessels might cause unreported fishing catch since it is not required to have fishing permits.

2.2. Institutional factors

Due to loopholes in the current international legal framework for the sea, some fishing activities are carried out outside the reach of national and international regulations. Unregulated fishing activities are those conducted by vessels with no nationality or those flying the flags of countries that are not members of relevant fisheries organizations and hence do not consider themselves bound by their regulations. Other institutional difficulties might exacerbate the IUU problem, with direct financial consequences for the IUU operator. First, it is commonly known that there is an insufficient level of Monitoring, Control, and Surveillance (MCS) both locally and in RFMOs (Regional Fisheries Management Organizations) [2].

A country with a low capability of MCS is indeed vulnerable to IUU fishing. In Indonesia, the enhancement of MCS gave a noticeable result in several cases of IUU fishing. Susi Pudjiastuti, the former Minister of Maritime Affairs and Fisheries of Indonesia under President Joko Widodo's 2014-2019 Working Cabinet has shown great work in handling IUU fishing in Indonesia. Her strategies by exploding the IUU fishing vessel have shown a positive impact on the reduction of IUU fishing in Indonesia. The fish stock also increases in some areas. So, in general, developing MCS can be highly efficient to reduce IUU fishing in Indonesia.

2.3. Organised IUU fishing operations

The establishment of organized IUU fishing operations has assisted and sped the development of IUU fishing by lowering the monetary and transaction expenses of engaging in IUU fishing [2]. IUU fishing is a transnational crime since the perpetrators frequently act within the territory of states without obtaining a permit or official authority to fish legitimately. It is believed that IUU fishing should be classified as a transnational organized crime due to the vast magnitude of the illegal fishing industry and its close ties to organized crime [5,6].

2.4. Social factors

Some of the variables that contribute to IUU fishing are social in nature. IUU/FOC (Flag of Convenience) vessels have access to an available and inexpensive labor pool due to the presence of dismal economic conditions and prospects in various developing nations (for example, many of the IUU crew observed are Indonesian, Chinese, or Filipino). Crew members arrested are commonly abandoned by their employers since they can be replaced at a minimal cost, therefore the presence of a ready and cheap labor pool reduces the true cost of risk for the vessel owner in such cases [2].

3. Addressing the IUU Fishing

Indonesia faces numerous issues in all facets of IUU fishing, and one of the government's top priorities is to overcome them. Indonesia contributes around 34% of the ASEAN region's fish products that reach the worldwide market, and IUU fishing is expected to cost the economy up to USD 3 billion per year [1].

According to a study conducted by Petrossian in 2015 [12], a country with strong patrol surveillance and good fisheries management experiences less IUU fishing problem within its sea territorial. It also mentioned that the number of species found in a territorial tends to attract the IUU fishing performer. Criminology research and theory can help identify the problem and suggest policy solutions to eliminate IUU fishing. Petrossian investigated the link between local situational elements and IUU fishing in 53 countries throughout the world in a study [12].

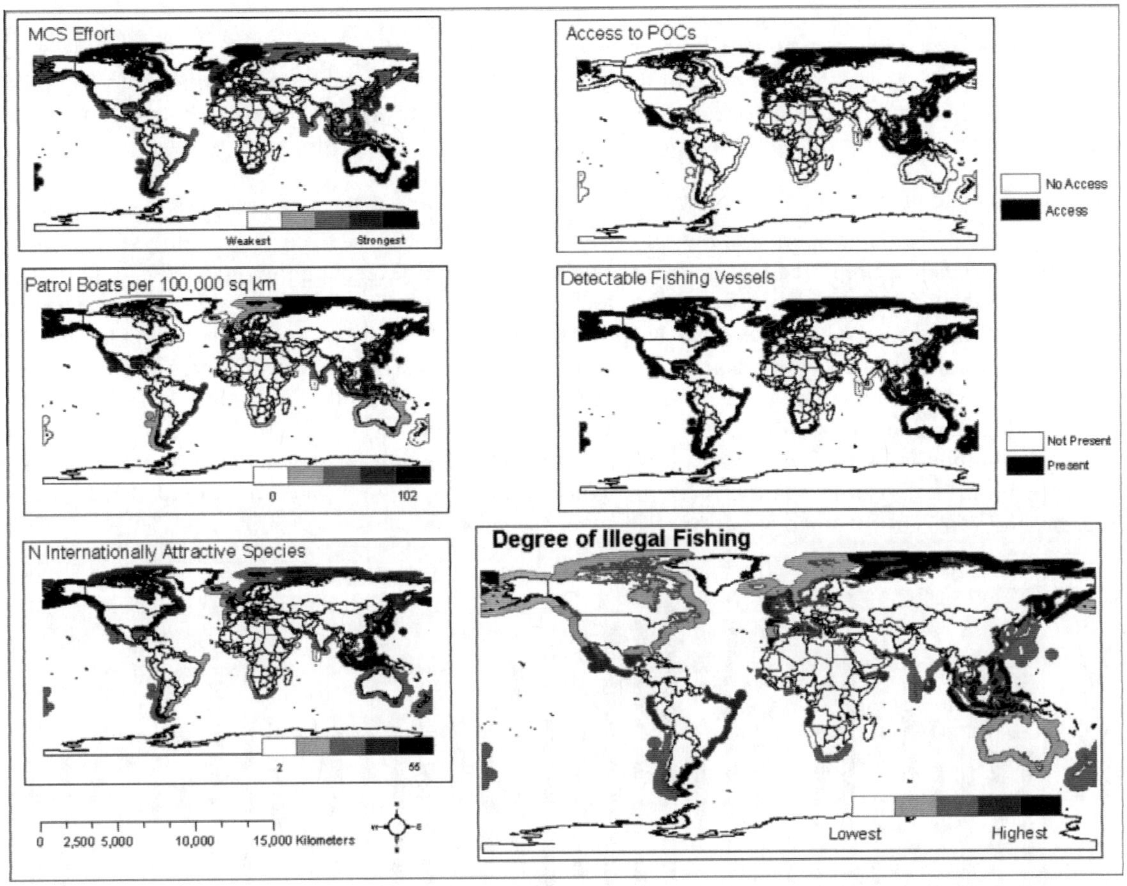

Figure 4. Spatial variations of the predictor variables and the results variable of "degree of illegal fishing" [12].

The method used by Petrossian [12] was using five predictor variables of each country that can contribute to the event of illegal fishing to find the degree of illegal fishing. Those five predictor variables are:

1. **The number of internationally attractive species**: It relates to how easily a fish can be traded in the global market.
2. **The number of patrol boats per 100.000 km^2**: It represents the information regarding the number of patrol boats owned by a country.
3. **Number of detectable fishing vessels**: Registered vessels from 1 January 2010-31 March 2010. The data included information about vessel type, traffic pattern, and navigation status.
4. **Monitoring, control, and surveillance (MCS) effort**: This represents ranking countries from 0-10 in terms of their performance on MCS.
5. **Access to ports of convenience**: It represents the distance from each country's centrally located to 10 ports of convenience (example: Port Singapore, Port Cartagena, Port Kaliningrad, etc). These

10 ports are chosen because they are expected to be the most frequently visited port by illegal fishing vessels during a six-year period (2004-2009).

It is reported that the number of species within sea territorial and access to a port of convenience can cause a country more vulnerable to IUU fishing. Countries with excellent MCS and strong patrol capacity are less likely to experience illegal fishing. Regarding the low MCS capacity of a country, it may be caused by an inability to invest in more fisheries management and patrol surveillance sector [12].

This tool or method is very essential in combating IUU fishing. Since this method shows the degree of illegal fishing in an area. It's very useful for a country to know first which areas are most vulnerable to IUU fishing. In a country like Indonesia, where almost 70% of the area is the ocean, it's very important to know where the best place to invest more MCS effort. So, the monitoring will be efficient and right on target.

In 2010,Varkey [13] conducted a study regarding IUU fishing in Raja Ampat Regency in Indonesia. In correlation with [12] argument that the number of attractive species can cause a region vulnerable to IUU fishing, it has been reported that illegal and unreported catch exceeded the reported catch by more than 40.000 tonnes in 2006. Where according to [13] Raja Ampat has high marine species diversity. This illegal and unreported catch is worth 40 million USD. It is correlated with Fishery Act no. 9/1985 and Fishery Act no.31/2004 which don't require small fishing vessels (less than 5 gross tonnage (GT) or vessels with engines less than 15 HP) to have fishing permits. This regulation leads to large unreported catch in Indonesia since the small-scale fishing vessel is dominant [13].

Just like Indonesia, Kuwait also has a problem with IUU. Kuwait is a country that could gain huge benefits from investing in its marine fisheries sector. Illegal, unreported, and unregulated fishing is a major problem for Kuwait since it reduces fish stock levels [4].

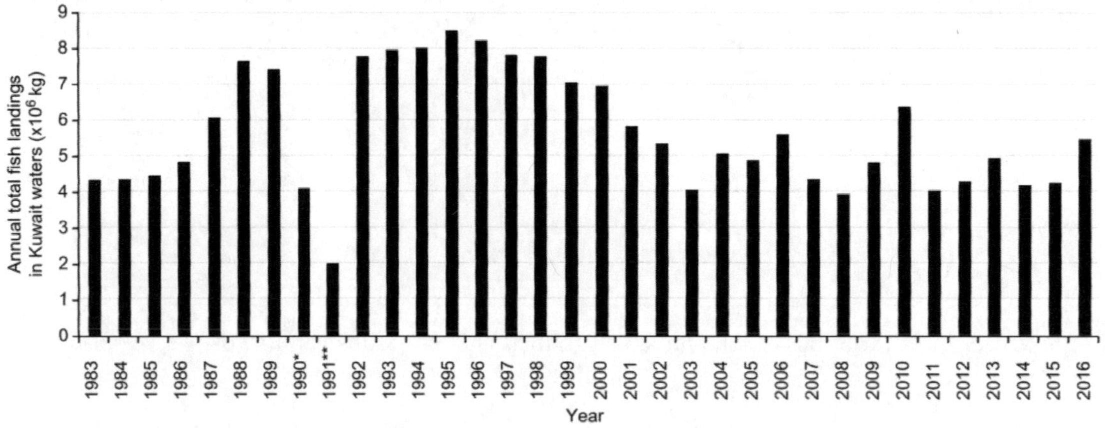

Figure 5. Yearly total fish landings in Kuwait sea from 1983-2016 in kg [4].

Fish production reached a peak in 1995 (8,5 million kgs). However, it gradually decreased to an average of 4 and 5 million kgs. The fisheries industry was stable between 2011-2013 with an annual production of 4 m-4,5 m kgs. The main cause of declining stocks is overfishing, and the second cause is pollution from factories and sea pollutants such as direct sewage discharge to the sea [4].

The main cause of IUU fishing in Kuwait is low enforcement of fisheries law; immigration monitoring failure (IUU fishing in Kuwait is dominated by immigrants, mostly from Egypt and India); and lack of control over the fish market which provides space for illegal fish. It is suggested that the government should take aggressive steps to punish sponsors of IUU fishers by blocking their business operations in Kuwaiti waters. Also, the government has to address the issue of alleged corruption in the judicial system, which allows IUU Kuwaiti fishers to escape punishment for harming fish stocks [4].

IUU fishing is also a concern in the United States. The United States has implemented the Seafood Import Monitoring Program (SIMP) to combat IUU fishing and seafood fraud by requiring chain-of-custody evidence for 13 species imported into the country. The SIMP asks that every seafood imported into the United States be traced and furnished with a complete production record. This will prevent the sale of IUU seafood in the United States. Exporters will be incentivized to enhance management in order to comply with the SIMP if the US has market strength as a result of considerable imports. If the US lacks market power, the SIMP will affect trade patterns, and the SIMP's costs will be passed on to US consumers in the form of higher prices and decreased seafood consumption. The SIMP will encourage importers to enhance their management procedures in the analyzed supply chains because the US has buying power for most items. Thus, this will make IUU fish not able to enter the US market [14].

This monitoring system can be very effective for a government to track the fish supply chains or seafood supply chain. By demanding documentation regarding the fish catch or seafood so the government can track the fish catch record. However, for scale small fisheries, for example, fishers in a rural area might not be able to follow this method. Preparing this kind of documentation could be impossible for them.

According to Vlad [18], to ensure the long-term viability of marine resources (for example fish stocks), more controls on foreign fleet activity and fishery resource management measures limiting destructive fishing practices, excessive by-catch, and the use of seasonal and area restrictions should be implemented, allowing resources to recover from overexploitation where it exists [18]. Controlling foreign fleet activity is a critical aspect to combat IUU fishing, implementing good MCS can effectively reduce the IUU fishing risk. And also implementing strict regulations regarding fishing equipment is also important. Using explosive equipment to catch fishing can rapidly reduce the fish stock because it kills the fish seeds which is endanger the fish stock level afterward.

Based on a study conducted by Gallic and Cox [2], there are some possible ways to fight IUU fishing, the first one is by reducing total revenues, which implies lowering the value of IUU catches. The second one is by increasing the operating cost of IUU/FOC vessels or even capital costs. Doing this will make IUU fishing more challenging, thus reducing the IUU fisher's tendency to do so. The third one is by raising the *cost* of risk associated with IUU fishing and also increasing the scope and level of sanction [2].

Increasing the cost of risk of evolving in IUU fishing might be one of the best solutions to prevent IUU fishing. For example, the sanction level issued by Susi Pudjiastuti, former Minister of Maritime Affairs and Fisheries of Indonesia during the 2014-2019 period. The IUU vessel will be exploded as the sanction of doing IUU fishing in Indonesia's territorial area. This strategy has been proven very effective to combat IUU fishing in Indonesia.

4. Strategies in Combating IUU Fishing in Indonesia

Indonesia has huge marine resources thus, utilizing those resources can help to enhance Indonesia's economic growth. IUU fishing is still happening in Indonesia and it can cause economic loss. So, appropriate strategies are needed. That is the reason why IUU Fishing is one of the issues that is prioritized to be solved by the government, see figure 3 below. Along with maritime defense and maritime security, the IUU Fishing problem sector is being prioritized by the Indonesian government to be solved.

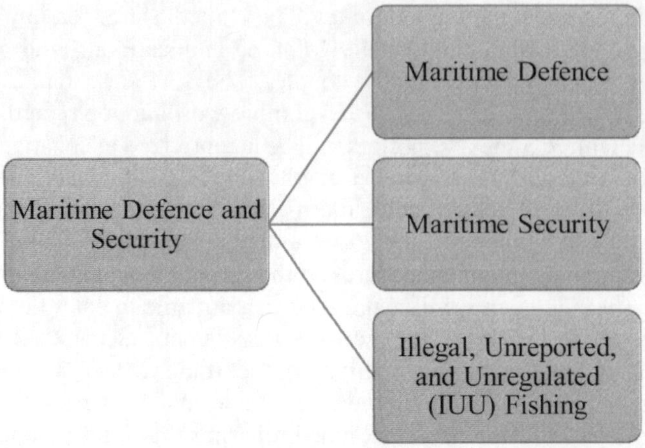

Figure 6. Indonesia maritime defense and security priority programs [1].

According to Chapsos [1], The local implications of IUU fishing are frequently overlooked. Local fishing communities are not only a part of the IUU problem but they should also be included in its solution. The government must devote greater resources to the creation of inclusive governance institutions, the formulation of policies, and the implementation of laws and regulations. Local fishing communities must be considered in this scenario. And also the cooperation between local and government need to be enhanced to increase trust between both sides [1,19].

Monitoring, Control, and Surveillance (MCS) are very important things in combating IUU fishing. MCS is all about ensuring that fisheries management procedures are followed. In practice, three elements play various roles in ensuring compliance. The first one is *monitoring*, it's all about gathering information, data collection, and reporting. The second one is *control* which means regulatory conditions which govern exploitation, control of fishing activities. The last one is *surveillance* which means observation required to maintain compliance [20,21]. MCS can help a country to reduce the effect of IUU fishing. Based on a study conducted by Doumbouya et al. 2017, MCS recovered the loss caused by IUU fishing (the case is in West Africa). It is reported that IUU fishing in West Africa is responsible for a loss of over USD 2,3 billion yearly, where USD 13,8 million are recovered through MCS. It may be concluded that higher fines contribute to reducing tendencies of IUU fishing through a better capability of catching offenders (IUU performers) by increasing MCS [22]. Rising sanction level is one of the best solutions in combating IUU fishing, it can reduce the encouragement of IUU fishers to do so.

According to a study conducted by Longépé [8], Since 2003, the Indonesia fisheries management is equipped with MCS technologies which are called Vessel Detection System (VDS). This technology can detect non-cooperative fishing vessels based on high-resolution radar images, which directly affects the performance of patrol vessels. This VDS is combined with Automation Identification System (Sat-AIS) data. In this scheme, VMS (vessel monitoring system) is attached to the legal ship, thus, the illegal ship is a ship that does not use VMS. Implementing this technology has reduced the total number of illegal fishing from 2015 to 2016 [8,23]. However, based on ministerial regulation number 42/PERMEN-KP/2015, attaching VMS is compulsory only for a fishing vessel with 30 GT or higher [23]. On the other side, this scheme could be less effective to be applied in Indonesia, since fishing vessel in Indonesia is dominated by small vessels as mentioned by [13]. This is also shown in figure 1. Small fishing vessel <5GT is still dominant.

In general, in order to overcome the IUU fishing issue, cooperation between locals and the government is needed especially in making fishing regulations such as fishing permits, etc. Raising the sanction level of IUU activities is efficient to reduce IUU fishing. And also, utilization of technology in terms of overcoming illegal fishing is needed. The government must be stricter in handling IUU fishing issues. Since IUU is a government priority program, the investment for this sector especially in patrol and surveillance should be viable. To summarize, the various methods to combat IUU fishing are presented below table.

Table 2. Summary of various solutions to address IUU Fishing

Solutions	References
- Enhancing MCS and patrol capacity	[8,12]
- Applying greater law enforcement	[4]
- Implementing Seafood Import Monitoring Program (SIMP)	[14]
- Controlling foreign fleet activity, implementing good MCS, and also implementing strict regulations regarding fishing equipment	[18]
- Reducing the value of IUU catches, raising the cost of risk associated with IUU fishing, and broadening the scope and intensity of sanctions	[2]
- Local fishing communities must be taken into account in the design and execution of policies and rules or regulations, as well as the establishment of inclusive governance institutions. And also the cooperation between local and government need to be enhanced to increase trust between both side	[1,19]
- Higher fines and better MCS effort	[22]

Table 2 shows the summary of various solutions based on the previous research. Monitoring, Control, and Surveillance are critical things that need to be improved to reduce IUU fishing. Applying more strict regulations and raising sanction levels may be effective to be implemented in Indonesia.

5. Conclusion and Recommendation

IUU fishing can result in a loss of revenue, environmental harm, and economic loss for coastal communities, as well as a reduction in fish stocks. It is important to overcome this IUU fishing issue. The number of cases and the motive of IUU fishing could be different. However, the causes of IUU fishing that happen in Indonesia and other countries are similar. Mostly, the main cause of IUU fishing is the weak capability of MCS (monitoring, control, and surveillance) of a country. And also laws and regulations regarding IUU fishing are not applied strictly. In general, any kind of method (as mentioned in table 2) to combat IUU fishing in another country may be applied in Indonesia but, there is some constraint. The method that needs a high level of technology might be harder to be applied in Indonesia since it cost a lot of money.

There are three key points that important in terms of overcoming IUU fishing. They are law and regulation, cooperation between local and Indonesian governments, and last but not least investment in technology. Thus, it is recommended for the Indonesia government to:

1. Perform strict enforcement regulations regarding IUU fishing issues. For example by implementing a program similar to Seafood Import Monitoring Program. Tracking the fish catch history can reveal how the fish were cached. And also, raising the sanction level of IUU fishing might be one of the best solutions to reduce IUU fishing in Indonesia.
2. Make sure that local fisher is not harmed by the regulation made by the government. It may lead to law violations performed by local fishermen. Communication and cooperation are needed between local fishers and the government to overcome IUU fishing.
3. Invest in technology, especially marine technology. With this ample sea area, it will be more efficient if automation technology is used to surveillance the sea. For example, underwater technologies such as SONARs detect illegal fishing vessels. Enhancing the MCS level may reduce IUU fishing in Indonesia significantly.

IOP Conf. Series: Earth and Environmental Science 1081 (2022) 012013 doi:10.1088/1755-1315/1081/1/012013

References

[1] Chapsos I, Koning J, Noortmann M. Involving local fishing communities in policy making: Addressing Illegal fishing in Indonesia. Marine Policy. 2019 Nov 1;109:103708.
[2] Gallic B le, Cox A. An economic analysis of illegal, unreported and unregulated (IUU) fishing: Key drivers and possible solutions. Marine Policy. 2006 Nov 1;30(6):689–95.
[3] Ma X. An economic and legal analysis of trade measures against illegal, unreported and unregulated fishing. Marine Policy. 2020 Jul 1;117:103980.
[4] Alqattan MEA, Gray TS, Stead SM. The illegal, unreported and unregulated fishing in Kuwait: Problems and solutions. Marine Policy. 2020 Jun 1;116:103775.
[5] Ardi Hendharto. Understanding Iuu Fishing As Transnational Organized Crime With Special Example of Benjina Case. Kajian. 2018 Jun;23(2):95–109.
[6] Chapsos I, Hamilton S. Illegal fishing and fisheries crime as a transnational organized crime in Indonesia. Trends in Organized Crime. 2019;22(3):255–73.
[7] Jan Lehmköster. World Ocean Review: The Future of Fish - the Fisheries of the Future. Tim Schröder, Dimitri Ladischensky, Dr. Jörn Schmidt, Dr. Rüdiger Voss, Dr. Kirsten Schäfer, editors. Hamburg: maribus gGmbH, Pickhuben 2, 20457 ; 2018.
[8] Longépé N, Hajduch G, Ardianto R, Joux R de, Nhunfat B, Marzuki MI, et al. Completing fishing monitoring with spaceborne Vessel Detection System (VDS) and Automatic Identification System (AIS) to assess illegal fishing in Indonesia. Marine Pollution Bulletin. 2018 Jun 1;131:33–9.
[9] Kementerian Kelautan dan Perikanan Indonesia. Siaran Pers Kementerian Kelautan Dan Perikanan: KKP Tangkap Kapal Illegal Fishing Asal Malaysia di Selat Malaka. Direktorat Jenderal Pengawasan Sumber Daya Kelautan Dan Perikanan. 2021 Sep 6;
[10] Hi Pontianak. Selama Pandemi, Kerugian Negara Akibat Illegal Fishing Capai Rp 30 Triliun. News. 2021.
[11] Kementerian Kelautan dan Perikanan. Siaran Pers Kementerian Kelautan dan Perikanan Nomor: SP.272/SJ.4/XII/2020: KKP Tangkap Kapal Ilegal, Menteri Kelautan dan Perikanan ad Interim: Pengawasan Tak Pernah Kendor. Direktorat Jenderal Pengawasan Sumber Daya Kelautan dan Perikanan. 2020 Dec;
[12] Petrossian GA. Preventing illegal, unreported and unregulated (IUU) fishing: A situational approach. Biological Conservation. 2015 Sep 1;189:39–48.
[13] Varkey DA, Ainsworth CH, Pitcher TJ, Goram Y, Sumaila R. Illegal, unreported and unregulated fisheries catch in Raja Ampat Regency, Eastern Indonesia. Marine Policy. 2010 Mar 1;34(2):228–36.
[14] Fang Y, Asche F. Can U.S. import regulations reduce IUU fishing and improve production practices in aquaculture? Ecological Economics. 2021 Sep 1;187:107084.
[15] KKP. Statistik-KKP. https://statistik.kkp.go.id/home.php?m=kapal&i=5#panel-footer. 2021.
[16] California Environmental Associates. Trends in Marine Resources and Fisheries Management in Indonesia: A 2018 Review. 2018.
[17] KKP. Analisis Data Pokok Kelautan dan Perikanan. Jakarta; 2015.
[18] Vlad M. Kaczynski SWL. Coastal Resources as an Engine of Economic Growth and Reduction of Poverty in West Africa: Policy Considerations. Coastal Management [Internet]. 2000 Jul 1;28(3):235–48. Available from: https://doi.org/10.1080/089207500408647
[19] Eriksson B, Johansson F, Blicharska M. Socio-economic impacts of marine conservation efforts in three Indonesian fishing communities. Marine Policy. 2019 May 1;103:59–67.
[20] Menyoli. Monitoring, Control and Surveillance (MCS) to combat IUU fishing. In: Fisheries Governance. 2019.
[21] Klaudija Cremers, Glen Wright, Julien Rochette. Strengthening Monitoring, Control and Surveillance in Areas Beyond National Jurisdiction. 2020.

[22] Doumbouya A, Camara OT, Mamie J, Intchama JF, Jarra A, Ceesay S, et al. Assessing the Effectiveness of Monitoring Control and Surveillance of Illegal Fishing: The Case of West Africa. Front Mar Sci. 2017;4:50.
[23] S. Budianto. Vessel Monitoring System Perkuat Pengawasan Illegal Fishing. Direktorat Jenderal Pengawasan Sumber Daya Kelautan dan Perikanan. 2017.

Ship to Ship Manoeuvring Simulation to Determine Elements of Tugboat Handling

I Putu Sindhu Asmara[1] and Adi Wirawan Husodo[2]

[1]Shipbuilding Department, Politeknik Perkapalan Negeri Surabaya
[2]Marine Engineering Department, Politeknik Perkapalan Negeri Surabaya

Abstract. The berthing and unberthing process of a shuttle tanker on the FSO unit has its difficulties if the FSO unit is between several mooring lines. This is an early study to simulate the berthing and unberthing of a 5000 DWT shuttle tanker at the FSO unit without going through the mooring ropes and still comply with the requirements of PIANC berthing speed. The manoeuvring simulation was based on a mathematical manoeuvring group model that used the effects of the propeller and rudder. The model simulated the percentage and direction of force and moment of tugboats to perform the manoeuvring properly. The model assumed the environmental disturbances of wind, current, and wave as the constant values and applied three scenarios of the current direction. As a case study, the berthing and unberthing manoeuvre of a 50000 DWT export tanker need two tugboats with the capacity of at least 2x42 tons bollard pull, exceptional for unberthing in scenario-2.

Keywords: tugboats; manoeuvring; MMG; environmental disturbances; FSO; shuttle tanker

1. Introduction

This paper provides a manoeuvring simulation using a simulator program developed in Matlab. Prediction of navigational safety in a restricted area such as port and channel use ship manoeuvring simulator besides for the training of navigation [1]. The experience of harbour pilot and tugboat captain is the main reference for towing operation of berthing and unberthing, and the effective operation method is unavailable [2]. The simulator can demonstrate the experience of the captains. Side berthing and unberthing manoeuvre of shuttle tanker to and from Floating Storage and Offloading (FSO) spread mooring also have a narrow path between the mooring lines and are considered acting as a restricted area. The simulation of vessel berthing and unberthing assisted by tugboats ensures the trajectory fits the planned route between the mooring lines and the berthing speed complies with the requirement.

The spread mooring has four (4x1) or eight (4x2) mooring lines attached at the corner of the FSO. The formation of mooring lines limits the FSO not moving freely [3]. This paper doesn't consider the surging, swaying, and yawing motions of the FSO and assumes their values are to be zero. This paper also doesn't concern the seakeeping analysis of heaving, rolling, and pitching motions. The authors only focus on the 3 DoF of manoeuvring simulation to ensure the shuttle tanker does not pass through the mooring lines and complies with the berthing speed requirement at the starboard of the FSO unit. The suggested approach velocity of a vessel from 10000 DWT to 50000 DWT for moderate conditions is 0.30 to 0.15 m/s [4]. The association recommends the berthing angle of those vessels be 10 to 15 degrees and a maximum of 6 degrees for ships of more than 50000 DWT. Gucma adopted the ship speed normal to the berthing line at the instant of contact as the design speed for fender and berthing structure [5].

This paper presents a case study in an offshore terminal in Indonesia planning to exchange the underwater pipe installation into a mother tanker acting as an FSO and a shuttle tanker to transport the oil from the FSO to the shore terminal. The study aims to determine the number and capacity of tugboats, the time series of tugs' power, and the time series of tugs' direction to achieve the allowable berthing speed and berthing angle. In addition, the study also intended to estimate the duration of berthing and unberthing manoeuvre.

2. Methodology

The manoeuvring simulator program developed in this study follows the Mathematical Manoeuvring Group (MMG) model and considers tugboat assistances and external disturbances of environmental conditions described below.

2.1. Motion equations of ship manoeuvring

Equations 1 to 3 express the surging, swaying, and yawing motion of vessels disturbed by the environment and assisted by tugboats.

$$(m + m_x)\dot{u} - (m + m_y)v_m r - x_G m r^2 = X_H + X_P + X_R + X_A + X_W + X_T \quad (1)$$

$$(m + m_y)\dot{v}_m + (m + m_x)ur + x_G m\dot{r} = Y_H + Y_R + Y_A + Y_W + Y_T \quad (2)$$

$$(I_{zG} + x_G^2 m + J_z)\dot{r} + x_G m(\dot{v}_m + ur) = N_H + N_R + N_A + N_W + N_T \quad (3)$$

where m ship's mass, m_x and m_y are added masses of x axis direction and y axis direction, respectively, u is surge velocity, v_m is the lateral velocity at midship, and r is the yaw rate. The x_G is the longitudinal coordinate of the centre of gravity of the ship, I_{zG} is the moment of inertia around the centre of gravity of the ship, and J_z is added moment of inertia. The X, Y, and N are the surge force, lateral force, and yaw moment around midship, respectively. Subscript H, P, R, A, W, and T represent hull, propeller, rudder, wind, wave, and tugs, respectively. Figure 1 shows the coordinate system of motion equation, where β is drift angle, δ is rudder angle, and ψ is the ship's true heading. The x_0 represents the North direction in the space fixed coordinate system and x denotes the surging direction in the body fixed.

2.2. Hydrodynamic forces and moment acting on ship hull

The shuttle tanker simulated in this study has a displacement of 50386 tons with a coefficient block of 0.82. The model used the resistance coefficient and hull hydrodynamics derivatives of the KVLCC2 model [6] for the shuttle ship by considering that the block coefficient is almost similar. The hull force

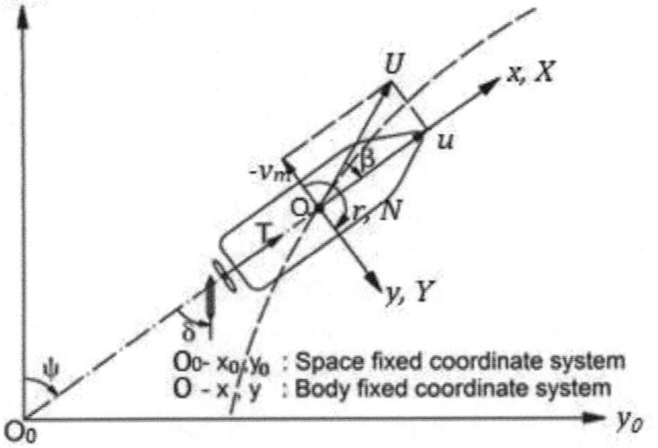

Figure 1. Coordinate System

and moment predicted by resistance coefficient and the hydrodynamics derivatives using prime ' putting in the symbol in Equations 4 to 6 are non-dimensional.

$$X_H = \frac{1}{2}\rho LdU^2\{-R'_0 + X'_{vv}v'^2 + X'_{vr}v'_m r' + X'_{rr}r'^2 + X'_{vvvv}v'^4_m\} \tag{4}$$

$$Y_H = \frac{1}{2}\rho LdU^2\{Y'_v v'_m + Y'_r r' + Y'_{vvv}v'^3_m + Y'_{vvr}v'^2_m r' + Y'_{vrr}v'_m r'^2 + Y'_{rrr}r'^3\} \tag{5}$$

$$N_H = \frac{1}{2}\rho L^2 dU^2\{N'_v v'_m + N'_r r' + N'_{vvv}v'^3_m + N'_{vvr}v'^2_m r' + N'_{vrr}v'_m r'^2 + N'_{rrr}r'^3\} \tag{6}$$

Where ρ is the water density, L is L_{pp}, d is ship draft and U is resultant speed. The forces and moment acting in the shuttle tanker were obtained by multiplying the non-dimensional force and moment by the $\frac{1}{2}\rho LdU^2$ and $\frac{1}{2}\rho L^2 dU^2$, respectively.

2.3. Hydrodynamic force due to the propeller

Surge force X_P due to the propeller is depending on propeller thrust T and the thrust deduction factor t_P. The force and thrust of the propeller are expressed as Equations 7 and 8, respectively. The thrust coefficient K_T expressed as the 2nd polynomial function of propeller advanced ratio J_P which is developed according to the K_T - J_P curve of the propeller and represented by Equation 9.

$$X_P = (1 - t_P)T \tag{7}$$

$$T = \rho n_P^2 D_P^4 K_T \tag{8}$$

$$K_T = a + bJ_P + cJ_P{}^2 \tag{9}$$

Where n_P and D_P are propeller revolution and propeller diameter, respectively. a, b, and c are constant.

2.4. Hydrodynamic forces and moment by the rudder

Equations 10 to 12 are the representation of effective rudder force in the longitudinal direction, rudder force in the lateral direction, and rudder yaw moment in the body-fixed coordinate system, respectively.

$$X_R = -(1 - t_R)(\frac{1}{2}\rho LdU^2)F'_N sin\delta \tag{10}$$

$$Y_R = -(1 - \alpha_H)(\frac{1}{2}\rho LdU^2)F'_N cos\delta \tag{11}$$

$$N_R = -(x'_R + \alpha_H x'_H)(\frac{1}{2}\rho LdU^2)F'_N cos\delta \tag{12}$$

Where x'_R denotes the dimensionless location of the rudder from the midship relative to ship length, δ is rudder angle, F'_N is non-dimensional rudder normal force. The forces are depending on the interaction coefficients between hull and rudder. The coefficients of t_R, α_H, x'_H are the steering resistance deduction factor, rudder force increase factor, and the non-dimensional longitudinal acting point the additional lateral force component, respectively. In this study, the simulation uses the coefficients of$t_R = 0.19$ [7] $x'_H = -0.45$ [8] and $\alpha_H = 3.6C_b/(L/B)$ [9].

2.5. Disturbance forces and moment by wind

Equations 13 to 15 express the external forces and moment due to a constant and uniform wind [10] acting on the ship and affecting the ship's manoeuvring.

$$X_A = \frac{1}{2}\rho_a V_A^2 A_T C_{XA}(\Theta_A) \tag{13}$$

$$Y_A = \frac{1}{2}\rho_a V_A^2 A_L C_{YA}(\Theta_A) f_A(\phi) \tag{14}$$

$$N_A = \frac{1}{2}\rho_a V_A^2 L A_L C_{NA}(\theta_A) f_A(\phi) \quad (15)$$

Where ρ_a denotes air density;$A_T$the transverse projected area; and A_L the lateral projected area. The constants of wind forces and moment are a function of relative wind direction θ_A. Fujiwara et al. [10] have published a detailed formulation of the aerodynamic force coefficients expressed as a function of the relative wind direction, ψ_A. Yasukawa (Yasukawa H and Sakuno Y, 2020) [11] wrote the coefficients as $C_{XA}(\theta_A), C_{YA}(\theta_A)$, and $C_{NA}(\theta_A)$ which are also a function of relative wind velocity θ_A. $f_A(\phi)$ is a correction coefficient due to the condition of ship heels but it isn't considered in this simulation. V_A is the relative wind velocity expressed by Equations 16 to 18.

$$V_A^2 = u_A^2 + vV_A^2 \quad (16)$$

$$u_A = u + U_W \cos(\theta_W - \psi) \quad (17)$$

$$v_A = v + U_W \sin(\theta_W - \psi) \quad (18)$$

Where U_W is absolute wind velocity; and θ_W the absolute wind direction using the North direction as 0 degrees.

2.6. Disturbance forces and moment of wave loads

The simulation used a non-dimensional time-averaged wave-induced steady forces and yaw moment ($\bar{R}, \bar{Y}$ and $\bar{N}$) of a vessel having a block coefficient C_B of 0.84 published by Wicaksono [12]. Equations 19 to 21 express the disturbance forces and moment of the wave.

$$X_W = \bar{R}\rho g \zeta_a^2 (B^2/L) \quad (19)$$

$$Y_W = \bar{Y}\rho g \zeta_a^2 (B^2/L) \quad (20)$$

$$N_W = \bar{N}\rho g \zeta_a^2 LB \quad (21)$$

Where ρ denotes water density;$g$the acceleration of gravity; ζ_a the wave amplitude; B the ship breadth; and L the ship length between perpendicular.

2.7. Assistance forces and moment by tugboats

Equations 22 to 24 are the expression of tug forces and moments applied in the simulation.

$$X_T = F_T \cos\theta_T \quad (22)$$

$$Y_T = F_T \sin\theta_T \quad (23)$$

$$N_A = Y_T (L/2) \quad (24)$$

Where F_T denotes the force of tug; and θ_T the relative direction of tug force. The arm of tug moment is half of the ship's length.

2.8. Current effects

The current effect is calculated based on the relative velocity concerning the moving frame. The U, u, v_m, and r are expressed as:

$$U = U_r = \sqrt{{u_r}^2 + {v_r}^2} \quad (25)$$

$$u = u_r = (u_b - u_c) = (u_b - U_c \cos\psi_{c_r}) \quad (26)$$

$$v_m = v_r = (v_b - v_c) = (v_b + U_c \sin \psi_{c_r}) \quad (27)$$

$$r = \dot{\psi}_{c_r} \text{ and } \psi_{c_r} = \psi - \psi_c \quad (28)$$

Subscript *b* and *c* represent the ship body and current, respectively.

3. Results and Discussions

3.1. Berthing and Unberthing Scenario

This berthing/unberthing simulation uses 3 different scenarios, which are a function of several environmental conditions such as maximum wind speed, significant wave height, significant wave period, and maximum current velocity. The parameters of the three scenarios can be seen in Table 1. The basic difference between the three scenarios lies in the direction of the current. The current directions are from the North West (NW) to South East (SE) for Scenario-1, from North East (NE) to South West (SW) for Scenario-2, and from South East (SE) to North West (NW) for Scenario-3.

Specific to berthing simulation, there are four detailed stages of shuttle tanker performing berthing for each scenario. Stage-1 is described by the maximum velocity of the shuttle tanker from 2 knots when in the anchor area then decreasing to 1 knot when the Tugboat starts to be attached to the shuttle tanker. At this stage, the shuttle tanker manoeuvre is still purely by the rudder, without the assistance of tugboats. Stage-2 is described by the maximum velocity of the shuttle tanker from 1 knot and then decreases to 0.5 knots. At this stage the tugboats have been already attached to the shuttle tanker, however, the manoeuvre of the shuttle tanker is still using the rudder. In Stage-3, the shuttle tanker starts from the position where the maximum velocity is 0.5 knots until the maximum velocity becomes 8 cm/s. The two Tugboats located at the stern side and forward side of the shuttle tanker assist the manoeuvre of the shuttle tanker. The direction of the shuttle tanker is still forward, in line with its direction at Stage-1 and Stage-2. In Stage-4, the shuttle tanker is directed parallelly close to the FSO with the assistance of Tugboats. The velocity of the shuttle tanker is kept at 8 cm/s maximum.

3.2. Berthing and Unberthing Simulation of Scenario-1

The simulation results of shuttle tankers berthing and unberthing against FSO for Scenario-1 in the form of trajectories are presented in Figure-2(a) and Figure-2(b) respectively. There are four stages of berthing simulation in Scenario-1. At Stage-1 the engine is in easy mode. The shuttle tanker manoeuvre uses the rudder without the assistance of tugboats. The rudder angle ranges from 0 to 5 degrees. The period of Stage-1 is 3600 sec (from t = 0 to t = 3600 sec). The shuttle tanker speed is reduced from 2 knots to 1.146 knots. The shuttle tanker's heading angle ranges between 315 degrees (at t = 0 sec) and 311.5 degrees (at t = 3600 sec). At Stage-2 the engine is in off mode, and the shuttle tanker manoeuvre only uses the rudder. The period of Stage-2 is 900 sec, from t = 3600 sec to t = 4500 sec. The speed of the shuttle tanker decreases from 1.146 knots to 0.4861 knots. The heading angle of the shuttle tanker is ranging from 311.5 degrees (at t = 3600 sec) and 320.3 degrees (at t = 4500 sec). At Stage-3, the engine and the rudder are in mode off. The manoeuvre of the shuttle tanker is assisted by two tugboats (stern tugboat and forward tugboat). The speed of the shuttle tanker decreases from 0.4861 knots to 0.3481 knots. The heading angle of the shuttle tanker is ranging from 320.3 degrees (at t = 4500 sec) and 313.9 degrees (at t = 8100 sec). The direction of the stern tugboat is –175 degrees (from t = 4500 sec to 6000 sec), –165 degrees (from t = 6000 sec to 7200 sec), and 175 degrees (from t = 7200 sec to 8100 sec). The direction of the forward tugboat is fixed at 0 degrees from t = 4500 sec to 8100 sec. At Stage-4, the

Table 1. Berthing/unberthing Simulation Scenarios

Environmental Condition	Scenario-1	Scenario-1	Scenario-1
Maximum wind speed (knots)	20	20	20
Significant wave height (m)	1.5	1.5	1.5
Significant wave period (sec)	6	6	6
Maximum current velocity (m/s)	0.3 (NW to SE)	0.3 (NE to SW)	0.3 (SE to NW)

engine and the rudder are also in mode off. The manoeuvre of the shuttle tanker is assisted by two tugboats (stern tugboat and forward tugboat). The speed of the shuttle tanker decreases from 0.3481 knots to 0.09294 knots. The heading angle of the shuttle tanker is ranging from 313.9 degrees (at t = 8100 sec) and 310.2 degrees (at t = 9200 sec). The directions of the stern tugboat are –120 degrees (from t = 8100 sec to 8900 sec), and 90 degrees (from t = 8900 sec to 9200 sec). While the directions of the forward tugboat are –60 degrees (from t = 8100 sec to 8900 sec), and 90 degrees (from t = 8900 sec to 9200 sec).

The time duration of unberthing of shuttle tanker is 3000 sec, which is described in five steps. At Step-1, the engine of the shuttle tanker is in off mode (t = 0 sec to t = 1400 sec), and the shuttle tanker manoeuvre is assisted by two tugboats, a stern tugboat and a forward tugboat. The direction of the stern tugboat is 80 degrees and the direction of the forward tugboat is 80 degrees. At Step-2, the engine of the shuttle tanker is in easy mode (t = 1400 sec to t = 2000 sec), with the rudder angle being 30 degrees. The manoeuvre of the shuttle tanker is also assisted by a stern tugboat and forward tugboat. The direction of the stern tugboat is –90 degrees and the direction of the forward tugboat is 90 degrees. At Step-3, the engine of the shuttle tanker is in easy mode (t = 2000 sec to t = 2400 sec), with the rudder being –10 degrees. The manoeuvre of the shuttle tanker is also assisted by a stern tugboat and forward tugboat. The direction of the stern tugboat is –90 degrees and the direction of the forward tugboat is 90 degrees.

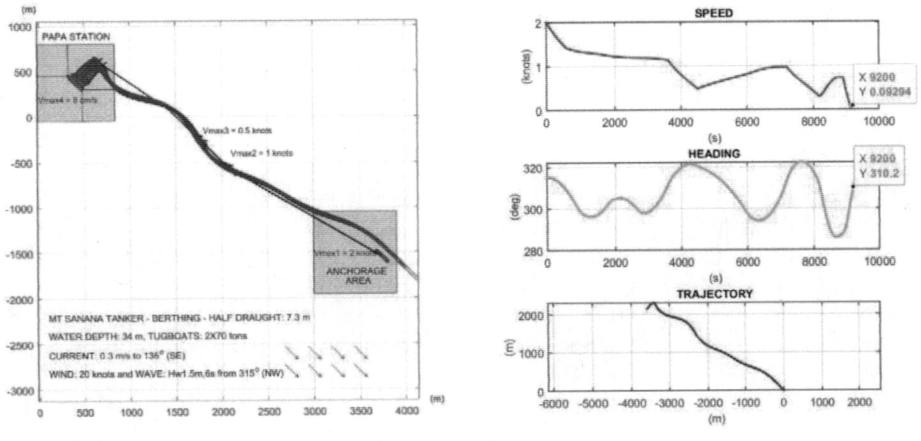

(a)

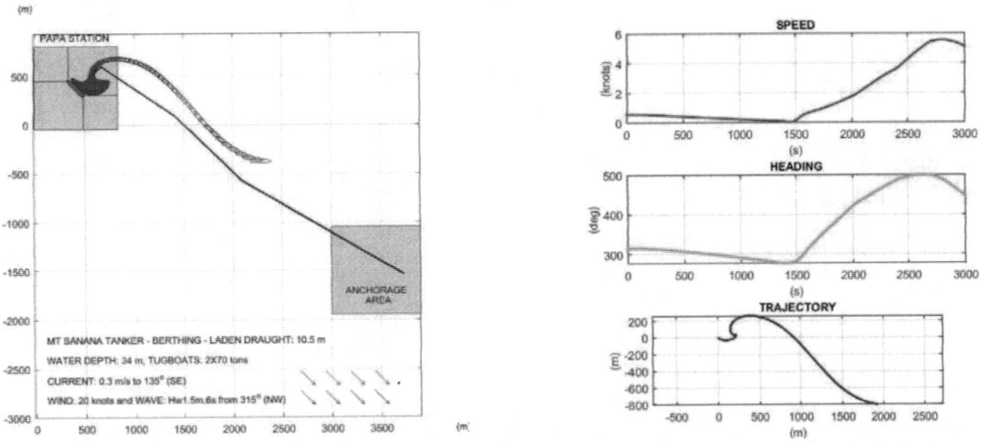

(b)

Figure 2. Full Trajectory Simulation of Scenario-1
(a) Berthing, (b) Unberthing

At Step-4, the engine of the shuttle tanker is in easy mode (t = 2400 sec to t = 2700 sec), with the rudder being –10 degrees. The manoeuvre of the shuttle tanker is also assisted by the Stern tugboat and forward tugboat. The direction of the stern tugboat is –30 degrees and the direction of the forward tugboat is 30 degrees. At Step-5, the engine of the shuttle tanker is in easy mode (t = 2700 sec to t = 3000 sec), with the rudder being –10 degrees. The manoeuvre of the shuttle tanker is carried out alone without the assistance of tugboats.

3.3. Berthing and Unberthing Simulation of Scenario-2

The simulation results of shuttle tankers berthing and unberthing against FSO for Scenario-2 in the form of trajectories are presented in Figure-3(a) and Figure-3(b) respectively. There are four stages of berthing simulation in Scenario-2. At Stage-1 the engine is in easy mode. The manoeuvre of the shuttle tanker uses the rudder without the assistance of tugboats. The rudder angle ranges from 0 to 10 degrees. The period of Stage-1 is 3000 sec. The shuttle tanker speed is reduced from 2 knots to 1.04 knots. The heading angle of the shuttle tanker ranges between 320 degrees (at t = 0 sec) and 336.7 degrees (at t = 3000 sec). At Stage-2 the engine is in off mode, and the shuttle tanker manoeuvre uses the rudder without tugboat assistance. The period of Stage-2 is 1100 sec, from t = 3000 sec to t = 4100 sec. The speed of the shuttle tanker decreases from 1.04 knots to 0.5271 knots. The heading angle of the shuttle tanker is ranging from 336.7 degrees (at t = 3000 sec) and 334.8 degrees (at t = 4100 sec). At Stage-3, the engine and the rudder are in mode off. The manoeuvre of the shuttle tanker is assisted by two tugboats (stern tugboat and forward tugboat). The speed of the shuttle tanker decreases from 0.5271 knots to 0.5166 knots. The heading angle of the shuttle tanker is ranging from

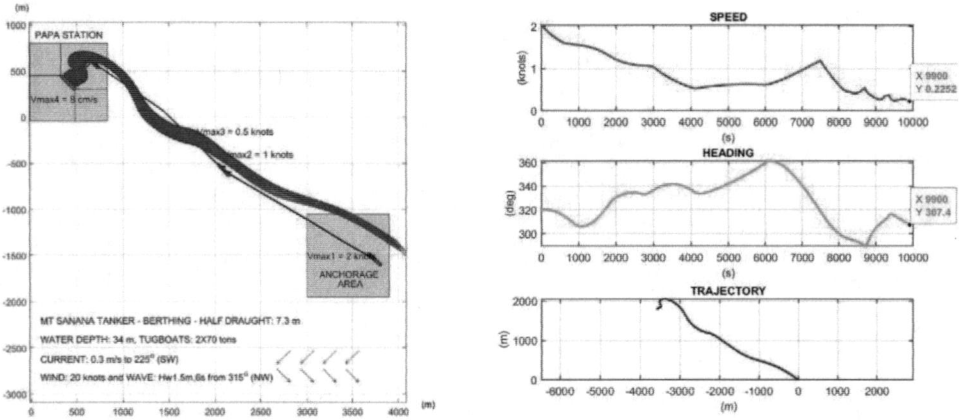

(a)

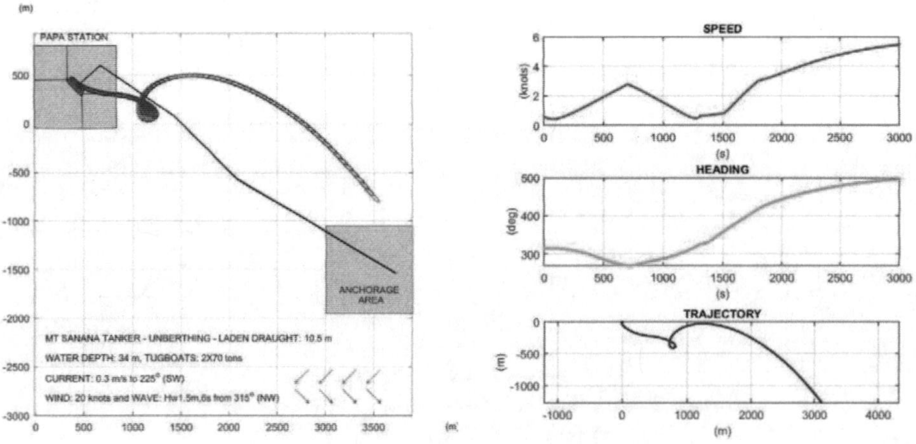

(b)

Figure 3. Full Trajectory Simulation of Scenario-2
(a) Berthing, (b) Unberthing

334.8 degrees (at t = 4100 sec) and 297.7 degrees (at t = 8700 sec). The direction of the stern tugboat is –175 degrees (from t = 4100 sec to 6000 sec), and 180 degrees (from t = 6000 sec to 8700 sec). The direction of the forward tugboat is fixed at 5degrees from t = 4100 sec to 6000 sec, and then it changes to 0 degrees from t = 0 sec to t = 8700 sec. At Stage-4, the engine and the rudder are in mode off. The manoeuvre of the shuttle tanker is assisted by two tugboats (stern tugboat and forward tugboat). The speed of the shuttle tanker decreases from 0.5166 knots to 0.2252 knots. The heading angle of the shuttle tanker is ranging from 297.7 degrees (at t = 8700 sec) and 304.4 degrees (at t = 9900 sec). The directions of the stern tugboat are 120 degrees (from t = 8700 sec to 9360 sec), and 103 degrees (from t = 9360 sec to 9900 sec). The directions of the forward tugboat are 60 degrees (from t = 8700 sec to 9360 sec), and 103 degrees (from t = 9360 sec to 9900 sec). The time duration of unberthing of shuttle tanker is 3000 sec, which is described in five steps. At Step-1, the engine of the shuttle tanker is in mode off (t = 0 sec to t = 700 sec), and the manoeuvre of the shuttle tanker is assisted by two tugboats, a stern tugboat and a forward tugboat. The direction of the stern tugboat is 120 degrees and the direction of the forward tugboat is 120 degrees. At Step-2, the engine of the shuttle tanker is in mode off (t = 700 sec to t = 1400 sec), and the manoeuvre of the shuttle tanker is assisted by two tugboats, a stern tugboat and a forward tugboat. The direction of the stern tugboat is –30 degrees and the direction of the forward tugboat is 30 degrees. At Step-3, the engine of the shuttle tanker is in easy mode (t = 1400 sec to t = 1500 sec), with the rudder being 30 degrees. The manoeuvre of the shuttle tanker is also assisted by a stern tugboat and forward tugboat. The direction of the stern tugboat is –30 degrees and the direction of the forward tugboat is 30 degrees. At Step-4, the engine of the shuttle tanker is in easy mode (t = 1500 sec to t = 1800 sec), with the rudder being 0 deg. The manoeuvre of the shuttle tanker is also assisted by a stern tugboat and forward tugboat. The direction of the stern tugboat is –30 degrees and the direction of the forward tugboat is 30 degrees. At Step-5, the engine of the shuttle tanker is in easy mode (t = 1800 sec to t = 3000 sec), with the rudder being 0 deg. The manoeuvre of the shuttle tanker is carried out alone without the assistance of tugboats.

3.4. Berthing and Unberthing Simulation of Scenario-3

The simulation results of shuttle tankers berthing and unberthing against FSO for Scenario-3 in the form of trajectories are presented in Figure-4(a) and Figure-4(b) respectively. There are four stages of berthing simulation in Scenario-3. At Stage-1 the engine is in easy mode. The shuttle tanker manoeuvre uses the rudder without the assistance of tugboats. The rudder angle ranges from 0 degrees, 15degrees, and 15 degrees. The period of Stage-1 is 2500 sec (from t = 0 to t = 2500 sec). The shuttle tanker speed is reduced from 2 knots to 0.974 knots. The shuttle tanker's heading angle ranges between 300 degrees (at t = 0 seconds) and 270.1 degrees (at t = 2500 seconds). At Stage-2 the engine is in off mode, and the shuttle tanker manoeuvre only uses the rudder. The period of Stage-2 is 1500 sec, from t = 2500 sec to t = 4000 sec. The speed of the shuttle tanker decreases from 0.974 knots to 0.610 knots. The heading angle of the shuttle tanker is ranging from 270.1degrees (at t = 2500 sec) and 261 degrees (at t = 4000 sec). At Stage-3, the engine and the rudder are in mode off. The manoeuvre of the shuttle tanker is assisted by two tugboats (stern tugboat and forward tugboat). The speed of the shuttle tanker decreases from 0.610 knots to 0.3835 knots. The heading angle of the shuttle tanker is ranging from 261 degrees (at t = 4000 sec) and 300.8 degrees (at t = 7200 sec). The direction of the stern tugboat is –170 degrees (from t = 4000 sec to 5000 sec), 170 degrees (from t = 5000 sec to 5200 sec), -10 degrees (from t = 5200 sec to 6000 sec), -90 degrees (from t = 6000 sec to 6600 sec), and -110 degrees (from t = 6600 sec to 7200 sec). The direction of the forward tugboat is 5 degrees (from t = 4000 sec to t = 5000 sec), 170 degrees (from t = 5000 sec to t = 5200 sec), -10 degrees (from t = 5200 sec to t = 6000 sec), -90 degrees (from t = 6000 sec to t = 6600 sec), and -110 degrees (from t = 6600 sec to t = 7200 sec). At Stage-4, the engine and the rudder are also in mode off. The manoeuvre of the shuttle tanker is assisted by two tugboats (stern tugboat and forward tugboat). The speed of the shuttle tanker decreases from 0.3835 knots to 0.11 knots. The heading angle of the shuttle tanker is ranging from 300.9 degrees (at t = 7200 sec) and 314 degrees (at t = 9000 sec). The directions of the stern tugboat are –130 degrees (from t = 7200 sec to 8900 sec), and 90 degrees (from t = 8900 sec to 9000 sec). The directions of the forward tugboat are –60 degrees (from t = 7200 sec to 8900 sec), and 90 degrees (from t = 8900 sec to 9000 sec).

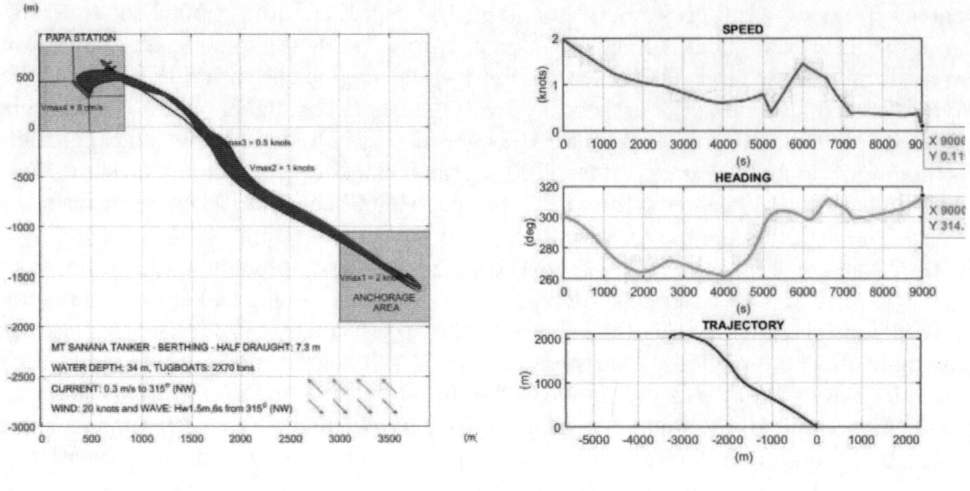

(a)

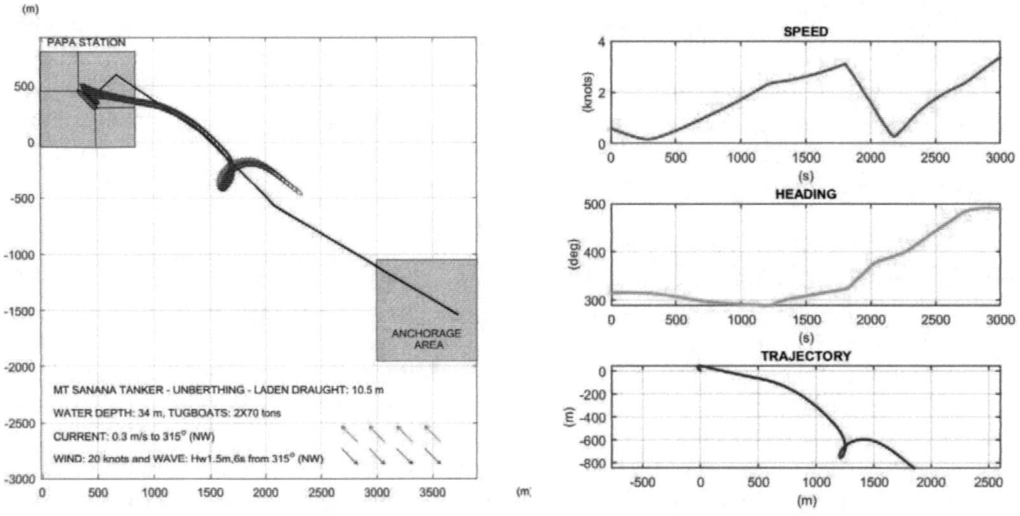

(b)

Figure 4. Full Trajectory Simulation of Scenario-3
(a) Berthing, (b) Unberthing

The time duration of unberthing of shuttle tanker is 3000 sec, which is described in 4 (four) steps. At Step-1, the engine of the shuttle tanker is in mode off (t = 0 sec to t = 1200 sec), and the manoeuvre of the shuttle tanker is assisted by two tugboats, a stern tugboat and a forward tugboat. The direction of the stern tugboat is 150 degrees and the direction of the forward tugboat is 150 degrees. At Step-2, the engine of the shuttle tanker is in mode off (t = 1200 sec to t = 1800 sec), and the manoeuvre of the shuttle tanker is assisted by two tugboats, a stern tugboat and a forward tugboat. The direction of the stern tugboat is –90 degrees and the direction of the forward tugboat is 90 degrees. At Step-3, the engine of the shuttle tanker is in easy mode (t = 1800 sec to t = 2700 sec), with the rudder being 30 degrees. The manoeuvre of the shuttle tanker is also assisted by a stern tugboat and forward tugboat. The direction of the stern tugboat is –90 degrees and the direction of the forward tugboat is 90 degrees. At Step-4, the engine of the shuttle tanker is in easy mode (t = 2700 sec to t = 3000 sec), with the rudder being –5 degrees. The manoeuvre of the shuttle tanker is carried out alone without the assistance of tugboats.

Table 2. Tugboats Operational Procedure of Berthing Scenarios-1

Time (s)	Stern Tugboat		Forward Tugboat	
	Bollard Pull (x70 t)	**Direction (deg)**	**Bollard Pull (x70 t)**	**Direction (deg)**
4500-6000	20%	-175	40%	0
6000-7200	20%	-165	40%	0
7200-8100	20%	175	20%	0
8100-8900	35%	-120	48%	-60
8900-9200	30%	90	40%	90

Table 3. Tugboats Operational Procedure of Berthing Scenarios-2

Time (s)	Stern Tugboat		Forward Tugboat	
	Bollard Pull (x70 t)	**Direction (deg)**	**Bollard Pull (x70 t)**	**Direction (deg)**
4100-6000	20%	-175	35%	5
6000-7500	20%	180	40%	0
7500-8700	40%	180	20%	0
8700-9360	20%	120	60%	60
9360-9900	60%	103	60%	103

Table 4. Tugboats Operational Procedure of Berthing Scenarios-3

Time (s)	Stern Tugboat		Forward Tugboat	
	Bollard Pull (x70 t)	**Direction (deg)**	**Bollard Pull (x70 t)**	**Direction (deg)**
4000-5000	20%	-175	35%	5
5000-5200	20%	-170	20%	170
5200-6000	20%	-10	20%	-10
6000-6600	40%	-90	40%	-90
6600-7200	30%	-110	60%	-110
7200-8900	50%	-130	60%	-60
8900-9000	60%	90	60%	90

3.5. Tugboat Standard Operational Procedure for Shuttle tanker Berthing Simulation

Two tugboats assist the shuttle tanker berthing operation for all scenarios; a stern tugboat and a forward tugboat. The bollard pull capacity of tugboats is 70 tons. Table-2 describes the tugboat operational procedure for berthing Scenario-1. Referring to Table-2, the stern tugboat and forward tugboat assist the berthing of the shuttle tanker at t = 4500 sec until 9200 sec. The bollard pull of a stern tugboat varies from 20% to 35%, while the bollard pull of a forward tugboat varies from 20% to 48%. The maximum capacity of the bollards for both tugboats occur at time t = 8100 sec until t = 8900 sec. The heading of tugboats relative to the heading of the shuttle tanker varies from -175 degrees to 175 degrees for the stern tugboat and -60 degrees to 90 degrees for the forward tugboat. The tugboat operational procedure for berthing Scenario-2 is described in Table-3. Referring to Table-3, the stern tugboat and forward tugboat assist the berthing of the shuttle tanker at t = 4100 sec until 9900 sec. The bollard pulls of stern tugboat and forward tugboat vary from 20% to 60%. The maximum capacity of the bollards for both tugboats occur at time t = 9360 sec until t = 9900 sec. The heading of tugboats relative to the heading of the shuttle tanker varies from -175 degrees to 180 degrees for the stern tugboat and 0 degrees to 103 degrees for the forward tugboat. The tugboat operational procedure for berthing Scenario-3 is described in Table-4. According to Table-4, the stern tugboat and forward tugboat assist the berthing of the shuttle tanker at t = 4000 sec until 9000 sec. The bollard pulls of stern tugboat and forward tugboat vary from 20% to 60%. The maximum capacity of the bollards for both tugboats occur at time t = 8900 sec until t

= 9000 sec. The heading of tugboats relative to the heading of the shuttle tanker varies from -175 degrees to 90 degrees for the stern tugboat and -110 degrees to 170 degrees for the forward tugboat.

3.6. Tugboat Standard Operational Procedure for Shuttle tanker Unberthing Simulation

Two tugboats assist the shuttle tanker unberthing operation for all scenarios; a stern tugboat and a forward tugboat. The bollard pull capacity of tugboats is 70 tons. The tugboat operational procedures of shuttle tanker unberthing operation Scenario-1 until Scenario-3 are presented in Table-5 until Table-7. The tugboats operational procedure for unberthing Scenario-1 is described in Table-5. The tugboats assist the unberthing of the shuttle tanker from t = 0 sec through 2700 sec. The bollard pull of the stern tugboat and the forward tugboats is 60% on average. The heading of the tugboats relative to the heading of the shuttle tanker varies from -30 degrees to 80 degrees for the stern tugboat and 30 degrees to 90 degrees for the forward tugboat. Referring to Table-6, the stern tugboat and forward tugboat assist the unberthing of the shuttle tanker Scenario-2 from t = 0 sec until 1800 sec. The bollard pulls of the stern tugboat and forward tugboat are 100%. The heading of tugboats relative to the heading of the shuttle tanker varies from -30 degrees to 120 degrees for the stern tugboat and 30 degrees to 120 degrees for the forward tugboat. The tugboat operational procedure for unberthing Scenario-3 is described in Table-7. The stern tugboat and forward tugboat assist the berthing of the shuttle tanker from t = 0 sec until 2700 sec. The bollard pulls of stern tugboat and forward tugboat vary from 40% to 60%. The maximum capacity of the bollards for both tugboats occur at time t = 1200 sec until t = 2700 sec. The heading of tugboats relative to the heading of the shuttle tanker varies from -90 degrees to 150 degrees for the stern tugboat and 90 degrees to 150 degrees for the forward tugboat.

Table 5. Tugboats Operational Procedure of Unberthing Scenarios-1

Time (s)	Stern Tugboat		Forward Tugboat	
	Bollard Pull (x70 t)	**Direction (deg)**	**Bollard Pull (x70 t)**	**Direction (deg)**
0-1400	60%	80	60%	80
1400-2000	60%	-90	60%	90
2000-2400	60%	-90	60%	90
2400-2700	60%	-30	60%	30
2700-3000	-	-	-	-

Table 6. Tugboats Operational Procedure of Unberthing Scenarios-2

Time (s)	Stern Tugboat		Forward Tugboat	
	Bollard Pull (x70 t)	**Direction (deg)**	**Bollard Pull (x70 t)**	**Direction (deg)**
0-700	100%	120	100%	120
700-1400	100%	-30	100%	30
1400-1500	100%	-30	100%	30
1500-1800	100%	-30	100%	30
1800-3000	-	-	-	-

Table 7. Tugboats Operational Procedure of Unberthing Scenarios-3

Time (s)	Stern Tugboat		Forward Tugboat	
	Bollard Pull (x70 t)	**Direction (deg)**	**Bollard Pull (x70 t)**	**Direction (deg)**
0-1200	40%	150	40%	150
1200-1800	60%	-90	60%	90
1800-2700	60%	-90	60%	90
2700-3000	-	-	-	-

4. Conclusions and Future Work

This case study simulated the berthing and unberthing activities of the shuttle tanker to the FSO unit based on three different berthing and unberthing scenarios. The performed simulation was aimed to determine the number and capacity of tugboats, time series of tugboat bollard pull, and time series of tugboat force direction to achieve the allowable berthing and unberthing manoeuvre, including the requirement of berthing speed and angle. For future work, the rule of thumb related to the optimum ratio between the shuttle tanker dimensions to the dimensions of FSO needs to be analyzed. The conclusions of this study are summarized as follows:

- The period of the berthing manoeuvres was 9000 seconds (2.5 hours) for Scenario-1 and Scenario-3, and 9900 seconds (2.75 hours) for Scenario-2. The period of the unberthing manoeuvres was 3000 seconds (50 minutes) for all scenarios.
- Two Tugboats with the capacity of 2x70 tons bollard pull could assist the berthing and unberthing manoeuvre of the shuttle tanker. The maximum power of tugboats used for the berthing manoeuvre is 60% (42 tons bollard pull), and the maximum power of tugboats used for Scenario-1 and Scenario-3 is 60% (42 tons bollard pull), and for Scenario-2 is 100% (70 tons bollard pull).

Acknowledgements

The authors wish to acknowledge PT Citra Mas, the Marine and Offshore Engineering consultant for the permission, encouragement, technical staff, and financial support to accomplish the project and publish this study.

References

[1] Tannuri E A and Martins G H A 2018 Application of a Manoeuvring Simulation Centre and Pilots Expertise to the Design of New Ports and Terminals and Infrastructure Optimization in Brazil (PIANC – World Congress Panama City).

[2] Gongxing W, Xiaolong Z, Yushan S, and Linling W 2021 Cooperative Manoeuvring Mathematical Modeling for Multi-Tugs Towing a Ship in the Port Environment *J. Mar. Sci. Eng.* 2021, 9, 384. https://doi.org/10.3390/jmse9040384 pp 1-23

[3] Murdjito, Pravitasari I Y, and Djatmiko E B 2020 An Analysis on the Spread Mooring of the Belida FSO Induced by Squall Loads *KAPAL :Jurnal Ilmu Pengetahuan dan Teknologi Kelautan* **17** 1

[4] PIANC 2002 Guidelines for the Design of Fenders System *Report of Working Group 33 of the Maritime Navigation Commission, International navigation Association.*

[5] Gucma S and Slaczka W 2018 Comprehensive method of formal safety assessment of ship manoeuvring in waterways *Scientific J. of the Maritime University of Szczecin* **54** 12

[6] Yasukawa H and Yoshimura Y, 2015, Introduction of MMG standard method for ship maneuvering predictions, *J. of Marine Science and Technology* **20** pp 37-52

[7] Maki A, Sakamoto, N, Akimoto Y, Nishikawa H, and Umeda N, 2020, Application of optimal control theory based on the evaluation strategy (CMA-ES) to automatic berthing *J. of Marine Science and Technology* **25** 1 pp 221-233

[8] Kobayasi E 2002, Basic Marine Vehicle Manoeuvring Mathematical Expression (Lecture Moduleof Kobe University) p 10

[9] Yoshimura Y and Masumoto Y 2011, Hydrodynamic Force Database with Medium High Speed Merchant Ships Including Fishing Vessels and Investigation into Manoeuvring Prediction Method *Translation from original paper Journal of JASNAOE* **14**

[10] Fujiwara T, Ueno M, and Ikeda Y 2006 *Proc. of the 16th Int. Offshore and Polar Conference*, Cruising Performance of a Large Passenger Ship in Heavy Sea (San Francisco, California, USA) pp 304-311

[11] Yasukawa H and Sakuno R 2020 Application of the MMG method for the prediction of steady sailing condition and course stability of ship under external disturbances *J. of Marine Science and Technology* **25** 1 pp 196-220

[12] Wicaksono A and Kashiwagi M 2017 Wave-induced steady forces and yaw moment or a ship advancing in oblique waves *J. of Marine Science and Technology* https://doi.org/10.1007/s00773-017-0510-6

Development of the Elements of Tugboat Handling for Berthing and Unberthing of Container Ships

I P S Asmara[1)], D Kristianto[2)], M A Mustaghfirin[3)], Y Praharsi[1)], Adianto[4)], A Z Arfianto[4)], C A Firmansyah[5)]

[1]Shipbuilding Department, Politeknik Perkapalan Negeri Surabaya
[2]PUI KEKAL, Institut Teknologi Sepuluh Nopember
[3]Marine Engineering Department, Politeknik Perkapalan Negeri Surabaya
[4]Marine Electrical Engineering Department, Politeknik Perkapalan Negeri Surabaya
[5]PT Terminal Petikemas Surabaya

Abstract. This paper provides a study on the procedure of tugboat handling for berthing and unberthing of container ships in the Port of Surabaya Container Terminal. The elements of the tugboat assistance consisted of time series of the percentage of tugboats' power, the direction of the tug's force, and the duration. This study developed a manoeuvring simulation program using the mathematical manoeuvring group (MMG) model to determine the handling elements to achieve the intended trajectory properly. The model considered the surging force, swaying force, and yawing moment of tugs, as well as the disturbances of current, wind, and wave. The planning trajectory is determined based on berthing and unberthing passages derived from the automatic identification system (AIS) data. The elements resulting in the berthing speed and berthing angle complied with the Permanent International Association of Navigation Congresses (PIANC).

Keywords: port; manoeuvring; MMG; AIS; tugboats; container ships; PIANC

1. Introduction

Senior pilots have experience in the critical task of harbour tug boats and teach their junior in the ship handling simulator. Unfortunately, they have no evaluation standard for tug boat handling management (Ishikura et al., 2014) [1]. The simulator could determine the optimum elements in terms of the number, capacity, duration, and time series of tugs direction, as well as the time series of its delivered working power along with the duration of the handling period. The senior navigators of the harbour pilot and captains of the vessel provide the direction to the tugboat operator based on their experience. In a condition that the ship's dimension is larger than the maximum dimension declared in the port entry manual, the element of tugboat handling for berthing and unberthing could be simulated in the simulator. Ishikura introduced three elements of tugboat handling for ship berthing. They are the propelling power of tug boats, their direction, and their duration. Another researcher has a similar opinion that the experience of harbour pilot and tugboats captain is the main reference for towing operation of berthing and unberthing, and the effective operation method is unavailable (Wu et al., 2020) [2]. Wu introduced different elements of handling. The elements are the time series of rope force and rope length of tugs. In addition, Wu recorded the time series of position, speed, and yaw angle of ships and tugboats.

The tugboats handling elements are the critical parts of the procedure to ensure the assistance of tugboats to achieve the standard of requirements for berthing and unberthing in any environmental condition. This study provides and agrees with the three elements proposed by Ishikura and also presents other evaluation factors of berthing and unberthing performance according to the work of Wu, namely the time series of position or trajectory, time series of speed, and time series of heading or yaw angle of the assisted vessel. In the simulation of manoeuvring prediction, the handling elements act as the input parameter. The performance indicators are the output elements of the program. This study is the initial part of a project to develop a Berthing Management System consisting of a manoeuvring simulation program, a monitoring system using an automatic identification system (AIS), and a berthing approach system (BAS). This paper aims to develop the tugboat handling simulator for berthing and unberthing of container ships.

2. Berthing and unberthing simulator

The tugboat handling simulator for berthing and unberthing operation program developed in this study refers to the 3 DOF Mathematical Manoeuvring Group (MMG) model (Yasukawa and Yoshimura, 2015) [3]. The model includes the forces of tugboats and the external disturbances of wind, and sea current.

2.1. The 3 DOF motion equations

The surging, swaying, and yawing motions of vessels expressed in Equations 1 to 3 consist of the forces of the environmental disturbance and tugboat assistance.

$$(m + m_x)\dot{u} - (m + m_y)v_m r - m x_G r^2 = X_H + X_P + X_R + X_A + X_T \quad (1)$$

$$(m + m_y)\dot{v}_m + (m + m_x)ur + m x_G \dot{r} = Y_H + Y_R + Y_A + Y_T \quad (2)$$

$$(I_{zz} + J_{zz} + m x_G^2)\dot{r} + m x_G(\dot{v}_m + ur) = N_H + N_R + N_A + N_T \quad (3)$$

Here, m and I_{zz} are the ship's mass and moment of inertia for yawing motion, respectively. m_x, m_y, and J_{zz} denote added mass for surging motion, added mass for swaying motion, and added moment of inertia, respectively. u, v_m, and r are surging velocity, swaying velocity at the mid-ship, and yawing rate, respectively. x_G is the longitudinal centre of gravity of the ship from mid-ship. X, Y, and N denote surging force, swaying force and the yawing moment around mid-ship, respectively. Subscript H, P, R, A, and T denote the ship's hull, propeller, rudder, wind, and tugs, respectively. Figure 1 shows the body-fixed and space-fixed coordinate systems. β, δ, and ψ denote drift angle, rudder angle, and the ship's true heading, respectively.

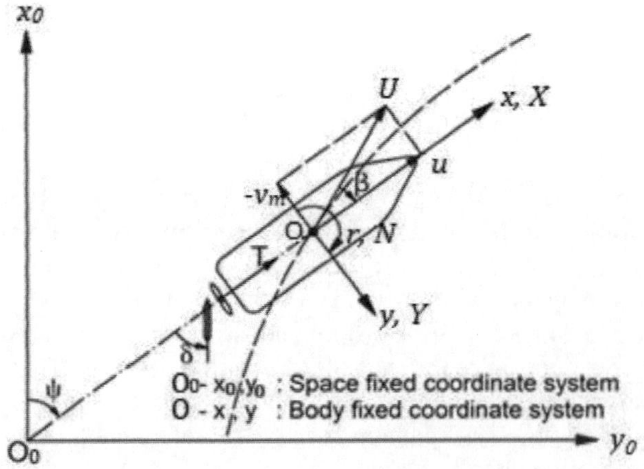

Figure 1. Coordinate system

2.2. *Hydrodynamic derivatives*

Yoshimura (Yoshimura Y and Masumoto Y, 2011) [4] published the hydrodynamic force database for manoeuvring prediction of vessels with the block coefficient (C_B) of between 0.51 and 0.65. The coefficient block of container ships is 0.57 to 0.66 and the recommended C_B is less than 0.65. (Charchalis A, 2018) [5]. The hull forces and moment predicted by Equations 4 to 6 based on resistance coefficient and the hydrodynamics derivatives using prime ' putting in the symbol are non-dimensional. This study adopts the approximation of the hydrodynamic derivatives based on the ship's main dimension published by (Taimuri G et al., 2020) [6]. Taimura et al. adopted the rapid estimation from several publications including (Norrbin, 1970) [7], (Kijima et al., 1990a) [8], (Brix, 1993) [9], (Yoshimura and Masumoto, 2011) [4], and other publications.

$$X_H = \frac{1}{2}(-R'_0 + X'_{vv}v'^2 + X'_{vr}v'_m r' + X'_{rr}r'^2 + X'_{vvvv}v'^4_m)\rho L d U^2 \quad (4)$$

$$Y_H = \frac{1}{2}(Y'_v v'_m + Y'_r r' + Y'_{vvv}v'^3_m + Y'_{vvr}v'^2_m r' + Y'_{vrr}v'_m r'^2 + Y'_{rrr}r'^3)\rho L d U^2 \quad (5)$$

$$N_H = \frac{1}{2}(N'_v v'_m + N'_r r' + N'_{vvv}v'^3_m + N'_{vvr}v'^2_m r' + N'_{vrr}v'_m r'^2 + N'_{rrr}r'^3)\rho L^2 d U^2 \quad (6)$$

Here, ρ is the water density, and L, d, and U are the ship's length between perpendiculars, ship draft, and resultant speed, respectively. $\frac{1}{2}\rho L d U^2$ and $\frac{1}{2}\rho L^2 d U^2$ express non-dimensional force and non-dimensional moment, respectively. Forces and moments due to propeller, rudder, tugboats, wind and current are described in another paper (Asmara and Husodo, 2022) [10].

2.3. *Acceptance criteria*

The PIANC requirement of berthing speed for ships with a displacement between 50.000 tons and 100.000 tons for moderate conditions is a maximum of 0.15 m/s, as shown in Table 1 [11].

Table 1. PIANC Requirement

Vessel Displacement	Favourable Condition (m/s)	Moderate Condition (m/s)	Unfavourable Condition (m/s)
Under 10,000	0.2 – 0.16	0.45 – 0.3	0.6 – 0.40
10,000 – 50,000	0.12 – 0.08	0.3 – 0.15	0.45 – 0.22
50,000 – 100,000	0.08	0.15	0.20
Over 100,000	0.08	0.15	0.20

3. Results and Discussions

3.1. *The subject ship*

The subject ship is a container ship with a capacity of 4300 TEUs, as seen in Table 2. The vessel berthed in Jetty 3 of the Surabaya Container Terminal on February 1, 2022, at 09:17:26 according to data derived from the PPNS AIS station. Figures 2 (a) and (b) show the traffic in the research area and the berthing-unberthing route of the vessel according to data derived from the PPNS-AIS monitoring system.

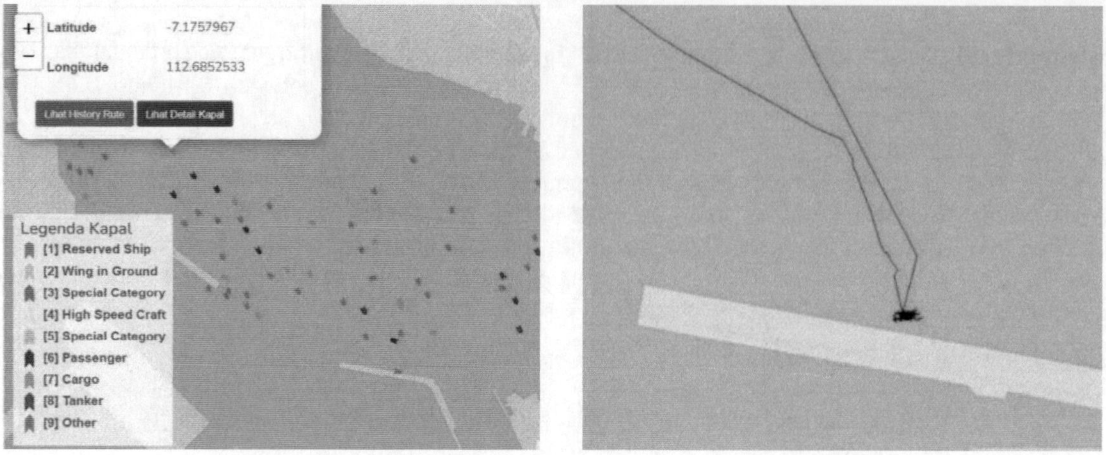

(a) (b)

Figure 2. PPNS-AIS monitoring system
(a) Traffic (b) Route

Table 2. Ship Dimensions

LOA (m)	DWT (m)	Capacity (TEUs	Beam (m)	Draught (m)	Block Coefficient
262.08	51693	4300	32.25	12	0.65

3.2. The environmental conditions

According to data from Meteorology, Climatology, and Geophysics Council in Tanjung Perak Station the environmental data on February 1, 2022, has a high current speed of up to 0.86 m/s as seen in Table 3. The vessel berthed on the starboard side in condition 1 and berthed on the port side in condition 2.

Table 3. Environmental Conditions

Environmental Condition	Condition-1	Condition-2
Maximum wind speed (knots)	12 (NW to SE)	12 (NW to SE)
Maximum current velocity (m/s)	0.87 (East to West)	0.87 (West to East)
Maximum tidal (m)	0.6 (High Tide)	-1.2 (Low Tide)
Water Depth (m)	15	15

3.3. Berthing and unberthing in condition-1

Figures 3 (a) and (b) present the berthing and unberthing simulation of the subject ship handled by 3 tugboats with a capacity of 3x45 tons bollard pull. Figure 3 shows the ship trajectory in blue and the simulation plotted the movement of the vessel every 30 seconds. The jetty of Surabaya Container Terminal has 4 docks as seen in Figure 3 (a) and the ship is berthed at dock-3 on the starboard. In the berthing simulation, according to a discussion with the port pilot, the initial speed of the vessel is not more than 3 knots in condition-1 and the current speed is about 1.67 knots. Figure 3 (b) shows that at the first stage of unberthing manoeuvre the bow moved and turned to follow the centre of the port channel.

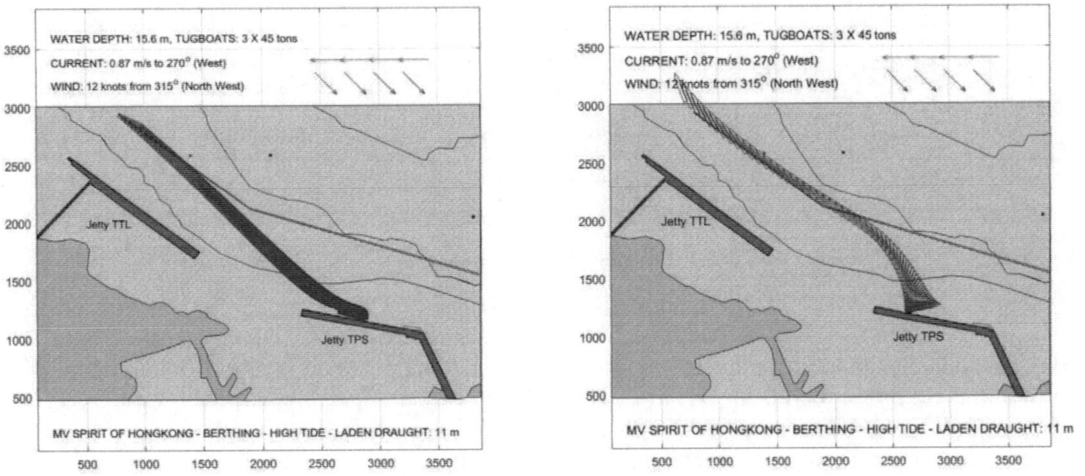

(a) (b)

Figure 3. Trajectory Simulation of Scenario-1
(a) Berthing (b) Unberthing

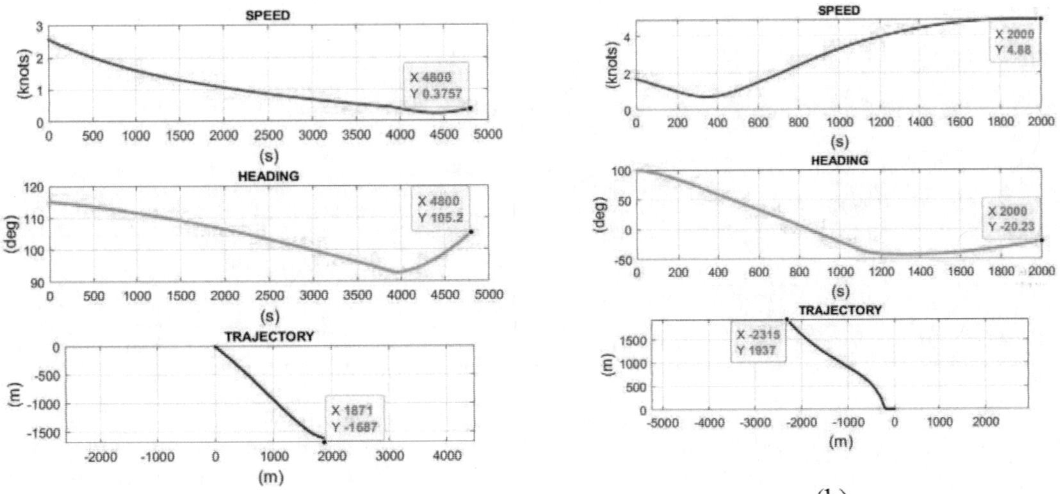

(b)

(a)

Figure 4. Berthing Speed, Berthing Heading and Trajectory of Simulation of Scenario-1
(b) Berthing (b) Unberthing

Figure 4 (a) shows that the berthing speed is 0.3757 knots or 0.19 cm/s. This doesn't comply with the moderate requirement of PIANC but it still complies with the unfavourable condition. Figure 4 (b) shows the time series of the ship's speed, heading, and trajectory. The initial speed for unberthing manoeuvre is the same as the current speed means that the ship drifted to the West but she is stopped by the dock structure.

Table 4. Berthing Procedure of Tugboats in Environmental Condition-1

Step	Time (s)	Engine Status	Rudder Angle	Astern Tug 45 tons Bollard Pull		Midship Tug 45 tons Bollard Pull		Forward Tug 45 tons Bollard Pull	
				Force	Direction	Force	Direction	Force	Direction
1	0 – 3900	Off	0 deg	20%x45 t	180 deg	-	-	60%x45 t	-5 deg
2	3900 – 4200	Off	0 deg	20%x45 t	-160 deg	-	-	20%x45 t	20 deg
3	4200 – 4600	Off	0 deg	60%x45 t	90 deg	-	-	60%x45 t	90 deg
4	4600 – 4800	Off	0 deg	60%x45 t	-90 deg	-	-	60%x45 t	-90 deg

According to the national requirement, the capacity of tugboats to handle a ship having a length of more than 250 m is a minimum 120-ton bollard pull. Table 4 presents 3 tugboats handling the vessel with a total capacity of 3x45 bollard pulls but only 2 tugs were working. The berthing manoeuvre has 4 steps, namely pulling, turning, pushing and stopping, respectively. In the first step, the forward tugs pulled the vessel from the initial position at the centre of the channel to approach the jetty. In the second step, the forward and astern tugs turned the ship into a heading of about 100 degrees to make the ship parallel with the dock. At the position about 100 m apart from the jetty, at step 3 both tugs pushed the vessel to the starboard and at step 4 the tugs took a correction by pulling to the portside to decrease the speed of stopping action. The tugs start the task from 3 nm at the jetty but in this simulation, the initial position is at the cannel with the same latitude as the Northside of the jetty of Terminal Teluk Lamong (TTL). The manoeuvre takes 4800 seconds or 1 hour and 20 minutes from the initial to the dock-3 of Surabaya Container Terminal or Terminal Petikemas Surabaya (TPS).

Table 5. Unberthing Procedure of Tugboats in Environmental Condition-1

Step	Time (s)	Engine Status	Rudder Angle	Astern Tug 45 tons Bollard Pull		Midship Tug 45 tons Bollard Pull		Forward Tug 45 tons Bollard Pull	
				Force	Direction	Force	Direction	Force	Direction
1	0 – 1100	Slow	-35 deg	85%x45 t	30 deg	-	-	85%x45 t	-90 deg
2	1100 – 2000	Slow	0 deg	60%x45 t	-30 deg	-	-	60%x45 t	60 eg

The procedure of unberthing manoeuvre in condition-1 consists of 2 steps, turning left and turning right as seen in Table 5. At the first step, the vessel uses the slow engine revolution and full rudder angel to the port side and is assisted by the full capacity of the tugboats to make a small turning radius entering the centre of the channel. In the second step, the ship heading is corrected by turning right using 60% of the bollard pull of the two tugboats. It takes 2000 seconds or about 33 minutes to move the vessel from dock-3 of the TPS jetty to out of the study area.

3.4. Berthing and unberthing in condition-2

In condition 2 the vessel is berthing following the current direction. Berthing manoeuvre in this condition is more difficult than in condition-1 due to the difficulty to control the manoeuvring speed. The port pilot suggested the speed at the initial condition is not more than 2 knots. Figure 5 (a) shows the initial position is moved closer to the TTL jetty to prevent the vessel from entering the anchorage area located in between the green and red buoys due to the difficulty to control the drifting motion. The speed decreased when the vessel was turned and pushed to the dock-3. Figure 6 (a) shows that the speed of the berthing manoeuvre decreased at the end stage of the manoeuvre. The berthing speed is 0.02 knots or about 0.01 m/s which is comply with the PIANC favourable condition. The berthing angle is 297.5 degrees and it takes 3700 seconds.

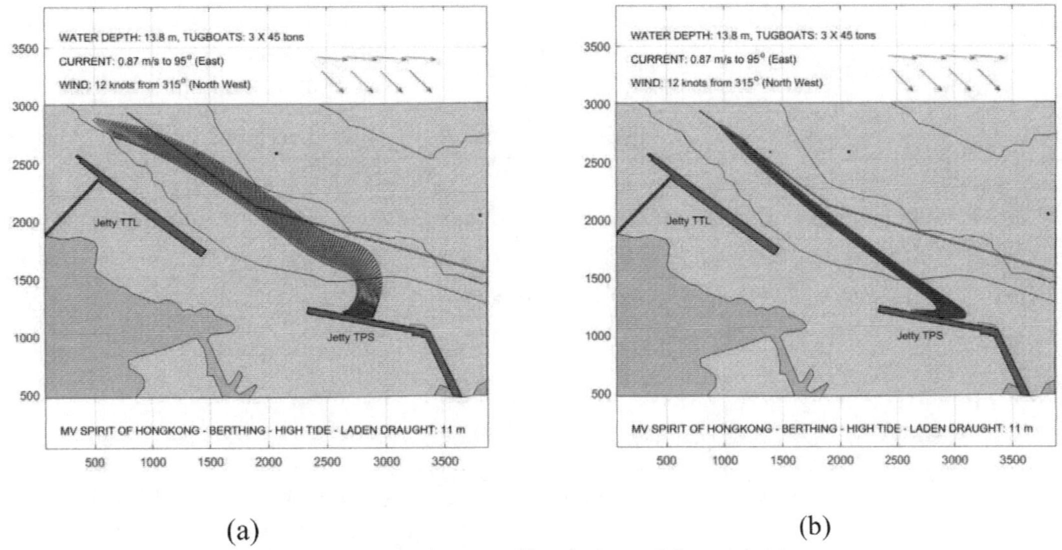

Figure 5. Trajectory Simulation of Scenario-2
(a) Berthing (b) Unberthing

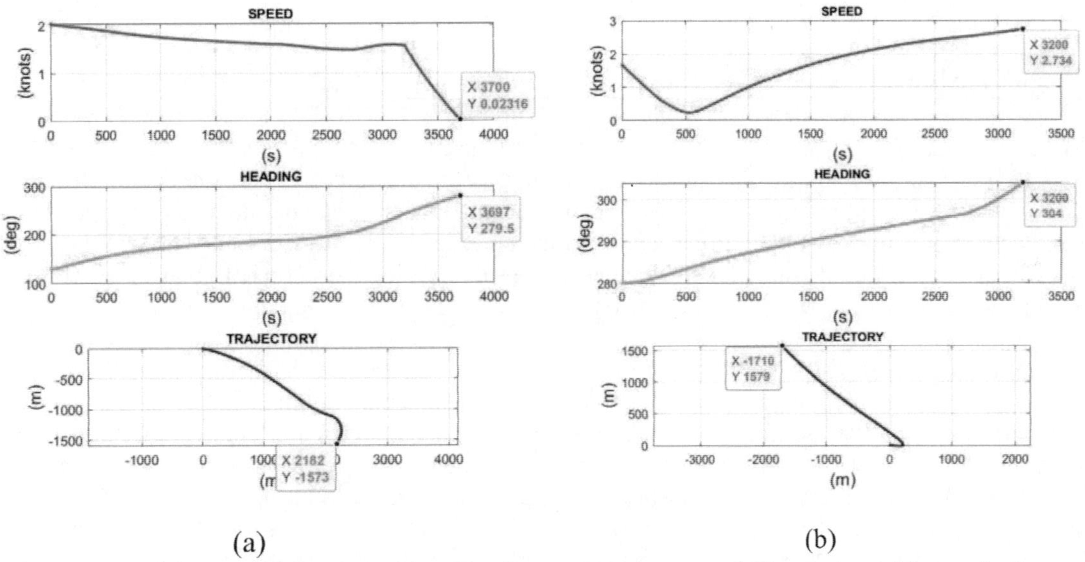

Figure 6. Berthing Speed, Berthing Heading and Trajectory of Simulation of Scenario-2
(a) Berthing (b) Unberthing

Figure 5 (b) and 6 (b) shows the vessel unberthing from the TPS jetty successfully without drifting motion in the channel area but the vessel drifted at the initial position of the dock-3. At the first step when the mooring line lose, the vessel drifted at the speed of 0.87 m/s the same as the current speed. The first procedure for berthing and unberthing the vessel to and from dock-3 is that dock-2 and dock-4 should be empty.

Table 6. Berthing Procedure of Tugboats in Environmental Condition-2

Step	Time (s)	Engine Status	Rudder Angle	Astern Tug 45 tons Bollard Pull		Midship Tug 45 tons Bollard Pull		Forward Tug 45 tons Bollard Pull	
				Force	Direction	Force	Direction	Force	Direction
1	0 – 2000	Off	0 deg	-	-	85%x45 t	90 deg	85%x45 t	75 deg
2	2000 – 2700	Off	0 deg	60%x45 t	180 deg	-	-	60%x45 t	90 deg
3	2700 – 3100	Slow	0 deg	-	-	-	-	-	-
4	3100 – 3200	Slow	0 deg	60%x45 t	90 deg	-	-	60%x45 t	90 deg
5	3200 – 3700	Off	0 deg	60%x45 t	-60 deg	-	-	60%x45 t	-60 deg

Table 7. Unberthing Procedure of Tugboats in Environmental Condition-2

Step	Time (s)	Engine Status	Rudder Angle	Astern Tug 45 tons Bollard Pull		Midship Tug 45 tons Bollard Pull		Forward Tug 45 tons Bollard Pull	
				Force	Direction	Force	Direction	Force	Direction
1	0 – 300	Slow	0 deg	20%x45 t	-15 deg	-	-	60%x45 t	0 deg
2	300 – 2700	Slow	0 deg	-	-	-	-	20%x45 t	20 deg
3	2700 – 3100	Slow	5 deg	-	-	-	-	20%x45 t	0 deg
4	3100 – 3200	Slow	5 deg	-	-	-	-	-	-

The berthing manoeuvre in condition-1 has 5 steps and uses three tugboats as seen in Table 6. In the first step, the forward and midship tugs use their full power (85%) to decrease the drifting motion. At the second step, the astern tug pulls backwards to decrease the speed. The manoeuvring speed can be decreased significantly by pushing the vessel to the starboard side at step 4 and at the final step the vessel is pushed to dock 3. Table 7 shows the vessel easily moving from the jetty by using 2 tugs in the first step and 1 tugboat in steps 2 and 3. In step 4 the vessel fully uses the engine power.

4. Conclusions and Future Work

This study simulated the elements of tugboat handling for the berthing and unberthing procedure of a container vessel in the Surabaya Container Terminal based on two scenarios of environmental conditions. The simulation confirms that the berthing manoeuvre in condition-1, against the current, has a better trajectory than condition-2. The simulation also determined the number and capacity of tugboats, time series of tugboat bollard pull, and time series of tugboat force direction to achieve the requirement of berthing speed and angle. For future work, the study will focus on the validity of the speed and angle of berthing using a monitoring system based on laser sensors called the Berthing Approach System (BAS) due to the low precision of monitoring the parameters using an Automatic Identification System (AIS).

Acknowledgements

The authors wish to acknowledge The Program of Domestic Applied Science Research for Vocational Lecturer funded by RISPRO LPDP for the encouragement and financial support with the Research Contract No. 0765/D6/KU.04.00/2021 to accomplish and publish this study.

References

[1] Ayumi I, Junichi S, Yuji H, and Koji M 2014 Evaluation on Elements of Tug Boat Handling during Ships' Berthing and Unberthing. The Journal of Japan Institute of Navigation 30 pp 129-134 https://doi.org/10.9749/jin.130.129

[2] Gongxing W, Xiaolong Z, Yushan S, and Linling W 2021 Cooperative Manoeuvring Mathematical Modeling for Multi-Tugs Towing a Ship in the Port Environment *J. Mar. Sci. Eng.* 2021, 9, 384. https://doi.org/10.3390/jmse9040384 pp 1-23
[3] Yasukawa H and Yoshimura Y 2015 Introduction of MMG standard method for ship manoeuvring predictions *J. of Marine Sci. and Tech.* **20** DOI 10.1007/s00773-014-0293-y pp 37-52
[4] Yoshimura Y and Masumoto Y 2011, Hydrodynamic Force Database with Medium-High Speed Merchant Ships Including Fishing Vessels and Investigation into Manoeuvring Prediction Method *Translation from original paper Journal of JASNAOE* **14** pp. 63-73
[5] Charchalis A 2018, Hydrodynamic Force Database with Medium-High Speed Merchant Ships Including Fishing Estimating the Main Dimensions of Ship's Hull *Journal of KONES Powertrain and Transport* **25** pp. 75-80
[6] Taimuri G, Matusiak J, Mikkola T, Kujala P, and Hirdaris S 2020, A 6-DoF manoeuvring model for rapid estimation of hydrodynamic actions in deep and shallow water *Ocean Engineering* **218** pp. 1-22
[7] Norrbin, N.H., 1970. Theory and observation on the use of a mathematical model for ship manoeuvring in deep and confined waters. Extended version of a paper presented at the 8th Symposium on Naval Hydrodynamics, Pasadena, 1970.
[8] Kijima, K., Katsuno, T., Nakiri, Y., Furukawa, Y., 1990. On the manoeuvring performance of a ship with the parameter of loading condition. J. Soc. Nav. Archit. Jpn. **168** pp. 141–148.
[9] Brix, J.E., 1993. Manoeuvring Technical Manual. Seehafen-Verl., Hamburg.
[10] Asmara I.P.S., and Husodo A.W., 2022 Berthing and Unberthing of Shuttle Tanker in FSO Unit, *Proceeding of MASTIC 2022.*
[11] PIANC 2002 Guidelines for the Design of Fenders System Report *of Working Group 33 of the Maritime Navigation Commission, International navigation Association.*

IOP Conf. Series: Earth and Environmental Science 1081 (2022) 012016 doi:10.1088/1755-1315/1081/1/012016

Experimental Study of the Effect of Bilge Keels Position as Roll Damping Devices on the Resistance of LCU Vessels

Sumarsono[1, 2], Beny Cahyono[1], Erwandi[2], B. Ali[2], A. Baidowi[1]
[1)] Department of Marine Engineering, Institut Teknologi Sepuluh Nopember
[2)] National Research and Innovation Agency - BRIN

Abstract. Landing Craft Utility (LCU) is one type of ship whose role is to transport combat equipment vehicles and amphibious troops to make landings on the beach. On the ship, bilge keels were installed. However, the installation of bilge keels on the surface of the ship's hull will add resistance to the ship's speed. To find out the effect of placing bilge keels on the additional ship speed resistance, a towing test experiment was carried out in the towing tank using the LCU ship model. Two bilge keels were manufactured in 4 (four) different positions, called A, B, C, and D positions. From the results of the study, it is known that the average value of the addition to the total resistance is; at A position there is an increase in ship speed resistance of 3.22%, at B position by 3.36%, at C position by 2.67%, and at D position by 1.83% of the total resistance in bare hull conditions. The results of this study indicate that the placement of bilge keels with the same shape and dimensions can add to different total resistances.

1. Introduction

The development of the anti-rolling components as a ship roll damper system is bilge keels. When the ship sails, the ship will face weather and water conditions that can change. This condition causes the ship to experience free movement in all directions including roll motion. Roll motion is an important mode of motion related to shipping stability. With the addition of a few other coefficients, the ship can capsize [1]. To overcome this problem, bilge keels were installed on the ship's hull. The bilge keel can function optimally on a stationary vessel or a vessel at low speed. Bilge keel at higher speeds will affect the increase in ship resistance [2]. Bilge is very easy to install and passive. Bilge keel is highly effective in dealing with roll motion and can function effectively in some marine conditions [3]. The installation of bilge keels on ships with low draft has a good effect on anti-rolling performance when the ship receives high waves [4]. Meanwhile, based on the length and width of the bilge keels, increasing the width is more effective than the length of the bilge keels [5]. On ships with a flat hull surface such as cargo ships, barges, and VLCC (Very Large Crude Carrier) by approaching and matching the angle with small roll amplitude areas and different vortices, for flat hull shapes, hull keels with high oscillation angles, large attenuation is an induced vortex phenomenon [6]. The roll attenuation approach on a flat hull ship with a large amplitude response area results in a larger attenuation which is possible because the large hull eddies are pulled under the flat hull [7]. Baskoro, et al [8] assessed the effect of bilge keel installation on resistance at various speed variations, where the ship has greater resistance than the ship without a bilge keel. Meanwhile, in terms of roll motion, ships with bilge keels have better maneuverability. A landing craft utility (LCU) is a ship used to transport vehicles, logistics, amphibious combat equipment, and troops. The LCU makes landings or anchorings at the beach or ESG jetty. LCU has a flat hull shape, especially in the middle part of the body. While the bow and stern have a streamlined hull shape. In loading and unloading vehicles on the beach, LCU requires calm water conditions so that loading and unloading do not endanger the safety of the vehicle or the personnel inside. The hull of the LCU is U-shaped. The use of a bilge keel in the U-shaped hull has the effect of reducing the rolling motion of the ship. Moreover, it also reduces the resistance (wave-making) and increases the frictional resistance [9]. Therefore, when the ship is operating in wavy water, the bilge keel can help reduce the roll motion of the ship. Besides, incidents of accidents due to shaky ships often occur on ships that have a hull shape similar to an LCU, such as container ships, bulk carriers, Landing Craft Tanks, and others. Bilge keel as a roll damping device serves to maintain the stability of the ship when it wobbles and reduce the influence of roll movement on the ship [10]. To reduce this motion, bilge keels are installed on the hull of the ship. The installation of bilge keels on the side of the ship has the effect of increasing the root mean square (RMS) value at low speeds and reducing the RMS value at high speeds [11]. On the Planning hull type of ships [12] conducted a study on the influence of the shape of the hull with several shapes and angles of the chain bilge on the total coefficient of resistance, where the results of this study provide an overview of the differences in the total coefficient of resistance in several forms of planning hull. The selection of the placement of the bilge keel in A, B, C, and D positions is following the hypothesis of the general location of the bilge keel area. So that the installation of a bilge keel in this area needs to be studied to determine its effect on resistance, and aspects of roll damping. Meanwhile, the study being conducted is on the effect on resistance, which can be continued by studying the effect of roll damping. The installation of bilge keels on the hull will affect the addition of the total resistance value of the ship. In this paper, a study is presented on the effect of the position of the bilge keels on the addition of total resistance, through a tensile test experiment in a towing tank test pond.

2. Model Test

The resistance test was carried out on the LCU ship model with the main sizes as presented in Table 1. And variations in the installation position of the bilge keels as shown in Figures 1a and 1b. The resistance test activity was carried out at the towing tank test pond facility of the Hydrodynamic Technology Laboratory – National Research and Innovation Agency (BRIN), Surabaya. Comparison of results is presented in the comparison of resistance tests with the speed of model tests, and comparisons of resistance values on the Froude number as has been done [13] in the resistance test of the catamaran model. In the resistance test, the model was pulled with the carriage and the drag force acting on the model ship was measured at various speeds. Measurements were carried out using a load cell resistance dynamometer which was connected to a data acquisition & analysis system recorder.

Table 1. Model Dimensions

Symbol Unit	Dimensions
LWL (m)	3.357
LPP (m)	3.285
B (m)	0.491
H (m)	0.329
T_F (m)	0.080
T_A (m)	0.098
Δ (kg)	121.0
WSA (m^2)	1.919
LCB from AP (m)	1.424

The test is carried out on the test model with variations in the installation of bilge keels in a predetermined position. In Figures 1 (a) and (b), the location of the variations in the installation of bilge keels is shown. This study is to find out the effect of bilge keel installation on technical aspects, in this case, its effect on resistance. The positions of the bilge keel which were installed at this location were not expected to have a huge effect on resistance but provide a large enough damping effect on the roll movement of the model. In this study, the study is conducted only on the technical aspect of its effect on resistance. A, B, C, and D positions are identified to make it easier to distinguish each position of the bilge keel installation location. A position was chosen to determine the effect of roll damping and the addition of the total resistance if the bilge keel is installed on the bottom. The position is installed in the area near the bilge. B position is installed right on the bilge. Where this position is the general position of the bilge keel installation. While C and D positions are installed on the sides. This is to determine the effect of roll damping and its effect on the addition of total resistance if the bilge keel is installed on the side of the ship. C position is installed near the bilge keel area. And D position is installed close to the line full of water.

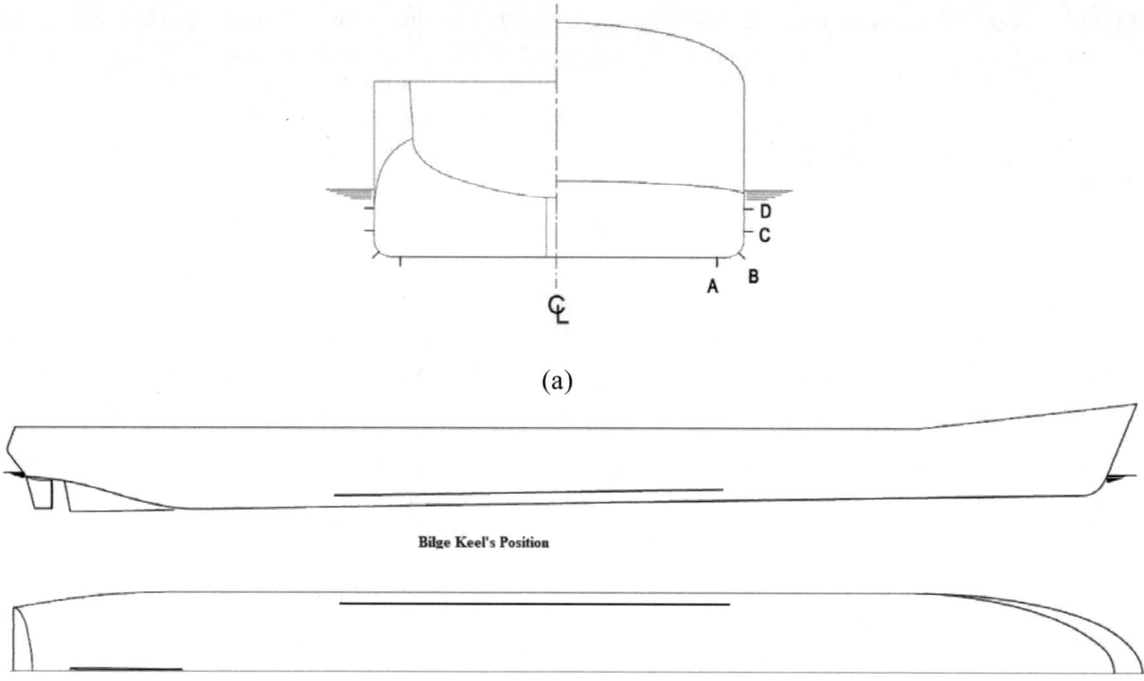

(a)

(b)

Figure 1. (a) & (b). Variations in the installation position of bilge keels.

The selection of A, B, C, and D positions in this experiment refers to several things. Among them is that generally, the installation of bilge keel is in that area. Dove [14], conducted a study where the bilge keel can reduce roll motion to the maximum when installed in a position one-fourth of the cross-section of the ship's baseline. The shape of the LCU hull mostly consists of a parallel middle body, where most of the hull is flat and elongated so that the flow of water around the hull tends to be straight. As has been stated by Bhattacharya [15], to achieve good performance the bilge keel is formed and installed following the current flow around the hull to minimize its effect on the addition of the total resistance. A variation bilge keel was installed in the bottom position of the ship model at a distance of 0.86 half the width of the model from the centerline. B variation bilge keel was installed in the round angle hull body position. C variation bilge keel was installed parallel to the bottom of the model ship at a distance of 0.035 m from the bottom. And the D variation was installed at a location close to the water-laden position (0.07 m from the baseline). Figure 2 shows the installation of a bilge keel on a model ship.

Bilge keels were installed in a parallel middle body position with a length dimension of 37%L and the width of the bilge keel 3% of the width of the ship [16]. Mounting in a position perpendicular to the surface of the model ship. With a thin dimension, it will have a small effect on the total resistance such as the shape of the bow of the ax bow type [17]. As it was also done by Sulisetyono [18] who studied the addition of the skeg component to the hull, where the addition of these components can significantly reduce the resistance wave-making on the hull, especially at high Froude number (Fn). The International Towing Tank Conference (ITTC) provides guidelines for conducting resistance tests [19], ITTC – Recommended Procedures, and Guidelines 7.5-02-02-01. The model testing process has been carried out following the ITTC procedure.

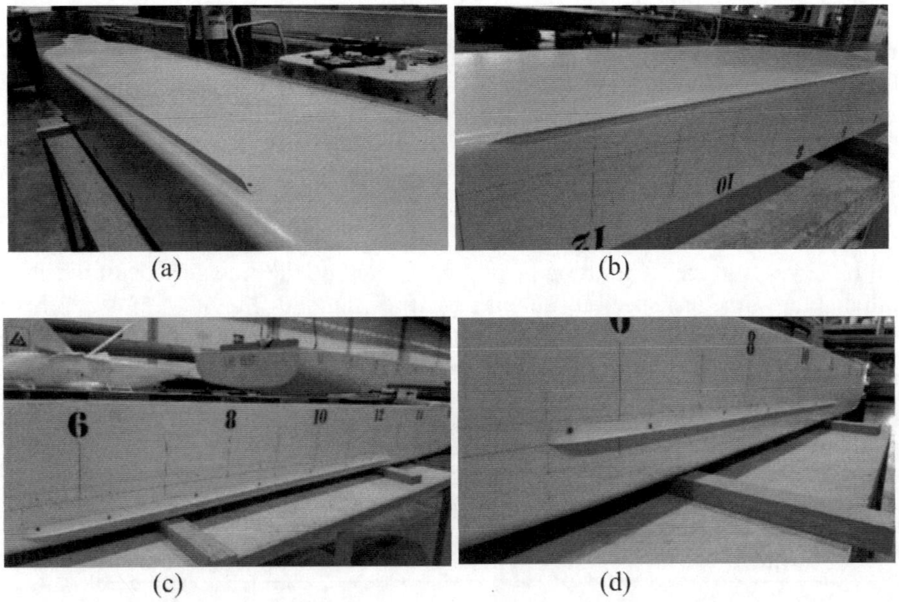

Figure 2. a. Bilge keel installation in A position; b. Bilge keel installation in B position; c..Bilge keel installation in C position; d. Bilge keel installation in D position.

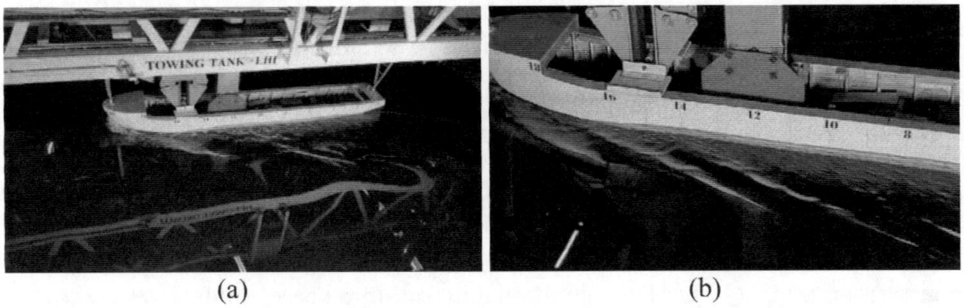

Figure 3. (a) & (b). Resistance test in Towing Tank.

The resistance test is carried out by pulling the ship model with a carriage in the test pool as shown in Figures 3 (a) and (b), at a speed of 0.9 m/s to 1.6 m/s. From this test, it can be obtained the relationship between the resistance of the model ship (in kg) and the speed of the model ship (m/s) on variations in the installation position of bilge keels can be obtained.

3. Result & Discussion

Several factors affect the ship resistance, including; ship length, water density, ship speed, water viscosity, gravity, and fluid pressure. The amount of resistance is stated in the calculation formula (1) below, where R_T represents the total resistance, ρ is the density of water, C_T is the coefficient of total resistance, S is the wet surface area, and V is the velocity. While the Froude number related to the wave resistance is stated in formula (2).

$$R_T = \frac{1}{2}\rho.C_T t.S.V^2 \quad (1)$$

$$Fn = \frac{V}{\sqrt{gLwl}} \quad (2)$$

In the resistance test of a model without the installation of bilge keels (bare hull), the resistance value was recorded at 1.245 kg at a model test speed of 1.424 m/s. Meanwhile, in the resistance test of the ship model that was installed with bilge keels in the A variation, bilge keels mounted on the bottom surface of the ship model produced a resistance value that was greater than the bare hull condition, which was at the same model test speed (1.424 m/s), the model resistance value ship of 1.289 kg. This is how it is shown in Table 2. Installation of bilge keels in the B position of variation, in the resistance test at that speed, the resistance value was recorded at 1.296 kg. In the ship model resistance test with the installation of bilge keels in the C variation position mounted on the side surface of the ship model, the resistance value at the same test speed results in a total resistance value of 1.259 kg. Whereas in the D variation where bilge keels were installed on the side of the ship model that is close to the water draft line position, the resistance test experiment produces a total resistance value of 1.278 kg at the test speed of the ship model.

Table 2. Result of bare hull resistance test & resistance variation of bilge keels position

V	Rm (barehull)	Rm "A"	Rm "B"	Rm "C"	Rm "D"
m/s	kg	kg	kg	kg	kg
0.890	0.373	0.381	0.383	0.380	0.377
1.068	0.541	0.567	0.561	0.578	0.550
1.246	0.804	0.825	0.833	0.829	0.829
1.424	1.245	1.289	1.296	1.259	1.278
1.602	1.764	1.820	1.812	1.775	1.779

Table 3. Show the addition of the total resistance value of the variation of the bilge keels installation to the total resistance value in bare hull conditions. In the A variation, the largest percentage addition to the total resistance value occurred at a test speed of 1.068 m/s, where the percentage in the resistance value was 4.75%. In the B variation there is an addition to the total resistance value which is the largest addition to the resistance value compared to the total resistance value for bare hull conditions of 4.10% occurring at a speed of 1.424 m/s. For the installation of bilge keels in the C variation is the largest percentage addition to the total resistance value occurs in the resistance test at a speed of 1.068 m/s, which is 6.80%. In the D variation, the largest increase in the resistance value occurred at the ship model test speed of 1.246 m/s with the percentage added to the total resistance value of 3.05%. The average increase in resistance value due to variations in the bilge keels installation to the total resistance value in bare hull conditions is; A variation is 3.22%, B variation is 3.36%, C variation is 2.67%, and D variation is 1.83%.

Table 3. Percentage of additional resistance variations in the position of bilge keels on bare hull conditions

V	Rm "A"	Rm "B"	Rm "C"	Rm "D"
m/s	%	%	%	%
0.890	2.04	2.72	1.77	0.94
1.068	4.75	3.68	6.80	1.61
1.246	2.60	3.59	3.02	3.05
1.424	3.54	4.10	1.16	2.71
1.602	3.20	2.70	0.61	0.87

The results of the resistance test on the ship model prove that there is additional resistance in the ship model that is installed with bilge keels. Analysis of the additional resistance as a result of installing a bilge keel in a variety of different positions gives different resistance values. Figure 4 shows a graph of the resistance test data for each variation of the bilge keels installation position, and resistance tests are carried out at speeds of 0.9 m/s to 1.6 m/s. The graph shows the comparison of total resistance between variations in the position of the bilge keel installation to form almost the same and coincident graph lines. The difference between each line is neither too big nor too significant. So this can make a recommendation to install bilge keels as roll damping devices in the area because it does not affect the addition of total resistance.

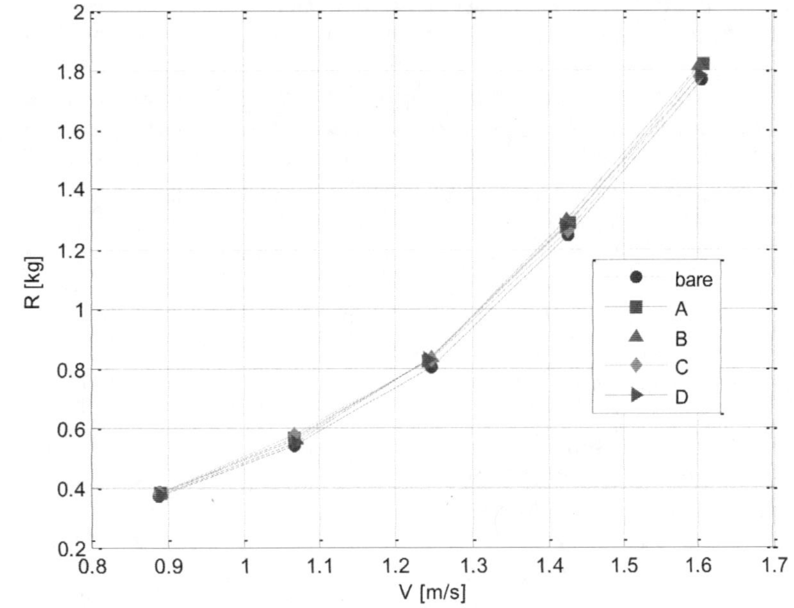

Figure 4. Total resistance on variations in test speed and variations in the installation position of bilge keels.

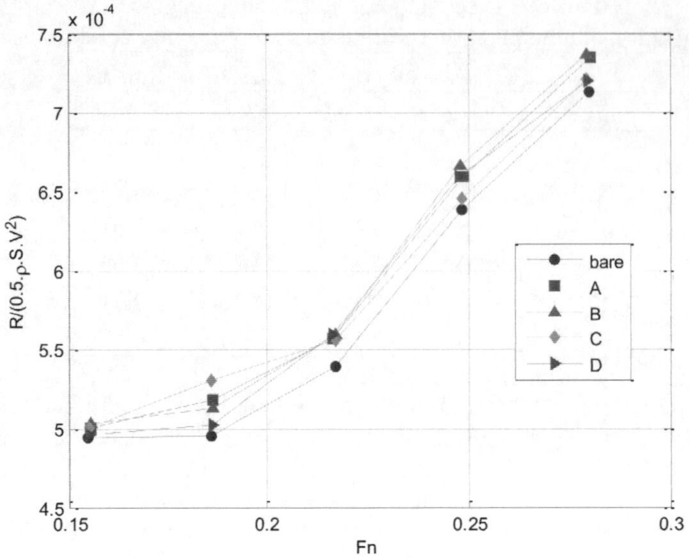

Figure 5. Comparison of the total coefficient of resistance at various Froude numbers.

The comparative analysis of the total resistance coefficient values on Froude numbers is presented in Figure 5. The total resistance coefficient values differ quite a bit at low Froude numbers. The Froude numbers value is higher in the variation of the installation position of the bilge keels C and D which is lower than the A and B variation. The bare hull condition is indeed the best because the R_T value is the smallest. At the initial speed of C_T in the C and D variations tend to be large, but at higher speeds, it tends to decrease and result below the C_T value of in the A and B variations. At Froude number 0.24 in the D variation has a high C_T value, but at a higher Froude number, C_T in the D variation is lower than the other variations. In another study Baskoro, et al [8], used Maxsurf software to analyze the effect of installing a bilge keel on total resistance, frictional resistance, viscous resistance, and wave resistance, where the results of the analysis showed an increase in total drag on the ship by 10.37% at speed 5 knots. Meanwhile, Liu W., et al [20], used Computational Fluid Dynamics (CFD) to simulate an increase in total resistance in the bilge keel design of fishing vessels, where the results showed an increase in the total resistance value of 1.17% on vessels where bilge keels were installed. However, several studies with the help of the CFD software need to be carried out further studies by conducting model test experiments. While the results of the experimental study in this study can also be carried out in further studies by conducting analysis using CFD software to validate the results of the model test.

4. Conclusion and Recommendation

From the technical aspect of using a bilge keel as a roll damping device, it is necessary to carry out a resistance test to determine the effect of the addition on the total resistance value. In LCU ships, the use of bilge keels is still needed to reduce the ship's roll motion. The positioning of the bilge keels needs to be considered because this is important in terms of the effect on the addition of total resistance. In the LCU model resistance test results, the average value of total resistance to bare hull conditions increased by 3.22% in the A variation, in the B variation by 3.36%, in the C variation by 2.67%, and in the D variation by 1.83%. In the C and D variations installed on the side of the model, the average percentage addition is smaller than for A and B variations installed in the bottom area. From the experimental results of the resistance test, the position of the bilge keel installation in D variation can be the best choice because it has the smallest effect of adding the total resistance value of 1.83%. To determine the effect of the installation position of bilge keels as a roll damping device, further studies can be carried out to determine the magnitude of the roll damping effect in each variation of the bilge keels position.

5. References

[1]. Kabylinski, L.K, and Kastner, "*Stability And Safety of Ships*," Elsevier, p.3 vol.1, 2003.

[2]. Ferrari, A. J, Jr, A. Ferreira, M. D, "*Assessment of the Effectiveness of the Bilge Keel as Anti Roll Device in VLCC*," IMO.*The Intact Stability Code MSC.267(85),*2008.

[3]. Ikeda, Y., "Roll Damping, Proceeding of Ship Motions," 1st Marine Dynamics Symposium, 1984.

[4]. Choiriya, Wasis D. Aryawan, "*Desain bilge keel* pada *shallow draft bulk carrier,"* Jurnal Teknik Pomits, 2013.

[5]. Hendratmoko, Hasanudin, I.K.A.P. Utama, "*Studi Eksperimen Pengaruh Lunas Bilga Terhadap Gerakan Rolling,"* Jurnal Teknik ITS, 2012.

[6]. Oliveira, Fernandes, "*Roll Damping of Flat-Bottom Hulls: Critique of the Quadratic Reason*," Marine Technology and Engineering, Taylor and Francis Group, London, UK, 2011.

[7]. Fernandes, Oliveira, "*The roll damping assessment via decay model testing (new ideas about an old subject),"* J. Marine. Sci. Appl, Researchgate.net, 2009.

[8]. Baskoro, G.A., B. Adietya, A., Sasmito, "*Analisa Pengaruh Pemasangan Bilge Keel dengan Variasi Kecepatan Terhadap Hambatan Gelombang,Viskos dan Gesek serta Olah Gerak dan Stabilitas pada Kapal Ikan Tradisional 70 GT*," Jurnal Teknik Perkapalan - Vol. 6, No. 1, 2018.

[9]. A. Saputra, Budiarto, G. Rindo, "*Analisa Pengaruh Bilge Keel Tipe Bulb Terhadap Lambung V Dan U Pada Kapal Patroli Dengan Menggunakan Metode Computational Fluid Dynamic (CFD)*", Jurnal Teknik Perkapalan, Vol. 7, No.1, 2019.

[10]. A. Syafiul M., "*Kajian Penambahan Bilge Keel Terhadap Gerak Oleng Kapal Kontainer Tipe Small Feeder*," Jurnal Inovtek Polbeng, vol. 8, no. 2, 2018.

[11]. Widyatmoko, Samuel,P. Manik, Andi T., "*Analisis Pengaruh Jumlah Bilge Keel terhadap Gerakan Rolling pada Kapal Patroli 14 m,"* Warta Penelitian Perhubungan, 2021.

[12]. I.K.A.P. Utama, Sutiyo, I K Suastika, A Sulisetyono, Hasanudin, Y A Hermawan, and W D Aryawan, "*Resistance Analysis of Rescue Boat in Calm Water Condition*," The 5th International Conference on Marine Technology (SENTA 2020), IOP Conf. Series: Materials Science and Engineering 1052, IOP Publishing, 2021.

[13]. E. Suwarni, I.K.A.P. Utama, "*Comparative Analysis of Catamaran Resistance in Shallow, Medium and Deep Waters,*" Jurnal Wave Volume 7 Nomor 2, 2013.

[14]. Dove, "*Effect of Bilge Keel Position on Roll Damping in Frigates*," Unpublished, MOD (PE) Report, 1958.

[15]. Bhattacharyya, "*Dynamics of Marine Vehicles*," Ocean Engineering, Wiley, 1978.

[16]. Sabuncu, T., "*Gemi Hareketleri*" ITU Published Book, 1983.

[17]. C. Kusuma, I.M Ariana, B. Ali, "*Redesign KCR 60m Bow with Axe Bow Type To Reduce Ship Resistance,*" Maritime Safety International Conference, IOP Conf. Series: Earth and Environmental Science 557, IOP Publishing, 2020.

[18]. A. Sulisetyono, "*The Wave Making Resistance Evaluation of The Fishing Ship With Skeg Using Tent Function Method*," Maritime Safety International Conference, IOP Conf. Series: Earth and Environmental Science 557, IOP Publishing, 2020.

[19]. International Towing Tank Conference, "ITTC - Recommended Procedures and Guidelines - Resistance Test (7.5-02-02-01)," 2011.

[20]. Liu W., Djatmiko, Eko Budi, et al, "*Bilge Keel Design for The Traditional Fishing Boats of Indonesia's East Java*," International Journal of Naval Architecture and Ocean Engineering, 1-16, 2018.

IOP Conf. Series: Earth and Environmental Science 1081 (2022) 012017 doi:10.1088/1755-1315/1081/1/012017

Form Factor Prediction Based on Ship Model Test Data by Statistical Method

Widodo[1,2], Agoes Santoso[1)*], Erwandi[2)], Achmad Baidowi[1)]

[1)] Department of Marine Engineering – Faculty of Marine Technology, Institute of Technology Sepuluh Nopember, Surabaya City-Indonesia
[2)] The National Research and Innovation Agency, Indonesia
*Corresponding Email : agoes295@gmail.com

Abstract. The ship model resistance test is carried out on the towing tank with a certain scale to get the total resistance coefficient of the ship model. The value of the ship's form factor is needed in extrapolation calculations to get the coefficient and total resistance value of a full-scale ship. The form factor can be obtained using the traditional method, namely testing the resistance of the ship model with a Froude Number between 0.1 – 0.2. This method is known as the Prohaska method. Another method or a more modern method is to use multiple linear regression statistics based on the ship model test. We used data from 41 displacement ship model tests conducted at the Hydrodynamics Technology Laboratory (LTH)-National Agency for Research & Innovation (BRIN). Data processing is done through statistical software using the OLS or Robust method which was previously done by transforming predictor variables to enlarge the variety of variables so that a good regression equation is obtained. The resulting form factor regression is a useful tool for predicting the ship's Effective Power value.

1. Introduction

Geographically, Indonesia is an archipelagic country with a large ocean area. In the distribution and equitable distribution of economic growth between islands, adequate transportation facilities are needed. The ship is one of the means of inter-island transportation, were in the design of the ship required stages, the first is the design and testing of designs both numerically and experimentally by making certain scale models. One of them is the prediction of the ship's thrust for the process of designing and operating the ship. Knowledge of the driving forces allows the size and mass of the propulsion engine to be determined and estimates of fuel consumption and operating costs. Prediction of the ship's thrust is carried out by experimental methods, numerical methods, and theoretical analysis on various aspects of the powering problem. The experimental method requires a ship test model at a certain scale to be tested in the Towing Tank (TT) basin in the hydrodynamics laboratory (LTH-BRIN). A ship moving in water exhibits two flow features, as shown in the figure below, [1].

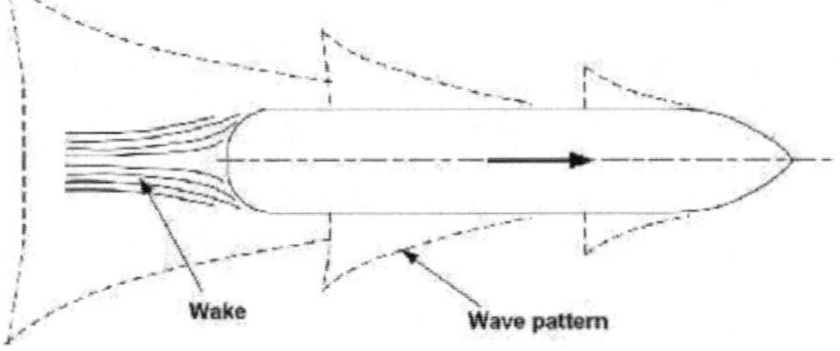

Figure 1. Waves and wake. [1]

Total resistance = Friction resistance + Viscous pressure resistance + Wave resistance

In predicting the full-scale ship resistance, the Form Factor value is needed for the extrapolation method for calculating the total resistance of the ship model into a full-scale ship. In most cases of low-speed resistance tests for form factor determination, the flow characteristics around the model vessel are not in full turbulence, and hence, differences in both draft and measured values are to be expected. [2].

The ship model testing carried out in the Towing Tank basin at LTH-BRIN, so far has used the Prohaska method to obtain the form factor (1+k). This value is important because the data obtained as a result of the test is the model-scale vessel resistance, so to obtain the full-scale vessel resistance value, it is necessary to enter the form factor value in the extrapolation process. Prediction of form factor values with regression equations provides alternative solutions in obtaining form factors by utilizing ship data, such as L_{WL}, B, C_B, C_M, C_P, W_{SA}, Δ & T which are more familiar, where ship data is transformed mathematically first. first according to the predictor variable (variable X) in the regression equation.

2. Background

The form factor is one of the important components in calculating the ship's resistance and the effective power of the ship. Extrapolation method for full-scale ship resistance calculation with model test data required ship form factor components. This component was first introduced by Froude, which was later continued by Hughes.

In 1978 the International Towing Tank Conference (ITTC) adopted the resistance test method by introducing the concept of form factor, as a performance prediction method for single-screw vessels. In conducting the ship test, a ship model test is carried out with a certain scale that can confirm the performance characteristics of a full-scale ship, wherein the model test, two types of dimensionless quantities are used, namely Froude Number (F_n) and Reynolds Number (R_n), however, the F_n and F_n scales are used. R_n between two "geosims" is practically impossible/ so that is conducting tests with the ship model, a form factor component is needed to find the coefficient of the total resistance of the ship. In getting the value of k, it can be done by measuring at a low speed, so that the C_W coefficient is close to zero and (form factor) = C_T/C_W. [1]

2.1. Prohaska Method.

The Prohaska method is based on the Hughes method, where the two methods differ in the definition of the shape factor. This method defines the form factor into three dimensions, where C_V is the viscous coefficient and C_F is the coefficient of frictional resistance in two-dimensional flow. The coefficient of total resistance can be written as: [3]

$$C_T = (1 + k).C_F + C_W \text{ , where } C_W = a.Fn^n \quad (1)$$

Where a is the coefficient and Fn is the Froude number. n varies depending on the C_B (between 4 and 6). The division of equation (2) with the coefficient of frictional resistance (C_W) results in the following equation:

$$C_T/C_F = (1+k) + C_W/C_F = (1+k) + a.(Fn^n/C_F) \quad (2)$$

To get the form factor can be done through experimental (with the Prohaska method) and statistical methods. As described above, the Prohaska method was carried out experimentally by pulling the test model in the range of $F_r < 0.2$. where at low speed the coefficient value is $C_W \sim 0$, (form factor) = C_T / C_F , so the Prohaska method can be used, where: C_T = (form factor) $C_F + aF_r^n$, it is assumed that $C_W = aF_r^n$ at low speed, ($F_r < 0.2$). The form factor is calculated through a straight-line plot of C_T/C_F vs F_r^4/C_F which intersects the ordinate ($F_r = 0$) at 1+k. [4].

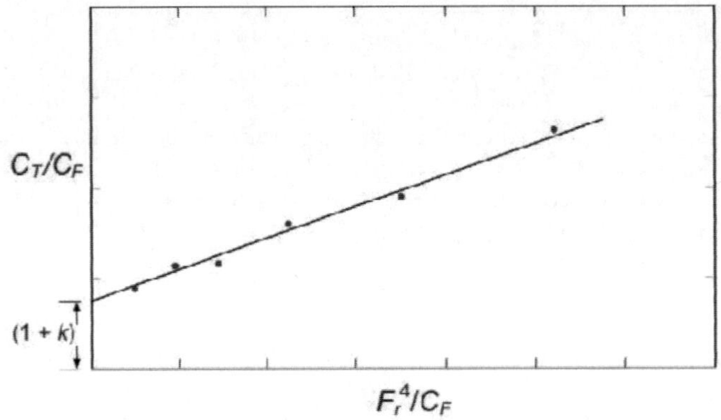

Graph 1. Plot Diagram Prohaska. [1]

From the actual construction case conditions, the following relationship is presented which can be applied to the prediction of full-scale ship resistance performance,

$$C_{TS} = (1+k)_S.C_{FS} + C_W + \Delta CF + C_{AA} \quad (3)$$

In almost all cases (conditions), the Reynolds number for full-scale ship designs is < 10^9, thus the appropriate form factor is $(1 + k)_S$ being the terminal form factor and can be considered constant [5].

2.2. Statistical Method

Calculation of form factor can also be done using statistics with linear regression method, the use of this method requires sufficient model test data so that the data is normally distributed. In predicting the shipshape factor, the linear regression method can be used, where various methods can generally be divided into two categories, namely simple linear regression and multiple linear regression. Linear regression methods such as Ordinary Least Squares (OLS), Robust Least Squares (RLS), Robust Least Trimmed (RST), and so on. [6].

In addition to the experimental method, the form factor can be obtained statistically with the linear regression method, where this method requires sufficient ship data so that data is normally distributed. LTH-BRIN has sufficient ship data and model test data to be used as predictor variables in conducting research. Linear regression method is one of the methods in statistics to model the approximation relationship between a scalar response or Y variable with one or more predictor variables X. In linear regression analysis, there are several methods whose use depends on the shape and condition of the data. Statistical application of linear regression method for ship form factor prediction has not been widely used. this method requires sufficient ship data and model test data, where the ship data used include L_{WL}, B, T, W_{SA}, C_M, C_P, C_B, & Δ, &, as X variable. Eight X variables from ship data will be transformed into 20 variations variables so that the resulting linear regression equation has a good statistical test value.

The use of the regression analysis method is carried out on fast boats [7], but not done on the displacement of the ship and the transformation of the variable X is not carried out optimally to get a better linear regression.

3. Basic Theory.

Prediction of ship shape factor can be done by using the experimental method known as the Prohaska method. In addition to this method, form factor prediction can be done statistically with the linear regression method. In general, this method can be divided into 2 types, namely simple and multiple linear regression methods. In multiple linear regression, the subject of the problem can be handled systematically with matrix rules, where linear regression of more than two independent variables X is easier to solve using the matrix method. [8]. The solution to the problem of linear regression is to analyze the relationship between one response variable and several predictor variables, which is known as multiple linear regression. [7]

One of the linear regression methods is the Ordinary Least Squares (OLS) method, where this method aims to minimize the sum of the squares of the residues. The use of statistics in prediction requires several statistical tests to obtain a good linear regression equation. Some statistical tests are Hypothesis Testing, namely:

3.1. F Test

This test is to determine the existence of a linear relationship between variable Y and variable X together (Montgomery and Runger, 2003). [8].

- Hypothesis

$H_0 : \beta_1 = \beta_2 = \ldots\ldots = \beta_k = 0$
H_1 : there is $\beta_j \neq 0$, with $j = 1,2,\ldots,k$

- Statistic Test.

$$F_{count} = (SS_R/k)/SS_E/(n-k-1) = MS\ \ /MSE \quad (4)$$

- Criteria Test.

H_0 rejected if $F_{count} > F_{(\alpha,k,n-k-1)}$ or $p\text{-}value < \alpha$

3.2. t Test

The t-test is a regression coefficient test to determine whether there is an effect of each X variable on the linear regression model, where according to Runger (2003), the step the t-test is : [8].

- Hypothesis

$H_0 : \beta_j = 0$
$H_1 : \beta_j \neq 0$

- Statistic Test.

$$t_{count} = \hat{\beta}_j/S_e(\hat{\beta}_j), with\ S_e(\hat{\beta}_j) = \sqrt{var\ (\hat{\beta}_j)} \quad (5)$$

- Criteria Test.

H_0 rejected if $t_{count} > t_{(\alpha/2,n-k-1)}$ or $p\text{-}value < \alpha$

3.3. Goodness of Fit

There are two kinds of this test, namely the determinant coefficient (Adjuster R^2), according to Ghozali (2011) R^2 is a tool to measure the model's ability to explain the Y & MSE variable (Mean Square Error).

Among these tests, there is an assumption test in multiple linear regression, such as the normality test, which is a test to determine whether the residuals in the regression model are normally distributed or not. Multicollinearity is an assumption test to find out that there is a fairly high correlation between X variables where multicollinearity is indicated by the Variance Inflation Factor (VIF) > 10. Autocorrelation is a test of this assumption using the Durbin Watson test. Autocorrelation test can use Breusch-Godfrey Serial Correlation LM Test on Eviews software. Heteroscedasticity is an assumption test to find out the existence of inequality in the variance from the residual of one observation to another observation. Outliers are an assumption test to state the existence of data points that are not of the same type as the others. [8]. In using the OLS method of linear regression, this test is required, to obtain a good linear regression equation.

In addition to the X variable, the Y variable is also needed. This variable is a form factor of data (1+k) obtained from the model test data using the Prohaska method which is technically explained in the introduction.

4. Related Theory

Statistics has many methods, one of which is the linear regression method. This method is used in the prediction of value or magnitude by using predictor variables. In this study, the method of multiple linear regression was used because the X variable was more than one. Some of the methods used in the research are:

4.1. Ordinary Least Squares Method.

Ordinary Least Squares (OLS) is one of the methods in linear regression that predicts the regression line by using the minimum value of the sum of the squares of error from the predicted value with the real value (Carl Friedrich Gaussimana). This method is known as Least Square, which is a multiple regression analysis method used to determine the relationship/influence of several variables X with variable Y (Henning, 2015 p. 7), this method will produce good predictions if all classical assumptions are met.

One of the problems in applying this method is the presence of outlier data which is an oddity and indicates different data when compared to other data.

A linear regression model with one independent variable can be expressed by,[9]

$$Y = a + \beta X + \varepsilon \tag{6}$$

The multiple regression equation can be described by the equation,

$$Y_i = \beta_0 + \beta_{i1} X_1 + \beta_2 X_{i2} + \cdots + \beta_n X_{in} + \varepsilon_i, \quad n = 1,2, \ldots \ldots, n \tag{7}$$

Where :
Y_i is the value of the dependent variable in the i-th observation
$\beta_0, \beta_1, \beta_2, \ldots \beta_n$ parameter to look for value
$X_{i1}, X_{i2}, \ldots, X_{in}$ value of the independent variable (predictor)
εi error value is random and normally distributed $N(0,\sigma 2)$. [10].

The application of the Ordinary Least Squares (OLS) method needs to be known about several influencing factors, where the OLS method is not feasible to be applied. Among the most contra and have a big influence on errors in the use and application of the OLS method are Multicollinearity and Missesification (that is, there is a high correlation between X variables) which can affect the correlation coefficient.

4.2. Stepwise Least Squares Method.

The stepwise method is one of the least-squares methods, this method is used to get the X variable (prediction variable).

In using the least-squares criterion (OLS) the relative mean error under certain circumstances represented in the two case variables, the stepwise estimator is better than the least-squares with direct prediction. (direct estimator).[11]. Implementation and Use of Stepwise LS in stages is to use the T (Student)-criterion to estimate the statistical significance of the regression coefficients and include in the model the variables that match the most significant coefficients that meet the conditions where the p-value < 0.5 while the variables that match the less significant coefficients are excluded. of the models. The use of Stepwise least squares is to get the optimal predictor variable with p-value and t -stat statistically qualified.

5. Results & Discussion.

Data collection carried out at LTH-BRIN is ship data and ship model test data. The data collected is displacement vessel data

5.1. Primary Data

Data collection is carried out at LTH-BRIN , where the data is ship data such as L_{WL}, B, C_B, C_M, C_P, Δ, W_{SA}, T, and data from ship model test results to get the form factor (1+k). This study used 41 ship data, with limitations on displacement ship data.

5.2. Secondary Data

Secondary data is data obtained from the transformation of primary data, wherein this secondary data will be obtained variations of the X variable that will be used in the data processing process with linear regression. The transformation of primary data will be developed into 20 variations of the X variable. In the secondary data, the X variable will be selected which gives the maximum R-squared value with statistical tests fulfilled, namely multicollinearity, heteroscedasticity, autocorrelation, and data distribution that is normally distributed.

The selection of the X variable in the secondary data was carried out using the Stepwise method on the Eviews software. The results of the selection with the Stepwise method obtained the variable X as follows: X_8; X_1; X_{15}; X_{17}; X_7; & X_{12}. were for :

$$X_8 = L_{WL}/\sqrt{W_{SA}};\ X_1 = B/L_{WL};\ X_{15} = \sqrt{W_{SA}};\ X_{17} = C_B/\sqrt{W_{SA}};\ X_7 = C_P;\ X_{12} = L_{WL}/\sqrt[3]{\Delta}$$

Variable Y or (1+k) was obtained from the data test model using the Prohaska method. Like Froude and Hughes, Prohaska's method uses a similar technique but applies more data points to the equation as follows:

$$C_T/C_F = (1+k) + AF_r^4/C_F \quad (8)$$

where (1+k) is the intersection point, and the slope is A. Prohaska's plot can be seen in Graph 1.

5.3. Eviews & Minitab Software Processing Results.

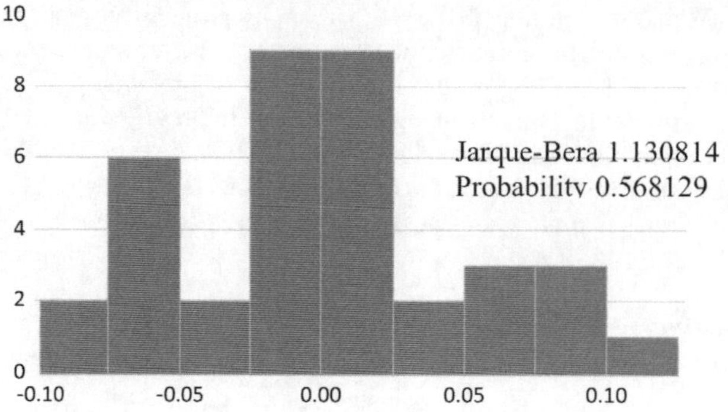

Graph 2. Normal Distribution, Data Processed by Eviews

Table 1. Descriptive Statistics

Variable	N	N*	Mean	SE Mean	StDev	Minimum	Q1	Median	Q3	Maximum
X_8	41	0	2,0079	0,0541	0,3464	0,3806	1,8974	2,0035	2,1893	2,7692
X_1	41	0	0,20518	0,00914	0,05850	0,01475	0,17477	0,20312	0,22843	0,40132
X_{15}	41	0	38,11	2,33	14,93	13,90	27,49	37,03	46,32	93,60
X_{17}	41	0	0,01972	0,00103	0,00660	0,00929	0,01579	0,01877	0,02339	0,03814
X_7	41	0	0,7315	0,0135	0,0867	0,6030	0,6390	0,7330	0,8165	0,8870
X_{12}	41	0	5,736	0,256	1,640	3,426	4,889	5,537	5,942	13,444

Data processed by Minitab

Table 2. The result of data processing with linear regression Eviews
Sample (adjusted) : 1 40
Included observations : 37 after adjustments

Variable	Coefficient	Std. Error	t-Statistic	Prob
C	1.216440	0.164676	7.386880	0.0000
X_8	-0.113445	0.031244	-3.630925	0.0009
X_1	0.710408	0.219491	3.236614	0.0027
X_{15}	0.004291	0.001235	3.474604	0.0014
X_{17}	5.207033	2.645475	1.968279	0.0572
X_7	-0.201389	0.138741	-1.451547	0.1558
X_{12}	0.008698	0.007259	1.198317	0.2391
R-Squared	0.492418	Mean dependent Var		1.303180
Adjusted R-Squared	0.402845	S.D. Dependent Var		0.080528
F-Statistic	5.497372	Durbin-Watson stat.		1.888711
Prob (F-Statistic)	0.000458			

Data processed by Eviews

Form factor prediction linear regression equation with OLS method :

$$Y = 1.216440 - 0.113445*X_8 + 0.710408*X_1 + 0.004291*X_{15} + 5.207033*X_{17} - 0.201389*X_7 + 0.008698*X_{12}$$

$X_8 = L_{WL}/\sqrt{W_{SA}}$; $X_1 = B/L_{WL}$; $X_{15} = \sqrt{W_{SA}}$; $X_{17} = C_B/\sqrt{W_{SA}}$; $X_7 = C_P$; $X_{12} = L_{WL}/\sqrt[3]{\Delta}$

A good linear regression equation requires several classical assumption tests to determine the feasibility of the data (variable X), where the tests are normal distribution, heteroscedasticity, autocorrelation, multicollinearity, and outlier data. Parameter test used Eviews software.

5.4. Statistical Test.

Standard deviation is the value of how much variation in the data, if the standard deviation is smaller than the average value, this indicates that the value can be used as a representation of the entire data (can be seen in Table 1).

From table 2, the R-Squared value is 49.2418%, which means that the X variable has an influence on the Y variable of 49.2418%. For Durbin Witson's value is 1.888711, this is it is necessary to test with the Breusch-Godfrey Serial Correlation LM test to determine whether there is an autocorrelation or not.

From table 2. Obtained the value of $X_{17} = C_B/\sqrt{W_{SA}}$; $X_7 = C_P$; $X_{12} = L_{WL}/\sqrt[3]{\Delta}$ probability value > 0.05, it can be concluded that the variable has a not too significant effect on Y, while $X_8 = L_{WL}/\sqrt{W_{SA}}$; $X_1 = B/L_{WL}$; $X_{15} = \sqrt{W_{SA}}$ probability value < 0.05, which means this variable has a significant effect on Y.

5.5. Autocorrelation Test.

In the autocorrelation test, the Breusch-Godfrey Serial Correlation LM Test was used.

Table 3. Autocorrelation Test.

Breusch – Godfrey Serial Correlation LM Test			
F-Statistic	0.049881	Prob. (2,32)	0.9514
Obs*R – Squared	0.127424	Prob. Chi-Squared (2)	0.9383

Data processed by Eviews

From the test results, obtained Prob. Obs*R squared is 0.9383 where this value is greater than 0.05 so it can be concluded that there is no autocorrelation symptom in the confounding error (residual value) of period t with period t-1. [12], [13].

5.6. Heteroscedasticity Test.

Heteroscedasticity testing is carried out to test whether in the regression model there is an inequality of variance from one residual to another observation [Ghozali, 2011]. In a good regression model, there should be no symptoms of heteroscedasticity. Below are the results of the heteroscedasticity test.

Table 4. Heteroskedasticity Test.

Heteroskedasticity test : Breusch-Pagan - Godfrey			
F-Statistic	0.845059	Prob. F(6.34)	0.5444
Obs*-squared	5.320774	Prob. Chi-Squared (6)	0.5034
Scaled explained SS	4.169132	Prob. Chi-Squared (6)	0.6538

Data processed by Eviews

From table 4, the F-Statistic value is > 0.05, so there is no problem with heteroscedasticity in the X variable.

5.7. Multikolinearitas Test.

Multicollinearity testing was conducted to determine the relationship between the X variables, where this test can be known from the value of the Variance Inflation Factor. One of the methods in determining multicollinearity problems in data that has independent variables [X_1, X_2,.....X_n) is the Pearson correlation matrix of predictor variables, eigenvalue matrix [X_0,X), or variance of inflation factor (VIF). [14]. The relationship between correlation and collinearity is not the same, so it can still be multicollinearity, even when all correlations are low.

The use of correlation matrix R is not enough. Another diagnostic is Determinant R (detR) for multicollinearity but there is the same problem with pairwise correlation. Even though the determinant is very close to one, there may still be multicollinearity among the X columns. The variance of the inflation factor (VIF) is very popular for its use as a diagnostic of multicollinearity. [15].

Table 5. Variance Inflation Factors.
(Sample : 1 : 41. Included observations : 41)

Variable	Coefficient Variance	Uncentered VIF	Centered VIF
C	0.027118	287.1207	NA
X_8	0.000976	42.87872	1.209809
X_1	0.048176	23.17769	1.702971
X_{15}	1.53E-06	26.96588	3.513539
X_{17}	6.998536	31.95545	3.153628
X_7	0.019249	110.5522	1.495237
X_{12}	5.27E-05	19.81511	1.464241

Data Processed by Eviews

VIF Equation

$$VIF_i = 1/(1-R_i^2) \quad for\ i = 1, 2, \ldots\ldots, k \tag{9}$$

From the table 5, the value of VIF < 10 is obtained, so that the X variable data does not have a problem with Multicollinearity.

5.8. Outlier.

The DfFITS and Leverage Value methods can detect the presence of outliers, where these two methods can be used to identify outliers in the Y variable and the X variable. A data variable Y can be detected. [16].

$$(DfFITS)_i = \hat{y}_i - \hat{y}_{i-1}/S_{i-1}^2 \sim \sqrt{h_{ii}} \tag{10}$$

$$h_{ii} = (X_i^T/(X^T.X)).X_i\ ; i = 1,2,\ldots.,n \tag{11}$$

$$|DfFITS_i| > 1.\ for\ n \leq 30$$

$$|DfFITS_i| > 2\sqrt{(p/n)}.\ for > 30 \tag{12}$$

Barnett and Lewis (1994) define outlier data as observations that appear inconsistent with other data. To overcome this problem, a new statistical technique has been developed which is not easily affected by the presence of data outliers. These robust methods are Least Median Squares (LMS), Least Trimmed Squares (LTS), Huber M Estimation, MM Estimation, Least Absolute Value (LAV), and S Estimation. [17].

In addition to the above method, outlier data can be obtained with statistical software, which in this study uses Minitab software, Outlier data is tested with Grubb's Test. From Minitab obtained outlier data, namely:

IOP Conf. Series: Earth and Environmental Science 1081 (2022) 012017 doi:10.1088/1755-1315/1081/1/012017

Table 6. Grubb's Test

Variable	N	Mean	StDev	Min	Max	G	P
X_8	41	2.0079	0.3464	0.3806	2.7692	4.70	0.000
X_1	41	0.20518	0.05850	0.01475	0.40132	3.35	0.012
X_{15}	41	38.11	14.93	13.90	93.60	3.72	0.002
X_{17}	41	0.01972	0.00660	0.00929	0.03814	2.79	0.140
X_7	41	0.7315	0.0867	0.6030	0.8870	1.79	1.000
X_{12}	41	5.736	1.640	3.426	13.444	4.70	0.000

Data processed by Minitab

Table 7. Variable outlier data X.

Variable	Row	Outlier
X_8	5	0.3806
X_1	36	0.4013
X_{15}	10	93.6002
X_{12}	4	13.4443

Data processed by Minitab

To achieve an effective forecasting target, it is necessary to identify outliers in the data. The presence of outliers can affect an effective prediction. Therefore, several tests have been applied to detect and remove outliers from the data. Among these tests is the Grubbs test introduced by (Grubbs, 1950) and recommended by ISO, which has been widely applied to detect outliers in data. The Grubbs test is easy to implement and operate using the mean and standard deviation of the data. [18].

The OLS method is less accurate because of the outlier data on variable X. Outlier data can affect data analysis, such as the average and standard deviation.

However, outlier data cannot be discarded because it will affect the regression model and result in inaccurate parameter estimates. The existence of data outliers requires a more robust method to overcome them. For this reason, a robust method is needed so that if there is a small change in the data it will not affect the estimated value.[6].

One of them is the Robust Least Squares (RLS) method which is very powerful in handling outlier data.

6. Conclusion

The calculation of ship resistance can be done through experiments with ship models at a certain scale, so a form factor is needed in calculating full-scale ship resistance. In addition to experiments, form factors can be obtained using statistical linear regression methods.

In statistics, it is necessary to take into account the correlation of the mean value and standard deviation, where standard deviation is the value of how much variation in the data, if the standard deviation is smaller than the average value, this indicates that the value can be used as a representation of the entire data (can be seen in Table 1).

The use of the linear regression method requires sufficient data to form a normally distributed data distribution. The test data of the LTH-BRIN ship model is primary data that requires a transformation process into secondary data or variable X. Eight X variables, namely L_{WL}, B, T, W_{SA}, Δ, C_P, C_B, C_M will be expanded into 20 variations of X. Secondary data will be selected using Stepwise Least Squares, 6 variables that meet the statistical test and R squares optimally obtained. The variables used in the linear regression equation are:

$$Y = 1.216440 - 0.113445*X_8 + 0.710408*X_1 + 0.004291*X_{15} + 5.207033*X_{17} - 0.201389*X_7 + 0.008698*X_{12}$$

In this study, the R-Squared value was 49.2418%. In predicting form factors from model test data, classical tests are needed such as autocorrelation, heteroscedasticity, multicollinearity, and outliers. From the classical test, outlier data were obtained (table 7), for that in the next research, predictions will be made with a stronger method of outlier data and the addition of a variation of the X variable to increase the value of R-Squared.

The regression equation to determine the form factor value has advantages when compared to the Prohaska method because it does not require a ship model experiment. Variable X in the regression equation uses ship data, such as L_{WL}, B, T, C_B, C_M, C_P, W_{SA} & Δ, which are familiar so that this equation is easy to understand.

References

[1] A. F. Molland, S. R. Turnock, and D. A. Hudson, Ship Resistance and Propulsion. 2011.

[2] N. Degiuli, N. Hadžić, M. Pedišić, and G. S. Buča, "Form factor determination of the full, large breadth and shallow draught ship series," Brodogradnja, vol. 58, no. 4, pp. 380–388, 2007.

[3] A. J. Jamaluddin and D. K. Kentjanawati, "Kajian Eksperimen Viscous Form Factor Pada Hambatan Kapal Katamaran," Wave J. Ilm. Teknol. Marit., vol. 7, no. 1, pp. 1–6, 2018, doi: 10.29122/jurnalwave.v7i1.3193.

[4] K. S. Min and S. H. Kang, "Study on the form factor and full-scale ship resistance prediction method," J. Mar. Sci. Technol., vol. 15, no. 2, pp. 108–118, 2010, doi: 10.1007/s00773-009 0077-y.

[5] Z. Aflakhah, J. Jajang, and A. T. Br. Sb., "KAJIAN METODE ORDINARY LEAST SQUARE DAN ROBUST ESTIMASI M PADA MODEL REGRESI LINIER SEDERHANA YANG MEMUAT OUTLIER," J. Ilm. Mat. dan Pendidik. Mat., vol. 11, no. 1, p. 21, 2020, doi: 10.20884/1.jmp.2020.12.1.1934.

[6] M. H. Alvin, M. Atok, and M. Indiaryanto, "Analisis Regresi untuk Memprediksi Tahanan Kapal Cepat," J. Sains dan Seni ITS, vol. 9, no. 1, 2020, doi: 10.12962/j23373520.v9i1.51386.

[7] A. D. Deria, A. Hoyyi, and M. Mustafid, "Regresi Robust Estimasi-M Dengan Pembobot Andrew, Pembobot Ramsay Dan Pembobot Welsch Menggunakan Software R," J. Gaussian, vol. 8, no. 3, pp. 377–388, 2019, doi: 10.14710/j.gauss.v8i3.26682.

[8] S. I. Bangdiwala, "Regression: simple linear," Int. J. Inj. Contr. Saf. Promot., vol. 25, no. 1, pp. 113–115, 2018, doi: 10.1080/17457300.2018.1426702.

[9] I. M. Yuliara, "Modul Regresi Linier Berganda," Regresi Linier Berganda, p. 18, 2016.

[10] T. D. Wallace, "Efficiencies for Stepwise Regressions," J. Am. Stat. Assoc., vol. 59, no. 308, pp. 1179–1182, 1964, doi: 10.1080/01621459.1964.10480758.

[11] J. Rowena and Hendra, "Earnings Volatility, Kebijakan Dividen, Dan Pertumbuhan Asset Berpengaruh Terhadap Volatilitas Harga Saham Pada Perusahaan Manufaktur Di BEI Periode," J. Adm. Kant., vol. 5, no. 2, pp. 231–242, 2017.

[12] R. Brüggemann, H. Lütkepohl, and P. Saikkonen, "Residual autocorrelation testing for vector error correction models," J. Econom., vol. 134, no. 2, pp. 579–604, 2006, doi: 10.1016/j.jeconom.2005.07.006.

[13] D. H. Vu, K. M. Muttaqi, and A. P. Agalgaonkar, "A variance inflation factor and backward elimination based robust regression model for forecasting monthly electricity demand using climatic variables," Appl. Energy, vol. 140, pp. 385–394, 2015, doi: 10.1016/j.apenergy.2014.12.011.

[14] A. Alin, "Multicollinearity," Wiley Interdiscip. Rev. Comput. Stat., vol. 2, no. 3, pp. 370–374, 2010, doi: 10.1002/wics.84.

[15] N. Nurdin and A. Islamiyati, "Penggunaan Regresi Robust Pada Data Metode Momen," vol. 10, no. 2, pp. 114–123, 2014.
[16] Ö. G. Alma, "Comparison of Robust Regression Methods in Linear Regression," Int. J. Contemp. Math. Sci., vol. 6, no. 9, pp. 409–421, 2011, [Online]. Available: http://m-hikari.com/ijcms-2011/9-12-2011/almaIJCMS9-12-2011.pdf.
[17] M. Aslam, "Introducing Grubbs's test for detecting outliers under neutrosophic statistics – An application to medical data," J. King Saud Univ. - Sci., vol. 32, no. 6, pp. 2696–2700, 2020, doi: 10.1016/j.jksus.2020.06.003.

Acknowledgements
This research was supported by the Hydrodynamics Technology Laboratory (LTH) - the National Research and Innovation Agency (BRIN) which has assisted in the availability and collection of data, and the Department of Marine Engineering - Faculty of Marine Technology ITS in terms of theoretical deepening.

IOP Conf. Series: Earth and Environmental Science 1081 (2022) 012018 doi:10.1088/1755-1315/1081/1/012018

The study of the Modification of the Ro-Ro Ferry's Angle of the Entrance Using Statistical Methods and Ship Model Resistance Tests

Cahyadi Sugeng JM[1,2], Semin[1,*], Erwandi[2]

[1]Department of Marine Engineering, Institut Teknologi Sepuluh Nopember, Surabaya Indonesia
[2]The National Research and Innovation Agency, Republic of Indonesia

*Email correspondent: semin@its.ac.id

Abstract. Aspects of speed and its correlation to ship resistance greatly affect the performance of ferry ro-ro ships. The fixing of the ship's resistance is very significant when related to the ship's economic costs. BTH-BRIN has carried out resistance tests and calculations using the Holtrop statistical method where the resistance result is considered too large. To reduce the resistance, modifications are made, one of which is by changing the angle of the entrance of the ship bow. Again, the Holtrop statistical calculations are then employed to determine the effective angle of the entrance after the ship is modified. Reducing the angle of the entrance of water at the ship's bow is expected to reduce the resistance in terms of the bow wave. It also needs paying attention to the amount of ship payload that has been planned. This paper describes the process of resistance reduction due to the change of the angle of the entrance of the ferry ro-ro ship. Firstly the resistance of the original ship is predicted using the statistical method and a scaled ship model is made and tested in BTH towing tank. Second, the angle of the entrance of the ferry ro-ro ship is modified. Finally, the modified one is predicted statistically based on the Holtrop method. The results of the statistical analysis of the modified angle of entrance showed that the resistance is smaller than the original one.

1. Introduction

In an archipelagic country like Indonesia, the ferry is the one which is most important and strategic mode of sea transportation, to connect one island to another [1]. The strategic role of the ferry ship is expected to support national economic growth. The main problem that is often experienced by ships is the decrease in speed due to the resistance of the ship. For ferry ships, especially in terms of reducing ship resistance, the EEDI (Energy Efficiency Design Index) aspect as required by the International Maritime Organization (IMO), is still considered very low [2].

The resistance is mostly affected by the main dimension and shape parameter of the ship, and of course, the dimension is determined in the design stage. Various shapes of the hull can be made from specific sizes and parameters [3], and all of them do not always give a good effect. For this reason, various considerations are needed in the calculation to obtain the size of the ship according to the speed and shape of the hull with minimal resistance. The common factor that has an important role when the ship sails is the value of the resistance experienced by the ship. The ship resistance is related to the Effective Horse Power (EHP) needed by the ship to reach the planned speed [4]. A complete and in-depth study is very much needed and is something that must be considered to determine the performance of the ship when carrying out hydrodynamic testing in the form of a ship resistance test.

The ship's resistance is influenced by several parameters that make up the hull of the ship, including block coefficient (Cb), prismatic coefficient (Cp), Wetted Surface Area (WSA), and Angle of entrance [5] and [6]. Between parameters are intercorrelated, so that if the size of one changes there will be a change in the other size [7]. The study of the influence of the shape of the hull is very important, many references are used in optimizing changes in the shape of the hull. The hull is useful for providing buoyancy and stability [8].

Various kinds of research efforts have been carried out to reduce the value of the ship's resistance. There are several ways to reduce ship resistance, one of which is optimizing the shape of the ship's hull. There are 3 (three) categories of ship hull shape optimization design process, namely: first, LCB position (midship 0%, FWD +, AFT-), second fore-body design, and third aft-body design [9]. In this case, the discussion is carried out by modifying the design of the ship's bow, especially on changes in the angle of the entrance. Modifications are made only by reducing the angle of the entrance of the waterline. The angle of the entrance is related to the wave resistance, as the angle of the entrance increases, the bow wave resistance also increases. Ships that have a good bow shape can provide efficiency the resistance so that ship operations can be better and more efficient [10]. The results of the change in the angle of the entrance can be obtained by comparing the design of the initial bow of the ship and the one after modifications have been made. The angle of the entrance is related to the wave resistance, the smaller the angle of the entrance, the smaller the effect of wave resistance, and vice versa [11].

This paper presents a study of the effect angle of the entrance on reducing ship resistance. The object used is the type of ferry ro-ro ship 1,500 GT which is a type of ferry that is usually used between islands in Indonesia. Modification of the angle of the entrance on the forebody is expected to reduce the wave resistance which in turn can reduce the total resistance when the ship is operating. The focus of the study in this research is to predict how big the impact of changes in the angle of the entrance on the total resistance experienced by the ship, which is one of the stages in ship design.

2. Lines plan

In the process of designing a ship, the lines plan is a very important design drawing. obtained based on the data of the ship that has been determined. Lines plan is needed to determine the shape of the ship's hull to be planned. And what is more important than the lines plan is the magnitude of the ship's resistance which is very dependent on the shape of the hull, especially those below the waterline. Lines plan is very influential on other design drawings such as general plans, profile construction, longitudinal and transverse construction, stability calculations, and others. The lines plan drawing is made in 2 dimensions so that to see the overall shape of the hull in 3 dimensions, it is divided into three parts, namely: body plan, sheer plan, and half breadth plan [10].

The body plan is a drawing of half of the entire vertical cross-section for each station, which consists of 2 sides, where the left side is the projection lines at the station behind and the right side is the projection lines at the station in front of the midship. In this picture, there are projection lines for each station across the ship in the form of curved lines, water lines in the form of horizontal lines, buttock lines in the form of vertical lines, and sent lines in the form of diagonal lines. The dimensions of this section are the height (H) and width (B) of the ship. The sheer plan is the appearance of the shape of the ship if the ship is cut vertically along the hull, which shows the shape of the bow and stern. In the sheer plan image, there are projection lines for each vertical buttock line extending the ship in the form of curved lines, body plan lines in the form of vertical lines, and half breadth plan lines in the form of horizontal lines. Therefore, in this section, there will be two dimensions, namely length (L) and height (H). Half breadth plan or the plan of half the width of the section viewed from the ship is obtained if the ship is cut horizontally along the hull. In this picture, there are projection lines for each waterline horizontally along with the ship in the form of curved lines, body plan lines in

IOP Conf. Series: Earth and Environmental Science 1081 (2022) 012018 doi:10.1088/1755-1315/1081/1/012018

the form of vertical lines, buttock lines in the form of horizontal lines, and sent lines in the form of curved lines. So this section will show the dimensions of length (L) and width (B).

The lines plan is the fundamental information used in the research of this paper. In Figure 1. the lines plan for the 1,500 gt ferry is shown as follows:

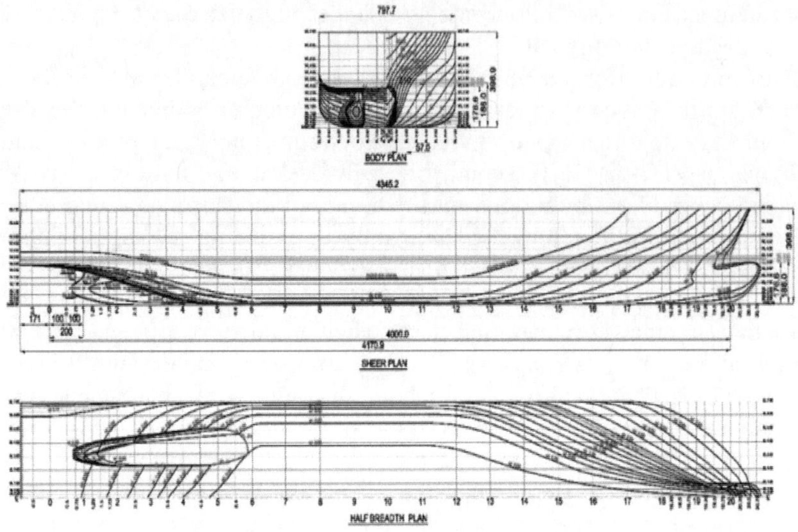

Figure 1. Lines plan of ferry ro-ro ship.

Lines plan of the ro-ro ferry ship 1,500 GT which includes the body plan, sheer plan, and half breath plan, while the material in the study is as follows :

- Primary data

  Primary data includes the principal dimensions of the ship, including coefficients obtained from the ship's planner or owner, as presented in table 1, namely:

Table 1. Principal dimensions of ship and ship model of ferry ro-ro ship

		Original Design		1st Modification		2nd Modification	
Parameter	Notation	Ship	Model	Ship	Model	Ship	Model
Length Over All (m)	LOA	76,720	4,371	76,720	4,371	76,720	4,371
Length Water Line (m)	LWL	73,200	4,171	73,200	4,171	76,600	4,365
Length Between Perpendiculars (m)	LBP	70,200	4,000	70,200	4,000	72,600	4,137
Breadth Moulded (m)	B	14,000	0,798	14,000	0,798	14,000	0,798
Depth Moulded (m)	D	4,600	0,262	4,600	0,262	4,600	0,262
Draught (m)	T	3,300	0,188	3,300	0,188	3,300	0,188
Displacement Volume Moulded (m^3)	Δ	2197	0,406	2197	0,406	2179	0,403
Wetted Surface Area (m^2)	WSA	1224	3,974	1224	3,974	1235	4,012
Angle of entrance (degrees)	I_E	21		3		3	
Scala	λ	17,550					

- Secondary Data
  The speed variation in the resistance test is between 10 knots to 18 knots, at a draft condition of 3.30 m, this is necessary to obtain an overview of the resistance experienced by each ship. The angle of entrance in the original design is 21 degrees and the design modification is 3 degrees.
- The ship model is made at the Ship Model Workshop (SMWS) with the main material of wood using a fiberglass laminate system.

3. Research methodology

This study aims to determine the effect of changes in the angle of the entrance on the total resistance value of the ferry ro-ro ship. Determination of ship resistance is carried out using the Holtrop statistical calculation method and the experimental method. In the Holtrop method, the prediction of the resistance is carried out based on statistical data for several hydrostatic parameters for the ship's hull. In the experiment, to get the actual total resistance of the ship, a physical model of the ship's resistance was tested, using the ship model extrapolation. The ship model resistance test was carried out at the Towing Tank, which is one of the Indonesian Hydrodynamics Laboratory (IHL), Balai Teknologi Hidrodinamika - Badan Riset dan Inovasi Nasional (BTH-BRIN), following the ITTC Standard Operational Procedure 1957, regarding the calm water resistance test.

The results of the statistical calculation of the Holtrop method of ship resistance in the initial design were verified by experimental tests, the comparison between the ship and the ship model was 1:17,55, while the dimensions of the ship and the ship model are shown in Table 1. The bow angle of the entrance of the ship's original design model is 21 degrees. This angle is then modified to 3 degrees, the speed of the ship being tested is between 10 knots to 18 knots or in the range of Fn 0.192 - 0.345.

To clarify the methodology of the research carried out, Figure 2. shows the flow chart used in this study.

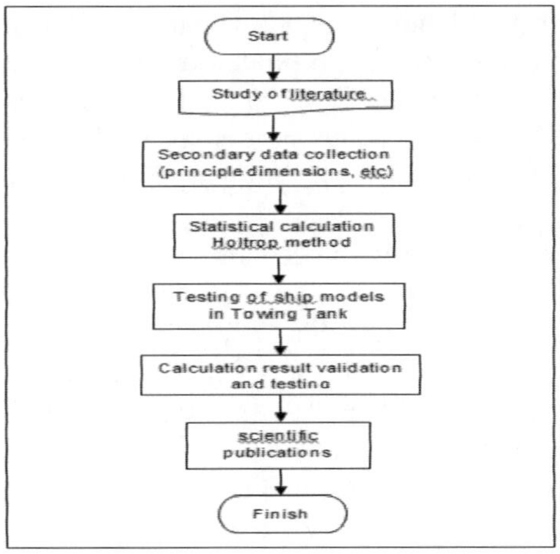

Figure 2. Flow chart Research method.

4. Ship resistance model test

To determine the hydrodynamic characteristics of a ship, one of the tests carried out is resistance testing. The resistance test is carried out in calm water conditions with variations in speed so that the characteristics of the ship's resistance in the form of a total resistance coefficient (CT) for the presence of hump and hollow phenomena can be observed and analyzed. The empty ship model should be scaled, and extrapolated from the model to a full-scale ship [12]. When predicting the magnitude of the ship's resistance using a physical model test, before testing it is determined the weight of the ballast will be used to regulate the draft conditions in the test. The model ship is towed by the carriage and the total longitudinal force acting on the model is measured by a resistance dynamometer, at various speeds as depicted in Figure 3. The model must be positioned so that it is in the centerline of the tank and parallel to the rail. The model should generally be as large as possible for the towing tank size concerning wall effects, shallow water, model mass, and the maximum speed of the towing carriage [13].

Figure 3. Resistance test of ferry ro-ro ship model

4.1. *Angle of Entrance*

As shown in Figure 4, when the ship moves in water, the ship's movement will cause waves. Wave is one of the main components of the total resistance of the ship. The wave resistance arising from the movement of the ship can occur both when the water is calm and when the water is wavy. In addition to the numerical model approach, several design guidelines can be used to predict wave resistance in the early stages of designing a ro-ro ferry, including model analysis that is calculated with many shape variants [14]. The percentage of wave resistance on low-speed ships is estimated to be around 25% of the total ship resistance. Meanwhile, for ships operating at high speed, the value of the wave resistance that occurs can reach 50% [15].

Efforts to reduce the amount of wave resistance can be attempted by making the shape of the bow of the ship as good as possible. The wave resistance is significantly related to the shape of the hull which is expressed in terms of principal dimensions and general coefficients such as block coefficients. The cause of the large value of the ship's resistance is influenced by several factors, one way to reduce the ship's resistance is to change the shape of the front hull (bow), namely by modifying the size of the angle of entrance [16].

The angle of the entrance (I_E) is the angle formed by the horizontal neutral axis of the ship with the bow tangent to the bow when the ship is fully loaded, as shown in the figure.

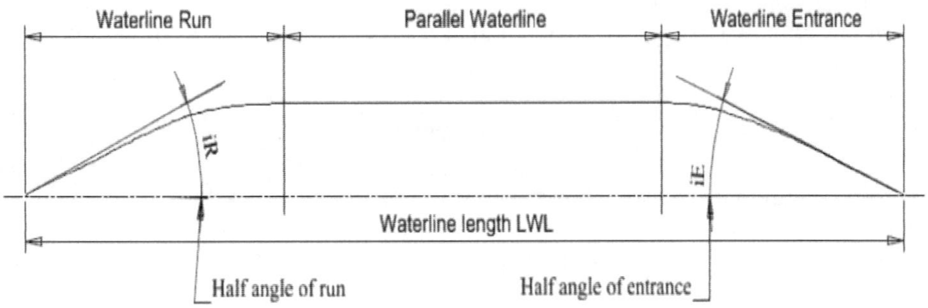

Figure 4. The angle of the entrance (I_E).

Half angle of the entrance (I_E) can be determined using the half breadth plan drawing on the ship's lines plan, where the half-angle of the entrance (I_E) is measured from the horizontal neutral axis of the ship (centerline) with the bow tangent to the bow waterline on the ship's draft which is determined. According to the resistance equation in Holtrop, the relationship between resistance and I_E can be seen from the following equation [17]:

$$C_1 = 2223105.C_4^{3,7861}(T/B)^{1,0796}(90-I_E)^{-1,3757} \tag{1}$$

$$\frac{R_w}{W} = C_1.C_2.C_3.e^{m1Fn^d + m_2\cos(\lambda - Fn^{-2})} \tag{2}$$

$$R_T = \frac{1}{2}.\rho.V^2.S_{tot}[C_F(1+k)+C_A]+\frac{R_w}{W}W \tag{3}$$

4.2. Ship resistance

In general, a ship moving in water at a certain speed will experience a resistance that is opposite to the direction of the ship's motion. The magnitude of the drag force that occurs must be able to be overcome by the ship's thrust generated from the work of the ship's propulsor. This drag force is what we usually call the ship's resistance [18]. The resistance on a ship moving on the water surface consists of 2 (two) main components, namely normal stress and shear stress (tangential stress), normal stress related to wave resistance, and viscous stress. While the shear stress is caused by the viscosity of the fluid [12].

The speed of the ship is strongly influenced by the amount of resistance, where the resistance of the ship at a speed is the fluid force acting on the ship in such a way that it opposes the movement of the ship [19]. The magnitude of the ship's resistance is influenced by the speed of the ship (Vs), the weight of the water displaced by the hull immersed in the water (displacement), and the shape of the hull (hull form).

The total resistance of the ship can be divided into several basic components, namely viscous resistance (R_V), wave resistance (R_W), and air resistance (R_A) [20] and [21]. Air resistance is often neglected, because its value is very small, only about 4% of the total resistance [21]. Because it occurs due to the effect of viscosity, viscous resistance can be further divided into frictional resistance and pressure resistance [22]. Meanwhile, the international standard from the International Towing Tank Conference (ITTC) is that the total resistance is a function of the Reynolds number and Froude number [13], which are :

$$R_{T(Rn,Fn)} = R_{V(Rn)} + R_{W(Fn)} = (1+k)_{(Fn)}.R_{F(\mathrm{Re})} + R_{W(Fn)} \tag{4}$$

4.3. Determination of ship resistance

According to ITTC 1957, the process of extrapolating the model resistance to full scale can follow the following equation:

$$R_T = \frac{1}{2}.\rho.V^2.S.C_T \tag{5}$$

The steps for calculating resistance (ship-model extrapolation) are as follows:

1. From the ship model test, the total resistance coefficient (C_T) of the model is obtained,

$$C_{T(model)} = \frac{R_{T(model)}}{\frac{1}{2}.\rho.WSA_{(model)}.V^2_{(model)}} \tag{6}$$

2. Determine the coefficient of frictional resistance (C_F), with the following equation

$$C_{F(model)} = \frac{0{,}075}{(\log_{10} R_n - 2)^2} \approx R_n = \frac{LWL_{(model)}.V_{(model)}}{\upsilon_{(Freshwater)}} \tag{7}$$

$$C_{F(ship)} = \frac{0{,}075}{(\log_{10} R_n - 2)^2} \approx R_n = \frac{LWL_{(ship)}.V_{(ship)}}{\upsilon_{(Seawater)}} \tag{8}$$

3. The form factor (1+k) was determined experimentally by applying the Prohaska Method at low speed (0.1<Fn<0.22 resistance data for the empty hull model). The Prohaska diagram as shown in Figure 5, shows the correlation between non-dimensional ship speed and total resistance coefficient in terms of frictional resistance coefficient (Prohaska, 1966) [19].

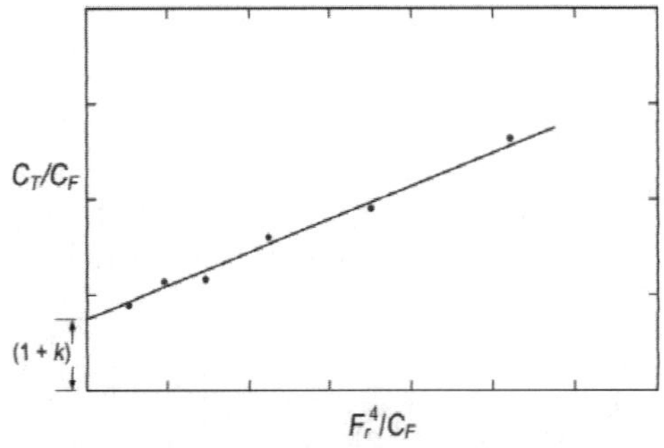

Figure 5. Prohaska diagram

4. Determining the coefficient of the total resistance of the ship, based on the "Modern MARIN" extrapolation procedure, the formula for the coefficient of frictional resistance formulated by ITTC-must be combined with the Correlation Allowance (C_A) between the model and the ship [17] with the Form Factor (1+k).

$$C_{T(ship)} = C_{T(model)} - (1+k)(C_{F(ship)} - C_{F(model)} + C_A) \tag{9}$$

Where :
C_A is the coefficient of addition of model-ship correlation resistance, is :

$$C_A = 0{,}006(L+100)^{-0{,}16} - 0{,}00205 \tag{10}$$

5. Getting the total resistance of the ship is :

$$R_{T(ship)} = \frac{1}{2}.\rho.V^2_{(ship)}.WSA_{(ship)}.C_{T(ship)} \tag{11}$$

5. Results and discussion

The change in the shape of the angle of the entrance on the ship's bow can have a significant effect on the amount of resistance experienced by the ship. To find out the shape of the ship's hull, especially those underwater, can be seen from the lines plan. The line plan depiction is based on the main size data made in 2 (two) dimensional to pay attention to all shapes of the ship's hull in 3 (three) dimensional, the drawing is divided into three parts, namely: body plan, sheer plan, and half breadth plan. Lines plan is very influential on other ship design drawings.

According to ITTC 1957, the amount of total resistance can use the equation, namely :

$$R_T = \frac{1}{2}.\rho.V^2.WSA\ .C_T \tag{12}$$

The results of the statistical calculation of total ship resistance using the Holtrop method with a change in the angle of entry from the original design (21 degrees) and after being modified to (3 degrees) and carried out with variations in speed (Vs) from 10 knots to 18 knots in draft conditions (T) 3, 30 meters, the results obtained are presented in Table 2. and a diagram of the total resistance curve at each speed can be shown in Figure 6.

The results of the total ship resistance value obtained based on the ship model resistance test in the Towing Tank can be presented in Table 3. Meanwhile, Figures 7. and 8. show the results of the total vessel resistance which are displayed in the form of a resistance curve diagram and coefficient of resistance.

5.1. The total resistance of ferry ro-ro ship from the calculation results

Table 2. Results of statistical calculation of the Holtrop method.

Speed (knots)	Total resistance (kN)		
	Original design	1st Modification	2nd Modification
10	57,0	55,9	55,5
11	71,6	69,2	68,8
12	90,7	85,9	85,5
13	114,5	106,3	106,0
14	145,3	131,8	132,5
15	192,3	169,6	170,2
16	248,5	214,1	212,8
17	293,3	250,4	248,9
18	330,0	280,8	283,0

The analysis and statistical calculation results of the Holtrop method, at speed (Vs = 18 knots) and draft conditions (T = 3.3 m) show that: the total resistance of the original design ship ($R_{T\text{-}O}$) is 330 KN, and the total resistance of the 1st modification ship ($R_{T\text{-}1st\ M}$) is 280.8 KN, the total resistance value is reduced to 17.521%. While the total resistance of the original ship ($R_{T\text{-}O}$) is 330 KN, the calculation of the total resistance of the 2nd modification design ship ($R_{T\text{-}2nd\ M}$) is 283 KN, able to reduce the total resistance by about 16.607%.

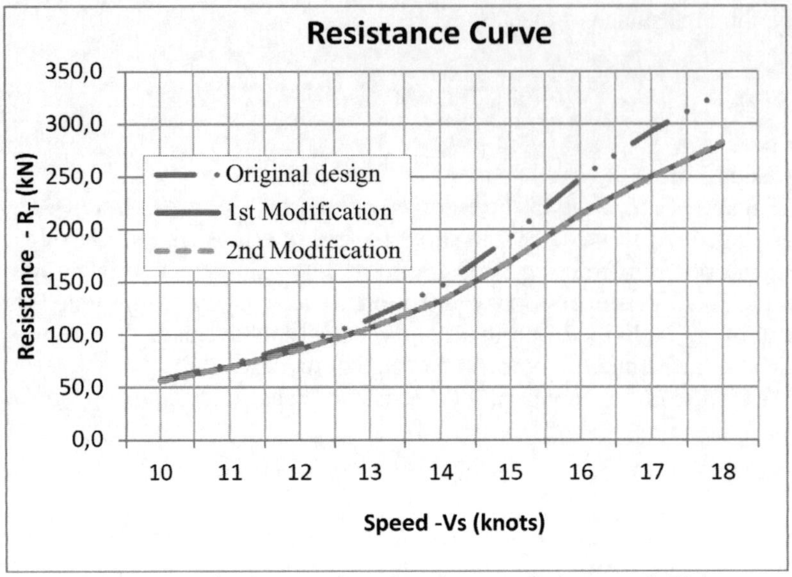

Figure 6. The Curve of total resistance (R_T) - speed (Vs).

5.2. The total resistance of ferry ro-ro ship from the model test results

Table 3. Result of resistance test in Towing Tank.

Speed (Vs) (knots)	Froude number (Fn)	Friction Resistance Coefficient (CFs)	Total Resistance Coefficient (CTs)	Total Resistance (R_T) (KN)
10	0,192	0,00170	0,00389	64,5
11	0,211	0,00168	0,00398	79,9
12	0,231	0,00166	0,00416	99,5
13	0,249	0,00164	0,00446	125
14	0,269	0,00163	0,00501	163
15	0,288	0,00161	0,00579	216
16	0,307	0,00160	0,00640	272
17	0,326	0,00159	0,00673	323
18	0,345	0,00157	0,00690	371

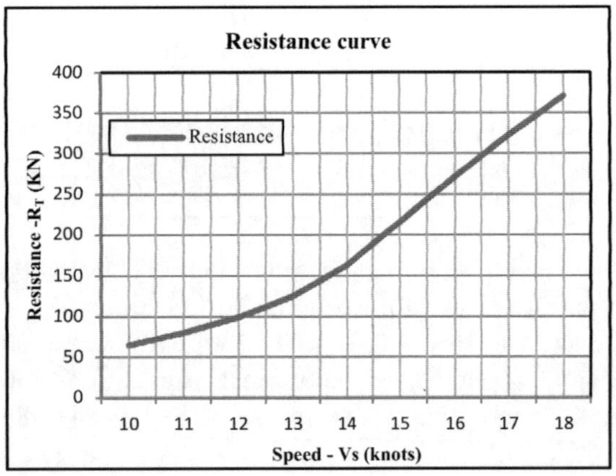

Figure 7. Diagram of resistance curve, at the draft (T) = 3.30 m.

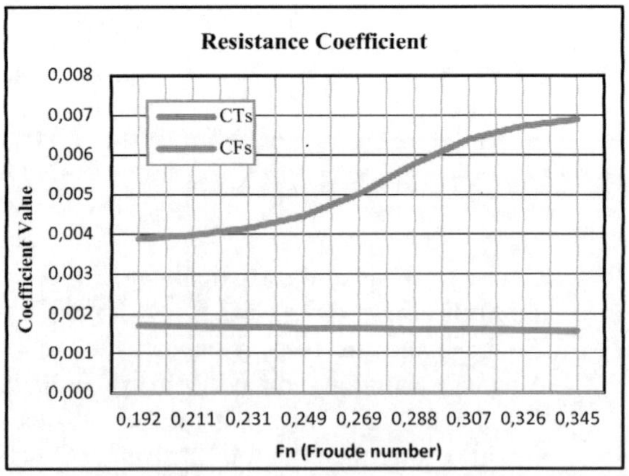

Figure 8. Diagram of resistance coefficient, at the draft (T) = 3.30 m.

Figure 9. below is a comparison of the results of the total resistance at a draft condition of 3.30 m obtained through statistical calculations of the Holtrop method and testing the total resistance of the ship model in the Towing Tank. The results show good agreement in the low speeds and the difference between them increases as the ship speed increases.

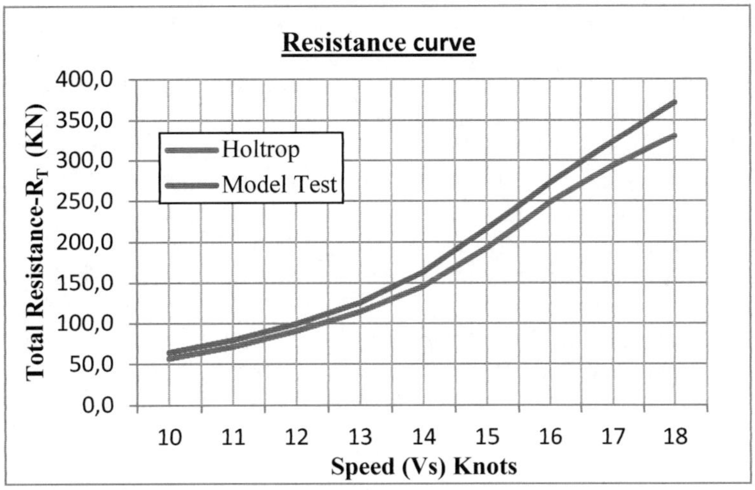

Figure 9. Diagram of resistance curve (Holtrop - Model test)

6. Conclusion

Some conclusions obtained from the analysis and calculation results using the Holtrop method statistical calculations are as follows:

The effect of changing the magnitude of the angle of the entrance on the bow of the ship from the initial design (21 degrees) for a 1500 GT ferry ro-ro ship to a modified design (3 degrees) was able to reduce the total value of the ship's resistance so that it had a positive effect on the speed of the ship. The decrease in the total resistance value of the ship occurs at each speed variation.

The results of the analysis and comparison between the statistical calculation of the Holtrop method and the model test on the total resistance of the 1,500 GT ferry ro-ro ship gave a difference of 13.16% at a speed of 10 knots and 9.17% at a speed of 13 knots, while at a speed of 18 knots is 12.42%.

Changes in the size of the angle of the entrance on the bow of the ship are closely related to the total resistance value of the ship, the increase in ship speed, and the need for effective power (EHP).

References

[1] Adam, L., 2016. Kebijakan Konektivitas Maritim di Indonesia. Jurnal Politica Dinamika Masalah Politik Dalam Negeri dan Hubungan Internasional, 6(1).
[2] Bazari, Z., 2016. IMO Train the Trainer Course on Energy Efficient Ship Operation. Ship Energy Efficiency Regulations and Related Guidelines, Module, 2, p.12.
[3] Utomo, B., 2010. Pengaruh ukuran utama kapal terhadap displacement kapal. Teknik, 31(1), pp.84-89.
[4] Rosmani, et al., "Pengaruh Bentuk Lambung Kapal terhadap Tahanan Kapal," Prosiding Hasil Penelitian Fakultas Teknik. 7, 2013.
[5] Holtrop J and Mennen G G J 1982, An Approximate Power Prediction Method, International Shipbuilding Progress, Vol 29.
[6] Holtrop J 1984, A Statistical Re-analysis of Resistance and Propulsion Data, International Shipbuilding Progress, Vol 31.

[7] Satoto, S.W., Perancangan Lambung Kapal Tanpa Awak Sebagai Alat Bantu Survei Di Kepulauan Riau. Kapal: Jurnal Ilmu Pengetahuan dan Teknologi Kelautan, 16(1), pp.9-15.

[8] Schneekluth, H. and Bertram, V., 1998. Ship design for efficiency and economy (Vol. 218). Oxford: Butterworth-Heinemann.

[9] Pak, K.R., Song, G.S., Kim, H.J., Son, H.J. and Park, H.G., 2020. Hull form design for resistance minimization of small-scale LNG bunkering vessels using numerical simulation. International Journal of Naval Architecture and Ocean Engineering, 12, pp.856-867.

[10] Hadi, E.S. and Zakki, A.F., 2017. Analisa Pengaruh Sudut Masuk Kapal Perintis 750 DWT Terhadap Resistance Kapal Dengan Mengunakan Metode Computational Fluid Dynamic (CFD). Jurnal Teknik Perkapalan, 5(2).

[11] Hadi, E.S., Manik, P. and Iqbal, M., 2018. Influence of hull entrance angle "Perintis 750 DWT", toward ship resistance: the case study for design development "Perintis 750 DWT.". In MATEC Web of Conferences (Vol. 159, p. 01057). EDP Sciences.

[12] Couser, P.R., Molland, A.F., Armstrong, N.A. and Utama, I.K.A.P., 1997. Calm water powering predictions for high-speed catamarans. Published in: FAST'97 proceedings, Sydney, Australia.

[13] ITTC (2002) Recommended Procedures and Guidelines, Testing and Extrapolation Methods Resistance Towing Tank Test, ITTC 7.5-02-02-02, 2002.

[14] Naukowe, Zeszyty, "Design guidelines for predicting wave resistance of ro-ro ferries at the initial designing stage," Scientific Journals Maritime University of Akademia Morska w Szczecinie, 22(94): 5–9, 2010.

[15] Muvariz, M.F., Wibowo, S., Abdurrahman, N., Saputra, H., Satoto, S.W., Yuniarsih, N. and Pamungkas, N., 2019. Studi Perhitungan Tahanan Kapal Kontainer 8500 DWT. Jurnal Teknologi Dan Riset Terapan (JATRA), 1(1), pp.14-19.

[16] Sijabat, Z. B., Hadi, E. S., & Rindo, G. (2018). Pengaruh Sudut Masuk Pada Kapal Perintis 750 Dwt Terhadap Resistance Kapal Dengan Penambahan Anti-Slamming Bulbous Bow Tipe Delta (Δ–Type). Jurnal Teknik Perkapalan, 6(1).

[17] Lewis, E.V., 1988. Principles of naval architecture second revision. Jersey: Sname, 2.

[18] Adji, S.W., 2009. Water Jet Propulsion System. Propulsion System, pp.1-21.

[19] Harvard, S.A., 1983. Resistance and Propulsion of Ships, John Willey & Sons. New York.

[20] Molland, A.F., Turnock, S.R. and Hudson, D.A., Ship resistance and propulsion: practical estimation of ship propulsive power, 2011.

[21] Carlton, J.S, 2010. Marine propellers and propulsion. Third edition Science and Technology 2010/11.

[22] Utama, I.K.A.P. and Molland, M.F., 2001. Experimental and numerical investigations into catamaran viscous resistance.

IOP Conf. Series: Earth and Environmental Science 1081 (2022) 012019 doi:10.1088/1755-1315/1081/1/012019

Planning an on-grid solar energy generation system for the Dwimatama pier at Tanjung Emas port

M R Arsalan, S Sarwito and E S Koenhardono

Department of Marine Engineering, Institut Teknologi Sepuluh Nopember, Jl. Teknik Kimia, 60111, Kota Surabaya, Indonesia

raihanarsalann@gmail.com

Abstract. Global warming is the gradual increase in the temperature of the earth's atmosphere caused by increasing levels of pollutant gases such as carbon dioxide. Natural characteristics are expected to change as the earth's surface temperature rises. By 2025, Indonesia's renewable energy use is expected to reach 23%. (Ministry of Energy and Mineral Resources, 2021). Existing critical sectors such as transportation and freight forwarding are being examined. For instance, vessels can reduce their carbon footprint by connecting to shore-connection facilities powering renewable energy sources in ports. The case study is the Dwimatama pier in Semarang's Tanjung Eams port. The city of Semarang receives approximately six hours of sunlight per day. The solar power plant could serve as a source of electrical energy in this region. The results of this study indicate that to generate electrical energy from 1 MWp of solar radiation, a panel area of approximately 1 hectare is required for a solar panel with a 20% efficiency.

1. Introduction

Due to the Indonesian government's ratification of the Paris Agreement, the Indonesian government must reduce greenhouse gas emissions and actively address climate change. The Paris Agreement limits global warming to well below 2 degrees Celsius and aims to keep it to 1.5 degrees Celsius. As a result, fossil fuels, which emit greenhouse gases, must be limited to clean and renewable energy sources. Indonesia has a plethora of clean energy sources that are extremely promising. Everything in Indonesia is based on water, geothermal energy, and sunlight as a source of electricity. Photovoltaic cell utilization in Indonesia is extremely promising, particularly in coastal areas. The promise of using PV is supported by the Ministry of Energy and Mineral Resources' renewable energy potential map, which indicates that Indonesia's coastal regions have the most significant potential compared to other areas. Photovoltaic cells can power port facilities in Indonesia, such as shore connections. Combining grid and photovoltaic energy is expected to reduce the electricity generated by conventional fossil fuel power plants. This research introduces novel features to the solar power generation system in the form of a solar panel system without a battery. This system has a much lower initial investment cost and is capable of optimally providing electricity needs under any conditions. With the publication of this paper through the ITS Surabaya-organized 2022 Matic Conference, it is expected that increased attention to renewable energy by utilizing it in the marine sector in Indonesia. Furthermore, this research can contribute to the development of the gren port concept in Indonesia, specifically in Tanjung Emas Semarang.

2. Method

This design planning requires data such as existing shore connection specifications, sunlight radiation in the region, data on supplied loads, location data that is considered appropriate for solar power plants, and some other supporting data.

2.1. Planned Location

In Indonesia, shore connection choices available are extremely limited, with the exception of Java, Kalimantan, and Bali. As a result, the authors chose the Tanjung Emas Semarang area of Java as the subject of their research. The author chooses Tanjung Emas port as a strategic location, but the land use around the port is quite poor. The appropriate solar energy generation area is depicted in Figure 1. Ten hectares of land are available. This area is two kilometers from the pier. This area is located in Semarang's Cipta Industrial Area. The location was chosen to maximize the area's land-use potential. Land in this area has a lower economic value due to the frequent occurrence of tidal floods. Numerous environmental issues have arisen as a result of the tidal flood in the North Semarang area (Agus Nur Shidik. 2019). Based on this information, it appears that land in this area would be better suited for a solar power plant that would supply electricity to Tanjung Emas Port. Due to its use as a solar power plant, it is relatively unaffected by tidal flooding. Tanjung Emas Port has Dwimatama Pier, which is known to operate a shore connection facility for ships unloading cargo at the dock. To carry out this planning, it is necessary to have information about the specifications of the land that will be used for the power plant. Solar arrays, inverters, and all other critical and supporting equipment for solar power plants can be laid out using this data.

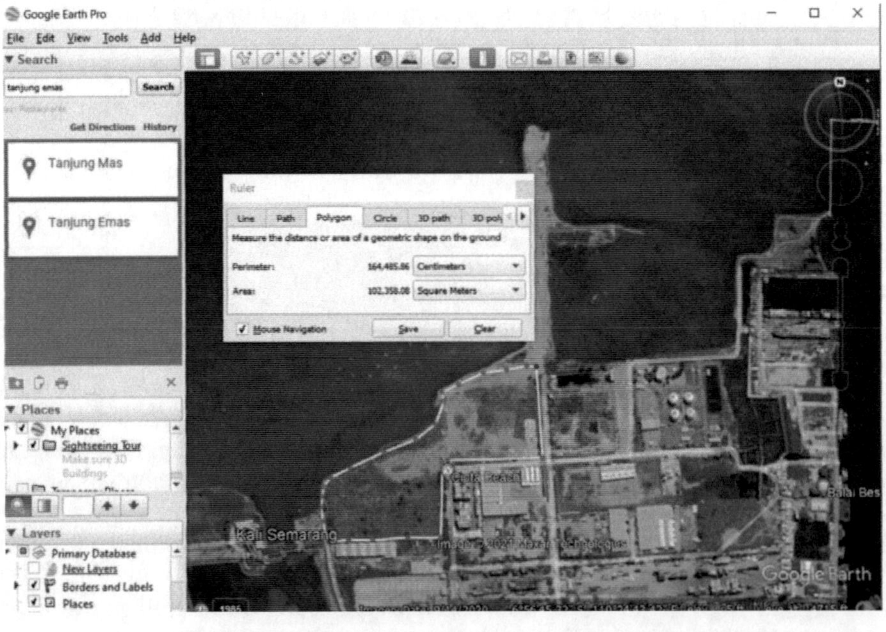

Figure 1. Planning Area Solar Power Plant.

2.2. Shore connection's specification

The shore connection specifications data is critical for designing a solar power plant capable of powering the facilities. Data Table 1 below pertains to the shore connection at Dermaga Dwimatama. Data on vessels that use the Dermaga Dwimatama shore connection facilities, in addition to shore connection data, are required to analyze the performance of solar power plants designed to meet the needs of vessels that lean against the Dermaga Dwimatama. According to the author's research, the Dwimatama jetty is a TUKS dock operated by Pupuk Indonesia, meaning that the vessels that dock here have similar specifications

Table 1. Specifications of shore connection Dwimatama

Voltage	400 V
Max Capacity	1000 KW
Frequency	50 Hz
Phase	3 Phase

2.3. Solar radiation data

The design of a solar power plant is inextricably linked to information about local solar radiation. The circumstance in which solar panels are installed significantly impacts their ability to capture solar radiation. The panel installation includes the direction and angle of inclination of installed solar panels. A fixed tilt angle of 10 degrees is recommended for long-term installation (Pangestuningtyas D.L, 2013). Table 2 displays the monthly slope angles of the solar panels. In addition to the slope angle, the azimuth angle is an essential angle for solar panel installation. The azimuth angle is the angle at which the direction of the solar panel is applied. The best azimuth angle for solar panels in Semarang is shown in Table 3. It is common knowledge that solar panels facing north are the most effective. The azimuth angle is consistent with Duffle and Beckmen's theory that the optimal azimuth angle for the southern hemisphere is 180°. The output of solar power plants is heavily influenced by the amount of sunlight that strikes their panels. Because the sun's intensity varies by hour, data on the sun's intensity in the area is required to calculate the output of solar panels every hour. Table 4 depicts the intensity of the hourly sunlight in the area. The data was gained from the journal " Analisis Potensi Energi Sinar Matahari Dan Energi Angin di Pusat Kota Semarang". Solar radiation data collected every hour in Semarang can support the planning of solar power plants. This relates to the design specifications for the equipment solar power plants used.

Table 2. Data of slop angle

Month	Angle	Sun Radiance ($MJ/m^2/day$)
January	1	11.017
February	1	12.933
March	2	12.773
April	16	13.477
May	27	13.676
June	34	15.639
July	33	17.855
August	23	16.378
September	8	15.511
October	1	16.274
November	1	14.903
December	1	10.846

Table 3. Data of azimuth angle

Month	90°	120°	150°	180°	210°	240°	270°
January	11	10.8	10.8	10.7	10.8	10.9	11
February	12.9	12.8	12.7	12.7	12.7	12.8	12.9
March	12.7	12.7	12.7	12.7	12.7	12.7	12.7
April	13	13.2	13.3	13.4	13.4	13.2	13.1
May	12.6	12.9	13.2	13.3	13.2	13	12.7
June	13.6	14.1	14.5	14.7	14.6	14.2	13.8
July	15.6	16.1	16.6	16.8	16.6	16.3	15.7
August	15.4	15.7	16	16.1	16	15.8	15.5
September	15.3	15.4	15.5	15.5	15.5	15.4	15.3
October	16.2	16.1	16	16	16	16.1	16.2
November	14.8	14.6	14.5	14.4	14.5	14.6	14.8
December	10.8	10.7	10.6	10.5	10.6	10.7	10.8

Table 4. Data of hourly radiation

Hour	lux	W/m2
7 - 8	5946	49.55
8 - 9	8050	67.0833333
9 - 10	10396	86.6333333
10 - 11	11168	93.0666667
11 - 12	26716	222.633333
12 - 13	54578	454.816667
13 - 14	56455	470.458333
14 - 15	54877	457.308333
15 - 16	18770	156.416667
16 - 17	7450	62.0833333

2.4. Vessel specification

Pupuk Indonesia operates the Dermaga Dwimatama special pier. As a result, the vessel leaning on the pier has been determined and has nearly identical specifications. Using a pier dedicated to unloading fertilizer cargo will make it easier to plan solar power plants to power the shore connection facilities found on this pier. Table 5 contains data on the power requirements of vessels moored at the Dermaga Dwimatama and the electricity bill data to the user's vessel to use from shore connection facilities from April to October 2021. Ships unloading at Dwimatama Pier require an average of 150 kW of electrical power. Pusri Indonesia is the ship that consumes the most electricity. Meanwhile, the Julianto consumes the least electricity compared to the others.

Table 5. Data of vessel

Vessel Name	Average Load (kW)
Ibrahim Zahier	156.9
Pusri Indonesia 1	174
Soemantri Brojonegoro	121.3
Abusamah	147.3
Julianto Md	135.5

3. Result and Discussion

3.1. Solar panel output per hour

To generate data on the hourly output of solar panels, we must first have information on the solar panels that will be installed. Table 6 contains specifications for solar panels intended for use in solar energy plants. After that, the author enters Table 6 into simulation software to obtain an accurate photovoltaic array calculation. In addition, the author enters the Semarang sun intensity table into the simulation software every hour to determine the solar array's output power. It is known that 5124 solar panels with the specifications listed in table 6 are needed to create a solar array MPP with a maximum solar radiation level of 470W/m2, which is close to the shore connection specifications at the Dermaga Dwimatama. Table 7 shows the various solar panel configurations used in solar power plants. Following establishing the proper format, the author enters the amount of solar radiation received per hour into the simulation software. Figure 2 depicts the hourly output of the solar array in comparison to the load on each vessel. With the graphic image, figure 2, it is clear that the solar array's electrical power output can fully meet the needs of several vessels beginning at 9 a.m. and continuing until 4 p.m., without requiring additional power from the PLN grid, thereby exacerbating the power crisis. For a short period, the output of these solar panels will exceed the total load required. According to the graph in Figure 2, electricity demand is constant or stable at all times. The fact that power requirements are relatively stable enables us to design solar power plants more easily. When solar power plants cannot supply vessel and shore connection facilities that fully utilize PLN electricity, excess energy generated by solar panels will be sold to PLN to generate additional revenue or reduce electricity bills.

Table 6. Specification of solar panel

STC	
VMP	31.5 V
IMP	12.98 A
VOC	37.45 V
ISC	13.88 A
Pmax	410 W
Module Eff	20 %
Operating Temp	-40°C + 85°C
NMOT	
Maximum Power NMOT	309.6 W
Optimum Operating Voltage	29.2 V
Optimum Operating Current	10.6 A
VOC	35.2 V
ISC	11.16 A

Table 7. Solar array configuration

Total Panel	5124 Panel
Series	21
Parallel	244

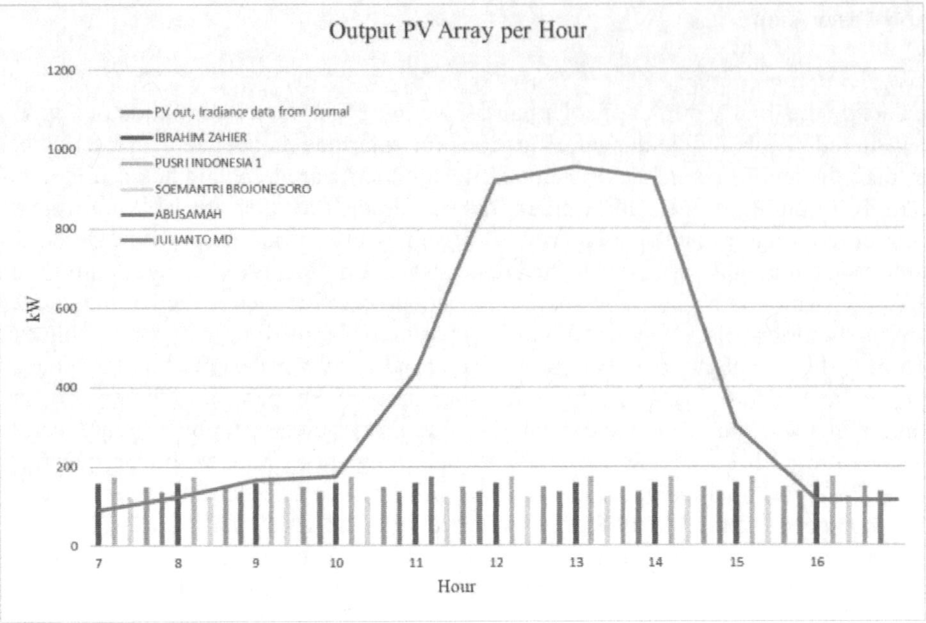

Figure 2. Hourly output solar array.

3.2. Solar power plant design

A wiring diagram and a single line diagram are required to design an electrical system and determine whether the electrical device circuits have been adequately prepared and understand the system globally or thoroughly to perform power flow analysis. The wiring diagram for the proposed solar power plant is shown in Figure 3. The solar power plant depicted in Figure 4 is a one-line diagram. Using the wiring diagram in Figure 3, the author creates a layout plan for a solar energy plant in an undeveloped area of the copyrighted beach area. Figure 5 depicts the proposed layout of a power plant in the copyrighted industrial area.

According to the plan, solar power plants require 3.8 hectares of the total available land area of 10 hectares. The land selected is in the easternmost section because of its proximity to the power plant. The power plant layout indicates that each array is 5 meters wide. This distance was chosen to account for the ease with which solar panels can be accessed for scheduled solar panel maintenance. As shown in Figure 5, the green line represents the cable pull from the solar energy plant to the shore connection facility. It is well known that the design of the cable through the seabed results in a cable transmission system that combines an air transmission system and an underwater transmission system in the design of this power plant. Based on the standards, this transmission system necessitates using a variety of cables.

As a result, marine-grade cables are required for air transmission systems and subsea cables for sea-borne transmission. Figure 6 depicts the proposed support structure for a solar panel. It is known that the image depicts a design drawing for a support structure for a solar array with a slope of 10 degrees and that the solar panel is mounted using a clamp. The purpose of using this clam is to facilitate maintenance on individual solar panels and to adhere to the installation procedures outlined in the product guide. When installing near the coast, various factors must be taken into account. The manufacturer depicted in Figures 7 and 8 provides instructions for installing solar panels near the beach. Insulation gaskets are known to be required to be made of mica lamination or other silicone or fluoride-based insulating materials. Figure 7 illustrates using gaskets to insulate the installation joint points, where mounting and grounding hardware is attached to the frame. Manufacturers provide recommendations for insulation gaskets using mica lamination materials, or other silicone, or fluoride-made insulating materials. The recommendation for using insulating gaskets is to prevent corrosion at installation joint points, where mounting and grounding hardware is attached to the frame. Figure 8 describes the manufacturer providing guidelines on the installation of installing grounding cables on solar panel installations near the coast. The general recommendation is to varnish the grounded part with a fluorocarbon varnish. Apply a fluorocarbon varnish to the grounding block to create an anti-corrosion coating. According to the manufacturer, coating the grounding part with this varnish can prevent corrosion.

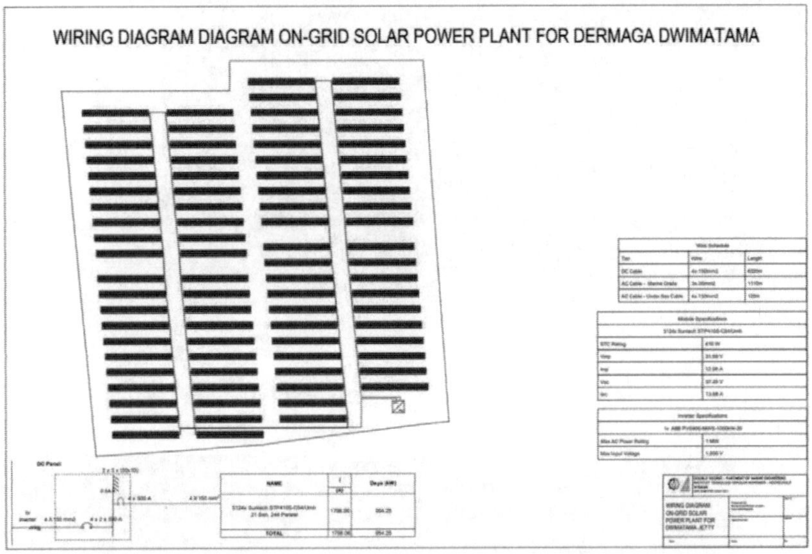

Figure 3. Proposed solar power plant's wiring diagram.

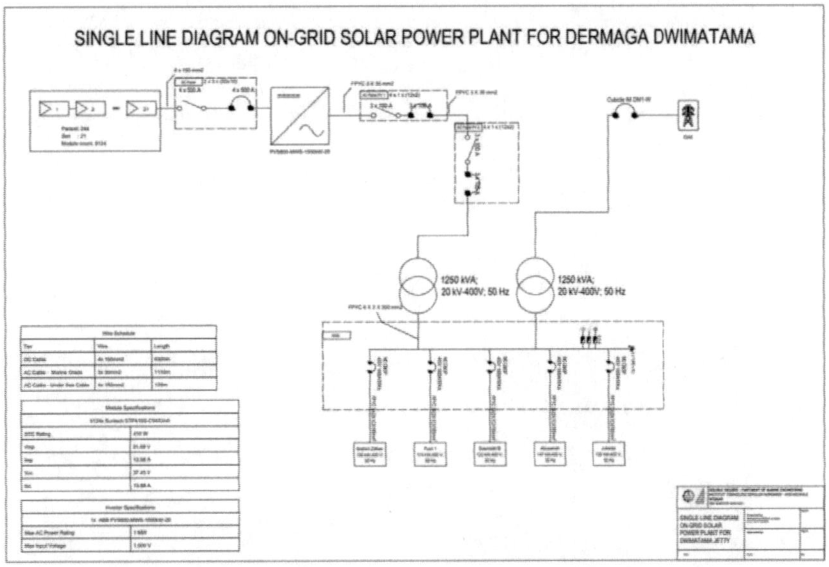

Figure 4. Proposed solar power plant's one line diagram.

IOP Conf. Series: Earth and Environmental Science 1081 (2022) 012019 doi:10.1088/1755-1315/1081/1/012019

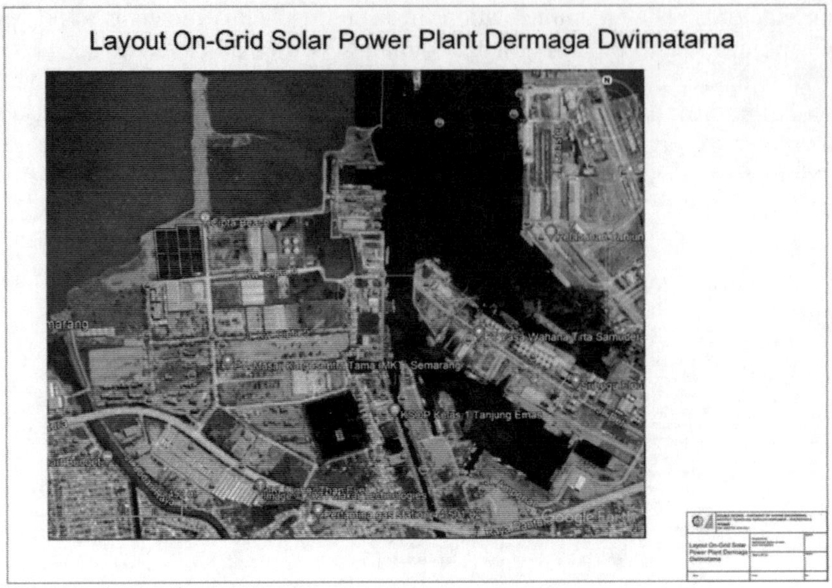

Figure 5. Proposed solar power plant's layout.

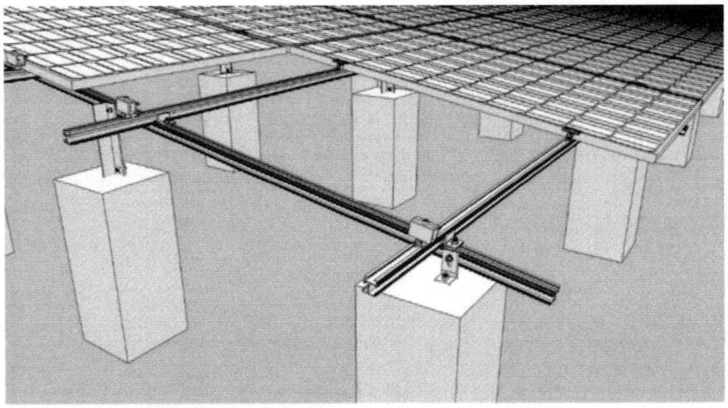

Figure 6. Proposed solar array's support structure.

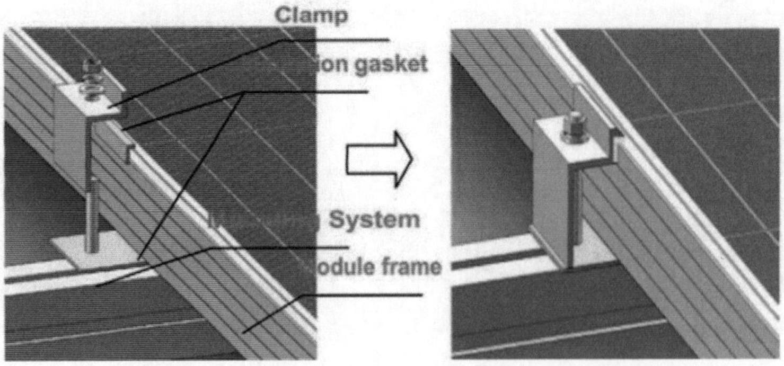

Figure 7. Guide lines installation near coast.

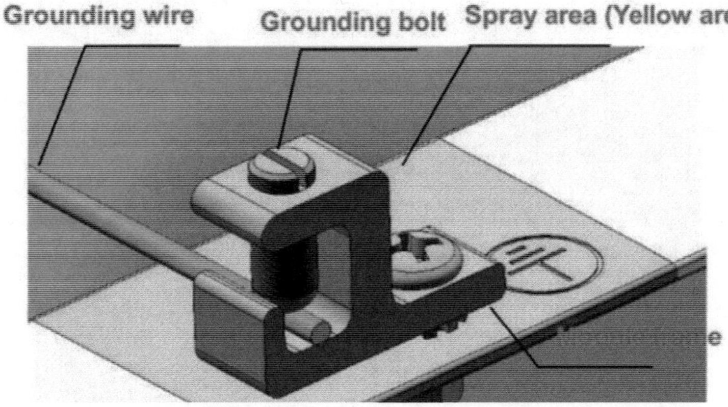

Figure 8. Guide lines grounding installation near coast.

3.3. Power flow analysis

When designing electronic systems, including power plants, a power flow analysis is required to determine the losses that occur in each component. These losses must be taken into account to comply with existing regulations and ensure workplace safety. Heat will accumulate in some areas of the electrical installation due to the excessive losses, resulting in material and non-material losses. A single-line diagram in a simulation application is depicted in Figure 9. How loads move through each bus in three different scenarios is shown in Tables 8, 9, and 10. The author illustrates the point by comparing three plans: 100% grid-supplied electricity, 50% PV and 50% grid-supplied electricity, and 100% grid-supplied electricity. The development of this scenario is based on its eventual application in the real world, where there is a scheme that uses solar panels and the grid to generate electrical energy. It is well known that the most significant power loss occurs in buses that are powered by transformers (Bus Load and Bus3). Power loss may occur due to transformer losses; the transformer can convert medium voltage to low voltage.

Furthermore, because the cable pull from the solar power plant to the shore connection facilities is quite long, the decrease in power for buses powered by solar power plants resulted in a reduction of medium voltage buses (Bus AC PV 2). Voltage drop analysis is performed in addition to load flow analysis. Voltage drop analysis is used to determine whether the power transmission system of a solar power plant meets applicable standards. Voltage drop data can be found in Tables 11, 12, and 13. A voltage drop analysis was also performed using three different scenarios. The first scenario is the grid supplies 100% of the energy, the second is the grid supplies 50% of the energy and PV supplies 50%, and the third is PV supplies 100% of the energy.

This analysis shows that most of the voltage drop occurs in a low voltage distribution system while remaining above the feasibility standard. Voltage drop occurs when carrying the same amount of power at a low voltage resulting in higher losses due to the higher current flowing than when carrying the same amount of power at a medium or high voltage. Furthermore, the medium voltage transmission system originating from solar power plants experiences a noticeable voltage drop while remaining well above the feasibility standard. Voltage drops occur in medium voltage transmission because the cable pulls are much further away from the solar power plant than the cable pulls used by PLN.

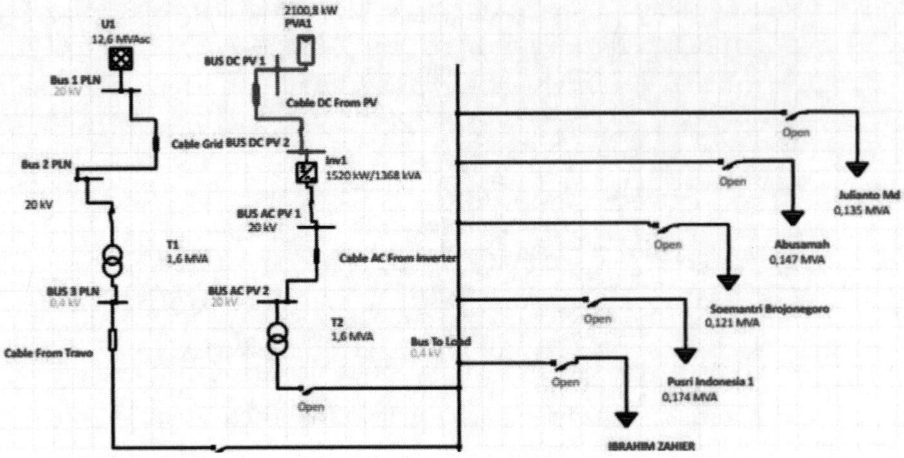

Figure 9. Single line diagram in a simulation application.

Table 8. Data of grid's load flow

Vessel	From	To	P (kW)	Q (kVAr)	S (kVa)	I (A)
Ibrahim	Grid	Bus 1	156.5	1.17	156.5	4.5
Zahier	Bus 1	Bus 2	156.5	1.17	156.5	4.5
	Bus 2	Bus 3	156.4	0.225	156.4	226
	Bus 3	Bus Load	156.2	0.084	156.2	226
Pusri	Grid	Bus 1	173.4	1.44	173.4	5
Indonesia	Bus 1	Bus 2	173.4	1.43	173.4	5
	Bus 2	Bus 3	173.2	0.276	173.2	250.3
	Bus 3	Bus Load	173	0.103	173	250.3
Sumantri B	Grid	Bus 1	120.7	0.696	120.7	3.5
	Bus 1	Bus 2	120.7	0.695	120.7	3.5
	Bus 2	Bus 3	120.6	0.134	120.6	174.3
	Bus 3	Bus Load	120.5	0.05	120.5	174.3
Abusamah	Grid	Bus 1	146.6	1.03	146.6	4.2
	Bus 1	Bus 2	146.6	1.03	146.4	4.2
	Bus 2	Bus 3	146.5	0.197	146.5	211.6
	Bus 3	Bus Load	146.3	0.074	146.3	211.6
Julianto MD	Grid	Bus 1	134.7	0.866	134.7	3.9
	Bus 1	Bus 2	134.7	0.865	134.7	3.9
	Bus 2	Bus 3	134.5	0.166	134.5	194.4
	Bus 3	Bus Load	134.4	0.062	134.4	194.4

Table 9. Data of Value Grid's + PV's Load Flow

Vessel	From	To	P (Kw)	Q (kVAr)	S (kVa)	I (A)
Ibrahim	Grid	Bus 1	74.3	-0.435	74.3	2.1
Zahier	Bus 1	Bus 2	74.3	-0.436	74.3	2.1
	Bus 2	Bus 3	74.3	-0.649	74.3	107.3
	Bus 3	Bus Load	156.6	0.084	156.6	226.2
	Inverter	Bus AC PV1	82.4	-29	87.4	2.5
	Bus AC PV1	BUS AC PV 2	82.4	1.03	82.4	2.4
	BUS AC PV2	Bus Load	82.3	0.76	82.3	189.9
Pusri	Grid	Bus 1	82.3	-0.407	82.3	2.4
Indoneisa 1	Bus 1	Bus 2	82.3	-0.407	82.3	2.4
	Bus 2	Bus 3	82.2	-0.668	82.2	118.8
	Bus 3	Bus Load	173.5	0.104	173.5	250.7
	Inverter	Bus AC PV1	91.4	-28.9	95.9	2.8
	Bus AC PV1	BUS AC PV 2	91.3	1.13	91.4	2.6
	BUS AC PV2	Bus Load	91.3	0.81	91.3	131.9
Sumantri B	Grid	Bus 1	57.5	-0.474	57.5	1.7
	Bus 1	Bus 2	57.5	-0.474	57.5	1.7
	Bus 2	Bus 3	57.5	-0.602	57.5	83
	Bus 3	Bus Load	120.7	0.05	120.7	174.4
	Inverter	Bus AC PV1	63.3	-29.2	69.8	2
	Bus AC PV1	BUS AC PV 2	63.3	0.825	63.3	1.8
	BUS AC PV2	Bus Load	63.3	0.671	63.3	91.4
Abusamah	Grid	Bus 1	69.7	-0.449	69.7	2
	Bus 1	Bus 2	69.7	-0.449	69.7	2
	Bus 2	Bus 3	69.6	-0.636	69.6	100.6
	Bus 3	Bus Load	146.6	0.074	146.6	211.8
	Inverter	Bus AC PV1	77.1	-29.1	82.4	2.4
	Bus AC PV1	BUS AC PV 2	77.1	0.967	77.1	2.2
	BUS AC PV2	Bus Load	77	0.738	77	111.3
Julianto MD	Grid	Bus 1	64.1	-0.463	64.1	1.8
	Bus 1	Bus 2	64.1	-0.463	64.1	1.8
	Bus 2	Bus 3	64	-0.621	64	92.5
	Bus 3	Bus Load	134.7	0.062	134.7	194.6
	Inverter	Bus AC PV1	70.7	-29.2	76.5	2.2
	Bus AC PV1	BUS AC PV 2	70.7	-29.2	76.5	2.2
	BUS AC PV2	Bus Load	70.7	0.9	70.7	2

IOP Conf. Series: Earth and Environmental Science 1081 (2022) 012019 doi:10.1088/1755-1315/1081/1/012019

Table 10. Data of PV load flow

Vessel	From	To	P (Kw)	Q (kVAr)	S (kVa)	I (A)
Ibrahim	Inverter	Bus AC PV1	156.6	-29	159.3	4.6
Zahier	Bus AC PV1	BUS AC PV 2	156.5	1.03	156.5	4.5
	BUS AC PV2	Bus Load	156.3	0.084	156.3	226
Pusri	Inverter	Bus AC PV1	173.5	-28.8	175.9	5.1
Indonesia	Bus AC PV1	BUS AC PV 2	173.3	1.26	173.3	5
	BUS AC PV2	Bus Load	173.1	0.104	173.1	250.4
Sumantri B	Inverter	Bus AC PV1	120.8	-29.4	124.3	3.6
	Bus AC PV1	BUS AC PV 2	120.7	0.612	120.7	3.5
	BUS AC PV2	Bus Load	120.6	0.05	120.6	174.3
Abusamah	Inverter	Bus AC PV1	146.4	-29.1	149.5	4.3
	Bus AC PV1	BUS AC PV 2	146.5	0.903	146.5	4.2
	BUS AC PV2	Bus Load	146.4	0.074	146.4	211.7
Julianto	Inverter	Bus AC PV1	134.7	-29.3	137.8	4
MD	Bus AC PV1	BUS AC PV 2	134.6	0.762	134.6	3.9
	BUS AC PV2	Bus Load	134.5	0.064	134.5	194.4

Table 11. Data of grid voltage drop

Vessel	Bus	kV	Vd	kV final	% V final
Ibrahim	BUS 1 PLN	20	0	20	100.00%
Zahier	BUS 2 PLN	20.0	0	20	100.00%
	BUS 3 PLN	0.4	0.8	0.3992	99.80%
	BUS Load	0.4	0.88	0.3991	99.78%
Pusri Ind	BUS 1 PLN	20.0	0	20	100.00%
	BUS 2 PLN	20.0	0	20	100.00%
	BUS 3 PLN	0.4	0.48	0.3995	99.88%
	BUS Load	0.4	1	0.3990	99.75%
Sumantri B	BUS 1 PLN	20.0	0	20	100%
	BUS 2 PLN	20.0	0	20	100%
	BUS 3 PLN	0.4	0.32	0.3997	99.92%
	BUS Load	0.4	0.68	0.3993	99.83%
Abusamah	BUS 1 PLN	20.0	0	20	100%
	BUS 2 PLN	20.0	0	20	100%
	BUS 3 PLN	0.4	0.4	0.3996	99.90%
	BUS Load	0.4	0.84	0.3992	99.79%
Julianto MD	BUS 1 PLN	20.0	0	20	100%
	BUS 2 PLN	20.0	0	20	100%
	BUS 3 PLN	0.4	0.36	0.3996	99.91%
	BUS Load	0.4	0.76	0.3992	99.81%

IOP Conf. Series: Earth and Environmental Science 1081 (2022) 012019 doi:10.1088/1755-1315/1081/1/012019

Table 12. Data of grid + PV voltage drop

Vessel	Bus	kV	Vd	kV final	% V final
Ibrahim Zahier	BUS 1 PLN	20	0	20	100%
	BUS 2 PLN	20	0	20	100%
	BUS 3 PLN	0.4	0.2	0.3998	99.95%
	BUS Load	0.4	0.4	0.3996	99.90%
	BUS AC PV 1	20	0	20	100%
	BUS AC PV 2	20	8	19.992	99.96%
Pusri Ind	BUS 1 PLN	20	0	20	100%
	BUS 2 PLN	20	0	20	100%
	BUS 3 PLN	0.4	0.2	0.3998	99.95%
	BUS Load	0.4	0.44	0.39956	99.89%
	BUS AC PV 1	20	0	20	100%
	BUS AC PV 2	20	10	19.99	99.95%
Sumantri B	BUS 1 PLN	20	0	20	100%
	BUS 2 PLN	2	1800	0.2	10%
	BUS 3 PLN	0.4	0.16	0.39984	99.96%
	BUS Load	0.4	0.32	0.39968	99.92%
	BUS AC PV 1	20	0	20	100%
	BUS AC PV 2	20	6	19.994	99.97%
Abusamah	BUS 1 PLN	20	0	20	100%
	BUS 2 PLN	20	0	20	100%
	BUS 3 PLN	0.4	0.16	0.39984	99.96%
	BUS Load	0.4	0.36	0.39964	99.91%
	BUS AC PV 1	20	0	20	100%
	BUS AC PV 2	20	8	19.992	99.96%
Julianto MD	BUS 1 PLN	20	0	20	100%
	BUS 2 PLN	20	0	20	100%
	BUS 3 PLN	0.4	0.16	0.39984	99.96%
	BUS Load	0.4	0.36	0.39964	99.91%
	BUS AC PV 1	20	0	20	100%
	BUS AC PV 2	20	8	19.992	100%

Table 13. Data of PV voltage drop

Vessel	Bus	Kv	Vd	V final	% V final
Ibrahim	BUS AC PV 1	20	0	20	100%
Zahier	BUS AC PV 2	0.4	0.36	0.39964	99.91%
	BUS Load	0.4	0.76	0.39924	99.81%
Pusri Ind	BUS AC PV 1	20	0	20	100%
	BUS AC PV 2	0.4	0.4	0.3996	99.9%
	BUS Load	0.4	0.84	0.39916	99.79%
Sumantri B	BUS AC PV 1	20	0	20	100%
	BUS AC PV 2	0.4	0.28	0.39972	99.93%
	BUS Load	0.4	0.6	0.3994	99.85%
Abusamah	BUS AC PV 1	20	0	20	100%
	BUS AC PV 2	0.4	0.32	0.39968	99.92%
	BUS Load	0.4	0.72	0.39928	99.82%
Julianto MD	BUS AC PV 1	20	0	20	100%
	BUS AC PV 2	0.4	0.28	0.39972	99.93%
	BUS Load	0.4	0.64	0.39936	99.84%

3.4. Economic analysis

Designing this type of solar energy system requires economic analysis. Due to the on-grid system model, the solar power plant's excess electricity generated cannot be stored. More power is sold to PLN to reduce electricity bills when solar power plants are not producing optimally. Table 14 compares the energy produced by solar panels with the energy required by each vessel. The table shows that solar power plants produce more energy per day than vessels need, but because energy is not stored, solar power plants can only participate at 37-40%. The participation of solar energy to fully supply the vessel's needs occurs from 9 a.m. to 3 p.m. In addition, the need for vessels is supplied by the PLN power plant. The other energy surplus is fed directly to the PLN grid. Table 14 states that the vessel that received the largest contribution from PV energy supply was the Soemantri Brojonegoro vessel. The vessel gets the largest supply of renewable energy from PV because it is a vessel that requires the least supply of electrical power compared to others.

Table 15 compares Pelindo's daily revenue when using hybrid PV and PLN electricity sources versus all PLN. As the data in table 15 shows, when Pelindo uses a mixture of PV and PLN electricity sources, its daily revenue increases by 30%. The revenue generated by Pelindo can increase because Pelindo has new revenue in the form of electricity sales to the PLN network. This sale can add to the company's revenue, which initially only sold electricity to vessels. This new economic activity triggers the potential for a drastic increase in Pelindo's revenue because the expenditure to buy electrical energy to PLN can be reduced by being replaced by using solar energy sources when the weather is sunny. This increased potential revenue is obtained because the total electrical energy produced by solar power plants is greater than the total demand. The results of the calculation of the power that the solar power plant can obtain can be seen in Figure 2 above.

In addition, this economic analysis also projects the BEP in this research. The authors conducted a thorough analysis of the BEP after establishing the basis for this revenue. BEP calculation begins with data on capital and operational costs that are expected to be incurred. Table 16 contains the calculation of capital and operational costs that are projected to be issued by Pelindo as the project maker. Table 17 contains the calculation of revenue earned by Pelindo every year. It can be seen in the table that there is a decrease in revenue every year. Revenue has decreased every year due to considering the decrease in the effectiveness of solar panels which are claimed by manufacturers to decrease by 2% in the second year and 0.55% in the following year. With this in mind, BEP can occur in 2028 after the operation of this solar power plant in 2021. The BEP graph can be seen in Figure 10.

Table 14. Comparison energy required by vessel and PV generated

Vessel name	Average load kw	Energy consumption in a day Kwh	Average energy produced by pv in a day Kwh	Solar energy sends to vessel kwh	Percentage vessel energy consumption supplied by pv
Ibrahim zahier	156.89	3765.44		1424.88	38%
Pusri indonesia 1	173.94	4174.47		1534.95	37%
Soemantri brojonegoro	121.28	2910.64	4202.42	1175.57	40%
Abusamah	147.30	3535.13		1357.71	38%
Julianto md	135.51	3252.16		1275.18	39%

Table 15. Comparison pelindo daily revenue

Vessel name	Total energy vessel need (Kwh)	PV Energy sends to vessel (Kwh)	Energy send to grid (Kwh)	Bill if 100% using energy grid (Rp)	Revenue from selling to vessel (Rp)	Revenue selling to PLN (Rp)	Total Revenue/day if using 100% PLN (Rp)	Total Revenue /day if Hybrid (Rp)
Ibrahim zahier	3765.44	1424.9	2777.54	3,900,165	7,741,739	2,876,918	7,741,739	10,618,657
Pusri indonesia 1	4174.47	1534.9	2667.47	4,323,829	8,582,703	2,762,916	8,582,703	11,345,619
Soemantri brojonegoro	2910.64	1175.6	3026.85	3,014,781	5,984,273	3,135,155	5,984,273	9,119,428
Abusamah	3535.13	1357.7	2844.71	3,661,618	7,268,230	2,946,494	7,268,230	10,214,723
Julianto md	3252.16	1275.2	2927.24	3,368,526	6,686,449	3,031,979	6,686,449	9,718,428
					Average revenue per day		7,252,679	10,203,371

Table 16. Projected capex and opex

Capex	Total Price	
Office n security	Rp	150,000,000
Planning Area	Rp	920,000,000
Price Solar Array Support Structure	Rp	2,629,032,000
Price Solar Panel	Rp	6,661,200,000
Price Inverter	Rp	1,000,000,000
Price Panel	Rp	500,000,000
Total Capex	Rp	11,860,232,000
Maintenance	Rp	261,000,000

Table 17. Projected Pelindo revenue

Vessel name	Revenue from selling to vessel	Revenue selling to PLN	Total Revenue each vessel
Ibrahim zahier	Rp 7,741,739	Rp 2,876,918	Rp 10,618,657
Pusri indonesia 1	Rp 8,582,703	Rp 2,762,916	Rp 11,345,619
Soemantri brojonegoro	Rp 5,984,273	Rp 3,135,155	Rp 9,119,428
Abusamah	Rp 7,268,230	Rp 2,946,494	Rp 10,214,723
Julianto md	Rp 6,686,449	Rp 3,031,979	Rp 9,718,428
Total average revenue in first year			Rp3,724,230,334.94
Total average revenue in second year (-2%)			Rp3,649,745,728.24
Total average revenue in third year (-0.05%)			Rp3,647,920,855.37

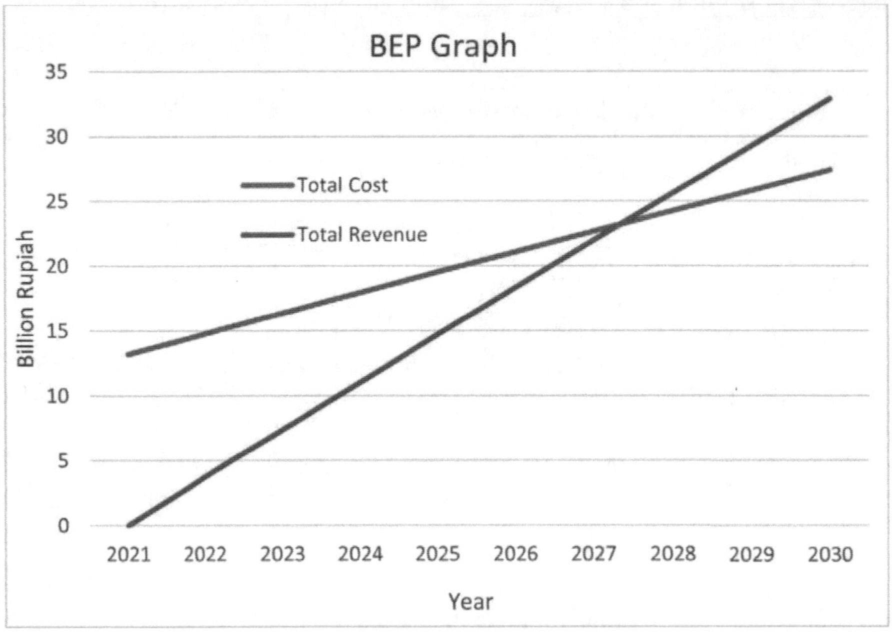

Figure 10. BEP Graph.

IOP Conf. Series: Earth and Environmental Science 1081 (2022) 012019 doi:10.1088/1755-1315/1081/1/012019

4. Conclusion

Solar power is likely to be used to run Semarang's Dermaga Dwimatama, according to this study of the Cipta industrial area (near the Cipta Beach area). According to the calculations, the required surface area is nearly 4 hectares when arrays are spaced 5 meters apart to generate 1000kWp from 410W panels. For the Semarang area, the panel's maximum output time is between 12 and 13 a.m. Sunlight from the solar panels powers the vessel from 9 a.m. until 4 p.m. Energy production is 4202.42 kWh/day at 954.25 kWp from this configuration's solar panels. Solar power can supply at least 37% to 40% of the vessel's energy needs.

In the range of 0.3 to 0.5 kW, the most power is lost. The bus output of the transformer is where this all happens. The transformer's bus outputs are PLN and Load Buses 3 and 4. These losses in the transformer result in a disproportionately large loss of power, as evidenced by the dropped voltage. Voltage drop in a 400V low voltage distribution system is the greatest. For a PLN distribution system alone, the voltage drop is between 0.02% and 1.25 %, whereas the voltage drop for PLN + PV systems is between 0.11% and 1.35 %. The nearly 2-kilometer-long pull cable of a complete solar panel system reduces medium and low voltage. Low-voltage savings can range from 0.15% to 0.21%. Furthermore, there is a tolerance of 0.7-0.1 percent for high voltage.

References

[1] Arifin, Z., Tamamy, A. J., & Amalia, A., Analisis Potensi Energi Sinar Matahari Dan Energi Angin di Pusat Kota Semarang. Setrum: Sistem Kendali-Tenaga-elektronika-telekomunikasi-komputer, 7(2), 296-304, 2019.

[2] Borkowski, T., & Tarnapowicz, D., "Shore to ship" system an alternative electric power supply in port. Journal of KONES, 19, 49-58, 2012.

[3] Ericsson, P., & Fazlagic, I., Shore-side power supply. Master of Science Thesis–Chalmers University of technology–Goteborg Sweden, 2008.

[4] Harahap, P., Pengaruh Temperatur Permukaan Panel Surya Terhadap Daya Yang Dihasilkan Dari Berbagai Jenis Sel Surya. RELE (Rekayasa Elektrikal dan Energi): Jurnal Teknik Elektro, 2(2), 73-80, 2020

[5] IEC 60364-5-52 Low-voltage electrical installations – Part 5-52: Selection and erection of electrical equipment – Wiring systems

[6] IEEE Std 141-1993 Recommended Practice for Electric Power Distribution for Industrial Plants

[7] Pangestuningtyas, D. L., Hermawan, H., & Karnoto, K., Analisis pengaruh sudut kemiringan panel surya terhadap radiasi matahari yang diterima oleh panel surya tipe larik tetap. Transient: Jurnal Ilmiah Teknik Elektro, 2(4), 930-937, 2014.

[8] Paul, D., & Haddian, V., Shore-to-ship power supply system for a cruise ship. In 2009 IEEE Industry Applications Society Annual Meeting (pp. 1-7). IEEE, 2009.

[9] Radwan, M. E., Chen, J., Wan, Z., Zheng, T., Hua, C., & Huang, X., Critical barriers to the introduction of shore power supply for green port development: case of Djibouti container terminals. Clean Technologies and Environmental Policy, 21(6), 1293-1306, 2019.

[10] Setyono, J. S., Mardiansjah, F. H., & Astuti, M. F. K., Potensi Pengembangan Energi Baru dan Energi Terbarukan di Kota Semarang. Jurnal Riptek, 13(2), 177-186, 2019.

[11] Silaksanti, D., & Sudarmo, S. T., COST AND ENVIRONMENTAL BENEFIT IN THE USE OF SHORE CONNECTION IN BJTI PORT. Advances in Transportation and Logistics Research, 3, 38-45, 2020.

[12] Widodo, D. A., & Andrasto, T., Pemberdayaan Energi Matahari Sebagai Energi Listrik Lampu Pengatur Lalu Lintas. Jurnal Teknik Elektro, 2(2), 6, 2010.

Study on the Application of Microcontroller-Based GPS Sensors on Floater Gliders to Reach the Specified Destination

Nanang J H Purnomo[1,2], A.A. Masroeri[1], Wibowo H Nugroho[2], Sahlan[2]

[1]Marine Engineering Department, Institut Teknologi Sepuluh Nopember, Surabaya, Indonesia
[2]National Research and Innovation Agency, Indonesia

Abstract. The purpose of the study is to determine the performance of the control system of the floater glider to a predetermined location. The method used was a combination between software simulation and an experimental free-running test. The Floater control system is equipped with GPS as input and output sensors to determine coordinates or waypoints. The microcontroller mounted was Pixhawk 2.4.8 based on ARM cortex M4 – 32 bits. The floater was driven by two Brushless DC motors and servo motors to drive the Z-Peller steering to the desired trajectory. The floater system applies the default application software from Pixhawk, namely mission planner to plan and also monitor the actual trajectory. To have a proper procedure for testing the designed control system, the three stages of the test were conducted, firstly numerical tuning test using step function on MATLAB simulation, then trajectory simulation test of waypoint using mission planner, and finally the field test in the ILH pond. From the three stages of the test, it was found that the selection of PID gain values Kp = 0.2, Ki = 0.3, and Kd = 0.02, indicating the performance of the floater glider control system to the planned location has a good fit between the trajectory simulation results and the waypoint testing in the pond. These results can be used as a reference in conducting further research so that the navigation accuracy of the control system will be improved.

Keywords: Floater Glider, GPS Sensor, Microcontroller, Trajectory

1. Introduction

Indonesia is located in the pacific ring of fire so there is the potential for earthquakes and volcanic eruptions in the Pacific Ocean area. The impact of the earthquake that occurred in the Pacific Ocean was a tsunami disaster. The tsunami disaster-caused damage to public infrastructure and caused enormous casualties. According to the map of the distribution of tsunami disaster data released by BNPB in 2018, it is necessary to get more attention so that mitigation can be carried out properly [1] [2].

Following the disaster mitigation policy launched by the Indonesian government to revitalize the Buoy-based early warning system, BRIN as one of the government institutions has succeeded in developing a design for tsunami early warning based on Buoy generation 3.1 which is already operating well and has been fixed placed in several Indonesian seawaters that are prone to tsunami disasters. Furthermore, it is necessary to develop a next-generation tsunami detection tool for monitoring changes in seawater elevation to maximize the tsunami early warning system in Indonesia.

The next-generation tsunami detection tool uses the concept of the Wave Glider System. The Wave Glider system itself has two parts. The first part is on the surface called the floater which floats on the surface of the seawater and the second part sinks which glides and attract the floater by moving the fins up and down due to ocean waves that occur on the surface. The basic concepts of floaters and gliders in the next-generation tsunami detection equipment are shown in Figure 1.

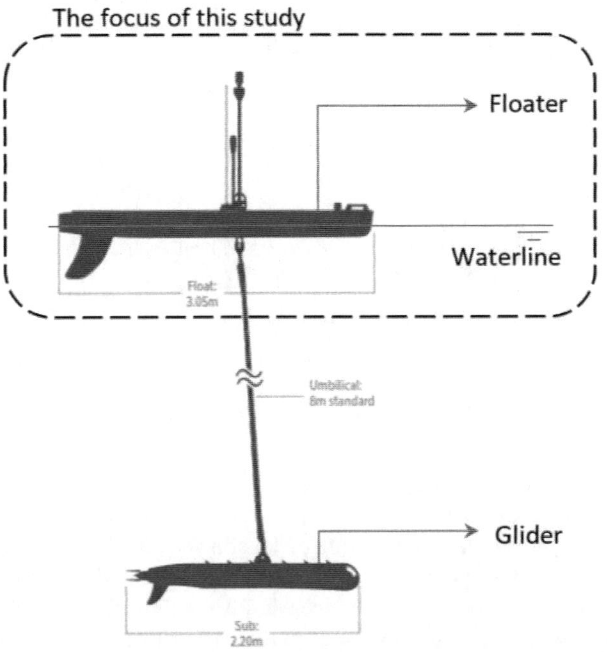

Figure 1. Basic Concepts of Floaters Gliders in the next Generation Tsunami Detection Equipment

The use of the Wave Glider makes it easier to explore in monitor the elevation of ocean waves [3]. The glider section itself has been studied by [1] on a numerical glider floater system for foil configurations in determining the optimal dimensions of the Wave Glider design [1], but the study is based only on numerical simulation. In addition, it was also studied by [4] the structural strength of the surface floater of wave glider using Eva Foam and Fiberglass material for skin and Aluminum material for frame and kell [4], but it has not yet studied setting the Autonomous control system on the floater glider.
Autonomous control system settings have been studied by [5] integrating APM (Ardupilot Mega) and mission planner applications in making the autonomous control system on the UAV (Unmanned Aerial Vehicle), the use of GPS and compass in setting the control direction can operate properly [5], but the use of APM makes data processing slow so it is necessary to make changes in the main control with a more reliable system. A reliable system is not only from the use of the control system but also from the firmware that is integrated with the control algorithm, the control algorithm itself has many types, one of which is using PID such as research conducted by [6] regarding the use of the PID method in regulating UAV wings. making it easier to maneuver [6], but the research conducted did not approach PID tuning.

But in this study, the propulsion system of the glider is not discussed in detail. The discussion is only focused on the surface floater in which there is an autonomous control system to reach the desired destination location. To make it easier to understand the surface floater is called a Floater Glider. On the back of this surface floater, a driving system in the form of 2 Z-Peller is attached as a substitute for the glider. This is intended as a backup in case of the glider propulsion system does not produce forward motion during calm wave conditions.

The type of microcontroller used in this study is the Pixhawk 2.4.8 based on ARM cortex M4 – 32 bits which is generally used for drones or UAVs. But in this study, try to apply it to the surface floater autonomous control system.

The purpose of the study is to determine the performance of the control system of the floater glider to a predetermined location. The method used was a combination between software simulation and an experimental free-running test.

2. Theoretical basis

In this study, the control system used is based on proportional integrator derivative (PID). The PID block diagram can be seen in Figure 2.

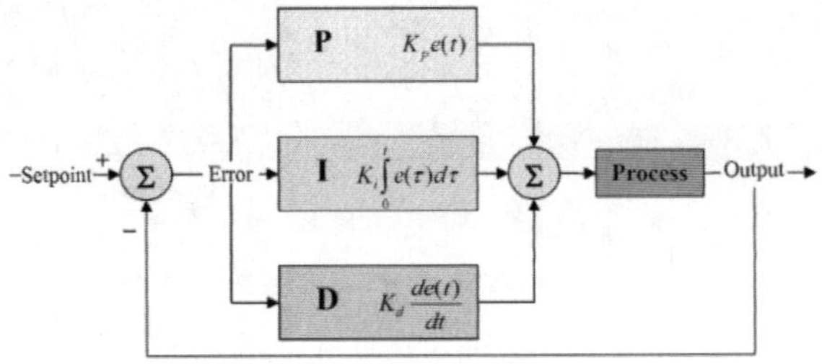

Figure 2. PID Block Diagram

From the PID block diagram, it has been explained that the setpoint setting is very important in adjusting the floater glider's tuning. Tuning control is needed in designing a floater glider with autonomous control systems to determine the performance of the control system of the floater glider to a predetermined location [7]. From the problems that have been described, it is necessary to develop the Floater section as anticipation if the movement of the floater glider system does not function properly by taking a PID control approach both in simulation and implementation for the optimization of the tsunami early warning system when heading to the specified location.

The design of a reliable autonomous control system in the Floater Glider design is very important in the performance of the control system of the floater glider, the selection of the use of instrumentation such as types of controls, sensors, and actuators is a way to increase reliability in tuning, in addition to integrating instrumentation with methods such as PID can maximize tuning accuracy on the Floater Glider.

Determination of the placement of the control, sensor, and actuator is also very important in addition to optimizing the work of the sensor as well as maximizing the placement of the actuator so that the Floater Glider can do a good tuning. The following is a picture of the placement of controls, sensors, and actuators shown in Figure 3, while the block diagram of the Autonomous system is shown in Figure 4.

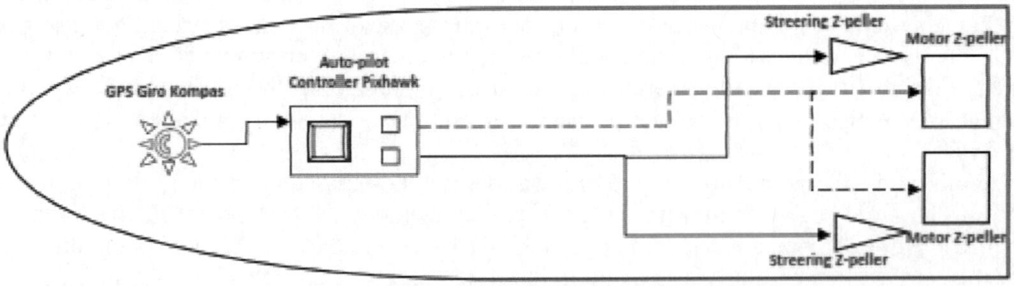

Figure 3. Placement of Components of the Control System of the Floater Glider

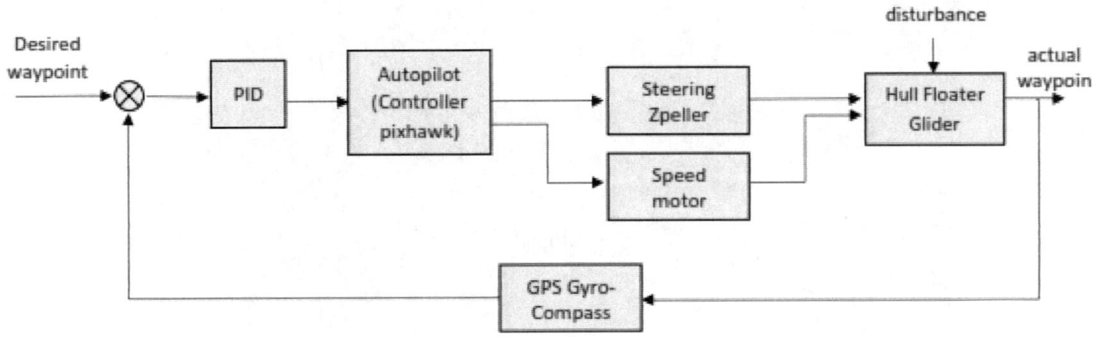

Figure 4. Autonomous Control System Block Diagram

Based on the block diagram in Figure 4, the selection of the coefficients Kp, Ki, and Kd, is carried out using the frequency tuning method in Equation 1

$$u(t) = K_p e(t) + K_i \int_0^t e(t)dt + K_d \frac{d}{dt} \quad (1)$$

Equation (1) describes the output value u(t), which is the sum of the proportional gain (Kp), integral gain (Ki), and derivative gain (Kd), each of which is changed by error (e) in a certain time interval (t) [8]. PID coefficient adjustment can also be based on the selection of Kp Ki Kd values through numerical simulations in MATLAB [8]. The selection of Kp Ki and Kd values, in numerical modeling "tuning", is the development of the default values of the mission planner software. The results of the numerical simulation of selecting the value of Kp Ki Kd for the "Step" function are shown in the form of a plot of the tuning factor against the time of each of these values.

3. Experimental Setup

The Floater control system is equipped with GPS as input and output sensors to determine coordinates or waypoints. The microcontroller mounted was Pixhawk 2.4.8 based on ARM cortex M4 – 32 bits. The floater was driven by two Brushless DC motors and servo motors to drive the Z-Peller steering to the desired trajectory. The floater system applies the default application software from Pixhawk, namely mission planner to plan and also monitor the actual trajectory.

This experimental setup consists of 2 steps, namely a simulation using mission planner software to determine the desired trajectory or waypoint and field testing of the floater glider in the pond. The value of Kp Ki Kd was obtained from the results of the numerical function test of the control system using a step function which is the input for the Floater Glider firmware simulation. This aims to determine the accuracy of the waypoint paths made. After the trajectory results from the mission planner are obtained, then a pool test is carried out as a validation of the selected Kp Ki Kd numerical simulation results.
The firmware used for simulation and implementation is Mission Planner because it is easy to use and has a good user interface. This Mission Planner firmware application is open source and can be downloaded on the web [9]. Mission Planner application settings are shown in Figure 5 and Figure 6

Figure 5. Mission Planner App View for Floater Glider Simulation on a Pond

Figure 6. PID Coefficient Setting Display of Mission Planner Floater Glider Software

Based on Figure 3. and Figure 4, this Floater Glider uses an autonomous system with a Pixhawk 2.4.8 microcontroller based on ARM cortex M4 – 32 bits, and a GPS-based compass sensor which is used to determine the direction and coordinates as well as the Floater Glider's destination point. While the actuators used are a pair of Z-Peller at the stern which function as the propulsion system of the Floater Glider. The Z-Peller propulsion system is also equipped with a servo motor as a substitute for the steering angle movement regulator. The use of two Z-Peller at the stern of the Floater Glider aims to make it more optimal when driving and maneuvering in the pool.

The position of GPS-based compass sensor is placed close to the controller component so that data transmission from the sensor is more optimal. Then the distance between the controller component and the Z-Peller is a bit far to reduce vibration due to motor rotation which can interfere with the performance of the controller. The dimensions of the Floater Glider are shown in Table 1.

Table 1. Properties of Floater Glider

Physical property	Units
LOA (length overall)	3.050 m
Breadth	0.800 m
Height	0.290 m
Draught	0.138 m
Displacement	175.1 kg

After getting the data from the mission planner simulation, the control system components are installed on the floater glider as shown in Figure 3. The components of the floater glider control system are shown in Table 2.

The next step is the calibration of the floater control system to synchronize the output value of the control system according to the input value. The calibration step taken is the calibration of the Z-Peller steering angle movement where the 0° point needs to be compared with the 90° value from the standard tool in the form of a protractor as shown in Figure 7.

After calibration of the Z-Peller angle, the floater glider (plant) electronic system function test is carried out as shown in Figure 8. If the function test is complete and successful then the floater glider is brought to the pool for testing the autonomous system as shown in Figure 9

The results of the simulation test used a mission planner in the form of a floater travel trajectory, while for testing in the pool in the form of GPS sensor reading data visualized in a graph.

Table 2. Components of the control system of the Floater Glider

No	Item	Specification
1.	Microcontroller	Pixhawk 2.4.8
2.	GPS Module Compass	U-Blox neo-m8n
3.	Z-peller	Z-Peller Azimuth Thruster (Z-Drive) Propulsion Combo Seri 120mm
4.	Brushless DC Motor	2 x 900kv Outrunner Brushless (Diameter 50mm)
5.	Electronic Speed Controller (ESC)	Hobbywing SeaKing V3 Waterproof 120A Brushless ESC for RC Racing Boat
6.	Servo Steering	Servo Hitec hs-645mg
7.	Telemetry	TTL 3D Robotics 3DR Radio 915Mhz Telemetry Kit for Pixhawk APM2.6
8.	Receiver RC	Remote Control Drone RC Transmitter 2,4GHz FlySky-FS-i6
9.	Batterie	Lithium 20 ah 24volt 4p8s lifepo4
10.	Software Aplikasi control	Mission Planner 1.3.75 ArduPilot
11.	Notebook PC	Asus A413E Core i7 8GB 512GB Nvidia MX330

Figure 7. Z-Peller Steering Angle Calibration

Figure 8. Function test of Pixhawk 2.4.8 Control System, GPS Sensor, and Z-Peller

Figure 9. Autonomous System Test of Floater Glider in Pond

4. Results & discussions

To have a proper procedure for testing the designed control system, the three stages of the test were conducted, firstly numerical tuning test using step function on MATLAB simulation, then trajectory simulation test of waypoint using mission planner, and finally the field test in the ILH pond. The field test was carried out in calm water conditions to only test the reliability of the autonomous system of the floater glider.

The results of the numerical simulation test tuning the value of the PID coefficient using the step function through MATLAB simulation are shown in Figure 10, Figure 11, and Figure 12.

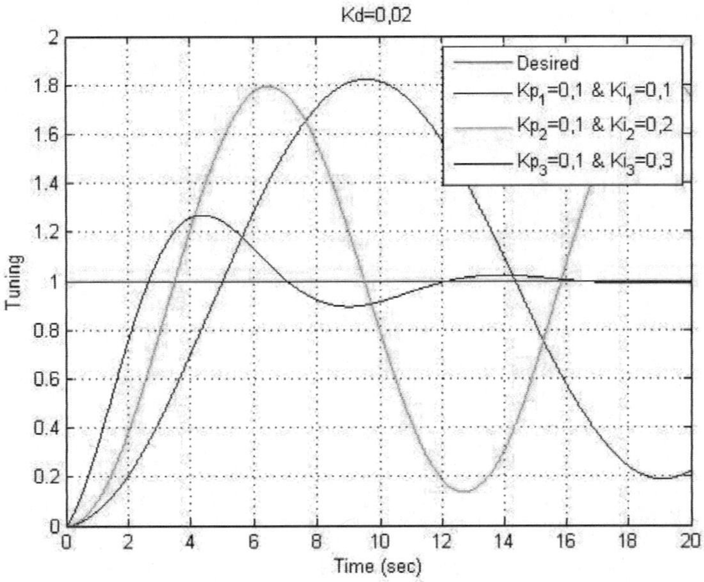

Figure 10. Graph of numerical test results PID coefficient value Kp_1=0.1, Ki_1=0.1, and Kd_1=0.02; Kp_2=0.1, Ki_2=0.2, and Kd_2=0.02; Kp_3=0.1, Ki_3=0.3, and Kd_3=0.02

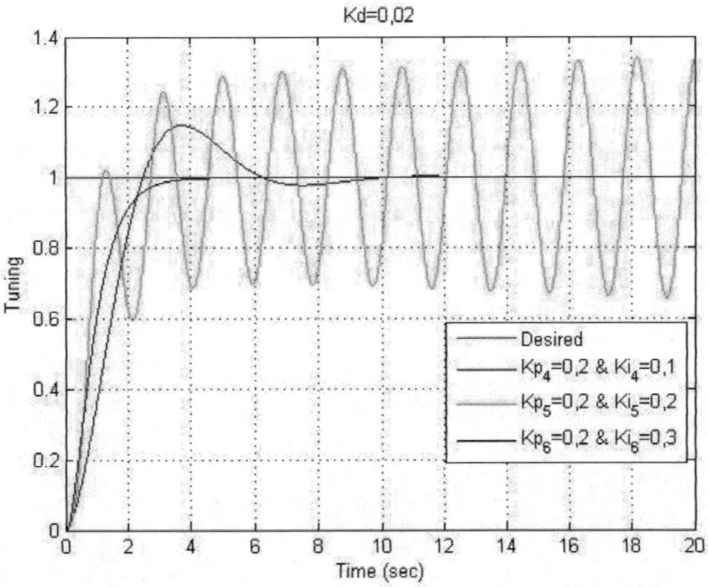

Figure 11. Graph of numerical test results PID coefficient value Kp_4=0.2, Ki_4=0.1, and Kd_4=0.02; Kp_5=0.2, Ki_5=0.2, and Kd_5=0.02; Kp_6=0.2, Ki_6=0.3, and Kd_6=0.02

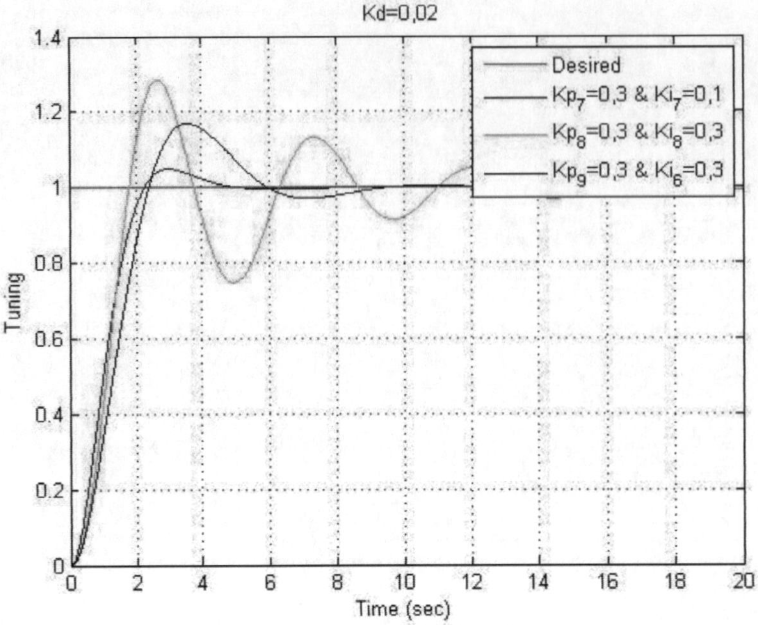

Figure 12. Graph of numerical test results PID coefficient value Kp_7=0.3, Ki_7=0.1, and Kd_7=0.02; Kp_8=0.3, Ki_8=0.2, and Kd_8=0.02; Kp_9=0.3, Ki_9=0.3, and Kd_9=0.02

As seen above, there are 5 combinations of Kp, Ki, and Kd values that can be used as simulation input values in the mission planner. These values are as follows; Kp_3=0.1 Ki_3=0.3 Kd_3=0.02; Kp_4=0.2 Ki_4=0.1 Kd_4=0.02; Kp_6=0.2 Ki_6=0.3; Kd_6=0.02; Kp_7=0.3 Ki_7=0.1 Kd_7=0.02; Kp_9=0.3 Ki_9=0.3 Kd_9=0.02 are values that are stable enough to be included in the mission planner's firmware settings. For this simulation, it was decided to use a speed of 1.0 m/s by following the speed capacity of the equipment for testing in the pond. Then the Kp Ki Kd values obtained from the numerical tuning simulation are used as input from the mission planner where the results are shown in Figures 13 to 17.

The results of the trajectory simulation test on the mission planner using the value of Kp=0.1; Ki=0.3; and Kd=0.02 at a speed of 1.0 m/s are shown in Figure 13.

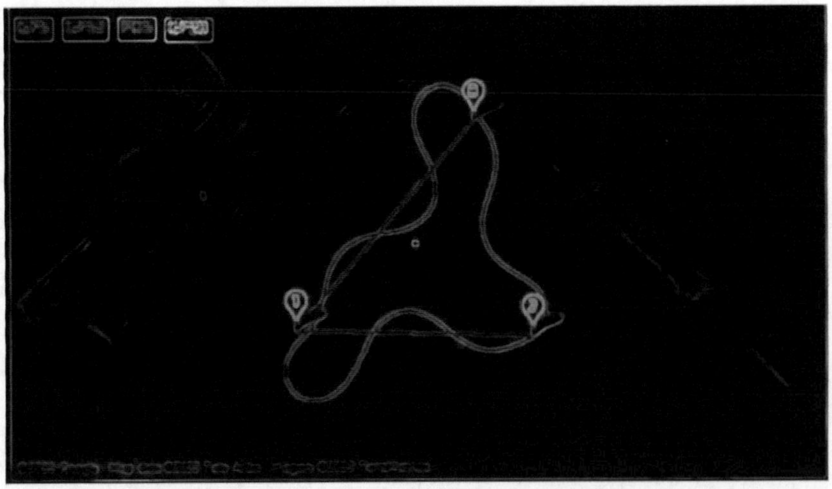

Figure 13. PID Trajectory Floater Glider Simulation Display For Kp=0.1; Ki=0.3; and Kd=0.02

The results of the trajectory simulation test on the mission planner using the value of Kp=0.2; Ki=0.1; and Kd=0.02 at a speed of 1.0 m/s are shown in Figure 14.

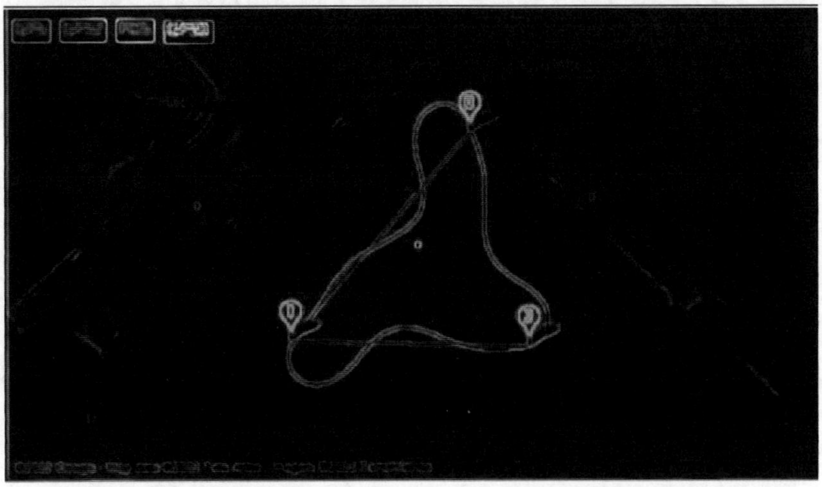

Figure 14. PID Trajectory Floater Glider Simulation Display For Kp=0.2; Ki =0.1; and Kd=0.02

The results of the trajectory simulation test on the mission planner using the value of Kp=0.2; Ki=0.3; and Kd=0.02 at a speed of 1.0 m/s are shown in Figure 15.

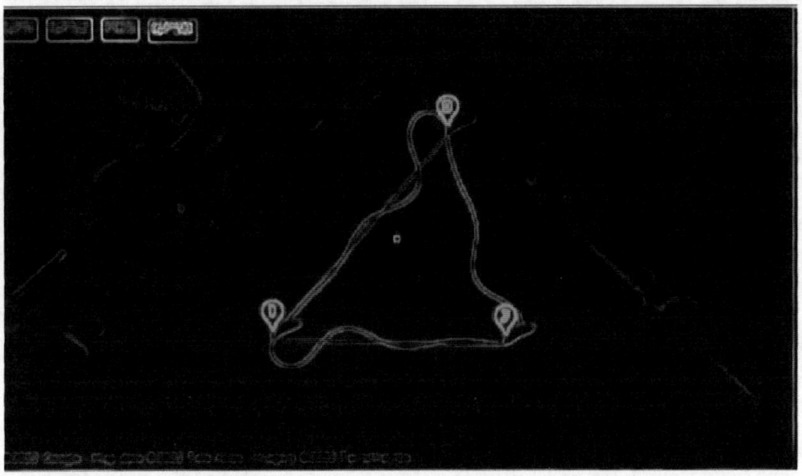

Figure 15. PID Trajectory Floater Glider Simulation Display For Kp=0.2; Ki=0.3; and Kd=0.02

The results of the trajectory simulation test on the mission planner using the value of Kp=0.3; Ki=0.1; and Kd=0.02 at a speed of 1.0 m/s are shown in Figure 16.

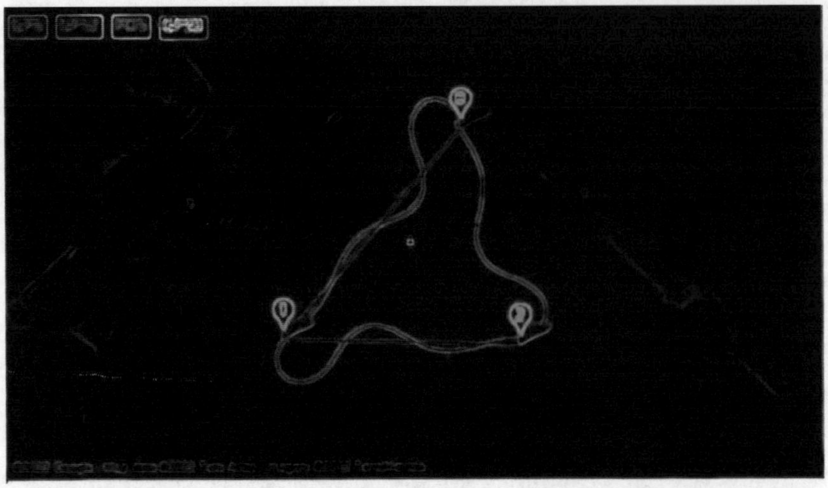

Figure 16. PID Trajectory Floater Glider Simulation Display For Kp=0.3; Ki=0.1; and Kd=0.02

The results of the trajectory simulation test on the mission planner using the value of Kp=0.3; Ki=0.3; and Kd=0.02 at a speed of 1.0 m/s are shown in Figure 17.

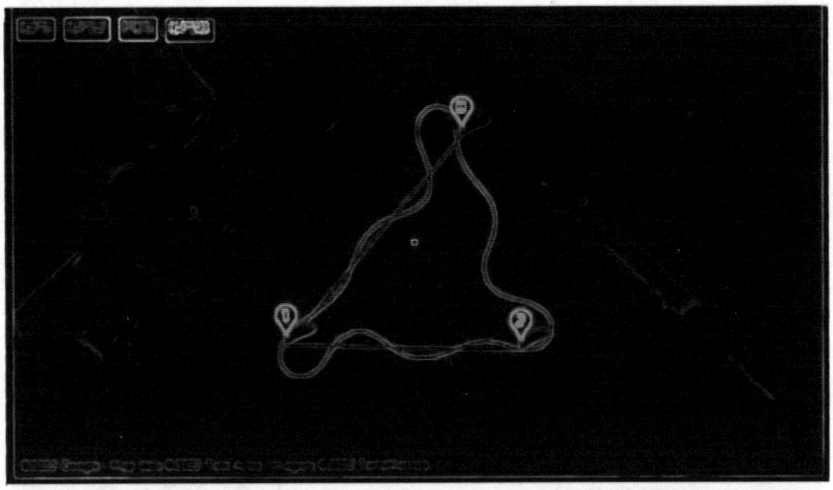

Figure 17. PID Trajectory Floater Glider Simulation Display For Kp=0.3; Ki=0.3; dan Kd=0.02

Due to the area limitations of the pond, it was decided to choose the 2 most stable values of Kp Ki Kd, which are the value of Kp = 0.2 Ki = 0.3 Kd = 0.02 and the value of Kp = 0.3 Ki = 0.1 Kd = 0.02 which shows good results also on the mission planner. The results of the pond test are shown in Figure 18 and Figure 19. Figure 18 shows that the Floater Glider moves in a slight circle but is stable enough to reach the planned location with the value of Kp = 0.2 Ki = 0.3 Kd = 0.02. It can be predicted from the tuning results in Figure 11 where the actual response of the control system is close to the desired response. It is different in Figure 19 where the Floater Glider move to the planned location has oscillation events on its trajectory as shown in the tuning results of the control system in Figure 12.

IOP Conf. Series: Earth and Environmental Science 1081 (2022) 012020 doi:10.1088/1755-1315/1081/1/012020

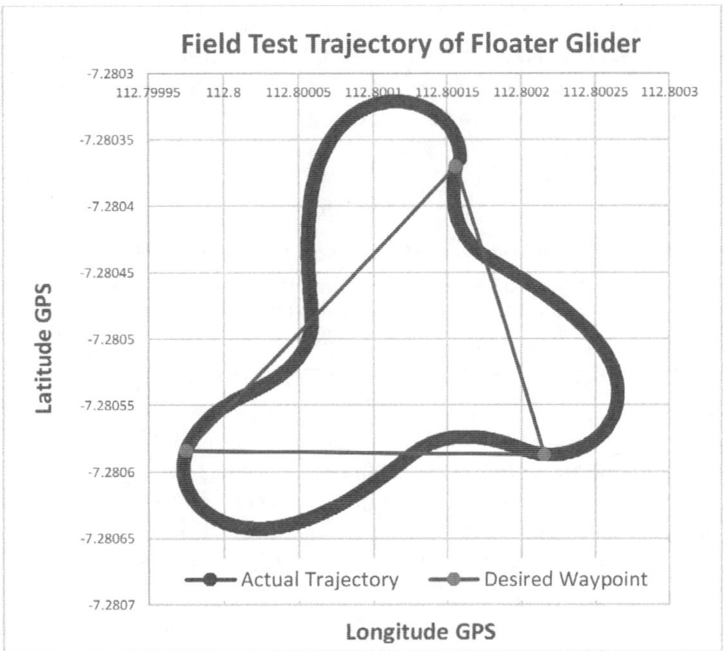

Figure 18. The trajectory of the field test results with the value of Kp=0.2, Ki=0.3, Kd=0.02

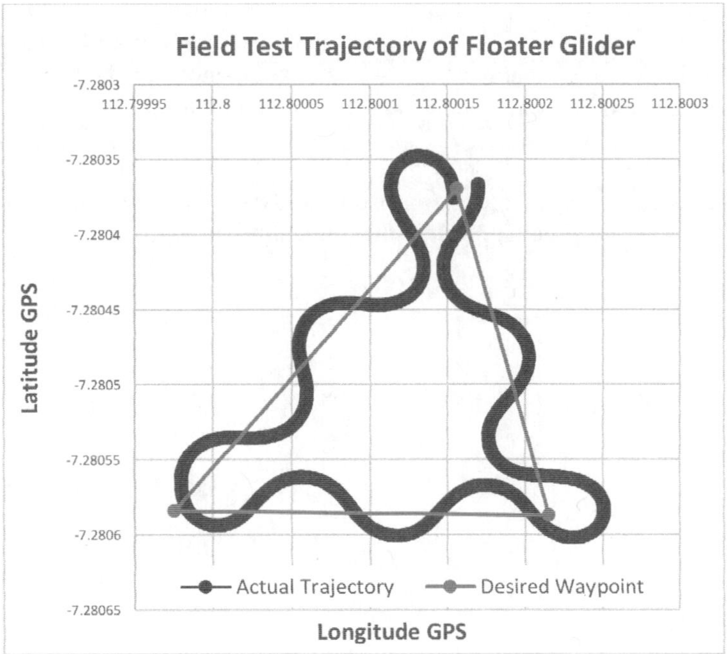

Figure 19. The trajectory of the field test results with the value of Kp=0.3 Ki=0.1 Kd=0.02

From the three stages of the test, it was found that the selection of PID gain values Kp = 0.2, Ki = 0.3, and Kd = 0.02 indicating the performance of the floater glider control system to the planned location has a good fit.

5. Conclusions

It can be concluded that the performance of the floater glider control system to the planned location has a good fit between the trajectory simulation results and the waypoint testing in the pond. These results can be used as a reference in conducting further research so that the accuracy of the navigation control system can be improved.

References

[1] Arifin, H. N. Wibowo, H. Buddin, and W. Bambang, "Numerical Prediction Of Foils Configuration In A Design of Buoy Glider System For Supporting Tsunami Early Warning," IOP Conf. Ser. Mater. Sci. Eng., vol. 1052, no. 1, p. 012017, 2021, DOI: 10.1088/1757-899x/1052/1/012017.

[2] D. P. Utomo and B. Purba, "Penerapan Datamining pada Data Gempa Bumi Terhadap Potensi Tsunami di Indonesia," Pros. Semin. Nas. Ris. Inf. Sci., vol. 1, no. September, p. 846, 2019, doi: 10.30645/senaris.v1i0.91.

[3] Liquid Robotics, "The Wave Glider Spec Sheet," pp. 3–4, [Online]. Available: https://www.info.liquid-robotics.com/wave-glider-spec-sheet.

[4] K. Priohutomo and W. H. Nugroho, "Strength Analysis and Assessment of Ina-TEWS Wave Glider," vol. 4, pp. 140–151, 2020.

[5] H. S. Saroinsong et al., "Rancang Bangun Wahana Pesawat Tanpa Awak (Fixed Wing) Berbasis Ardupilot," vol. 7, no. 1, pp. 73–84, 2018.

[6] F. T. Industri, "Rancang Bangun Dan Analisa Sistem Kendali Pid Pada Unmanned Aerial Vehicle (UAV) Fixed Wing Pid Pada Unmanned Aerial Vehicle (UAV)," 2016.

[7] "Advanced Control Engineering," Adv. Control Eng., 2001, doi: 10.1016/b978-0-7506-5100-4.x5000-1.

[8] "MATLAB based PID controller - File Exchange - MATLAB Central." https://www.mathworks.com/matlabcentral/fileexchange/35163-matlab-based-pid-controller.

[9] "Installing Mission Planner — Mission Planner documentation." https://ardupilot.org/planner/docs/mission-planner-installation.html.

Acknowledgements

The authors highly appreciated a research grant from the National Research and Innovation Agency, Indonesia, through a degree by research program to conduct this study.

Uncertainty Analysis Study on Seakeeping Tests of Benchmark Model

Nandiko Rizal[1,2], Dian Purnama Sari[2], Beny Cahyono[1], Dedy Dwi Prastyo[3], Baharuddin Ali[2] and Erdina Arianti[2]

1. Marine Engineering Department, Institut Teknologi Sepuluh Nopember, Kampus ITS - Sukolilo, Indonesia.
2. Indonesian Hydrodynamic Laboratory-BRIN.
3. Departement of Statistics, Institut Teknologi Sepuluh Nopember, Kampus ITS - Sukolilo, Indonesia.

Abstract. Experimental uncertainty and measurement, particularly in engineering, have been increasingly emphasized over the last 20 years. Uncertainty analysis of ship model testing in seakeeping experiments based on International Organization for Standardization, Guide for Uncertainty of Measurements (ISO-GUM), report with recommendations published by ITTC (2008). In this paper, the uncertainties for free-running tests were performed for a model scale 1:62 as a benchmark model test at the Indonesian Hydrodynamic Laboratory (IHL). The benchmark model is representative of a full scale of 186 meters. The model has full appendages as well as a rudder and propellers to allow for free-running tests, natural roll period, free roll decay, and swing tests. Factors of basic uncertainty (type A and type B) are estimated using model tests. The standard deviation of the mean value of a repeated measurement is uncertainty type A. However, Type B uncertainty is estimated from manufacturing specifications. The uncertainty components resulting from both types are quantified by standard deviations. The uncertainty value of Type B is approximated by a corresponding variance. This study helps to understand the underlying uncertainty associated with this research, which resulted in a total geometry uncertainty value (Lpp, B, T, H, $\overline{KG}$, and k_{yy}) of 0.17598% and the total uncertainty value of the instrument calibration results (wave height sensor and qualysis motion tracking) is 0.01161%. The results of this study are expected to minimize the impact of sources of uncertainty and allow the potential for underlying uncertainty to be corrected for the next seaworthiness test.

Keywords: ISO-GUM, ITTC, Seakeeping Test, Uncertainty.

1. Introduction

To increase the safety of a ship in transit, the ship must be calculated and tested with different wave movements before construction or during the design process. Therefore, the ship's model was tested. Measurements include 6 DoF (degrees of freedom), and motions that are a surge, sway, heave, roll pitch, and yaw. The quality of ship movement is known from the presentation of statistical analysis measurements. Measurement results are displayed in the response amplitude operator (RAO), to support analysis. Nandiko Rizal (2018).

Uncertainty measurement is an essential factor in assessing the accuracy of any experiment or measurement. Seakeeping testing is one of them. Uncertainty analysis aims to determine the level of confidence in the measurement data and can be used to assess the error rate as an objective indicator.

Furthermore, maximizing the quality of the experiment is obtained from the results of the analysis of uncertainty and insight. Uncertainty analysis over the years has observed confidence levels in both experimental and computational studies. However, shipbuilding and offshore structures have not applied uncertainty analysis. Concerns about the uncertainty of the seakeeping test were explained at the 15th and 16th ITTC at the seaworthiness committee, and the report (ITTC, 1978, 1981) reported that seakeeping has shown that the experimental data for the tests is significant. ITTC (1990). The use of the ANSI and ASME (1985) approaches takes into account the uncertainties in hydrodynamic vessel experiments and computational studies. Conference ITTC at 2008, the 1995 Guidelines (ISO-GUM) declares measurement uncertainty was recommended by the International Organization for Standardization (ISO) for the analysis of uncertainty in the ITTC community and its member organizations. Uncertainty is classified into two categories, type A and type B according to the method recommended by ISO-GUM, and how to evaluate uncertainty. Random uncertainties evaluated by statistical methods based on repeated measurements are represented by type A. Methods other than repeated observations, i.e. through manual information, manufacturer information, or data provided as certificates are represented by type B. ISO-GUM procedures referring to ITTC 7.5-02 -07-02.1 Rev. 4 (2011) regarding the measurement of seakeeping test uncertainty approved by the Maritime Commission. ISO-GUM, uncertainty analysis includes five principles, namely: Principle 1. Uncertainty can be divided into 2 categories, Type A and Type B; Principle 2. An element of the estimated variance is defined as a type A uncertainty, which includes the degree of freedom (DOF) effect; Principle 3. The elements of the appropriate variance are approximated by type B uncertainty where they are expected to exist; Principle 4. Combining the variance with the usual method of calculating it using Combined Uncertainty; Principle 5. For some applications, the combined uncertainty must be multiplied by the coverage factor to obtain the total uncertainty value. The 95% confidence level of the extended uncertainty is called the overall uncertainty.

Uncertainty analysis of ship model testing in seakeeping experiments based on International Organization for Standardization, Guide for Uncertainty of Measurements (ISO-GUM), report with recommendations published by ITTC (2008). IHL has applied a method of measurement uncertainty in accordance with ITTC 7.5-02-07-02.1. In some cases, the type of test procedure can interfere with the calculation of statistically and meteorologically valid measurement uncertainty. IHL identifies all factors of uncertainty, makes reasonable estimates, and ensures that the method of reporting test results does not give a false impression of uncertainty. Past experience serves as an estimate of knowledge of the method's performance and range of measurements. When estimating measurement uncertainty, all components of uncertainty that are important analysis used appropriately in certain situations are taken into account.

Experimenter Klein (1985) defined uncertainty analysis as follows: Uncertainty analysis is an important part of experiment planning, management, and reporting. Proper uncertainty analysis is important, but during the planning and testing stages, applying uncertainty analysis is more important. Coleman and Steele. (1998, 2018) state that uncertainty analysis is a process that needs to be followed during all experimental stages of any research program, from the program planning stage, through design and instrumentation, installation and test set-up, actual testing, data acquisition and data reduction analysis, all the way to final reporting. Hermanski et al. (2001) Experiments with regular and irregular waves are carried out on still water. Self-propelled ship models have been tested at various speeds and bow orientations. The experimental results of the ship must be performed by an uncertainty analysis (UA). The total calculated uncertainty is less than 2%. An overall uncertainty value of 2% is considered acceptable and satisfactory. For a large total uncertainty (> 2%), the value of U is dominated by the random uncertainty P component. In experiments with irregular waves, both systematic and random uncertainties are equivalent. Tests run on irregular waves have lower random uncertainty than the other two test types. Irvin et al. (2008) Normal head waves introduced an uncertainty assessment of the vertical movement of surface fighters. Normal head waves in uncertainty assessment of the vertical movement of surface fighters provided bias estimates and random accuracy limits. The overall uncertainty of root thumb squares (RSS) provided bias estimates and random accuracy limits. also researched ballast parameters, time course, Fourier series amplitude, transfer function, and heave and pitch phase uncertainty. According to Kim et al. (2011) and Park et al. (2013), the short simulation time is the different simulation code of the main difference between parametric rolling simulations. Most of the existing studies are too short and inaccurate, with nonlinear roll motion based on lengthy and multiple simulations. Parametric rolling of prediction of the statistical characteristics. In addition, parametric roll simulation of a large container ship at the GM metacentric height error sensitivity was observed by running a Monte Carlo simulation. Yonghwan Kim and Greg Hermansky (2014) some of the technical issues entered into numerical analysis and seakeeping load analysis on loads and response to waves and presented uncertainties in seakeeping analysis and ITTC procedures. Durante et al. (2020) expressed that regular wave uncertainty quantification models are used to estimate the relevant statistical estimators. The statistical uncertainty of the expected value and standard deviation were evaluated by autocovariance and block-bootstrap analysis. A statistically converged experimental benchmark study of the catamaran model in irregular waves was carried out in conjunction with the regular wave quantification uncertainty (UQ) model, which was used to approximate the relevant statistical estimator. The regular wave quantification uncertainty (UQ) model and irregular waves were carried out in conjunction with the use of approximate the relevant statistical estimator at in statistically converged experimental benchmark study of the catamaran model. Autocovariance analysis and the block-bootstrap method evaluate statistical uncertainty related to expected values (EV) and standard deviation (SD) of the main variables. D. Purnamasari et al. (2020) presented quantify the uncertainty of the drag data of the model tanker obtained from the towing tank test. Drag in the velocity range (1029-1.543 m / s). By post-processing, the raw resistance test data using a standard Kalman filter algorithm, the standard deviation of the measurement is minimized. Parametric models have been proposed in the form of Kalman filters, and simulation studies have presented useful techniques for adjusting the noise covariance matrix. D. Purnamasari et al. (2020) studies for verification and validation were carried out using the International Towing Tank Conference's methodology and guidelines. The accuracy of the resistance results was dependent on the mesh's grid quality. The CFD findings for the smallest mesh were fairly similar to the outcomes of the experiment. I K A P Utama et al. (2021) carry out estimation uncertainty benchmark testing in order to enhance the accuracy of resistance. From 2010 through 2018, the Indonesian Hydrodynamic Laboratory used the IHL 007 Ro-Ro ferry model supplied by the Maritime Research Institute Netherlands (MARIN) for benchmark testing. There was reasonable agreement in case studies with speed ranges of 1.24 m/s to 2.47 m/s. Resistance calculations have proven to be far more accurate when using KF-based state estimates than when using LPF-based state estimations, hence KF has been suggested as the preferred choice.

The purpose of this study is to identify the geometry uncertainty and mass distribution and to calculate the calibration uncertainty of measuring instruments for seakeeping tests, using the ITTC 7.5-02-07-02.1 procedure based on ISO-GUM.

2. Method

The seakeeping experiment of the benchmark model was conducted at the MOB-IHL. The testing will take place under full load conditions with regular wave conditions. The model has full appendages as well as a rudder and propellers to allow for free-running tests so that it can move within 6 DoF (surge, sway, heave, roll, pitch, and yaw). Motion data is obtained, when the wireless optical tracking system records the model being tested. The target sensor installed in the model will be detected by motion tracking equipment for further recording in the form of model movement data, which is expected to better describe the phenomenon of ship motion while sailing. The model test has a scale ratio of 1:62. Statistical analysis of measurement results is presented to determine the quality of the experiment data. Seakeeping testing is usually used to describe all aspects of ship performance in undulating waters that are closely related to the field of hydrodynamics, to see the ship's ability to sail safely at sea for a certain period, especially in bad weather conditions, or to see the ship's response in undulating waters, which is mainly affected by 6 degrees of freedom of movement, related to passenger safety and comfort, load movement, bending moment, and others.

The model used in this test of seakeeping is shown in Figure 1. The material for making the model is made of wood covered with glass fibre and finished with paint for a smooth surface. In the process of modelling, the dimensions have been checked on a 3-axis measuring table. Table 1 below explains the main sizes of the models. Before seakeeping testing, the static similarity of the model was adjusted to approximate real conditions.

Figure 1. Model of benchmark seakeeping test.

Table 1. Main sizes of models.

Items	Symbol	Unit	Value	
			Model	Prototype
Scale			1:62	1:1
Length Over All	LOA	m	3,000	186,000
Length of Water Line	LWL	m	2,930	181,673
Length Between Perpendiculars	Lpp	m	2,891	179,270
Breadth	B	m	0,500	31,000
Depth	H	m	0,119	7,400
Draught	T	m	0,077	4,800
Vertical Centre of Gravity	$KG_{(from\ base)}$	m	0,102	6,900
Longitudinal Cent of Gravity	$LCG_{(from\ AP)}$	m	1,398	86,700

In this test, the factors in question include:

- The first calculation of the geometric uncertainty of the model.
- The second calculation of the uncertainty of the free-running system.
- The third calculation of the uncertainty of the model test.
- The fourth calculation of the uncertainty of the combination results.

2.1. The ISO-GUM approach
The method recommended by ISO-GUM is to classify the elements of uncertainty, type A and type B.

2.2. Sources of uncertainty.
2.2.1. Type A uncertainty. Type A uncertainty is evaluated by statistical methods based on repeated experiments i.e:

- Measurement dimensions of the model at the marking table.
- Measurement of swing table.
- Measurements in the MOB test pool before the seakeeping test is carried out.
- Measurements in the MOB test pool during the seakeeping test.

The measurement of seakeeping performance is planned to be carried out 3 times with experimental repetitions on regular wave conditions, with a significant wave height of Hw = 3.5 meters and variations in wave period. Tw = 8.3 sec., 10.7 sec., and 12.5 sec. At headings 180 and 90 degrees, the ship's speed Vs = 12 knots. The dominant motion measurements are heave and pitch.

The factors that affect the type A uncertainty in seakeeping testing include:

- Model heading direction.
- Model speed.
- Elevation of water waves.
- Wavelength.
- Heave motion.
- Pitch motion.

2.2.2. Type B uncertainty. Type B uncertainty is an uncertainty value obtained by means other than statistics, using available information, including model geometry, measurement tools, and calibration. The cause of type B uncertainty in seakeeping testing at IHL. In general, there is uncertainty in the calculation of geometry uncertainty in the model and mass distribution and calculation of calibration uncertainty on measuring instruments. The respective uncertainties must be defined and presented.

Table 2 below, explains the classification of uncertainty types type A and type B, in seakeeping experiments.

Table 2. Uncertainty classification.

No.	Items	Uncertainty Type A	Uncertainty Type B
1	Model geometry	Measurement dimensions of the model at Marking table.	Manufacture specification of Marking table.
2	Model mass distribution	Measurement in swing table.	Manufacture specification of stopwatch, and electric hanging scale.
3	Calibration	Measurements in the MOB test pool before the seakeeping test is carried out.	Manufacture specification of sensors: wave height sensor, qualisys surface motion, and the marker (wand).

Sensors and equipment used consist of:

- Marking table type AKL 16, serial number 4683/1Bj.01/92, manufactured by Zett Mess.
- Stopwatch type SW-02, manufactured by Casio.
- Electric Hanging Scale type CCS 600 K serial number NSC20036, manufactured by UWE.
- Wave Height Sensor type HRIA-1014 (600mm), manufactured by HR Wallingford.
- Wave Height Sensor type HRIA-1018, manufactured by HR Wallingford.
- Qualisys Surface Motion Camera type Oqus 300+, produced by Qualisys AB.
- Qualisys Surface Motion Camera type Oqus 700+, produced by Qualisys AB.
- The marker (wand) used is a long-range active marker, produced by Qualisys.

3. Results and Discussion

ITTC procedure 7.5-02-07-02.1 as a reference basis for uncertainty analysis of seakeeping tests carried out at the MOB facility at IHL. Each experiment has an uncertainty (u), and the values obtained have an effect on the experiment results. As depicted in figure 2, the factors in question can be grouped into several parts, i.e: uncertainty in terms of geometry, calibration, free-running system, and repeated measurements.

Then, having obtained this uncertainty, the combined uncertainty can be calculated and the uncertainty expanded. A further explanation of the factors that affect the uncertainty of seakeeping testing is presented below:

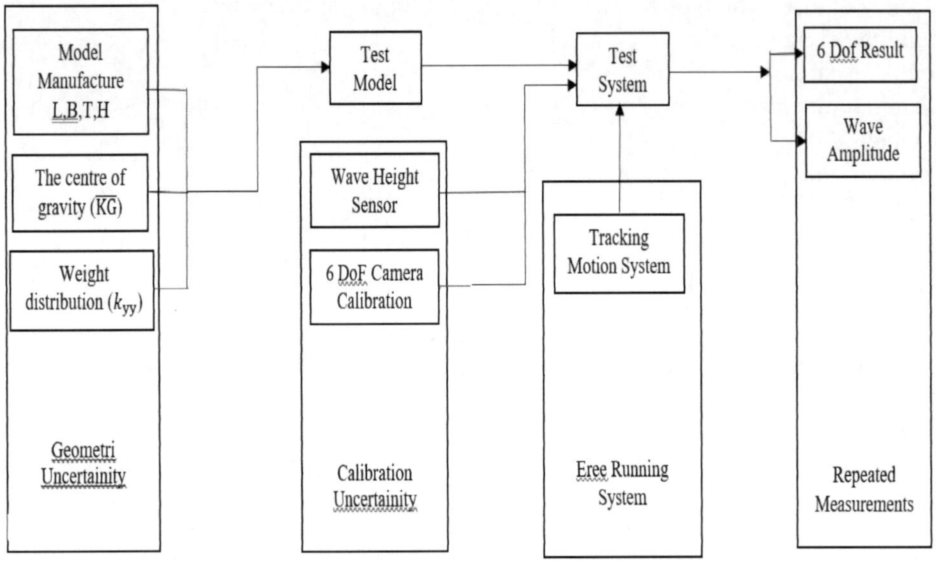

Figure 2. Factors that affect the uncertainty of seakeeping testing

3.1. Sources of uncertainty in seakeeping testing

Essential body movement in seakeeping experiments is a minimum requirement. Factors that affect seakeeping uncertainty consist of the calculation of the geometry uncertainty in the model and mass distribution, the calculation of the calibration uncertainty on the measuring instrument, the direction of the model heading, the velocity of the model, the elevation of the water wave, the wavelength, the heave movement, and the pitch motion. All parameters of the test procedure are subject to type A and type B uncertainties.

Based on model testing, type A uncertainty is evaluated. The benchmark model seakeeping test was carried out 3 times with repeated experiments on regular wave conditions and a significant wave height of Hw = 3.5 meters, with variations in wave periods of Tw = 8.3 seconds, 10.7 seconds, and 12.5 seconds. At heading 90 and 180 degrees with the ship's speed of Vs = 12 knots. model test measurement data in the form of: model speed (V (m/s)), wave elevation (amplitude za (m), period Tw (s)), model heading direction (yaw motion (deg)), heave and pitch values displayed in the form of a non-dimensional response amplitude operator (RAO) of heave and pitch motion.

Type B uncertainty are the geometry represented by the dimensions of the model (length, breadth, draught and depth), mass distribution represented by $\overline{KG}$ and k_{yy}, and preparing data for the uncertainty value of the measuring instrument used, namely the uncertainty of the wave height sensor (WHS) U_W is obtained from a report on the calibration results of the equipment. The uncertainty of qualisys motion tracking U_Q is obtained from the calibration of the qualysis system carried out for each test heading. The marker (wand) length of 1000 mm is obtained from the certificate of calibration of the equipment, the uncertainty of electric hanging scales is obtained from the certificate of calibration of the equipment, the uncertainty of the marking table U_{mt} is obtained from the certificate of calibration of the equipment, and the uncertainty of the stopwatch U_{st} is obtained from the certificate of calibration of the equipment.

In this study, only type B uncertainty affecting seakeeping uncertainty is discussed, the uncertainty of each of the model geometry, $\overline{KG}$, k_{yy}. The uncertainty of the calibration of the surface motion camera quality, the 90 and 180 deg headings and the uncertainty of the wave height sensor calibration, and instrument equipment uncertainty, means the uncertainty must be determined. The progress of this research ends here because this study has been achieved.

3.2. Geometry Uncertainity

3.2.1. Calculation of geometry uncertainty in the model and mass distribution. The sources of the geometry represented by the dimensions of the model (length, breadth, draught and depth), mass distribution represented by $\overline{KG}$ and k_{yy} should be determined as uncertainty. In the process of making the model, the dimensions have been checked through a 3-axis measuring table/marking table, and data from the measurement results of the model in the form of length Lpp (m), breadth B (m), draught T (m) and depth H (m) from the measurement results in the marking table is presented in table 3, ITTC (2017, Rev. 04), and the measurement process is presented as shown in Figure 3.

To calculate the uncertainty of the dimensions (length, breadth, draught and depth) it takes the deviation value of each parameter and the main size of the model as in the equation (1).

$$U_{dimensions} = \frac{\text{deviation}}{\text{main dimensions}} 100\% \quad (1)$$

From each component, the uncertainty is calculated.
U_{Lpp} = uncertainty for length Lpp.
U_B = uncertainty for width B.
U_T = uncertainty for draught T.
U_H = uncertainty for depth H.

Model main general, i.e. Model of Width (Y) and Depth (Z). The recommended acceptance error for the model length is 1.0 mm, and the model length tolerance is 0.05% LPP or 1.0 mm. The uncertainty of the model geometry (length, breadth, and draught) is determined based on experience in modelling. If with a 95% confidence level, then the extended uncertainty value in the model length is +/-, using the confidence factor in table 4. it is possible to determine the representative value using the normal distribution.

Table 3. The measurement results of benchmark the ship model in the marking table

No.	Items	Symbol	Unit	Size (m)			Tolerance
				Design	Actual	Deviation	
1	Length Between Perpendiculars	Lpp	m	2,891	2,891.34	+0,00034	± 0,001 m
2	Breadth	B	m	0,500	0,499.79	- 0,00021	± 0,001 m
3	Draught	T	m	0,077	0,077.01	+0,00001	± 0,001 m
4	Depth	H	m	0,119	0,119.03	+0,00013	± 0,001 m

Figures 3. Main dimension check of benchmark ship model in marking table.

Table 4. Factors for each confidence level.

Level of confidence (%)	Factor
50.00	0.6757
68.27	1.0000
90.00	1.6450
95.00	1.9600
99.00	2.5760
99.73	3.0000

Meanwhile, to estimate the uncertainty of $\overline{KG}$, and k_{yy}, the formula used to calculate each value uses uncertainty propagation. The uncertainty in the $\overline{KG}$ model is as in equation (2), while the k_{yy} uncertainty is obtained from several measurements as in equation (4), the k_{yy}, measurement data is presented in table 5, and the measurement process is presented as shown in Figures 4 and 5. the formula used to calculate each value using uncertainty propagation. The following formula is used: $\overline{KG} = \overline{KB} + \overline{BM} - \overline{GM}$. The vertical centre of buoyancy $\overline{KB}$ and the transverse metacentre ($\overline{GM}$) depend on the geometry, while the metacentric height ($\overline{GM}$) is from the inclining test, the results obtained.

Using the known standard uncertainties of the main parameters and assuming a simple hull shape geometry evaluation of the $\overline{BM}$ standard uncertainty. For example, calculating the triangular prism of the transverse $\overline{BM}$. The aggregate uncertainty KG can be calculated using the following formula:

$$u_c(KG) = \left[\left(\frac{\partial KG}{\partial KB}\right)^2 u^2(\overline{KB}) + \left(\frac{\partial KG}{\partial BM}\right)^2 u^2(\overline{BM}) + \left(\frac{\partial KG}{\partial GM}\right)^2 u^2(\overline{GM})\right]^{1/2} \quad (2)$$

Each component, the uncertainty is calculated.
U_{KG} = uncertainty for the position of the centre of gravity $\overline{KG}$

In general, the uncertainty value is obtained by equation (3) below.

$$U = \sqrt{\frac{1}{N-1} \sum_{i=1}^{N} (x_i - \bar{\bar{x}})^2} \quad (3)$$

From each component, the uncertainty is calculated.
$x_i =$ measurement value to -i.
N = the number of measurements taken.
U = measurement uncertainty.

$$U_{k_{yy}} = \frac{\sqrt{\frac{1}{N-1}\sum_{i=1}^{N}(x_i-\bar{\bar{x}})^2}}{\bar{\bar{x}}} \quad (4)$$

Each component, the uncertainty is calculated.
U_{kyy} = uncertainty for the weight distribution represented by the radius of gyration k_{yy}.

Table 5. The results of the measurement of the weight distribution of the model (k_{yy}) on the swing table

No.	Items	Symbol	Actual Data	Unit
1	First measurement	x_1	20.430	second
2	Second measurement	x_2	20.328	second
3	Third measurement	x_3	20.301	second
	Measuring average	$\bar{x}$	20.353	second

Measured natural period Tn = 4.071 sec.
Measured natural frequency Wn = 1.543 rad/sec
Measured mass moment of inertia Iym = 48.104 kg m^2
Measured radius of gyration kym = 0.713 m

Figure 4. Setting the weight distribution of the model (k_{yy}) in the swing table.

Figure 5. Measurement of the weight distribution of the model (k_{yy}) on the swing table

3.3. Calibration Uncertainty

3.3.1. Calculation of calibration uncertainty on measuring instruments. Prepare data for measuring instrument uncertainty values, wave height sensor instruments, qualisys surface motion cameras, long-range active markers, stopwatch, marking table and electric hanging scales must be calibrated before seakeeping testing. The calibration of the wave height sensor and qualisys surface motion camera instruments were carried out by IHL internal parties. In the qualisys surface motion camera calibration, the marker used is a long-range active marker with a size of 1 meter. The qualisys surface motion camera calibration process is carried out in 2 positions, namely the 180-degree position and the 90-degree position. The results of this calibration can be used to measure 6 DoF in the seakeeping free running test at both 90 and 180 deg. The marker (wand) is the manufacturer's calibrate, while the calibration of the electric hanging scales and stopwatch are carried out by an external party, namely Caltesys. The calibration of the marking table is carried out by an external party, namely metal industries development centre (MIDC) Indonesia. End-to-end in situ calibration procedures are applied to all instruments. before seakeeping experiments, instrument calibration must be carried out first, as calibration minimizes instrument errors. The instrument calibration process is subject to uncertainty.

The data for the uncertainty value of the measuring instrument used, namely the uncertainty of the wave height sensor (WHS) U_{whs} is obtained from the report on the calibration results of the equipment, as shown in Tables 8 and the graph is presented as shown in figure 6. The uncertainty of qualisys motion tracking U_{qmt} is obtained from the calibration of the qualysis system carried out for each test heading, as shown in tables 6 and 7. The marker (wand) length of 1000 mm is obtained from the certificate of calibration of the equipment with a result calibration is 1001 mm and relative accuracy ± 0,1 mm (2σ). The uncertainty of electric hanging scales is obtained from the certificate of calibration of the equipment. Calibration results are traceable to the international system of units (SI) through LK-023-IDN. The uncertainty of the marking table U_{mt} is obtained from the certificate of calibration of the equipment that is 15 µm with a confidence level of approximately 95%, with a coverage factor k=2, and the uncertainty of the stopwatch U_{st} is obtained from the certificate of calibration of the equipment that is ± 0,65 second with a confidence level of approximately 95%, with a coverage factor k=2.
The instrument calibration results are presented below:

- Qualisys surface motion camera.

Table 6. Calibration result data qualisys surface motion camera, heading: 90 deg.

Camera result					
Camera	X (mm)	Y (mm)	Z (mm)	Points	Avg.residual (mm)
01	-12332.80	-18922.58	2398.35	454	1.28968
02	-13473.60	-12608.62	2414.33	912	0.70859
03	-13945.39	- 6929.84	3101.01	896	0.93020
04	Unused camera	-	-	-	-
05	Unused camera	-	-	-	-
06	-16185.77	12593.68	2514.32	310	2.70550

IOP Conf. Series: Earth and Environmental Science 1081 (2022) 012021 doi:10.1088/1755-1315/1081/1/012021

Table 7. Calibration result data qualisys surface motion camera, heading: 180 deg.

Camera result					
Camera	X (mm)	Y (mm)	Z (mm)	Points	Avg.residual (mm)
01	15656.67	-13020.54	2680.12	1377	0.29480
02	9284.11	-13426.15	2680.18	2387	0.29693
03	3585.96	-13450.28	3234.37	2571	0.20314
04	- 2948.09	-13920.05	3368.38	2562	0.20220
05	-11014.59	-14255.94	2512.41	2405	0.27367
06	-16041.78	-14124.28	2411.18	1056	0.28774

- Wave height sensor.

Table 8. Calibration result data wave height sensor.

Output Voltage (volt)	Water Depth (cm)
8.151	26
7.544	24
6.934	22
6.309	20
5.687	18
5.041	16
4.431	14
3.800	12
3.174	10
2.548	8
1.905	6
1.281	4
0.657	2
-0.003	0
-0.636	-2
-1.298	-4
-1.947	-6
-2.575	-8
-3.184	-10
-3.880	-12
-4.512	-14
-5.178	-16
-5.828	-18
-6.469	-20
-7.104	-22
-7.763	-24
-8.405	-26

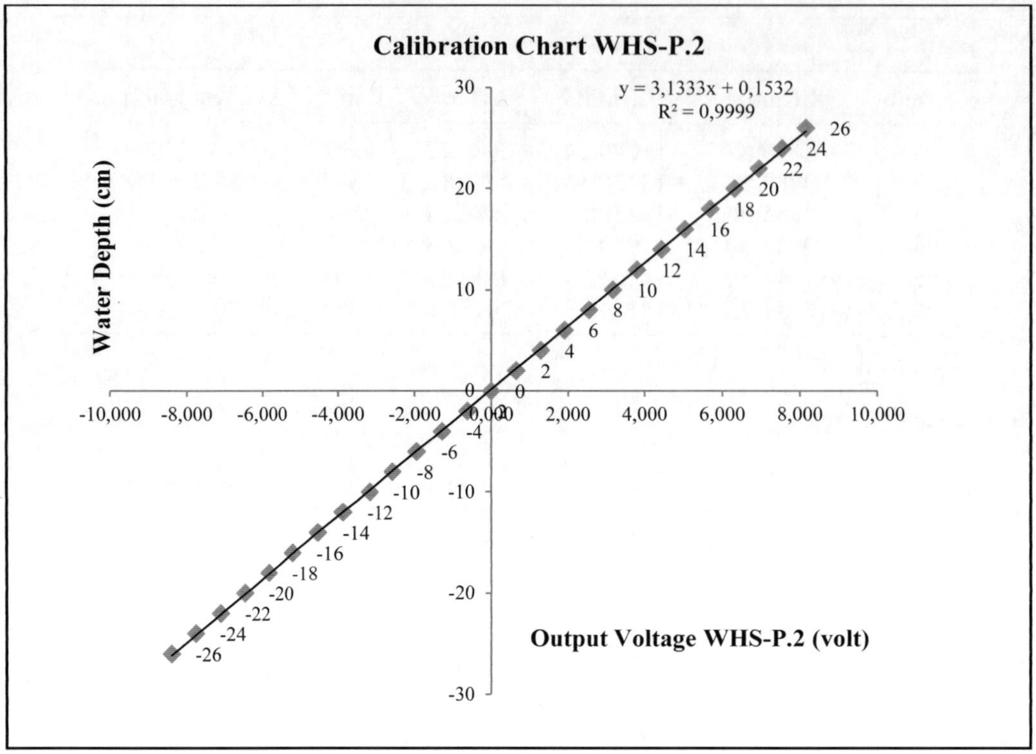

Figure 6. Calibration result chart wave height sensor.

Calculation of uncertainty in the benchmark model seakeeping test is carried out according to ITTC procedure 7.5-02-07-02.1. The data from the uncertainty calculation of the geometry model, the mass distribution, and the uncertainty calculation of the equipment used is the wave probe and qualisys motion tracking obtained from the calibration results of each device performed internally (IHL), shown in table 9. For qualisys motion tracking, calibration is carried out for each test heading.

In calculating the uncertainty in the seakeeping test of the benchmark ship model, a 95% confidence level is used. Therefore, the value of the confidence level factor is 1.96 according to the ITTC - procedure 7.5-02-07-02.1 procedure which can be seen in Table 4.

Table 9. The results of the calculation of the uncertainty of geometry, mass distribution and uncertainty of the results of the calibration of the instrument.

Source of Uncertainty	Value		Unit	Type A Uncertainty (%)	Type B Uncertainty (%)
	Prototype	Model			
Lpp	179.27	2.891	m	-	0.01176
B	31.00	0.500	m	-	0.042
T	4.80	0.077	m	-	0.01298
H	7.40	0.119	m	-	0.10924
KG	6.90	0.102	m	-	0.08026
k_{yy}	42.966	0.693	m	-	0.5108

Qualysis Motion Tracking 180 deg	-	-	m	-	0.00032
Qualysis Motion Tracking 90 deg	-	-	m	-	0.00859
Wave Height Sensor	-	-	m	-	0.0027

Conclusions

The purpose of the uncertainty analysis research is a seakeeping benchmark model testing study, namely to determine the measurement value of geometry uncertainty, mass distribution and instrument calibration results. Measurement of geometry uncertainty and mass distribution consists of model dimensions (length, width, draught), $\overline{KG}$, and k_{yy}. Measurement of the uncertainty of the results of the calibration of the tool consisting of a wave height sensor (WHS) and qualisys motion tracking and the objectives of this study have been achieved.

The method used in this study types A and type B uncertainty as recommended by ISO-GUM and in accordance with ITTC-Recommended Procedures and Guidelines 7.5-02-07-02.1.

The results of the uncertainty analysis research of this benchmark seakeeping model test study yielded the uncertainty values obtained from:

- The total geometric uncertainty value (Lpp, B, T, H, $\overline{KG}$, and k_{yy}) is 0.17598 %.
- The total uncertainty value of instrument calibration results (wave height sensor and qualisys motion tracking) is 0.01161%.

References

[1] D. Durante, R. Broglia, M. Diez, A. Olivieri, E.F. Campana and F.Stern 2020 Accurate experimental benchmark study of a catamaran in regular and irregular head waves including uncertainty quantification *Jurnal Ocean Engineering* Volume **195**

[2] Greg Hermanski, Ahmed Derradji-Aouat and Peter Hackett May 23 – 26, 2001 Uncertainty analysis - preliminary data error estimation for ship model experiments, *6th Canadian Marine Hydro-mechanics and Structures Conference* (Vancouver, BC, Canada)

[3] Hugh W. Coleman and W. Glenn Steele 1998 Experimentation and uncertainty analysis for engineers Second Edition Book *John Wiley & Sons, Inc*

[4] Hugh W. Coleman and W. Glenn Steele 2018 Experimentation, validation and uncertainty analysis for engineers Fourth Edition Book *John Wiley & Sons, Inc*

[5] Dian Purnamasari, I Ketut Aria Pria Utama, I Ketut Suastika and Giles Anthony Thomas 2020 Application of kalman filter to the uncertainty of model resistance data obtained from experiment *Journal of Engineering Science and Technology* Volume **15** No. 2 1455 - 1465

[6] Dian Purnamasari, I Ketut Aria Pria Utama and I Ketut Suastika 2020 Verification and validation of a resistance model for tanker 17.500 DWT *Journal of Marine Science and Technology* Volume **28** Iss. 1 Article 3.DOI: 10.6119/JMST.202002_28(1).0003

[7] I Ketut Aria Pria Utama, Dian Purnamasari, I Ketut Suastika, Nurhadi and Giles Anthony Thomas 2021 Toward improvement of resistance testing reliability *Journal of Engineering and Technological Sciences* Volume **53** No. 2

[8] Irvine, M., Longo, J. and Stern, F 2008 Pitch and heave tests and uncertainty assessment for a surface combatant in regular head waves *Journal Ship Res.* 52 (2), 146–163

[9] ITTC 1978 Report of Seakeeping Committee *In: Proceedings of the 15th ITTC*

[10] ITTC 1981 Report of Seakeeping Committee *In: Proceedings of the 16th ITTC*

ITTC 1990 Report of the panel on validation procedure *In: Proceedings of the 19th ITTC* (Madrid)

[11] ITTC 2008 Guide in the Expression of Uncertainty in Experimental Hydrodynamic ITTC Procedure 7.5-02- 01-01, Revision 01.

[12] ITTC 2014 Recommended Procedures and Guidelines 7.5-02-05-04 Rev. 01
[13] ITTC 2017 Recommended Procedures and Guidelines 7.5-02-07-02.1 Rev. 06
[14] JCGM 100: 2008 GUM 1995 with Minor Corrections September 2008 Evaluation of measurement data-guide to the expression of uncertainty in measurement First edition
[15] Y. Kim, D.M. Park and T. Kim 2011 On uncertainty in numerical analysis of parametric roll *In: Proceedings of the 12th International Ship Stability Workshop* (Washington DC, USA)
[16] Kline, S. J 1985 The purpose of uncertainty analysis *Journal of Fluids Engineering* Vol.**10**/153
[17] Lisa M, Minnick, Lauren, W. Hanyok, Heather A. Tomaszek, Mark P. Melendez, Charles R. Turner, Joel T. Park, Vadim Belenky, Christopher C. Bassler *July 2012 Model-scale experiment of the seakeeping performance for R/V melville, Model 5720 Hydromechanics Department Report* NSWCCD-50-TR-2012/035
[18] Nandiko Rizal 2018 Seakeeping test of ship perambuan model at maneuovering and ocean engineering basin (MOB) Laboratory for Hydrodynamics Technology *Journal ISOMAse/JOMAse*/A.L/30.9.18/P.**59.3**
[19] Dong-Min Park, Yonghwan Kim and Kang-Hyun Song 15 July 2013 Sensitivity in numerical analysis of parametric roll *Journal Ocean Engineering* Volume **67**, pages 1–12
[20] Yonghwan Kim and Greg Hermansky 1 August 2014 Uncertainties in seakeeping analysis and related loads and response procedures *Journal Ocean Engineering* Volume **86**, Pages 68-81

Acknowledgements
The first author expressed her gratitude to The National Research and Innovation Agency (BRIN) which funded her Master's research-based education program at ITS Surabaya under contract number: SP/110/BPPT/08/2021. The author thanks Dr. Dian Purnama Sari, ST., MT., Beny Cahyono, ST., MT., Ph.D., Dr. rer.pol. Dedy Dwi Prastyo, S.Si, M.S., Erdina Arianti, ST., MT., and Baharuddin Ali, ST., MT., who always provide guidance so that this paper can be produced properly. The authors would like to thank the staff members of the Indonesian Hydrodynamic Laboratory for their support during the experiment.

IOP Conf. Series: Earth and Environmental Science 1081 (2022) 012022 doi:10.1088/1755-1315/1081/1/012022

Analysis of Uncertainties in Open Water Test of B-Series Propeller at Indonesian Hydrodynamic Laboratory

A Machfudin[1,2], A A B Dinariyana[1], D Purnamasari[2]

[1] Department of Marine Engineering, Institut Teknologi Sepuluh Nopember, Surabaya - Indonesia

[2] The National Research and Innovation Agency (BRIN), Surabaya – Indonesia

Abstract. A propeller's performance characteristics in open water typically refer to the variation of its thrust, torque, and efficiency with speed and revolution rate. This investigation provides a method for gaining a better understanding of the achievable accuracy. of the open water propeller test and presents an example of how to conduct a comprehensive uncertainty analysis of prediction models4 blades B-series model propeller, diameter 157.59 mm, pitch ratio 1.133 and Blade Area Ratio (BAR) 0.65 was carried out for this study. The methodologies presented using the ITTC recommended procedure to obtain individual measurement systems, measurement of individual variables, data reduction and experimental results. The bias limits for the individual measurement systems are estimated: propeller geometry, speed, thrust, torque, propeller rate of revolution, and temperature/density. The precision limits for single and multiple runs have been determined for the thrust coefficient (K_T) and torque coefficient (K_Q) for one advance coefficient (J), 5 sets of tests with 3-speed measurements in each set are conducted, totaling 15 test runs. The experimental test results have represented as relative numbers the bias for K_T represents 98.1 % and the bias for K_Q represents 99.4 % of the total uncertainty for the mean value of 15 tests. The uncertainty analysis results of the open water propeller were recommended in an acceptable range of the 95% level of confidence.

1. Introduction

The open water propeller test is used in analyzing the propulsion capability and estimating the required power, ITTC (*International Towing Tank Conference*) [1]. The Open Water Test is carried out by rotating the propeller at a certain speed by varying the speed to get the measured values for speed, rpm, thrust and torque. It was performed with the propeller alone in uniform flow to establish basic thrust, torque and propeller efficiency. A towing tank is used to conduct methodical propeller testing. The propeller is driven by a streamlined housing that receives power from the carriage. It is moved along the tank with the propeller in front of the housing, effectively immersing the propeller in uninterrupted water. For a variety of carriage speeds and propeller revolutions (advance coefficient)., thrust and torque measurements are taken. These tests remove cavitation and give information on the propeller inflow conditions. The earliest modern uncertainty analysis concept was introduced by Kline and McClintock [2]. They introduced notions like single-sample and multiple-sample uncertainty, uncertainty interval and the law of uncertainty propagation. Since the 24th ITTC in 2005, ITTC member organizations have demanded the use of ISO [3] rather than AIAA [4] or ASME [5]. AIAA and ASME are both American organizations, and ISO was regarded as the legitimate international organization for the establishment of guidelines and standards. At the ITTC 2008 conference, the International Organization for Standardization (ISO) Guide to the Expression of Uncertainty in Measurements (ISO-GUM) was approved for uncertainty analysis by the ITTC community and its members' organizations. Based on the technique through which they are evaluated, the ISO-GUM recommends categorizing uncertainty components into two types: type A and type B., Giacomo [6] and ISO [3]

Become the basic guideline that there is inherent uncertainty in all measured results when conducting all types of physical experiments. A test result is just one example of the experiment's various possible outcomes. It is rather unlikely that the exact same measurement result will be obtained if the experiment is repeated. It's crucial to know how far the measured value ranges from the unknown, true value. It must perform to determine such an uncertainty analysis, Steen [7]. According to Kline [8], Uncertainty analysis is a necessary element of planning, controlling, and reporting experiments. The most crucial thing is to perform an acceptable uncertainty analysis. Using an uncertainty analysis at the planning and checkout stages of an experiment is particularly essential. In experimental work, uncertainty analysis is at the essence of quality control. Measurement uncertainty is the difference between the measurement results and the actual value of the object being measured. The nature of uncertainty is that it is possible to occur with a maximum difference between the measured and ac. Uncertainty is a parameter that determines the range of values in which the estimated correct value being measured is Kantasubrata[9]; KAN [10].

Experimental uncertainty assessments are critical for identifying risks in design when using data directly or calibration and or qualifying simulation methods**,** Longo and Stern [11]. As a research institution that demands high accuracy results and low error rates, the Indonesian Hydrodynamic Laboratory (IHL) as a member of ITTC has investigated the uncertainty of the resistance test, Utama et al [12]; Purnamasari et al [13]; [14]. Taylor [15] assessed the uncertainties in measurements using equipment to measure podded propulsor performance parameters using the methodology presented by Bose and Luznik [16]. the uncertainty in propeller performance and pod unit thrust in the analysis without resolving the uncertainty in unit forces and moments measurements. Islam et al [17] provided the results of uncertainty analysis for measuring the forces and moments of a podded propulsor but did not go into details about the analysis.

Wei et al [18] research the latest open water propeller uncertainty analysis. The researcher performed an uncertainty analysis using ISO GUM's standards for determining the uncertainty value are compared it to the results for uncertainty analysis using the Monte Carlo method. It was recommended the uncertainty analysis using the ISO GUM method is proven to be applicable and its reliability can be accounted for. Choi et al [19] analyze the uncertainty of the open water propeller test and concluded that uncertainty analysis can be effective in discovering dominant error sources and therefore improving experimental measurement accuracy.

The main objective of this study is to provide uncertainty analysis for standard open water propeller performance coefficients using a systematic methodology. The experiment was conducted at the Laboratory Hydrodynamic Indonesia. Geometry propeller, revolution, speed, and multiple tests are used to determine open water propeller performance parameters. For the precision limits, 5 sets of experiments with 3-speed measurements in each set have been undertaken to determine the thrust coefficient (KT) and torque coefficient (KQ) for three advance coefficients (J). The paper is recommended that 15 tests points with three speeds be performed to obtain acceptable uncertainty values.

This paper is organized in the following the uncertainty analysis methodology of the open water experiments, as well as data reduction equations and definitions, and also measuring systems (instrumentation and data acquisition and reduction). It described discussions of bias and precision limits, total uncertainties and methods for uncertainty. Finally, conclusions and suggestions are presented. Ideally, the availability of such extensive analysis and benchmark values will encourage and enable the application of uncertainty analysis in a wider range of situations.

2. Method uncertainty

2.1 Open water propeller method

The focus of this research is to obtain a better knowledge of the open water propeller test's achievable accuracy and provides an example of how a thorough uncertainty analysis of prediction models can be performed. 4 blades B-series model propeller, diameter 157.59 mm, pitch ratio 1.133 and Blade Area Ratio (BAR) 0.65 was carried out for this research. The propeller is a standardized propeller that was frequently used for open water testing in towing tank at the Indonesia Hydrodynamic Laboratory and it was created by a CNC machine with a high precision margin value. Experimental uncertainty analysis was carried out using the ITTC 2014 method of uncertainty analysis for open water test [20] following the ITTC Procedures 7.5-02-01-01, Rev.00 [21] and 7.5-02-01-02, Rev.00[22]. The methodologies presented using the ITTC recommended procedure, the total uncertainty will be obtained for the individual measurement systems, measurement of individual variables, data reduction and experimental results. The Biases and precision limits are combined using a root mean square, end-to-end and standard error estimate (SEE) methods to determine the total uncertainty estimates for each of variable interest. For each of the following measurement systems, bias limits are estimated: propeller geometry, speed, thrust, torque, propeller rate of revolution, and temperature/density. The beginning stages in uncertainty analysis are determined bias limits of geometry B_D, speed advance B_v, rate of revolution B_n, density $B\rho$, thrust B_T and torque B_Q as well as K_T, K_Q, and J 's partial derivatives. The thrust coefficient (K_T), torque coefficient (K_Q), and advance coefficient (J) are calculated using the standard error estimate (SEE) method. The precision limits are determined to provide estimates mean, standard deviation of the thrust coefficient K_T, the torque coefficient K_Q and the advance coefficient J for single and multiple tests. The Biases and precision uncertainty are combined using a root mean square method to determine the total uncertainty.

2.2 Equation of Measurement

The quantity Y being measured, defined as the measurand, is not measured directly, but it is determined from N other measured quantities $X_1, X_2, ...X_N$. Thus, the measurement equation:

$$Y = f(X_1, X_2, X_3...X_N) \tag{1}$$

The corrections as well as values that account for additional causes of variability, such as various observers, instrument calibrations, different laboratories, and times at which observations were made, are all included in the function f, along with the quantities Xi. (y) is an estimate of the measurand (Y) produced from equation (1) using the estimates xi for the values of the N quantities Xi. As a result, the result of the measurements is the estimate (y):

$$y = f(x_1, x_2, x_3, ...x_N) \tag{2}$$

The following are examples of typical measurement functions or data reduction equations of the propulsion performance functions of ITTC [23].
Advance ratio:

$$J = f(V, n, D) = \mathrm{V}/(nD) \tag{3}$$

Thrust coefficient:

$$K_T = f(T, \rho, D, n) = T/(\rho, D^4, n^2) \tag{4}$$

Torque coefficient:

$$K_Q = f(Q, \rho, D, n) = Q/(\rho, D^5, n^2) \tag{5}$$

From the equations shows that D is the diameter of the propeller (m), T is the torque (N.m), ρ referring to the density of water (kg/m3), n is the rate of rotation (1/s), and Q is the value of the thrust (N). As a result, K_Q values are derived from assessments of the variables D, Q, ρ and n, whereas KT assessments are created from variables D, T, ρ and n. The estimates for each quantity Q, T, and D are obtained from direct measurements while ρ is computed as a function of temperature, t in °C. The values for each parameter Q, T, and D come from direct measurements, whereas ρ is calculated as a function of temperature, t in degrees Celsius. The uncertainty in the measurement of y, which is symbolized by u(y), is obtained from the uncertainties of u(xi) through the predictions of the input xi in the measurement function, equation (2).

2.3 Uncertainty Classification

Standard Uncertainty, Combined Uncertainty, and Expanded Uncertainty are the three (3) kinds of uncertainty defined by ISO [2.] Standard uncertainty which is symbolized by a lowercase letter (u) is obtained by calculating the positive square root of the variation predicted. The standard uncertainty of the test measurement results is divided into 2 types: type A uncertainty and type B uncertainty. According to ISO GUM, the sources of measurement uncertainty are classified into two types: type A uncertainty and type B uncertainty. Type A uncertainty is assessed by statistical analysis of a set of data, while type B uncertainty is assessed by various sources of information other than statistical analysis. This information can come from calibration certificates or reference material certificates, tool specifications, or previous measurement data. In this experiment, for every three speed measurements and advance coefficients, type A uncertainty is calculated by repeating the test as many as 15 times. Type B uncertainty is calculated using data from geometric measurements, H-39 dynamometer (figure 1) calibration, carriage speed calibration, equipment setup and propeller model in the towing tank, and historical open water measurements.

Figure 1. H-39 Dynamometer.

The Combined Standard Uncertainty symbolized by (UC) is the value of the measurement function, f, which is derived from the uncertainty of several individual measurements, xi. The rule of uncertainty propagation from the measurement function or data reduction equation is used to calculate the combined uncertainty, equation (1). The following is the outcome for standalone or unrelated measurements:

$$u_c^2(y) = \sum_{i=1}^{N} c_i^2 u^2(x_i) \quad (6)$$

where c_i is the susceptibility coefficient, $\partial f/\partial x_i$. The cumulative uncertainty for properly related measurements is thus:

$$u_c(y) = \sum_{i=1}^{N} c_i u(x_i) \quad (7)$$

2.4 The rules in ITTC [20].
The main principle of Extended uncertainty (U) is obtained by computing the multiplication value of the combined uncertainty and scope factor (k). The scope factor (k) refers to a range of values around a measurement's result that may be predicted to cover a significant portion of a valuation distribution that could be assigned to the measurand. As a result, the numerical value for the scope factor (k) must be adjusted thus an interval $Y = y \pm U$ matching a specific level of confidence can be obtained. In laboratory hydrodynamics, k generally leads to a 95% confidence level. At the 95 percent confidence level, all ITTC conclusions will be provided with an extended uncertainty. ISO [2] denotes that a simpler method is more likely to be acceptable in measurement conditions, in which the measurement probability distribution is nearly normal or Gaussian. The central limit theory in statistics is efficiently used here. The distribution can be considered to be Gaussian if the degrees of freedom value is considerable ($v >$ 30) and for the 95 % confidence level, k will be given a score of 2. The inverse student (t) distribution at the 95 % level of confidence is the result of k with a lesser sample data. ITTC [24] contains comprehensive information and procedures for conducting uncertainty analysis calculations, containing standard, combined, and extended uncertainty.

3 System and Procedure for Measurement
Figure 2 illustrates the individual measurement systems, individual variable measurements, data reduction, and experimental data for the propeller open water tests.

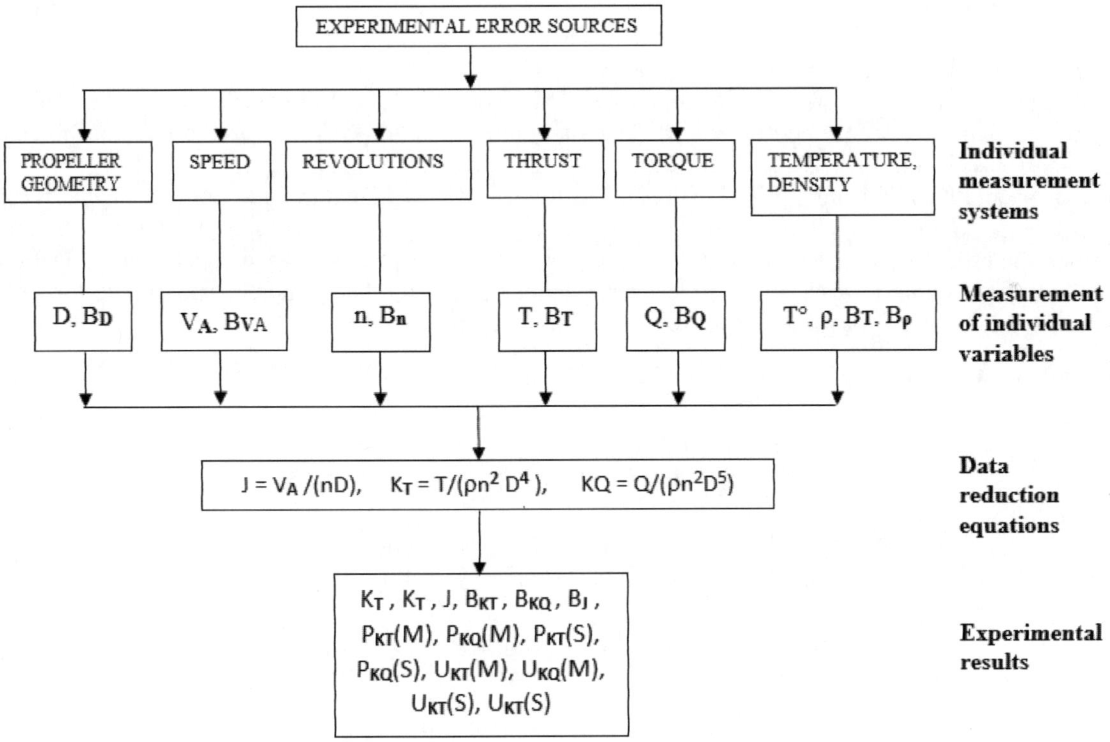

Figure 2 Propeller open water tests block diagram.

For each system of measuring, temperature/density, propeller geometry, speed, thrust, torque, and propeller rate of revolution will all be approximated as bias limits that contribute to the total uncertainty. For each measuring system, the elementary bias limitations are evaluated for the following categories: calibration, data acquisition, data reduction, and conceptual bias. The bias limits are instead decreased into B_{KT}, B_{Kq} and B_J, using the dimension reduction equations (3), (4), and (5). An end-to-end technique for multiple tests (M) and a single run (s) is used to determine the precision limits for the thrust coefficient (P_{KT}) and torque coefficient (P_{KQ}) in the model scale. The main characteristics of the model propeller, as well as the constants employed in the experiment, are described in Table 1.

Table 1. Specifications and constants for propellers.

Definition	Symbols	Values (unit)
Blade	Z	4 (-)
Propeller Diameter	D	0.1576 (m)
Propeller Pitch ratio	P/D	1.133 (-)
Propeller Blade Area Ration	BAR	0.65 (-)
Gravity	g	9.81 (m/s^2)
Density	ρ	1000 (kg/m^3)
Water temperature	$t°$	27 (degrees)

The ITTC [25] additionally specifies the propeller model's permitted precision. To estimate the open water performance of the propeller at full scale, the diameter (D), pitch ratio (P/D), and chord length (c) are used to assess the systematic standard error in these propeller design parameters.

4. Uncertainty analysis experimental of open water

The total uncertainty of experimental testing to determine the value of the thrust coefficient (K_T), torque coefficient (K_Q) and advance coefficient (J). The total uncertainty for the thrust and torque coefficients is calculated as the root sum square of the uncertainties of the total bias and precision limit. The precision limit for K_T and K_Q will be determined using an end-to-end procedure that includes all precise errors for speed, thrust, torque, propeller revolution and temperature /density. To account for random errors, whether you establish the precision limit of single or several runs should be based on standard deviation calculated from multiple tests. If the repeat test cannot be run, the experimenter should use the best information available at the time to estimate the value of the precision error.

$$(U_{K_T})^2 = (B_{K_T})^2 + (P_{K_T})^2 \quad (8)$$

$$(U_{K_Q})^2 = (B_{K_Q})^2 + (P_{K_Q})^2 \quad (9)$$

$$(B_{K_T})^2 = \left[\frac{\partial K_T}{\partial D} B_D\right]^2 + \left[\frac{\partial K_T}{\partial n} B_n\right]^2 + \left[\frac{\partial K_T}{\partial T} B_T\right]^2 + \left[\frac{\partial K_T}{\partial \rho} B_\rho\right]^2 \quad (10)$$

$$(B_{K_Q})^2 = \left[\frac{\partial K_Q}{\partial D} B_D\right]^2 + \left[\frac{\partial K_Q}{\partial n} B_n\right]^2 + \left[\frac{\partial K_Q}{\partial T} B_Q\right]^2 + \left[\frac{\partial K_Q}{\partial \rho} B_\rho\right]^2 \quad (11)$$

$$(B_J)^2 = \left[\frac{\partial J}{\partial V_A} B_{V_A}\right]^2 + \left[\frac{\partial J}{\partial n} B_n\right]^2 + \left[\frac{\partial J}{\partial D} B_D\right]^2 \quad (12)$$

For multiple tests, the precision limit is derived using the formula:

$$P(M) = \frac{K\, S_{Dev}}{\sqrt{M}} \quad (13)$$

Wherein M is the range of runs for which the precision limit is to be found, S_{Dev} is the standard deviation calculated from multiple runs, and $K=2$ is the frequency of runs employed in the strategy. The formula can be used to establish the precision limit for a single run:

$$P(S) = K.S_{Dev} \tag{14}$$

4.1 The geometry of the propeller

The model is constructed to be mathematically comparable to the real propeller geometry. In spite of the fact that endeavors are made to deliver a precise propeller show, mistakes in measurements and off-sets can occur when using CNC processing machines, causing errors in diameter, chord length, pitch, and edge area shape. Flow characteristics around the propeller blades as well as the measured thrust and torque can be affected by errors in measurement and geometry. As previously stated, systematic variations in propeller geometry and offsets can be used to assess these mistakes. The affectability of pitch setting can be evaluated by conducting repeated open water experiments with the pitch re-set between each set of testing. Errors resulting from mistakes in blade shape can be investigated by running several experiments propeller models made from the same surface description.

Figure 3. Propeller Model.

In this scenario, the manufacturing error in model propeller diameter is evaluated using the restrictions set forth in ITTC Procedure 7.5-01-01-01 Rev 01[26]. When the model diameter error is expected to be within 0.1 mm, the bias errors are calculated as BD=0.1 mm when compared to the percent value of the propeller model's nominal diameter. In this instance, the propeller diameter bias mistake manifests itself as it were influence data reduction equations for thrust and torque coefficients.

4.2. The advanced speed of the propeller

Individual measuring systems for pulse count (*c*), wheel diameter (*D*), and 12-bit DA and AD card time($\otimes t$) basis make up the carriage speed measurement system. The speed is calculated by using an optical encoder to track the rotations of one of the wheels. The speed circuit has a time base of 100 milliseconds, allowing the pulse to be updated every 10th of a second. The maximum number of pulses in 100 milliseconds is limited to 4096 by a 12-bit DA conversion in the pulse count. The speed circuit's output is 0-10 V, therefore 4096 counts in 100 milliseconds equals 10 V output. The encoder output is determined using the following equation, which is based on the equation in ITTC procedure (20):

$$V = \frac{c\pi D}{8000\Delta t} \tag{15}$$

where *D* is the diameter of the carriage wheel and c is the number of counted pulses in $\otimes t$=100 ms.

When translating the analogue voltage to a frequency that indicates the pulse count across 10-time bases or one second, the final bias can occur. If the specified frequency is correlated to a comparable voltage output, this is permitted. The bias limit is determined by fitting a linear regression curve to a calibration (set of data). Standard error estimate (SEE) is a statistic defined by Coleman and Steele [27] as follows:

$$SEE = \sqrt{\frac{\sum_{i=1}^{N} \quad (Y_i - (aX_i + b))^2}{N-2}} \tag{16}$$

Coleman and Steele [24] argue that a ± 2(SEE) band around the regression curve contains about 95% of the data sets and that this band provides a confidence interval on the curve fit. The following is the total bias limit for pulse number:

$$B_c = \sqrt{B_{c1}^2 + B_{c2}^2 + B_{c3}^2 + B_{c4}^2} \tag{17}$$

By measuring against a distance away and a recorded transit time, the bias limit for the speed could be calculated using end-to-end method.

4.3 Revolutional Speed of Propeller

Individual detection systems for pulse count (c) and 12-bit AD conversion at a time base (t) make up the Revolutionary Speed of Propeller measuring system. The rate of revolution is determined by using an optical encoder to track the propeller shaft's movement. Due to identically spread and measured windows at one rotation, the encoder has a pulse capacity of 600.

A 100 ms time basis is built into the circuit of speed, allowing for signal upgrades every 10th of a second. The maximum number of beats in 100 milliseconds is limited to 4096 by a 12-bit DA transformation within the pulse count. The yield of the speed circuit is 0-10 V so 4096 tallied in 100 ms compares to 10 V yield. The yield of the encoder is obtained through the condition:

$$n = \frac{c}{No_{Pulse} * \Delta t} \tag{18}$$

The total bias limit for rate of revolution and pulse is calculated using the following equation:

$$B_n = \left(\left(\frac{\partial n}{\partial c} B_C\right)^2 + \left(\frac{\partial n}{\partial \Delta t} B_{\Delta t}\right)^2\right)^{\frac{1}{2}} \tag{19}$$

$$B_C = (B_{C1}^2 + B_{C2}^2 + B_{C3}^2)^{\frac{1}{2}} \tag{20}$$

$$\frac{\partial n}{\partial c} = \frac{1}{600\Delta t} \tag{21}$$

$$\frac{\partial n}{\partial \Delta t} = \frac{c}{600}\left(-\frac{1}{\Delta t^2}\right) \tag{22}$$

4.4 Torque and Thrust.

When the model propeller is dragged through the water and turns, it is necessary to monitor the axial force (thrust) and the rotational moment of a propeller (torque). Weights are used to calibrate the thrust transducer.

The weights serve as a a reference for calibrating load cells and a cause of inaccuracy, depending on the level of excellence of the standard. The individual weight used has a tolerance of ± 0.005 percent, according to certification. A certificate certifying the weights' calibration to a certain class is included. Weights are put at a predetermined distance from the shaft line to calibrate the torque dynamometer. The weight has the same ±0.005 percent accuracy as such the thrust calibration. The moment arm has a length of 0.5 m and a precision of ±0.2 mm. The equation is used to calculate the moment:

$$Q = F.ma \tag{23}$$

The bias errors can be expressed as follows:

$$\left(B_{Q_w}\right)^2 = \left(\frac{\partial Q_w}{\partial F} B_F\right)^2 + \left(\frac{\partial Q_w}{\partial ma} B_{ma}\right)^2 \tag{24}$$

Or

$$\left(B_{Q_w}\right)^2 = (ma.B_F)^2 + (F.B_{ma}) \tag{25}$$

As a result, the limit thrust, and torque bias can be calculated as follows:

$$B_T = \sqrt{\left((B_{Tw})^2 + (B_{T_{AD}})^2 + (B_{T_{cf}})^2\right)} \tag{26}$$

$$B_Q = \sqrt{\left((B_{Qw})^2 + (B_{Q_{AD}})^2 + (B_{Q_{cf}})^2\right)} \tag{27}$$

4.5 Temperature/Density.

Water characteristics were determined using water temperature readings under the test settings. The value of ± 0.1 °C is the acceptable water accuracy limit [28]. The matching water density and viscosity were determined using a water property chart [29]. The manufacturer calibrates the thermometer with a guaranteed accuracy of ±0.30 °C within the temperature range from -5 °C until +50 °C. For g=9.81m/s^2, the relationship between density and temperature according to ITTC [30] is as follows:

$$\rho\text{=}1000.1\text{+}0.0552\cdot t^{\circ}\text{+}0.0077\cdot t^{\circ 2}\text{+}0.00004t^{\circ 3} \tag{28}$$

$$B_{\rho 1} = \left|\frac{\partial \rho}{\partial t^{\circ}}\right| 0.0552 - 0.0154t^{\circ} + 0.000120t^{\circ 2} \tag{29}$$

$$B_{\rho 1} = \left|\frac{\partial \rho}{\partial t^{\circ}}\right| B_t \tag{30}$$

The bias for ρ can then be determined using the following formula:

$$B_\rho = \sqrt{\left(B_{\rho 1}\right)^2 + \left(B_{\rho 2}\right)^2 + \left(B_{\rho 3}\right)^2} \tag{31}$$

4.6 Precision Limit

The standard error for multiple testing, with the model disassembled and restored among each series of measurements, should be established in order to establish the precision limits. Five sets of testing with three speed measurements in each set were used in this investigation, totaling 15 test points. The measurements were carried out for three advanced coefficients. Repeating tests with several settings is an effective method to incorporate unintentional fallacies into the setup.

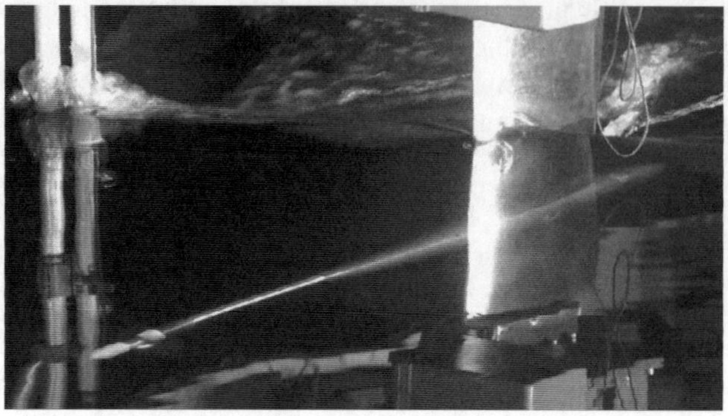

Figure 4. Open Water Test Running.

When the measurements are repeated, the observed quantities are likely to be carried out at different speeds and rpm for the several trials, as well as utilizing various water temperatures. Due to the challenges in precisely selecting the correct speed and rpm, deviations in speed and rpm will occur. For each point of measurement, the thrust and torque coefficients can be calculated and then the data was then corrected to the correct J value by matching the slopes of the K_T and K_Q curves. As input parameters, differences in speed and rpm should not be construed as precision errors. The K_T and K_Q measurements are just implemented at various J grades, i.e., under various circumstances. There is no compensation for temperature differences between the various test sets. The temperature of the open water test is not taken into account when extrapolating to full scale using the ITTC-78 method, therefore such a modification would inaccurately affect the uncertainty.

5. Result& Discussion

5.1 Bias Limit

5.1.1 Propeller geometry

The bias error is obtained to be B_D=0.1 mm, which is a value equivalent to 0.063 % of the diameter length of 157.6 mm, providing the model diameter accuracy is within 0.1 mm. As a result, the bias inaccuracy in propeller diameter solely affects this example's data reduction formulae for thrust and torque coefficients.

5.1.2 The advanced speed of the propeller

Taking into account the pulse bias, wheel diameter bias, and time base bias, the carriage speed bias value may be calculated. Based on measurements against a predetermined distance and a recorded stop time, the bias limit for the speed might be calculated with end-to-end method for this experiment as Table 2. The speed bias can be calculated using the end-to-end method and is equal to 0.007556 m/s, or 0.756 percent of the measured speed value.

Table 2. Speed calibration result data.

v set point (m/s)	v Measurable (m/s)	v Calibration (m/s)
0.500	0.503	0.500

1.000	1.003	1.000
2.000	2.005	1.998
3.000	3.009	2.999
4.000	4.010	3.997
5.000	5.016	5.002
6.000	5.997	5.993
7.000	6.952	6.947
8.000	7.967	7.958

5.1.3 Speed of advance of the propeller

Individual measuring devices for pulse count(*c*) and 12-bit AD converter on a temporal basis (*t*) make up the propeller revolution measurement system. An optical encoder tracks the rotation of the propeller shaft to determine the rate of rotation. In consequence of equally spaced and sized slots at single rotation, the encoder has a pulse capacity of 600. The oscillator module has already been calibrated by the manufacturer, its quantified accuracy on every update is 1.025 10^{-5} seconds, providing $B_{\Delta t}$= 1.025 10^{-5} seconds.

Based on the value obtained of c = 840, *Bc* = 2.345 and Δt=0.1 s and the average speed of n=22.016 rps, the partial derivatives can be computed according to equations (21) and (22) as:

$$\frac{\partial n}{\partial c} = \frac{1}{600\Delta t} = 0.01667$$

$$\frac{\partial n}{\partial \Delta t} = \frac{c}{600}\left(-\frac{1}{\Delta t^2}\right) = -140$$

The total bias limit of propeller rate revolution can then be calculated from equation (20) as:

$$B_n = ((0.01667 \cdot 2.345)^2 + (140 \cdot 1.025\ 10^{-5})^2)^{\frac{1}{2}} = 0.0391$$

Bn=0.0391 rps, which is a percent 0.178 of the actual propeller revolution of 22.016 rps, is the overall bias limit associated with the propeller rotation.

5.1.4 Torque, thrust and density

Table 3. The total bias limit of thrust, torque and density.

	Elementary bias error			Bias limit
B_T	B_{Tw}	$B_{T_{AD}}$	B_{Tcf}	B_T
	0.0050	0.0240	0.4839	0.4845
B_Q	B_{Qw}	$B_{Q_{AD}}$	B_{Qcf}	B_Q
	0.0022	0.0119	0.0037	0.0127
B_ρ	$B_{\rho 1}$	$B_{\rho 2}$	$B_{\rho 3}$	B_ρ
	-0.0787	0.0830	3.2059	3.2079

According to equations (26), (27) and (31) the bias for *T, Q* and ρ can be determined in Table 3 The bias limit for thrust is B_T = 0.4845 corresponding to 0.00485 % of *T* = 100N and for B_Q = 0.0127 corresponding to 0.00231% of Q = 5.5 Nm. The density of bias limit for is Bρ =3.2079 kg/m³, which corresponds to 0.321 % of ρ = 1000 kg/m³.

5.1.5 Total Bias Limits

The total bias and precision limits must be determined using partial derivatives applying input data of T = 100 N; Q = 5.5 Nm; n = 22.02 rps; v = 2.429 m/s; D = 0.156 m, and ρ =1000 kg/m³.
The partial derivatives of K_T, K_Q and J for this condition can be determined as Table 4.

Table 4. The partial derivatives of K_T, K_Q and J.

The Partial derivatives of K_T		The partial derivatives of K_Q		The Partial derivatives of J	
$\frac{\partial K_T}{\partial D}$	0.8378	$\frac{\partial K_Q}{\partial D}$	3.6468	$\frac{\partial J}{\partial V}$	0.2875
$\frac{\partial K_T}{\partial n}$	0.0301	$\frac{\partial K_Q}{\partial n}$	0.0105	$\frac{\partial J}{\partial n}$	0.00317
$\frac{\partial K_T}{\partial T}$	0.0033	$\frac{\partial K_Q}{\partial Q}$	0.0210	$\frac{\partial J}{\partial \rho}$	4.4195
$\frac{\partial K_T}{\partial \rho}$	0.00033	$\frac{\partial K_Q}{\partial \rho}$	0.00012		

Based on the calculation value from Table 4, the total bias limit (Table 5) for the thrust coefficient (K_T), torque coefficient (K_Q), and advance coefficient (J) may be calculated.

Table 5. The total bias limit of K_T, K_Q and J.

	Elementery Bias Error				Bias Limit	
B_{K_T}	$\frac{\partial K_T}{\partial D}B_D$	$\frac{\partial K_T}{\partial n}B_n$	$\frac{\partial K_T}{\partial T}B_T$	$\frac{\partial K_T}{\partial \rho}B_\rho$	B_{K_T}	B_{K_T}(%)
	0.0052780	5.356E-05	1.599E-07	1.059E-06	0.00527 8	0.0194
B_{K_Q}	$\frac{\partial K_Q}{\partial D}B_D$	$\frac{\partial K_Q}{\partial n}B_n$	$\frac{\partial K_Q}{\partial T}B_Q$	$\frac{\partial K_Q}{\partial \rho}B_\rho$	B_{K_Q}	B_{K_Q}(%)
	0.002297	1.869E-05	4.851E-07	3.852E-07	0.00229 6	0.04319
B_J	$\frac{\partial J}{\partial V_A}B_{V_A}$	$\frac{\partial J}{\partial n}B_n$	$\frac{\partial J}{\partial D}B_D$		B_J	B_J(%)
	0.002174	5.643E-05	0.002784		0.00353 3	0.0059

The total bias limit can be computed using the following formula equations (10), (11) and (12) as Table 5 are B_{K_T}= 0.005278 ; B_{K_Q}= 0.002296 ; B_J=0,003533. It correlates to 0.0194 % of the thrust coefficient

(K_T) = 0.27202 , 0. 04319 % of the torque coefficient (K_Q) = 0.0532, and 0.0059 % of the advance coefficient (J) = 0.60.

5.2 Precision Limit

The standard deviation for multiple tests, with the model taken and reinstated between every series of measurements, must be obtained to estimate the precision limits. Within the experiment, 15 testing with 3 speed measurements have been performed. The measurements have been performed for 3 advanced coefficients (J = 0.50, J = 0.60 and J = 0.70) as Table 6.

Table 6. The data of precision limits for thrust and torque coefficients.

Run/ Series	Thrust and torque coefficient J=0.50		Thrust and torque coefficient J=0.60		Thrust and torque coefficient J=0.70	
	K_T	K_Q	K_T	K_Q	K_T	K_Q
1	0,32693	0,06214	0,27110	0,05327	0,23411	0,04942
2	0,31557	0,05721	0,27472	0,05386	0,22510	0,04880
3	0,31605	0,05878	0,28202	0,05537	0,22212	0,04835
4	0,32148	0,05975	0,26690	0,05233	0,22078	0,04471
5	0,31377	0,05991	0,27179	0,05211	0,22645	0,04781
6	0,31472	0,06118	0,27424	0,05193	0,22453	0,04505
7	0,31012	0,05975	0,27238	0,05193	0,22436	0,04901
8	0,31280	0,06063	0,27499	0,05319	0,22931	0,04908
9	0,31082	0,06008	0,27203	0,05284	0,22931	0,05035
10	0,31507	0,06132	0,27338	0,05414	0,22579	0,04857
11	0,31097	0,05862	0,27040	0,05266	0,22831	0,04913
12	0,31296	0,05927	0,27131	0,05259	0,22795	0,05036
13	0,31526	0,06020	0,27306	0,05371	0,22582	0,04883
14	0,31645	0,06429	0,26808	0,05422	0,22905	0,04911
15	0,32513	0,06175	0,26395	0,05384	0,22342	0,04826
Mean	0,31587	0,06033	0,27202	0,05320	0,22643	0,04846
SDev	0,00498	0,00169	0,00410	0,00099	0,00336	0,00161
PS	0,00996	0,00338	0,00819	0,00198	0,00672	0,00321
PS (%)	3,15320	5,59553	3,01221	3,71994	2,97000	6,62707
PM	0,00129	0,00044	0,00106	0,00026	0,00087	0,00041
PM (%)	0,40708	0,72238	0,38887	0,48024	0,38343	0,85555

Table 6 shows the mean results, standard deviation (S_{Dev}), and precision limits for individual tests equation (14) and multiple tests equation (13) respectively. Within brackets, the corresponding percentage values have also been given. Because of the challenges of precisely selecting the proper speed and rpm, deviations of speed and rpm will occur. The thrust and torque coefficients for every measurement value may be obtained, and the J value can then be rectified utilizing the slope of the KT and KQ curves to compare the data. Because the speed and rpm variances are input parameters, they should not be construed as precision errors. The K_T and K_Q measurements are only carried out under different circumstances at various predetermined J values.

5.3 Total Uncertainties

Total uncertainty can be determined using equations (8) and (9) for J = 0.60 through integrating the bias limits with repeated and individual tests of precision limits. Table 7 determined the total uncertainty for the average value of 15 experiments.

Total uncertainty for multiple runs calculated as $U_{\underline{K_T}}$ = 0.005383 corresponding to 0.0198 % of the thrust coefficient K_T= 0.27202 and $U_{\underline{K_Q}}$= 0.002312 Correlating to0. 04346 % of the torque coefficient K_Q = 0.0532. For a single run, the total uncertainty was estimated as U_{K_T} = 0.009744 Correlating to0.0358 % of the thrust coefficient K_T= 0.27202 and U_{K_Q} = 0.003033 corresponding to 0.0570 % of the torque coefficient K_Q= 0.0532.

Table 7. The total uncertainty for multiple and single tests.

Elementary total uncertainty			Multiple tests		Single test	
B_{K_T}	$P_{\underline{K_T}}$	P_{K_T}	$U_{\underline{K_T}}$	$U_{\underline{K_T}}(\%)$	U_{K_T}	$U_{K_T}(\%)$
0.005278	0.00106	0.00819	0.005383	0.0198	0.009744	0.0358
B_{K_Q}	$P_{\underline{K_Q}}$	P_{K_Q}	$U_{\underline{K_Q}}$	$U_{\underline{K_Q}}(\%)$	U_{K_Q}	$U_{K_Q}(\%)$
0.002296	0.00026	0.00198	0.002312	0.04346	0.003033	0.0570

In terms of relative numbers for a single run, the bias for K_T accounts for 54.8 percent of the total uncertainty, and It accounts about 98.1 percent of the total uncertainty for the average value of 15 experiments. For a single run, the bias for K_Q accounts for 75.8 percent of the entire uncertainty and for the mean value of 15 tests, it accounts about 99.4 percent of the total uncertainty. The relative contribution of each term can be estimated through analyzing the precision limits and bias limit, as well as the uncertainty. This allows researchers to determine the sections where a measurement system upgrade will have the most impact.

6. Conclusions

The results of open water propeller uncertainty analysis by the ITTC 2014 method are presented in this paper. The Biases and precision limits are combined using a root mean square, end-to-end and standard error estimate (SEE) methods to determine the total uncertainty estimates for each variable interest. The bias error is the primary source of uncertainty in performance coefficients in the majority of cases (90 percent or more on the total uncertainty). The principal interest must be on reducing the bias limit in the instruments to minimize overall uncertainty in the final results. The high degree of accuracy of bias limit is found in the temperature/ density, propeller diameter, propeller rate of revolution and advanced speed. According to methodology of the ITTC 2014 , the uncertainty is assumed to be influenced by environmental situations and other factors, thus the concatenated standard uncertainty result is acquired and then compared to the confidence level. Based on experimental test results, descriptives that the bias value for K_T accounts for 54.8 % of the total uncertainty for a single run, while for multiple tests, it accounts for 98.1 % of the total uncertainty. For a single test, the bias for KQ accounts for 75.8% of the total uncertainty, and multiple tests calculate 99.4% of the total uncertainty. . It recommends that the uncertainty analysis results for open water propellers are still within the acceptable range of 95% level of confidence. This shows that the open water test conducted at the Indonesian Hydrodynamic Laboratory is still acceptable because it has a value above 95% confidence level in multiple tests.

References

[1] ITTC 2014 *Open Water Test* ITTC Recommended Procedures and Guidelines Procedure 7.5 – 02-03-02.1 Rev 03

[2] Kline S J and McClintock F A 1953 *Describing Uncertainties in Single-Sample Experiments* ASME Mechanical Engineering **75** No 1 pp 3-8

[3] ISO 1995 *Guide to the Expression of Uncertainty in Measurement* International Organization for Standardization Genève Switzerland
[4] AIAA S 071A 1999 *Assessment of Experimental Uncertainty With Application to Wind Tunnel Testing* American Institute of Aeronautics and Astronautics Reston Virginia USA
[5] ASME PTC 19.1 2005 *Test Uncertainty* American Society of Mechanical Engineers New York
[6] Giacomo P 1981 *News from the BIPM* Metrologia **17** No 2 pp 69-74
[7] S Steen 2014 *Lecture Note Experimental Methods in Marine Hydrodynamics Faculty of Engineering Science and Technology* – NTNU Trondheim Norwegian University of Science and Technology
[8] Kline S J 1985 *The Purpose of Uncertainty Analysis* Journal of Fluids Engineering **107** No 2 pp 153-160
[9] Kantasubrata, 2014 Diklat teknik estimasi ketidakpastian pengukuran kimia: *Dasar Ketidakpastian pada Pengujian* Jakarta Badan Standardisasi Nasional
[10] KAN 2002 *Pedoman Evaluasi Pelaporan Ketidakpastian Pengukuran* (DP.01.23) Jakarta: Komite Akreditasi Nasional
[11] Joe, Longo, Fred and Stern 2005 *Uncertainty Assessment for Towing Tank Tests with the Example for Surface Combatant DTMB Model 5415* J. Ship Res. 49 March 1 55-68
[12] I K A P Utama, D Purnamasari, I K Suastika, Nurhadi and G A Thomas 2021 *Toward Improvement of Resistance Testing Reliability* Journal of Engineering and Technological Sciences Vol. 53, No. 2, 2021, 210201
[13] D Purnamasari, I K A P Utama, I K Suastika 2020 *Application of Kalman Filter to the Uncertainty of Model Resistance Data Obtained from Experiment* Journal of Marine Science and Technology 28 No 1 pp. 18-24 2020
[14] D Purnamasari, I K A P Utama, I K Suastika and G Thomas 2020 *Uncertainty Assessment for Towing Tank Tests with the Example for Surface Combatant DTMB Model 5415* Journal of Engineering Science and Technology 15 No 2 2020 1455 – 1465
[15] Taylor R 2006 *Experimental Investigation of the influence of Hub Taper Angle on the Performance of Push and Pull Configuration Podded Propellers* Master in Engineering. Memorial University of Newfoundland Canada 150p
[16] Bose N and Luznik L 1996 *Uncertainty analysis in propeller open water tests* Int. Shipbuilding. Prog 43 (435) 237-246
[17] Islam F M 2007 *Uncertainty Analysis of NSERC-NRC Pod Dynamometer System* National Research Council Institute for Ocean Technology Arctic Avenue Canada
[18] Wei L, Wang W, Sun L and Mao Z 2021 *Research on Uncertainty Analysis of Propeller Open Water Test Based on the ISO GUM Method* Journal of Jiangsu University of Science and Technology 35 No 1 February 2021
[19] Choi G I, Chun H H, Kim J S and Lee C M 1994 *Uncertainty Analysis for the Propeller Open Water Test* Transactions of the Society of Naval Architects of Korea Vol 1 No 1 Feb 1994
[20] ITTC 2014 *Uncertainty Analysis, Example for Open Water Test* ITTC Recommended Procedures and Guidelines Procedure 7.5-02-03-02.2 Rev 01
[21] ITTC 1999a *Uncertainty Analysis in EFD, Uncertainty Assessment Methodology* 22nd International Towing Tank Conference Seoul/Shanghai ITTC Recommended Procedures and Guidelines 7.5-02-01-01 Rev 00
[22] ITTC 1999b *Uncertainty Analysis in EFD, Guidelines for Towing Tank Tests* 22nd International Towing Tank Conference Seoul/Shanghai ITTC Recommended Procedures and Guidelines 7.5-02-01-02 Rev 00
[23] ITTC 2002 *Propulsion, Performance Podded Propulsor Tests and Extrapolation* ITTC Recommended Procedures and Guidelines Procedure 7.5 - 02-03-01.3 Rev 00
[24] ITTC 2008a *Guide to the Expression of Uncertainty in Experimental Hydrodynamics* ITTC Procedure 7.5-02-01-01 Rev 01

[25] ITTC 2017d *Propeller model accuracy* In: ITTC – Recommended Procedure and Guidelines 75-01-02-02 International Towing Tank Conference Zurich Switzerland
[26] ITTC 2008 *Model Manufacture Ship Models* ITTC Recommended Procedures and Guidelines 7.5-01-01-01 Rev 02
[27] Coleman H W and Steele Jr W G 1999 *Experimentation and Uncertainty Analysis for Engineers* John Wiley and Sons Inc New York
[28] ITTC 2014a *General guidelines for uncertainty analysis in resistance test* In: ITTC – R.P & G 75-02-02-02 International Towing Tank Conference Zurich Switzerland
[29] ITTC 2011 *Fresh water and seawater properties* In: ITTC – Recommended Procedure and Guidelines 5-02-01-03 Vol 7 International Towing Tank Conference Zurich Switzerland.
[30] ITTC 2006 *Testing and Extrapolation Method, General Density and Viscosityof Water* ITTC Recommended Procedures and Guidelines 7.5-02-01-03 Rev 01

Acknowledgements
This research is supported by The National Research and Innovation Agency (BRIN) as a sponsor of research-based education programs by Nomor: SP/112/BPPT/08/2021. The author thanks Dr. Dian Purnamasari, ST., MT., A.A.B Dinariyana, ST., MES., Ph.D., Mahendra, ST., MT., and Vido, ST for their guidance and support in this research. The authors also would like to thank the staff members of The Indonesian Hydrodynamic Laboratory for their support during the experiment.

The effect of safety climate on safety behavior with intervention

Rahajeng Pradeti[1*], Yuning Widiarti[1], Indri Santiasih[1]

[1] Shipbuilding Institute of Polytechnic Surabaya, Surabaya, Indonesia 60111

*rahajengpradeti@student.ppns.ac.id

Abstract. Safety climate influences safety behavior and tends to unsafe behavior. Accidents caused by unsafe behavior are quite high, 80%-85% than unsafe conditions. Previous research has addressed the safety climate to safety behavior by looking at the impact and involvement of individuals and the result is controlled safety climate to increase in safety behavior. Changes in unsafe behavior need to be made by applying intervention methods. Based on this, research is carried out on the influence of safety climate on safety behavior by providing interventions to change behavior. The study was conducted on 198 production workers in manufacturing companies. The safety climate influence analysis was conducted using the NOSACQ-50 questionnaire and multivariate linear regression test, Structural Equation Model (SEM) to see who influenced worker behavior and T-Paired test to see results of the given intervention. Implementation of behavior change is done by intervention. Intervention is provided by installing special safety signs at work and training for workers. The results of the intervention were analyzed using the Critical Behavior Checklist (CBC). The results of the influence test showed that management had a large influence on worker behavior with 5 variables. Testing with SEM showed that there were 3 variables with significant values so that worker behavior was influenced by management opinions and attitudes. The T-Paired test showed that unsafe behavior decreased and safe behavior increased after the intervention was administered. Thus, management has a considerable influence on the formation of worker behavior so management needs to participate in realizing work safety.

Keywords: Behavior Based Safety, Multivariate Linear Regression Test, SEM, CBC

1. Introduction

Work accidents in the industry are still very high. The cause is a work accident caused by unsafe action of 80% – 85% higher than unsafe conditions [1]. In one of the manufacturing companies in the Tangerang area that in the process uses large and heavy materials, several types of work accidents occured in the production process from light to heavy over the past 5 years as in **Figure 1**. Accidents that often occur are plate scratches and material being hit.

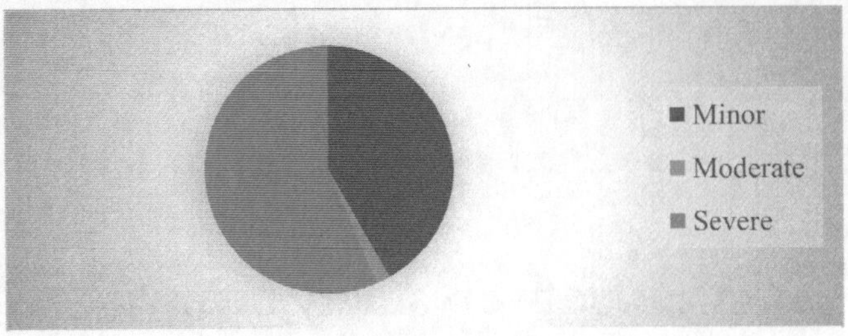

Figure 1. Tangerang Manufacturing Company Work Accident

The cause of frequent work accidents occurs because workers are too focused on work deadlines without regard to the elements of work safety and also the lack of attention from management about the fulfillment of work safety in the workplace such as the installation of safety signs tailored to the type of work and the provision of training internally and externally to improve the ability of workers to do their work. But unsafe behavior often occurs due to the influence of the knowledge, abilities, and motivations of the worker itself, while the safety climate consists of a safe and comfortable work environment. Then the success of the company can be realized if the work environment is safe and comfortable it will create safe behavior [2].

Previous research has addressed the safety climate to safety behavior by looking at the impact and involvement of individuals. As a result, if there is a controlled safety climate, there will be an increase in safety behavior [3]. But research on the effect of safety climate on safety behavior is still lacking in its application. In another study, there was the use of a behavior based safety (BBS) program which is a program to identify and change worker behavior. This BBS is getting significant attention in the industry. BBS is done by providing special treatment tailored to the conditions of workers and the workplace to change unsafe behavior to be safe[4]

This study, testing on the effects of climate and determines the interventions to be given with multivariate linear regression testing and Structural Equation Modeling (SEM). Furthermore, tests were conducted to assess the effectiveness of interventions administered with the Critical Behavior Checklist (CBC) questionnaire and the t-paired test. The results of this study are expected to be used by companies as a consideration in improving workers' safe behavior based on the safety climate and providing interventions to change unsafe worker behavior.

2. Method

The study was conducted using the NOSACQ-50 questionnaire to look at the effect of the safety climate on safety behavior and the CBC questionnaire to look at the effectiveness of providing interventions to workers in specific units. This questionnaire was distributed to 198 respondents who were workers in the production section with 2 shifts of work, namely morning shift and night shift. Furthermore, tests with multivariate linear regression to see the magnitude of the influence resulting from the 7 variables present in the **Table 1**.

Table 1. Variable NOSACQ-50 Questionnaire.

Variable	Description
X_1	Management safety priority, commitment and competence
X_2	Management safety empowerment
X_3	Management safety justice
X_4	Workers' safety commitment
X_5	Workers' safety priority and risk non-acceptance
X_6	Safety communication, learning, and trust in co-worker safety competence
X_7	Workers' trust in the efficacy of safety systems

Another test is SEM to see indicators of each of the most influential variables from previous test results. SEM is done with AMOS software that is tested on each variable and indicator, to see the significant influence on the indicators in each variable. And, t-paired tests were conducted to see the effectiveness of interventions given before and after the intervention with the CBC questionnaire. Questionnaires are in the work section that has been designated from the results of multivariate linear regression at the beginning which then the work unit fills out questionnaires at the time before being given intervention and after being given intervention. The intervention to be selected is tailored to the company's conditions and the outcome of discussions with the HSE team.

3. Discussion

3.1. The impact of safety climate on safety behavior

This influence test is done with SPSS software by entering data from the spread of questionnaires that are converted to interval/ratio data, then from 7 variables in NOSACQ-50, there are 3 variables (X_3, X_5. X_6) which has a value of significance that is close to zero. So, these variables are the most influential because workers' perceptions of work safety obtained from management and between workers affect the behavior of workers and the lack of contribution from workers regarding work safety. Therefore, the influence of management leadership on the formation of worker behavior is very high because the more positive the influence given, the more positive the worker's behavior. The importance of behavioral influence is from management as a leader in an organization.

3.2. Determining the type of intervention

After obtaining the results of multivariate linear regression testing for the influence of the safety climate on safety behavior, it further determines the type of intervention to be used to change worker behavior. The determination is by looking at the indicators of the most frequently asked questions about the management or workers of the NOSACQ-50 questionnaire with 7 variables in it. Determination is done with SEM with several steps in it. The result of the previous multivariate linear regression of the most significant variables is 3, of the 3 variables are tested with all indicators in each variable. The indicator in question is a question on the questionnaire.

The result of SEM is that X3 (Management Safety Justice) obtained 2 significant indicators in questions number 19 and 22 (A19 and A22) can be seen in **Figure 2** as the test result and results of the goodness of fit in **Table 2**. Both of these questions are the perception of workers towards management, workers feel if management pays less attention and is fair to workers. In X5 (Workers' safety priority and risk non-acceptance) there are 4 significant question indicators in questions number 30, 32, 34, and 35 that can be seen in **Figure 3** which is the test result and results of the goodness of fit in **Table 3**. These four questions are workers' perceptions of themselves and between workers that they tend to underestimate minor injuries or minor accidents that occur and tend to have the job done immediately rather than report the work. In X6 (Safety communication, learning, and trust in co-worker safety competence) there are 4 significant question indicators in questions number 38, 39, 40, and 43 that can be seen in **Figure 4** which is the test result and results of the goodness of fit in **Table 4**. The four questions are workers' perceptions of themselves, between workers, and management that they have no attachment to realizing a safe and comfortable workplace and workers feel they do not get the freedom of opinion, especially regarding work safety to management.

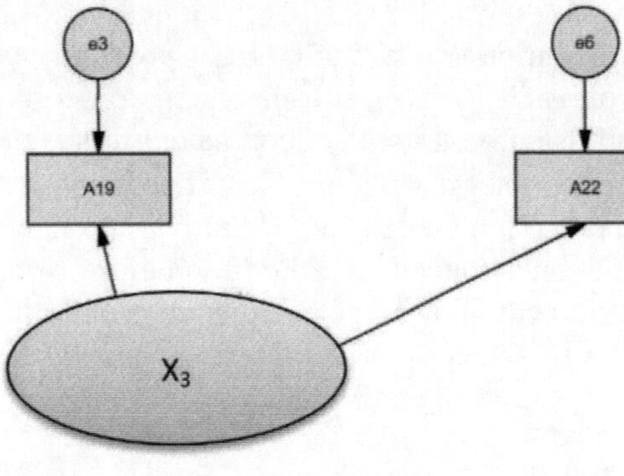

Figure 2. X_3 Line Modification Model

Table 2. Results X_3

Goodness of fit Index	Cut Off Value	Results
Chi-Square	Smallest value	2,5
Probability	≥0,05	0,28
CMIN/DF	≤ 2,00	1,27
GFI	≥0,90	0,99
TLI	≥0,90	0,98
CFI	≥0,90	0,99
RMSEA	≤ 0,08	0,04

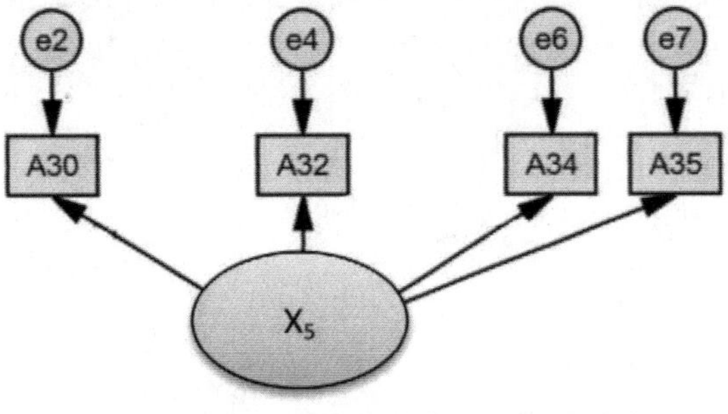

Figure 3. X_5 Path Modification Model

Table 3. Results X_5

Goodness of fit Index	Cut Off Value	Results
Chi-Square	Smallest value	11,19
Probability	≥0,05	0,26
CMIN/DF	≤ 2,00	0,12
GFI	≥0,90	0,98
TLI	≥0,90	0,97
CFI	≥0,90	0,98
RMSEA	≤ 0,08	0,03

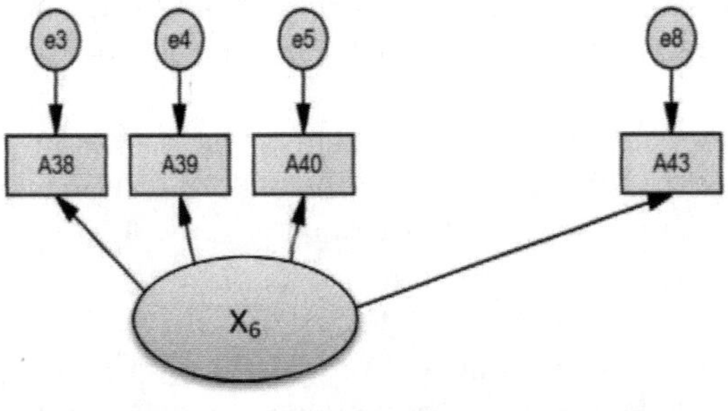

Figure 4. X_6 Path Modification Model

Table 4. Results X_6

Goodness of fit Index	Cut Off Value	Results
Chi-Square	Smallest value	4,10
Probability	≥0,05	0,54
CMIN/DF	≤ 2,00	0,82
GFI	≥0,90	0,99
TLI	≥0,90	1,01
CFI	≥0,90	1,00
RMSEA	≤ 0,08	0,00

Test results with SEM showed that management affects workers' perception of work safety as well as workers' perceptions of themselves and between workers. So from these results selected interventions that can involve both to play an active role in work safety to realize a safe and comfortable workplace. The intervention provided is expected to change the behavior of the worker little by little.

3.3. Intervention

Intervention is one part of the BBS program. The provision of these interventions needs to be done to see the extent to which worker behavior can change for the better or workers' safe behavior can increase compared to unsafe behavior. The intervention technique used in this study was an intervention with activators i.e. the installation of safety signs using prohibition signs and warning signs [6]. The installation of error signs is adjusted to each work unit and uses existing safety signs in the company and prioritizes prohibition and warning signs. The introduction of this safety sign is followed by the explanation of each toolbox meeting so that it can remind workers and help workers to avoid possible dangers that can occur.

Other interventions with intervention as a behavior change agent, namely the provision of internal company training. The training provided is simple training on each job such as forklift operators getting training on forklift use, things that can be done and should not be done during the operation and work safety at the time before, during, and after driving the forklift. This training is provided by their respective supervisors. This is expected to remind workers in their respective work sections to be able to cooperate in realizing a safe and comfortable workplace.

Intervention with safety signs is done by appealing to toolbox meetings that are expected to provide workers with an understanding of prohibition signs or warning signs installed around their workplaces. The provision of intervention with training is carried out by a team that has competence in their field to be able to provide understanding to workers about the work they do so that there are no work accidents. After being given both interventions, then an assessment of worker behavior by comparing before and after the intervention was carried out. The results of these interventions showed an increase in worker behavior in each work unit and a decrease in workers' unsafe behavior. It can be concluded that the interventions provided can change workers' behavior for the better. This assessment is carried out by distributing CBC questionnaires to each responsible supervisor in their respective work section. Questionnaires were distributed before and after the intervention.

After the administration of the intervention is further assessed before and after the intervention. This assessment is conducted with CBC questionnaires distributed to each supervisor of the work section to assess changes in worker behavior.

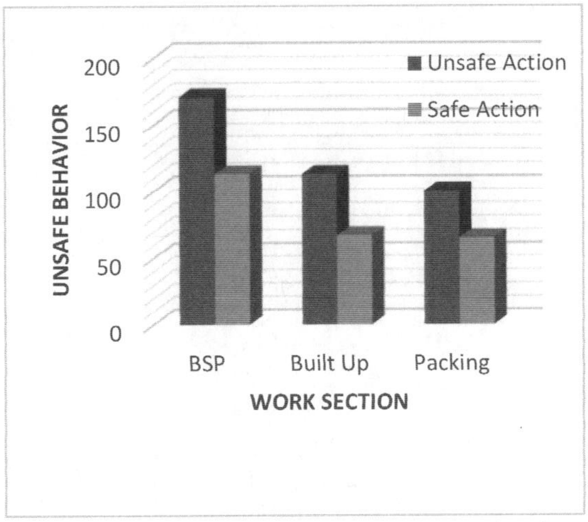

Figure 5. Before Change in Worker Behavior

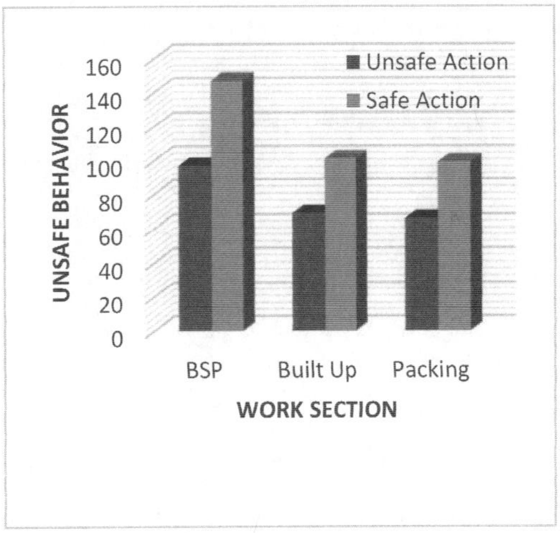

Figure 6. After Change in Worker Behavior

The behavior of workers before the intervention is applied can be seen in **Figure 5**, the diagram shows if the behavior of workers for unsafe behavior is higher than the safe behavior they do. So that the application of interventions to workers to change unsafe behavior needs to be done and after 4 weeks of implementation, the changes can be seen in **Figure 6** that the worker's unsafe behavior was high enough before the intervention was given and the low safe behavior of the worker also before the intervention was given. But after the administration of these two interventions, there was a decrease in worker behavior, especially in unsafe behavior and there was an increase in the safe behavior of workers in all three work sections (BSP, Built-Up, and Packing).

4. Conclusion

The results of testing and application of interventions found that the perception of workers influenced by management so that worker behavior becomes unsafe because workers feel less support from management, but after intervention changes behavior, it is found that interventions succeed in changing worker behavior to be safer.

5. Reference

[1] Maslahat W 2020 Relationship of Worker Conditions with Work Accidents in Workers In Building Project A by PT. XYZ Year 2020 p 250–262

[2] Neal A and Griffin MA 2002 Safety climate and safety behavior *Australian. J. Management* **27** p 67–75

[3] Renecle M, Curcuruto M, Gracia Lerín FJ & Tomás Marco I 2021 Improving safety in highrisk operations: A multilevel analysis of the role of mindful organizing in translating the safety climate into individual safety behavior. *J. Safety Science* **138**

[4] Guo BHW, Goh YM, & Le Xin Wong K 2018 A view of system dynamics from a behaviorbased safety program in the construction industry *J Safety Science* **104** p 202–215.

[5] Chen H, Li H, and Goh YM 2021 Construction safety climate review: Definition, factor, relationship to safety behavior and research agend *J. Safety Science* **142**

[6] Geller ES 2016 *The Psychology of Safety Handbook* (*second edition*) (CRC Pers) p 560.

[7] Bian J, Fu H and Jin J 2020 Are we sensitive to different types of safety signs? Proof of ERP p 495–505.

Design of Real-Time Monitoring System for Performance Parameter and Alarm of Main Engine RV. Baruna Jaya III

T Handoyo[1], D W Handani[1], A Sudaryanto[2], and T H Nugroho[2]

[1]Department of Marine Engineering, Institut Teknologi Sepuluh Nopember, Surabaya, 60111, Indonesia
[2]Research and Innovation Agency (BRIN), Jakarta, 10340, Indonesia

tris014@brin.go.id[1];dhimas@ne.its.ac.id[2];agus.sudaryanto@bppt.go.id[3]; tsan001@brin.go.id[4]

Abstract. With the increasing age of RV. Baruna Jaya III (30 years), the ships often face various problems. The continuous operation of the engine causes a decrease in its performance which eventually results in the failure of the main engine. For instance, damage to the main engine can stop operating of the ship at sea which eventually have various impacts on ship and life safety as well as operational cost. In this context, efforts to manage ships especially for the operation of main engine and its prevention from errors through real time monitoring and control system have become a major issue. This study aims to design a monitoring and alarm system for the main engine performance on the RV. Baruna Jaya III as the preliminary step for the development of the overall monitoring system. As a result, a design for data acquisition of the main engine performance parameters in real time and the design of a ship's main engine failure alarm system have been developed using PLC-based design technology and a SCADA-based monitoring system. This system can be thus further built for the performance of the ship's main engine, as part of management ship operations and maintenance policies.

1. Introduction

The research vessel of RV. Baruna Jaya III is now one of the ships belonging to the National Research and Innovation Agency (BRIN) which has oceanographic and geological functions. The ship, which was launched in 1989, has two similar main engines, namely the Niigata SEMT Pielstick 5PA5L which is connected to a Leroy Somer 625 KVA alternator. The main engine on the ship is operated 24 hours a day for 1 week to a month during ship operation, so that the operating conditions of the engine that are carried out continuously will cause a decrease in engine performance [1]. Continuous decline in engine performance can result in failure of the main engine operation. Some of the fatal major engine failures were incomplete combustion which caused the ship to stop operating at sea. Main engine failure that occurs when operating at sea can cause fatal problems for ships and ship members [2]. The operation of the main engine causes the need for special treatment of the components on the main engine when operating at sea [3].

Monitoring the performance of ship engines is very important to know the safety of engine operation. Many studies on ship engine monitoring have been carried out, with several concepts and methods developed [4, 5, 6, 7, 8]. Currently, the main engine of the RV. Baruna Jaya III is still using visual (analog) measuring instruments to see the parameters of engine condition and performance. The manual measuring tool requires the crew to directly check the engine area. Routine observation of the parameters of temperature, pressure, viscosity and flow oil and fuel is one method that is quite good for knowing the condition of the engine. With today's technological advances, these parameters can be seen easily on the monitor screen if the machine is fitted with the right sensors. Some of the sensors that can be installed are Pressure Transmitter to measure pressure, Flow Meter to measure flow rate, Temperature Sensor to measure temperature and Magnetic Pick Up to measure Engine Revolutions Per Minute (RPM). The data from sensor readings can be processed through the Programmable Logic Controller (PLC) and displayed on the monitor in the room that requires the data, namely the Engine Control Room (ECR) and the Steering Room through a Supervisory Control and Data Acquisition (SCADA) based system. In this regard, real time supervision and control of processes in a machining system can be carried out through an automatic control system [4], where the control system has hardware and software components that allow for supervision and control [5]. Thus, the parameter data of the machining system can be monitored more easily and quickly for further processing and provides a communication interface between the user and the machine [6].

Observation and supervision of a machining system process is generally used as a reference in making decisions on maintenance activities that aim to maintain system function [7]. The SCADA control system is the basis for controlling, monitoring and data acquisition and it is also possible to manage and control the system locally and remotely through a graphical interface that communicates the user with the system [8]. SCADA, a system (arguably) a continuation of an automated system built with PLC is the basis for a distribution automation system consisting of I/O signaling hardware, controllers, software, network & communications [9]. This system has the advantage of saving technical costs, time, and small risks due to easy integration with all devices in the machining system [10]. In addition to monitoring the main engine parameters, it is also continued with the development of an alarm system design that functions as an indication of the extent to which the ship's operations are under management control. Efficient and reliable alarm design in conducting onboard monitoring systems is essential for safe and sound operation. One of the reliable and efficient alarm design methods for monitoring in the engine room is the design method using PLC and SCADA-based systems, where this design technique is reliable, efficient and effective for online monitoring of machine control room devices [11]. The purpose of this research is to develop a design that will later be used to help take action as preventive measures and can be used as an analytical medium to determine the reliability of the system and as a reference for carrying out appropriate maintenance strategy policies.

In a previous study, an information technology application for the acquisition of machine performance data using General Packet Radio Service (GPRS) technology has been built in a simple way at RV. Baruna Jaya [12]. GPRS-based communication system architecture is used for monitoring the performance of ship machinery and collecting field data, such as temperature (manual input) and pressure (sensor pressure). By using a GPRS-based Information System, the data is sent to the Ship Operator's office [12]. Research that has been done previously is able to perform data acquisition of the ship's main engine performance parameters [12], it's just that the sensor is limited to only 1 pressure sensor which results in most of the monitoring of the main engine parameters being visual or analog, data transmission is still GPRS-based which has a limited operating area as well as only uses a microcontroller which makes the number of sensors designed is limited. Current technological developments are capable of monitoring and controlling industrial system parameters from anywhere, anytime in real time using the internet [13]. For this reason, it is necessary to develop an existing design by developing a monitoring system or monitoring the performance of the ship's engine on the research vessels of RV. Baruna Jaya III in real time. This research will use technology based on Internet of Things (IoT) with PLC design method and monitoring system based on SCADA. The reason of utiliziation of PLC is because PLC is compatible and recommended for industrial environmental standards, especially PLC specifications can enter the standard environment for ships, so it is very suitable for use in the engine room area which is a "hazard environment. PLC is also designed in such a way for high reliability and protection and usage. in the long term for the industrial environment, besides that PLCs are lower in complexity, cost effective and robust, that is the reason why PLCs are preferred for real time monitoring systems over using other devices such as microcontrollers or computers (single board computers).Furthermore,data transmission from the ship to the ship operator's office is planned to use information technology based on the Very Small Aperture Terminal (VSAT) satellite.

IOP Conf. Series: Earth and Environmental Science 1081 (2022) 012024 doi:10.1088/1755-1315/1081/1/012024

2. Methodology

This research was conducted as one of the steps for optimizing the ship with the application of information technology which is used as a monitoring system in the scope of the ship's engine so that the engine can be monitored in real-time by the ship's management. The scope of activities carried out in this study include determining the parameters for monitoring the performance of the ship's main engine based on previous research literature, critical components of the ship and the Niigata engine manual; and development of monitoring and alarm system design which is carried out with a modular system so that it can be realized in stages. Furthermore, PLC programming is also carried out for data collection and processing systems from sensors for the purpose of monitoring the condition of the ship and controlling several critical actuators on the ship, making a SCADA system for monitoring purposes by ship management in the ECR room and steering room, making a reporting system for data of the ship condition, and then the creation of a real-time monitoring system for ship conditions that is displayed at the ship operator's office. For this research, a survey has been carried out on the current condition of the RV. Baruna Jaya III. Figures 1 and 2 show the main engine and condition of the wheelhouse of RV. Baruna Jaya III, respectively.

Figure 1. M/E Niigata SEMT Pielstick 5PA5L.

Figure 2. Condition of the wheelhouse.

In this study, the components of monitoring and alarm system were placed in the engine room with a remote I/O system to minimize wiring and also make the system modular. Figure 3 shows the layout of the remote I/O system placement in the engine room. While Figure 4 which is the architectural design of the monitoring and alarm system based on the survey data and discussions with the RV. Baruna Jaya team shows the flow and placement of the system as a whole.

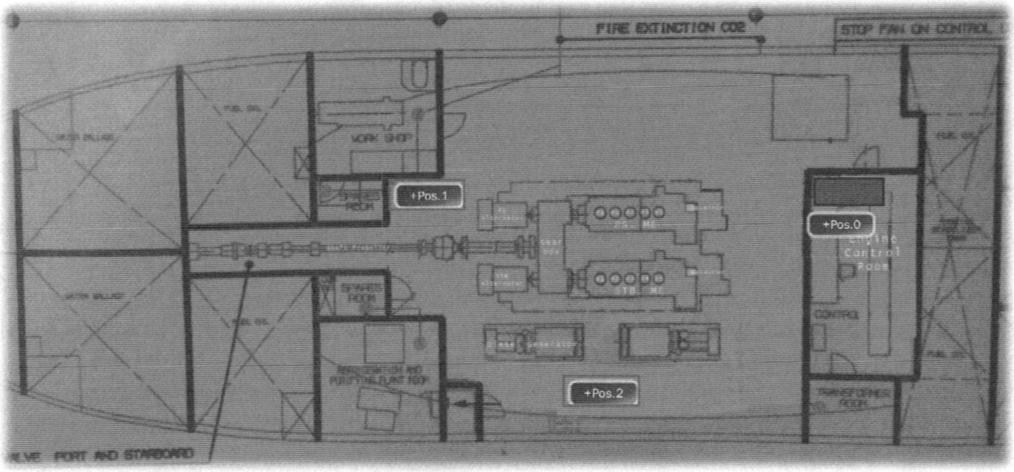

= Location of remote I/O = Locaion of PLC

Figure 3. Location of PLC panels and remote I/O in the engine room.

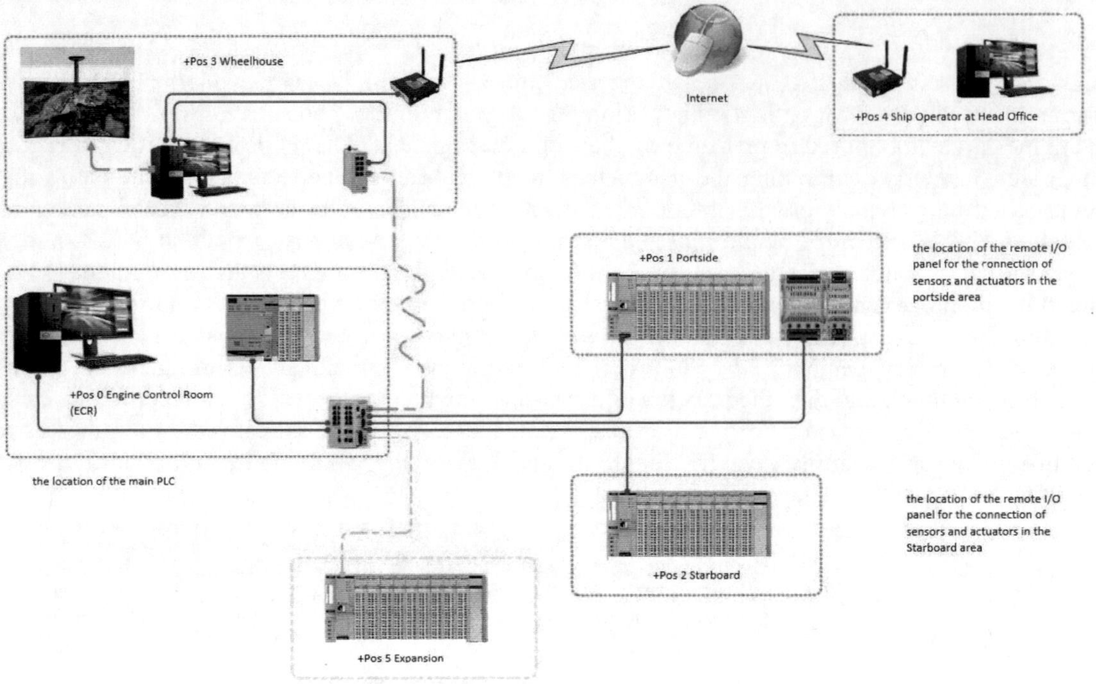

Figure 4. Architecture of monitoring performance and alarm system for RV. Baruna Jaya III.

+Pos.0 is the ECR space that will be used as the location of the main PLC. This location will be filled with the main controller of the PLC system, network switch, and a computer that functions as a SCADA Server. In this case, the Allen-Bradley PLC series compactlogix is used because of its good quality and specifications according to the conditions of use on the ship (marine-grade). Then, +Pos.1 is the location of the remote I/O panel for the connection of sensors and actuators in the portside area. Meanwhile, +Pos.2 is the location of the remote I/O panel for sensor and actuator connections in the starboard area.

In addition to the engine room, the location of the placement of the components of this automation system is also in the wheelhouse (bridge) in the form of a dashboard that displays a resume of engine condition information that can be used by the ship captain in operating the ship (Figure 4). In addition to the dashboard, the wheelhouse is also provided with a computer that functions as a SCADA client as well as a reporting system for this monitoring and alarm system. The system components in the wheelhouse are then identified by location ID +Pos.3. Monitoring and alarm system data from the RV. Baruna Jaya III is designed to be specifically accessed via the internet network from the ship operator's office. For this purpose, an internet modem will be installed in the RV. Baruna Jaya III connected to a computer at the +Pos.3 location in the wheelhouse. Furthermore, a computer with internet access capability is required in the office to be able to display the RV main engine performance parameter data of RV. Baruna Jaya III in real time. The location of the ship operator at the head office is assigned a location ID +Pos.4.

Furthermore, +Pos.5 is a remote I/O module expansion template for future development. If it is necessary to add a remote I/O location to accommodate the placement of sensors and actuators that are relatively far from the remote I/O panel location on the starboard or portside, it is possible to add a remote I/O panel with the component template. The main component is an Ethernet/IP adapter module with part number 5069-AENTR. As for the number and type of input modules or output modules will adjust to the type and number of sensors and actuators to be used.

If the location of the placement of the new sensor or actuator is still possible to be connected to the remote I/O panel on the existing starboard or portside, it can be directly connected to the remote I/O panel without the need for expansion of the remote I/O. If the module on the remote I/O panel, starboard or portside is insufficient, it is enough to add the missing module. On one remote I/O panel can be added up to a maximum of 31 input or output modules.

The configuration list of components for each location ID above is a minimal configuration (back-bone) of the monitoring performance and alarm system that will be implemented on the RV. Baruna Jaya III. The number of input and output modules can change according to the number of parameters/sensors and actuators that will be connected to the configuration of this PLC system.

3. Result and discussion

3.1. Determination of Main Engine Performance Parameters

Determination of parameters for monitoring main engine performance is based on available previous research literature studies, discussions with ship personnel, engine manuals and historical failures of ship main engine operations as well as critical ship components. Determination of parameters for the development of the design of main engine performance monitoring system and alarm is carried out simultaneously with the manufacture of the PLC system backbone. In addition to meet the urgency, the determination of parameters also looks at several aspects including cost, component availability and processing time. If time, cost and availability of the components are not possible, then additional parameters will be carried out in future research activities.

As the initial stage of making this backbone, herewith the details of the parameters which have been defined based on above determination:

✓ Main engine on Starboard
- Engine Exhaust Temp per-cylinder (5 cylinders)
- Engine Shaft RPM

These two types of parameters are used to determine the performance of the Main Engine when operating. The anomaly of this value (indicated by an alarm/trip) to the reference value is an indicator to the operator that there is a problem with the Main Engine so that an inspection action should taken immediately.

✓ Engine Oil: Pressure, Temp In/Out, Level

Engine Oil parameters will be monitored to determine the condition of Engine Oil when the Main Engine is operating.

✓ Fuel Oil: Flow, Pressure

This data will be used primarily to determine the trend of fuel consumption on the Main Engine easy to remove / install so that in case of trouble, it can be easily returned to the manual system.

✓ Alternator on Starboard
- Power Meters: V, I, F, φ.

A power meter will be installed to determine all electrical parameters output from the alternator as part of monitoring.
- Body Temp.
- Shaft RPM.

These two parameters are monitored if an anomaly occurs in the alternator performance.

✓ Main engine on Portside
- Engine Exhaust Temp per cylinder (5 cylinders)
- Engine Shaft RPM

Similar to the main engine on the Starboard, these two types of parameters are also used to determine the performance of the Main Engine while operating. As stated above, this value anomaly (indicated by an alarm/trip) to the reference value is an indicator to the operator that there is a problem with the Main Engine so that an inspection action should taken immediately.

✓ Engine Oil: Pressure, Temp In/Out, Level

Engine Oil parameters will be monitored to determine the condition of Engine Oil when the Main Engine is operating.

✓ Fuel Oil: Flow, Pressure

This data will be used primarily to determine the trend of fuel consumption on the Main Engine

✓ Alternator on Portside

- Power Meters: V, I, F, φ

The alternator at Portside will also be installed with a power meter to determine all electrical parameters output from the alternator for monitoring purposes.

- Body Temp.
- Shaft RPM.

Like the alternator on the starboard, the two parameters of the alternator on the portside are also used as a monitor if an anomaly occurs in the alternator's performance.

✓ Gearbox

✓ Propeller Shaft Vibration

✓ Eccentricity Shaft Propeller

3.2. Design of Wiring Panel System

Based on the ease and uniformity of installation, as well as maintenance and operation for the selection of instruments, the type of instrument with type 2 wire – differential current was chosen according to the manual wiring of the analog input PLC module (5069-IF8). The results of the design of the wiring diagram for the assembly of the remote I/O panel and the installation of the monitoring system as a whole will be attached and presented in a separate wiring diagram document (not shown here). However, some screenshots of the wiring diagram design for the purposes of making the engine performance monitoring system wiring and alarms are as follows can be seen in Figures 5 - 9.

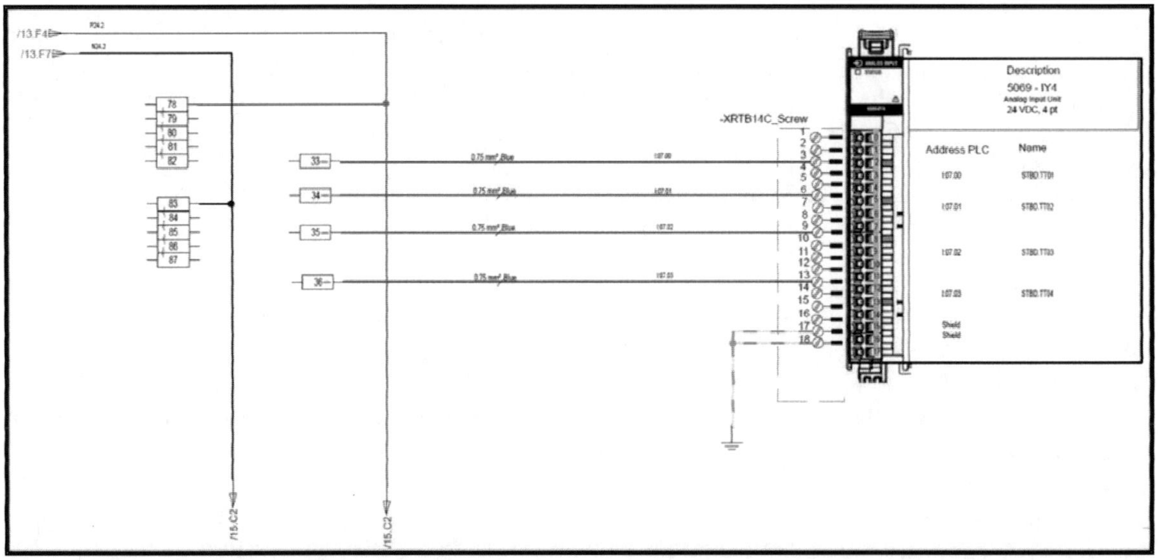

Figure 5. Wiring diagram AI (Analog Input).

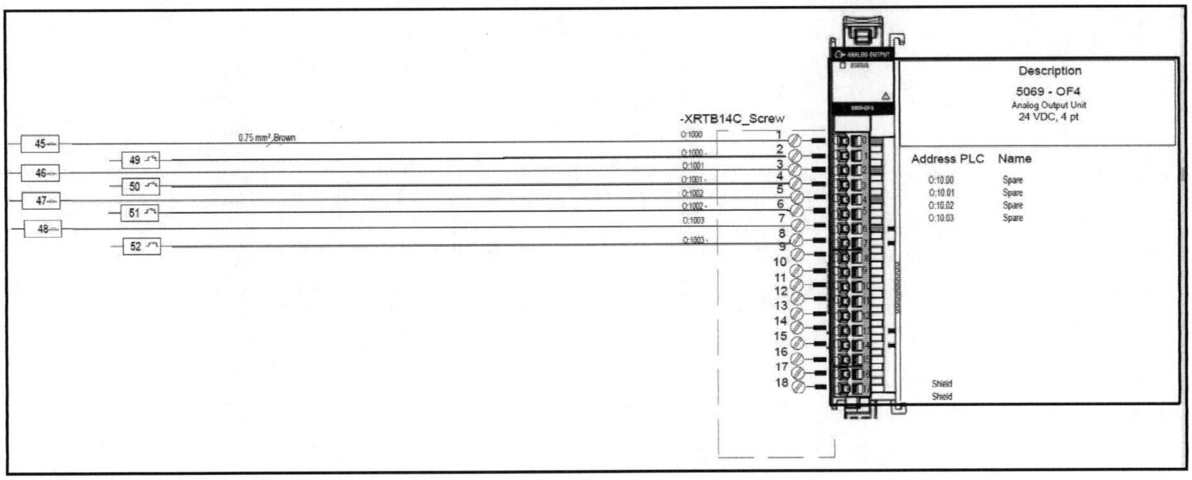

Figure 6. Wiring diagram AO (Analog Output).

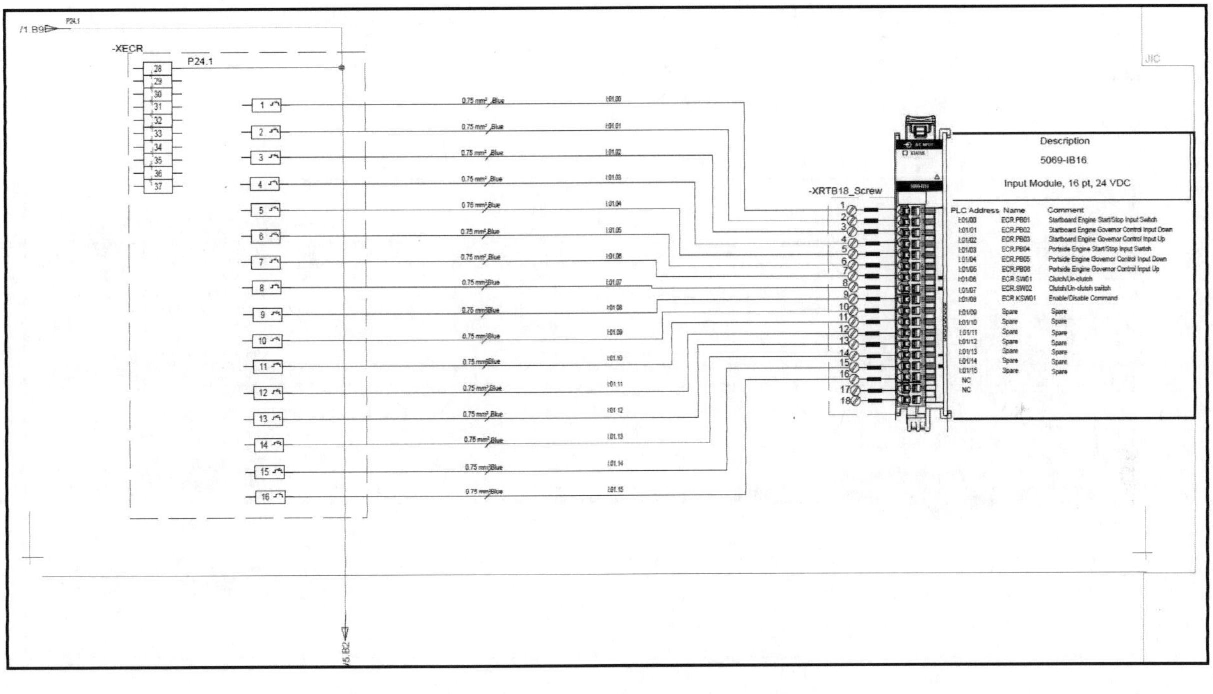

Figure 7. Wiring diagram DI (Digital Input)

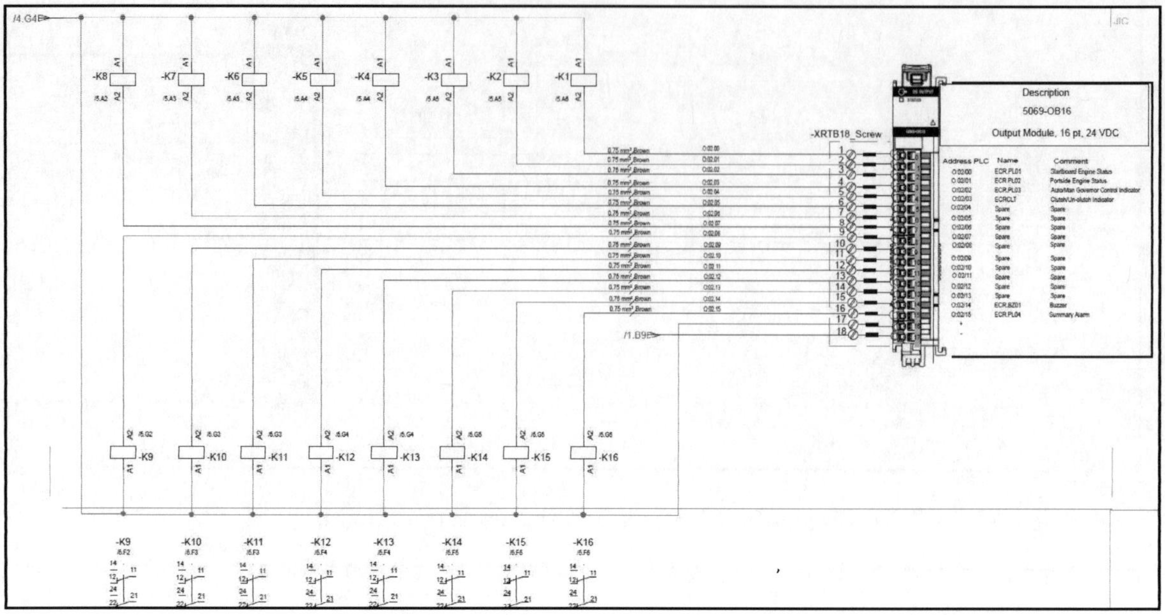

Figure 8. Wiring diagram DO (Digital Output)

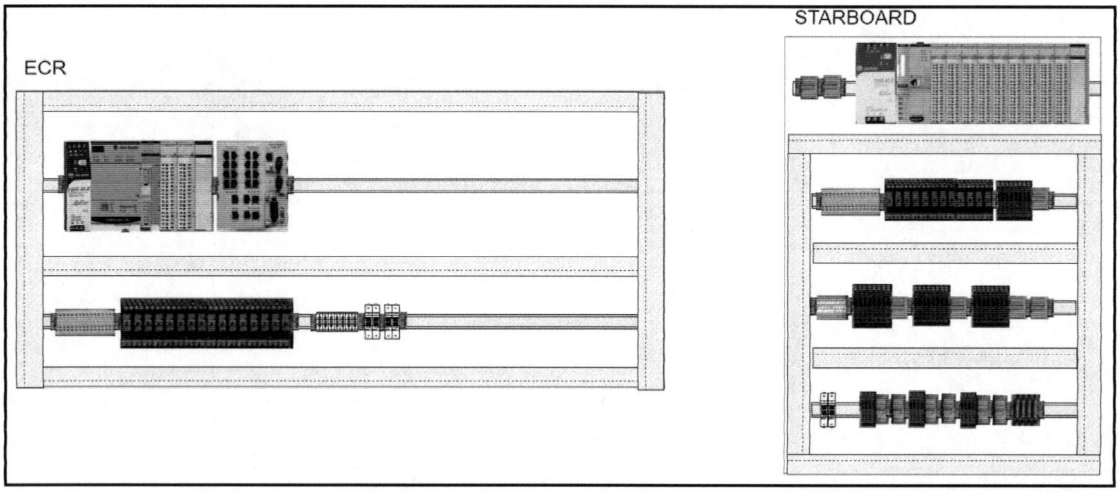

Figure 9. Main PLC panel design and remote I/O.

This design of wiring diagram will be used for the assembly of the remote I/O panel and the installation of the monitoring system according to the needs of the instrument that has been set. Furthermore, the design will be used as a reference for the implementation of system installation and PLC programming.

3.3. PLC Programming

The PLC program was created using software studio 5000 from Allen-Bradley. The steps taken in making this program include project creation and target configuration, making the main program for the monitoring and alarm system of RV. Baruna Jaya III, and simulations to check the programs that have been made.

IOP Conf. Series: Earth and Environmental Science 1081 (2022) 012024 doi:10.1088/1755-1315/1081/1/012024

3.3.1 *Project creation and target configuration.* From the startup screen software studio 5000, there is an option for creating a new project. By clicking on this option, a new project creation wizard will appear by first selecting the type of PLC to be used. In accordance with the list of components that have been made above, the type of PLC selected is 5069-L320ER as shown in Figure 10.

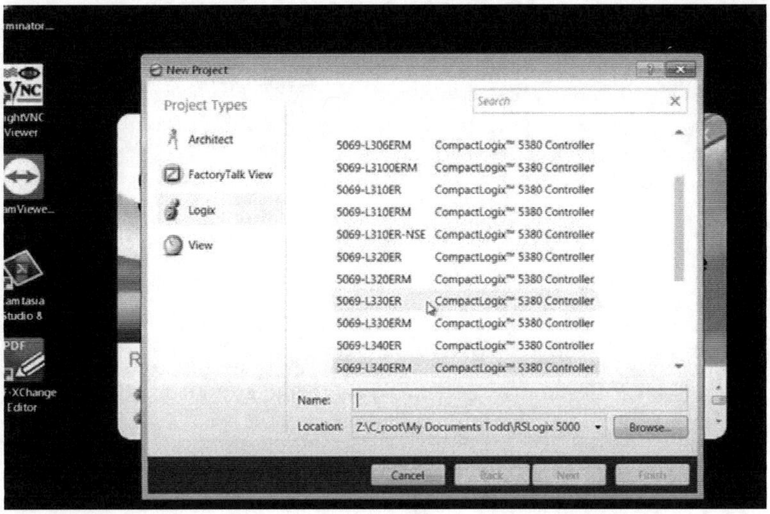

Figure 10. Project creation wizard in software studio 5000.

After that, the workspace for programming the PLC will be displayed. The next step is to configure the PLC according to the above architecture in the "I/O Configuration" folder menu section as shown in Figure 11. This configuration is made as the same as the original including the complete remote I / O configuration used.

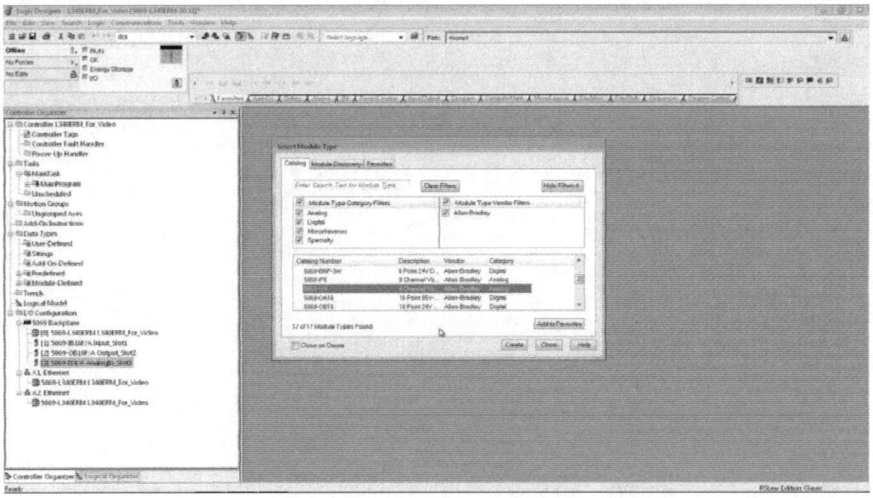

Figure 11. PLC target configuration.

3.3.2 *Making the main program for the monitoring and alarm system of the RV. Baruna Jaya III.* The manufacture of software for monitoring performance and alarm system of RV. Baruna Jaya III is done through the "Task" folder menu. The form of program code can be in the form of ladder diagrams, function blocks, instruction lists, or a combination of several forms of code. Studio 5000 software has flexibility in the type of code used. Some of the menus in programming are shown in Figures 12, 12, 13, 14 and 15.

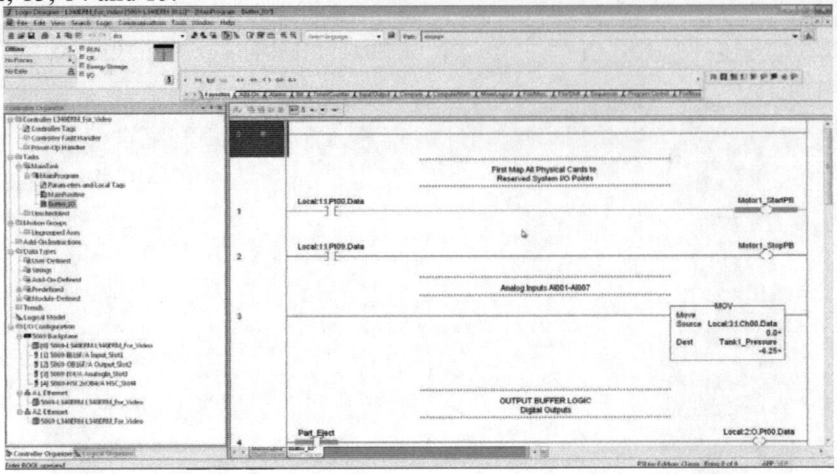

Figure 12. Coding of software monitoring and alarm system.

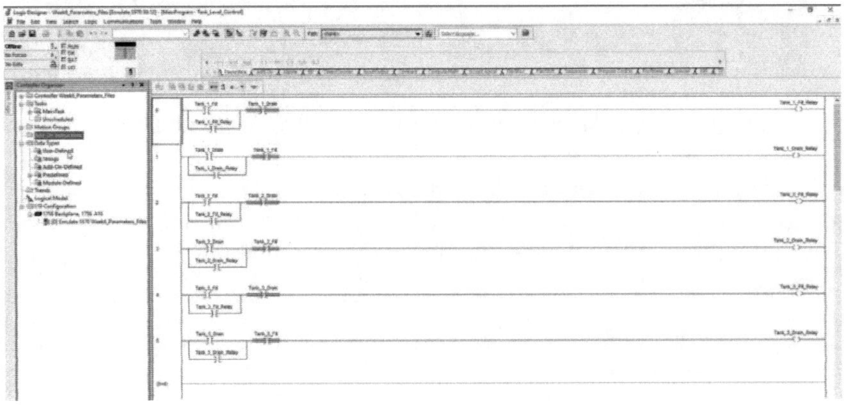

Figure 13. Coding software in the form of a ladder diagram.

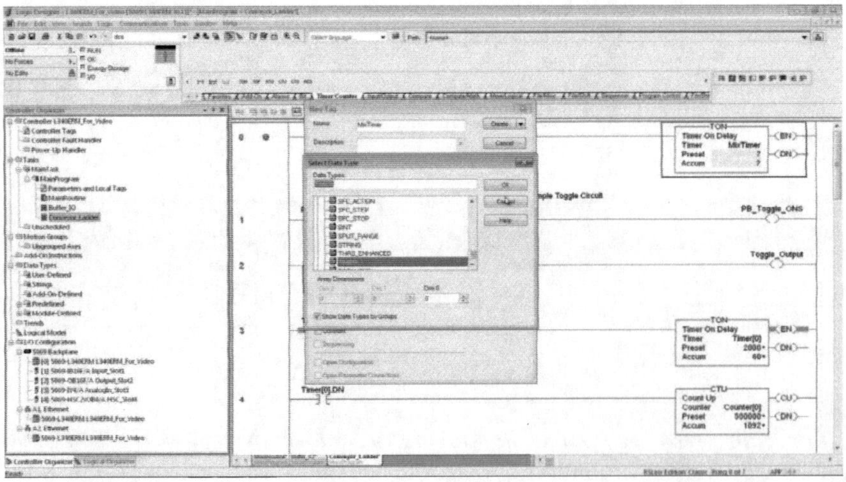

Figure 14. The internal function facilities of the studio 5000 software which are ready to used.

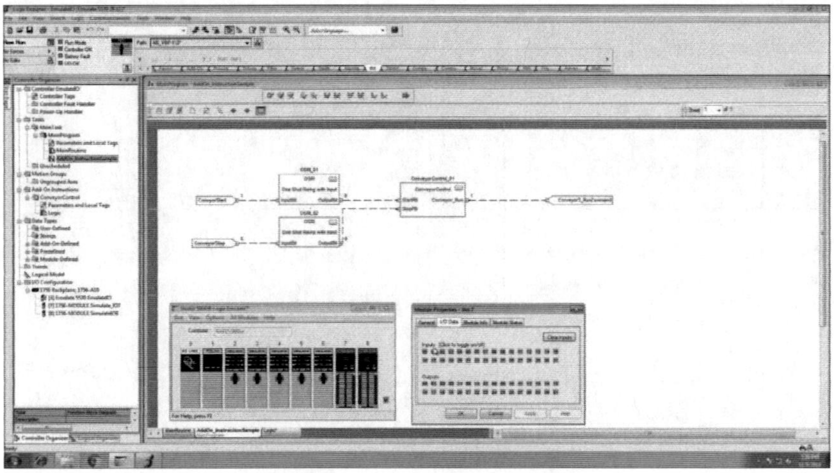

Figure 15. Simulation to check PLC Program.

3.3.3 Simulation for checking the program that has been made. If the studio 5000 software has been connected to the PLC, the program can be transferred to the PLC and can then be run. Studio 5000 has a debugger facility with animations to see the program running on the PLC (Figure 16). If there is a register that is in the ON status, the color will change to green. This animation makes it easier for programmers to check whether there are errors in the program that has been made.

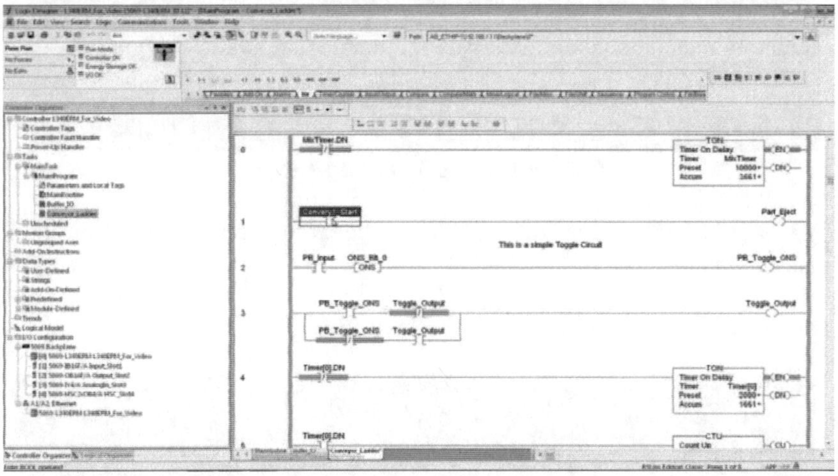

Figure 16. Animation of the debugging process for checking the program.

3.4. SCADA and Dashboard Programming

The SCADA software used for the automation system on the RV. Baruna Jaya III research vessel is Indusoft Web Studio. *Dashboard.* Used to view an overview and resume of ship conditions. This SCADA cannot be operated and only functions as monitoring. This dashboard will be displayed using an LED Monitor mounted on the wheelhouse (bridge) so that it can be seen by the captain of the ship for the needs of the ship's operation. The dashboard display is as shown in Figures 17 and 18 below.

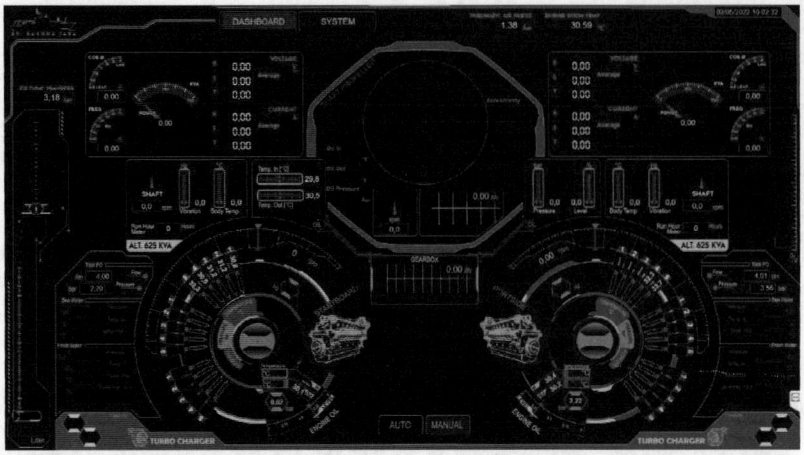

Figure 17. Dashboard display 1.

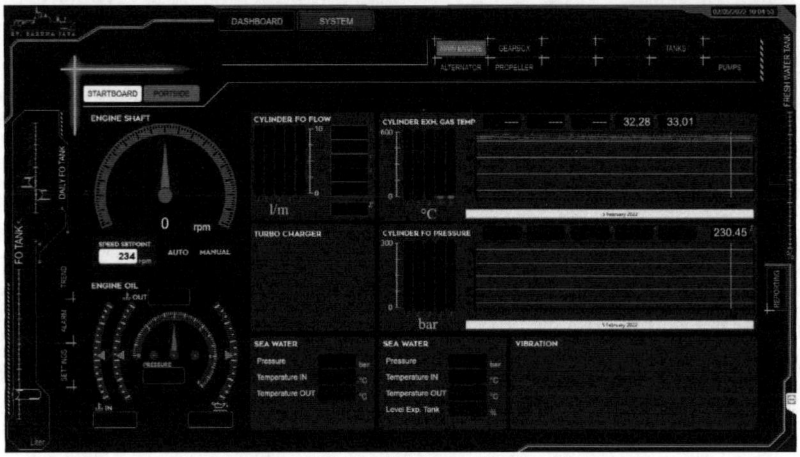

Figure 18. Dashboard display 2.

SCADA ECR. This SCADA is a master that fully describes the automation system on the RV. Baruna Jaya III. Ship managers can operate this SCADA from the ECR room for monitoring and controlling components on the ship. Figure 19 shows the SCADA ECR display while Figure 20 shows the design view of the alarm system.

Figure 19. Display of SCADA ECR.

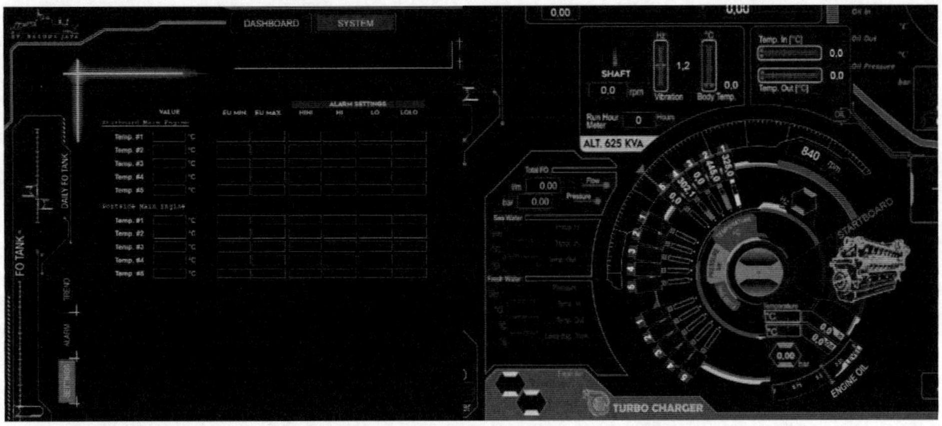

Figure 20. Display of alarm system

3.5. Reporting System

The reporting system can be accessed by ship management using a computer located at location of ID + Pos.3 in the wheelhouse (bridge). To make a report, the steps taken are:

- Activate the system reporting display (Figure 21) and enter a user-name and password
- Choose the start date and start of the reporting period
- Enter some additional notes in the field if needed
- Clicking the "generate" button. Preview reporting will be displayed on the monitor screen to be reviewed by ship management
- From the preview view, there is an option to directly print the report using a printer or export it as a pdf file.

Figure 21. Display of reporting system.

3.6. Web SCADA

Web SCADA is part of the SCADA software features, namely Indusoft web studio. To activate the SCADA web feature can be done by setting the parameters in the web dialog in the project settings as shown in the following Figure 22.

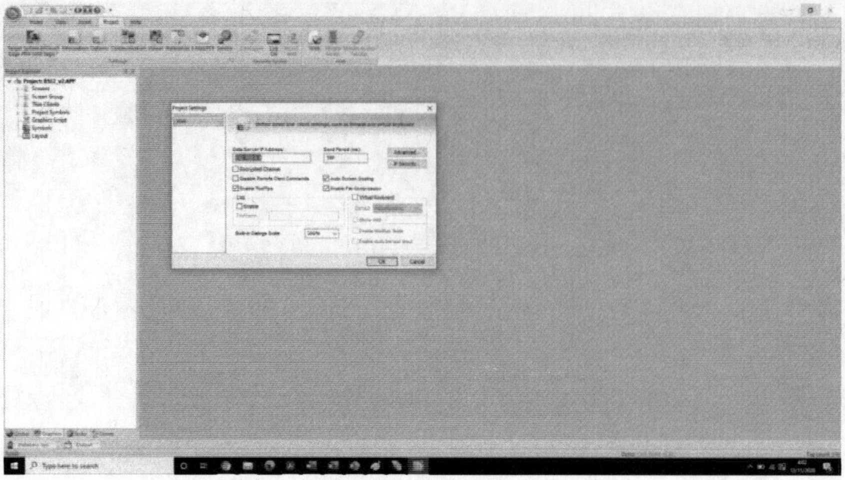

Figure 22. SCADA web settings.

After setting the SCADA software, the IIS service is also set on the Windows operating system for the web-server (Figure 23). Settings include specifying the physical path folder according to the saved SCADA project, MIME-Type registration, firewall settings, and permission settings.

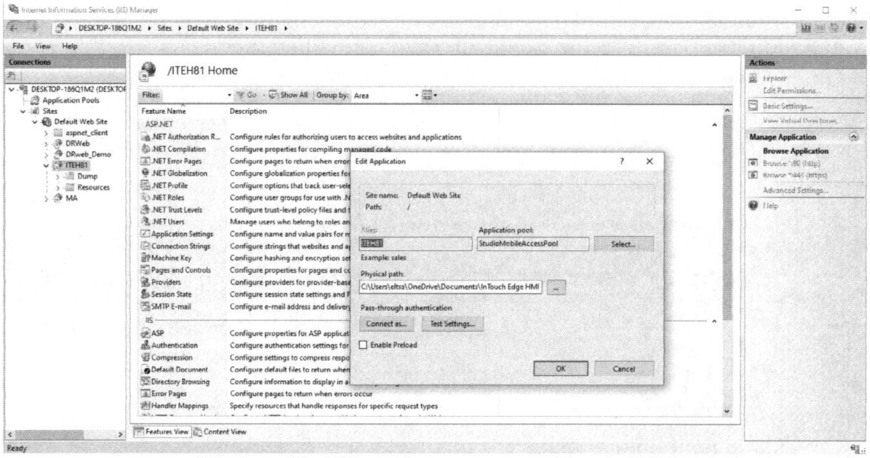

Figure 23. IIS settings to support web SCADA.

After the settings are complete and SCADA is activated, the SCADA display can be accessed using internet explorer by entering the SCADA server address. If successful, a user-name and password dialog will appear to gain access to the main SCADA display (Figure 24).

IOP Conf. Series: Earth and Environmental Science 1081 (2022) 012024 doi:10.1088/1755-1315/1081/1/012024

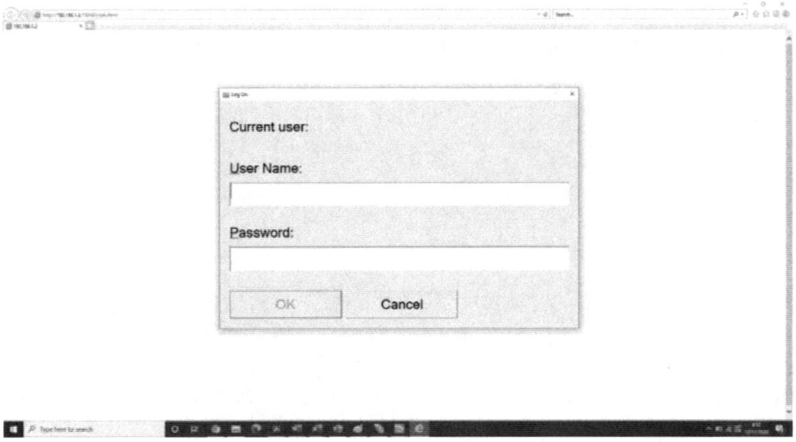

Figure 24. SCADA login accessed using internet explorer.

After successful login, internet explorer will display a SCADA mimic which can be operated by the user like accessing SCADA on the server computer. Figure 25 shows a description of the performance of ship equipment via SCADA which is accessed via the internet explorer network.

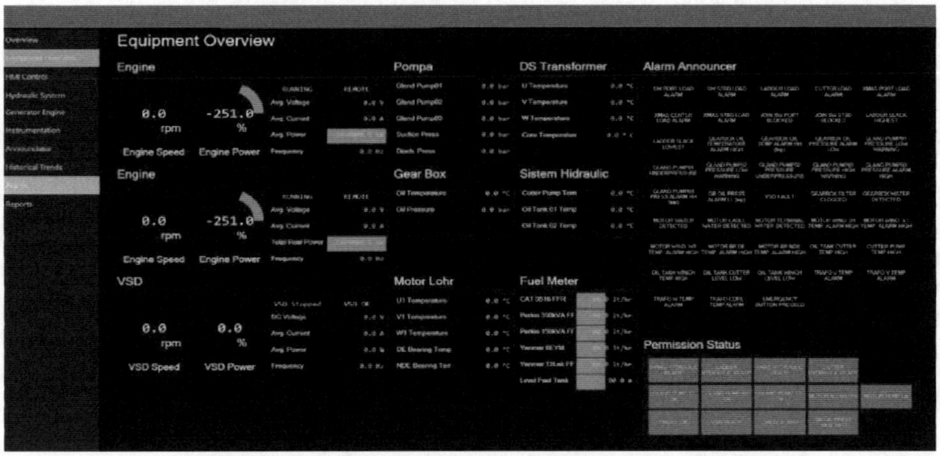

Figure 25. SCADA accessed using internet explorer.

There are several notes that must be considered when using this SCADA web feature, namely:

- Browsers can only use internet explorer
- If you want to access via the internet network, the SCADA computer must be connected to an internet provider that provides a static IP feature.
- It is necessary to analyze the data that can be accessed through this SCADA web and the direction of the flow of information for security reasons.

4. Conclusions

Based on literature references review, it is known that the technology for monitoring of the ship's main engine can be done using internet of things-based technology. This study has produced a design of real time monitoring performance and alarm system for main engine of RV. Baruna Jaya III research vessel. This system will collect data from sensors and make it available to users from certain locations using PLC design methods and SCADA based systems. The development of an IoT-based real-time monitoring system on the ship's main engine is based on determining the appropriate measurement data parameters and depending on the critical components to be measured. These developments are include designing a modular monitoring and alarm system, designing a reporting system which will then be continued in further research to carry out system development by installing hardware and sensors, calibration, data acquisition, data transmission and data analysis. This study could be a reference for recommendations for selecting the right maintenance strategy through the Condition-based Maintenance (CBM) method.

References

[1] Priharanto Y E, Latif M Z, & Saputra R 2017 Risk Assessment of Cooling Engines on Fishing Vessels Using the FMEA Approach *Jurnal Airaha* ***6*(1)** 24-32

[2] Latif M Z, Priharanto Y E, Prasetyo D, and Muhfizar 2018 Preliminary Hazard Analysis and Fault Tree Analysis to Identify Causes of Failure of the Main Engine Lubricant System for Fishing Vessels *Jurnal Airaha* ***7*(02)** 077-087

[3] Yaqin R I, Zamri, Siahaan J P 2020 The FMEA Approach in Risk Analysis of Main Engine Fuel System Maintenance: A Case Study in KM. Sidomulyo *Jurnal Rekayasa Sistem Industri* **9(3)** 189-200

[4] Flores J 2017 Supervision, Control and Integration by PLC of a Process that Simulates the Selection and Drilling of Parts *M.S. thesis, Polytechnic University of Catalonia, Catalunya, España*

[5] Andrade R 2018 Teaching module to control water level and flow using SCADA system, PLC and PID algorithm *Journal RIEMAT* **4** 50-62

[6] Sánchez G and Custodio A 2017 Development of SCADA system for flow control based on LINUX University *Science and Technology* **11(44)** 121-8

[7] Zapata G et al 2010 Aportes Metodológicos para el Diseño de Sistemas de Supervisión de Procesos Continuos *Journal Technological information* **22(3)** 97-114

[8] Lopez E 2017 SCADA Systems in Industrial Automation *Journal Technology in Progress* **28(4)** 3-14

[9] Chauhan R K, Chauhan K, Dewal M L 2010 Intelligent SCADA *System International Journal on power system optimization and Control* **2(1)** 143-9

[10] Rodríguez A et al 2019 Synchronization and Integration in SCADA of a process desalination *Journal Electronic Engineering, Automation and Communications* 334-54

[11] Tawiah I, Ashraf U, Song Y, and Akhtar A 2018 Marine Engine Room Alarm Monitoring System, Fault Detection and Monitoring Technique via PLCs and SCADA-based System *International Journal of Advanced Computer and Applications* **9(6)** 433-41

[12] Prasetyo R B et al 2011 Aplikasi Teknologi Informasi Berbasis GSM SMS Untuk Akuisisi Data Unjuk Kerja Permesinan Secara Real Time *Jurnal Teknoin*

[13] Shah S et al 2017 IOT Based Industrial SCADA *System International Research Journal of Engineering and Technology (IRJET)* **4(4)** 2432-35

Acknowledgments

This study partly supported by former Center of Technology for Marine Survey (Balai Teksurla). We acknowledge the encouragement and assistance from leaders and staffs of former Balai Teksurla, Center of Technology for Machine Tools, Production & Automation (BTMEPPO) and PT. CBI who have provided opportunities, facilities, and support to us in carrying out this research until the preparation of this paper.

A Computational of benchmark model of resistance test at Indonesian Hydrodynamic Laboratory

D Purnamasari[1], M Nasir[1], Rina[1]

[1]Indonesian Hydrodynamic Laboratory

Abstract. Benchmarking are the activity of comparison of different methods, procedures, physical models to obtain the same target in order to achieve a better standard of performance and exercises provide a common basis for validation of numerical methods. The work develops on prior studies into benchmarks and uncertainty analysis for ship model resistance tests of LHI-007 was used for benchmark testing from 2010 to 2020 at the Indonesian Hydrodynamic Laboratory. A computational study based on the free-surface capturing approach performed with the FINE/Marine flow solver to predicted total resistance in calm water of four speed (1.63 m/s - 2.47 m/s). Verification and validation method studied uncertainty analysis for both for experimental and computational following the ITTC guidelines. The results show comparison between computed results and previous published experimental tests of total resistance (R_T) and coefficient total resistance (C_T) are done to validated CFD simulations. The result of this study concludes that predicted resistance of the benchmark model has reasonable accuracy that the present solver.

Keywords: Ferry Ro-Ro ship; CFD; calm water; resistance

1. Introduction

The objective of the benchmark tests is to look up common knowledge or performance standard, the chance of the benchmark test represents learning something, using the prime competence, and great loyalty in order to serve direction for best implementation. Benchmarking is the process of comparasion different of methods and procedures, and physical models serve as a common basis for the validation of numerical methods. Types of benchmarking can be defined as internal benchmarks, it is an appeal between processes executed within one organization.

The flow around a ship in steady motion can be realistically simulated by methods belonging to the wide field of computational fluid dynamics (CFD). The comprehensive CFD validation is mostly accepted as provided that the assessment of the confidence of a CFD solution through comparisons with a benchmark in general provided by experimental data. The first meaningfull test under an international collaborative study on EFD/CFD and uncertainty estimate between Iowa Institute of Hydraulic Research (IIHR), Istituto Nazionale per Studi ed Esperienze di Architettura Navale (INSEAN) in Rome, Italy, and David Taylor Model Basin, USA Navy as a preliminary design for a surface combatant DTMB 5415 with a sonar dome bow and transom stern [1]

ITTC has recommended the benchmark database for CFD validation for resistance and propulsion focus on modern hull forms specifically as test cases for ship hydrodynamics there have been reported by the Resistance Committee of the 22nd International Towing Tank Conference in 1999. Three ships for Tanker (KVLCC), container ship (KCS), and surface combatant (DTMB 5415) hull forms were recommended for practice and are presently availability used as test cases in the Gothenburg 2010 Workshop on CFD for Ship Hydrodynamics [2]. KVLCC and KCS have bulbous bows and transom sterns. The KVLCC and KCS data were procured by Korean Institute of Ships & Ocean Engineering (KRISO) in collaboration with Pohang University of Science and Technology, Korea and Ship Research Institute, Japan, respectively.

Verification and validation (V&V) studies for the improvement of the quality assurance of the CFD methods is essential for the CFD applications. By reason of the derived uncertainty levels are only valid for a unique case and condition of each test. The fundamental strategy of validation is to assess how accurately the computational results compare with the experimental data, with quantified error and uncertainty estimates for both. According to the resistance statistics presented in [3], the mean comparison error of the measured data value is 1.7% and 1.3% for KVLCC2 and KCS in fixed trim and sinkage condition while the standard deviations are 1.3% and 1.2% of the data value, respectively. Recently, researchers perfomed simulations for the KCS model at three Froude number 0.22,0.26,0.28, investigation for different turbulence models *SST* and Standard K-ω, the results are generally in good agreement with the data. Error in C_T for *SST* are 3.2%, 3.78%, and 4.97%, and for Standard K-ω were 4.75% 5.08% and 6.03% respectively [4]. Uncertainty analysis implement for resistance, sinkage and trim of KCS, DTC, KVLCC2, and JBC ship models using factor of safety (*Fs*) and correction factor (*Ci*) method. The result showed the total predicted uncertainties for all cases have a significant difference, *Ci* based method estimated higher level of uncertainty comparing factor of safety method [5]. The numerical simulations for the DTMB 5415 case, mesh-independence study was performed at *Fr* 0.28 to compare with the experimental data, relative error on the resistance coefficient in different mesh resolutions is less than 4.13% [6]. KCS containership model scale is simulated using the unsteady RANSE method from coarse, medium, and fine grids with the Realizable k-ε turbulence model to estimate total resistance, dynamic trim, and sinkage. The total resistance is about 2.50% of the measured value and 7.74% for trim and sinkage on all mesh sizes. The discrepancy reduces from 2.366% to 0.903% with grid refinement [7].

The present study a continuing effort and updates towards improvement of Indonesian Hydrodynamics Laboratory (IHL) uncertainty estimates [8,9] with specific focus on the surface ship model scale database of resistance. IHL was initiate benchmark testing in 2010 and repeated it annually up to 2020, LHI-007 Ro-Ro Ferry ship model was from the commencement used as a benchmark in 1994 when the facility was in commission with identical tests conducted at both the Maritime Research Institute Netherlands (MARIN) and IHL [10]. This study was deliberate to mandate a variety of dominant physics and develop the surface-ship database with quality datasets and uncertainty assessment.

The paper is structured in the following: the background of international benchmark test case determination by experimental methods and validation CFD presented in Section 1, specified informatin of the flow solver, numerical methods, boundary conditions, grid generation and supplementary grid refinements, computational conditions are described in the research method in Section 2. Results and discussions are presented in Section 3 and the paper ends with conclusions are summarized in section 4.

2. Method

2.1 *Ship Model and particulars*

The LHI-007 Ro-Ro Ferry ship model was selected as the target of computational, the scale ratio of the model was 1:20.97. The larger model and the larger forces are both features leading to greater accuracy in the measurement of resistance. The parameter of the ship and the model are presented in Table 1. Body plan of Ro-Ro Ferry is shown in Figure 1.

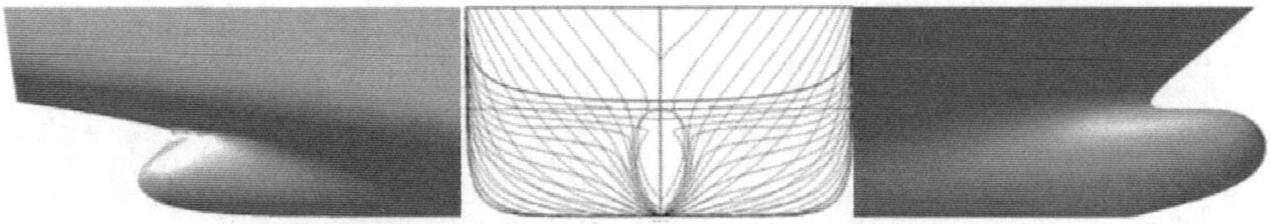

Figure 1. Body Plan of LHI 007 Ferry Roro

Table 1 Particulars of ship and model LHI-007 Ro-Ro Ferry

Parameter	Symbol	Prototype	Model	Unit
Length on waterline	*Lwl*	146.94	7.007	m
Breadth	B	24.00	1.150	m
Draft	*T*	6.50	0.310	m
Wetted surface area	*S*	4042.00	9.192	m^2
Displacement Volume	∇	14.21	1.579	m^3
Block coefficient	C_b	0.80	0.80	

The calm water resistance tests for routine tests and IHL database of the repeatability have been periodically. Table 2 shows the water temperature for the first test in 1994 and during 2010-2018 presented in [9] and continued until 2020, it is to be noted that experimental results are given diagrammatically for wide range of temperature. Temperature 27° (2010) and 28°(2018) at four speed (1.63 m/s, 1.81 m/s, 2.25 m/s, 2.47 m/s) were chosen with concern to employ of data for CFD validation.

Table 2 Variation of water temperature in Towing Tank IHL for model benchmark LHI 007

Year	Temperature (°C)	Density (kg/m^3)	Kinematic Viscosity (m^2/s)
1994	28	996.236	$8.35.10^{-7}$
2010	27	996.515	$8.54.10^{-7}$
2011	27.1	996.488	$8.52.10^{-7}$
2012	27.8	996.292	$8.39.10^{-7}$
2013	28.3	996.150	$8.30.10^{-7}$
2014	29	995.947	$8.18.10^{-7}$
2015	27.8	996.292	$8.39.10^{-7}$
2016	28.2	996.178	$8.32.10^{-7}$
2017	27.5	996.377	$8.45.10^{-7}$
2018	28	996.236	$8.35.10^{-7}$
2019	26.9	996.543	$8.56.10^{-7}$
2020	27.3	996.433	$8.48.10^{-7}$

2.2 Computational

The calm water resistance computations were resolved using CFD software produced by NUMECA with volume of fluid (VOF) method (free surface capturing approach). NUMECA has product for mesh generator, flow solver and post processing [11]. The present work focuses on the k-ω Shear-Stress Transport turbulence model was first proposed by Menter [12]. The CFD simulations output is the total resistance, which is calculated by incorporate the shear and pressure force on bare hull model. Total Resistance of four speed were computed in calm water conditions.

Coordinate systems, domain and boundary conditions have been adopted as described in Figure 2. The global coordinate system (x, y, z) is clearly defined to reflect the flow patterns close to the hull. The x-axis is positive from the bow to the stern and the z-axis is negative in the gravity direction. The boundary conditions for a solid wall for the hull and deck are wall function and slip respectively, the top and bottom are prescribed pressure, the inlet is far-field with 0 m/s velocities, the outlet, side, and mirror is the symmetry plane. The velocity is assigned to the ship center of gravity.

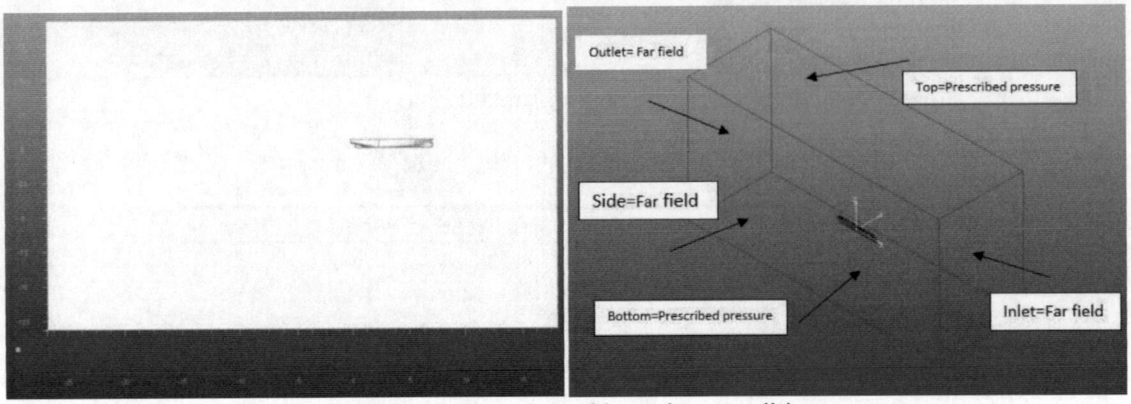

Figure 2. Domain and boundary conditions

FINE/Marine's plugin called C-Wizard used to mesh generation, three different (coarse, medium, fine) are generated of the mesh. The case setup a plugin allowing the whole process of setup of mesh and solver parameters based on the geometry of model, patch names and physical inputs. The geometry is symmetric and the flow around the model hull can be assumed symmetry, the simulation performed in half domain as shown in Figure 3. The last step of the meshing process is to control the grid quality, verify that there are no Negative, Concave or Twisted cells. The Orthogonality must be checked, computation should 1 deg < Minimal orthogonality < 5 deg.

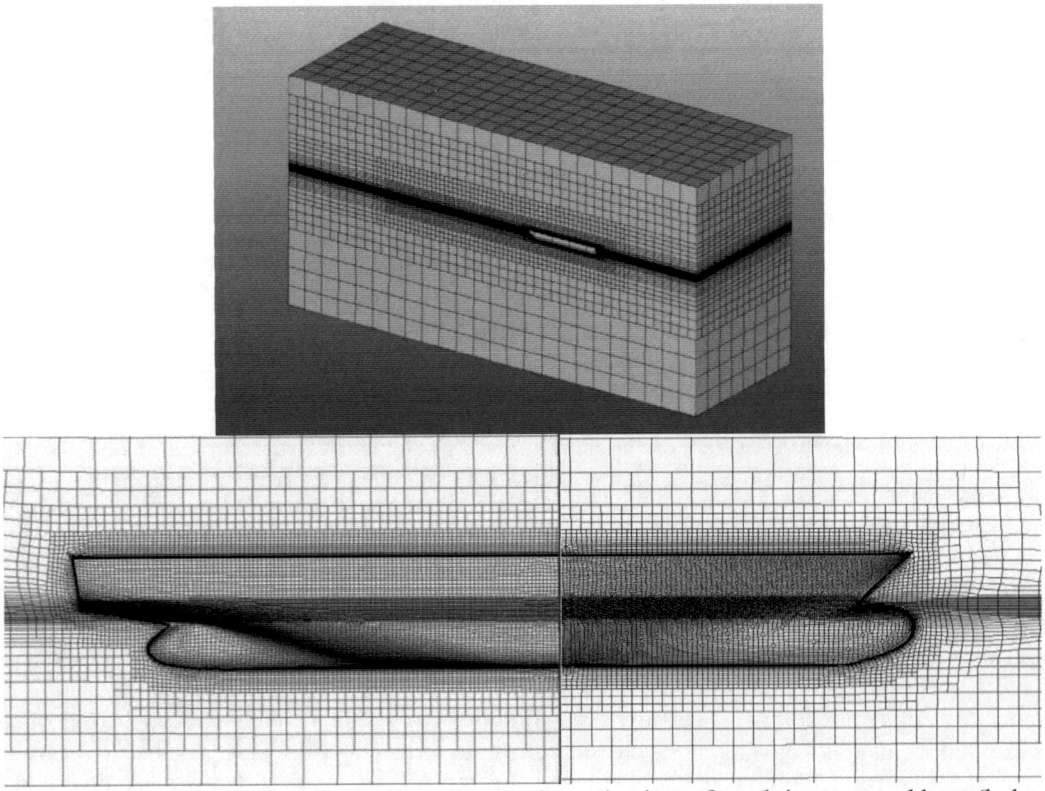

Figure 3. View of mesh LHI 007 Ferry Ro-Ro (upper), view of mesh in stern and bow (below)

2.3 Verification and Validation

Verification is done for three different grid size from fine (S_1), medium (S_2) and coarse (S_3). Convergence studies of the 3 solutions were used to define the convergence ratio R_G by following equation:

$$R_G = \varepsilon_{21}/\varepsilon_{32} \qquad (1)$$

Using generalized Richardson Extrapolation (*RE*), the order of accuracy p_G and the error in the solution δ can be estimated as follow:

$$\delta^*_{RE_{G,1}} = \frac{\varepsilon_{21}}{r_G^{p_G} - 1} \qquad (2)$$

$$p_G = \frac{ln(\varepsilon_{32}/\varepsilon_{21})}{ln(r_G)} \qquad (3)$$

Estimate error from Richardson Extrapolation is multiplied by with factor of safety *Fs* value (1.25) to definitive simulation error U_G followed in [13] recommendation:

$$U_G = F_S|\delta^*_{RE_{G1}}| \qquad (4)$$

Validation is defined by assessing simulation uncertainty U_G by using experimental data uncertainty U_D. Experimental data were obtained through periodically towing tank test of LHI 007 Ferry Ro-Ro at IHL, thorough uncertainty analysis according to ITTC [13] presented in [9). The validation uncertainty is calculated as:

$$U_V^2 = U_D^2 + U_G^2 \qquad (5)$$

3. Result and Discussions

Grid dependency analysis for prediction resistance in calm water condition, three different mesh were tested. The mesh resolve applied are present in Figure 4. The different numbers of cells are given in Table 3. The quality plays a direct proportional for the quality of the analysis, furthermore, the solver will be more robust and efficient when using a well-constructed mesh.

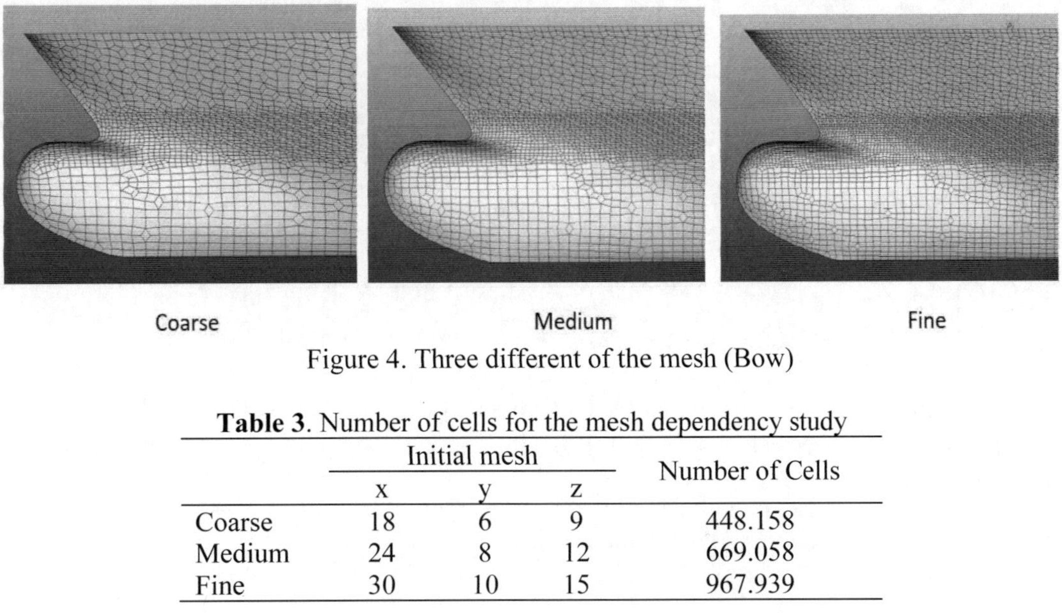

Figure 4. Three different of the mesh (Bow)

Table 3. Number of cells for the mesh dependency study

	Initial mesh			Number of Cells
	x	y	z	
Coarse	18	6	9	448.158
Medium	24	8	12	669.058
Fine	30	10	15	967.939

The coefficient total resistance using fine mesh of the benchmark model was calculated for four different values of speed in the range and the obtained results are compared with the experimental results [9]. For the highest speed 2.47 m/s, the coefficient total resistance has a significantly difference, in other hand, another speed has smallest difference.

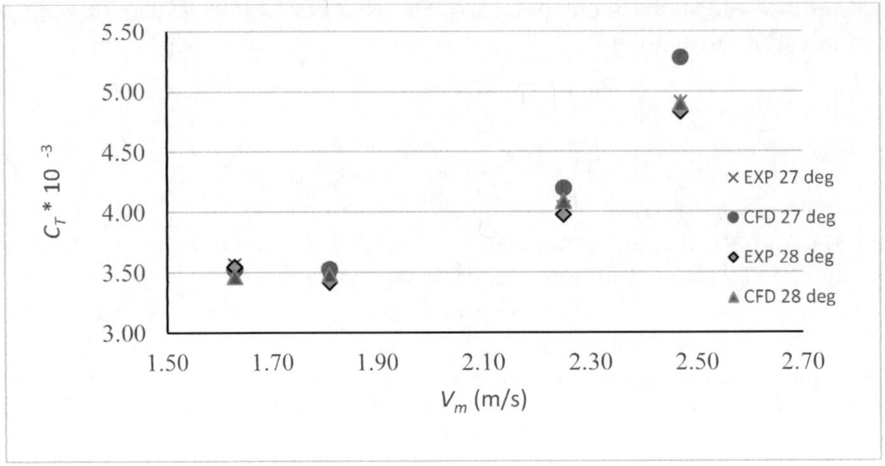

Figure 5. The total coefficient resistance of LHI 007

The first point from the wall, y^+ on the submerged model hull identical to 60 as shown in Figure 6. Extreme refinements near the wall are eases when using wall functions. In the condition of y + value, it is adequate to have the first grid node in the range 30<y+<100 as described in [13].

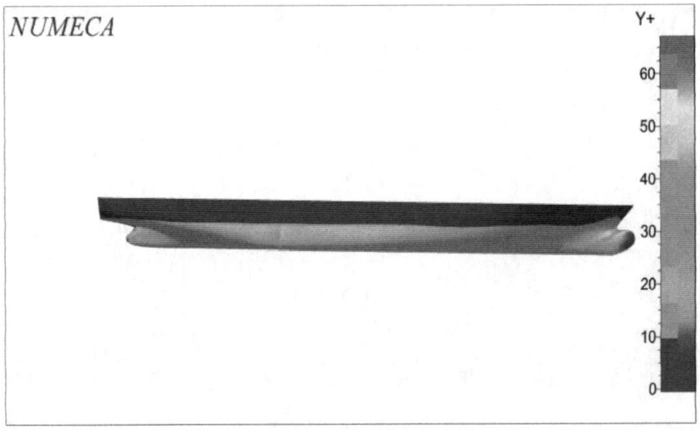

Figure 6. The average y+ of LHI 007 Ferry Ro-Ro

To further illustrate the results, the hydrodynamic pressure distribution on LHI 007 hull surface during calm water motion has been shown for simulation in Figure 7, variant of the pressure around the bulb section is natural for this ship type of ferry roro. The pressure near the model is projected by means of entire computational volume through the inlet toward the outlet. The total pressure accommodates both the effect of static pressure and hydrodynamic.

IOP Conf. Series: Earth and Environmental Science 1081 (2022) 012025 doi:10.1088/1755-1315/1081/1/012025

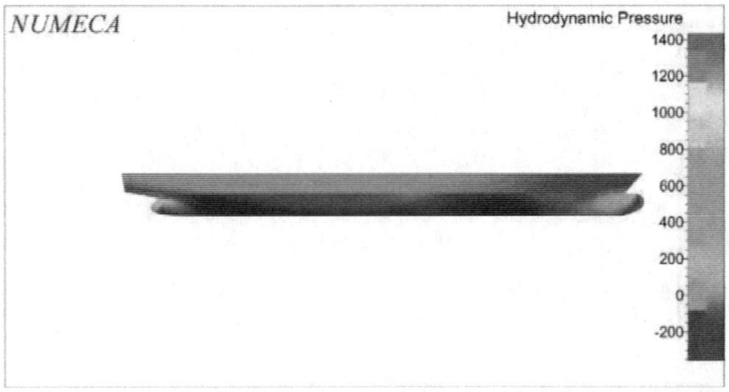

Figure 7. Hydrodynamic pressure distribution on the hull surface of LHI 007 ferry Ro-Ro

When ship moves forward, create waves on the free surface around benchmark model, the wave making resistance encountered by model. Thus, higher resolution at the bow, stern and at the free surface near hull of ship model are importance to obtain a good prediction. Figure 8 and 9 show wave elevation patterns close to the model of speed 1.63 m/s, 1.81 m/s, 2.25 m/s, and 2.47 m/s. It can be observed, wave elevation patterns appear differently at different speed.

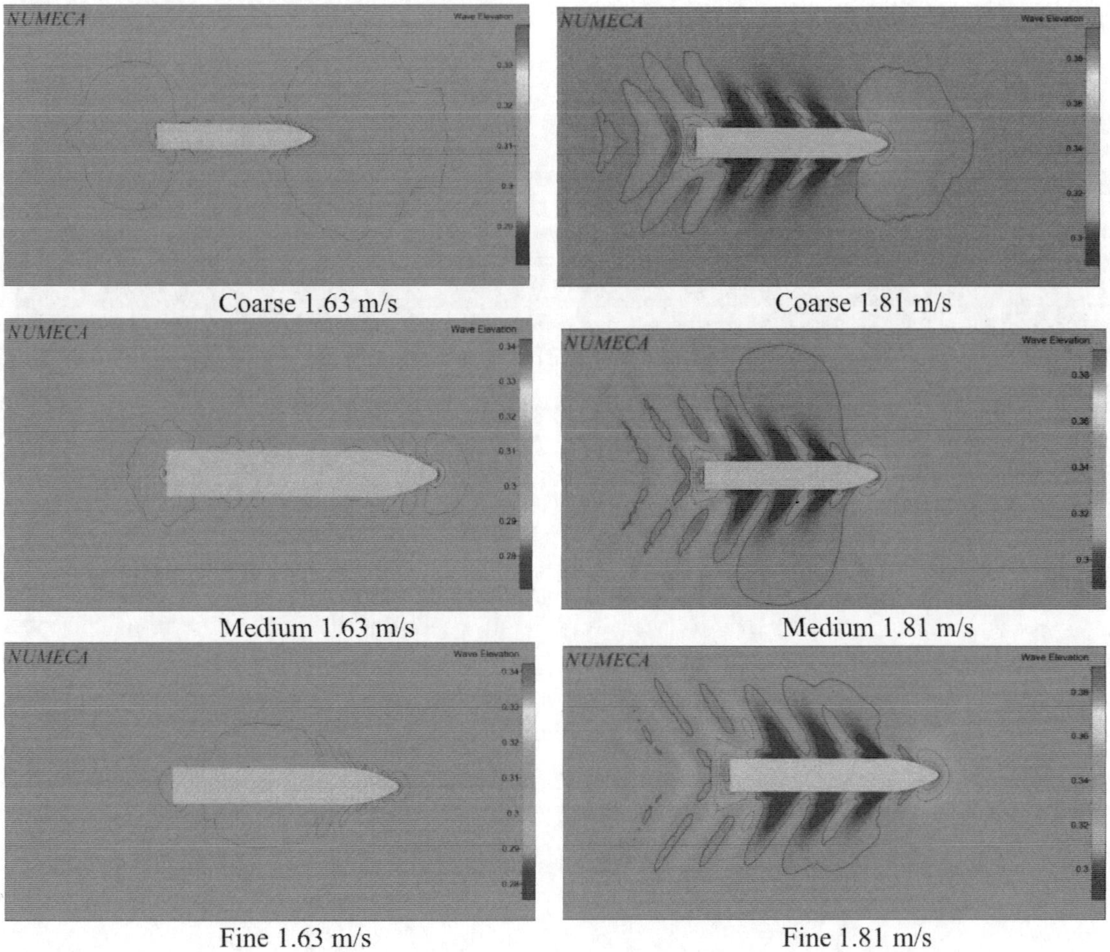

Figure 8. Free surface elevation at speed 1.63 m/s and 1.81 m/s

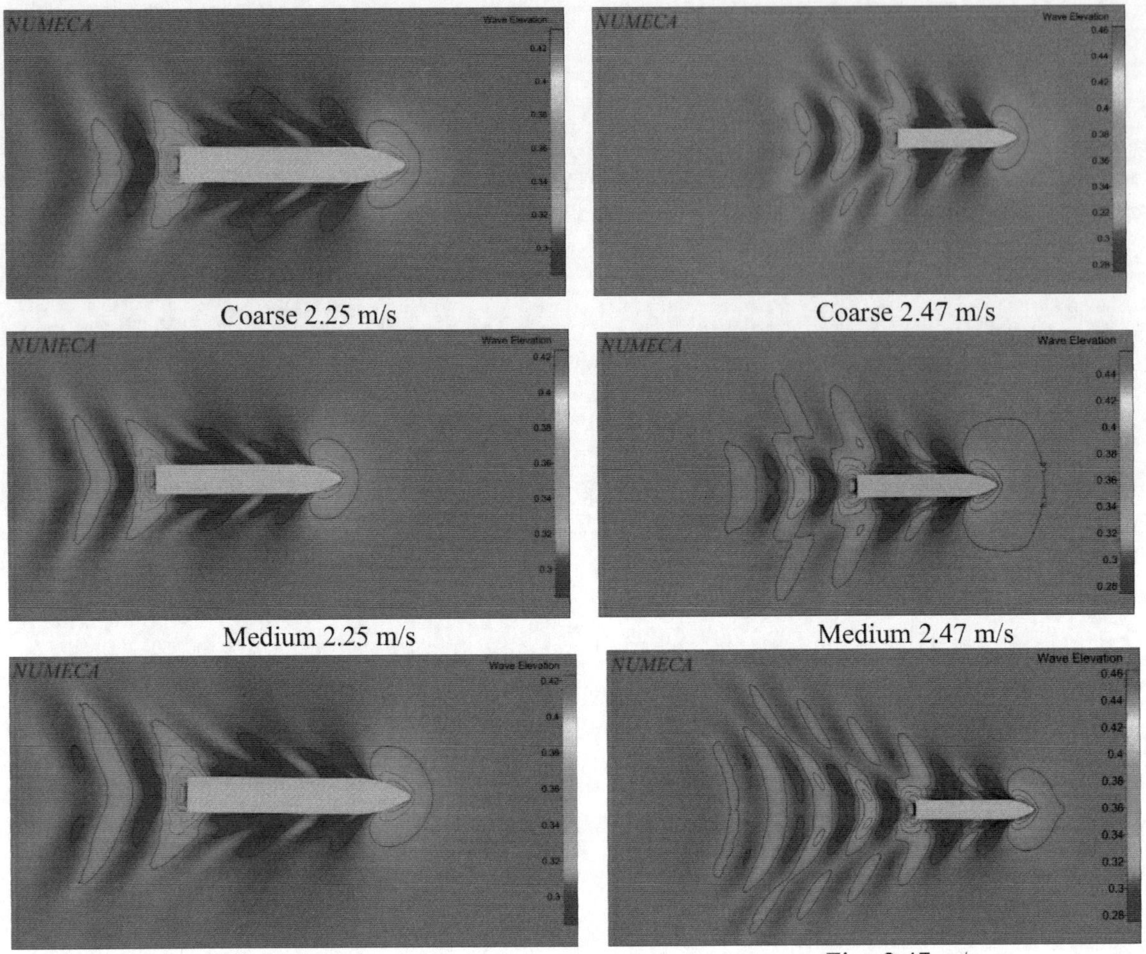

Figure 9 Free surface elevation at speed 2.25 m/s and 2.47 m/s

A systematic verification and validation show in Table 4 and 5. The uncertainty analysis, shows convergence for all speed $R_G < 1$ that confidence in solutions exist, however assessing variability for order of accuracy. A factor of safety approach used to assign the uncertainty wherein an error calculation from RE to bound simulation error, the uncertainties are obtained from Equation (4). The factor safety based approach provides a small value of uncertainty for all speed. Result of validation using corrected solution is effective at four speed set condition.

Table 4. Verification and validation of the coefficient total resistance C_T of LHI 007 (27°)

V_m (m/s)	ε_{21}	ε_{32}	R_G	p_G	δ^*_{RE}	U_G	U_D	U_{val}
1.63	0.104	1.704	0.061	8.069	0.007	0.008	0.784	0.784
1.81	0.654	1.854	0.353	3.007	0.356	0.446	0.828	0.940
2.25	1.354	2.904	0.466	2.202	1.183	1.478	0.737	1.652
2.47	1.048	3.090	0.339	3.120	0.538	0.672	0.916	1.136

Table 5. Verification and validation of the coefficient total resistance C_T of LHI 007 (28°)

V_m (m/s)	ε_{21}	ε_{32}	R_G	P_G	δ^*_{RE}	U_G	U_D	U_{val}
1.63	0.344	1.044	0.330	3.203	0.169	0.211	0.654	0.687
1.81	0.472	0.966	0.489	2.067	0.451	0.564	0.772	0.956
2.25	0.632	3.360	0.188	4.821	0.146	0.183	0.791	0.812
2.47	1.178	11.066	0.106	6.464	0.140	0.175	0.861	0.879

Photograph of experimental and computational of benchmark model of resistance LHI 007 Ro-Ro model as shown in Figure 10. Predicted simulation for resistance were compared against experimental, showing good agreement.

Figure 10. Photograph of experimental and computational of LHI 007 Ro-Ro model

4. Conclusions

The present study, systematic computations for four different speeds, were carried out to determine the model resistance of LHI-007 model ship, a benchmark Ferry Ro-Ro utilized for validation of numerical methods of resistance tesi in the towing test at Indonesian Hydrodynamic Laboratory. It can be summary from the results, the method of safety factor provides a small value in uncertainty prediction. Furthermore, verification and validation study shows monotonic convergence for all speed. Further study is recommended for the benchmark of seakeeping test.

References

[1] Olivieri A, Pistani F, Avanaini A, Stern F, and Penna R 2001 Towing Tank Experiments of Resistance, Sinkage and Trim, Boundary Layer, Wake, and Free Surface Flow Around a Naval Combatant INSEAN 2340 Model Iowa Institute of Hydraulic Research The University of Iowa, IIHR *Report* No. 421.

[2] L Larsson, F. Stern, and M. Vsonneau 2010 Gothenburg 2010 A Workshop on Numerical Ship Hydrodynamics Editors. II.

[3] Larsson, L. Stern, F. Visonneau, M 2014 Numerical Ship Hydrodynamics: An Assessment of the Gothenburg 2010 Workshop. Springer Verlag, Germany. Chapter 5.

[4] Muhammad Tufail Shahzad 2017 Ship Resistance Computations using OpenFOAM, Master Thesis University of Liege

[5] Hafizul Islam and C Guedes Soares 2019 Uncertainty analysis in ship resistance prediction using OpenFOAM", *J. Ocean. Eng.* **191** p 105805.

[6] Chunhui Liu 2019 Prediction of Forces and Moments in Multiphase Naval Simulations, A thesis Master of Applied Science in Mechanical & Mechatronics Engineering the University of Waterloo

[7] Ngoc Kien Vu and Hong Quang Nguyen 2020 Numerical Simulation of Flow Around the Ship Using CFD Method *J of Mech Eng Res and Dev* **43** p 75-86 ISSN: 1024-1752

[8] Purnamasari, D., Utama, I.K.A.P. & Suastika, I.K 2018 Benchmark Study of Ship Model Resistance Test, *Applied Mechanics and Materials* p 114-120 ISSN: 1662-7482

[9] I K A Utama, D Purnamasari, I K Suastika, Nurhadi, G Thomas 2021 Toward Improvement of Resistance Testing Reliability *J of Eng and Tech Sci* **5** (2) DOI: 10.5614/j.eng.technol.sci.2021.53.2.1

[10] Indonesian Hydrodynamic Laboratory (IHL) 1994 Combined Performance Test in Towing Tank, *Report* No. IHL 001-2-TT

[11] Numeca International, FINE/Marine 7.2 2018 Theory Guide (NUMECA online documentation platform).

[12] F. R. Menter, M. Kuntz, and R. Langtry 2003 Ten Years of Industrial Experience with the SST Turbulence Model," *Turbul. Heat Mass Transf.* 4

[13] ITTC 2017 Uncertainty Analysis in CFD Verification and Validation Methodology and Procedures 7.5-03-01-01.

IOP Conf. Series: Earth and Environmental Science 1081 (2022) 012026 doi:10.1088/1755-1315/1081/1/012026

Hydraulic Stability of BPPT – lock on Breakwater Head

[1]Salestiano Cuimbra, [1]Haryo Dwito Armono, [1]Wahyudi

[1]Department of Ocean Engineering, Institut Teknologi Sepuluh Nopember

Abstract. Breakwater is one of the coastal buildings built by humans to protect coastal areas such as ports, settlements, and tourist attractions from wave attacks. This test aims to determine the condition damage of BPPT – lock in one- and two-layers placements in the head of the rubble mound breakwater. The method used to conduct this research was physical model test and carried out at laboratory of *Balai Teknologi Infrastruktur Pelabuhan dan Dinamika Pantai* (BTIPDP) Yogyakarta – Indonesia. The armour unit model used BPPT–locks were 66.048 grams or equivalent to 2.268 tons in prototype and tested in various wave heights, starting from 7.5 cm, 9.2 cm, and 11.4 cm with period 1.5 S or equal to 4 meters of wave height in the prototype. Based on the laboratory results of testing, installing a double armour layer was recommended because the percentage of damage of the double armour layer was lower, and the coefficient of damage was higher.

Keywords: BPPT- lock, single and double layer, breakwater, irregular waves.

1. Introduction

Breakwater is one of the artificial coastal structures to protect coastal areas such as ports, residential areas, and tourist attractions from wave attacks. The breakwater will help reduce wave energy so that the waters behind the breakwater become calm (Jauzi et al., 2020). The activities behind the breakwater, such as activities in the port (loading, unloading goods, and passengers), residential areas on the beach, and tourist places, will not be disturbed by incoming waves. Before 1933 there was no method for calculating breakwater type rubble mounds but only based on experience and qualitative criteria such as the influence of wave height, angle of inclination, and the armour units' weight. Castro (1933) was the first to publish a formula for calculating the weight of rock material from rubble mound breakwater (Vanhoutte, 2008). After 1950 various types of armour units were designed and have wave stability interlock with adjacent units, one of the breakwater units designed with interlocking capacity is Tetrapod (de Rover et al., 2009). Based on the historical review of the previous armour units produced in various countries (Jauzi et al., 2020), *Balai Teknologi Infrastruktur Pelabuhan dan Dinamika Pantai (BTIPDP / BPPT)* has been developed a new concept of the armour units by considering technical and economic advantages. The new armour unit was named BPPT–lock, developed and designed to be placed randomly, single layer and interlocked between armour units. The patent for BPPT – lock has been submitted to the Ministry of Law and Human Rights of the Republic of Indonesia. The patent certificate was issued in August 2012 with ID P0031532 (Balai Teknologi Infrastruktur Pelabuhan dan Dinamika Pantai, 2021). Figure 1. Prototype of BPPT - lock below show the prototype of BPPT – lock at Laboratory of *Balai Teknologi Infrastruktur Pelabuhan dan Dinamika Pantai* (BTIPDP).

Figure 1. Prototype of BPPT - lock

This study was conducted to assess the stability of BPPT – lock on the head of rubble mound breakwater with the physical model test. The armour layer installed on rubble mound breakwater was one, and two layers were built based on bathymetric data, water depth, tide, and wave height. Figure 2. Structure rubble mound breakwater, shows the head of the rubble mound breakwater model.

Figure 2. Structure rubble mound breakwater

The purpose of this test is i) to evaluate the damage of BPPT – lock in one layer and two layers placement on the head of rubble mound breakwater, ii) to determine the percentage damage of BPPT – lock unit on the head of rubble mound breakwater, and iii) evaluate the hydraulic parameters that affect the stability of BPPT – lock unit on the head of rubble mound breakwater. This paper describes the material preparation, process construction, testing with different wave heights, and interpreting results to determine the stability of BPPT – lock on the head of rubble mound breakwater. Physical modelling and testing in the laboratory were performed to observe the failure mechanism, which was directly influenced by hydraulic parameters such as wave variation and water elevation (Triatmadja & Sulistyowati, 1996) and (Briggs, 2013). The physical model test is chosen as the mathematical model (Latham, Xiang, & Higuera, 2015) will not show the physical displacement or armour damage and does not entirely describe the interaction process between the structure and wave. The armour unit's stability analysis is usually evaluated based on the Hudson formula (Hudson, 1958) and Van der Meer (Meer, 1988).

There are several physical model testing activities to determine the hydraulic stability of BPPT – lock on the head of rubble mound breakwater, namely model preparation, testing, and interpretation of the result. Physical modeling is a process of replication cases in the field or prototype that will be reduced according to the law scale, similarity criteria, and laboratory capacity to get a good result and prevent the problem of scale effects (de Rover et al., 2009). If the scale factor is not used correctly, the scale effect will occur and lead to inaccurate or unrealistic results from the model test. There are three similarities in the similarity scaling criteria: dynamic similarity, kinematic similarity, and geometry similarity. In the dynamic similarity, the force acting in the model must be the same as the force in the field or prototype (friction force, surface tension, pressure, and gravitational force) because the model represents the actual shape or prototype. The kinematic similarity refers to the fluid motion similarity between the model and the prototype particle. Geometry similarity is needed to ensure that all the dimensions in the model have the same ratio as the prototype's dimension. The model's shape must represent the prototype to show that the model is geometrically the same.

2. Research Methodology

The scaling model used in the physical model test on the head of structure rubble mound breakwater is based on the Froude number. The Froude number is the main scaling criterion for modelling coastal and inland processes with the free surface flow (Briggs, 2013). In this study, since gravity wave is the dominant driving force, the similarity was based on the Froude number was expressed as:

$$F_r = \sqrt{\frac{F_I}{F_g}} = \sqrt{\frac{\rho L^2 V^2}{\rho g^3 g}} = \frac{V}{\sqrt{gl}} \quad (1)$$

Therefore, the Froude similarity between model and prototypes was expressed as:

$$F_r = (\frac{V}{\sqrt{gL}})_m = (\frac{V}{\sqrt{gL}})_p \quad (2)$$

where F_r is Froude number, g is gravity (m / s^2), V is the velocity (m/s), L is length (m), and subscript p and m refers to prototype and model. In the rubble mound stability test, there are driving forces that will affect the movement of the unit of BPPT – lock, such as drag force, lift force, share force, and gravity force (van Buchem, 2009). Drag force comes from water movement, causing waves and hits BPPT – lock units, expressed as:

$$F_D = \frac{1}{2} C_D . \rho_\omega . A_f . u. |u|. \quad (3)$$

where F_D is the drag force, C_D is the drag coefficient, ρ_ω is water density, A_f is the cross-sectional area, and u is the object's speed relative to the fluid. The lift force comes from the differential pressure between the top and bottom of the armour unit. It was expressed as:

$$F_L = (p_{under} - p_{above}) . A_L \quad (4)$$

where F_L is the lift force, p is the pressure, and A_L is the subjected area.
Shear force (Fs) is caused by water flowing along with the armour unit. This force was expressed as:

$$F_S = C_f . \frac{1}{2} . p_\omega . u. |u|. A_s \quad (5)$$

where F_S is shear force, C_F is share coefficient, ρ_ω is water density, A_S is the cross-sectional area, and u is the object's speed relative to the fluid. Physical model tests are affected by gravity force because the test is carried out in the laboratory.

$$F_G = = (\rho_c - \rho_w) . V . g \quad (6)$$

Since the gravity force is dominant, the model is based on the Froude similarity presented above. However, the Reynolds number for physical model testing on the head of the rubble mound breakwater is maintained above 3 x 10^4 to ensure that the viscosity force does not affect the experiments.

$$Re = \frac{l_a\sqrt{gH}}{\upsilon} \tag{7}$$

where Re is Reynolds number, l_a is the dimension of the stone (m), g is the acceleration of gravity force (m/s^2), H is wave height in front of the model (m), υ is the kinematic viscosity of water (at 28°C, the value is 1 x 10 $^{-6}$ m/s^2).

The quasi three dimensional of physical model test on the head of rubble mound breakwater was tested in 3 scenarios, where tests 1 used a single armour layer and tests 2 and 3 used double armour layers and the test was executed in the wave flume at the Laboratory of *Balai Teknologi Infrastruktur Pelabuhan dan Dinamika Pantai* (BTIPDP) and *Badan Pengkajian dan Penerapan Teknologi* (BPPT), which is under of *Kedeputian Teknologi Industri Rancang Bangun dan Rekayasa* (TIRBR), Yogyakarta – Indonesia. Specifications of wave flume facilities available in the laboratory to test physical models on the head of rubble mound breakwater are follows; (Basin et al., n.d.). Figure 3. Condition of wave flume, wave generator, wave probe, and control room shows the wave flume, wave generator, wave probe, and control room condition. The following features are available in the BTIPDP laboratory.

- Piston paddles: Equipped with a single or dual electric actuator.
- Active wave absorber: To reduce waves reflected from the paddles and calm the basin or flume after the test.
- Wave absorbing beach: To prevent splashing from the rearward paddle's movements.
- Flume: width 1.9 meters, length 40 meters, height 1.6 meters.
- Water depth operation: 1 meter (maximum).
- Wave periods: 0.4 – 0.5 seconds (wave frequency of 0.2 Hz to 2.5 Hz)
- Significant wave height: 0.25 meters.
- Maximum regular wave high (H): 0.52 meters.
- Maximum irregular wave high (H): 0.25 meters.

Figure 3. Condition of wave flume, wave generator, wave probe, and control room

Variation parameters used to test the structure of rubble mound breakwater were single and double armour layer, water deep, wave type, wave height, slope, and duration of testing. Each test was repeated three times. Table 1. The testing parameters, shows the various parameters that were used to test the structure.

IOP Conf. Series: Earth and Environmental Science 1081 (2022) 012026 doi:10.1088/1755-1315/1081/1/012026

Table 1. The testing parameters

Testing number	Armour layer type	Wave input			Water deep (cm)	Inclination	Duration of testing (Min)
		Wave height (cm)	Period (sec)	Wave type			
1	Single layer	9,12,17,20	1,5	Irregular	24,7	1: 2	60
2	Double layer						
3	Double layer						

In this test, the weight and volume of 20 BPPT–lock model units were measured to estimate their densities ($\text{BPPT} - \text{lock density} = \frac{\text{BPPT - lock weight}}{\text{BPPT - lock volume}}$). The measurement process is shown in Figure 4. Weighing BPPT - lock and measurement density in the laboratory and in the Table 2. The average value of BPPT - lock weight, volume BPPT - lock, and BPPT - lock Density show the average value of BPPT – lock weight, volume BPPT – lock, and BPPT – lock density.

Figure 4. Weighing BPPT - lock and measurement density in the laboratory

Table 2. The average value of BPPT - lock weight, volume BPPT - lock, and BPPT - lock Density

No	1	2	3	4	5	6	7	8	9	10	11	12	13	14	15	16	17	18	19	20	Average Value
BPPT - lock weight (gram)	68,36	62,5	67,06	70,2	69	67,86	61,77	70,26	63,25	63,98	67,43	65,9	67,2	65,49	66,5	67,6	66,48	63,1	63,21	64,16	66,048
Volume BPPT - lock (ml)	34	34	34	34	34	34	34	34	34	34	34	34	34	34	34	34	34	34	34	34	34
Density of BPPT - lock (gram/Cm^3)	2,011	1,84	1,972	2,06	2	1,996	1,817	2,066	1,86	1,882	1,983	1,94	1,976	1,926	1,95	1,99	1,955	1,86	1,859	1,887	1,94259

The armour units used in testing numbers 1,2, and 3 are BPPT – locks, with an average weight of 66,048 grams. The total armour units used in testing 1 are 2145 units and installed single layer, while the total armour units used in testing 2 and 3 are 2902 units and installed in a double layer. Table 3. Availability of BPPT - lock units at BTIPDP - BPPT shows the type of armour unit, weight average of armour unit, average density, and total armour unit.

Table 3. Availability of BPPT - lock units at BTIPDP - BPPT

Testing number	Total armour unit	Armour unit	Average armour unit weight (gram)	Average density (gram/cm^3)
1	2145			
2	2902	BPPT - lock	66,048	1,943
3	2902			

The process of converting the armour unit model or BPPT – lock in the model to armour unit prototype or BPPT – lock in the prototype use the Hudson equation, and the calculation process can be seen in Table 4. Calculation process and Table 5. Conversion model to prototype

$$N_W = \frac{N_\rho (N_H)^3}{N_K (N_\Delta)^3 N_{cot\alpha}} \quad (8)$$

where : $N_w = \frac{W_p}{W_m}$ is weight scale, $N_\rho = \frac{\rho_p}{\rho_m}$ is the density scale, ρ_p is the prototype density of BPPT – lock (ton / m^3), ρ_m is the density model of BPPT – lock (gram/m^3), $N_H = \frac{\rho_p}{\rho_m}$ is scale wave height, H_p is wave height in prototype (m), H_m is wave height in the model (cm), $N_K = \frac{\rho_p}{\rho_m}$ is a scale of coefficient damage, $K_{DP} = \frac{\rho_p}{\rho_m}$ is coefficient damage of prototype, K_{Dm} is coefficient damage of model, $N_\Delta = \frac{\rho_p}{\rho_m}$ is a scale of relative density, Δ_P is the relative density of BPPT – lock prototype, Δ_m is the relative density of BPPT – lock model, $N_{cot\alpha} = \frac{\rho_p}{\rho_m}$ is the scale of structure slope, $Cot\alpha_p$ is structure slope in prototype and $Cot\alpha_m$ is structure slope in the model.

Table 4. Calculation process

No	Prototype density of BPPT – lock (ρ_p)	Density model of BPPT – lock (ρ_m)	$N_p = \frac{\rho_p}{\rho_m}$	Sea water density (ρ_r)	Water density (pw)	$\Delta_p = \frac{\rho_r}{\rho_w} - 1$	$\Delta_m = \frac{\rho_r}{\rho_w} - 1$	$N_\Delta = \frac{\Delta_p}{\Delta_m}$	$N_p = \frac{\rho_p}{\rho_M}$	Hp (cm)	Hm (cm)	$N_H = \frac{H_p}{H_m}$	Kdp	Kdm	$N_K = \frac{K_{DP}}{K_{Dm}}$	$Cot\alpha_p$	$Cot\alpha_M$	$N_{Cot\alpha} = \frac{Cot\alpha_p}{Cot\alpha_M}$
1	2,2	1,94	1,134021	1,025	1	1,1463415	0,94	1,219512	1,13402	380	10	38	17	17	1	2	3	1

Table 5. Conversion model to prototype

No	Nw	Nρ	NH	NK	NΔ	Ncotα	Weight of BPPT - lock model (gram)	Nw x weight of BPPT - lock model (gram)	BPPT - lock (ton)
1	34309,41	1,13	38	1	1,21951	1	66,048	2266068,16	2,266

The rubble mound breakwater model consists of sand, gravel, BPPT – lock, plywood, and spraying paint. Sand is used as the core layer; gravel is used as the second layer; BPPT–lock is used as the armour layer; plywood is used as a template for the breakwater structure; and finally, the amour units were colored using spraying paint. The models have a structural height of 37.83 cm; bottom structure diameter is 234.61 cm, head structure diameter is 56.17 cm, parapet diameter is 22.24 cm, and incline angle is 1:2. Figure 5. Dimension of rubble mound breakwater shows the dimension of the rubble mound breakwater model, and Figure 6. process construction of rubble mound breakwater shows the construction process of the structure rubble mound breakwater.

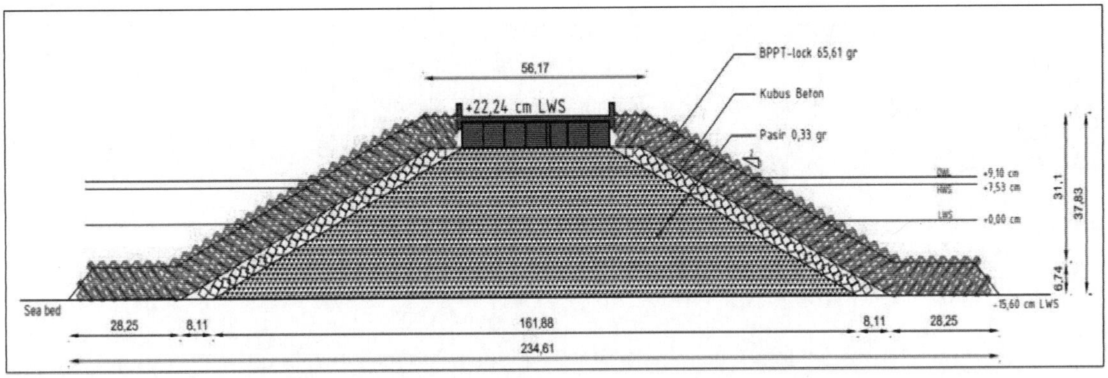

Figure 5. Dimension of rubble mound breakwater model

Figure 6. process construction of rubble mound breakwater

The structure testing of rubble mound breakwater is built in the middle of the wave flume, with a minimum distance of the bathymetric slope apron to the wave generator are 3 – 5 meters. The distance from the edged apron to the model is between 3 – 5 times wavelength, and the slope of the bathymetric foot apron to the bottom of the flume tank is recommended maximum of 1:10. Figure 7. Position rubble mound breakwater in the wave flume shows the rubble mound breakwater position in the wave flume.

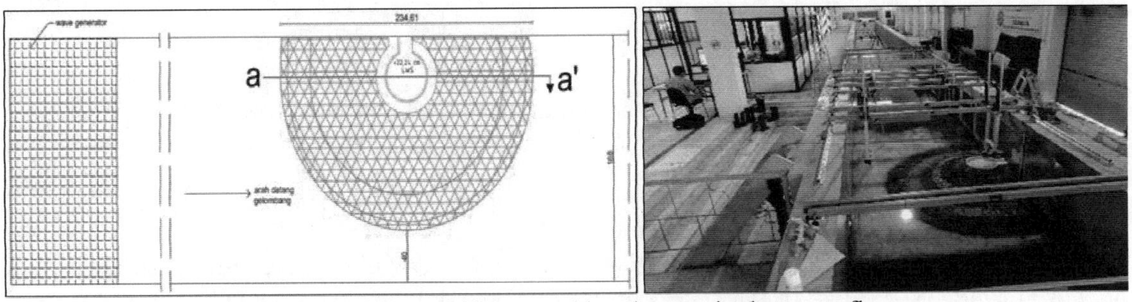

Figure 7. Position rubble mound breakwater in the wave flume

Sensors placed in the wave flume to measure water level elevation and wave high during physical model testing on the head of rubble mound breakwater totaled 11 sensors. Sensors 1, 2, and 3 were placed in front of the wave generator and at the center of the platform. In front of the model, sensors 4, 5, 6, and 7 were placed to record incoming wave height. Furthermore, sensors 8 and 9 were placed on the top and behind the model, while sensors 10 and 11 were placed in the left and right of the model to measure run-up. Figure 9 shows the position of the sensor in the wave flume.

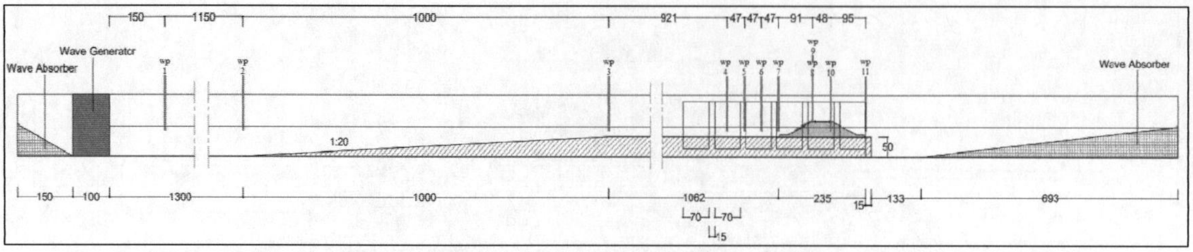

Figure 8. Position sensor in the wave flume

After preparation of the model and placement of the sensor has been completed, calibration and setting of the equipment process called basic research test (BRT) were needed to check the ability of the wave generator and the measurements process of wave probes. The wave generator had to generate the wave design / Hd (Hd is the wave with a return period of 100 years / H_{100} years). The wave probes measured the water level fluctuation, and with further analysis, the significant wave height and periods were obtained. This process is needed to ensure that the data obtained from the physical model test can be accurate and follow the predetermined design wave. Physical model testing was carried out at a high water level (HWL) with water deep of 24.7 centimeters, and the test duration was 60 minutes for each test. According to the Froude scale, 60 minutes of testing in the laboratory is the same as storm conditions with a return period of 100 years in the prototype. The test was carried out three times, namely testing 1, 2, and 3. Each test starts by generating wave heights of 9 cm, 12 cm, 17 cm, and 20 cm with a duration of 60 minutes.

To calculate the structural stability on the head of rubble mound breakwater using an average of BPPT – lock weight, BPPT – lock volume, average density, wave height, percentage of damage, and coefficient of damage (KD). The percentage of damage (% damage) is the ratio of amour unit number that move from their position to the total amour units that use in the test times 100 percent or damage ratio or ($D_i = \frac{Number\ of\ displaced\ amor\ units}{Total\ number\ of\ amor\ unit}\ x\ 100\ \%$). The coefficient damage (KD) was evaluated based on armour unit density, wave height, armour unit weight, relative density, and structure slope using Hudson's formula (Hudson, 1958). This formula is shown in the following equation :

$$W = \frac{\rho_a H^3}{K_D(S_r-1)^3 Cot\alpha} \quad \text{or} \quad K_D = \frac{\rho_a H^3}{W\ (S_r-1)^3\ Cot\ \alpha} \tag{9}$$

Where: W is weight average of armour unit (gram), H is wave hight in front of the model (cm), α is structure slope, S_r is relative density (ρ_a / ρ_w), ρ_a is BPPT - lock density (gram / cm^3), ρ_w is water density (gram / ml = gram / cm^3).

3. Results and Discussion

The quasi-three-dimensional physical model test on the head of rubble mound breakwater obtains data on wave height, armour unit damage, armour unit average volume data, armour unit average weight data, and average density data. The percentage and coefficient damage to each test can be calculated. The structure is considered stable when the damage is below 0.05%. In this test, a 2% damage is set as the upper limit and 0.5% as the lower damage limit. If the percentage damage to the armour unit is above 2%, the structure is considered unstable. The damage of the armour unit below 0.5% means that there is no damage to the structure. Table 6. The percentage damage in the testing 1, Table **7.** The coefficient damage in testing 1, Table 8. The percentage damage in the testing 2, Table 9. The coefficient damage in testing 2, and Table 10. The percentage damage in the testing 3, shows the percentage damage and coefficient of damage on testing 1,2, and 3.

Table 6. The percentage damage in the testing 1

Testing 1						
No	Minutes	Wave height in front of the structure (cm)	Total BPPT – lock on breakwater head	Number of damages	Accumulation	Damage
1	60'	7,5	2145	5	5	0,23%
2	60'	9,2	2145	5	10	0,47%
3	60'	11,2	2145	28	38	1,77%
4	60'	11,4	2145	93	131	6,11%

Table 7. The coefficient damage in testing 1

Testing 1									
No	Wave height in front of the structure (cm)	Damage	Weight average of amour unit (gram)	Cot α	Average density of BPPT – lock unit (gram/cm^3)	Density of water at 25°C (gram/cm^3)	Sr	(Sr - 1)^3	Coefficient of damage (KD)
1	7,5	0,23%	66,048	2	1,942	1	1,942	0,837	7,408
2	9,2	0,47%	66,048	2	1,942	1	1,942	0,837	13,674
3	11,2	1,77%	66,048	2	1,942	1	1,942	0,837	24,671
4	11,4	6,11%	66,048	2	1,942	1	1,942	0,837	26,016

Table 8. The percentage damage in the testing 2

Testing 2						
No	Minutes	Wave hight in front of the structure (cm)	Total BPPT – lock on breakwater head	Number of damages	Accumulation	Damage
1	60'	7,3	2902	1	1	0,03%
2	60'	9,1	2902	3	4	0,14%
3	60'	10,4	2902	1	5	0,17%
4	60'	11,1	2902	0	5	0,17%

Table 9. The coefficient damage in testing 2

Testing 2									
No	Wave hight in front of the structure (cm)	% Of damage	Weight average of amour unit (gram)	Cot α	Average density of BPPT – lock unit (gram/cm^3)	Density of water at 25°C (gram/cm^3)	Sr	(Sr - 1)^3	Coefficient of damage (KD)
1	7,3	0,03%	66,048	2	1,942	1	1,942	0,837	6,831
2	9,1	0,14%	66,048	2	1,942	1	1,942	0,837	13,233
3	10,4	0,17%	66,048	2	1,942	1	1,942	0,837	19,753
4	11,1	0,17%	66,048	2	1,942	1	1,942	0,837	24,016

Table 10. The percentage damage in the testing 3

Testing 3						
No	Minutes	Wave hight in front of the structure (cm)	Total BPPT – lock on breakwater head	Number of damages	Accumulation	Damage
1	60'	7,3	2902	0	0	0,00%
2	60'	8,9	2902	0	0	0,00%
3	60'	10,3	2902	1	1	0,03%
4	60'	10,8	2902	0	1	0,03%

Table 11. The coefficient damage in testing 3

Testing 3									
No	Wave height in front of the structure (cm)	% Of damage	Weight average of amour unit (gram)	Cot α	Average density of BPPT - lock unit (gram/cm^3)	Density of water at 25°C (gram/cm^3)	Sr	(Sr - 1)^3	Coefficient of damage (KD)
1	7,3	0,00%	66,048	2	1,942	1	1,942	0,837	6,831
2	8,9	0,00%	66,048	2	1,942	1	1,942	0,837	12,379
3	10,3	0,03%	66,048	2	1,942	1	1,942	0,837	19,188
4	10,8	0,03%	66,048	2	1,942	1	1,942	0,837	22,121

The results of tests 1, 2, and 3, which are shown in tables 6, 7, 8, 9, 10, and 11, will be displayed in graphical form to see the relationship between the percentage of damage (% damage), wave height (Hd) and the coefficient of damage (KD) in each test. In the graph, the Y-axis direction represents the percentage of damage (% damage), and the X-axis direction indicates the wave height (Hd) and coefficient of damage (KD). The blue line in the graph shows the results of the percentage damage and the coefficient of damage for the different waves. Figure 9. The effect of wave height on the percentage of damage and coefficient of damage in testing 1, show the relationship between wave height (Hd), percentage of damage, and coefficient damage.

Figure 9. The effect of wave height on the percentage of damage and coefficient of damage in testing 1

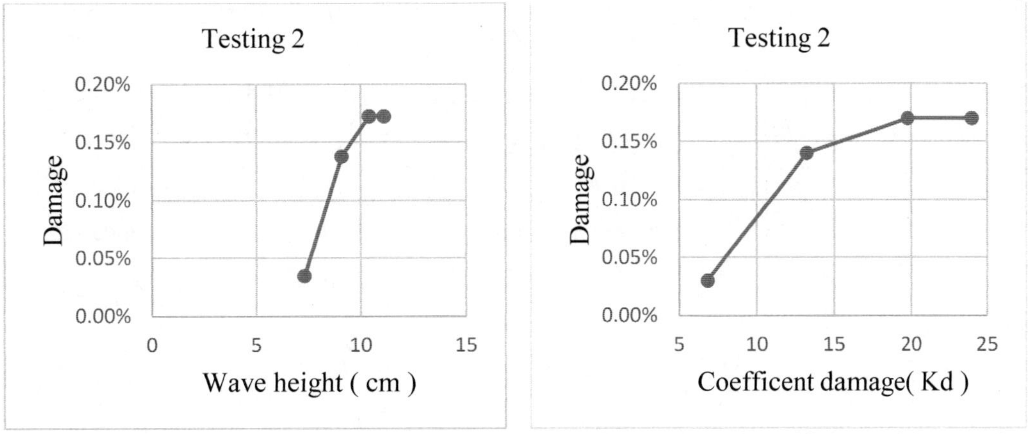

Figure 10. The effect of wave height on the percentage of damage and coefficient of damage in testing 2

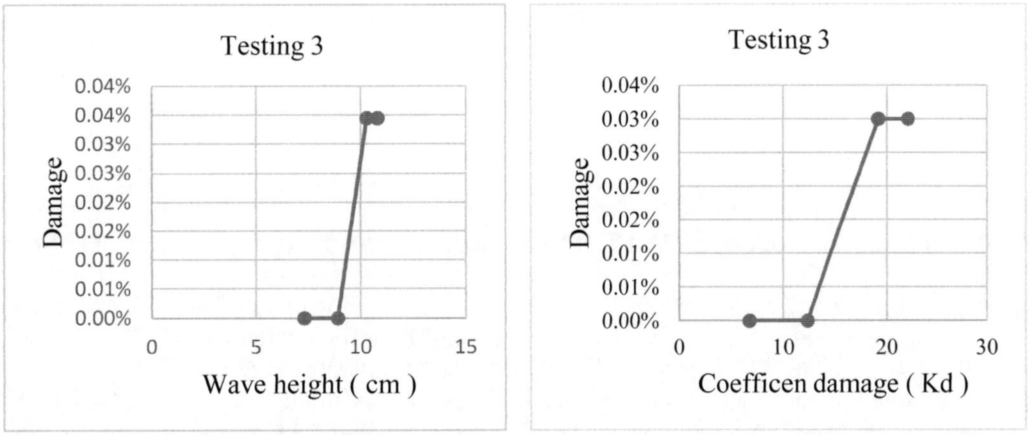

Figure 11. The effect of wave height on the percentage of damage and coefficient of damage in testing 3

The result of tests 1,2 and 3 in tables 6, 7, 8, 9, 10 and 11 will be summarized in tables 12, 13 and 14 to see the various damage and coefficient damage that occurs.

Table 12. Recapitulation of the percentage damage and coefficient of damage on test 1

	Testing 1		
	Single layer		
Average of armour unit weight (gram)	Wave height (cm)	Damage	Coefficient damage
66,048	7,5	0,23%	7,408
	9,2	0,47%	13,674
	11,2	1,77%	24,671
	11,4	6,11%	26,016

Table 13. Recapitulation of the percentage damage and coefficient of damage on test 2

Testing 2			
	Double layer		
Average of armour unit weight (gram)	Wave hight (cm)	Damage	Coefficient damage
66,048	7,3	0,03%	6,831
	9,1	0,14%	13,233
	10,4	0,17%	19,753
	11,1	0,17%	24,016

Table 14. Recapitulation of the percentage damage and coefficient of damage on test 3

Testing 3			
	Double layer		
Average of armour unit weight (gram)	Wave hight (cm)	Damage	Coefficient damage
66,048	7,3	0,00%	6,831
	8,9	0,00%	12,379
	10,3	0,03%	19,188
	10,8	0,03%	22,121

The test results for a single armour layer show that the damage and coefficient of damage increase when the wave height increases. The 7.5 cm wave height caused 0.23% damage with a value of coefficient damage (KD) are 7.408, for 9.2 cm wave height caused 0.24 % damage with 13.674 coefficient damage (KD), for 11.2 cm wave height caused 1.77% damage with 24.671 coefficient damage (KD), and finally, 11.4 cm wave height caused 6.11% damage with a 26.016 coefficient damage (KD).

The double armour layer test shows the percentage of damage is smaller while the coefficient of damage increases. The 7.3 cm wave height caused 0.03% damage with 6.831 coefficient damage (KD). The 9.1 cm wave height caused 0.14% damage with 13.233 coefficient damage (KD). The 10.4 cm wave height caused 0.17% damage with 19.753 coefficient damage (KD), and the 11.1 cm wave height caused 0.17% damage with 24.016 coefficient damage (KD). In the last test, a repeating test for the double armour layer, the percentage of damage is minimal. No damage was found at 7.3 cm and 8.9 cm wave height. The coefficient damage (KD) for both tests are 6.831and 12.379. At 10.3 cm wave height, the 0.03% damage was observed with a 19.188 coefficient of damage (KD), and for 10.8 cm wave height, it caused 0,03% damage with a coefficient of damage (KD) 22.121.

The results of testing in the laboratory show that two armour layers are more stable than one armour layer for a breakwater head. Therefore, installing a single armour layer on the structure breakwater head is not recommended. BPPT – lock units' average weight was 66.048 grams or equivalent to 2.268 tons in the prototype, and the wave height was 11.4 cm or equal to 4 meters of wave height in the prototype. As the damage reaches 6.11%, however, it is recommended to install two armour layers because tests 2 and 3 show that the value of percentage damage is 0.17%, with coefficient damage (KD) being 24.016 at 11.1 cm of wave height or equivalent to 4 meters of wave height in the prototype.

4. CONCLUSION

The results of tests 1, 2, and 3 show that the damage condition for the single armour layer (test 1) is higher than the double armour layer because the single armour layer shows the damage is 0.23% at wave heights 7.5 cm, 0.24% at wave height 9.2 cm, 1.77% at wave height 11.2 cm and 6.11% at wave height 11.4 cm. For the double armour layer (test 2), the percentage of damage is 0.03% at wave height 7.3 cm,

0.14% at wave height 9.1 cm, 0.17% at wave height 10.4 cm, and 0.17% at wave height 11.1 cm. Test 3 (double armour layer) shows the percentage of damage is 0.00% at wave height 7.3 cm, 0.00% at wave height 9.9 cm, 0.03% at wave height 10.3 cm, and 0.03% at wave height 10.8 cm.
The test results show that the structure is considered stable if the armour layer is double because the percentage of damage is below 2%. The single armour layer is considered unstable since the results of the damage is above 2%. The hydraulic parameters that primarily affects the stability of the BPPT – lock unit is the wave height, as the higher wave generates increment of the percentage of damage.

References

[1] Basin, W., Basin, W., Flume, W., & Flume, W. (n.d.). *Spesifikasi Wave maker dan Kolam Gelombang*. 2(m), 1–5.
[2] Bppt-lock, I., Di, U., Sanur, P., Bppt-lock, P., Aplikasi, C., Di, X., Dunia, S., & Lokasi, D. I. (2021). *Implementasi BPPT-lock untuk di Pelabuhan* (SANUR, BALI).
[3] Briggs, M. J. (2013). *Basics of Physical Modeling in Coastal and Hydraulic Engineering. US*
[4] *Army Corps of Engineers*, 1(September), 11. https://www.researchgate.net/publication/258440862*_Basics_of_Physical_Modeling_In_Coastal_and_Hydraulic_Engineering*
[5] de Rover, R., Verhagen, H. J., van den Berge, A., & Reedijk, B. (2009). *Breakwater Stability With Damaged Single Layer Armor Units*. 3368–3380. https://doi.org/10.1142/9789814277426_0279
[6] Jauzi, M. Z., Istiyanto, D. C., Subarkah, A., & Suranto. (2020). *BPPT-lock Armor Unit Implementation for Improving Breakwater Sustainability at the Pacitan Power Plant. Journal of Physics*: Conference Series, 1625(1). https://doi.org/10.1088/1742-6596/1625/1/012047
[7] Triatmadja, R., & Sulistyowati, S. (1996). *gamapod stability.pdf. Media Teknik*, 2.
[8] van Buchem, R. V. (2009). *Stability of a single top layer of cubes*. September.
[9] Vanhoutte, L. (2008). *Hydraulic stability of Cubipod armor units in Breaking conditions*.

Meshing generation strategy for prediction of ship resistance using CFD approach

Serliana Yulianti[1], S Samuel[1], T S Nainggolan[1] and Muhammad Iqbal[2]

[1]Department of Naval Architecture, Faculty of Engineering, Diponegoro University, Semarang, 50275, Indonesia
[2]Department of Naval Architecture, Ocean, and Marine Engineering, University of Strathclyde, Glasgow, UK.

Coresponding author: samuel@ft.undip.ac.id

Abstract. CFD is a numerical approach used to solve fluid problems. In the CFD simulation process, the meshing stage is crucial to produce high accuracy. Meshing is a process where the geometric space of an object is broken down into many nodes to translate the physical components that occur while representing the object's physical shape. The research objective was to analyze the characteristics of the mesh technique in the *Finite Volume Method* (FVM) using the RANS (*Reynolds - Averaged Navier - Stokes*) equation. The numerical simulation approach used three mesh techniques, namely overset mesh, morphing mesh, and moving mesh. The k-ε turbulent model and VOF (*Volume of Fluid*) were used to model the water and air phases. The mesh technique approach in CFD simulation showed a pattern under experimental testing. This research showed the difference in value to the experimental results, namely by using the moving mesh method, the difference in resistance difference was 8% at high-speed conditions, the difference in trim value at overset mesh was 11%, and the difference in heave value with the moving mesh method was 14% at low speed. The conclusion reported that overset mesh had better than other mesh methods.

Keywords: CFD, Fridsma hull, morphing mesh, moving mesh, overset mesh

1. Introduction

An experiment conducted by G. Fridsma in 1969 has sparked many researchers to conduct similar research related to planing hulls. Fridsma conducted experimental analysis on ships with the planing type, hereinafter known as Fridsma ship, with several L/B configurations, displacement, deadrise angle, LCG (Longitudinal Center of Gravity), and so on [1]. Supported by the ship's simple geometry, until now, there has been much research discussing the Fridsma ship.

The rapid development of technology makes research more effective to do. One technology supporting the research of ships is a numerical simulation method based on Computational Fluid Dynamics (CFD). CFD is a system program that can plan and analyze an engineering product using mathematical solutions. In the analysis using CFD, especially ship type planing, the methods used to predict the resistance and movement of the planing hull included FVM (Finite Volume Method), FEM (Finite Element Method), FDM (Finite Difference Method), and analytical-experimental. According to Yousefi in 2013, the most appropriate method used to predict drag, trim, and heave on ships was FVM because

IOP Conf. Series: Earth and Environmental Science 1081 (2022) 012027 doi:10.1088/1755-1315/1081/1/012027

it produced better calculation accuracy than other methods [2]. In 2019, a numerical simulation using the FVM method was carried out on the Fridsma ship. The simulation explained that CFD could be used for ships with low speeds (Fr 0.59). However, at high speed (Fr 1.78), there was an inability to calculate numerically, requiring evaluation [3]. Research on the spray strips application has also been carried out on the Fridsma to reduce ship resistance. The innovation from this research has succeeded in increasing the performance of the Fridsma ship by reducing drag by up to 6% at high speed [4].
As for the meshing method, it is crucial to pay attention to the simulation process using CFD. The meshing techniques that researchers often use are the overset mesh[5], morphing mesh [6], and moving mesh[7]. In 2021, research was carried out on fridsma hull ships using an overset mesh system to reduce numerical ventilation problems [8]. Previously, research was conducted on the RBF (radial basis functions)-based grid morphing mesh system[9]. Furthermore, research was conducted on the characteristics of the moving mesh method [10]. In 2017, Agustino De Marco analyzed the hydrodynamics of the storied hull type using overset mesh and morphing mesh techniques to determine the two mesh techniques' characteristics[11]. Therefore, it is essential to study each mesh method on vessel as one of the parameters to improve calculation accuracy. In this research, the Fridsma hull ship was used to analyze each mesh technique characteristic.

2. Method

2.1 Research objects

Early research on the hydrodynamics of planing hulls has been started in the United States since 40 years ago [12]. Based on the Froude number value, ships can be classified based on the range, as shown in figure 1. In a planing type ship, the balance of the ship's weight was caused by the pressure acting on a wetted surface area. The pressure was composed of two components, namely hydrostatic, which related to buoyancy, and hydrodynamics relating to the ship's speed[13]. Therefore, a ship can be classified based on the pressure acting when moving as follows:

1. Displacement vessel, if the hydrostatic pressure was more significant than the hydrodynamic pressure.
2. Semi-displacement vessel, if the hydrostatic and hydrodynamic pressures were the same.
3. Planing vessel, if the hydrostatic pressure was less than the hydrodynamic pressure.

This research used experimental data Fridsma hull form as the research objects as described in Table 1 in calm water condition. The lines plan drawing of the Fridsma ship was described in Figure 2. In this research, the movement of the ship using DFBI used two degrees of freedom to simulate trim and heave.

Table 1. Experiment data by Gerard Fridsma [1]

Dimension	Unit	Value
L/B	-	5,00
L	m	1,143
B	m	0,229
TAP	m	0,081
LCG by AP	m	0,457
VCG by keel	m	0,067
τo	Degree	1,569
β	Degree	20,00
Δ	Kg	107,67
Iyy = Izz	Kg.m2	0,235

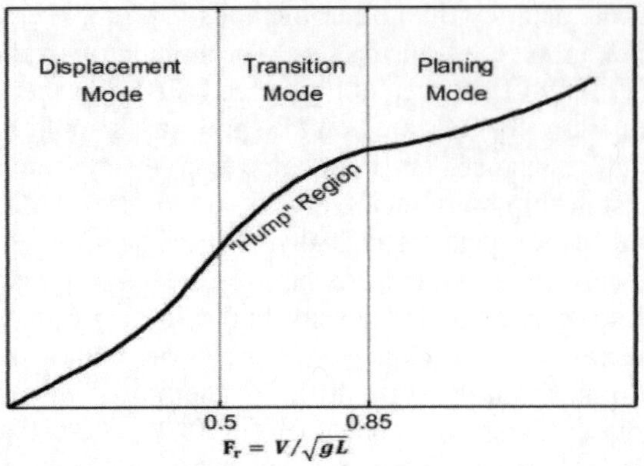

Figure 1. Ship classification based on froude number [14]

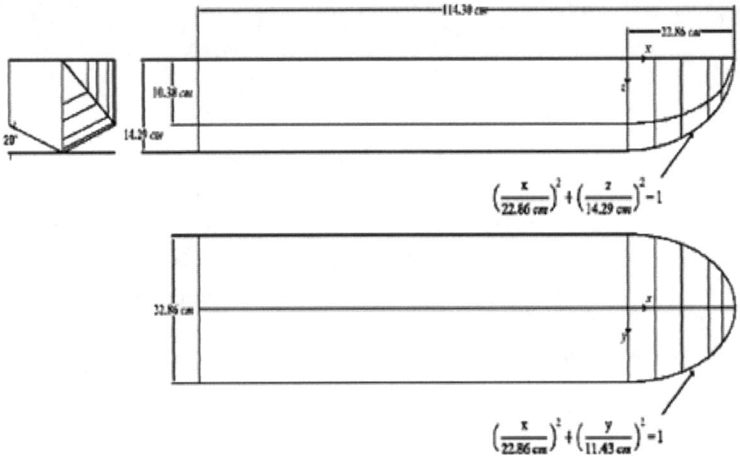

Figure 2. Lines plan of Fridsma hull form

2.2. Numerical method

CFD was developed to predict the shape of the flow without compromising the accuracy of the calculation. Every CFD program used mathematical equations to solve fluid flows. FVM is a CFD that represented and evaluated partial differential equations using RANS. The Reynolds averaged Navier-Stokes equation (RANS) defined the conservation laws of mass and momentum. The first discretization stage was to divide the computational domain into a finite number of volumes, forming a mesh. Next, the partial differential equations were integrated into each volume using the divergence theorem and producing an algebraic equation for each cell. In the cell center, the cell meant the variable flow value was stored at that node [15]. An essential goal of any CFD program was to solve equations using boundary and initial conditions. The RANS equation as in equation 1, was developed based on the concept that the speed and length of the ship described the turbulence effect around the hull. In the calculation, the k- ε turbulence model, which served as a wall, was used to describe the effect of turbulence on the flow[16].

$$\frac{\delta F}{\delta t} + \frac{\delta uF}{\delta x} + \frac{\delta vF}{\delta y} + \frac{\delta wF}{\delta z} = 0 \quad (1)$$

Wall function (y+) divided the wall and the fluid flow. It served to capture the boundary layer, which played a vital role in calculating resistance. The value of y+ played a vital role in reducing calculation inaccuracies. Lotfi and Ahmed Gultekin researched to obtain accurate results. Lotfi used the value of y+

50 – 150 [17], and Ahmed Gultekin used a value of y+ 45-60 [18]. Meanwhile, in this research, the results obtained y+ 45-70. The calculation of y+ value was based on ITTC [19], which is as described in equation 2 as follow:

$$\frac{y}{L} = \frac{y^+}{Re\sqrt{Cf/_2}} \tag{2}$$

To determine the time-step in this research, the *Courant-Friedrichs-Lewy* (CFL) number was used. The CFL number indicated the number of points traveled by a fluid particle in a time interval. The faster the ship, the smaller the time-step that would be used. In this research, the time-step used was 0.009 s. The time-step referred to the ITTC calculation in equation 3, with *L* as the ship's length and *U* as the ship's speed.

$$\Delta t\ ITTC\ =\ 0{,}005 \backsim\ 0{,}01\ \frac{L}{U} \tag{3}$$

The dimensions of the towing tank in the CFD simulation followed the ITTC recommendations described in Figure 3, consisting of (a) overset mesh, (b) morphing mesh and moving mesh[19].

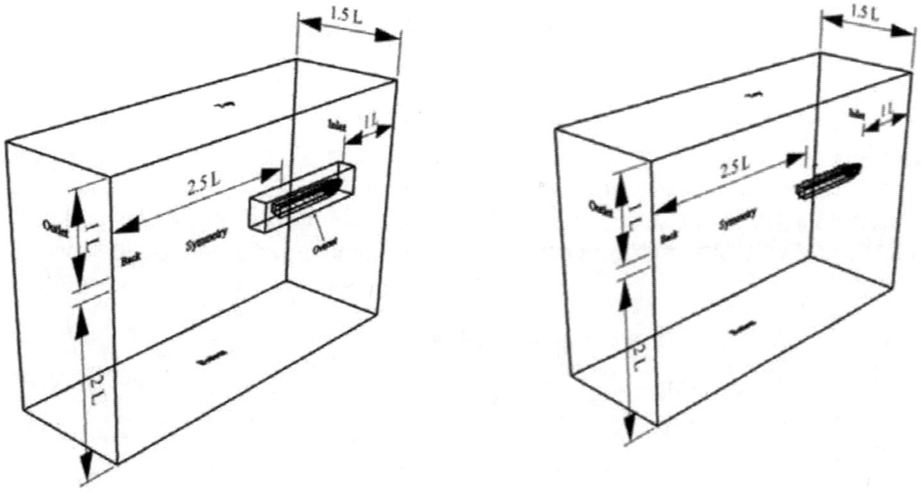

Figure 3. Fluid domain (a) overset mesh (b) morphing mesh and moving mesh.

The mesh density was centered on the hull and water surface so that the results remained accurate in a relatively faster computation time. In addition, refinement in the bottom area was needed for more accurate results. The local meshing was performed using the anisotropic mesh to focus on the x, y, or z ordinates. In this simulation, the mesh density was divided into several parts, as in Table 2 and Figure 4 visualize the surface mesh domain density.

Table 2. Mesh density

Parts	Type of Refinements	Size
Tank	Surface	0,7874 L
Water Surface Fine	Volume	0,0062 L
Water Surface Medium	Volume	0,0123 L
Water Surface Coarse	Volume	0,0246 L
Hull Box 1	Volume	0,0984 L
Hull Box 2	Volume	0,0492 L
Hull Box 3	Volume	0,0246 L
Hull Box 4	Volume	0,0123 L
Wake Fine	Volume	0,0492 L

Wake Medium	Volume	0,0984 L
Wake Coarse	Volume	0,1969 L
Hull	Surface	0,0030 L

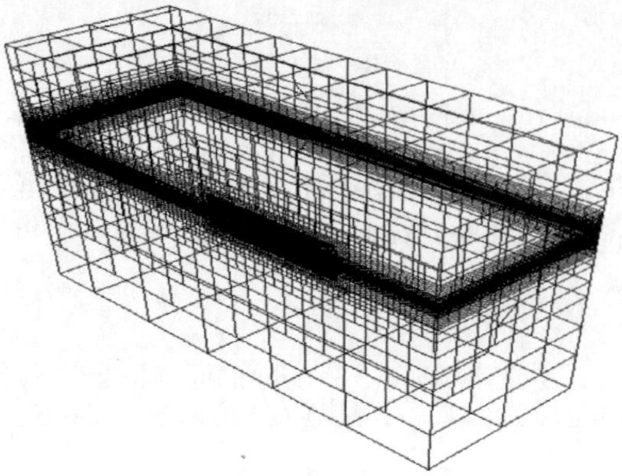

Figure 4. Mesh configuration

2.3 Overset mesh

Overset mesh was a mesh method using donor-acceptor cells, so there was an overset box and background, as shown in figure 5. This method required more than one geometry with a background as a donor and an overset as a donor-recipient. Active cells were found at each end of the geometry in the overset as an intermediary for donor-acceptor cells. There were passive cells in the background, replaced by overset cells. According to research conducted by Simon Manchini, the donor-cells variable can be calculated as in equation 4 [20]:

$$\varphi_{acceptor} = \sum\alpha_1\varphi_1 \quad (4)$$

Where α_1 was the interpolation weighting factor, φ_1 was the dependent variable, and φ were acceptor donor cells. The overset mesh consisted of the background and the overset box, so the set-up for each geometry is shown in Table 3.

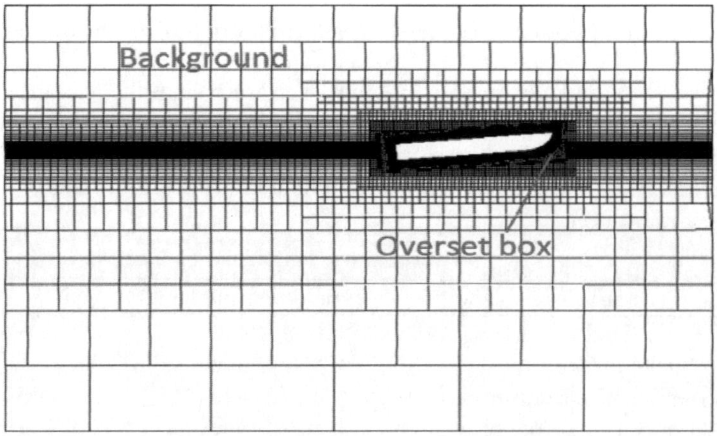

Figure 5. Background and overset box

IOP Conf. Series: Earth and Environmental Science 1081 (2022) 012027 doi:10.1088/1755-1315/1081/1/012027

Table 3. Setup on overset mesh

Part	Node	Properties	Set up
Background	Region, background, physic values	Motion specification	Stasioner
Overset box	Region, substract, physic values	Motion specification	DFBI rotation & translation

2.4 Morphing mesh

Morphing mesh was an interpolation technique that allowed a single object to change shape [21]. By creating an intermediate frame to the model, the intermediate form was produced indefinitely. In the ship cases, the entire hull could be transformed parametrically using the principal dimensions. Morphing mesh was suitable for complex relative motions, while more significant deformations might require new cells to maintain a high-quality mesh. Figure 6 is a visualization of the morphing mesh technique before deforming. The mesh morphing technique only used a towing tank geometry, different with overset mesh technique.

The morphing mesh technique required special treatment of moving nodes to control the accuracy of space derivatives and time-stepping schemes. It was done by interpolating the fluid flow variables precisely [22]. According to research conducted by Agustino De Marco, an interpolated field was used to replace mesh nodes based on the *Radial Basis Functions* (RBF) method. To generate the interpolated field, a system of equations was solved as in equation 5, using the control vertices and their defined displacements:

$$d'_i = \sum_{j=0}^{n} \lambda_j \sqrt{r_{ij}^2 + c_j^2} + \alpha \tag{5}$$

Where d'_i was the mesh displacement, i was every vertex of the mesh, λ_j was the coefficient of expansion, n was the number of cells, c_j was a fundamental constant, α was a constant value, and r_{ij} was the distance between two vertices. To get r_{ij} value, Agustino De Marco used the following equation 6:

$$r_{ij} = [\, x_i - x_j] \tag{6}$$

x_i was the difference in vertex distance x, and x_j was the vertex distance y. Table 4 was a set-up of the morphing mesh method using Star CCM software.

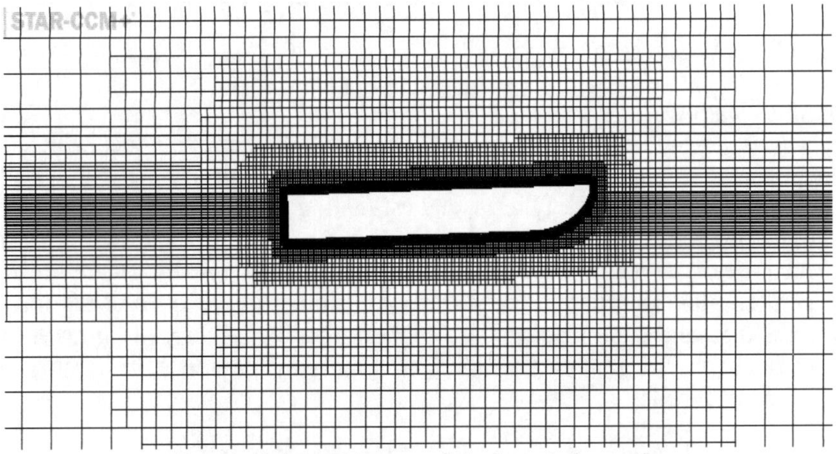

Figure 6. Morphing mesh before deformation

Table 4. Setup on morphing mesh.

Part	Node	Property	Set up
Background	Region, substract, physic values	Motion specification	DFBI morphing

2.5 Moving mesh
The moving mesh technique moved the entire grid according to the rigid motion of the ship. The grid itself was not altered but remained rigid. This approach had several advantages because only the flow variable must be corrected according to the ship motion, the method was excellent, and the computational error was small. This moving mesh technique differed from the morphing or overset mesh methods. The primary difference in this mesh technique was shown on the background movement and the ship as described in Table 5 and the visualization of Figure 7.

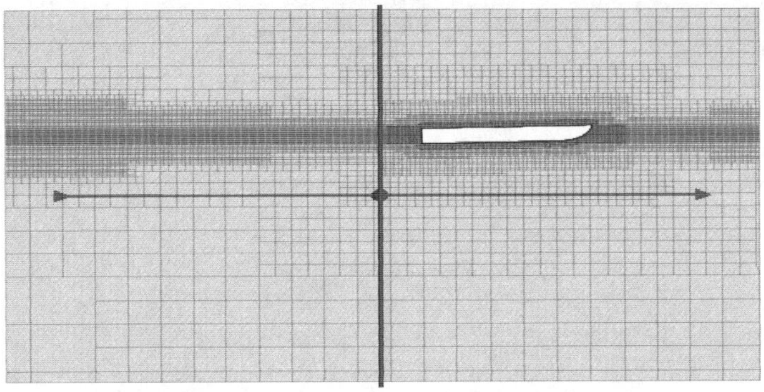

Figure 7. Moving mesh before analyzed

Table 5. Setup on moving mesh.

Bagian	Node	Properti	Set up
Background	Region, substract, physic values	Motion specification	DFBI rotation & translation

3. Result and Discussion

3.1 Overset mesh technique
Overset mesh was used to solve the equations of motion and rotation of objects, which occurred in solid objects interacting with fluids [21]. The overset mesh method was commonly used for numerical simulation of maneuver tests, roll decay tests, and estimating the ship's response to waves. In Figure 4, it can be seen how data transfer works from the overset mesh method. The meshing visualization can be seen in figure 8a, the overset mesh condition before running. Additionally, figure 8b shows the overset mesh condition after running. There were different positions in the overset section to translate the physical properties that worked, including the 2DOF motion of the ship. In figure 9, there were active cells at each end of the overset geometry as donor recipients and passive cells in the background, which were replaced by overset cells.

IOP Conf. Series: Earth and Environmental Science 1081 (2022) 012027 doi:10.1088/1755-1315/1081/1/012027

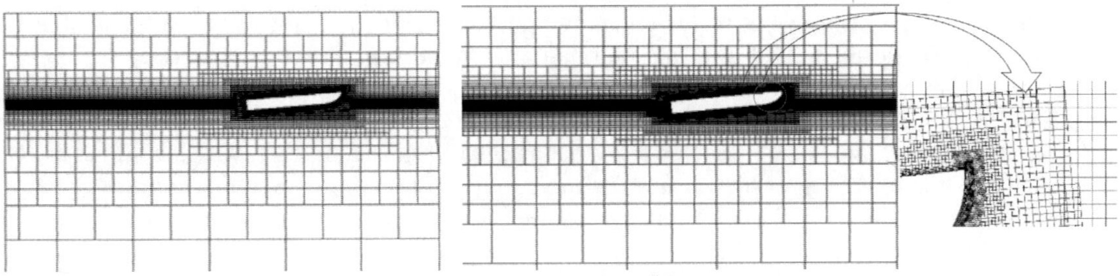

(a) (b)

Figure 8. Overset mesh (a) before analyzed (b) after analyzed

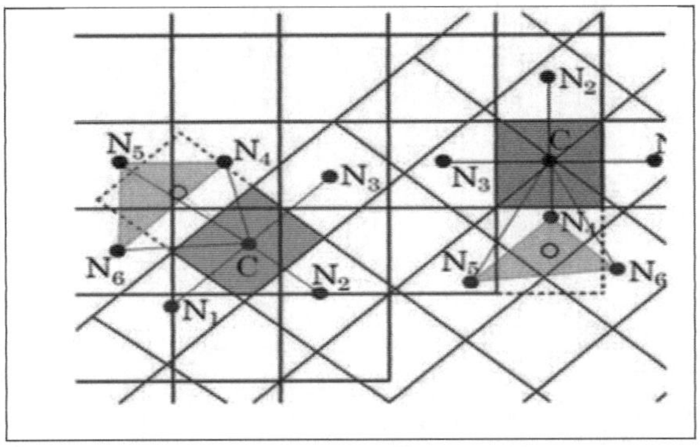

Figure 9. Data transfer visualization between overset mesh (red) and background mesh (blue), where acceptor cell (orange) received data from donor cell (green) [20].

3.2 Morphing mesh technique

Morphing mesh was a technique without movement of the ship's domain and background. Based on interpolation between two forms, morphing would require two model forms consisting of a source and a target so that the transformation could occur from the source to the target. No target model was available. Furthermore, one of the extreme models could be generated by applying Laplace coordinates. Morphing mesh was suitable for complex relative movements. The morphing mesh method was often used in the marine sector to analyze offshore or offshore buildings. Figure 10a is an image of the morphing mesh before deforming. Figure 10b shows that it has been deformed.

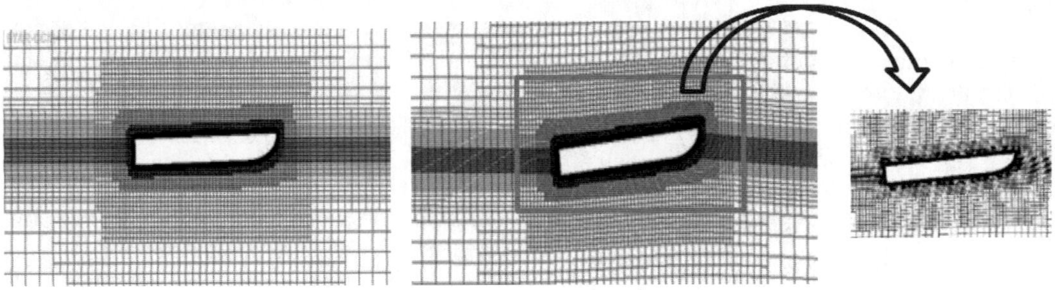

a b

Figure 10. Morphing mesh (a) before deformation (b) after deformation

3.3 Moving mesh technique

In this method, the entire grid was moved according to the ship's motion. This approach had several advantages because only the flow variable had to be corrected according to the ship's motion. The method was excellent, and the computational error was small. On the other hand, this method only

IOP Conf. Series: Earth and Environmental Science 1081 (2022) 012027 doi:10.1088/1755-1315/1081/1/012027

applied to the motion of one rigid body. The moving mesh technique had two primary and significant weaknesses in testing the durability of the hull, namely the need to create a set-up-free surface where the background moves freely, and the object under investigation did not move. Other research on moving mesh has also been carried out for applications on CFD[22] . The second drawback was that it could cause additional oscillations or spurious waves that affected trim and sinkage, which could cause errors in the analysis. Figure 11 a is a visualization of the moving mesh before analysis. Figure 11b has a red box where the red box shows the occurrence of oscillations or false waves.

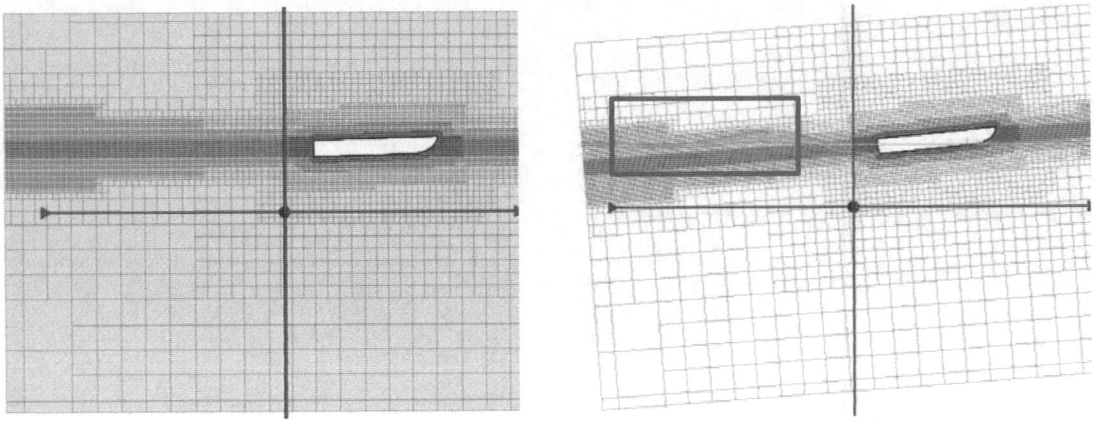

(a) (b)

Figure 11. Moving mesh (a) before analyzed (b) after analyzed

3.4 Benchmark

To compare the results of the mesh method, Fridsma's experimental data were used as validation with values of L/B = 5 and LCG = 0,6 L from AP 5 speeds were used based on the Froude number in the Fridsma experiment, namely 0,6; 0,9; 1,2; 1,5 and 1,8 (Table 1). Simulations were carried out with the same density so that the number of overset mesh was 1.462K, morphing mesh 684K, moving mesh 1.557K concentrated on the water surface and hull. The resistance was described by the non-dimensional unit R/Δ, where R was the drag and Δ was the displacement/ weight of the ship. The trim graph was expressed in units of degrees. Furthermore, the heave graph with non-dimensional units s/B with s was the displacement between the starting and ending points on the z-axis of the ship and B as the ship's width.

The resistance results obtained in this research were the difference between the CFD results and the Fridsma experiment, as shown in Figure 12. At low speeds, the morphing mesh method was closer to the Fridsma experiment, with an error of less than 4,5%. In the state of the hump overset mesh, it was closer to the experiment, with an error of less than 0,5%. Meanwhile, at high speed, the moving mesh method was close to the experiment, with an error of 8%.

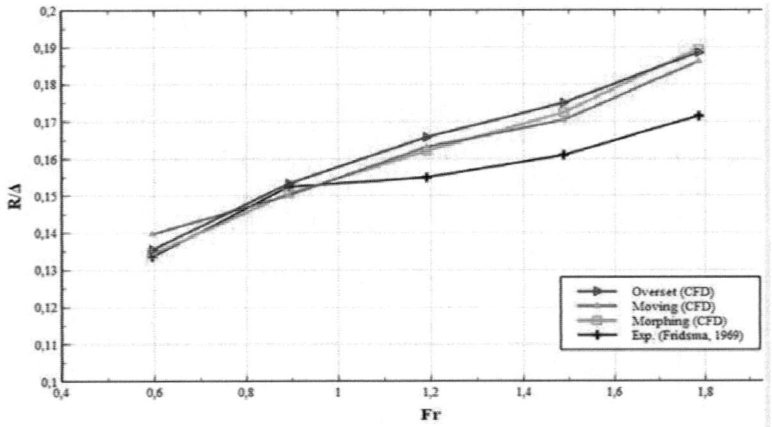

Figure 12. Resistance

From the trim graph in figure 13, there was a difference between the CFD results and the Fridsma experiment. In this research, the overset mesh results tended to approach the Fridsma experiment with a maximum error of 11%. At the same time, the morphing mesh or moving mesh tended to be less close to the Fridsma experiment, with a maximum error of 15%.

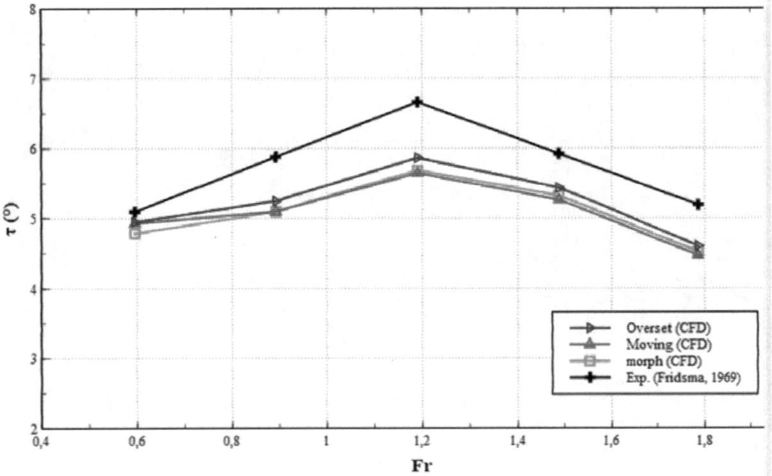

Figure 13. Trim

Based on the heave graph in figure 14, there was a difference between the CFD results and the Fridsma experiment. In this time, the moving mesh method at low speed obtained results that tended to approach the Fridsma experiment, with an error of 14%. In the hump state, the morphing mesh method approached the Fridsma experiment with an error of 11%. At high speed, the overset mesh method tended to approach the Fridsma experiment, with an error of 6%.

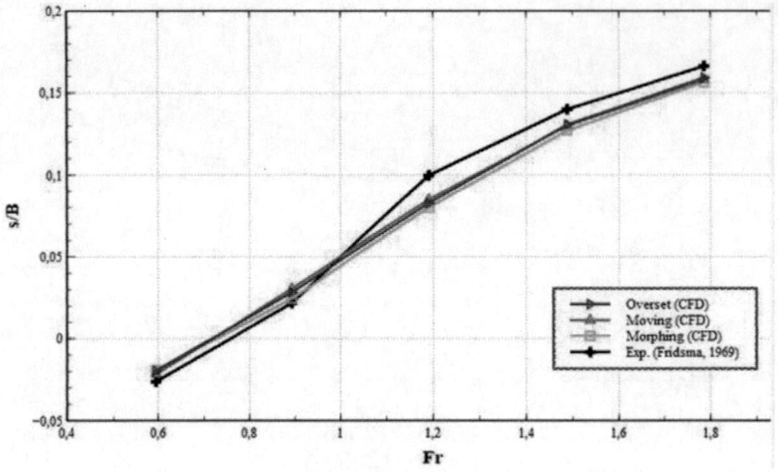

Figure 14. Heave

The results of the numerical simulation showed positive results when compared to the results of the Fridsma experiment. It can be seen from figure 12, figure 13, and figure 14. Although on the resistance and heave graph, there were a few inaccuracies in Fr 1,2 – Fr 1,8. In contrast, inaccuracies also occurred in the trim graph, namely at Fr 0,6 – Fr 1,8.
Research by Wheeler et al. in figure 15 performed a numerical simulation with the fridsma hull form. Wheeler's research had an error of 17,26% at Fr 1,78. Meanwhile, this research obtained an error of 15% at the same Fr. Hence, the results of this numerical simulation could be accepted.

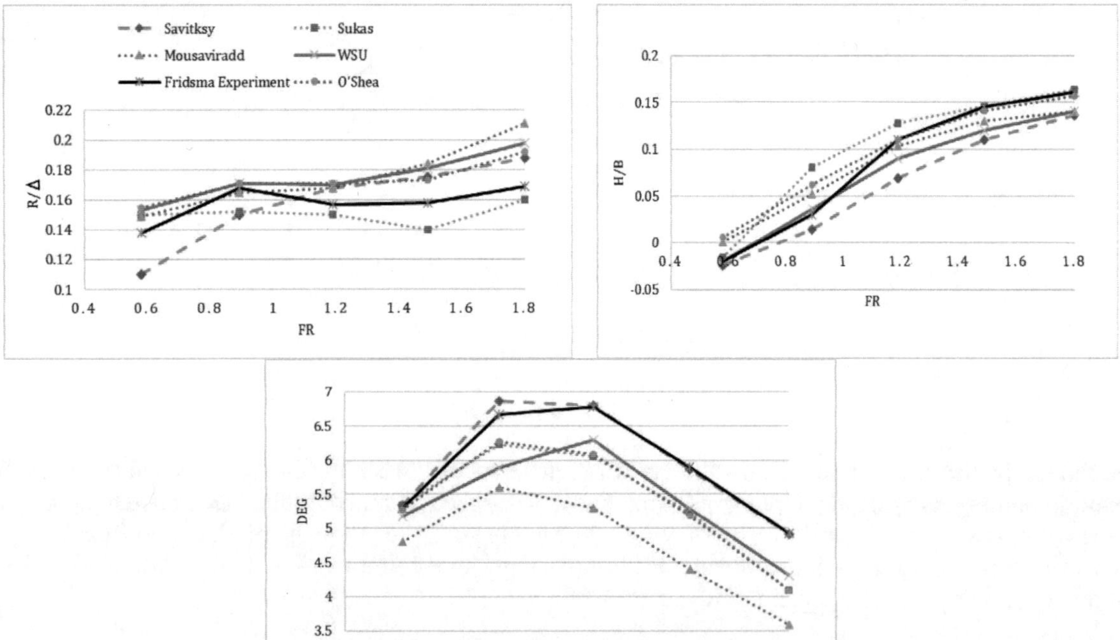

Figure 15. Comparison of drag, trim and heave [23]

In this research, the time-step used was 0.009 s. The time-step referred to the ITTC calculation. In addition, the function of the time-step was to speed up the convergence of a simulation. The simulation could be ended when the graph had converged or had a relatively small difference in value (stagnant). Furthermore, the value taken was based on the average of the converged results. The overset mesh, morphing mesh, and moving mesh methods had different processing times to achieve convergence. Table 6 explains the time required to obtain convergent results from the three methods with the number of grids and times steps that were not too different. In Figure 16, it can be seen the difference in the processing time of each mesh method.

In this research obtained several characteristics of the overset mesh, morphing mesh, and moving mesh methods which were listed in the form of the advantages and disadvantages of the three methods, which are reported in Table 7. The advantages and disadvantages of the table did not guarantee accuracy due to differences in the ship characteristics.

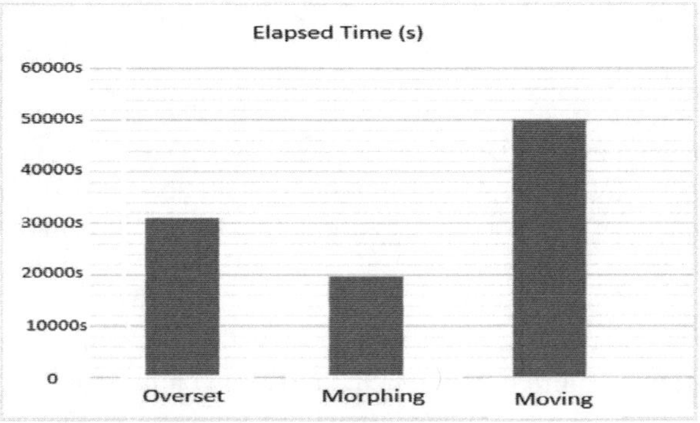

Figure 16. Comparison of processing time from each kind of mesh

Table 6. Time description

Parameters	Overset mesh	Morphing mesh	Moving mesh
Time of mesh (s)	31080	19821	50077
Solution Time	10	10	10
Sum of mesh	1.457.318	1.367.837	1.557.979

Table 7. Strategy mesh characteristics

Method	Advantages	Disadvantages
Overset mesh	• Tends to approach validation when analyzing at a minimum and maximum speed • When analyzing the trim, the results obtained are slightly closer to the validation or experiment.	• Take a long time to get analysis results • In this study, when analyzing the constraints of the overset mesh method, it was not so close to the validation or fridsma experiment.
Morphing mesh	• Tends to approach experimentation when analyzing resistance, trim, and trim at low speeds. • It does not take too long time to get the results of the analysis	• At low speeds the results obtained tend to approach the validation of the experiment, but at high speeds obtained are less close to the validation of the experiment.

Method	Advantages	Disadvantages
Moving mesh	• When analyzing resistance, trim, and heave at high speeds, the results obtained tend to be close to validation or experimentation.	• It takes a little longer to get the analysis results • At high speeds the results obtained tend to be close to validation or experimentation, while at low speeds the results are less close to validation or experimental

4. Conclusion

Based on the results of numerical analysis using the Computational Fluid Dynamics (CFD) method, some conclusions are obtained as follows:

1. The numerical computational validation based on CFD had relatively good accuracy compared to the experimental results of Fridsma in predicting the drag, trim, and heave of fast boats. The inaccuracy of the results occurred at Fr > 1.
2. Based on the time efficiency point of view, moving mesh took more time to get the analysis results, while overset mesh and morphing mesh took a long time in one analysis.
3. In this research, it was reported that overset mesh had better results than other mesh techniques.

References

[1] G. Fridsma, 1969, *A Systematic Study of The Rough-water Performance of Planing Boat* (Hoboken, New Jersey)

[2] R. Yousefi, R. Shafaghat, and M. Shakeri, 2013, Hy*drodynamic analysis techniques for high-speed planing hulls*, *Appl. Ocean Res*, **42**, pp. 105–113, doi: 10.1016/j.apor.2013.05.004.

[3] S. Samuel, A. Trimulyono, and A. W. B. Santosa, 2019, *Simulasi CFD pada kapal planing hull, Kapal J. Ilmu Pengetah. dan Teknol. Kelaut.*, **16**, pp. 123–128, doi: 10.14710/kapal.v16i3.26397.

[4] S. Samuel, A. Trimulyono, P. Manik, and D. Chrismianto, 2021, *A numerical study of spray strips analysis on fridsma hull form, Fluids*, **6**, p. 420, doi: 10.3390/fluids6110420.

[5] O. F. Sukas, O. K. Kinaci, F. Cakici, and M. K. Gokce, 2017, *Hydrodynamic assessment of planing hulls using overset grids," Appl. Ocean Res,* **65**, pp. 35–46,doi: 10.1016/j.apor.2017.03.015.

[6] M. E. Biancolini, I. M. Viola, and M. Riotte, 2014, *Sails trim optimisation using CFD and RBF mesh morphing, Comput. Fluids*, **93**, pp. 46–60, doi: 10.1016/j.compfluid.2014.01.007.

[7] S. Mancini, 2015, *The Problem of Verification and Validation Processes of CFD Simulations of Planing Hulls*,(Università Degli Studi Di Napoli Federico II)

[8] S. Samuel, D. J. Kim, A. Fathuddiin, and A. F. Zakki, 2021, *A numerical ventilation problem on fridsma hull form using an overset grid system, IOP Conf. Ser. Mater. Sci. Eng,* **1096**, no. 1, p. 012041, doi: 10.1088/1757-899x/1096/1/012041.

[9] M. E. Biancolini, 2011, *Mesh Morphing And Smoothing By Means Of Radial Basis Functions (RBF): A Practical Example Using Fluent And RBF Morph*, (Handbook. Res. Comput. Sci. Eng. Theory Pract) pp. 347–380, doi: 10.4018/978-1-61350-116-0.ch015.

[10] I. Viola, R. Flay, and R. Ponzini, 2012, *Edinburgh research explorer CFD analysis of the hydrodynamic performance of two candidate America ' s Cup AC33 hulls,*" *Int. J. Small Cr. Technol,* **154**

[11] A. De Marco, S. Mancini, S. Miranda, R. Scognamiglio, and L. Vitiello, 2017, *Experimental and numerical hydrodynamic analysis of a stepped planing hull, Appl. Ocean Res,* **64**. pp. 135–154, doi: 10.1016/j.apor.2017.02.004.

[12] Savitsky, 1964, *Hydrodynamic design of planing hulls, Mar. Technol. SNAME*, **1**, pp. 71–95,
[13] O. M. Faltinsen, 2006, *Hydrodynamics of High-Speed Marine Vehicles,* (New York: Cambridge University Press)
[14] R. Marshall, 2002, *All about powerboats: Understanding Design and Performance,* (McGraw Hill Professional)
[15] F. Moukalled, L. Mangani, and M. Darwish, 2016, *The Finite volume method in computational fluid dynamics in fluid mechanics and its applications,* **113**, (Springer)
[16] B. Launder and D. Spalding, 1974, *The numerical computation of turbulent flows*, *Comput. Methods Appl. Mech. Eng*, **3**, pp. 269–289.
[17] P. Lotfi, M. Ashrafizaadeh, and R. K. Esfahan, 2015, N*umerical investigation of a stepped planing hull in calm water*, *Ocean Eng,* **94**, pp. 103–110, doi: 10.1016/j.oceaneng.2014.11.022.
[18] A. G. Avci and B. Barlas, *"An experimental and numerical study of a high speed planing craft with full-scale validation," J. Mar. Sci. Technol.*, vol. 26, no. 5, pp. 617–628, 2018, doi: 10.6119/JMST.201810_26(5).0001.
[19] ITTC, 2011, *Practical Guidelines for Ship CFD Applications*, *ITTC – Recomm. Proced. Guidel.* ITTC, pp. 1–8.
[20] S. Mancini, 2015, *The Problem of Verification and Validation Processes of CFD Simulations of Planing Hulls*, (Università Degli Studi Di Napoli Federico II)
[21] J. Y. Kang and B. S. Lee, 2010, *Mesh-based morphing method for rapid hull form generation CAD Comput. Aided Des,* **42**, no. 11, pp. 970–976, doi: 10.1016/j.cad.2009.07.001.
[22] A. Fathuddiin, S. Samuel, K. Kiryanto, and A. Widyandari, 2020, *Prediksi hambatan kapal dengan menggunakan metode overset mesh pada kapal planing hull*, *J. Rekayasa Hijau*, **4**, no. 1, pp. 24–34, doi: 10.26760/jrh.v4i1.24-34.
[23] C. P. Fabio De Luca, Simone Mancini, Salvatore Miranda, *"An Extended Verification and Validation Study of CFD Simulations for Planing Hulls," J. Sh. Res*, **60**, no. 2, pp. 101–108, 2016.
[24] T. Tang, 2005, *Moving mesh methods for computational fluid dynamics*, pp. 141–173, doi: 10.1090/conm/383/07162.
[25] M. P. Wheeler, K. I. Matveev, and T. Xing, 2018, *Validation study of compact planing hulls at pre-planing speeds*, *Am. Soc. Mech. Eng. Fluids Eng. Div. FEDSM*, **2**, pp. 1–8, doi: 10.1115/FEDSM2018-83091.

Virtual private network (VPN) model for AIS real time monitoring

A Maulidi[1], M Abdullah[1], DW Handani[2]

[1] Lecturer at Politeknik Negeri Madura, [2] PUI-KEKAL ITS
Jl. Raya Taddan KM. 4 Camplong Sampang, Jawa Timur, Indonesia.

amd.2201@gmail.com

Abstract. The need for a ship monitoring system based on Automatic Identification System (AIS) has been developed and provides many broad benefits for the maritime world in Indonesia. The system can become an early warning system (EWS) for ship operators, port operators, and even ship owners for tracking a movement of their ships. By internet connection based Virtual Private Network (VPN) can connect several remote base stations (RBS), VPN technology can integrate many RBS from several remote places into one AIS monitoring system. VPN-based internet connection is used to guarantee a secure network, private internet connection, and can access a network remotely. VPN technology encrypts data exchange even before data transmission reaches the public network. Connecting to the internet using a VPN connection is like accessing the internet using a special tunnel, not in the main internet network. Thus the use of a VPN-based internet connection in the AIS monitoring system can ensure the security of data transmission from RBS to the AIS monitoring server.

Keywords: AIS, EWS, VPN, Monitoring.

1. Introduction

The regulations for ship traffic monitoring equipment are contained in the International Maritime Organization (IMO) Convention on the Safety of Life at Sea (SOLAS) Chapter V, which organize the safety of shipping for all ships (1). Research and development of monitoring system based on automatic identification system (AIS) has succeeded in creating a ship monitoring system in real time (2). ITS Center of Excellence of Maritime Safety and Marine Installation (PUI-KEKAL) released several AIS-based safety monitoring development products such as AISITS Early Warning System, AISITS Ship Tracking, AISITS Ship Inspection Score (3).

The EWS system early identify conditions that can be dangerous and cause accidents in the area around the subsea pipeline installation due to the fall of the ship's anchor or the sinking of the ship. Early warning can be used by pipe operators or ship captains for areas that have a risk of accidents (4). AISITS is also a platform that can detect, and track vessels based on MMSI or IMO numbers. By this tracking system, it will be able to track vessels that have been proven to have committed violations (5). The basic data on the AISITS platform are ship static data, ship name, flag, year of manufacture, engine type, ship dimensions, etc. This data will make it easier for the government to monitor the condition of ships in

Indonesian sea (6). The development of more real time AIS data communication based on IoT has also been carried out to create a ship monitoring system in the Madura Strait (7).

However, the monitoring system still requires a connectivity support system from one remote base station (RBS) to another RBS and also to the server computer. This paper will explain the connectivity system model using a Virtual Private Network (VPN).

VPN technology also has a high level of security, therefore VPN can be used as an alternative technology to build an integrated network at a relatively low cost. The advantages of virtual private networks in terms of security are no longer in doubt. The virtual private network security method is capable of resisting attacking DoS (denial-of-service) and hacking. The way virtual private networks work, and network security is the reason many people use this technology to meet the needs of data communication over public networks. VPN implementation consists of remote access VPN, and site-to-site VPN. Remote access, also known as a virtual private dial-up network (VPDN), connects mobile users to a local area network (LAN).

2. Methods

2.1 *Remote Access* VPN

Remote access VPN can be used by web base client monitoring who want to connect to a special network from various remote locations from the network center. To access the central network, the user must be connected to the NAS by dialing the specified phone number. Then by using the client software, the user can connect to the destination network center (8).

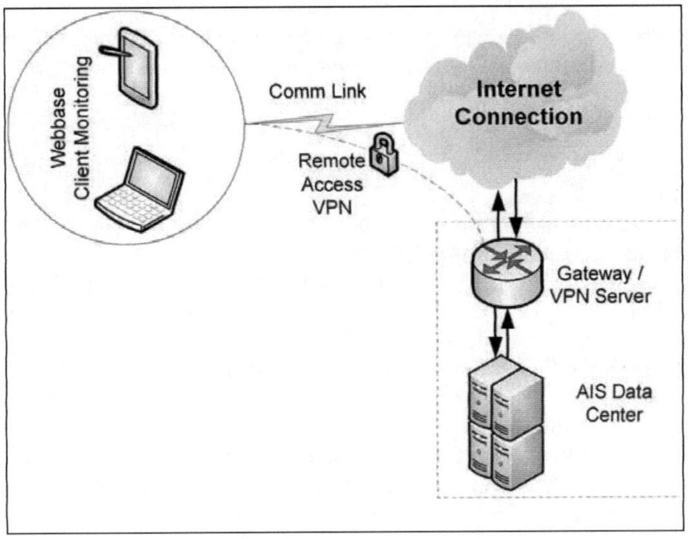

Figure 1 Remote Access VPN

2.2 *Site-to-site VPN*

Site-to-site VPN is used to connect between two or more RBS that are located far apart, both RBS owned by the institution itself and partner institutions in the consortium. A VPN that is used to connect an RBS to another RBS is called an extranet. This implementation belongs to the type of intranet site-to-site VPN, as shown in figure 2 (8).

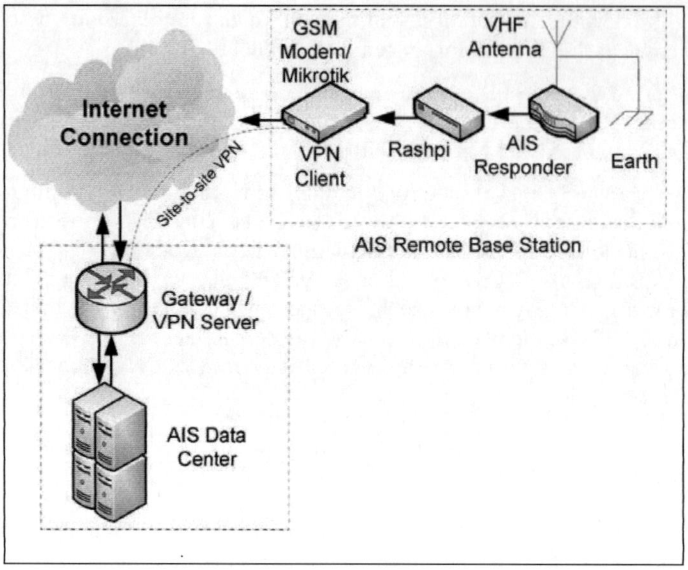

Figure 2 Site-to-site VPN

2.3 VPN Topology of AIS Monitoring System

In principle, the VPN network in the AIS monitoring system consists of three main devices as Remote Access VPN or Site-to-site VPN, namely client monitoring devices, routers or GSM modems as VPN clients and Mikrotik routers as VPN servers.

Communication flow of VPN Topology of AIS Monitoring System (7):

1. VHF antenna receiving signal from ships (AIS Transponder) (9) and then forwarding to AIS Responder.
2. NMEA data receiving in module AIS Responder such as ship identity, ordinates and directions are received like a data logger and then forwarded to Raspberry (10).
3. Raspberry Pi with the algorithm is converts logger data into binary data and sends it to the GSM Modem.
4. GSM modem is used to connect to internet and then by host-to-host connection configuration, RBS can connect to the Gateway/VPN in the AISITS server room.
5. In the room server, host to host connection accepted by Gateway/VPN Server and data forwarded to AISITS server.
6. The data received by the AISITS Server is in the form of binary data and then stored in the MySQL database. Then data is generated using PHP Programming and displayed in HTML format according to the client's request (11).
7. By IoT-based data communication, user can use a browser to access the web monitoring by using a PC or Smartphone. Thus, the client can see within the RBS reception radius how many ships are within the RBS reception radius. For special purposes, clients can connect remotely using Secure Shell (SSH) (12).

Figure 3 below is a complete topology of the AIS Monitoring System that connects RBS, User Client and AIS Data Center.

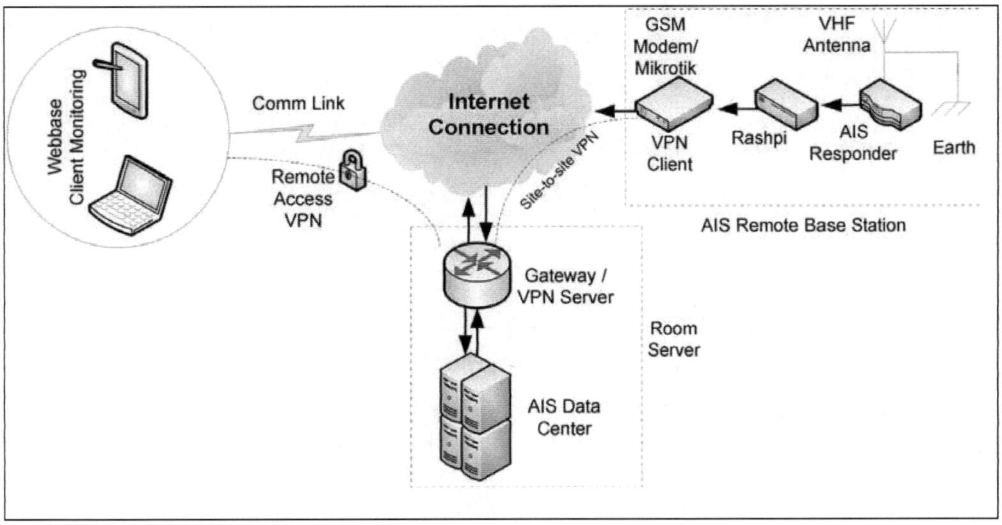

Figure 3 VPN Topology of AIS Monitoring System

2.4 *VPN Configuration*

The devices used in create a VPN using a Mikrotik RB750 are as follows (13) (14):

1. Winbox is used to configure Mikrotik, this will be easier than using command line.
2. Activating server PPTP in Menu PPP → PPTP Server → checked enabled status.
3. Create username and password in client Menu PPP → Secrets → create new (+) → entry
   a. name: name of client (RBS1 and RBS2)
   b. Password: user password RBS
   c. Profile: chose default encryption means VPN that has been built already applies encryption methods for network security.
   d. Local Address: IP address stored in the mikrotik to connect with RBS using remote address
   e. Remote address: IP address used by the RBS to connect to the mikrotik via the local address
4. Make connection client Setup in connection or network → choose a connection option → connect a workplace
   a. Internet Address: entry IP Public from ISP
   b. Destination Name: entry VPN name.
5. To test connection, entry username and password in the client.

2.5 *Integration Data Architecture*

Infrastructure analysis at the application level aims to obtain an integrated database. Distributed database, for database integration for each site, using full replication. The analysis steps at the application integration infrastructure level are choosing a Database Management System (DBMS), building a distributed database architecture, and installing and configuring software (14).

The integration application infrastructure is required to integrate the database on each RBS (node). Distributed database integration using gallery replication, as shown in the figure 4 (8).

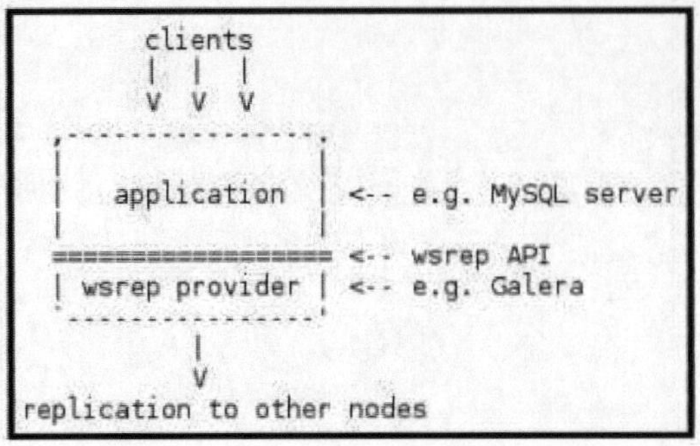

Figure 4 Galera Replication Concept

There are various ways to transfer state from one node to another, namely using mysqldump. A weakness of mysqldump is it works slowly because mysqldump blocks the server table from changes that might occur during the transfer. The method used in this research is rsync.

Rsync is a tool for one-way transfer and synchronization of files or trees, both local transfers (in the same system) or remotely (network/internet). The rsync function is similar to the cp, mv, scp, FTP clients. If the transfer breaks in the middle, Rsync will continue the synchronization process from the last point before it broke.

Table 1 Database integration requirements

No.	Applications/OS	Specification	Function
1.	Windows OS	Windows XP / 7 / 8	System Monitoring
2.	Ubuntu	Ubuntu 10.4	MySQL Server
3.	MySQL Server		Database
4.	Galera		Database Replication

The tools for building a distributed database with full replication method are as presented in table 1. Table 1. Specification of integration requirements Building a distributed database using galera requires mySQL with wsrep API. The state of wsrep API, Galera replication and MySQL in figure 5 (8).

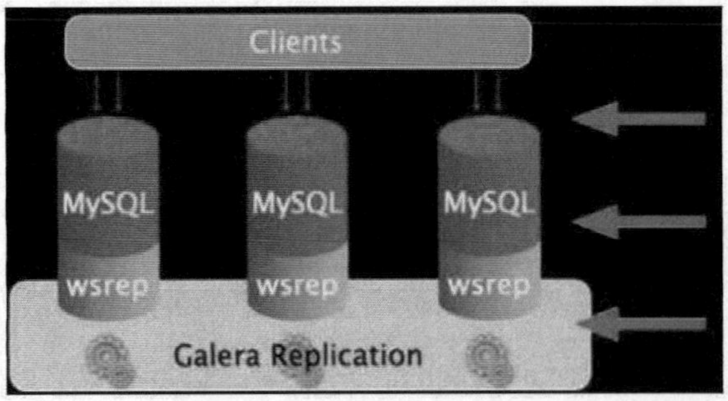

Figure 5 Structure of database

IOP Conf. Series: Earth and Environmental Science 1081 (2022) 012028 doi:10.1088/1755-1315/1081/1/012028

2.6 Database Integration Configuration

Step manual installation galera and mySQL with wsrep package

1. wget http://launchpad.net/galera/0.8/0.8.1/+download/galera-0.8.1-amd64.deb
2. wget http://launchpad.net/codershipmysql/0.8/0.8.1/+download/mysql-serverwsrep-5.1.57-0.8.1-amd64.deb
3. dpkg -i galera-23.2.2-I386.deb mysql-serverwsrep-5.5.28-23.7-I386.deb
4. entry user root and others configuration “/usr/bin/mysql_secure_installation”
5. configuration wsrep for RBS1
   a. “wsrep_provider=/usr/lib/galera/libgalera_s mm.so”
   b. “wsrep_cluster_address="gcomm://"
   c. “wsrep_sst_auth=wsrep_sst:wspass”
6. configuration wsrep for RBS2
   a. “wsrep_provider=/usr/lib/galera/libgalera_s mm.so”
   b. “wsrep_cluster_address="gcomm://192.168.2.122"”
   c. “wsrep_sst_auth=wsrep_sst:wspass”
7. Restart Service MySQL “service mysql restart”
8. User Privileges
   a. mysql -u root -p -e "SET wsrep_on=OFF; DELETE FROM mysql.user WHERE user='';";
   b. mysql -u root -p -e "SET wsrep_on=OFF; GRANT ALL ON *.* TO 'wsrep_sst'@'%' IDENTIFIED BY 'wspass'";
   c. mysql -u root -p -e "GRANT ALL PRIVILEGES ON *.* TO 'root'@'%' IDENTIFIED BY 'password' WITH GRANT OPTION; FLUSH PRIVILEGES;";
9. Bootstrap cluster to connected between RBS
   a. RBS11 : mysql -u root -p -e "SET GLOBAL wsrep_cluster_address='gcomm://';";
   b. RBS22 : mysql -u root -p -e "SET GLOBAL wsrep_cluster_address='gcomm://192.168.2.122:4567';";
10. Verification status “mmysql -u root -p -e "SHOW STATUS LIKE 'wsrep%';"; “

3. Result and Discussion

3.1 Testing Connection

After configuring RBS, we can test the connectivity of the VPN network that has been created. This test can be done on a remote user client using the Command Prompt application, before testing make sure the device used is already on the same VPN network. Open the application by typing CMD in the windows dialog box. Enter the 'PING' command followed by the destination RBS IP. Figure 6 below shows the results of the success test carried out with a testing time of less than 1 second. Data statistics for IP 192.168.2.122, Packets sent 4, Packets received 4, Packets lost 0.

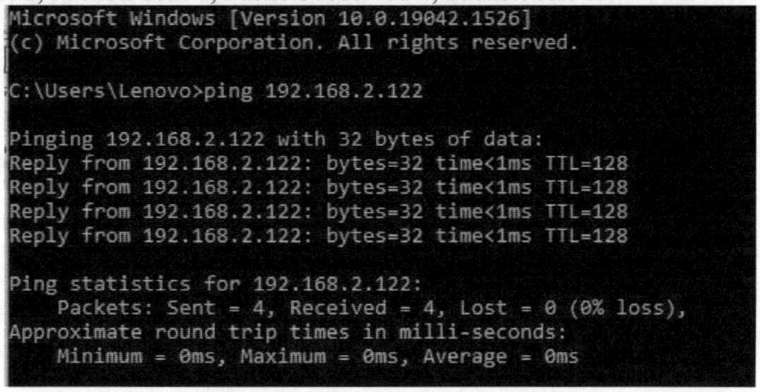

Figure 6 Result Test Connection

IOP Conf. Series: Earth and Environmental Science 1081 (2022) 012028 doi:10.1088/1755-1315/1081/1/012028

3.2 Security Testing using Pingflood

One of the attacks which is widely used on computer networks is to use Denial of Service (DoS). This attack method aims to stop or turn off the service on the VPN server. In this DoS test using the pingflood.exe application, the target IP address to be attacked is the public VPN IP server (202.155.xx.xx). Command prompt is activated by start > run > cmd. The next step is type command: pingflood 202.155.xx.xx –s 65000 –n 100000, as shown in Figure 7.

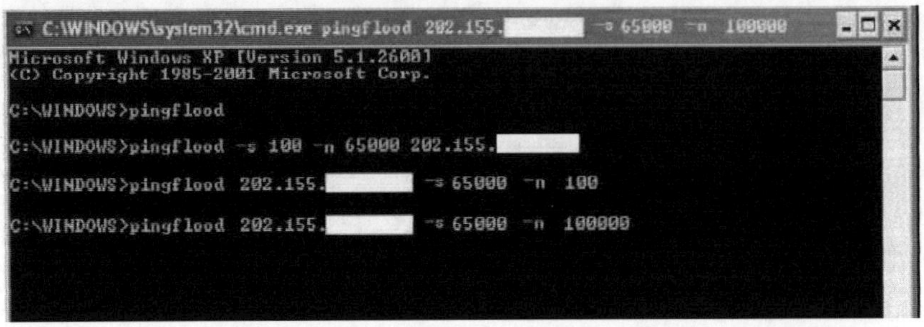

Figure 7 Pingflood using Command Prompt

Pingflood command above will send a number of packets to the VPN Server IP, then analyzed the effect of sending the packet whether the IP Server is affected or not. The impact of this pingflood attack certainly affects the connection between the VPN client and the VPN server, which was connection normally but suddenly went down with the Request Time Out display shown in Figure 8.

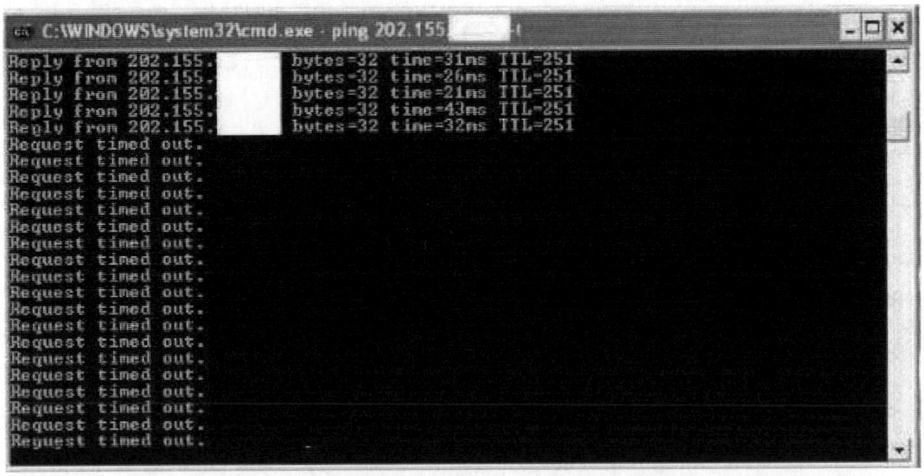

Figure 8 Pingflood attack effect

Pingflood attack process will stop after the number of pingflood packets sent has been complete. Then the connection to the target IP will return to normal. Shown in Figure 9.

Denial of Service (DoS) attack on the target IP using pingflood have an effect compared to the usual ping command. So it can be concluded that the experimental Denial of Service (DoS) attack, especially pingflood, was to disrupt network activity on the VPN server even though temporarily. Thus, it is necessary to add security at the server and gateway or router levels, especially at the VPN Server.

IOP Conf. Series: Earth and Environmental Science 1081 (2022) 012028 doi:10.1088/1755-1315/1081/1/012028

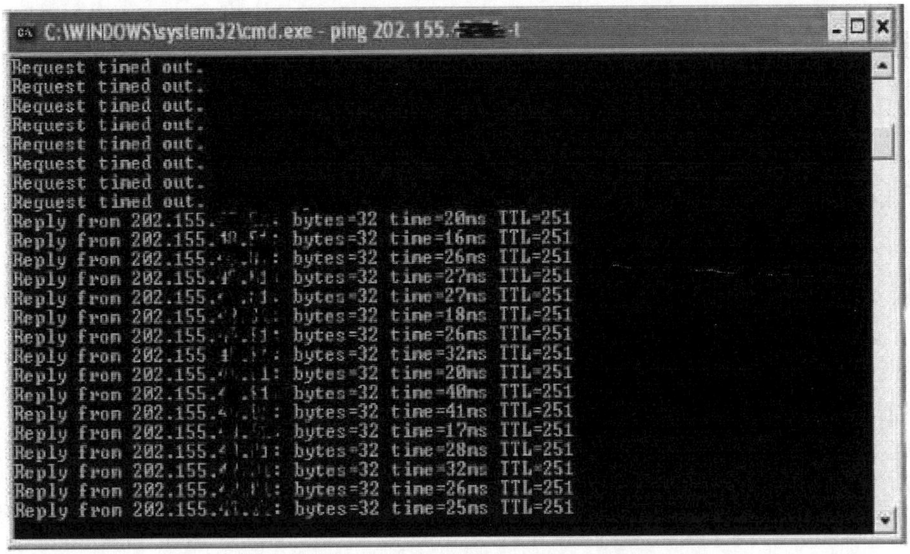

Figure 9 End off pingflood attack

3.3 Data Record from VPN Client

Figure 10 shows the raw data in the Interpreter/AIS Decoder received from the AIS Responder device. This data will then be sent to the data server via the VPN server according to the IP address that has been entered. Connectivity between the VPN client and the VPN server is safe from attacking DoS (denial-of-service) and hacking because it is in a tunnel that is formed only for the IP address that has been previously set.

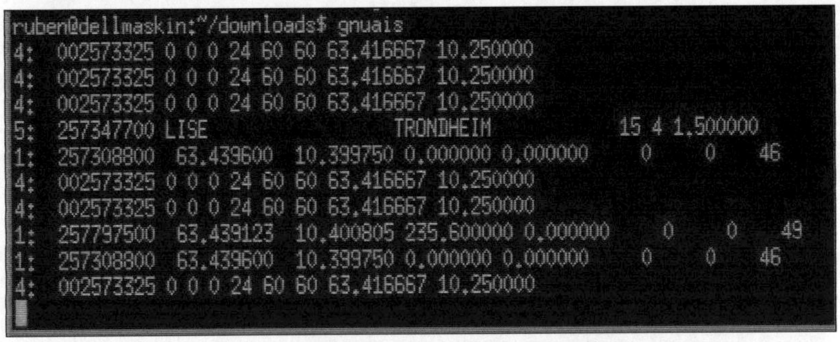

Figure 10 Row Data From VPN Client

The data sent from the VPN client on the image above can only be read by machine because it is raw data consisting of numbers that still need to be generate become information of the ship condition, for example name of the ship, MMSI number, direction, speed, and destination of the ship.

3.4 Data query from Data Center

In the data center, there is a database that accommodates all ship data sent from the VPN Client. This data can be read manually and become information on a ship in an area. By using the mysql command prompt can be given a command to query the mysql database to view the data at a certain time. Figure 11 is a display of query result information like time receive, MMSI, ship position latitude, longitude, heading, course and speed. This data is increase as long as any data is received from the RBS client.

IOP Conf. Series: Earth and Environmental Science 1081 (2022) 012028 doi:10.1088/1755-1315/1081/1/012028

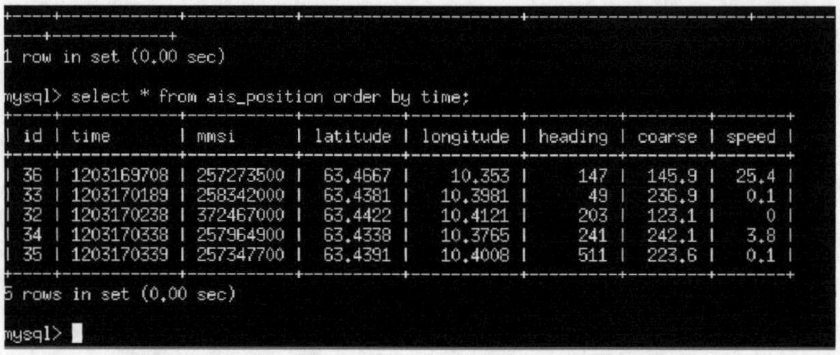

```
1 row in set (0.00 sec)

mysql> select * from ais_position order by time;
+----+------------+-----------+----------+-----------+---------+--------+-------+
| id | time       | mmsi      | latitude | longitude | heading | coarse | speed |
+----+------------+-----------+----------+-----------+---------+--------+-------+
| 36 | 1203169708 | 257273500 | 63.4667  |    10.353 |     147 |  145.9 |  25.4 |
| 33 | 1203170189 | 258342000 | 63.4381  |   10.3981 |      49 |  236.9 |   0.1 |
| 32 | 1203170238 | 372467000 | 63.4422  |   10.4121 |     203 |  123.1 |     0 |
| 34 | 1203170338 | 257964900 | 63.4338  |   10.3765 |     241 |  242.1 |   3.8 |
| 35 | 1203170339 | 257347700 | 63.4391  |   10.4008 |     511 |  223.6 |   0.1 |
+----+------------+-----------+----------+-----------+---------+--------+-------+
5 rows in set (0.00 sec)

mysql>
```

Figure 11 Display query ships database in data center

3.5 Display in web browser

Figure 12 The system output is a web browser display with information. name of ship, flag, type of ship, engine used, MMSI, and Speed at the time the data is received. The web page also provides information on categories of vessels that are safe or that require further inspection of vessels in the monitoring area.

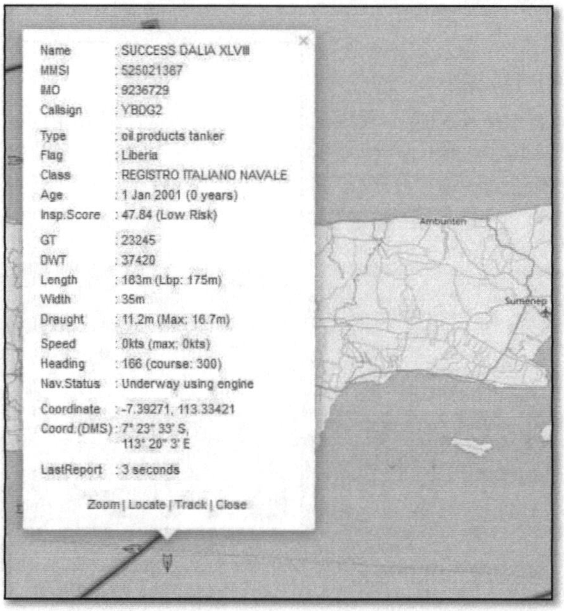

Figure 12 ships information in the web browser

4. Conclusion

Based on the results obtained from research on the Implementation of Virtual Private Network (VPN) on the AIS Monitoring System in real time, as follows:

1. The virtual private network (VPN) server design implemented in the system is capable of supporting the integration of the AIS monitoring system in real time.
2. VPN network able to defend DoS attacks and restore the network to normal conditions so that it can anticipate data loss.

3. A distributed database with a full replication approach using Galera Replication is able to meet the needs of data integration in the AIS monitoring system.
4. Galera replication is very good for building distributed databases with full replication method and is very suitable for use with large capacities.

5. References

[1]. SOLAS, 1. SOLAS, Safety Of Life At Sea. *Carriage Requirements for Shipborne Navigational Systems and Equipment.* s.l. : SOLAS, 2000. Vol. Chapter V.

[2]. Dinariyana, AA B, et al. 2, *Pengembangan Prototype Software Real Time Monitoring Berbasis Data Automatic Identification System (AIS).* Surabaya : IPTEK Journal of Proceedings Series, 2018, Vol. 10.

[3]. *PUI KEKAL.* Product Developments. [Online] The Center for Excellence of Maritime Safety and Marine Installation (PUI KEKAL), 2018. [Cited: July 2, 2021.] https://riset.its.ac.id/pui-kekal/products/.

[4]. Artana, Ketut Buda, et al. *Penlilaian risiko pipa gas bawah laut.* Surabaya : Guna Wijaya, 2013.

[5]. Artana, Ketut Buda, et al. *Milestone For E-Navigation: A Concept for Navigational Safety Platform by Means of AISITS.* Makassar : POLITEKNIK ILMU PELAYARAN MAKASSAR, 2021.

[6]. Artana, K.B., Pitana, T., Dinariyana, D.P. et al. *Real-time Monitoring of Subsea Gas Pipelines, Offshore Platforms, and Ship Inspection Scores Using an Automatic Identification System.* March 2018, Switzerland : Journal of Marine Science and Application (JMSA), 2018, Vol. 17. 1993-5048.

[7]. Maulidi, A and Irmiyana, T. *The development of internet-of-things (IoT)-based ship monitoring system in madura strait.* Surabaya : IOP Publishing Ltd, 2021. 1052 012052.

[8]. MariaDB. About Galera Replication. *MariaDB.* [Online] [Cited: June 12, 2021.] https://mariadb.com/kb/en/about-galera-replication/.

[9]. Maulidi, Akh., Prasetyo, Taufan and Irmiyana, Triyanti. 01,*Disain Sistem Navigasi Automatic Identification System (Ais) Transceiver Berbasis Mini Computer Pada Kapal Nelayan Tradisional Di Madura.* Bengkalis : INOVTEK, 2019, INOVTEK POLBENG, Vol. IX, pp. 12-17. 2580-2798 .

[10]. Maulidi, Akh., Prasetyo, Taufan and LAR., Anauta. *Sistem Penerima (Receiver) Automatic Identification System (AIS) Berbasis Mini Computer Pada Kapal Nelayan Tradisional Di Madura.* 2019, INOVTEK POLBENG, pp. 344-349.

[11]. Maulidi, Akh., Pitana, Trika and Buda, Ketut. *Pengembangan Sistem Monitoring Automatic Identification System (AIS) Berbasis Website Secara Real Time.* 2017, INOVTEK BOLBENG, pp. 153-166.

[12]. Maulidi, A, et al. *The development of internet-of-things (IoT)-based safety monitoring system in north sea madura.* Surabaya : IOP Publishing Ltd, 2021. 972 012013.

[13]. Mikrotik, VPN Overview. *Mikrotik.* [Online] February 3, 2011. [Cited: January 22, 2022.] https://wiki.mikrotik.com/wiki/VPN_Overview.

[14]. Masykuri, A, Utami, E and Sudarmawan. 2, *Implementasi VPN Server dalam Sistem Informasi Apotek (Studi Kasus Integrasi Sistem Informasi Apotek Santi Pontianak).* Yogyakarta : STIMIK AMIKOM, 2016, Vol. 17. 1411-3201.

[15]. Maulidi, Akh, et al. *Integrasi AIS dan Shipping Database Sebagai Dasar Pengembangan Metode Keselamatan Kelautan.* Surabaya : FTK ITS, 2013. SENTA. pp. D45-D49.

Acknowledgments
We would like to thank AISITS Research Consortium, PUI KEKAL ITS, Director of Politeknik Negeri Madura (POLTERA), Colleagues of Poltera, All Committee Maritime Safety International Conference 2022 (MASTIC 2022) which has been approve and publish the results of this research, and my beloved family always support.

Risk Assessment of Ship Collision on FSO Abherka and Oil Spill Modelling Due to Structural Damage

Rafiuddin Adyaksa Sukma[1], Dhimas Widhi Handani[1,2], Taufik Fajar Nugroho[1,2], Widiastuti Tyasayumranani[3]

[1]Department of Marine Engineering, Institut Teknologi Sepuluh Nopember, Surabaya, Indonesia
[2]Center for Excellence of Maritime Safety and Marine Installation (PUI KEKAL), Institut Teknologi Sepuluh Nopember, Surabaya, 60111, Indonesia
[3]Department of Maritime Transportation System, Mokpo National Maritime University

[1]rafisukma04@gmail.com

Abstract. Floating storage and offloading (FSO) is an important vessel in offshore as shuttle storage of crude oil before it will transfer to the onshore facility located near the drilling platform. One of the locations that use FSO is Poleng Field in West Madura Offshore by operating FSO Abherka. Poleng field is supplying for two regions in Java and Kalimantan, it can affect the energy supply of the regions if there is an accident. The possible accident is a ship collision with FSO Abherka in addition the location is near Surabaya West Access Channel (SWAC). It is necessary to conduct a risk assessment of ship collisions on FSO Abherka and a consequences analysis of oil spills due to structural damage. Calculating ship collision frequency using CRASH method using two traffic scenarios found the result of ship collision frequency for external and internal powered collision: 4,20E-03 and 1,28E-03, and external and internal drifting collision: 4,51E-03 and 4,01E-03. Consequences of Collision simulation is conducted in three variations of collision scenario and impacting to the hull damage. Total asset & equipment loss due to collision in various degree of 30°, 60°, and 60° as follows: \$23.462.196,03, \$23.666.532,98, and \$23.921.683,88. Those total asset & equipment loss and environmental impact from the oil spill is on the category of unacceptable. Applying a traffic monitoring system is effective to reduce the frequencies by about 71-93%.

Keywords: CRASH Model, FSO, Oil Spill Modelling, Ship Collision Simulation, SWAC

1. Introduction

The existence of FSO becomes vital in offshore operation as floating storage and offloading of oil & gas products before it will be sent to the onshore facility. Offshore exploration will become more significant to support oil & gas domestic needs. In 2023 SKK Migas targeted to increase the national oil and gas lifting into 1 million barrels per day (BOPD). To achieve the target, it needs to increase the production of existing wells and conduct new explorations. Based on data from the Ministry of Energy and Mineral Resources, Indonesia has 128 basins which consist of 20 active production basins, 27 new reserves found, and 13 without reserves. The remaining 68 basins have not been explored yet and those are mostly located in the offshore area. With the remaining potential of national reserves, exploration offshore will become the focus of the government. Exploration in the offshore area is facing some difficulties such as the area that is located far access in the middle of the ocean and needs high capital cost to start the project. The use of FSO becomes important to help in the process of distributing the lifting products to the onshore facility as floating storage. In economics, the usage of FSO can be more beneficial and press the cost than making a subsea pipeline. By the targets of the government, the usage of FSO will increase in line with the development of offshore production in Indonesia. One of the locations that already used FSO as their floating storage terminal is Poleng field under Pertamina Hulu Energi West Madura Offshore (PHE WMO) by operating FSO Abherka.

FSO Pertamina Abherka is a conversion vessel of oil tanker MT. Geudongdong has converted into floating storage offloading. FSO Abherka has officially operated as a shuttle storage terminal in Poleng field in August 2012. FSO Abherka can store crude oil products from Poleng Processing Platform (PPP) with a capacity of up to 600,000 barrels by utilizing 9 cargo oil tanks. Within a year, FSO Abherka will transfer crude oil to shuttle tankers approximately 10-12 times per year. The product of crude oil from FSO Abherka will be carried out to two main refinery units in RU-IV Cilacap and RU-V Balikapan.

The usage of FSO in offshore facility especially FSO Abherka has high risk due to its ability to store in high amount of crude oil at one place. If there is leak or incident that happen in FSO Abherka, the crude oil is possible to spill and spread into the sea. In addition, location of FSO Abherka in west Madura offshore is located about 1,8 km from PPP which connect to all wells in PHE WMO. The occurrence of oil spill can be affected by various factors such as the probability of failure in the system or possibility of collision. From the collision can result in the damage of the FSO hull and it will trigger into leaks and resulting of oil spills.

The location of Poleng field itself is close to the shipping line of Surabaya West Access Channel (SWAC). The distance from the Buoy No.1 before getting into the lane is 24,4 km. There is also ships inside of the Poleng field for shuttle activity that approaching FSO Abherka to load crude oil. With those two traffics at any time could be possibly to have failure that can cause a ship to deviate from their lane and endanger to FSO Abherka. If a collision happens, it will make a big risk for the safety of the crew and environment due to oil spills. Also, it can affect to the supply of crude oil for Kalimantan and Java because both regions are supplied by Poleng Field. It needs to conduct risk assessment based on the possibility of collision and oil spill as the consequences due to structural damage.

2. Method

The location of FSO Abherka in West Madura Offshore is close to the Surabaya West Access Channel (SWAC) which has high intensity of ship traffic. If an accident of ship collision happen, it will impact to the oil spill due to the structural damage of FSO Abherka. It needs to held risk assessment based on the risk that already identified.

There are some literatures that can be used as guidelines for this research such as Quantitative Risk Assessment for Offshore Installation [1] , Pedoman Pengelolaan HSSE Pertamina Berbasis Risiko [2], and GNOME Guidelines book. There are several related research that can be considered as comparison regarding the risk assessment methodology: comparing CRASH Model and Traffic Based Model to calcualte the frequency of collision [3]; using CRASH model for frequency analysis in the wellhead platform PHE-12 [4]; and others related studies to analyze the consequences [5], [6].

This research is using Computerized Risk Assessment of Shipping Hazard (CRASH) for the frequency analysis. This method is possibly to calculate the collision of ship into two scenarios collision: powered collision and drifting collision. The result of the frequency analysis will be analyzed further for the consequences. There are two possible consequences that may come up impacted of collision: oil spill due to the structural damage and loss in asset & equipment due to the collision. For the oil spill it will be simulated the dispersion directly after the collision and can calculate the amount of crude oil that spread in the sea. The impact of collision could give structural damage to the ship hull. cost of repairment and loss of some equipment and asset of FSO can be analyzed.

From both analysis result of frequency and consequences will determine the risk level of the risks. If the level of risks is unacceptable, it needs to do mitigation immediately to prevent the wider risks. Risk control options and cost-benefit analysis could be possible to arrange the suitable mitigation actions regarding the risk level of each risk.

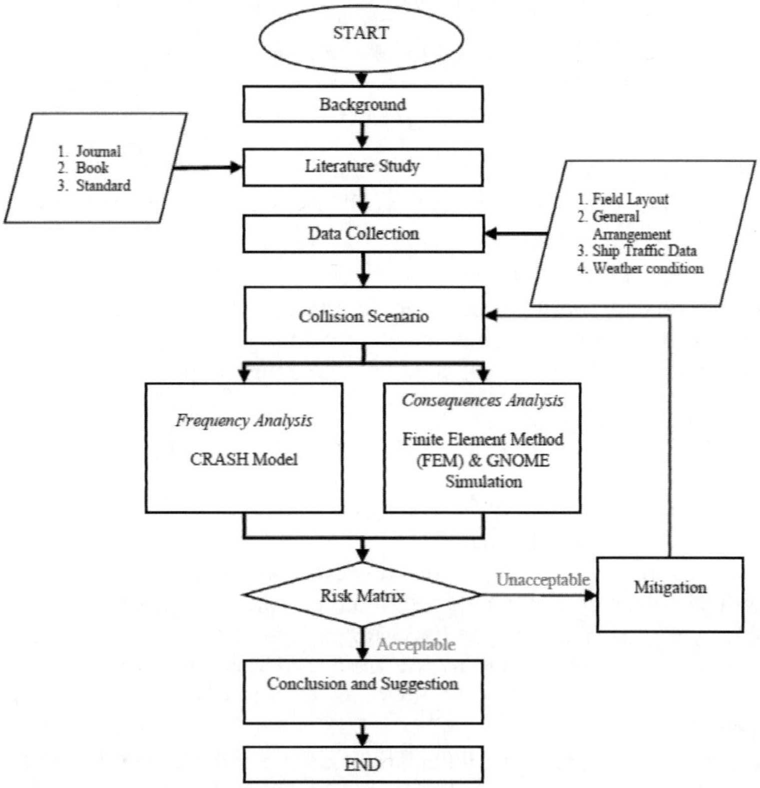

Figure 1. Flow Chart Methodology

2.1. Data Collection

There are several data that need to be fulfilled to support this research such as: Poleng field layout, number of ship traffic, FSO Abherka general arrangement, and area characteristic. Those data are collected as an input and consideration for this research.

2.2. Collision Scenario

This research is considering two traffic scenarios of ship collision on FSO Abherka, as follows:

1) External traffic
   All vessel passing SWAC that measured from Buoy No.1 that possibly to hit FSO Abherka. The distribution of this traffic is considering all vessel that using SWAC.
2) Internal traffic
   All vessel that related to offshore activity inside of the area and all vessel that passing close to the FSO Abherka by considering historical of traffic lane. This traffic scenario will be measured from centerline of shipping route to Kalimantan and/ East Indonesia.

Calculation of frequency of ship collision using CRASH model, needs to define the scenario of collision. In CRASH model, to calculate frequency of collision there are two methods: powered collision and drifting collision. From both methods will be determine for two traffic scenario that is external traffic and internal traffic. There are four collision scenarios which are represented by the following codes: A1, A2, B1, and B2. Description for each code shows in the table 1 as follows:

Table 1. Collision scenario code description.

Code	Description
A1	External Traffic Powered Collision
A2	External Traffic Drifting Collision
B1	Internal Traffic Powered Collision
B2	Internal Traffic Drifting Collision

2.3. Frequency Analysis

In the calculation of the frequency of ship collision on FSO Abherka is using quantitative risk assessment (QRA) method of CRASH modeling. There are two collision scenarios that will calculate that is powered collision and drifting collision.

1) Powered Collision

A type of ship collision that vessel hit the object in high speed due to various causes. Several factors causing the collision that related to each other: failure during voyage, failure of watchkeeping by the crew, and failure of alerting to avoid collision with the object or platform. Powered collision can be calculated using this formula:

$$FCP = N \; x \; Fd \; x \; P1 \; x \; P2 \; x \; P3 \tag{1}$$

F_{CP} = Collision powered frequency with the object
N = Number of ships through the lane
F_d = Proportion of ships coming out of the servant groove towards the object area
P_1 = Probability of the ship experiencing failure in sailing
P_2 = Probability of ship failure in watchkeeping
P_3 = Probability for the object to fail to provide a sign to prevent collisions

2) Drifting Collision

Type of collision caused by a ship experiencing breakdown in the engine and carried by the direction of the wind towards to the platform or object. Drifting collisions occur at a relatively slow speed. There are two causes that related each other that causing drifting collision: engine breakdown and weather condition of wind blowing. Drifting collision can be calculated using following formula:

$$\mathbf{F_{CD} = N_b \; x \; P_B \; x \; P_W \; x \; D/BL} \tag{2}$$

F_{CD} = Collision drifting frequency (per year)
N_b = Number of ships in the box (ships/year)
PB = Breakdown probability in the box
PW = wind blowing probability from the box to the object
D = Collision diameter
BL = The length of the box perpendicular to the wind direction

2.4. Consequence Analysis

There are two consequences to be considered: oil spill and structural damage in asset & equipment loss. Oil spill is affected by structural damage due to the collision and will be simulated using GNOME oil spill modeling. For the structural damage will be simulated using Finite Element Method to simulate the collision using various scenarios of collision degree: 30°, 60°, and 90°.

2.5. Risk Acceptance Criteria

After frequency and consequences has already analyzed, will be inputted to the risk matrix. Risk matrix is defined to determine the level of risk for both oil spill and structural damage. This research is using risk matrix to determine the level of risks from both oil spill and loss in asset & equipment of FSO Abherka.

2.6. Mitigation
The mitigation is arranged based on the risk matrix result. If the result shows the risk on category or level unacceptable, means it need to do mitigation immediately to prevent the risks getting bigger. The mitigation method that used is using Risk Control Options and Cost-Benefit Analysis.

3. Results and Discussion
3.1. Ship Traffic Data Collection
The data be used on this research is the total amount of ship traffic through Surabaya West Access Channel (SWAC) per year and forecasting for the next 10 years. In addition to support this research, specifications, and detail data of destination are also needed. Author collected data with referring from two sources: Kantor Otoritas Pelabuhan Tanjung Perak and Kantor Syahbandar dan Otoritas Pelabuhan Gresik. The data of ship traffic passing SWAC shows in the table 2 as follows:

Table 2. Ship traffic data passing Surabaya West Access Channel (SWAC).

Year	Type of Vessel						Total
	Bulk Carrier	General Cargo	Tanker	Passenger	Container	Others	
2018	712	2.519	1.501	2.064	3.910	6.209	16.915
2019	721	2.461	1.773	2.410	3.931	6.638	17.934
2020	670	2.722	1.577	1.820	3.935	6.716	17.440

3.2. Powered Collision
A type of collision when a ship hits a platform at high speed caused by several factors causing the collision that related to each other: failure during voyage, failure of watchkeeping by the crew, and failure of alerting to avoid collision with the object or platform. Powered collision can be calculated using this formula:

3.2.1. Fault Tree Analysis (FTA) of Powered Collision. Before calculating the frequency of powered collision, it is necessary to determine the probabilities value of the failure according to the existing scenario of powered collision using fault tree analysis (FTA). Probability failure based on casualty of powered collision using fault tree analysis (FTA) can be seen in figure 2 as follows:

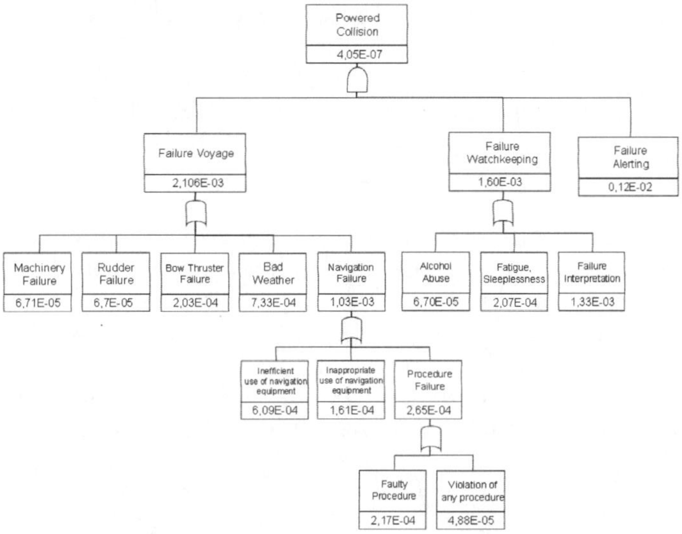

Figure 2. FTA of powered collision.

Failure voyage (P1) is a condition in when ship fails in voyage due to several factors that cause the ship deviate from its lane. The causes of failure could be possible in voyage such as: machinery failure, rudder failure, bow thruster failure, bad weather, and navigation failure [7]. The probability of ship fails in voyage from its lane is determined by 0,002106.

Failure watchkeeping (P2) is a condition of a collision that caused by crew fault or human error. Some of the ship's crew fault or errors that can result in watchkeeping failures are due to under alcohol abuse, fatigue and sleeplessness, and interpretation failure [7]. The probability of failure watchkeeping after calculated using fault tree analysis is determine by 0,001604.

Failure alerting (P3) is a condition in which hazard sign fail to inform the ship to prevent collision. The probability of this failure is assuming by 0,12 of failure in watchkeeping referring to the previous research. The probability of failure that causing powered collision is determine by 0,000000405536.

3.2.2. Powered Collision Calculation Result.

Table 3. External powered collision calculation.

Formula	Annual Frequency			
	2020	2022	2027	2032
N	17.440	18.217	19.530	20.842
Fd	5,69E-01	5,69E-01	5,69E-01	5,69E-01
P1	2,11E-03	2,11E-03	2,11E-03	2,11E-03
P2	1,60E-03	1,60E-03	1,60E-03	1,60E-03
P3	1,20E-01	1,20E-01	1,20E-01	1,20E-01
P1 x P2 x P3	4,06E-07	4,06E-07	4,06E-07	4,06E-07
Total	**4,02E-03**	**4,20E-03**	**4,50E-03**	**4,81E-03**

Table 4. Internal powered collision calculation.

Formula	Annual Frequency			
	2020	2022	2027	2032
N	5.406	5.647	6.054	6.461
Fd	5,61E-01	5,61E-01	5,61E-01	5,61E-01
P1	2,11E-03	2,11E-03	2,11E-03	2,11E-03
P2	1,60E-03	1,60E-03	1,60E-03	1,60E-03
P3	1,20E-01	1,20E-01	1,20E-01	1,20E-01
P1 x P2 x P3	4,06E-07	4,06E-07	4,06E-07	4,06E-07
Total	**1,23E-03**	**1,28E-03**	**1,38E-03**	**1,47E-03**

The table 3 and 4 shows that the frequency of ship collisions with FSO Abherka for both external and internal scenario with the highest frequency is on container vessels. For external powered collision the highest frequency of container vessel is 8,02E-04 and internal powered collision of container vessel is 5,22E-04. Container vessel has the highest value of collision probability due to its high number of vessels around the location. Can be analyzed that frequency of powered collision is affected by causation factor which is probability and number of ship traffic that considered on the calculation. Total frequency for external powered collision for 2022, 2027, and 2032 respectively are 4,20E-03; 4,50E-03; and 4,81E-03. And for total frequency of internal powered collision for 2022, 2027, and 2032 respectively are 1,28E-03; 1,38E-03; and 1,47E-03.

3.3. Drifting Collision

Collision caused by a ship experiencing breakdown in the engine and carried by the direction of the wind towards to the platform or object. Drifting collisions occur at a relatively slow speed. There are two causes that related each other that causing drifting collision: engine breakdown and weather condition of wind blowing. Drifting collision can be calculated using following formula:

3.3.1 Fault Tree Analysis (FTA) of Drifting Collision. From the cause of the possibility of the ship drifting to crashing into the platform can be determined the probability value by using fault tree analysis.

IOP Conf. Series: Earth and Environmental Science 1081 (2022) 012029 doi:10.1088/1755-1315/1081/1/012029

Probability failure based on casualty of drifting collision using fault tree analysis (FTA) can be seen in figure 3 as follows:

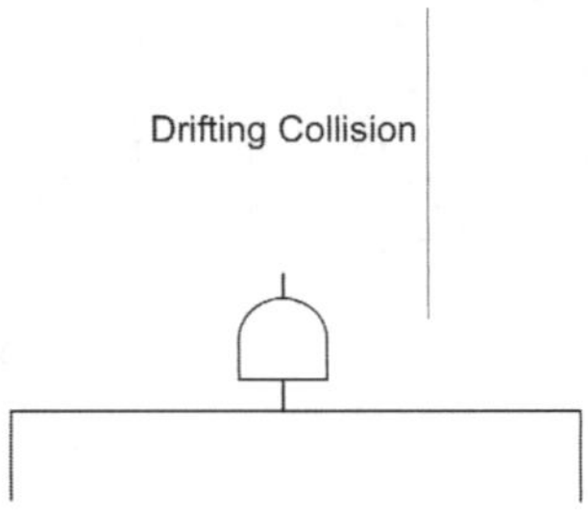

Figure 3. FTA of drifting collision.

Engine breakdown failure is a failure during the repair process within a certain time so that the ship can drifted out of from the lane. Another probability that can cause the drifting collision is wind blowing. Wind blowing factor can cause drifting collision on the ship that will be carried away by the wind out of its lane and drifted towards to the platform or object.

The engine breakdown factor on a tanker is generally worth for 2 x10^{-5} per hour in 6 nm [1]. The assessment is based on the assessment of merchant ships and shuttle tankers considering the factor of repair time within a few hours.

Wind blowing is estimated that the probability of wind blowing is 0.302 towards to the east based on wind rose data in Surabaya. Which assuming this calculation is conducted during west wind season due to the probability of wind blowing towards to the east from calculated point is higher than east wind season. West wind means the wind will blow to the east, FSO Abherka is located towards to the east from SWAC.

3.3.2 Drifting Collision Calculation Result. The total ship traffic is known and refers to the use of the FSO Abherka which will operate for approximately 10 years, so it is necessary to forecast the ship traffic for the next 10 years. With extrapolation method, the calculation of drifting collision will be considered for 2027 and 2032 until the operation of FSO Abherka is end. The result of calculations for the frequency of drifting collision respectively for external traffic scenario and internal traffic scenario as follows:

Formula	2020	2027
Pb0	2,00E-05	2,00E-05
Pb1		
Pw		
Total	**4,32E-03**	**4,83E-03**

Formula	2022		2032
Pb0	2,00E-05	2,00E-05	2,00E-05
Pb1	4,20E-06	4,20E-06	1,46E-06
Pw	3,02E-01	3,02E-01	3,02E-01
Total	**3,84E-03**	**4,01E-03**	**4,30E-03**

IOP Conf. Series: Earth and Environmental Science 1081 (2022) 012029 doi:10.1088/1755-1315/1081/1/012029

Calculation result in table 5 and 6 is the frequency of ship collision on FSO Abherka obtained in drifting collision. In the external traffic scenario obtained the highest frequency value of drifting collision is 8,61E-04 by container. For internal traffic scenario obtained the highest frequency value of drifting collision is 1,63E-03 with the highest probability on container vessel. Total frequency for external drifting collision for 2022, 2027, and 2032 respectively are 4,76E-03; 6,08E-03; and 7,75E-03. And for total frequency of internal drifting collision for 2022, 2027, and 2032 respectively are 4,15E-03; 4,45E-03; and 4,76E-03.

3.4. Structural Damage Analysis

This research is used analysis of the collision scenario using finite element method (FEM). Then continued to analyze the consequences of the impact on oil spill.

3.4.1. Collision Simulation Model. In the process of FEM simulation is using ANSYS, it requires modeling of two ships, namely: Striking Ship and Stucking Ship. Striking ship is a ship that will be modified as a impactor, while Stucking ship is a ship that will be modeled as being hit. The detail specification of the ship that will be simulated shows in table 7 as follows:

Table 7. Specification of simulation model.

Principle Dimension	Striking Ship	Stucking Ship
Name	CMA CGM	FSO Abherka
LoA	294.1	258
Breadth	32.2	42
DWT	54,309	86,866

Striking ship is chosen as the striking ship model due it is the highest frequency of ship that calculated based on CRASH model calculation. CMA CGM Container Vessel is the highest ship that recorded based on ship traffic data after considering from various data from Kantor Otoritas Pelabuhan Tanjung Perak, KSOP Gresik, and approaching vessel data of FSO Abherka.

3.4.2. Collision Simulation Assumption and Variation. The analysis of the ship collision with FSO Abherka will be conducted to find out the damage to the ship's hull of FSO Abherka. To help this analysis process will be used ANSYS Workbench explicit dynamic simulation software. Using the software will be able to analyze the total deformation of the simulated ship collision. There are several data assumptions to conduct the simulation, such as:

- ✓ Velocity (initial condition): 11,83 m/s
- ✓ Density: 1.025 kg/m^3
- ✓ Thickness: 8mm
- ✓ Time frame: 2 second

The simulation scenario of ship collision is carried out in several variations based on vessel collision degree such as 30°, 60°, and 90°. Those scenarios are determined based on many conditions that possible to strike FSO Abherka in relatively from various collision degree. Variation of *90°* collision simulation result, the variation where the container ship will hit FSO Abherka in a perpendicular state to FSO Abherka

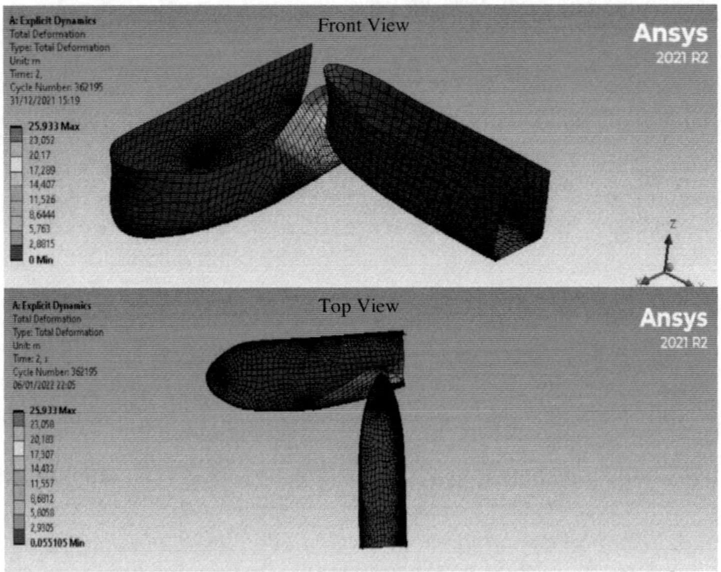

Figure 4. 90° Collision Simulation Result

3.5. Calculation of Asset & Equipment Loss

After conducting the simulation of ship collision on FSO Abherka, it can be calculated the total loss in asset & equipment due to the collision. It can be known the area of damage or total deformation resulting from the collision. Based on data of total damage area of the ship, it is necessary to repair the damages by replacing the plate on the hull with the new one. Also, needs to calculate the total loss in asset & equipment by considering the cost for ship chartering, oil spill loss, and clean-up cost. The calculations may differ following the recent price changes.

Table 8. Calculation of total asset & equipment loss for 90° collision.

No	Item	Quantity	Unit	Price	Total Price
1	Plate Changing 8mm x 6' x 20'	292.185	kg	Rp30.000,00	Rp8.765.555.555,56
2	FSO Chartering	60	day	Rp350.779.440,00	Rp21.046.766.400,00
3	Docking	54.326	GT	Rp20.000,00	Rp1.086.520.000,00
4	Oil Loss	56.476,2	Barrel	Rp1.149.313,44	Rp64.908.855.700,13
5	Oil Spill Clean-up	2.502	tons	Rp99.277.200,00	Rp248.377.489.982,54
		Total (Rp)			Rp344.185.187.638,22
		Total ($)			**$23.921.683,88**

The example of total asset & equipment loss calculation shows the table 8 which for 90° collision, can be concluded that oil spill clean-up gives the biggest amount of cost from all the items. The cost for oil spill cleaning is spent for about $6900 per tons assumption refer to journal from (Etkins,2000) for average clean-up cost per ton in several regions. For this case, the total cost for oil spill cleaning for 2,502 tons crude oil is about Rp248.377.489.982,54 or $17.262.822 in USD. From three variations in collision degree, 90° collision degree has the biggest amount of loss. This is in accordance with the area of hull damage in 90° degree collision have biggest area coverage of damage. With total 4.508 m² hull damage, the total loss in asset & equipment in 90° degree of collision is $23.921.683,88. Those calculation can be even bigger if the loss in operational and benefit is being considered. The calculation summary for each variation can be seen on the table 9 as follows:

Table 9. Summary of total asset & equipment loss.

Collision Degree	Area Damage	Total Asset & Equipment Loss
30°	1,108 m^2	$23.462.196,03
60°	2,620 m^2	$23.666.532,98
90°	4,508 m^2	$23.921.683,88

3.6. Oil Spill Modelling

The simulation of oil spill modelling is using GNOME Oil Spill Modelling. Before conducting oil spill simulation, it needs some supporting data as input for simulation and the data input such as: volume of oil spills, location map of oil spill, and wind & waves data.

After all data required has inputted to GNOME, the next step is running the simulation of oil spill based on data. The simulation of this research is run in between time of 8 hours, 16 hours, and 24 hours to determine the oil distribution due to the accident. After running process following those simulation time, here are the result of the simulation shows in the figure 5 as follows:

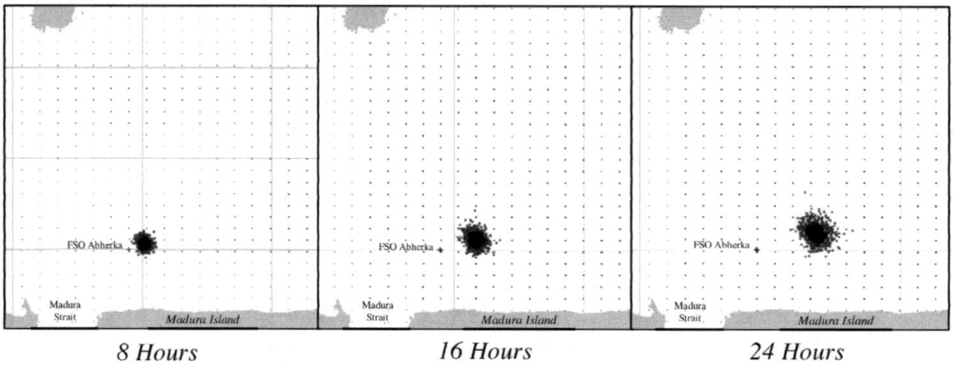

Figure 5. Oil Spill Modelling Result

Within 24 hours, there are 44.560 barrels or 79% of crude oil volume released in the sea. And there will be 11.916 barrels or 21% of cargo oil tank released is evaporated. The result of the modelling can be seen through the following table 10 below. Based on the modelling maps, we can see that the location of spill is located in the middle of the sea and far enough from coastal area. Time setting of the simulation is set in period of west wind that wind blowing tend to blow to the east. This is affecting to the oil spill heading which towards to the east and keep clear from the other platform near FSO Abherka.

Table 10. Result of oil spill modelling.

Condition	Oil Volume					
	8 Hours		16 Hours		24 Hours	
	Barrels	%	Barrels	%	Barrels	%
Evaporated	5.365	9%	9.319	17%	11.916	21%
Beached	-	0%	-	0%	-	0%
Floating	51.111	91%	47.158	84%	44.560	79%

3.7. Risk Acceptance Criteria

Risk can be identified by analysing both frequency and consequences of an accident. The frequency and consequences have already defined the result based on each methodology. The next step is interpreting the result of both frequency and consequences into risk matrix to get to know the level of risk from ship collision on FSO Pertamina Abherka. Risk matrix standard that be used on this research is refer to Pertamina Risk Assessment Matrix (RAM) HSSE Pertamina.

3.7.1 Frequency Interpretation. Previously, has been analyzed the probability of ship collision on FSO Abherka by using several scenarios that have been described in previous discussions with code A1, B1, A2, and B3 to represent the scenario of powered collision and drifting collision. For the frequency has been calculated and interpretated to the risk criteria shows in the table 11 as follows:

Table 11. Risk interpretation of ship collision frequency.

Scenario		Frequency	Risk Level	Risk Interpretation
A1	*External Traffic Powered Collision*	4,20E-03	3	Moderate
B1	*Internal Traffic Powered Collision*	1,28E-03	3	Moderate
A2	*External Traffic Drifting Collision*	4,51E-03	3	Moderate
B2	*Internal Traffic Drifting Collision*	4,01E-03	3	Moderate

3.7.2 Consequences of Oil Spill. The number of oil spills spilled due to the collision of the ship with FSO Abherka has been simulated and has obtained the results of calculating the probability of the frequency of the collision. From both data can be interpreted to be entered into the risk matrix to find out the severity of the risk that may occur. Based on the results of the simulation obtained the amount of oil spilled due to the impact amounted to 44.560 barrels, which include into the category of risk level 5 or catastrophic. According to the amount of crude oil spilled to the ocean is 44.560 barrels and interpretated with the frequency of various scenarios of collision, the risk is on the level of unacceptable. The risk of this case is unacceptable means need to be taken of immediately mitigation action to reduce the risk condition. From that information can be inputted into risk matrix as shows in the figure 6 as follows:

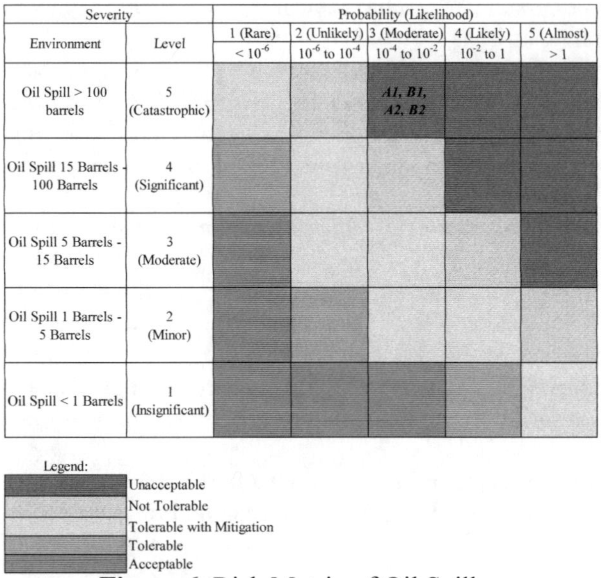

Figure 6. Risk Matrix of Oil Spill

3.7.3 Consequences of Asset & Equipment Loss. The consequences of total asset & equipment loss in three collision degree are in the risk level of 5 or catastrophic. The range of level 5 or catastrophic is total loss with value greater than \$5,000,000. And the interpretation result of frequency and consequences of total asset & equipment loss are in unacceptable level. The risk of unacceptable cannot be accepted and need to do immediate mitigation to reduce the risks with some of specific condition and step. From that interpretation can be inputted into risk matrix as shows in the figure 7 as follows:

Severity		Probability (Likelihood)				
Asset & Equipment	Level	1 (Rare) $< 10^{-6}$	2 (Unlikely) 10^{-6} to 10^{-4}	3 (Moderate) 10^{-4} to 10^{-2}	4 (Likely) 10^{-2} to 1	5 (Almost) > 1
\$5,000,000	5 (Catastrophic)			***A1, B1, A2, B2***		
\$1,000,000 - \$5,000,000	4 (Significant)					
\$100,000 - \$1,000,000	3 (Moderate)					
\$10,000 - \$100,000	2 (Minor)					
< \$10,000	1 (Insignificant)					

Legend:
Unacceptable
Not Tolerable
Tolerable with Mitigation
Tolerable
Acceptable

Collision Scenario:
A1: 30°, 60°, 90°
B1: 30°, 60°, 90°
A2: 30°, 60°, 90°
B2: 30°, 60°, 90°

Figure 7. Risk Matrix of Total Asset & Equipment Loss

3.8. Risk Control Options and Cost-Benefit Analysis

In determining the risk control option (RCO) starts from identifying the areas that require risk control the most. Areas that require control are those that include into the unacceptable, not tolerable, and tolerable with mitigation on risk matrix. Risk control is done to reduce the risk to the stage of tolerable. The category in the risk matrix is obtained by two variables, that is frequency (likelihood) and consequences, so it is necessary to control both variables to lower the risk level. Based on Guidelines for Formal Safety Assessment (FSO) For Use in The IMO Rule-Making Process [8], RCO is considered to evaluate in several area categories as follows: (1) Technical/engineering sub-system, (2) Working environment, (3) Personnel sub-system, and (4) Organizational/management sub-system.

3.8.1 Control options consideration. For this case of ship collision on FSO Abherka, there are some categories that van be considered as the control options of the risk. The consequences in losses and environment are very high, the mitigation would not be effective to mitigate the risk if it already happened. The effective mitigation is to prevent and reducing the frequency of ship collision on FSO Abherka. There are several options maybe consider for reducing the frequency of collision. To reduce the frequency there are several options can be implemented such as: Radar Beacon (Racon), Stand-by patrol vessel, Traffic Monitoring System, and emergency response plan training.

3.8.2 Cost-Benefit Analysis. Cost-Benefit Assessment (CBA) is to assess the total cost for every risk control options that already determined before. After calculating the total cost, then will be calculated the benefit in form of cost. So, there will be a difference between cost and benefit. From the differences will be chosen the most beneficial options to be recommended.

Table 12. Cost-Benefit analysis of control options.

Risk Control Option		Cost	Reduction
Radar Beacon	$	21.500,00	5 – 20%
Stand-by Guard Vessel	$	207.003,36	57%
Traffic Monitoring System	$	101.493,85	71 – 93%

3.9. Recommendation for Decision Making
Recommendations for decision making based on RCOs that have been spelled out before is implementing Traffic Monitoring System. Here are some of recommendation that can be implemented to reduce the risks of ship collision.

3.9.1 Traffic Monitoring System. After doing the cost-benefit analysis process by considering the cost of capital costs and the benefits in reducing risk. Considering the economic factors and traffic density of ships such as FSO Abherka, the application of traffic monitoring system becomes effective. The installation cost of the traffic monitoring system is $101,493.85. Where the installation cost is cheaper than having to add a fleet for stand-by guard vessels. And these costs can be used in the long term not only for one year but several years ahead with maintenance. Based on John Spouge (1999), traffic monitoring system can reduce for several condition as shows in the figure 8.

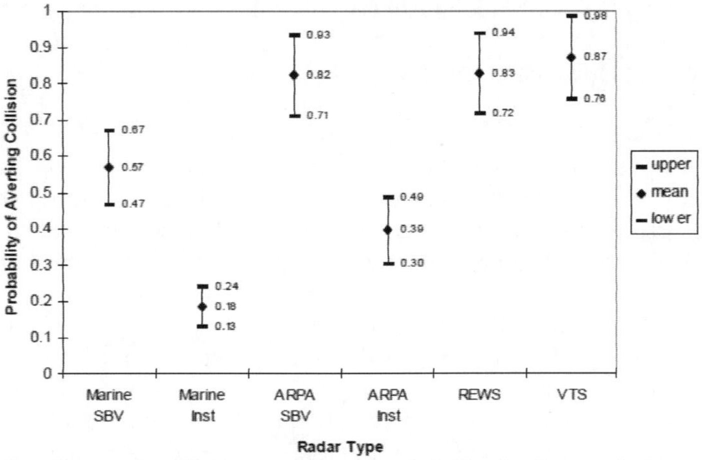

Figure 8. Effects of collision avoidance probability by Dovre Safetec (1999).

Traffic monitoring system can help to monitor all passing vessels that sail close to the area of FSO Abherka. If there is a ship that enters the area that has been set up, the system will give a notification signal to the ship to avoid the area and can also do physical preventing by sending a guard vessel that connect with the system to avert the ship. With the ability to reduce of ship collision, if it implemented to the frequency result can be seen the reduction as shown in the table 13.

Table13. Traffic Monitoring System reduction after implementation.

Scenario		Before Reduction		After Reduction	
		Frequency	Risk Level	Frequency	Risk Level
A1	*External Traffic Powered Collision*	4,20E-03	3 (Moderate)	2,94E-04	2 (Unlikely)
B1	*Internal Traffic Powered Collision*	1,28E-03	3 (Moderate)	8,96E-05	2 (Unlikely)
A2	*External Traffic Drifting Collision*	4,51E-03	3 (Moderate)	3,16E-04	2 (Unlikely)
B2	*Internal Traffic Drifting Collision*	4,01E-03	3 (Moderate)	2,81E-04	2 (Unlikely)

3.9.2 Mandatory Tugboat Assistance in Restricted Area. In the offshore area, it is necessary to apply mandatory assistance for each ship that will approach, especially for tankers that will load for crude oil. So, one of the recommendations given to mitigate the occurrence of ship collision on FSO Abherka is to apply mandatory assistance to all shuttle tanker activities that enter the restricted area. For every shuttle tanker that would hold loading activity should be assisted by tugboat to get into area of Poleng field and get out from Poleng field. This is to minimize the possibility incident while loading crude oil between shuttle tanker and FSO Abherka.

3.9.3 Emergency Response Plan (ERP). In overcoming the occurrence of incidents in FSO Abherka needs to be applied also mitigation efforts in the form of emergency response plan (ERP). An emergency response plan is a step to address an event through several standard activities in a fast and responsive time. ERP must exist in every condition that has a risk of a danger to anticipate safety hazards and widening losses. Implementing ERP in FSO Abherka becomes very important because it has a high risk of danger. The application needs to be accompanied by the readiness and ability of each crew to carry out established procedures to overcome every emergency condition.

4. Conclusion

Based on the result of this research can be obtained that the frequency of ship collision on FSO Abherka until the next 10 years (2023) has low frequency of collision. But, after analyzing the impact of the collision is resulting great impact in asset & equipment loss also the environment. The highest losses that can be resulted due to the collision is for about $23.921.683,88. For the environmental impact there will be possibly 44.560 barrels of crude oil spill to the sea. Both of the consequences of collision are included to the risk category of level 5 or catastrophic and needs to be mitigated immediately. However, due to the high impact of the consequences, it will not be effective to mitigate in the consequences of collision. It would be more effective to mitigate the possibility of collision that resulting great impact of consequences. By applying traffic monitoring system could be possibly to reduce for about 71%-93% of collision with several integrated system. Also implementing mandatory tugboat assistance inside the area of FSO Abherka can prevent the collision.

Acknowledgments

All gratitude and thanks to God or the grace that allows me to finish this research. Gratitude for my parents, supervisor, mentors, friends, and everyone that I can not to mention it who always supporting and sending love during the completion of this research. Hopefully with the completion of this research, author can gain more knowledge, understanding, and perspective that can be useful for the future career.

References

[1] J. Spouge, A Guide To Quantitative Risk Assessment fro Offshore Installations, DNV Technica, 1999.

[2] Pertamina, Pedoman Pengelolaan HSSE Berbasis Risiko, Pertamina HSE Management System, 2019.

[3] D. W. Handani, T. Pitana, A. A. B. Dinariyana and R. L. Pratama, "Risk assessment of ships collision during installation of anode protection on subsea gas pipeline located in Surabaya West Access Channel," *Maritime Safety International Conference,* 2004.

[4] F. Farhansyah, Y. Mulyadi and A. B. Dinariyana, "Penilaian Risiko Kuantitatif Tubrukan Kapal dengan Platform: Studi Kasus Tubrukan Kapal dengan Wellehead Platform PHE-12," *Jurnal Teknik ITS Vol. 5,* 2016.

[5] A. A. B. Dinariyana, M. A. Avesina, K. Sambhodo and F. Fahrirozan, "Collision Risk Analysis and Oil Spill Dispersion Modelling in," *Maritime Safety International Conference,* 2020.

[6] J. Sutrisno, A. Dinariyana and D. Priyanta, "RISK ASSESSMENT OF SHIP COLLISION AND GROUNDING IN SURABAYA WEST ACCESS CHANNEL DUE TO THE EXISTENCE OF SHIPWRECKS," *Jurnal IPTEK ITS,* 2018.

[7] Ö. Uğurlu, U. Yildirim, E. Köse and E. Yuksekyildiz, "Marine accident analysis for collision and grounding in oil tanker using FTA method," *Maritime Policy & Management,* 2013.

[8] IMO, GUIDELINES FOR FORMAL SAFETY ASSESSMENT (FSA), International Maritime Organization, 2002.

Feasibility Study of 7.8 m Fiberglass Boat Using Longitudinal Hollow Steel on Sagging and Hogging Condition

Hartono Yudo [1], Ocid Mursid [1)*], Wilma Amiruddin [1], Imam Pujo Mulyatno[1]

[1]Department of Naval Architecture, Diponegoro University, Semarang, Indonesia

Corresponding author : ocidmursid@gmail.com

Abstract. The longitudinal structure is needed to hold the boat, especially in sagging and hogging conditions. Modifying a longitudinal system using laminating steel and fiber is required to increase strength and reduce weight. This study uses a 7.8 m ship modified from conventional to hybrid construction. The finite element method compared the conventional and hybrid ship structures under sagging and hogging conditions to evaluate ship strength. In sagging conditions, the maximum stress in conventional boats is 16.68 MPa, reduced up to 13.65 MPa in hybrid boats. In hogging condition indicate that conventional has maximum stress is 11.18, which is higher than in conventional structure, as much as 10.38 MPa. The keel stress distribution in a couple of conditions of sagging and hogging indicates that the hybrid structure has lower stress than conventional vessels. The simulation shows that a longitudinal hybrid frame makes the ship stiffer in sagging and hogging conditions. The weight of conventional and hybrid boats is similar and affordable to more developing hybrid ships.

1. Introduction

Figure 1 illustrates the ship under sagging and hogging conditions. Sagging and hogging is a common condition to evaluate ship longitudinal strength. Ship stiffener designs to hold shipload on the ship structure consisting of the longitudinal and transversal frame. The design of longitudinal stiffener is to hold longitudinal strength caused by sagging and hogging conditions.

Fiber glass ships has lower ship resistance which compare to wooden boat [1]. Modification of the longitudinal frame is the most effective modifications of the longitudinal strength of boats is using hybrid materials system. Hybrid hull material combines two or more materials of different mechanical properties hybrid steel and fiber are used in this study. Additional materials and configurations are combined to reach stronger and lighter ships.

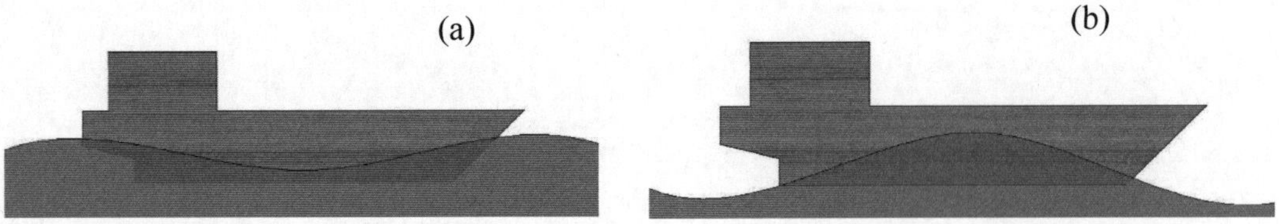

Figure 1. Ship Sagging (a) and Hogging (b) Condition

Replacing the frame with steel construction on a 35 GT fishing wooden boat reduces deflection up to 35%, making the ship hull more stiffened [2]. Steel construction has higher mechanical strength that makes the hull stiffened. Combining steel and the composite hull are used in 50 meters of the ship, the results show reducing the overall failure mode from 4,09 to 3.79 [3]. In other research, covering steel truss on fiberglass in ship stiffener suggests making the ship stronger and lighter [4]. Application sandwich steel and FRPU (fiberglass reinforced polyurethane elastomer) show lighter and stronger boats than conventional ships [5]. Another research Finite element method (FEM) used to calculate the strength of composite hatch cover indicates reducing 44.32% of the weight and reduced maintenance cost [6]. In 38.5 m of yacht super structure indicate that composite material fiberglass and steel has lighter of the ship weight than aluminium [7].

This study aims to predict the effect of modifying longitudinal fiber stiffener to steel frame laminate. This study compares the sagging and hogging conditions between the conventional and hybrid structures of 7.8 m boats affecting strength and weight. This study's benefit is to gift an alternative design for building stronger ships in traditional shipyards which increase life time of the boat.

2. Methods

Table 1. Ship Main Dimension

Description	**value**	**unit**
Length Over all	7.8	m
Breadth moulded	1.85	m
Draft Design	0.3	m

Table 1 shows the primary dimension of the 7.8 m boat made in the traditional boat industry. Reverse engineering is used to model ship construction based on fabricated ships. In the modeling and fabrication of vessels, both conventional and hybrid boats use identical hull shapes and structures.

Figure 2. Conventional Boat Construction

IOP Conf. Series: Earth and Environmental Science 1081 (2022) 012030 doi:10.1088/1755-1315/1081/1/012030

Figure 2 shows the conventional structure of the boat, which builds in a traditional shipyard using fiberglass material only. Conventional boat is the most common type of structure in the fiberglass ship. The longitudinal and transversal frame uses a C angle.

Figure 3. Hybrid Boat Construction

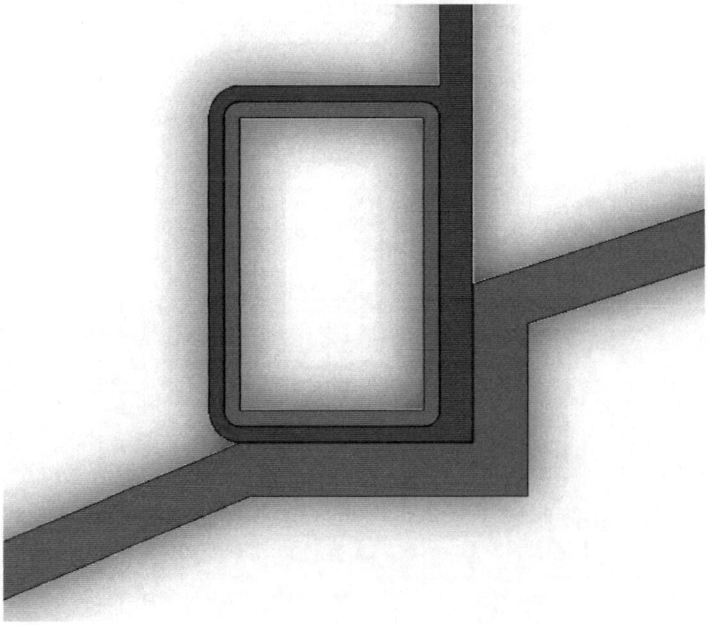

Figure 4. Hollow Laminate Detail

Figure 3 shows the developed ship, which has hybrid stiffened. Details of the developed stiffener are shown in figure 4, the hollow steel (grey color) covered by fiberglass (blue). Covering of hollow steel is needed to bind hollow steel to the ship's bottom plate (orange color). Transversal frame space is equal between conventional and hybrid systems.

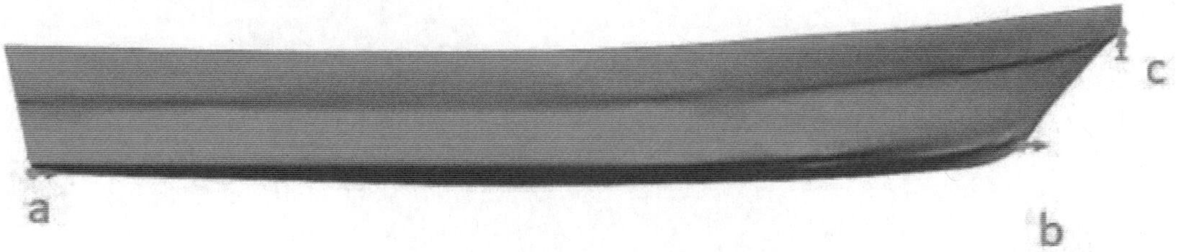

Figure 5. Simulation Constraint

Figure 5 shows simulation constraints in this study which on three nodal are a constraint. The coordinate system in this study consists of the X-axis being forward of the ship, the Y-axis being starboard, and the Z-axis being upward of the boat. Detail constraint of each node describes in table 2 according to DNV GL [8] .

Table 2. Displacement Detail

Nodal	**Direction of support**
a	X, Y, Z
b	Y
c	Y, Z

On sagging and hogging conditions, the wavelength in this study is 7.8 m, equal to ship length, and wave height is 0.39 m (similar to L/20). Load in this study consists of LWT, DWT, and buoyancy load. The buoyancy load follows the wave pattern for sagging and hogging conditions.

Table 3. Material Properties

Parameter	**Unit**	**Fiber Laminate**		**ASTM A36**
		Woven roving	Chopped strand mats	
Density	kg/m^3	1613	1428	7850
Modulus Elastisitas	MPa	15101	8316	200000
Poison ratio	-	0.11	0.3	0.29
Tensile	MPa	217	103	250

Table 2 shows the material properties of this study. Some laminate woven roving and chopped strand mats are used to build the 7.8 m, while hollow steel is inserted into a hybrid frame.

3. Result and Discussion

In this study, von misses criteria evaluate the ship's strength. The hybrid hollow and the rest of the ship structure describe the hybrid ship's strength in a separable figure.

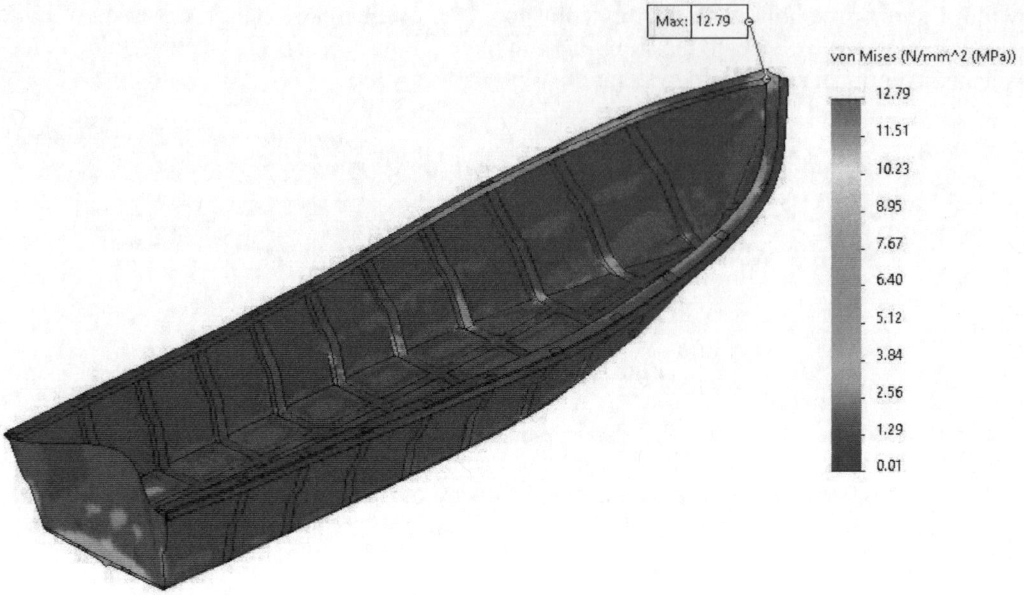

Figure 6. Stress of Conventional Boat on Sagging Condition

Figure 6 and 7 shows the comparison of stress on the ship in conventional and hybrid in sagging condition. In conventional ships, sagging conditions show that maximum stress is 12.79 MPa. The maximum stress on the conventional structure is near the "C" constraint, which indicates that the upper hull holds hull pressure on the hogging condition.

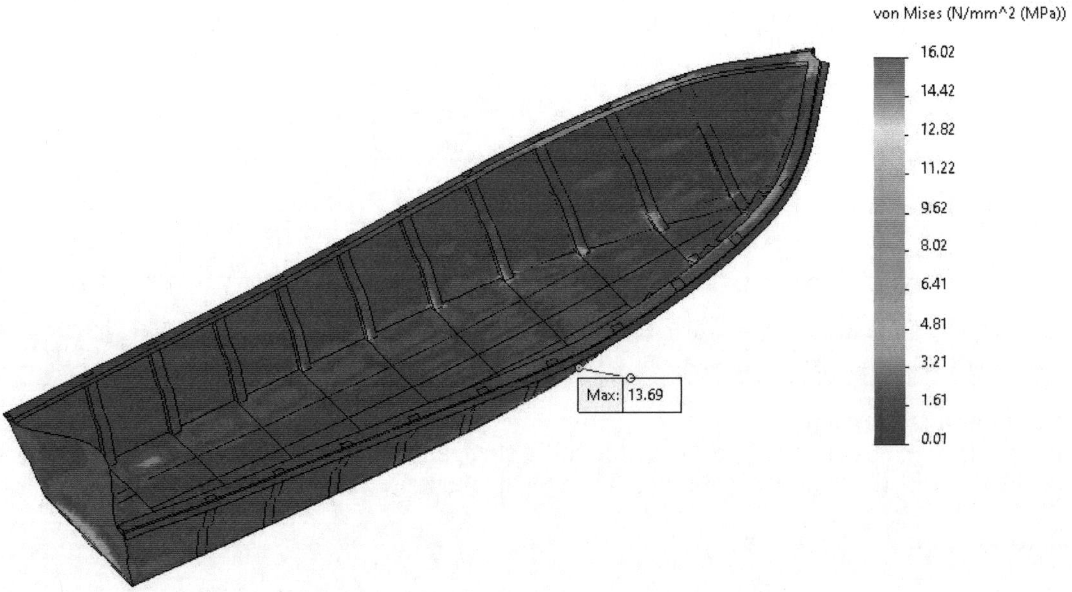

Figure 7. Stress of Hybrid Boat on Sagging Condition

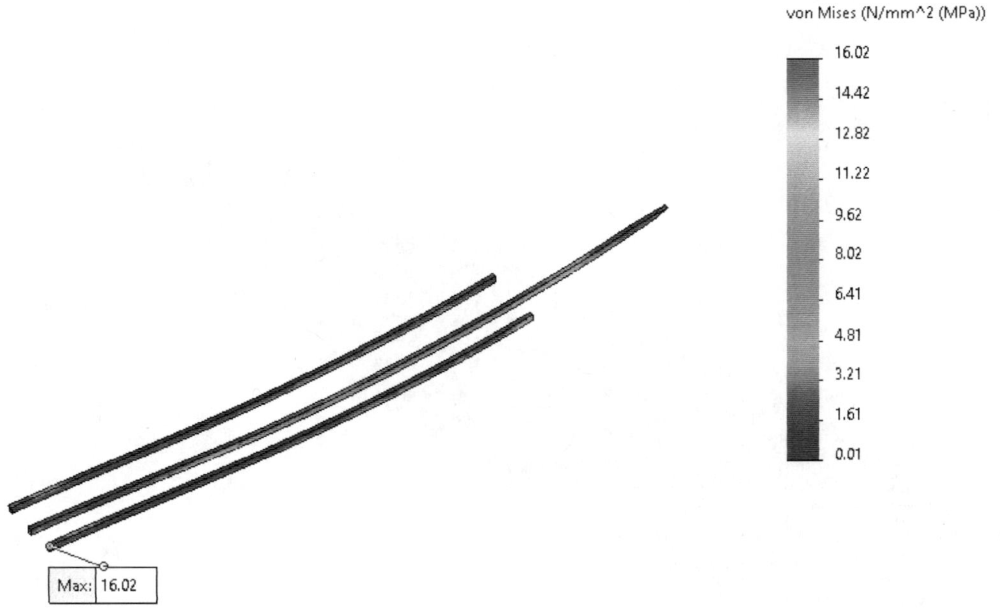

Figure 8. Stress of Hollow in Sagging Condition

Figure 8 shows maximum stress on the hollow hybrid structure under sagging conditions. It shows that the hollow holds a higher load in the hybrid structure than in other fiber construction. The hollow structure near to midship deforms downward caused by loading distribution in the sagging condition. The maximum stress of the hollow hybrid frame is 16.02 MPa.

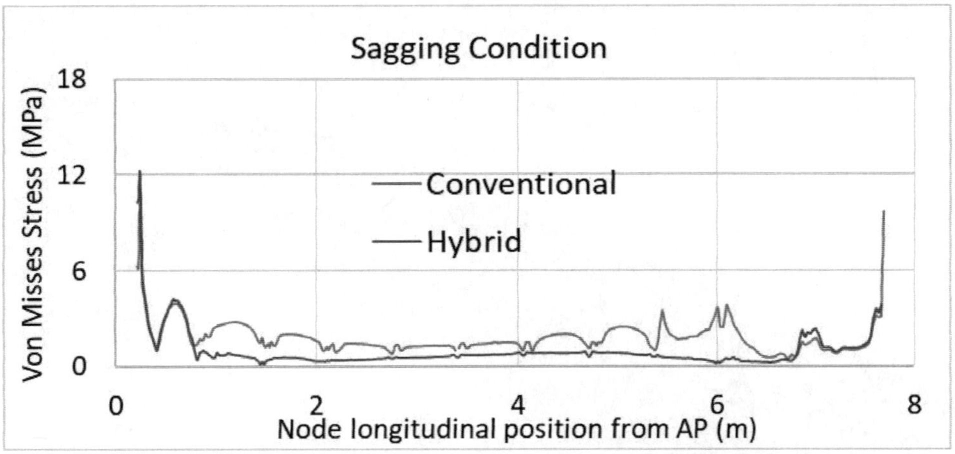

Figure 9. Keel Stress Distribution in Sagging Condition

Stress on the ship's keel is needed to show the comparison of the longitudinal frame to the ship's longitudinal strength. A longitudinal stiffener structure is designed to reduce the stress on the keel of the boat; therefore, the ship can hold more payload. The line graph in figure 9 shows a comparison of stress on the keel ship to hold the load in sagging conditions. The hybrid ship has lower stress than conventional in each position, and it shows that the steel structure makes the ship stiffer. The stress peak is located near the A and C nodes where its nodes are constrained in the z-axis. The trough on the graphic figure indicates transversal frame location.

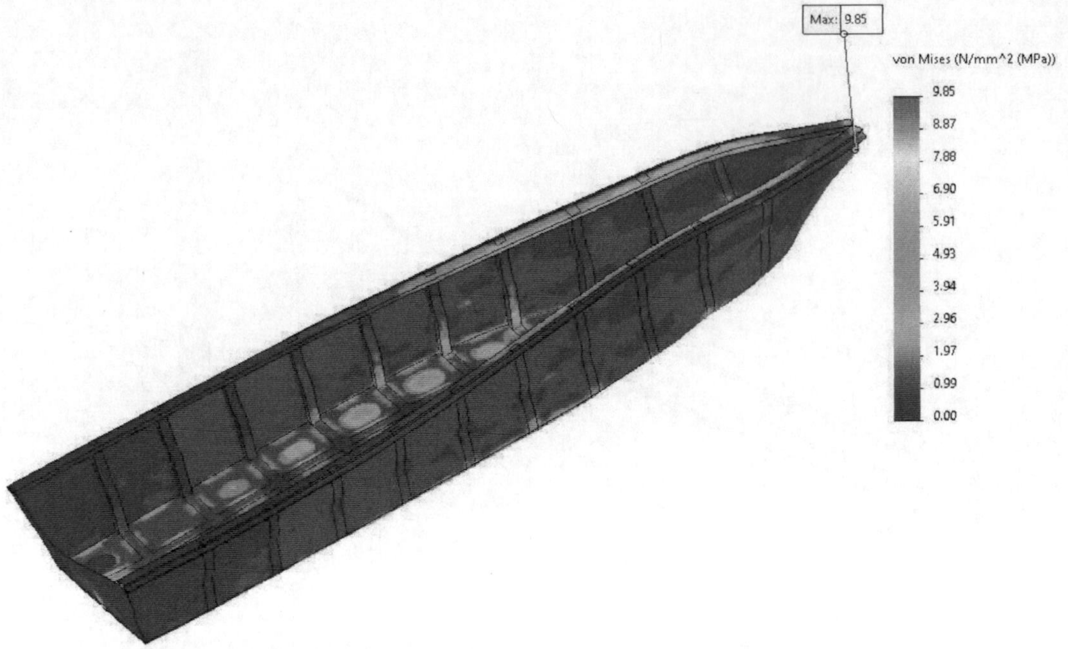

Figure 10. Stress of Conventional Boat on Hogging Condition

In Figures 10 and 11, a couple of conventional and hybrid boat stress under hogging conditions is shown in Figures 10 and figure 11. Both hogging and sagging load have an identical location of the maximum stress. The maximum stress indicates a critical point of the ship structure. The maximum stress of conventional and hybrid structure are similar.

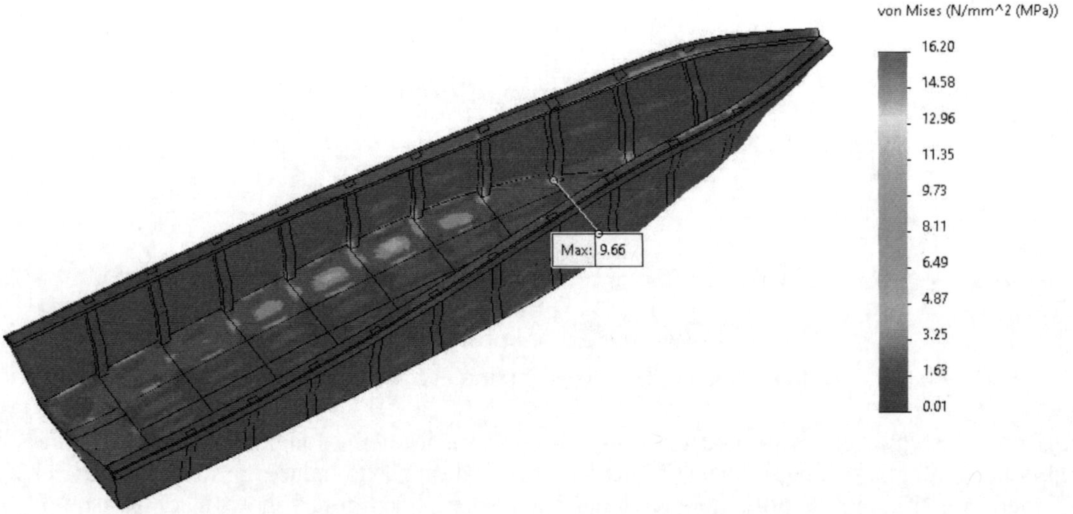

Figure 11. Stress of Hybrid Boat on Hogging Condition

Figure 12 shows distribution stress on the hollow hybrid structure. The location of maximum stress in the hogging condition is in the same location as the sagging condition. The wave on hogging condition makes the mid of hollow deform upward.

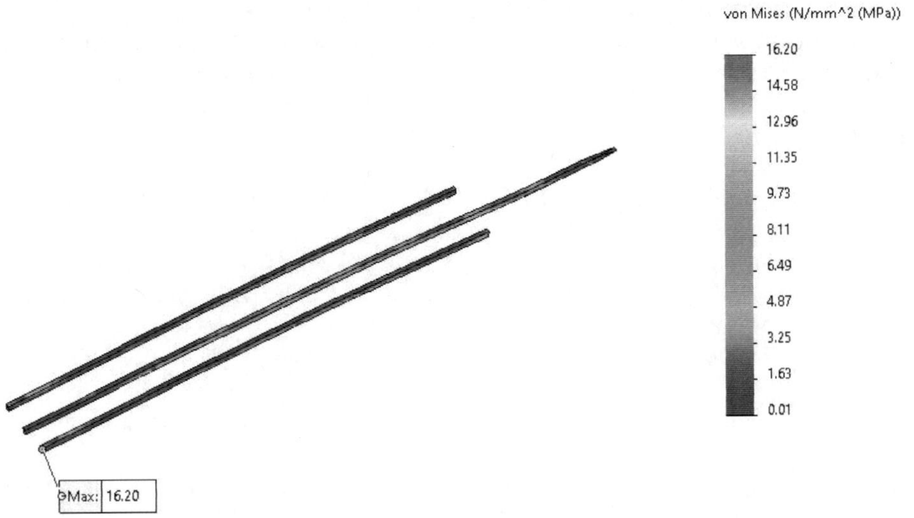

Figure 12. Stress of Hollow in Hogging Condition

Figure 13 show the comparison of stress distribution on the keel of the conventional and hybrid boat. The trend stress in the hybrid boat is lower than in a conventional boat. The sagging and hogging conditions show a similar trend line, although the hogging shows wavier. In the hogging condition, the amplitude of the stress distribution on the midship is higher than the sagging load, which causes in hogging condition the peak of the wave load located in the midship.

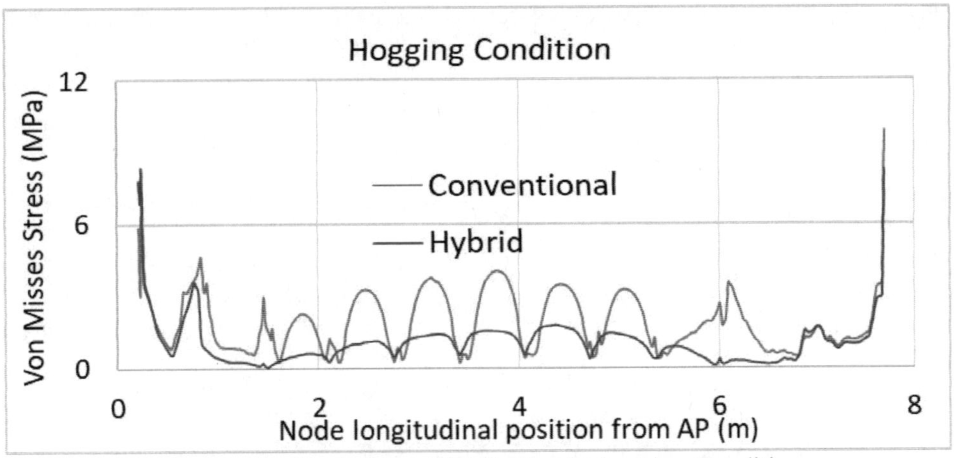

Figure 13. Keel Stress Distribution in Sagging Condition

The weight of the hull structure on the conventional is 417 kg, and the hybrid condition is 456 kg. The conventional and hybrid boat's weight is identical but, on another side, increases the ship's longitudinal strength. The benefit of the laminating longitudinal is increasing longitudinal strength significantly; otherwise, weight is similar.

4. Conclusion
Both sagging and hogging conditions indicate that hybrid structure is stronger than conventional. Simulation shows that conventional structure is needed to develop into a hybrid structure to reduce stress on ship construction. The maximum stress on conventional and hybrid in sagging and hogging conditions is under tensile stress of chopped strand mat (103 MPa), which indicates that the ship is eligible to voyage in the sagging and hogging condition in this research.
The weight in conventional and hybrid is similar. It indicates that a hybrid structure is appropriate to apply. Development of hybrid ship is needed in deeper aspect to reach extra benefit.

References

[1] Utama D, Nasirudin A, Iqbal M. *Design of River Tour Boat's Hull For Taman Nasional Tanjung Puting,* Central Borneo. Kapal J Ilmu Pengetah dan Teknol Kelaut. 17(1):28–39.
[2] Bochary L, Firmansyah MR, Sitepu G, Asri S. *A Study for the Application of Steel Frames on a Traditional Wooden Fishing Boat.* In: IOP Conference Series: Materials Science and Engineering. IOP Publishing; 2019. p. 12016.
[3] Mushtaq SM. *Optimized design and reliability of a hybrid ship hull.* University of California, Berkeley; 2007.
[4] Cao J, Grenestedt JL, Maroun WJ. Testing and analysis of a 6-m steel truss/composite skin hybrid ship hull model. Mar Struct. 2006;19(1):23–32.
[5] Ismail A, Zubaydi A, Piscesa B, Tuswan T, Chandra AR. *Study of sandwich panel application on side hull of crude oil tanker.* J Appl Eng Sci. 2021;19(4):1090–8.
[6] Tawfik BE, Leheta H, Elhewy A, Elsayed T. *Weight reduction and strengthening of marine hatch covers by using composite materials.* Int J Nav Archit Ocean Eng. 2017;9(2):185–98.
[7] Dominguez Ruiz F, Carral Couce LM. *Superstructure design: combination of fiberglass panel and tubular structure with naval steel hull.* In: Pan-American Conference of Naval Engineering. Springer; 2017. p. 81–91.
[8] DNVGL-CG-0127. Class guideline—finite element analysis. 2016;

Acknowledgements
We Ackowledge to DRPM of Kemenristek / BRIN to founding this research We Acknowledge to DRPM Kemenristek / BRIN to support this research.

Stability Assessments of RoPax Open Car Deck on Longitudinal Wave

Hasanudin Hasanudin [1], Achmad Zubaydi[1], Wasis Dwi Aryawan[1]

[1] **Departement of Naval Architecture, Faculty of Marine Technology, ITS**

Abstract. This paper discusses RoPax Open Car Deck (OCD) without side casings (wing tanks) on longitudinal wave water. The purpose is to know the critical intact stability of RoPax OCD on the longitudinal wave so that the other ship can avoid the condition. A case study takes a RoPax OCD, which capsized on Halmahera seawater. The stability calculation uses the energy balance method on calm water and longitudinal waves with various amplitude, phases, and lengths of waves. Survival stability assessments use the lever righting arm (GZ) of intact stability criteria IMO (IS Code 2008). Results show RoPax OCD full load on calm water has failed to fulfil several of the stability criteria of IS Code 2008, especially on area 30°-40° of GZ curve 1.221 meter-degree ≤ 1.7189 meter-degree, and GZ maximum 15° ≤ 25°. Hogging and sagging waves reduce the GZ height compared to calm water, but the length of the GZ range on calm water increases 0°~29° increase becomes 0°~32° on the sagging wave and decreases 0°~13° on the hogging wave. With varying wave phases 0-0.8, the highest GZ curve occurs in phase 0.2, and the lowest GZ curve occurs in phase 0.6. With varying wavelength, 10-50 meter, shows highest GZ curve occur on wavelength 10 m, and the lowest GZ curve occurs on wavelength 40 m, the same as the ship length (Lpp). All GZ stability curves have positive values, but the longitudinal wave reduces the GZ heigh and causes the water to ingress onto the car deck. Hence, the effect on the water-free surface influences ship capsizes. The authors suggest that RoPax with opened car deck without side casings should not be operated in the open sea or the strait bordered by the open sea to avoid the same accident.

1. Introduction

Marine transportation covers 90% of world trade compared to land and air transportation. RoPax (Roll-on Roll-off Passenger) ship represents marine transportation that can carry multipurpose [1]. The ship could simultaneously load various cargos, cars, and passengers to load and unload the ship using a ramp door installed in front of the stem, behind the stern, and inside the hull. RoPax is becoming popular globally because of its flexibility in carrying multipurpose [2]. The ship is usually operated in the rivers, lakes, straits, and sea-spread islands. Most RaPax is used in developing archipelagos and tropical countries with fewer land infrastructures [3], [4].

A single cause rarely causes ship accidents but multi-combination (human error, accident nature/damage, natural hazards, negligence / poor maintenance, failure to provide service, and others) [5]. Data from 2011- 2015 increased significant ship accidents from 1750 cases to 3300 cases; the trend slightly decreased in 2018 became 2200 cases [6]. Every year around 2000, causality happens on the passenger ship, 91% causality contributing on Ferry / RoPax. More ironic, 76% of RoPax accidents and 90% of causalities have happened in tropical developing countries like Bangladesh, the Philippines, and

Indonesia [4]. Cargo ship has the highest rank of ship accident, but RoPax has the highest rank of causalities. Records of ship accident cause in Indonesia water of National Transportation Safety Committee (NSC): 35% fire, 31% collision, 24% sink, 6% grounded, and 4% others [7]. The data indirectly reveal that stability becomes a severe problem that needs solving to minimize the risk of ship accidents.

This paper discusses RoPax OCD, which sunk on Halmahera water Indonesia, the ship's poor survival intact stability, and a longitudinal wave's reducing lever arm stability GZ effect. This incident was experienced by RoPAx OCD when exposed to a 4-meter height wave with an astern direction [8]. Wave influences buoyancy distribution on underwater ship hull and reduces high lever stability arm GZ. This paper is expected to get a critical wave that needs to be avoided of existing RoPax OCD with the same type operating today and improvement in RoPax OCD design.

2. Effect of Wave on Stability

When ships operate on the sea often face waves with astern, quarter, beam, bow, and ahead seas directions. The ship can face on wave crest (the maximum value or upward displacement within a cycle) or wave trough (the opposite of a crest, so the minimum or lowest point in a cycle). The wave changes the distribution of buoyancy underwater of the ship's hull and influences the righting lever arm stability GZ. International Maritime Organization (IMO) is developing a realistic criterion to measure ship safety on the wave using dynamic motion that is called Second Generation Intact Stability Code (SGIS) [9]. New criteria of intact dynamic stability are developed for some capsized ships in the wave even though the ship has fulfilled the criteria of static intact stability (IS Code 2008) [10]. The development of computer technology can calculate SGIS quickly, complicatedly, and complexly. A vulnerability assessment of dynamic wave stability use: parametric roll, pure loss of stability, and surf-riding / broaching. The assessment use three-tier, the first simple vulnerability, the second medium vulnerability, and the third complex vulnerability [11].

Longitudinal wave / pure loss stability is part of SGIS analyses that are very important and should be considered. Pure loss stability use wave exposes with astern direction. A ship that floats on the water follows the equilibrium force and moment law of weight (W) and buoyancy (B). The value of buoyancy depends on the float pressure distribution of water. Wave also influences the distribution and ship stability. Left Figure 1.a, the white line shows the ship's water line on calm water while the shipping's yellow line shows the line on wave trough. The wave makes the water plan area wider than calm water or a more comprehensive blue line than red. The wave trough also changes the righting lever arm stability GZ on calm water or the dast rad line lesser than the straight blue line. Otherwise, Figure 1.b on the right-side white line shows the ship's water line on calm water while the ship's yellow line shows the line on the wave crest. The wave changes the waterplane area smaller than calm water or the blue line more minor than the red line. The wave crest also changes the righting lever arm stability GZ on calm water to become higher or dast rad line higher than the straight blue line.

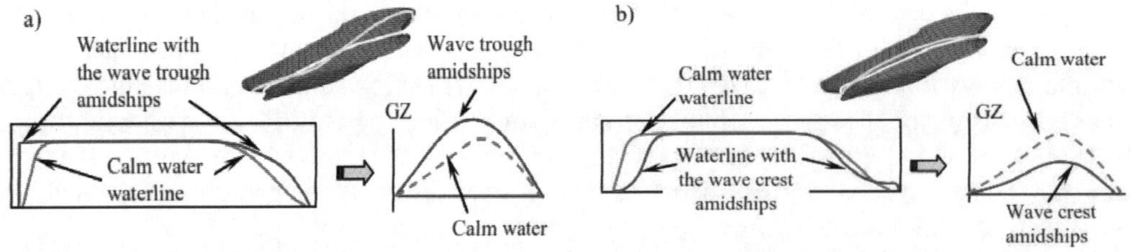

Figure 1. Longitudinal Wave / Pure Loss of Stability [10] [11]

Wave in the sea happens randomly with various waves: wave phase, wavelength, and height. Wave profiles depend on gravity that pulls wave mass down, sea topography, and wind blowing on the sea. Although the wave has a random profile but has a specific formula ratio. US Navy can calculate the formula wave height and wavelength as follows:

IOP Conf. Series: Earth and Environmental Science 1081 (2022) 012031 doi:10.1088/1755-1315/1081/1/012031

$$Wave\ height = 0.607\sqrt{Wave\ length}\ [m] \quad [1]$$

Wave happens on the surface sea that influences ship hull pressure distribution. It increases or decreases the height of center gravity to the metacenter (GM) and lever righting stability (GZ). The effect of wave crest and wave trough on righting arm stability (GZ) is formulated as follows [12]:

$$GZ(\varphi, t) = GM_0\left(\frac{GM_{crest} + GM_{trough}}{2 \cdot GM_0} + \frac{GM_{crest} - GM_{trough}}{2 \cdot GM_0}\right)\varphi \quad [2]$$

3. Case Study

This paper used a case study RoPax that sunk on the Halmahera water effect hit with waves with astern direction on 18 August 2018 [8]. First, the ship departure from Tobelo-Ternate to Bitung-North Sulawesi Utara Sea in **Figure 2** below. The ship at Laloda Water hit with 4-meters wave height on head seas direction. After that captain decides to return to Tabelo Port; in the way the ship sunk on the water with astern longitudinal wave direction, the incident very likely involves pure loss stability / longitudinal sagging and hogging waves to reduce writhing arm stability (GZ). After that, the water enters the car deck, influences the free surface, capsizes, and the ship sinks to undersea bad.

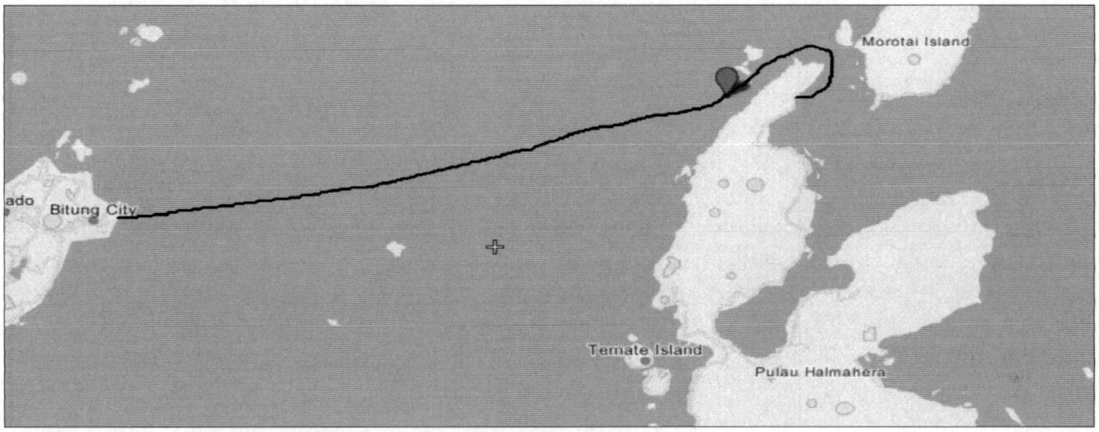

Figure 2. Route operation and location RoPax accident case study

RoPax that sunk is the oldest that operating at Nort Maluku Sea. The RoPax OCD before operated at Bastion-Sidangoli-Bastion-Sofifi and then switched Tobelo-North Halmahera Utara to Bitung-Nort Sulawesi that route further and waved higher. At the incident, RoPax carries 12-unit vehicles consisting of 8 medium cars and four trucks that sunk together with the ship. RaPax carried 34 passengers and 17 crews, and five passengers were missing. Main dimension RoPax that use case study show **Table 1** as follows:

Table 1. Main dimensions of case study RoPax OCD

Item	Symbol	Unit	Ship Dimensions
Length Overall	Loa	meter	45.00
Length Per Perpendicular	LPP	meter	40.00
Breadth Moulded	B	meter	11.0
Height Moulded	H	meter	3.20
Summer Draft	T	meter	2.2
Freeboard	Fb	millimetre	1000
Gross Tonnage	GT	Tonnage	401
Net Tonnage	NT	Tonnage	120
Machinery		HP	2 x 650

The side view of the RoPax general arrangement becomes a case study shown in **Figure 3** left below. The RoPax has a maximum capacity of nine trucks, four rows, two columns at three rows three columns. The RoPax is an open car deck (OCD) with small partial side casings. The purpose reduces GT is to reduce taxes and reduce safety regulations. The left and right-side hull have a wide window behind the car deck without a rooftop. The ship only has some small partial side casings at the stern and stem to install a mooring system. There are expansive windows so small side casings can be obeyed or the ship without side casings, so the body plan models of RoPax OCD are as follows in **Figure 3** right below.

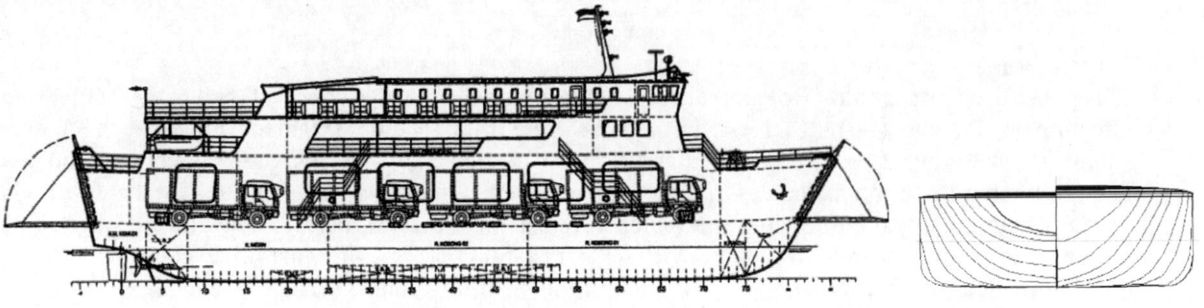

Figure 3. Side view and hull form without side casings of RoPax OCD case study

4. . Discussions

4.1. Ship Stability Analysis at Calm Water Assessed with IS Code 2008

Before calculating stability, the ship was modelled: the hull, compartments, tanks, and load case (weight and center gravity). All data to modelling and load case can be got from stability booklet and loading condition report before ship accident with source data from NSC. In this case, stability calculation assumes the full fuel oil, fresh water, and lubrication oil because the ship departed recently. Result stability calculation of the departure condition can be shown as a curve in left **Figure 4** and numerical in the right side **Figure 4** below.

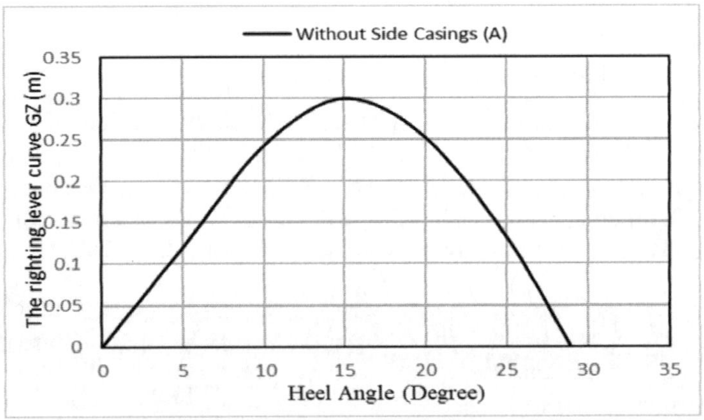

Heel Angle (degree)	GZ (meter) Calm Watear
0	0
5	0.12
10	0.243
15	0.299
20	0.252
25	0.132
30	-0.039
35	-0.245
40	-0.471
45	-0.707
50	-0.946
55	-1.184
60	-1.416
70	-1.641
75	-1.856

Figure 4. The righting lever curve GZ Stability of RoPax at full load departure condition on calm water

Figure 4 at the right side above the x-axis is heel angle, and on the y-axis, the righting lever curve GZ stability of RoPax OCD on calm water with full load departure condition. A table on the left shows the angled heel with result calculation of GZ at calm water with heel angle range 0°-75° and heel interval 5°. A curve has a positive range heel at the angle of 0 degrees until 29 degrees. The maximum stability lever of GZ on calm water occurs at 0.299 meters and has a heel angle of 15°. Although the righting arm

IOP Conf. Series: Earth and Environmental Science 1081 (2022) 012031 doi:10.1088/1755-1315/1081/1/012031

positive is needed to assess with intact stability ISC 2008 to know whether the ship is safe or unsafe. The result comparison of RoPax stability values with ISC 2008 is shown in **Table 2** as follows:

Table 2. Stability assessment of RoPax OCD based on IMO's criteria

IMO's criteria	Value of IMO's Criteria	Unit	Value of GZ on calm water		Comparison with IMO's standard
Area 0° to 30°	3.1513	meter-degree	8.8637	Pass	181.3%
Area 0° to 40°	5.1566	meter-degree	10.0847	Pass	95.6%
Area 30° to 40°	1.7189	meter-degree	1.221	Fail	-29.0%
Max GZ at 30°	0.2	meter	-0.039	Fail	-47.0%
The angle of maximum GZ	25	degree	20.9	Fail	-16.4%
Initial GMt	0.15	meter	1.358	Pass	805.3%
Passenger crowding	10	degree	0.9	Pass	1111%
Turn: angle of equilibrium	10	degree	1.7	Pass	588%

Table 2 above shows: that the first column is the number item IMO's criteria of Intact Stability Code 2008; the second column is the limit value of intact stability criteria; the third column is unit; the fourth column is the value of RoPax stability on calm water (with its status pass or fail), and the end is comparison percentage between ship stability criteria with and actual ship stability. Result calculation RoPax OCDR fails at three criteria to satisfy ISC 2008: area curve stability GZ with heel angle 30°~40° is 1.221 meters-degree less than 1.7189 IMO's IS Code 2008 criteria; The maximum GZ curve at 30° degree is -0.039 meters less than 0.2 meters IMO's IS Code 2008 criteria; the maximum curve stability GZ is 20.9 degrees less than 25 degrees IMO's IS Code 2008 criteria.

4.2. Ship Stability Assessment at Hogging Wave and Sagging Wave with Variation Wave Along with Ship

RoPax OCD accident was exposed with the longitudinal waves. There are two conventional extreme longitudinal waves: hogging and sagging. The wavelength is the same as the ship's waterline length Lw=40 meters, and waves high h_w = 4 meters. Wave profile used sinusoidal that show **Figure 5** as following:

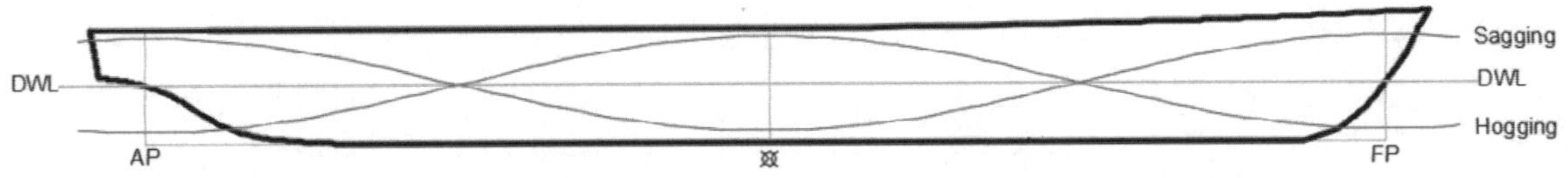

Figure 5. RoPax side profile on calm water, a hogging wave, and a sagging wave.

The wave profiles above were used to calculate the lever-righting arm stability (GZ), and the stability value would be compared. Green-line is calm water, the red line is a sagging wave, and the magenta line is a hogging wave. The result of comparing lever arm stability (GZ) of calm water, hogging, and sagging can be shown in left **Figure 6,** and the numeric result on the right is as follows:

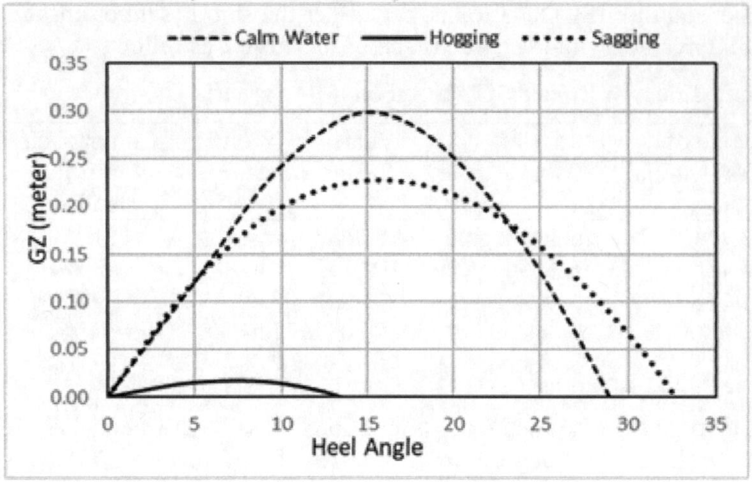

Heel Angle (degree)	GZ (meter) Hogging	Sagging
0	0.00	0.00
5	0.02	0.12
10	0.01	0.20
15	-0.01	0.23
20	-0.06	0.21
25	-0.13	0.16
30	-0.22	0.07
35	-0.34	-0.06
40	-0.48	-0.24
45	-0.64	-0.43
50	-0.82	-0.64
55	-1.02	-0.87
60	-1.23	-1.09
65	-1.44	-1.31
70	-1.64	-1.53

Figure 6. Comparison of stability righting arm GZ on calm water, sagging wave, and hogging wave of RoPax OCD. Figure 6 shows the comparison of the stability righting lever GZ on calm water (straight line), sagging (dot line), and hogging (step line), in which all the stability righting lever GZ is positive. The lowest curve of the stability righting lever GZ happens on the hogging wave. The maximum stability righting lever GZ has a value of 0.299 meters in calm water and becomes 0.23 meters at the sagging wave and 0.01 meters at the hogging wave. The positive range of the stability righting lever GZ is 0°~29° at calm water, becomes 0°~32° at the sagging wave, and 0°~13° at the hogging wave. The trend of the stability righting lever GZ in Figure 6 differs from **Figure 1** because ship dimensions and hull form characteristics are different on sagging waves, and sagging waves influence the stability righting lever GZ.

4.3. Ship Stability Assessment at Hogging Wave and Sagging Wave with Variation Phase Wave

Wave on the sea continuously oscillates and translates, generating different phases. The varying phase impacts the buoyancy distribution and influences righting stability lever arm GZ. This sub-chapter discusses the analysis of a RoPax OCD with a 4-meter wavelength, the same with accident incident, phase variation use: phase 0, phase 0.3, phase 0.4, phase 0.6, phase 0.8, and phase 1. Longitudinal wave profile with varying phases with ship hull can be shown in **Figure 7** as follows:

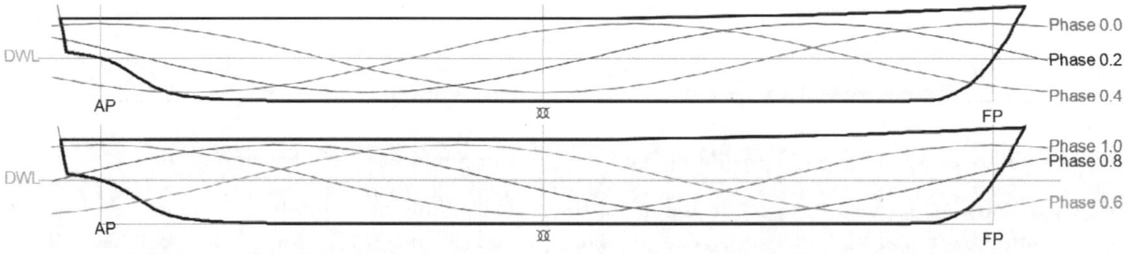

Figure 7. Shipside view with calm water and wave water with phase variation

Figure 7 above shows waves profile with different phase variations. Wave phase variation is created with interval phase 0.2 and range start phase 0 until phase 0.8. To show the wave pattern clearly, divide two sections: phase 0-0.4 **Figure 7** above and phase 0.6-1.0 **Figure 7** below. **Figure 7** above red colour wave shows phase 0, black wave colour shows phase 0.2, and magenta wave colour shows phase 0.4. While **Figure 7** below, the green wave colour shows phase 0.6, the blue wave colour shows phase 0.8, and the red wave colour shows phase 1.0. Results wave phase calculation of righting lever arm stability GZ shown in Table 3 as follows:

Table 3. Lever arm stability GZ with varying wave phase

Angle (deg.)	Phase 0	Phase 0.2	GZ (m) Phase 0.4	Phase 0.6	Phase 0.8
0	0.00	0.00	0.00	0	0
5	0.12	0.12	0.08	0.049	0.135
10	0.20	0.23	0.12	0.064	0.242
15	0.23	0.32	0.13	0.052	0.297
20	0.21	0.34	0.11	0.014	0.295
25	0.16	0.30	0.06	-0.051	0.244
30	0.07	0.19	-0.03	-0.143	0.143
35	-0.06	0.03	-0.17	-0.263	-0.007
40	-0.24	-0.16	-0.34	-0.411	-0.192
45	-0.43	-0.38	-0.53	-0.584	-0.399
50	-0.64	-0.61	-0.73	-0.777	-0.621
55	-0.87	-0.84	-0.95	-0.983	-0.85
60	-1.09	-1.07	-1.17	-1.196	-1.082
65	-1.31	-1.31	-1.39	-1.41	-1.312
70	-1.53	-1.53	-1.61	-1.622	-1.537

Table 3 result calculations lever righting stability (GZ) with varying phase 0-0.8 with interval 0.2 and with varying heel angle 0° ~ 70° with interval 5°. The calculation also shows in figure lever righting stability GZ `**Figure 8** as follows:

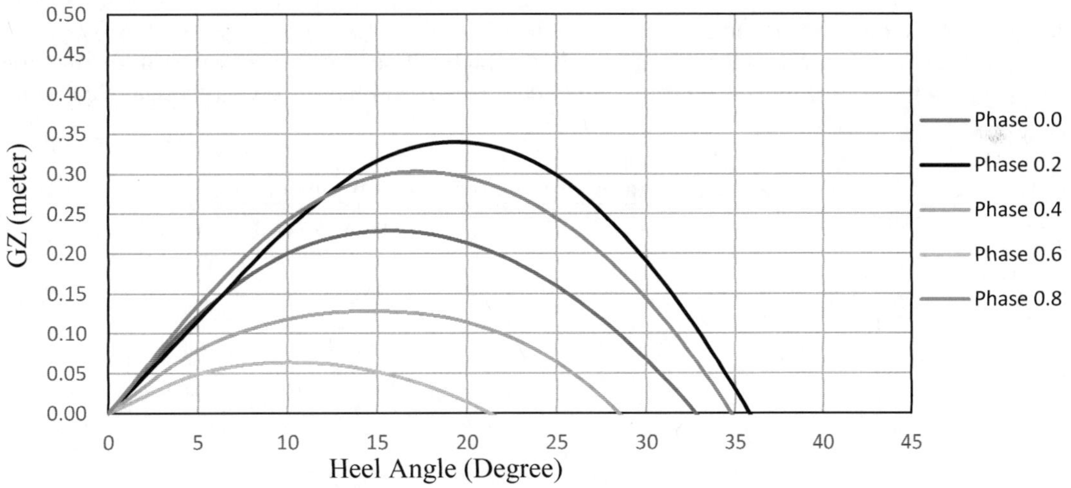

Figure 8. Comparison of stability righting arm GZ with varying phase

Figure 8 shows that all curve has a positive value. The highest curve of lever righting stability GZ at wave phase 0.2 near sagging wave. The lower lever righting stability (GZ) at wave phase 0.6 near hogging wave. At wave phase 0.2 has a maximum high of lever righting stability GZ 0.34 meter with a positive range of lever righting stability GZ 0-35°. At wave phase 0.6, there is a top lever righting stability of 0.064 meters with a positive range of lever righting stability GZ 0°-21°. Suppose lever righting stability GZ sorted from most minor to highest obtained: phase 0.2, phase 0.8, phase 0.0, phase 0.4, and phase 0.6. It concludes that it was decreasing/increasing, not along decreasing/increasing with the curve of lever righting stability GZ.

4.4. Ship Stability Assessment at Hogging Wave and Sagging Wave with Variation Wavelength

Wavelengths and wave heights on the sea occur randomly with specific patterns. This paper calculates righting lever arm GZ with varying wavelengths of 10 m, 20 m, 30 m, 40 m, and 50 m. Formula correlation a wavelength and a wave height uses the US Navy Standard in formula no [2] above to get variations in wavelength and wave height, such as **Table 4.** is as follows:

Table 4. Variation of wavelength and wave height

Wavelength (L_w)	10 m	20 m	30 m	40 m	50 m
Wave height (h_w)	1.92 m	2.71 m	3.32 m	3.84 m	4.29 m

Table 4 above shows that the higher the wavelength, the higher the wave, which has a harmonious relationship between increases and decreases. The type of wave used is a sinusoidal wave. Then the wavelength, wave height, and wave pattern are plotted into the RoPax hull case study, modeled in computer software. The results of the wave graph plot are shown in **Figure 9** below.

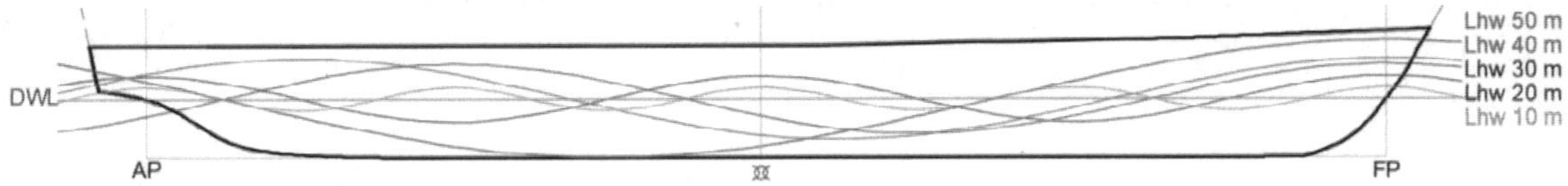

Figure 9. Wavelength and wave height variation profile

Figure 9 above shows variations in wavelength and variations in wave height. Initial wave modelling starts at the FP with the most significant wave amplitude. From the picture above, the red line is a wavelength of 50 meters, the magenta line is a wavelength of 40 meters, the black line is a wavelength of 30 meters, the blue line is a wavelength of 20 meters, and the green line is a wavelength of 10 meters. Then the righting lever stability (GZ) calculation is performed, and the results on the wavelength data and heeling angle are as follows:

Table 5. Righting stability GZ with variation wave height

Angle (deg.)	Lhw 10 m	Lhw 20 m	GZ (m) Lhw 30 m	Lhw 40 m	Lhw 50 m
0	0	0	0	0	0
5	0.198	0.198	0.198	0.198	0.198
10	0.403	0.402	0.402	0.402	0.402
15	0.595	0.614	0.614	0.614	0.614
20	0.673	0.82	0.82	0.82	0.82
25	0.64	0.984	0.984	0.984	0.984
30	0.522	1.067	1.067	1.067	1.067
35	0.355	1.087	1.087	1.087	1.087
40	0.159	1.019	1.019	1.019	1.019
45	-0.055	0.864	0.864	0.864	0.864
50	-0.28	0.668	0.668	0.668	0.668
55	-0.509	0.465	0.465	0.465	0.465
60	-0.74	0.251	0.251	0.251	0.251
65	-0.968	0.022	0.022	0.022	0.022
70	-1.192	-0.223	-0.223	-0.223	-0.223

Table 5 shows the righting stability (GZ) result with a variation wavelength between 10~50 meters. GZ is calculated at variation heel 0°~70° with interval heel 5°. Calculation results can also be displayed as a graph of the stability arm GZ` as follows:

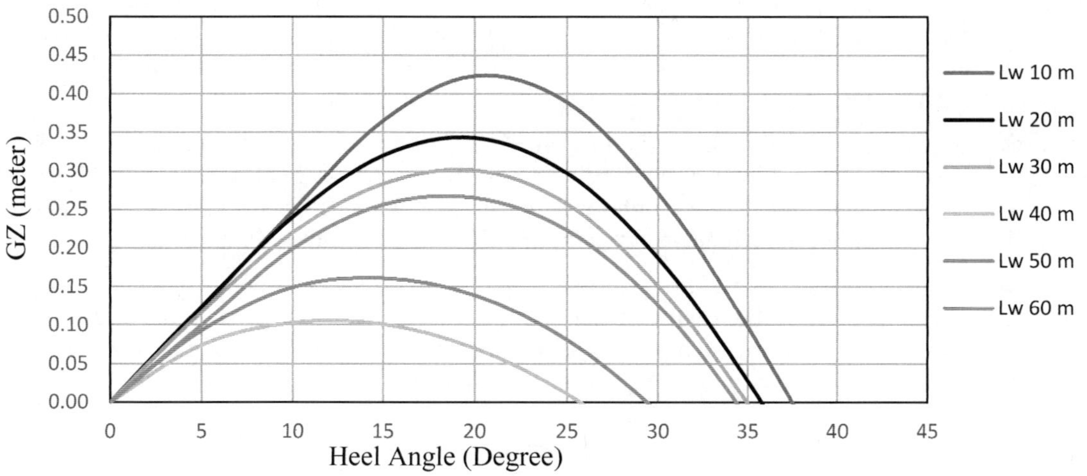

Figure 10. Curves of Righting stability GZ with variation wave height

Figure 10 shows the righting stability curve (GZ) with a variation wavelength between 10~50 meters. The curve in red is the stability arm curve of GZ for a wavelength of 10 meters: a black curve is GZ at the wavelength of 20 meters, a blue curve is GZ at the wavelength of 30 meters, a yellow curve is GZ at the wavelength 40 meters, a blue curve is GZ at wavelength 50 meter, and a green curve is GZ at wavelength 60 meter. The highest GZ curve is obtained at a wavelength of 10 meters, then the longer the wavelength of the GZ curve decreases until a wavelength of 40 m. The curve goes up again at 50 meters and 60 meters in length. The critical wavelength occurs at a wavelength equal to the length of the ship's Lpp. The wavelength is smaller than the length of the ship's Lpp. The greater the stability arm curve of GZ, and vice versa, the greater the wavelength of the Lpp length, the greater the stability curve of GZ.

5. Conclusions

From analyzing and discussing the case study above can be concluded that some results stability assessments of righting lever stability GZ on calm water and wave water are as follows:

- The curve of lever righting stability of case study RoPax does not satisfy all intact stability criteria (IS code 2008). Especially at criteria area heel angle 30° - 40° has value 1.221 meter-degree less than IS Code criteria 1.7189 meter-degree, and heel angle maximum of lever righting stability GZ maximum is 15° less than requirement IS Code 25°. So, the case study ship was improper to operate on Halmahera water.
- With use analysis, conventional hogging and sagging with wavelength 40 m and wave height 4 meters the same condition ship accident compares calm water. The maximum GZ curve on the hogging and the sagging waves is more than in calm water. Curve maximum height of lever righting stability GZ at calm water is 0.299 meters reduce become 0.23 meter at the sagging wave and reduce to be 0.01-meter pada at hogging wave. The positive range of GZ at calm water is 0°~29°, increases to 0o~32o at the sagging wave, and reduces to 0°~13° at the hogging wave.
- Wave on the sea occurs with random phase in this paper simulates at varying phase 0, 0.2, 0.4, 0.6, 0.8 at wavelength 40 m and wave height 4 m. The highest curve GZ occurs at phase 0.2 near the sagging wave, and the lowest curve GZ at phase 0.6 near the hogging wave. At wave phase 0.2, maximum GZ occurs at 0.34 meters and positive range GZ at heel 0°-35°. At wave phase 0.6, maximum GZ occurs at 0.064 m and positive range GZ at heel 0°-22°.

- Wave on the sea occurs with the random wavelength in this paper simulates at varying phases 10 meters until 60 meters. The highest GZ occurs at wavelenght10 meters, decreases until wavelength 40 meters, and increases until wavelength 60 meters. Critical wavelength occurs at almost the same wavelength at a ship length of 40 meters.
- This case study is positive for all lever righting stability GZ curves at calm and waves water. Wave reduces righting stability of the GZ curve. Their side openings on the car deck generate water ingress and free surface moment, reducing the righting stability GZ curve becomes negative. This condition makes the ship have high-risk accidents, proven the ship capsize on the Halmahera Sea.

Ocean waves affect decreasing the stability of survival integrity in OCD RoPax. The authors suggest that the RoPax with an open car deck without side casings should not be operated in high-wave seas or straits bordering the high seas to avoid the same accident. If the RoPax cannot be replaced with another ship, then the RoPax OCD must be modified by closing the car deck or installing a side casing/wing tank on the car deck.

References

[1] A. Christodoulou, Z. Raza, and J. Woxenius, *"The integration of RoRo shipping in sustainable intermodal transport chains: The case of a North European RoRo service,"* Sustainability, vol. 11, no. 8, p. 2422, 2019.

[2] T. Lamb, "*Ship Design And Construction*," Ed. Thomas Lamb Publ. SNAME ISBN 0-939773-40-6, 2003.

[3] A. Nurwahyudy, "*Contemporary Issues In Domestic RO-RO Passenger Ferry Operation In Developing Countries*" World Marit. Univ., 2014.

[4] N. Baird, "*Fatal Ferry Accidents, Their Causes, And How To Prevent Them*" 2018.

[5] H. Kim, S. Haugen, and I. B. Utne, "*Assessment of Accident Theories for Major Accidents Focusing on The MV SEWOL Disaster: Similarities, Differences, And Discussion for A Combined Approach*," Saf. Sci., vol. 82, pp. 410–420, 2016.

[6] EMSA, *Annual Overview of Marine Casualties and Incidents* 2017. European Maritime Safety Agency Lisbon, Portugal, 2017.

[7] K. KNKT, "*Data Investigasi Kecelakaan Pelayaran Tahun 2010 – 2016,*" http://knkt.dephub.go.id/knkt/ntsc_home/Media_Release/Media%20Release%20KNKT%202016/Media%20Release%202016%20-%20IK%20Pelayaran%2020161130.pdf, Nov. 25, 2016. http://knkt.dephub.go.id/knkt/ntsc_home/Media_Release/Media%20Release%20KNKT%202016/Media%20Release%202016%20-%20IK%20Pelayaran%2020161130.pdf (accessed Jun. 23, 2020).

[8] Hasanudin, A. Zubaydi, W. D. Aryawan, and I. K. A. P. Utama, "*The Effects of Side-Casings Balance on Open-Deck-Ro-Ro Vessels in Term of Intact Stability*," IOP Conf. Ser. Earth Environ. Sci., vol. 557, p. 012052, Sep. 2020, DOI: 10.1088/1755-1315/557/1/012052.

[9] S. Krueger, H. Hatecke, P. Gualeni, and L. Di Donato, "*On The Application of the 2nd Generation Intact Stability Criteria to Ro-Pax and Container Vessels,*" Proc. STAB, pp. 453–463, 2015.

[10] K. Oh, J. Ok, and B. Moon, "*Research and Preparation Status on IMO Second-Generation Stability Criteria,*" Presented at the 2018 International Conference on Mathematics, Modelling, Simulation and Statistics Application (MMSSA 2018), 2019.

[11] W. Peters et al., "*The Second-Generation Intact Stability Criteria: An Overview of Developmen*t," 2011.

[12] C. Andrei, "*A Proposed New Generation of Intact Stability Criteria for Assessment of Ship Stability In Longitudinal Waves*," MS&E, vol. 227, no. 1, p. 012005, 2017.

Acknowledgements
The authors would like to express their sincere gratitude to Mr Aleik Nurwahyudi from the National Committee for Safety Transportation (KNKT) for providing valuable data and constructive discussion during the paper preparation.

IOP Conf. Series: Earth and Environmental Science 1081 (2022) 012032 doi:10.1088/1755-1315/1081/1/012032

Development of the Hub and Spoke network model in natural disaster management

E Rasyid[1], K B Artana[2] and K Sambodho[2]

[1]Graduate Student Department of Management Technology, Institut Teknologi Sepuluh Nopember, Surabaya
[2]Department of Marine Engineering, Institut Teknologi Sepuluh Nopember, Surabaya, 60111, Indonesia

Abstract. Indonesia is a country with high disaster vulnerability. Disasters have a huge impact on a nation's economic development. According to the National Agency for Disaster Management, a disaster is an event or a series of events that threatens and disrupts the lives and livelihoods of people, caused by either natural factors and/or non-natural factors, as well as human factors and resulting in human casualties to lead. Environmental damage, property damage and psychological impact. This study is limited to hazards caused by natural factors, including nine types of threats, namely: earthquakes, tsunamis, volcanic eruptions, landslides, floods, flash floods, drought, forest and land fires, and extreme waves and abrasion. As an archipelago country, Indonesia has challenges in disaster management. Another thing behind this study is that there is a discrepancy between disaster maps and the location (distribution) of ships that have the potential to be deployed in disaster management, which will affect the duration of the emergency response time and the high cost of disaster management. What's more, there are conditions in which ships cannot perform certain coping functions, affecting the inability to effectively treat disaster victims, potentially increasing fatalities. Not all ships are suitable for disaster response operations in specific areas, resulting in the emergency response time being too long and the cost of disaster response being high. The purpose of this study is to evaluate the effectiveness of the existing ship distribution state that can be used in disaster management versus the existing disaster map. The research method used is to collect data and identify potential disasters in Indonesia. Then collect data and map the type and number of ships and the distribution of ship bases that can be used in disaster management. Using qualitative analysis, assess the discrepancy between vessel availability and distribution and the potential location and nature of the disaster in Indonesia. The expected result is that there is a match between the availability and distribution of vessels and the potential of the location and nature of the disaster in Indonesia.

1. Introduction

Geographically, the Indonesian archipelago is located between mainland Asia and mainland Australia, and between the Indian Ocean and the Pacific Ocean. Geographic locality is the location of an area or region as seen from reality on the earth's surface. Therefore, the territory of Indonesia is in a cross position and has significance in the situation and economy of Indonesia through international trade routes.

In terms of geology, some parts of Indonesia are areas prone to natural disasters. This is because the territory of Indonesia is the meeting point of two young mountain ranges in the world, the Pacific Rim (mountains that surround the Pacific Ocean) and the Mediterranean Rim (mountains) The meeting point of three major tectonic plates, namely the Indo - Australian Plate, Eurasian Plate and Pacific Plate is also causing Indonesia to be included in the Ring of Fire of the World or Pacific Ring of Fire, an area that frequently experiences earthquakes and volcanic eruptions around the Pacific Ocean basin [12] According to Sutopo Purwo Nugroho, head of the Data Information Center and Public Relations of the National Disaster Management Agency, the trend of natural disasters has increased by 350% worldwide in the last three decades, especially hydro-meteorological disasters, or disasters influenced by weather aspects, such as floods, landslides, hurricanes and droughts 80% of these Kind of disasters happen in In Indonesia. This is caused by environmental changes and regional factors in the form of global climate change.

Disasters have a significant effect on the economic development of a nation. The potential for disruption to socio-economic life has been experienced for residents living in disaster-prone areas such as Indonesia. The risk of natural disasters hurts development, especially economic development.

According to the National Agency for Disaster Management [4] [5] [6], a disaster is an event or series of events that threatens and disrupts people's lives and livelihoods caused, either by natural factors and/or non-natural factors as well as human factors, resulting in human casualties, environmental damage, property losses. objects, and psychological effects.

The definition states that disasters are caused by natural, non-natural, and human factors. Therefore, Law Number 24 of 2007 also defines natural disasters, non-natural disasters, and social disasters. Natural disasters are disasters caused by events or a series of events caused by nature, including earthquakes, tsunamis, volcanic eruptions, floods, droughts, hurricanes, and landslides. Non-natural disasters are disasters caused by non-natural events or a series of events, which include technological failures, failed modernization, epidemics, and disease outbreaks. While social disasters are disasters caused by events or series of events caused by humans which include social conflicts between groups or between communities, and terror [5].

This study is limited to hazards caused by natural factors, which include nine types of threats, namely: Earthquakes, Tsunamis, Volcanic Eruptions, Landslides, Floods, Flash Floods, Drought, Forest, and Land Fires, and Extreme Waves and Abrasion.

Natural conditions that are prone to disasters, Indonesia as an archipelagic country has challenges in disaster management. For this reason, the government needs to anticipate disasters with disaster mitigation, emergency response, and reconstruction rehabilitation, this is following Article 33 of Law Number 24 of 2007 concerning Disaster Management, which explains that the Central Government and Regional Governments are responsible for implementing disaster management. The task is handled by the National Agency for Disaster Management for the national level and the Regional Disaster Management Agency for the regional level.

Natural disaster management is not a matter of National Agency for Disaster Management alone but is the shared responsibility of all parties, known as the Penta helix, namely the government, the community, the business world, academia, and the media.

Just like National Agency for Disaster Management, National Agency for Search and Rescue (BASARNAS) is also a non-ministerial government agency that is under and directly responsible to the President. The main task is to assist the president in carrying out government affairs in the field of search and rescue. In terms of disaster management, BASARNAS is usually tasked with evacuating or providing assistance to victims of natural disasters. In addition, BASARNAS is also tasked with finding victims who have disappeared due to natural disasters.

The Ministry of Transport also plays an important role in natural disaster management in Indonesia. Several disaster preparedness measures implemented by the Ministry of Transport in ports include the installation of a new generation warning receiver system (WRSnGen) to support the process of early detection of earthquakes and tsunamis in Indonesia, which has been installed at 316 sites.

In addition, the Ministry of Transportation has also prepared approximately 100 human resources and patrol boats consisting of 75 navigation ships and 41 ships from the Marine and Coast Guard Unit

(KPLP) of the Directorate General of Sea Transportation spread across several parts of Indonesia to support disaster management.

After the passage of Law No. 32 of 2014 on Maritime Affairs, the function was expanded. Bakorkamla. The functions of Bakorkamla are defined in Article 4 of Presidential Decree No. 81 of 2005 on Bakorkamla. The most visible extension of the function from Bakorkamla to Bakamla is the implementation of the early warning system, which the Indonesian state, which is geographically and climatologically an area at risk of (natural) disasters, urgently needs. Bakamla is defined in Article 3 of Presidential Order No. 178 of 2014 on Bakamla.

In addition to non-ministerial governmental institutions, the potential for natural disaster management in Indonesia can be implemented through the deployment of vessels of the Indonesian National Army and Police of the Republic of Indonesia, Basarnas, Bakamla and others in Indonesian waters and offices of the Harbor Master and Port Authority spread throughout Indonesia. Indonesia is also a potential that can be used in natural disaster management.

To date, for ships used for natural disaster management, there is still a discrepancy in the location (distribution) of the disaster map and the ships that may be used for disaster management. This idea will affect the length of emergency response time and the high cost of disaster management. Another thing is that there are still conditions under which certain mitigation functions cannot be performed by ships, effecting that disaster victims cannot be treated effectively, so they have the potential to increase fatalities. Not all ships are suitable for disaster management in certain areas. This will affect the length of emergency response time and the high cost of disaster management.

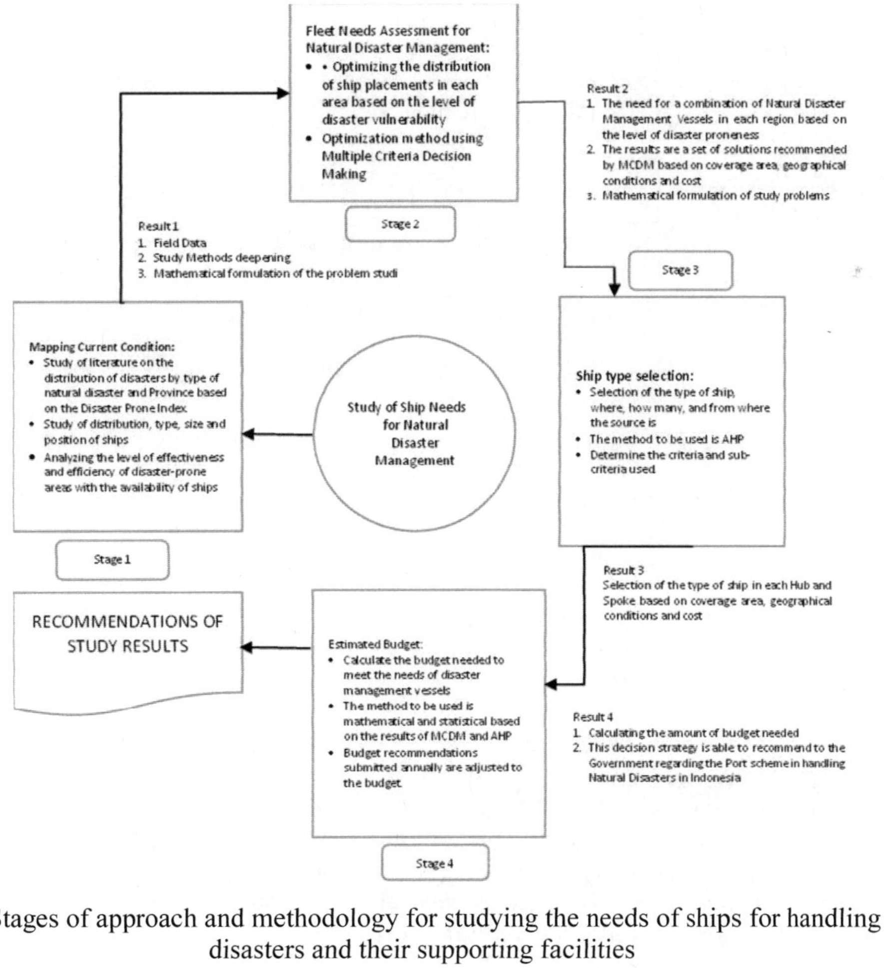

Figure 1. Stages of approach and methodology for studying the needs of ships for handling natural disasters and their supporting facilities

IOP Conf. Series: Earth and Environmental Science 1081 (2022) 012032 doi:10.1088/1755-1315/1081/1/012032

2. Methodoogy

In general, the approaches and methods that will be used in this study are as shown in Figure 1. Several stages in this study process can be carried out in parallel, both from the data search process and model making.

2.1. Optimization of the distribution of disaster management vessels using the Multiple Criteria Decision Making Method

The main goal of this research is to shape the distribution of ships in number and type can be used in disaster management and to find a deployment pattern that will ensure a brief emergency response time, minimal operating costs and low investment in deploying vessels utilizing the Multiple Criteria Decision-Making method.

There are several general functions used in MCDM, namely [7]:

- Alternatives are items that are different and have the same possibility of being chosen by the decision
- Manufacturer
- Attributes, often referred to as decision criteria.
- Conflicts between criteria, some criteria are usually in conflict with each other, for example profit criterion will conflict with the cost criteria.
- Decision Weight, Decision Weight shows the relative importance of each criterion, = (1, 2, 3, ...).5. The decision matrix of size x, contains elements that represent the ratings of the alternatives; = 1, 2, 3, …, against the criteria; = 1,2,3, …

Multi-Criteria Decision Making (MCDM) is one of the most widely used decision-making methods. The purpose of MCDM is to select the best alternative from multiple mutually exclusive alternatives based on the overall performance of various criteria (or attributes) determined by the decision maker. There are four decision steps in MCDM, namely [11]:

- Identify the problem
- Develop preferences.
- Evaluate alternatives.
- Determine the best alternative. Suppose there are m criteria (C1, …, Cm) and n alternatives (a1, …, an).

MCDM problems are usually represented in the form of a decision table as shown in Table 1.

Table 1. Decision Table [3]

		a_1	a_2	---	a_n
w_1	C_1	a_{11}	a_{12}	---	a_{1n}
w_2	C_2	a_{21}	a_{22}	---	a_{2n}
---	---	---	---	---	---
w_m	C_m	a_{m1}	---	---	a_{mn}

The aij value shows the alternative performance score Aj on the Ci criteria which is the preference of the decision maker. Each criterion has a weight of wi which indicates the level of importance of the Ci criteria in the decision-making process.

2.2. Designing a Decision Support System (DSS) based on The Technique for Order of Preference by Similarity to Ideal Solution (TOPSIS)

TOPSIS is a multi-criteria decision method or alternative choice, which is an alternative that has the smallest distance from the positive ideal solution, and the largest distance from the negative ideal solution from a geometric point of view using Euclidean distance. In order to achieve the ideal solution with this method, it can be assumed that all factors increase and decrease the scatter. In general, the TOPSIS procedure follows the following steps [7]:

- Compiling the normalized value of the decision matrix
- Calculate the weighting of the normalized decision matrix;
- Arrange the matrix values of positive and negative ideal solutions;
- Calculating the measurement/assessment for each alternative to the positive ideal solution or negative;
- Determine the ranking by calculating the relative closeness to the positive ideal solution A+ or negative A-

It is hoped that this study can obtain a special ship network model for disaster management based on the shortest response time, the largest coverage area and the smallest operating cost, using existing ships.

3. Mapping of Disaster-Prone Areas and Current Ship Distribution

- Earthquake and Tsunami

Indonesia lies at the confluence of the world's major plates and several smaller plates, making Indonesia an area prone to earthquakes and tsunamis. Indonesia is surrounded by major plates, namely the Eurasian Plate, the Indo-Australian Plate, the Philippine Marine Plate, the Pacific Plate and the Sunda Plate. The meeting of the Eurasian tectonic plates created a volcanic arc in western Indonesia, one of the most seismically active regions in the world. In addition, from the research results, it is knowing that 295 sources of active fault earthquakes have been identified on the territory of Indonesia at the moment, affecting the territories of the island of Sumatra, the island of Java, the island of Kalimantan, the island of Sulawesi, Maluku, and Papua, Bali and Nusa Tenggara.

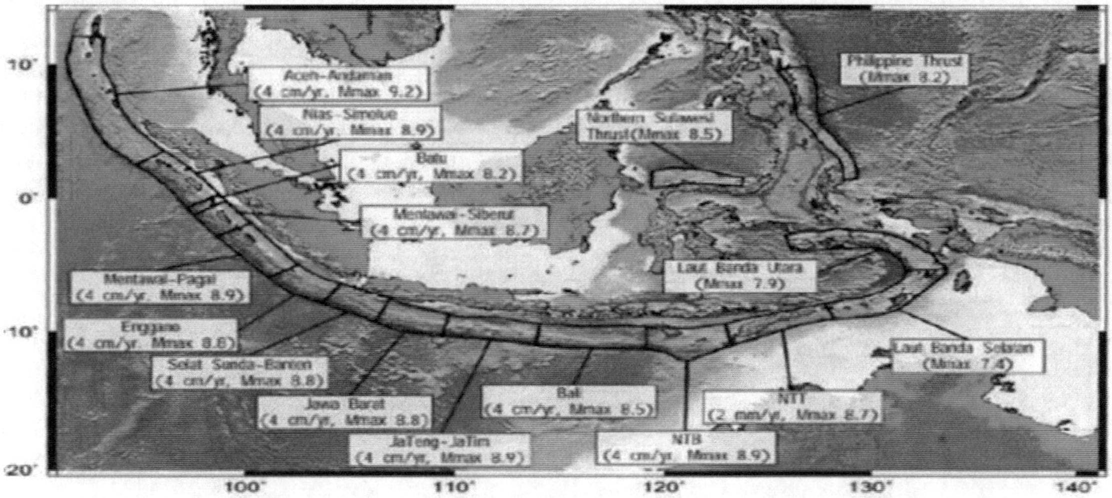

Figure 2. Earthquake and Tsunami Disaster Events [5]

There are 346 regencies/cities in the high earthquake risk level with a population at risk of 60.9 million people and 165 regencies/cities in the moderate earthquake hazard zone with a population at risk of 142.1 million people. The trend of earthquakes in Indonesia from January 2008 to December 2018 showed an increase, especially in the 2013–2018 period, for more details, it can be seen in the following figure:

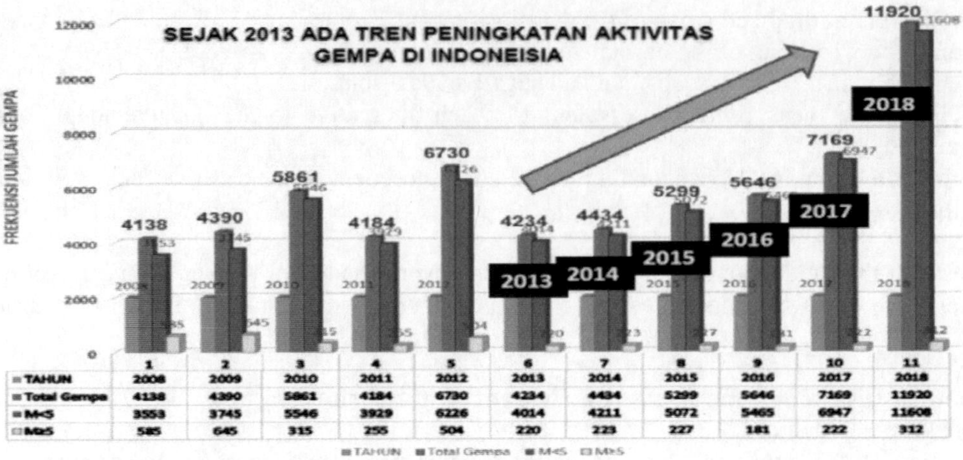

Figure 3. Trends in Earthquake Occurrence in Indonesia (January 2008 – December 2018 Period) [5]

- Volcanic Eruption

Indonesia is a country that has the most volcanoes in the world, so geographically it is prone to volcanic eruptions. Volcanoes in Indonesia are part of the Pacific ring of fire formed by the subduction zone between the Eurasian and Indo-Australian plates. Currently, Indonesia has 129 active volcanoes with approximately 3.9 million people living in the surrounding area. Of the 129 active volcanoes that exist today, 68 are the most active type A volcanoes, 29 type B volcanoes, and 31 type C volcanoes in Indonesia. Type A volcanoes have been recorded to have experienced magmatic eruptions at least once after 1600.

East Nusa Tenggara Province is the area with the most active volcanoes, with 16 active volcanoes. Meanwhile, the regency/city area that has the most active volcanoes in East Flores Regency with 4 volcanoes.

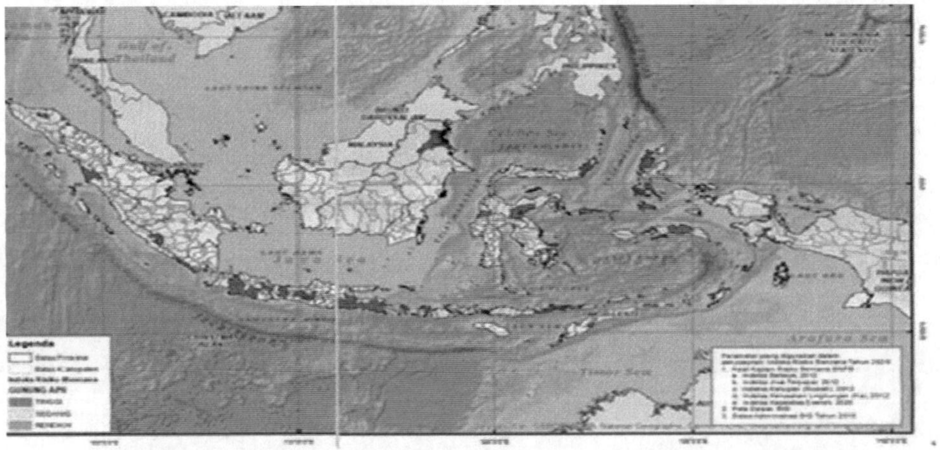

Figure 4. Map of the distribution of volcanic eruption risk areas [6]

- Flood

Flooding is caused by rainfall that is well above normal, so the existing water drainage system cannot accommodate the accumulation of stormwater and overflows. The water capacity of the drainage system is not always the same, but changes due to sedimentation, river narrowing due to natural and man-made phenomena, clogged garbage and other obstacles. Deforestation in stormwater catchments also leads to an increase in flood runoff due to the discharge of water entering the sewer system and other water reservoirs. What's more, reduced water catchment areas also contribute to increased flood discharge. In densely populated residential areas, water infiltration into the soil reducing. Hence, when it rains with heavy rains, most of the water becomes surface water, which enters the water drainage system directly, exceeding its capacity and causing flooding Figure 5. Map of distribution of flood-prone disaster risk areas

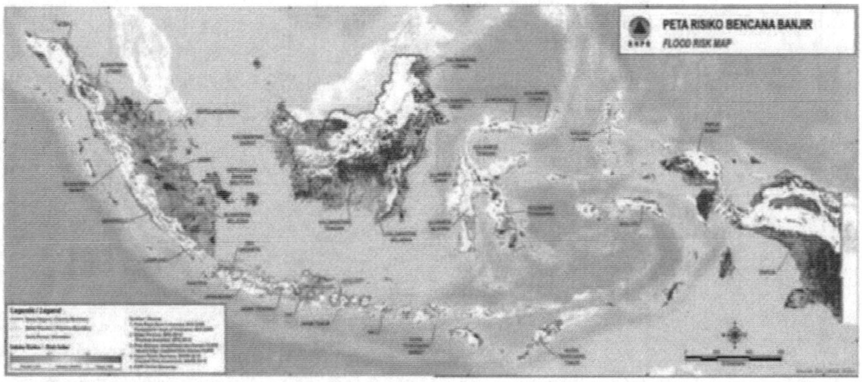

Figure 5. Map of distribution of flood-prone disaster risk areas [5]

Areas prone to flooding are the coastal and low-lying parts of the large islands of Indonesia. Based on data on flood events in Indonesia recorded at National Agency for Disaster Management in the 2015–2019 period, it appears that every year there are always provinces whose areas are affected by floods.

- Land Movement/Landslide

The landforms and geological conditions in Indonesia are mountainous or mountainous, which can lead to landslides or ground movement due to disturbances in the stability of the soil or the rocks that make up the slopes. Disturbances in slope stability controlling by morphological conditions (particularly the slope), rock or soil conditions that make up the slopes, and hydrological conditions or water management on the slopes. Even if an embankment is vulnerable or has the potential for landslides due to the nature of the embankment, rock/soil and water systems, the embankment will not slide or its stability will be disturbed without triggering a tripping event such as a gradient in the water content in the slope stretching the bonds between soil grains, so it can slide easily, vibration. B. from earthquakes or explosions, especially on sandy soils with water content, often result in liquefaction, increased loads that exceed the bearing capacity of the soil, and indiscriminate cutting of the slope foot, causing the slope to lose its bearing capacity.

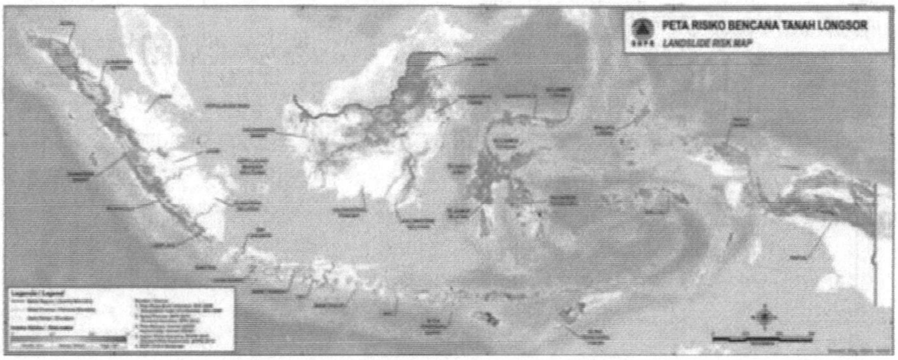

Figure 6. Landslide disaster risk map [5]

Areas prone to landslides/soil movements in Indonesia extend from Aceh, North Sumatra, West Sumatra, Bengkulu, Lampung, Banten, West Java, Central Java, DIY, East Java, Bali, NTB, NTT, parts of Kalimantan, especially in the East and North, most of Sulawesi, Maluku and North Maluku and parts of Papua and West Papua.

- Drought

Indonesia is prone to drought disasters, from historical data, drought in Indonesia is closely related to the ENSO (El-Nino Southern Oscillation) phenomenon. The influence of El-Nino on rain diversity has several patterns, namely: (1) the end of the dry season is back from normal, (2) the start of the rainy season is back from normal, (3) the dry season rainfall drops sharply compared to normal, (4) the day series The dry season is getting longer, especially in the eastern part of Indonesia. The impact of the danger of drought is often gradual/slow, so that if there is no good and continuous monitoring it will result in disasters in the form of reduced water sources, reduced food production which can trigger hunger and can also cause land and forest fires.

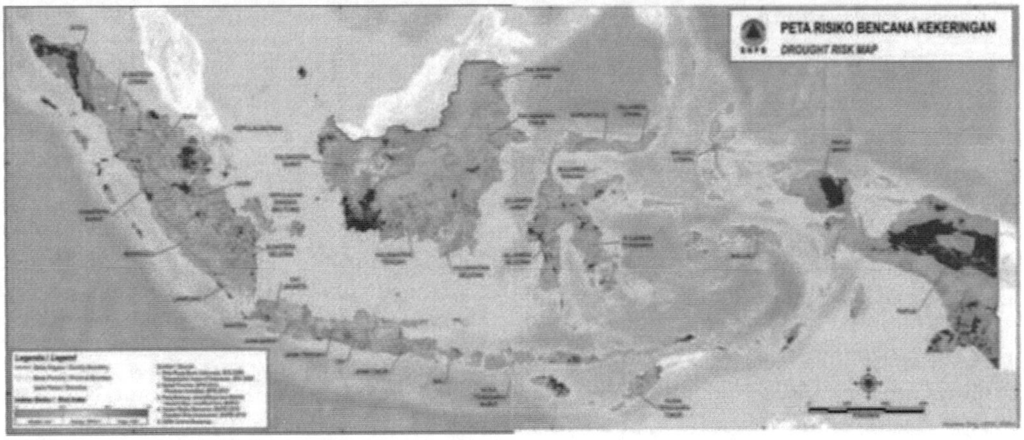

Figure 7. Drought risk map [5]

Drought-prone areas in Indonesia extend from Aceh, to Lampung, Banten, West Java, Central Java, DIY, East Java, Bali, NTB, NTT, parts of Kalimantan, Sulawesi, Maluku and North Maluku.

- Land and Forest Fires

The incidence of land and forest fires in Indonesia is not only influenced by natural factors but also largely by human factors who intentionally burn and neglect to prepare land for agriculture and plantations. The largest land and forest fires were in Riau Province and almost all provinces in Indonesia in 2013 and 2015. This disaster resulted in the transboundary haze to neighboring countries, namely Singapore and Malaysia.

The extent of land and forest area burned in Indonesia to date is also influenced by the biophysical characteristics of the existing land. Most of the fires in the last 10 years occurred on peatlands. This land is naturally a non-combustible wetland, but if the peat is dry due to excessive drainage it is highly susceptible to fire. Dry peatlands can also change their nature and cannot return to their original form of wetlands, so the level of vulnerability to fire is higher. Thus, aspects of land and climatic conditions become important aspects that affect the incidence of land and forest fires.

IOP Conf. Series: Earth and Environmental Science 1081 (2022) 012032 doi:10.1088/1755-1315/1081/1/012032

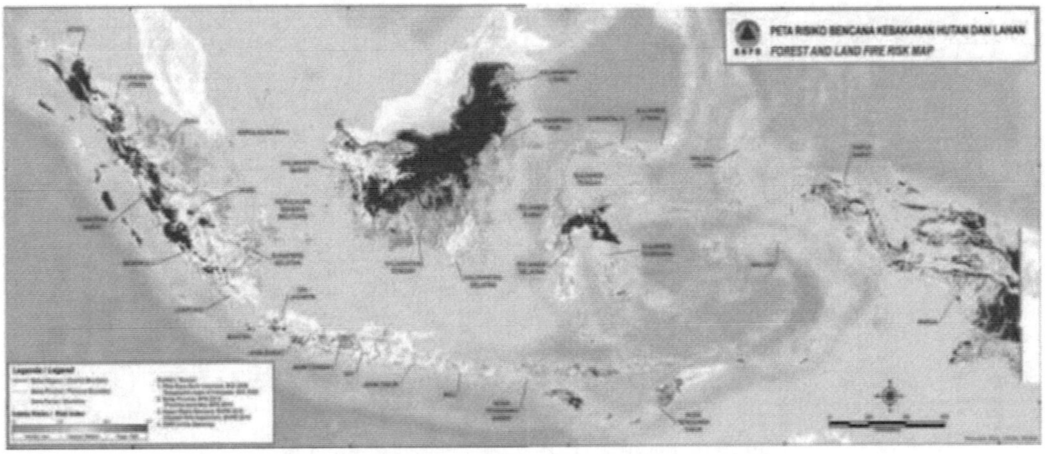

Figure 8. Forest and land fire disaster risk map [5]

Most parts of Indonesia are prone to land and forest fires. Since 2015, almost every year, land and forest fires have occurred in the provinces of Riau, Jambi, South Sumatra, West Kalimantan, Central Kalimantan, South Kalimantan and East Kalimantan.

- Tornado/Strong Wind

Extreme weather is strong winds with speeds of 120 km / h or more often occur in tropical areas between the agrarian turning north and south, except in areas very close to the equator. This wind is caused by a pressure difference in a weather system such as an increase in air pressure or the apparent movement of the sun perpendicular to the earth's surface (culmination).

Tornadoes/strong winds can occur anytime and anywhere, both on land and at sea and if they occur at sea, the duration is longer than on land. Tornadoes / strong winds generally occur in the afternoon or evening, sometimes at night, and more often during the transition of seasons (transitional). Until now, there have been many reports of hurricane/strong wind disasters in many places. Tornadoes/strong winds are quite large which can cause a lot of damage and at the same time cause significant losses, such as destroying people's houses, trees, and means of transportation.

Almost all areas in Indonesia are prone to tornadoes. For more details, see the following image.

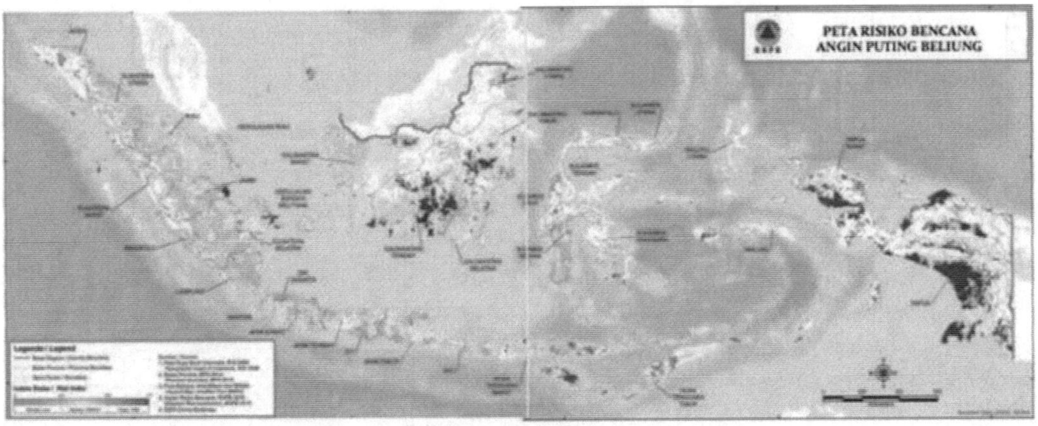

Figure 9. Tornado risk map [5]

The distribution of the Disaster Risk Index in Indonesia by province can be seen in Figure 10 below.

IOP Conf. Series: Earth and Environmental Science 1081 (2022) 012032 doi:10.1088/1755-1315/1081/1/012032

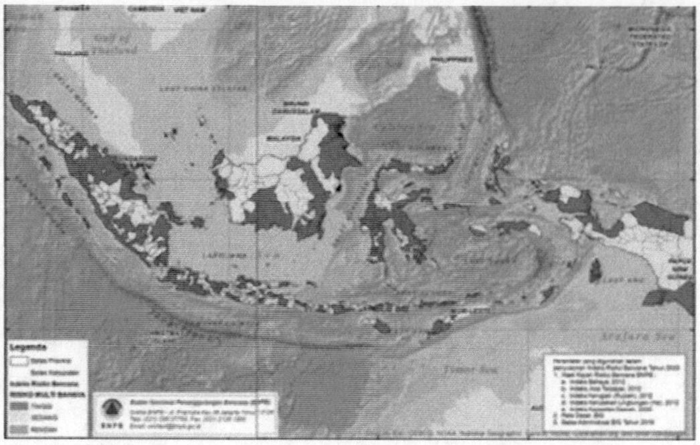

Figure 10. Indonesia's 2020 Disaster Hazard Map [6]

Most of the islands of Sumatra have a high category of disaster risk, starting with the highest being Bengkulu, Bangka Belitung Islands, Aceh, West Sumatra, Riau, Lampung, and North Sumatra. South Sumatra, Jambi, and Riau Islands are in the medium category. Java Island, Banten, and West Java provinces are in the high category, while the Special Region of Yogyakarta, East Java, Central Java, and DKI Jakarta are in the medium category. Bali and Nusa Tenggara are in the medium category. East Kalimantan, North Kalimantan, and South Kalimantan are included in the high category on the island of Kalimantan. For the island of Sulawesi which is in the high category, the provinces of West Sulawesi, South Sulawesi, Southeast Sulawesi, and Central Sulawesi are in the high category. The next provinces that fall into the high category are North Maluku and West Papua. The rest fall into the medium category. None of the provinces in Indonesia are in a low category.

3.1. Distribution of Indonesian Navy Ships

The Indonesian Navy has 14 Lantamal spread throughout Indonesia. With the distribution: Lantamal I in Belawan, Lantamal II in Padang, Lantamal III in DKI Jakarta, Lantamal IV in Tanjung Pinang, Lantamal V in Surabaya, Lantamal VI in Makassar, Lantamal VII in Kupang, Lantamal VIII in Manado, Lantamal IX in Ambon, Lantamal X in Jayapura, Lantamal XI in Merauke, Lantamal XII in Pontianak, Lantamal XIII in Tarakan, North Kalimantan and Lantamal XIV in Sorong. Where each Lantamal has a ship that can be used as a disaster management ship [10] [14].

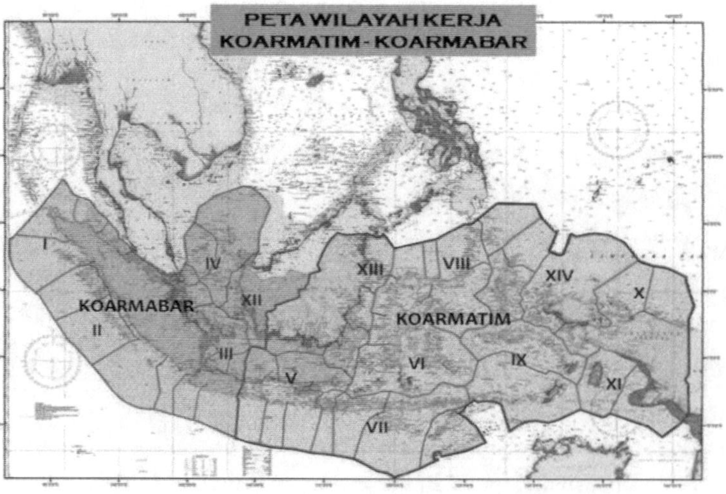

Figure 11. Distribution of Lantamal in Indonesia [10]

3.2. Syahbandar's Office Distribution
Indonesia currently has 100 harbormaster offices spread throughout Indonesia. Consists of 4 (four) main class offices, namely in Belawan Medan, Tanjung Priuk Jakarta, Tanjung Perak Surabaya and Makassar South Sulawesi. Nine Class 1 (one) offices, 15 Class 2 (two) offices, 16 Class 3 (three) offices, 16 Class 4 offices, and 40 Class 5 offices.

3.3. Distribution of BASARNAS Offices
In carrying out its main tasks, the National Search and Rescue Agency has a Technical Implementation Unit (UPT) in the area called the SAR Office and a Technical Implementation Unit (UPT) from the Training Center. Currently, there are 43 SAR offices consisting of 21 Class A SAR offices and 22 Class B SAR offices.
To speed up the location of the disaster which is spread over a fairly wide area, the SAR Office places a rescue team at the SAR Post. SAR posts are placed in the SAR office area in two places with regional priorities that have a high level of vulnerability to disasters/disasters. In accordance with the Minister of Transportation Regulation no: KM 40 of 2006, concerning Search and Rescue Posts (POS SAR) there are 65 SAR posts, namely: Sibolga, Tanjung Balai Asahan, Nias, Cirebon, Pasaman, Jember, Banyuwangi, Pagar Alam, Tulung agung, Raja Ampat, Amurang, Tual, Saumlaki, Sukabumi, Belitung, Banten, Jembrana, Karangasem, Buleleng, Solo, Bone, Selayar, Nabire, Serui, Lhokseumawe, Meulaboh, Simeuleu, Bengkalis, P. Natuna Besar, Tanjung Balai Karimun, Tahuna, Anambas, Cilacap, Trenggalek, Kayangan, Manggarai, Waingapu, Labuan Bajo, Ketapang, Sintete, Kotabaru, Sampit, Tarakan, East Kutai, Bau-Bau/ Buton, Wakatobi, Kolaka, North Gorontalo, Luwuk Banggai, Namlea, Tobelo, Banda, Raja Ampat, Fakfak, Wamena, Sarmi, Bintang Mountains, Mamuju, Agats, Bima, Batam, Jepara, Kaimana, Kimam/ P. Dolak, Pangkalan Bun.

3.4. BAKAMLA Ship Distribution
BAKAMLA ships totaled 31 (thirty one) units spread over five bases, namely Tanjung Priok Base as many as 12 (twelve) units, Tanjung Uban base recorded 8 (eight) units, Tanjung Perak 3 (three) units, Bitung Base 4 (four) units and the Tual Base as many as 4 (four) ships.

3.5. Functions and Suitability of Ships in Disaster Management
Based on National Agency for Disaster Management's decision Number 173 of 2015 concerning National Clusters for Disaster Management, the concept of how disaster management ships work consists of 8 clusters in response to disaster emergencies, which include: (1) health; (2) search and rescue; (3) logistics; (4) refuge and protection; (5) education; (6) facilities and infrastructure; (7) economy; and (8) early recovery [1] [2].
From National Agency for Disaster Management Decree Number 173 of 2015 concerning National Clusters for Disaster Management concerning 8 (eight) disaster emergency response clusters, it can be explained:

Table 2. Functions and Suitability of Types of Ships in Disaster Management

Description	SHIP CLASS					
	I - A	I - B	II	III	IV	V
Operating Area	All of Indonesia	All of Indonesia	Indonesian waters are limited	Scope of interest in port waters	Scope of interest in port waters	Port
cruising range (NM)	≥ 4.000	≥ 2.000	≥	≥ 300	≥ 150	≤ 50
Length (Meter)	≥ 70	60 - 70	35 - 60	20 - 35	10 – 20	≤ 10
Width	≥ 9	7 - 9	6 - 7	5 - 6	3 - 5	≤ 3
Height	≥ 5	4 - 5	3 - 4	2 - 3	1 - 2	≤ 1
Velocity	15 - 20	15 - 20	15 - 25	20 - 25	25 - 30	25 - 30
Crew	55	50	20	16	10	5

Table 3. Functions and Suitability of Types of Ships in Disaster Management

Function	**Patrol Ship**				**LCU**	**LCT**	**Tanker**	**Hospital ship**	**Passenger ship**	**Survey Ship**	**Cargo Ship**	**etc**
	Kls IA	Kls IB	Kls II	Kls III								
Victim Evacuation	√	√	√	√				√	√			
Distribution Logistics					√	√			√		√	
Heavy Equipment Distribution					√	√						
Hospital								√	√			
Personnel transportation	√	√	√	√					√			
Weather Detection Equipment	√	√								√		
Transportation of reconstruction materials					√	√	√					
Temporary Community Shelter								√	√			
Fuel Supply							√					
etc												

4. Result and Discussion

Disaster management in Indonesia always involves the Indonesian National Armed Forces Ship. From several existing KRIs, several types of KRIs can be optimized to transport aid materials, such as food, medicines, laboratory equipment, vehicles, and Indonesian National Army personnel. This has been done in the handling of the Wasior flood disaster, namely by deploying five warships from the ranks of the Indonesian Eastern Fleet Command (Koarmatim), such as KRI Hasanuddin-366, KRI Fatahillah-361, KRI Kalakay-818, KRI Dr. Soeharso-990, and KRI Ahmad Yani-351. Near the disaster site, several ships are belonging to BASARNAS which are in Biak, Sorong, Jayapura, Timika, and Merauke. Meanwhile, the BAKAMLA office located around the location can also be used to assist the disaster management process.

In the Mentawai Tsunami disaster, the Indonesian Navy deployed five warships from the ranks of the Indonesian Western Fleet Command (Koarmabar), such as KRI Imam Bonjol-383, KRI Teluk Gilimanuk-531, KRI Teluk Peleng-535, KRI Teluk Sabang-544, and KRI Teluk Cirebon-543, and an element of the Military Cross-Sea Command (Kolinlamil) namely KRI Teluk Manado-537.

To support the handling of victims of natural disasters due to cyclone Seroja that occurred in 18 districts/cities in East Nusa Tenggara (NTT). One of them is a hospital ship. The five KRIs seconded for handling natural disasters in NTT include KRI Ahmad Yani, KRI Oswald Siahaan, KRI Makassar which is a hospital KRI, KRI Tanjung Kambani 971, and KRI Escolar.

The Indonesian National Army also has 3 Hospital Assistance Vessels (BRS), namely KRI dr Wahidin, KRI dr. Soeharso (operated in earthquake response in Mamuju, carried out humanitarian operations in Timor Leste in early 2016, Mamuju and Majene disaster management in 2020) and KRI Semarang-594 (picked up 188 Indonesian citizens of the World Dream Cruise Ship crew who had finished observing the Sebaru Island, Thousand Islands, heading to Kolinlamil Pier, Tanjung Priok, Jakarta, on March 14, 2020. In addition, KRI Semarang-594 had a "pick-up" mission of 2,100 litres of hand sanitizer concentrate assisted by the Singapore Government on April 8, 2020). Based on the Indonesian National Army Law No. 34 of 2004, in the OMSP mission, Hospital Assistance Ships can carry out operational tasks to help cope with the consequences of natural disasters, evacuation, and provide humanitarian assistance as well as assist in search and rescue in accidents (search and rescue).

In addition to the KRI ships mentioned above, in dealing with disasters in Indonesia, the Indonesian National Army also has the KRI Tarakan-905 which measures 122.40 meters long and 16.50 meters wide, is capable of transporting 5,500 metric tons of oil, and has RAS capability. which is able to carry out the process of refuelling to other ships while continuing to sail. In addition, it can serve the delivery or transfer of goods from ship to ship in a position both are still sailing.

5. Conclusions

As an archipelagic country, most of which is water, Indonesia has challenges in disaster management. In addition, Indonesia has the potential to overcome natural disasters from the sea. This can be seen from the many governments and non-governmental institutions that can function in tackling the threat of disasters, such as National Agency for Disaster Management, Indonesian National Army and Republic of Indonesia Police, BASARNAS, BAKAMLA, volunteers, and so on. There is still a mismatch between the disaster map and the location (distribution) of ships that have the potential to be used in disaster management which has an impact on the length of emergency response time and the high cost of disaster management.

6. References

[1] Ade Prasetia, 2014, *Optimalisasi Peran Kapal Republik Indonesia (KRI) Dalam Penanggulangan Bencana di Inonesia (Sebuah Sinergitas Alutsista* Indonesian National Army *AL Dalam Operasi Militer Selain Perang).*

[2] Adril. 2011. *Peran TNI Dalam Penanggulangan Bencana Alam (Studi Kasus Peran Korem 032/Wirabraja Dalam Penanggulangan Bencana Alam di Sumatera Barat).* [Thesis]. Jakarta: Program Pascasarjana Universitas Indonesia.

[3] Agung N. Pramudhita, Hadi Suyono, Erni Yudaningtyas, 2015. Penggunaan Algoritma Multi Criteria Decision Making dengan Metode Topsis dalam Penempatan Karyawan. Jurnal EECCIS Vol. 9, No. 1, Juni 2015

[4] Badan Naional Penanggulangan Bencana, 2011, Jurnal Penanggulangan Bencana, Volume 2 no 1 Juni tahun 2011, ISSN 2087636X, Jakarta

[5] Badan Naional Penanggulangan Bencana 2021. Peraturan- Badan Naional Penanggulangan Bencana -no-5-Tahun-2020

[6] Badan Naional Penanggulangan Bencana, 2020. Indeks Risiko Bencana Indonesia

[7] Falahah, S. R., 2016, Penerapan Metoda TOPSIS pada Analisis Penentuan Posisi Ideal Pemain Sepak Bola (Studi Kasus Pembinaan Pemain Usia Dini pada SSB), Seminar Nasional Aplikasi Teknologi Informasi (SNATi), Yogyakarta, 6 Agustus

[8] L. Tri Setyawanto R., Soekotjo Hardiwinoto, Nazili Abdul Azis. 2016. Kewenangan Badan Keamanan Laut (Bakamla) dalam Pelaksanaan Pengamanan di Wilayah Perairan Indonesia. Diponegoro Law Review, Volume 5 No. 4, 2016.

[9] Maarif, S, 2011, Kapasitas Masyarakat dan Sistem Kerjanya Hubungannya dengan Indek Ketahanan dalam Mengurangi Resiko Bencana Alam.

[10] Okol Sri Suharyo, 2017. *Model Penentuan Lokasi Pangkalan Angkatan Laut Berbasis Sustainabilitas*. Disertasi ITS.

[11] Pratyush Sen and Jian-Bo Yang, 1998. Multiple Criteria Decision Support in Engineering Design. Springer

[12] Surono, 2010. Perspektif Analisis Resiko Bencana dariAspek Geologis dan Pengembangannya; Workshop Pengembangan Analisis Resiko Bencana untuk Pembangunan Nasional; Hotel Millenium 25 Oktober 2010, Badan Naional Penanggulangan Bencana Jakarta

[13] Widodo, B. (2014). Konsepsi Tugas TNI dalam Penanganan Bencana. Retrieved from Kompasiana:https://www.kompasiana.com/bambangsixnal/54f9304ca333116f068b485a/konsepsi-tugas-Indonesian National Army-dalam-penanganan-bencana#:~:targetText=Hukumsebagailandasanyangdapat%20mendasari%20adalah%20Undang%E2%80%93undang,terhadap%20keutuhan%20bangsa%20dan%20negara

The operational concept of Mini LNG Carrier: Preventing sedimentation on the seabed

Kadir, Abdul [1,3*]; Istadi, I. [2]; Iskendar, I. [3]; Subagio, Agus [4]; Ali, Baharuddin [3]; Nurcholis, N.[3]; Waluyo, W.[3]

[1] School of Postgraduate Studies, Universitas Diponegoro, Semarang, Central Java 50275, Indonesia
[2] Department of Chemical Engineering, Faculty of Engineering, Universitas Diponegoro, Semarang, Central Java 50275, Indonesia
[3] National Research and Innovation Agency (BRIN), Jakarta 10340, Indonesia
[4] Department of Physics, Faculty of Science and Mathematics, Universitas Diponegoro, Semarang, Central Java 50275, Indonesia

* Correspondence: abdkptrim@students.undip.ac.id; abdu014@brin.go.id

Abstract. Utilization Mini LNG Carrier is an alternative solution to support the supply chain for the distribution of liquefied natural gas (LNG) to gas power plants, especially in Eastern Indonesia. Mini LNG carriers must have adequate seakeeping and manoeuvrability in carrying out their missions. The biggest challenge faced when manoeuvring in shallow water where the rotation of the propeller can produce high-velocity water flows that cause sedimentation on the seabed. The distance between the tip of the propeller to the seabed greatly determines the level of ease of erosion and sedimentation that can occur. This study aims to provide an alternative ships' operational concept by mapping the suitability between the depth of the water and the rpm of the propeller that does not cause sedimentation on the seabed. This concept determines the velocity of the water flow at which sedimentation begins to occur as the propeller rpm threshold. The final result is a depth division map where the propeller can be used at maximum rpm, the propeller is used with a limited rpm and the depth at which the propeller is not recommended to be used.

1. Introduction

The Mini LNG Carrier is the right choice to support the supply chain of LNG distribution to a small capacity of gas power plant (PLTG) spread in several remote areas in Eastern Indonesia [1]. For this purpose, the Agency for the Assessment and Application of Technology of the Republic of Indonesia (BPPT) developed a Landing Craft Tank type design for a Mini LNG Carrier with a capacity of 36 ISO Tanks Containers of 20 feet.

Landing Craft Tank (LCT) or Landing Ship Tank (LST) type vessel can dock directly on open shores, small ports, manoeuvre in shallow waters, along rivers and bays [2]. This type is suitable for coastal and inland areas or areas that do not have adequate seaport infrastructure. To ensure that the Mini LNG Carrier can distribute LNG to the destination, it is necessary to carry out a series of experiments such as powering, seakeeping and manoeuvring which are carried out on a model scale in the hydrodynamics laboratory. All tests are assumed to be in deep sea conditions at maximum propeller rpm.

The manoeuvring behaviour of a ship becomes even more important when approaching a harbour and shallow water [3] [4]. The water flow produced by the rotation of the propeller is high-velocity flow,

IOP Conf. Series: Earth and Environmental Science 1081 (2022) 012033 doi:10.1088/1755-1315/1081/1/012033

which is capable of scratching the seabed if it is not controlled. The closer the distance of the ship's propeller to the seabed, the easier it is for erosion and sedimentation to occur [5]. Sedimentation can cause siltation of navigation channel and ports basin causes delays in shipping activities and the process of loading and unloading ships cannot run normally [6]. Dredging is a method that is widely used to overcome the effects of siltation due to sedimentation even though it has environmental impacts on marine biota and the diffusion of contaminants and pollutants on the seabed [7] [8]. However, this method only overcomes the impact of sedimentation but cannot prevent the occurrence of sedimentation itself. There needs to be another solution to prevent sedimentation from the source, one of which is through the ship's operational system. Under these conditions, the speed of the Mini LNG Carrier's manoeuvring needs to be adjusted to the water depth to avoid damaging the seabed. This paper explains how to provide an alternative ships' operational concept with an environmental approach, by mapping the suitability between the depth of the water and the rpm of the propeller that does not cause sedimentation on the seabed due to the manoeuvring of the Mini LNG Carrier.

2. Material and Methods

This study uses an inductive approach by investigating the characteristics of the Mini LNG carrier Landing Craft Tank type with a capacity of 36 ISO Tanks Containers of 20 feet by experimental models and literature references. Due to ITTC (International Towing Tank Conference), Mini LNG Carrier was tested at the Indonesian Hydrodynamic Laboratory in free-sailing mode. The model used in the experiment is shown in Figure 1. The material for the model was made of wood covered with fibreglass and finished with a model scale of 1: 19.8312. The main dimension of the model and prototype are presented in Table 1.

Figure 1. Model of Mini LNG Carrier

Before testing, it was necessary to adjust the static similarity of the model to approach the real conditions.

Table 1. Main dimension of Prototype and Model of Mini LNG Carrier

Item	Model	Prototype
Scale	1: 19.8312	1:1
LOA (m)	2.702	53.584
LWL (m)	2.550	50.574
Lpp (m)	2.500	49.578
B (m)	0.625	12.400
H (m)	0.313	6.20
D (m)	0.126	2.50
Displacement (kg)	157.12	1,256,000
KG (m)	0.24	4.83
LCG (m)	1.25	24.693

IOP Conf. Series: Earth and Environmental Science 1081 (2022) 012033 doi:10.1088/1755-1315/1081/1/012033

2.1. Seakeeping test

Seakeeping represents the strength, stability and endurance of the ship to operate at sea in all conditions under the influence of waves [9]. Seakeeping testing was carried out using regular waves with variations in the period of the waves and the direction of the ship's heading to the waves. The test was carried out at forwarding speed conditions with an initial speed of 12 knots. The seakeeping test program was conducted in 2 m Hw, 10 times measurement for 180° as shown in Figure 2 and 90° heading as shown in Figure 3, on a range between 3 until 12 Tw (s), where Hw is wave heigh and Tw is wave period. All measurements were carried out by defining according to the righthand axis system.

Figure 2. Seakeeping model test at heading 180 deg

Figure 3. Seakeeping model test at heading 90 deg

2.2. Manoeuvring test (turning circle and zig-zag)

Manoeuvring test aims to see the ship's ability to perform a circular motion (turning circle) and zig-zag motion. The turning maneuver test was carried out in the starboard direction as shown in Figure 4, and port side direction as shown in Figure 5 with a steering angle of 35 deg each, while the zig-zag test was carried out at a steering angle of 20/20 deg. In the turning maneuver test, a wireless optical motion tracking system is used to measure the model's motion, while in the zig-zag maneuver test, an automatic control system is used which is connected wirelessly to the controlling computer.

Figure 4. Starboard direction

Figure 5. Portside direction

All tests are intended to ensure the ship can operate properly according to existing regulatory standards.

2.3. Equation for velocity analysis

Determining the equation for the analysis of the velocity of the water flow from the propeller is one of the steps in this research. Identification and analysis of water flow velocity is carried out in Zone of Flow Establishment (ZFE) at a distance of 0 - 3.25Dp axial direction behind the propeller and Zone of Established Flow (ZEF) for distances > 3.25Dp [10]. For twin propeller, limitations of (ZFE) and (ZEF) refer to the proposed by Jiang et al [11] [12].

2.4. Operational concept of Mini LNG Carrier
The operational concept is based on the hydrodynamic characteristics of the Mini LNG Carrier, water depth and maximum water flow velocity as the propeller rpm threshold to prevent erosion and sedimentation on the seabed.

3. Result and discussion

3.1. Seakeeping test result
The test results are given in graphical form, RAO-Heave, RAO-Roll, and RAO-Pitch respectively, as shown in Figures 6-11.

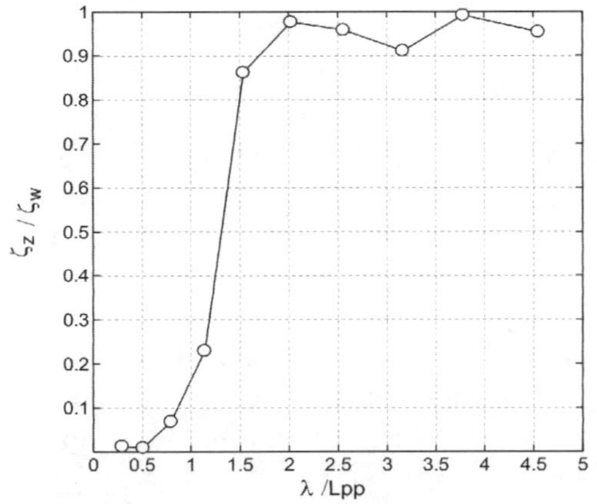

Figure 6. RAO-heave at heading 180 deg

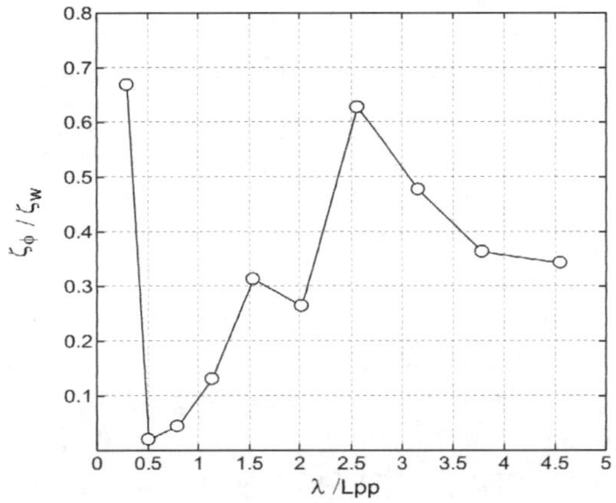

Figure 7. RAO-roll at heading 180 deg

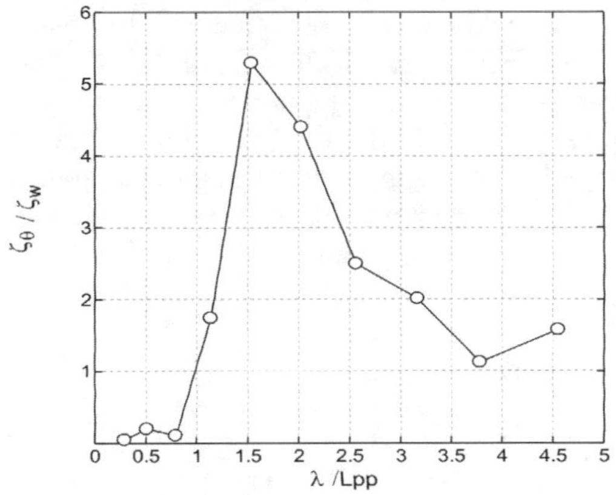

Figure 8. RAO-pitch at heading 180 deg

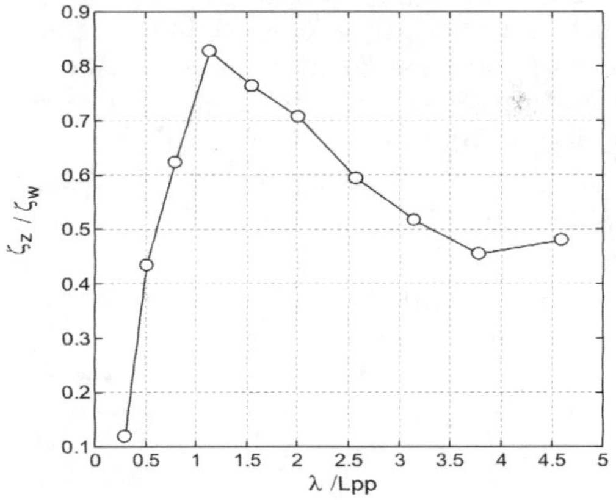

Figure 9. RAO-heave at heading 90 deg

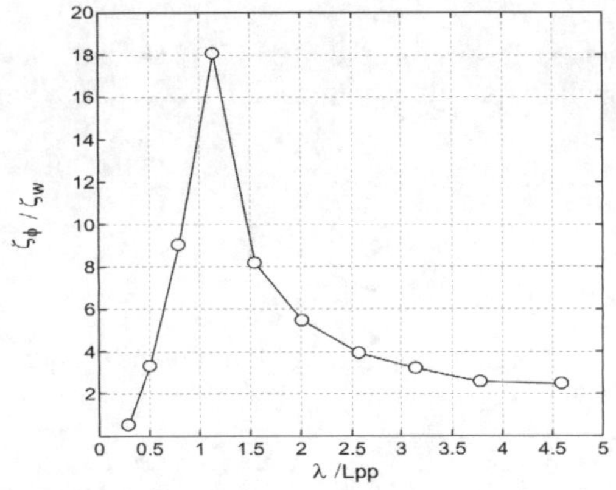

Figure 10. RAO-roll at heading 90 deg

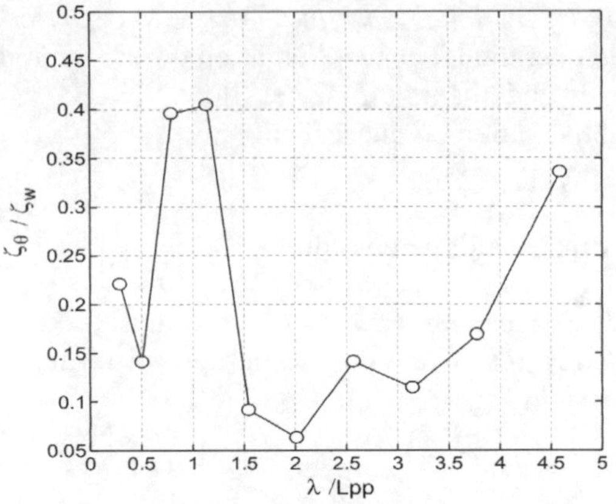

Figure 11. RAO-pitch at heading 90 deg

From the results of the seakeeping test at 180 deg heading (see Figure 6-8), the dominant motions that occur are heave and pitch modes. The heave motion response increases when $\lambda/Lpp \geq 1.5$, where the heave motion amplitude is approximately the same as the wave amplitude. While the largest pitch response that occurs is 5.2888 deg at λ/Lpp around 1.5. From the results of the seakeeping test at 90 deg heading (see Figure 9-11), the largest roll angle response value of 18.051 deg occurs at a wave length condition equal to the ship's natural roll period of 5.99 sec, or in this test it occurs at λ/Lpp around 1.1394. The largest heave response also occurs in the roll resonance condition, namely when λ/Lpp is around 1.1394 with a heave response of 0.827 m for a wave height of 2 m, where λ is wave length, Lpp is ship length, ς_ϕ is motion amplitude, ς_ω is wave amplitude.

3.2. *Manoeuvring test (turning circle and zig-zag)*

From the analysis of turning circle test data, manoeuver parameters will be obtained which consist of advanced, transfer, tactical, and turning diameter as the main parameters of turning circle movement. While the parameters considered in the zig-zag maneuver test are steering angle and overshoot angle. The results of the 20/20 deg turning circle and zig-zag test show that the ship's turning and zig-zag ability meets the IMO criteria [13]. The results of the turning and zig-zag manoeuvring tracking results are shown in Figures 12 ~ 14. Some of the parameter values for turning circle and zig-zag test results are shown in Table 2 and Table 3. The simulation results show that all parameter sizes are still below the maximum criteria threshold IMO criteria.

Table 2. Turning circle result.

Parameter	Turning circle Rudder angle = 35 deg				IMO
	Port direction result		Starboard direction result		
	(m)	Non dimensional	(m)	Non dimensional	
Advance	84.64	1.71 Lpp	111.15	2.24 Lpp	4.50 Lpp
Transfer	62.38	1.26 Lpp	70.77	1.43 Lpp	
Tactical diameter	110.90	2.24 Lpp	130.77	2.64 Lpp	5.00 Lpp
Turning diameter	99.55	2.01 Lpp	120.38	2.43 Lpp	

Table 3. Zig-zag test result.

Parameter	Zig-zag 20/20 deg	
	Result (deg)	IMO (deg)
First over shoot angle	7.3615	25

From the test results, it is known that the ship has the ability to make circular movements in the port direction with an advance length of 84.64 m (1.71 Lpp) and tactical 110.90 m (2.24 Lpp), while for the starboard direction the advance length is 111.50 m (2.24 Lpp) and tactical 130.77 m (2.64 Lpp) and this value is still safe in the IMO determination where the maximum value for advance length is 4.50 Lpp and tactical 5.00 Lpp. This ship also has the ability to perform zigzag movements of 20/20 deg with a first overshoot angle of 9,023 deg, this value is still within the safe limit set by IMO where the maximum first over shoot value is 25 deg.

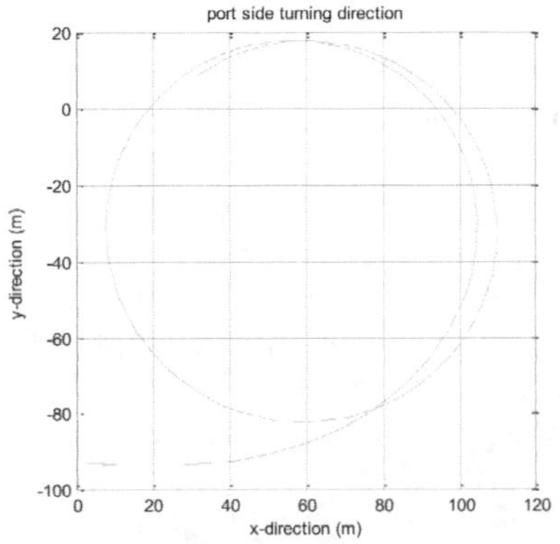

Figure 12. Tracking port side turning circle manoeuver Mini LNG carrier

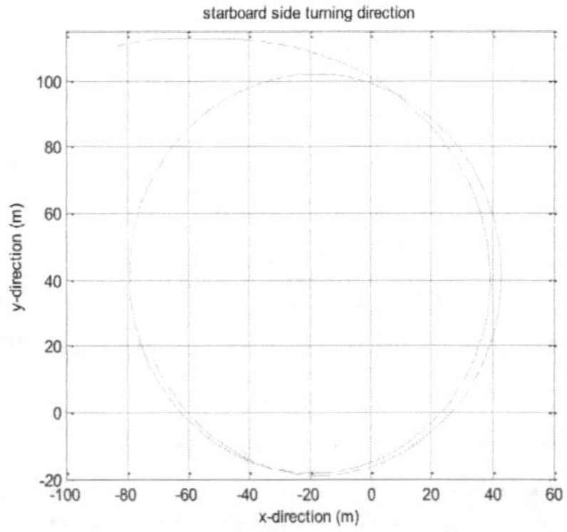

Figure 13. Tracking starboard side turning circle manoeuver Mini LNG carrier

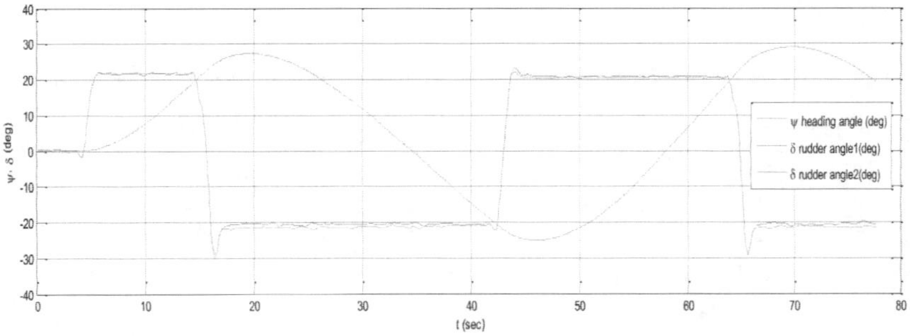

Figure 14. Movement Tracking 20/20 deg zig-zag manoeuver Mini LNG Carrier

3.3. Design analysis of water flow velocity of propeller

The important thing expected from the manoeuvrability of the ship is how to maintain the desired direction at all water depth conditions. In fact, the seakeeping and manoeuvring capabilities of ships will change when sailing in shallow waters. In shallow water, the ship will be slower and the heeling is significantly reduced and the turning diameter becomes larger so that it is not sufficient to assess the ship's turning ability based solely on the IMO criteria.[4]. Based on PIANC criteria, for shallow water if 1.2<h/t<1.5 and very shallow for h/T<1.2, where h is water depth and T is draught of ship [14].

Manoeuvring characteristics were significantly affected by under keel clearance (UKC) [15]. The smaller the UKC the larger the turning diameter of the ship as shown in Figure 15.

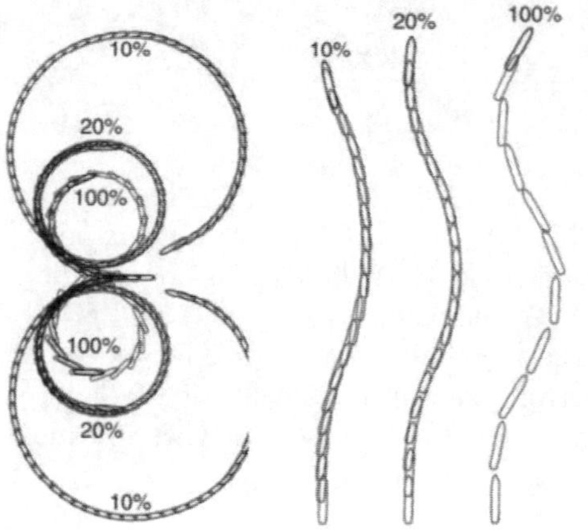

Figure 15. Ship's turning and zig-zag capability in UKC variations (100%, 20% and 10%) [16]

Manoeuvring activities in shallow water are riskier because they have the potential to cause sedimentation on the seabed. The relationship between the performance of the propeller with the occurrence of erosion and sedimentation at the bottom of the water can be explained in the form of a mathematical equation. Prediction of the speed of the ship's propeller jet is the first step to investigate the scour carried out by the propeller jet [17]. The sedimentation process begins to occur when the velocity of the current flowing over the bottom of the water has crossed the threshold of about 0.15 m/s and 0.30 meters above the bed [18]. Another study concluded that the bottom layer shift will occur if the shear stress θ is greater than the threshold θ_{cro}, $\theta > \theta_{cr}$ where $\theta = \tau_{cro},o$ / [(ps – pw)gd_{50}] where τ_{cro},o = critical shear stress for particles without cohesion, ps = sediment density, pw = fluid density, s = ps/pw = relative density, and d_{50} = median diameter of sediment [19] [20].
The velocity of the water Vo generated from the rotation of the propeller can reach 8 m/s and will scratch the bottom of the water [21]. The equation for the maximum velocity in front of the propeller is expressed by Fuehrer and Romisch in [21] by:

$$Vo = nD_p\sqrt{\frac{2C_tD_p{}^2}{A}} \tag{1}$$

where n is the rotation of propeller (1/sec), D_p is the diameter of propeller (m), A is the projection plane of the propeller (m2), and C_t is the coefficient of thrust. By subtitution $= A = \pi D_p{}^2/4$ we get the equation;

$$Vo = 1.59nD_p\sqrt{C_t} \tag{2}$$

the value of C_t is not constant but depends on the characteristics of the propeller so that in the form of an equation it can be given as [17];

$$Vo = \varsigma nD\sqrt{C_t} \tag{3}$$

where

$$\varsigma = D^{-0.0686}\left(\frac{P}{D}\right)^{1.519-0.403} \quad -\grave{0}.323 \tag{4}$$

BAR is the blade area ratio (projected area of all blades related to the propeller disc area) and P/D is the pitch ratio of the propeller (quotient of a pitch and the propeller diameter), D is the propeller diameter. The propeller's water flow velocity distribution pattern can be divided into two zones in Figure 16.

IOP Conf. Series: Earth and Environmental Science 1081 (2022) 012033 doi:10.1088/1755-1315/1081/1/012033

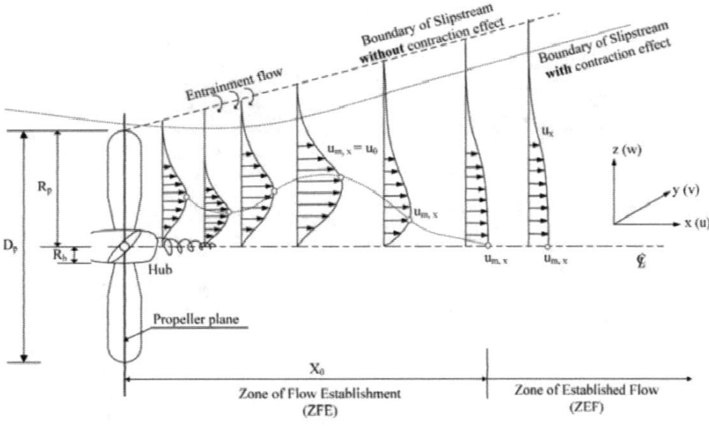

Figure 16. Water flow pattern of single propeller [22]

Zone of flow establishment is the area of flow formation starting from the rotating propeller. The maximum speed in this area can be calculated based on the following equation [23].

$$V_{x,max} = V_o \quad (5)$$

Velocity changes with increasing flow distance traveled by the following equation:

$$\frac{Vx,max}{Vo} = 1.107 - 0.184.\left(\frac{x}{Dp}\right) \quad (6)$$

where D_p is the diameter of the propeller, x distance from the center of the propeller shaft.
Zone of established flow is an area where the water flow is stable so that the calculated flow velocity is radial with the equation:

$$\frac{Vx,r}{Vx,max} = e^{\left[-22.2.\left(\frac{r}{x}\right)^2\right]} \text{, where r is the flow radius} \quad (7)$$

A similar approach can be given for the maximum flow axial velocity V_o based on the following formula [24]:

$$Vo = 1.60 f_n n_d D \sqrt{K_t} \text{ , for silent ship (J=0)} \quad (8)$$

$$V_{oj} = \sqrt{\;}\ V_o \text{ , for moving ship (J} \neq 0 \quad (9)$$

where n_d [1/sec] is the propeller design rotation rate; f_n is the propeller application rotation rate factor is between 0.7 – 0.8 during the manoeuver; D is the propeller diameter [m]; P is the propeller design pitch [m]. P/D value can be assumed 0.7. and K_t or K_{tj} propeller thrust coefficient whose magnitude is determined by the equation:

$$K_t = 0.55 \frac{P}{D} \quad (10)$$

$$K_{tj} = 0.55 \frac{P}{D} - 0.46J \quad (11)$$

J is the propeller ratio of 0.2 to 4.5 depending on the wave factor w. For a moving ship J is determined by the equation::

$$J = \frac{vs(1-w)}{nD} \quad (12)$$

The eroded material will be carried by the current to move places and at a certain distance from a mound of material. The description of erosion and material mounds due to the flow of water in the rotation of the propeller is in Figure 17.

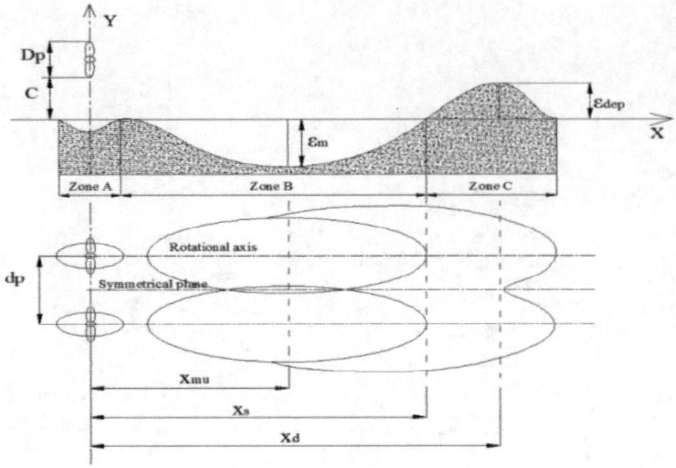

Figure 17. The size of the abrasion posture due to the twin propeller [12]

From Figure 17, it was found that the maximum abrasion position occurs at a distance of Xm, twin = 2.1 Dp, where Dp is the diameter of the propeller. For further calculations that are suitable for Mini LNG Carrier, the propeller data needs to be adjusted to the data in Figure 18 and Table 4 below;

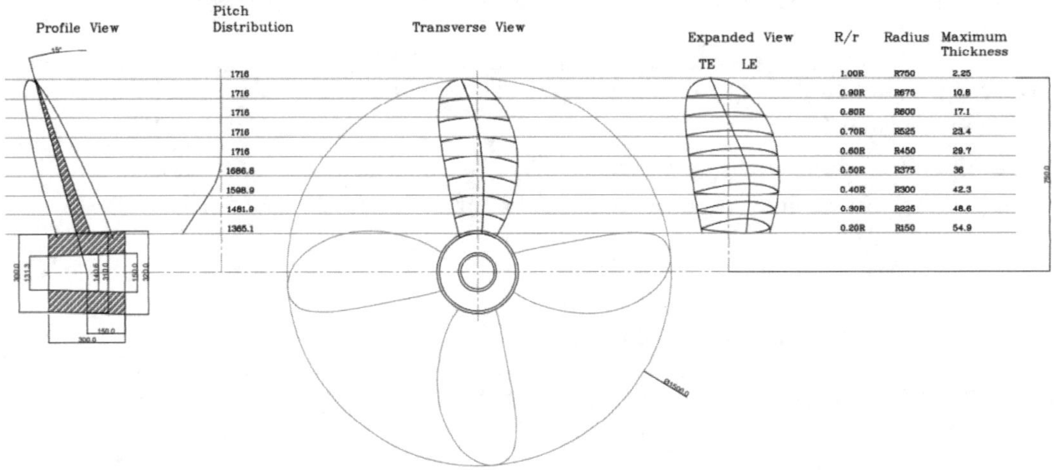

Figure 18. Propeller projection drawing of Mini LNG carrier Iso Tank Container

Table 4. Scaling factor

Parameter	Symbol	Full scale	Model scale	unit
Diameter	D	1.499	0.131	m
Pitch at Diameter Ratio	P/D	1.144	1.144	
Expanded blade area ratio	A_E / A_0	0.400	0.400	
Numbers of blade	Z	4	4	blade

3.4. Environmental approach for operational concept of Mini LNG Carrier

The coastal environment is very vulnerable to damage because it is an area where there are many activities, one of which is shipping activities. The increase in Mini LNG Carrier activities in the coastal area has the potential to hurt the maritime environment, one of which is the occurrence of erosion and sedimentation which can cause silting of the seabed. Sedimentation is also an ecological problem where organisms that depend on the interior of the water cannot survive if their habitat is buried by sediment[25]. Sedimentation caused by shipping activities causes ecological changes where the ratio of the composition of the chemical elements of the basic structure of the water changes. The ecological response becomes slower than the sedimentation process so it affects the sedimentation date records on a hundred-year scale[26].

The scheme for the impact of environmental damage due to shipping activities through the sedimentation process can be given as shown in Figure 19 [27]. To avoid some of these impacts, the ship's operational concept needs to be made environmentally friendly.

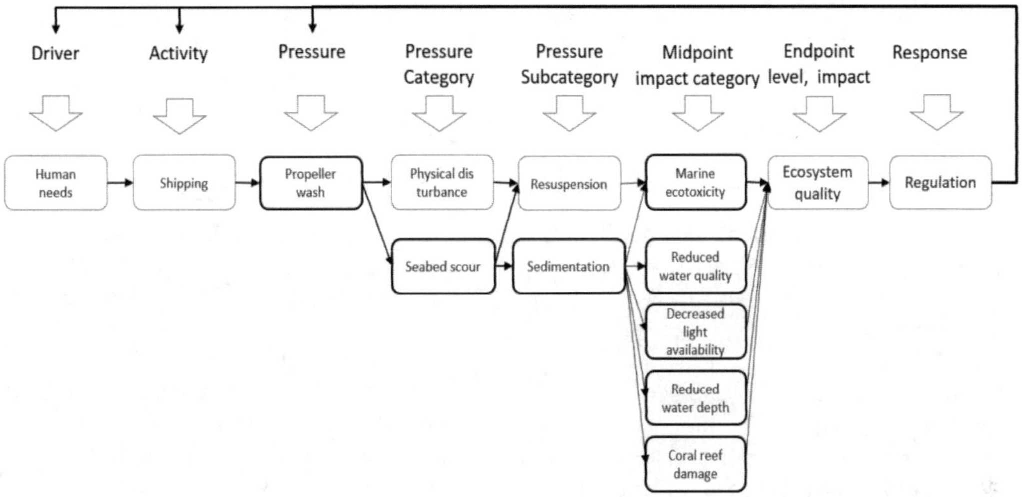

Figure 19. The scheme for the impact of sedimentation due to shipping activities

The operational concept framework proposed by this study is shown in Figure 20, with sequential stages; determine the depth of water in the shallow water category with a ratio value of water depth to draught ($h/T < 1.5$), identify the value of the velocity of the water flow from the ship's propeller at each rpm variation, determine the critical velocity of water flow at each selected water depth, make a graph of the velocity of the water and the rpm of the propeller, tabulate safe propeller rpm data for each depth of water, graphing the water depth zone with the appropriate propeller rpm, create an automated system for preventing sedimentation.

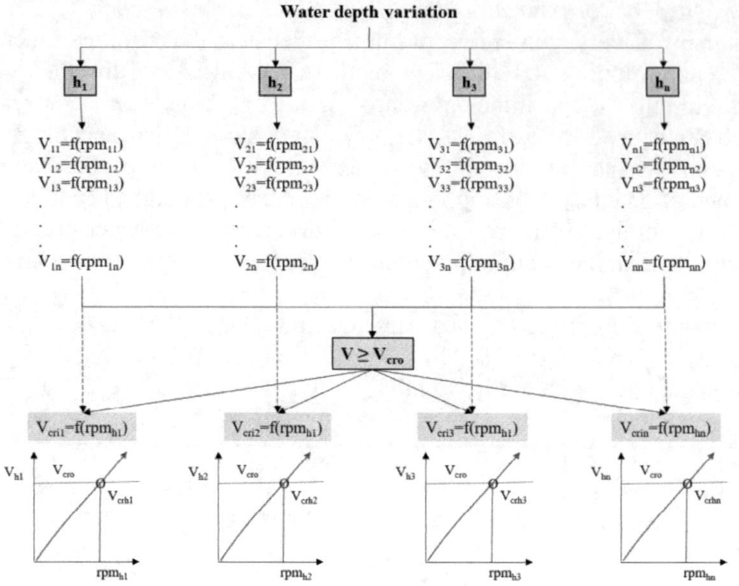

Figure 20. Scenario of determining critical velocity value

Where h_1, h_2, h_3, h_n are variations of water depth, V_{1n}, V_{2n}, V_{3n}, V_{mn} are variations of water flow velocity, rpm_{1n}, rpm_{2n}, rpm_{2n}, rpm_{mn} are variations of propeller rotation, V_{cro} is critical velocity of water flow where sedimentation begins to occur, V_{crh1}, V_{crh2}, V_{crh3}, V_{crnm} are variations of critical velocity at h_1, h_2, h_3, h_n respectively.
The velocity of water flow at each selected depth is calculated at each specified rpm, if the velocity value is equal to or greater than the critical velocity, the rpm at that time is considered the maximum rpm at that depth. Furthermore, the maximum rpm value of the inward variation is transformed into a graph to form a map of the depth and maximum rpm of the vessel under study. Example of a map based on the concept of determining the critical velocity value in Figure 21.

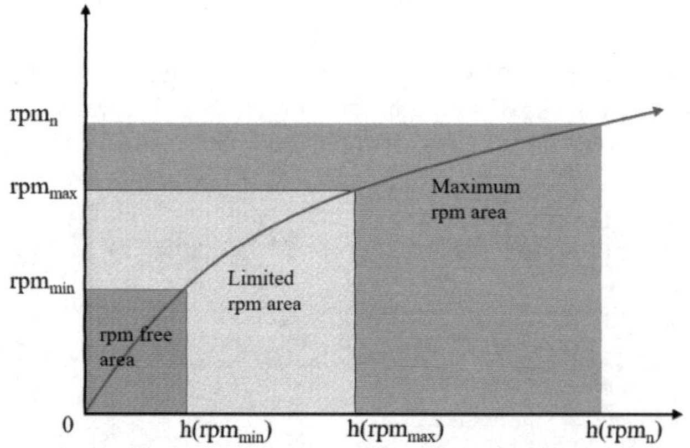

Figure 21. Schematic chart of the division of the operational depth of the ship

Figure 21 is the division of the ship's operational area zone based on the depth and rpm of the propeller that is safe for the environment where; The red zone indicates the depth at which the propeller should not be used; The yellow zone indicates the depth at which the propeller can be used with a limited rpm; The green zone is the depth at which the ship can use the propeller with the maximum rpm.

The data from the graph in Figure 19 becomes a reference for making an automation system where the ship can adjust the propeller rotation based on the water depth data obtained through the inward detection sensor mounted on the bottom of the ship's bow. The concept of the automation system as shown in Figure 22.

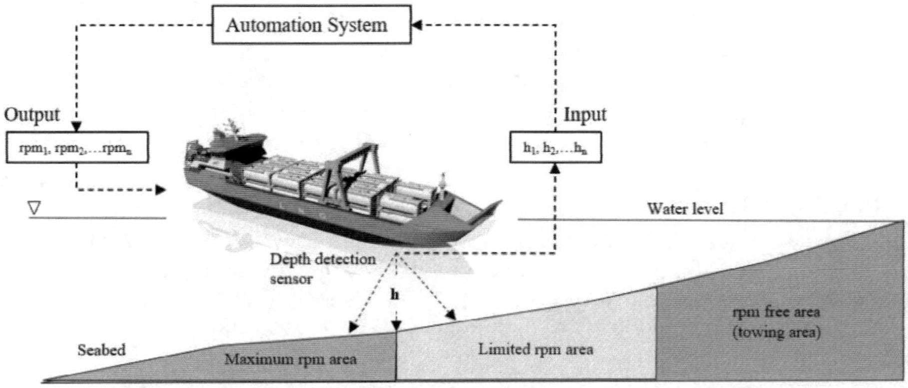

Figure 22. Ship sedimentation prevention automation system

The use of a depth detection system with automatic propeller rpm adjustment is an early warning of the possibility of erosion and sedimentation so that it can be anticipated earlier. This system can be used for several variables other than depth so that it supports the shipping system automation system in the current digital era.

4. Conclusion

The results of the model scale test on the Mini LNG Carrier show that the criteria for seakeeping and manoeuvring capabilities are by IMO (International Maritime Organization) requirements for deep sea conditions. To operate in shallow water, the Mini LNG Carrier needs to adjust the rotation of the propeller to the water depth so as not to cause sedimentation to the seabed. The manoeuvrability of the Mini LNG Carrier in shallow water can be optimized by adjusting the rotation of the propeller to the depth of the water with a water flow velocity threshold from the propeller of 0.15 m/s as the maximum water flow velocity to prevent sedimentation on the seabed. The development of an automation system between the propeller rpm and water depth will help a digital-based marine transportation system prevent damage to the coastal environment due to sedimentation caused by ships. The implementation of this concept on a wide scale needs to be supported by policies and regulations from relevant government agencies and good cooperation with ship owners.

References

[1] M. A. Budiyanto, I. K. Singgih, A. Riadi, and G. L. Putra, "Study on the LNG distribution to Mobile Power Plants using a Small-Scale LNG Carrier for the case of the Sulawesi region of Indonesia," *Energy Reports*, vol. 8, pp. 374–380, 2022, doi: 10.1016/j.egyr.2021.11.211.
[2] L. Brandon C. Montanye, "Analysis of the Landing Ship Tank (LST) and its influence on amphibious warfare during world war two," Iowa State University, 1997.
[3] P. M. Carrica, A. Mofidi, K. Eloot, and G. Delefortrie, "Direct simulation and experimental study of zigzag maneuver of KCS in shallow water," *Ocean Eng.*, vol. 112, pp. 117–133, 2016, doi: 10.1016/j.oceaneng.2015.12.008.
[4] E. M. Milanov and V. Y. Chotukova, "Roll motion of containership in shallow water," *RINA, R.*

Inst. Nav. Archit. - Int. Conf. Sh. Manoeuvering Shallow Confin. Water Bank Eff. 2009, no. May 2009, pp. 368–399, 2009, doi: 10.3940/rina.bank.2009.10.

[5] M. Wei and Y.-M. Chiew, "Characteristics of Propeller Jet Flow within Developing Scour Holes around an Open Quay," *J. Hydraul. Eng.*, vol. 144, no. 7, p. 04018040, 2018, doi: 10.1061/(asce)hy.1943-7900.0001470.

[6] P. Dinar Da and B. Erika, "Effect of sedimentation volume on productivity and dwelling time of ships at bom baru port," *MATEC Web Conf.*, vol. 138, 2017, doi: 10.1051/matecconf/201713807006.

[7] A. Bianchini, F. Cento, A. Guzzini, M. Pellegrini, and C. Saccani, "Sediment management in coastal infrastructures: Techno-economic and environmental impact assessment of alternative technologies to dredging," *J. Environ. Manage.*, vol. 248, no. January, p. 109332, 2019, doi: 10.1016/j.jenvman.2019.109332.

[8] T. V. Rehitha, N. Ullas, G. Vineetha, P. Y. Benny, N. V. Madhu, and C. Revichandran, "Impact of maintenance dredging on macrobenthic community structure of a tropical estuary," *Ocean Coast. Manag.*, vol. 144, pp. 71–82, 2017, doi: 10.1016/j.ocecoaman.2017.04.020.

[9] A. F. Molland, *The maritime Engineering Reference Book*. 2008.

[10] G. A. Hamill and C. Kee, "Predicting axial velocity profiles within a diffusing marine propeller jet," *Ocean Eng.*, vol. 124, pp. 104–112, 2016, doi: 10.1016/j.oceaneng.2016.07.061.

[11] J. Jiang *et al.*, "Ship Twin-propeller Jet Model used to Predict the Initial Velocity and Velocity Distribution within Diffusing Jet," *KSCE J. Civ. Eng.*, vol. 23, no. 3, pp. 1118–1131, 2019, doi: 10.1007/s12205-019-1370-x.

[12] Y. Cui, W. H. Lam, T. Zhang, C. Sun, and G. Hamill, "Scour induced by single and twin propeller jets," *Water (Switzerland)*, vol. 11, no. 5, 2019, doi: 10.3390/w11051097.

[13] 28th_ITTC_Manoeuvring_Committee, "Full Scale Manoeuvring Trials - ITTC," *28th ITTC Conf.*, pp. 1–18, 2017.

[14] T. M. Committee, "Final Report and Recommendations to the 23rd ITTC," in *The 23rd ITTC*, 2002, vol. I, pp. 153–687.

[15] G. Delefortrie and M. Vantorre, "Modeling the maneuvering behavior of container carriers in shallow water," *J. Sh. Res.*, vol. 51, no. 4, pp. 287–296, 2007, doi: 10.5957/jsr.2007.51.4.287.

[16] M. Vantorre, K. Eloot, G. Delefortrie, E. Lataire, M. Candries, and J. Verwilligen, "Maneuvering in Shallow and Confined Water," *Encycl. Marit. Offshore Eng.*, pp. 1–17, 2017, doi: 10.1002/9781118476406.emoe006.

[17] W. Lam, G. A. Hamil, Y. C. Song, D. J. Robinson, and S. Raghunathan, "A review of the equations used to predict the velocity distribution within a ship's propeller jet," *Ocean Eng.*, vol. 38, no. 1, pp. 1–10, 2011, doi: 10.1016/j.oceaneng.2010.10.016.

[18] British Standard, "Maritime Structures - Part 1 : Code of practice for general criteria," in *Code of practice for general criteria.*, vol. BS 6349-1:, no. 196, 2003, p. 254.

[19] L. C. van Rijn, "Unified View of Sediment Transport by Currents and Waves. II: Suspended Transport," *J. Hydraul. Eng.*, vol. 133, no. 6, pp. 668–689, 2007, doi: 10.1061/(asce)0733-9429(2007)133:6(668).

[20] S. Alhaddad, R. J. Labeur, and W. Uijttewaal, "Breaching flow slides and the associated turbidity current," *J. Mar. Sci. Eng.*, vol. 8, no. 2, 2020, doi: 10.3390/jmse8020067.

[21] G. A. Hamill, J. A. McGarvey, and D. A. B. Hughes, "Determination of the efflux velocity from a ship's propeller," *Proc. Inst. Civ. Eng. Marit. Eng.*, vol. 157, no. 2, pp. 83–91, 2004, doi: 10.1680/maen.2004.157.2.83.

[22] H. Shih-chun, "3-D Flow Measurements of a Swirling Jet Induced by a Propeller by Using Piv," no. 1967, 2013.

[23] O. Stoschek, E. Precht, O. Larsen, M. Jain, L. Yde, and N. Ohle, "Sediment Resuspension and Seabed Scour Induced By Ship-Propeller Wash," *PIANC World Congr.*, pp. 1–18, 2014.

[24] A. Guarnieri *et al.*, "Effects of marine traffic on sediment erosion and accumulation in ports: A new model-based methodology," *Ocean Sci.*, vol. 17, no. 2, pp. 411–430, 2021, doi: 10.5194/os-

17-411-2021.

[25] B. J. O'Neill, J. Steffen, and M. Jakubauskas, "Effects of Sedimentation on the Profundal Benthic Macroinvertebrates of a Great Plains Reservoir," *Trans. Kansas Acad. Sci.*, vol. 117, no. 1–2, pp. 61–68, 2014, doi: 10.1660/062.117.0108.

[26] C. Xue *et al.*, "Impact of Ship Traffic on the Characteristics of Shelf Sediments: An Anthropocene Prospective," *Front. Mar. Sci.*, vol. 8, no. August, pp. 1–16, 2021, doi: 10.3389/fmars.2021.678845.

[27] E. Ytreberg, S. Åström, and E. Fridell, "Valuating environmental impacts from ship emissions – The marine perspective," *J. Environ. Manage.*, vol. 282, no. January, 2021, doi: 10.1016/j.jenvman.2021.111958.

IOP Conf. Series: Earth and Environmental Science 1081 (2022) 012034 doi:10.1088/1755-1315/1081/1/012034

The Performance of KH and KG Flexible Riser with Distributed Buoyancy Modules Configuration

Muhammad Pratikto Priyanto[(a)], Danasesha Paradinda Putra[(a)], Iis Iskandar[(a,b)]

[a]Star Energy (Kakap) Ltd., Indonesia ; [b]Corresponding Author.

Muhammad.Pratikto@starenergy.co.id, Danasesha.Paradindaputra@starenergy.co.id, Iis.Iskandar@starenergy.co.id

Abstract. As a relatively new technology in the riser structure, Distributed Buoyancy Modules (DBM) are seldom to use on flexible risers in Indonesia Oil and Gas Fields which are dominated by brown fields. DBM have several advantages compared with conventional subsea structures such as Mid-Water Arches (MWA) in installation phase. DBM provides lower budget spending due to the shorter period and easier method in installation phase. The paper presents engineering design of DBM operation and installation in Kakap Field, Natuna Sea, Indonesia to replace the existing KH and KG Risers. Several cases was evaluated in this paper to perform the behaviour of risers using DBM. Based on the static analysis, the maximum effective tension for KH Riser and KG Riser are 100.7 KN and 83.4 KN respectively during their operation with allowable value of less than 1530 KN. Meanwhile, the minimum bending radius for the risers when they are being operated are above the allowable value of 3.926 m which are 6.6 m and 6.7 m for KH and KG Riser in turn. The analysis indicates the both of risers are safe to operate during their life period.

1. Introduction

In the offshore oil and gas production process, exploited hydrocarbon resources are delivered from sea well to the offshore floating storage station called Floating Production Storage and Offloading (FPSO) [1]. This storage critically connected with flexible risers equipped with buoyancy module, which highly vulnerable and complex in terms of technical or even environmental risks which may lead to higher economic cost.

The configuration of flexible riser could be seriously determined by the selection of its buoyancy system that maintain riser to be “float” at mid-water level [1,2]. Utilization of Distributed Buoyancy Module (DBM) becomes one of the alternatives to generate efficient risers yet simple configuration with simple installation procedure and simple equipment involved [2,3]. The use of buoyancy modules not only to assist in achieving desired configurations and shape but also to achieve optimal performance of structural strength against topside or tension loads of the deepwater flexible risers [1,4]. One of common preferred configuration is Lazy Wave riser configuration, which is known to its better resistance against dynamic condition in comparison with conventional Steel Catenary Riser (SCR)[1]. The high stress and fatigue damage responses induced by dynamic condition of particular riser configuration could be determined and analysed by an aid by several design engineering simulation.[4]

In this paper, a performance of flexible risers after equipped with DBMs in order to achieve desired Lazy Wave riser configuration is discussed based on the actual case. The 2 (two) sets of flexible risers,

KH & KG risers were equipped with new risers and DBM buoyancy system. In the case of assessment on the installed risers, Interference assessment, static & dynamic condition analysis, are carried out and discussed. It is shown via dynamic simulation and analysis, that these flexible risers with DBM have desirable structural performance against possible force due to bending and tension, showing that these risers' configuration could carry operation safely. This installation carried out at Kakap Natuna Sea, Indonesia.

2. Data and Configuration

2.1. Riser Parameters

A lazy-wave configuration will be used for both KG and KH 10" risers with the length of 199 m and 175 m respectively. This configuration utilizes distributed buoyancy modules along a determine section of the riser length to provide a multi-curve/wavy shape to the catenary. This aims at reducing the riser weight hanged from the platform, minimize bending moment at the sag bend near the seabed and stress along the riser length and connections.

A section of the flexible will be stationary on the seabed in depth of 92.52 m (Mean Sea Level), followed by a buoyant section where the buoyancy modules are distributed. This induces a hog bend into the flexibles, followed by a sag bend and with the final hanging catenary terminating into a flange at the SPM. Bend stiffeners for both risers are located at the topside SPM hang-off point. The general riser configuration along with the SPM mooring lines which have 6 chains are shown in Figure 1. The chains are divided into 2 sections which are Segment A (at SPM) and Segment B (at Anchor) with the length of 165 m and 289 m as well as R4 and ORQ for their grade respectively. In addition, detailed parameters of the flexible pipe for both the KG & KH riser are shown in Table 1. Note that KG and KH risers are identical in characteristic and properties. The nomenclatures of KG and KH Flexible Risers are described in Figure 2. The other environment parameters with respect to coming direction (wave origin direction) are shown in Table 2.

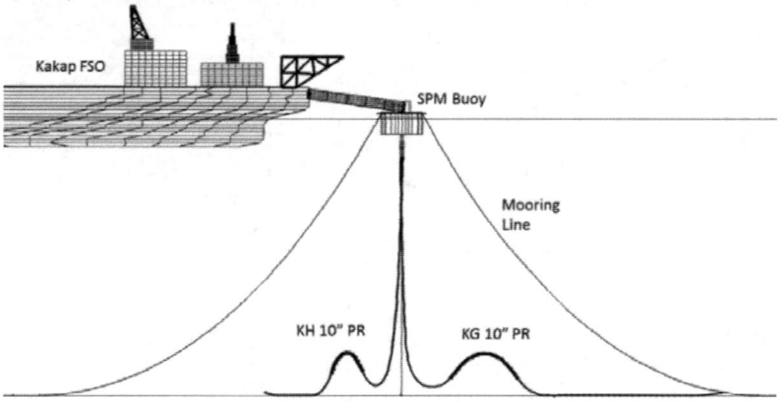

Figure 1. Flexible risers & mooring line configuration elevation view.

Table 1. KH & KG Flexible Riser Configuration.

Flexible Riser	SBM to PLEM horizontal distance	Lower Catenary Length	Buoyant Section Length	Upper Catenary Length	Number of Buoyancy Modules	Buoyancy Module Centre to Centre Spacing
KH	88.5 m	44 m	30 m	101 m	14 m	2.19 m
KG	106 m	60 m	46 m	93 m	21 m	2.19 m

Table 2. (a) Extreme wave 100yr RP sea state parameters.

Parameter	Unit	N (0°)	NE (45°)	E (90°)	SE (135°)	S (180°)	SW (225°)	W (270°)	NW (315°)
H_{max}	[m]	9.5	12.5	9.3	6.0	7.6	8.5	7.1	8.7
T_{lower}	[s]	8.2	9.4	8.1	6.5	7.3	7.7	7.0	7.8
T_{mean}	[s]	9.1	10.4	9.0	7.2	8.1	8.6	7.8	8.7
T_{upper}	[s]	10.0	11.4	9.9	7.9	8.9	9.5	8.6	9.6
H_s	[m]	5.1	6.7	5.0	3.2	4.1	4.6	3.8	4.7
T_p	[s]	9.9	11.3	9.8	7.9	8.8	9.3	8.5	9.5
T_z	[s]	7.7	8.8	7.6	6.1	6.9	7.3	6.6	7.4

Table 2. (b) Extreme wave 1yr RP sea state parameters.

Parameter	Unit	N	NE	E	SE	S	SW	W	NW
		0°	45°	90°	135°	180°	225°	270°	315°
H_{max}	[m]	7.2	9.1	6.7	4.2	5.4	6.4	5.2	6.4
T_{lower}	[s]	7.1	8.0	6.8	5.5	6.2	6.8	6.0	6.8
T_{mean}	[s]	7.9	8.9	7.6	6.1	6.9	7.5	6.7	7.5
T_{upper}	[s]	8.7	9.8	8.4	6.7	7.6	8.3	7.4	8.3
H_s	[m]	3.9	4.9	3.6	2.3	2.9	3.4	2.8	3.5
T_p	[s]	8.6	9.7	8.3	6.6	7.5	8.1	7.3	8.1
T_z	[s]	6.7	7.5	6.4	5.1	5.8	6.3	5.7	6.3

Table 2. (c) Extreme 100yr RP current parameters.

Depth [m]	N	NE	E	SE	S	SW	W	NW
	0°	45°	90°	135°	180°	225°	270°	315°
0.0	0.60	0.60	0.60	1.00	1.10	0.90	0.40	0.50
23.1	0.40	0.40	0.40	0.80	0.90	0.80	0.30	0.30
46.3	0.20	0.20	0.30	0.70	0.80	0.60	0.20	0.20
69.4	0.20	0.10	0.20	0.60	0.70	0.50	0.10	0.10
91.5	0.10	0.10	0.10	0.40	0.40	0.30	0.10	0.10

Table 2. (d) Extreme 1yr RP current parameters.

Depth [m]	N	NE	E	SE	S	SW	W	NW
	0°	45°	90°	135°	180°	225°	270°	315°
0.0	0.50	0.50	0.50	0.90	1.00	0.80	0.40	0.40
23.1	0.40	0.30	0.40	0.80	0.90	0.70	0.30	0.30
46.3	0.20	0.20	0.30	0.70	0.80	0.60	0.20	0.20
69.4	0.20	0.10	0.20	0.60	0.70	0.50	0.10	0.10
91.5	0.10	0.10	0.10	0.40	0.40	0.50	0.10	0.10

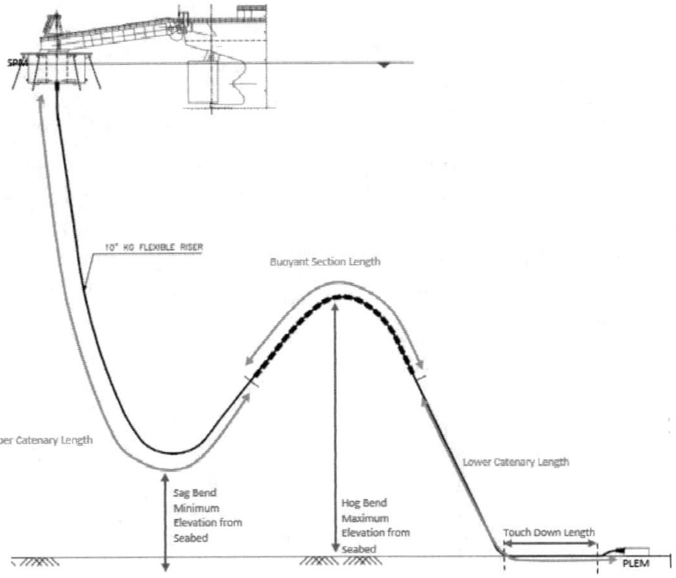

Figure 2. KH & KG Flexible Riser Lazy Wave Configuration Nomenclature.

2.2. Vessel Parameters

Kakap Natuna Floating Storage Offloading (KN FSO) Vessel is the vessel that will be used during the operation of risers. The vessel has an adequate stability due to large size with the length of 314.64 m, 43.40 m width and 8.49 m ballasted draught. The vessel Response Amplitude Operator (RAO) for Kakap FSO are extracted from analyses performed in Moses of the vessel, for both ballasted and loaded.

3. Methodology

The design methodology for this paper will be based on applicable standard and company specification. Figure 3 shows the workflow adopted for the KG & KH riser replacement design. The workflow is primarily driven by the requirement to provide regular interface deliverables for detailed design engineering phase. Analysis will be performed by Analysis was performed using Moses and Orcina's OrcaflexTM Version 10.1C and it was assisted by Third-party.

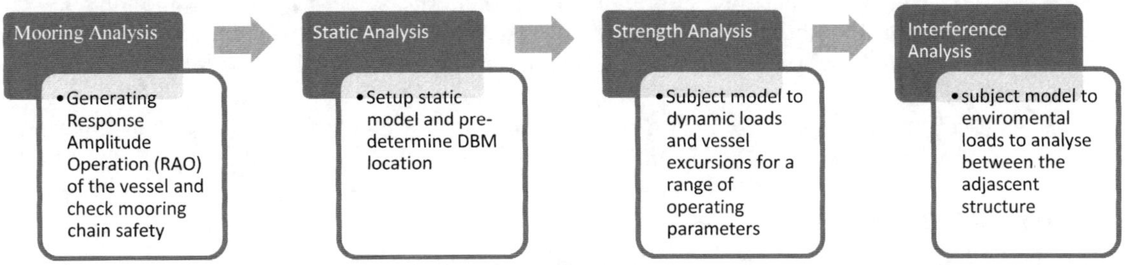

Figure 3. Workflow.

3.1 Mooring Analysis

There are two stages to carry out mooring analysis. The first stage is to get the motion response of the FSO. The analysis will be performed by employing diffraction approach to obtain the first-order (high frequency) motion responses under the excitation of regular waves, the second-order (low frequency) motion responses and motion statistics of the vessel under the excitation of random waves. Further, the hydrodynamic motion properties of the FSO will be obtained based on the result of environmental loading. This analysis will be carried out in MOSES.

The second stage is the computation of mooring line tension and buoy excursion. A moored structure can be characterized as a dynamical system subject to the various environmental conditions with low and high frequency excitation as result of the wind, waves and current acting on the system. The reaction forces are a combination of added mass forces, wave drift damping forces, viscous forces and combined restoring and damping forces interacting with mooring lines. The buoy and mooring line tension are simulated using ORCAFLEX based on the excitation and reaction for simulating the buoy excursion and mooring line tension.

3.2 Static Analysis

The static analysis is the analysis step to ensure the correct design of the DBM. The basics of the static analysis are to:

- Insert input data from basis of design document into finite element analysis software;
- Re-adjust and iterate the location, spread, etc of the DBMs along the flexible riser to ensure visibly correct lazy-wave configuration of both risers.
- Perform static / quasi-static analysis, and re-adjust DBMs as per above point.

3.3 Strength Analysis

The dynamic analysis is performed as time domain analysis. Each of load case is analysed based on the current profile and regular waves. The relevant extreme regular wave height Hmax and associated period Tass is derived from significant wave height, peak period and spectral shape. Morrison's equation is used to calculate the hydrodynamic action on the riser and distributed buoyancy modules. The extreme load case matrix considers the following parameters. If any items in the load case are shown to be non-critical, these may be excluded during detailed design.

1. System State. Two states are considered in the analysis. Start of Life (SOL) is characterised by initial as-built conditions of riser system. End of Life (EOL) is based on the changes occurred in the riser system (e.g. Marine Growth development and water ingress).
2. Contents Density. Minimum and maximum density during operation are considered in the analysis.
3. Environment (Wave-Current). The directionality of the environmental conditions is applied in accordance with the associated vessel offset and vessel loading condition. Note that the environmental conditions are given for the 8 compass directions N, NE, E, SE, S, SW, W and NW. 100 year and 1 year environmental conditions are considered in the analysis.
4. Riser Temperature and Pressure. Critical combinations of riser pressure and temperature will be analysed. In general, a pressurized cold state will give the highest interface loads, while a depressurized hot state will give the highest utilization of minimum bending radius of the flexible riser.
5. Vessel Offset. Vessel excursion is considered in the analysis to ensure that all aspects of interference and structural design criteria are sufficiently investigated.

3.4 Interference Analysis

During the interference analysis, the extreme load case matrix (Table 17) is investigated by extracting relevant clearance between risers and adjacent structures. Furthermore, visual inspection of critical load cases is carried out.

The following cases are investigated:

- Mooring line interference: Clearances between risers and mooring lines are extracted for all load cases.

- Seabed interference: Clearances between risers and mooring lines are extracted for all load cases.
- SPM buoy body interference: Clearances between risers and buoy body are extracted for all load cases. Interference between risers and the buoy body is not allowed at any times.

4. Analysis and Discussion

4.1. Mooring Analysis

4.1.1 *FSO RAOs and Combined System Motions.* The Kakap Natuna FSO vessel Response Amplitude Operators (RAOs) have been generated based on the vessel characteristics (geometry and dead weight) and body linesplan to accurately model the hull shape. A set of environmental loads were applied to the model and the dynamic responses of the vessel body have been calculated in all six (6) degrees of freedom. The analysis and motions calculations were performed using the commercial software MOSES. As the FSO is attached to a Single Point Mooring structure it was necessary to create a combined model of FSO and SPM linked by a rigid structure with hinged connection to generate the motions of FSO plus SPM.

4.1.2 *FSO and SPM Mooring Analysis.* The analysis study was carried out in two phases. Phase one relates to 3D diffraction analysis performed in MOSES to obtain the motion responses of the FSO in free floating condition. Phase two relates to the coupled time domain analyses performed using ORCAFLEX, with the motion responses obtained from Phase one.

Two mooring configurations were applied (in-line and between-line) as per Figure 4(a) and 4(b) for two FSO draughts. For each analysis condition, Ultimate Limit State (ULS) with "all line intact" and Accidental Limit State (ALS) with "one line damaged case" are considered as per statutory requirement. Intact indicates that the mooring system is operating normally (vessel excursions are controlled), whilst Damaged means that one (1) mooring leg has failed (vessel excursions are more prominent). The survival condition with 100 years return period of environmental load and operating condition with 1 year return period of environmental load have been considered. All environmental elements (wave, current, and wind) are assumed collinear as conservative approach.
From the analysis results, maximum mooring lines tension and maximum excursion of the FSO plus SBM are summarized in the Table 3 and Table 4.

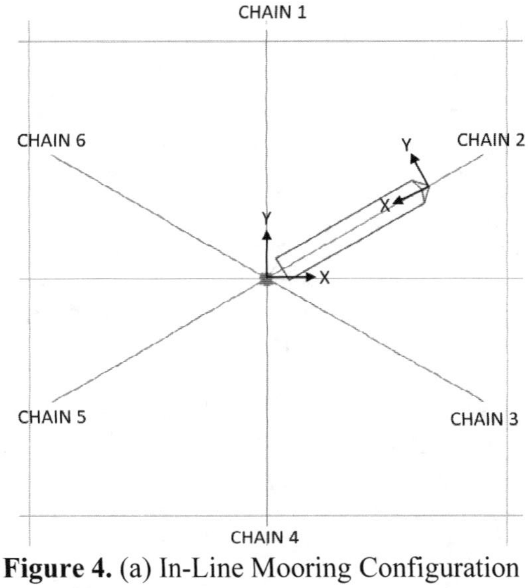

Figure 4. (a) In-Line Mooring Configuration

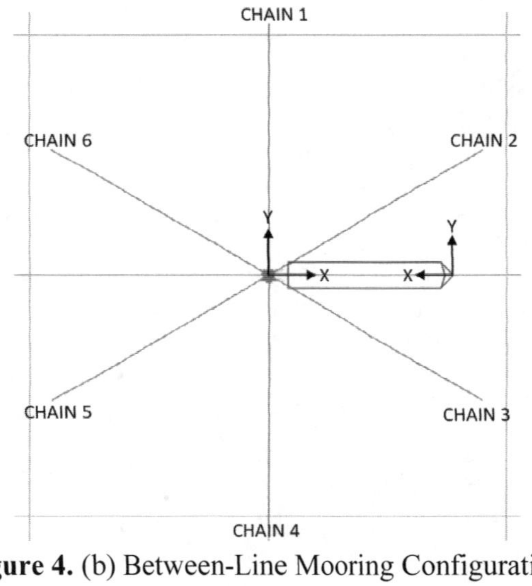

Figure 4. (b) Between-Line Mooring Configuration

Table 3. Mooring Analysis Results Summary for Intact Condition

Environmental Load	Vessel Condition	Condition	Mooring Line Tensions			Maximum SBM Excursion
			Max.	**SF**	**Remark(1)**	
Survival Condition (1 YRP)	Ballast	In-Line	59.33 MT	8.00	OK	6.16 m
		Between-Line	53.87 MT	8.81	OK	7.47 m
	Fully Loaded	In-Line	83.51 MT	5.68	OK	15.65 m
		Between-Line	93.58 MT	5.07	OK	14.33 m
Operating Condition (100 YRP)	Ballast	In-Line	101.71 MT	4.67	OK	8.39 m
		Between-Line	84.71 MT	5.60	OK	10.60 m
	Fully Loaded	In-Line	281.72 MT	1.68	OK	20.99 m
		Between-Line	228.82 MT	2.07	OK	22.66 m

Notes:

1. The mooring lines strength assessment criteria are 60% and 80% of minimum breaking load for intact and on-line damage case [11]. SF for Safety Factor.
2. The minimum breaking load calculation is based on de-rated capacity considering 38 years of service life.

Table 4. Mooring Analysis Results Summary for One-Line Damage Condition.

Environmental Load	Vessel Condition	Condition	Mooring Line Tensions			Maximum SBM Excursion
			Max.	**SF**	**Remark(1)**	
Survival Condition (1 YRP)	Ballast	In-Line	80.16 MT	5.92	OK	8.52 m
		Between-Line	85.50 MT	5.55	OK	8.96 m
	Fully Loaded	In-Line	108.55 MT	4.37	OK	21.39 m
		Between-Line	136.24 MT	3.48	OK	14.82 m
Operating Condition (100 YRP)	Ballast	In-Line	112.12 MT	4.23	OK	13.77 m
		Between-Line	125.39 MT	3.78	OK	13.79 m
	Fully Loaded	In-Line	252.09 MT	1.88	OK	31.97 m
		Between-Line	346.20 MT	1.37	OK	19.07 m

4.2. Static Analysis

The static analysis [1] is considered to be the foundation of subsequent dynamic analyses and aims to establish the equilibrium configuration of the riser system due to weight, buoyancy, tension and current. In this analysis, Finite Element Analysis was conducted in Orcaflex based on provided flexible data and hang-off / tie-in locations which were modelled as fix. Vessel excursions (herein called "Offsets") of the SPM & FSO are categorized into three (3) types: Nominal offset, Far Offset and Near Offset which are depicted at Figure 5. The Far Offset is categorized into two (2) subsets: Intact and Damaged. The load case matrix for the static analysis is shown in Table 5.

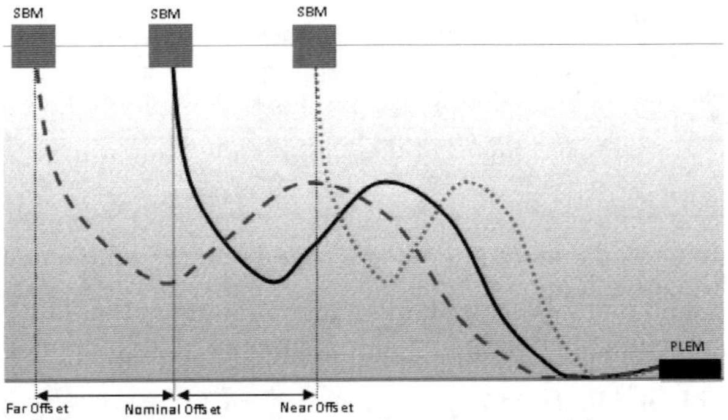

Figure 5. Nominal, Far & Near Offset Descriptions

Table 5. Flexible Riser Configuration Analysis Load Case Matrix.

Flexible Riser	Service	Riser Content Density [kg/m3]	Riser Content Pressure [barg]	Vessel Excursion	Current Loading(3)
KH & KG	Start of Life	Operational Mean (796)	Operational Maximum (28.3)	- None(1) - Damaged Near & Far(2)	100yr RP

Analysis results for the KH and KG flexible riser configuration static assessment are presented in Table 6 and Table 8, respectively, with interface loads at the hang-off position at the SPM and subsea tie-in at the PLEM presented in Table 7 and 9 in turn. For all considered flexible riser integrity critical results, the KH and KG flexible riser in its current configuration is acceptable.

Table 6. KH Flexible Riser Results (Static).

Position	Offset [m]	Maximum Effective Tension [kN]	Minimum Effective Tension [kN]	Minimum Bending Radius [m]	Touch Down Length [m]	Sag Bend Minimum Seabed Clearance [m]	Hog Bend Maximum Elevation from Seabed [m]
Nominal	0	96.7	5.8	8.4	14.9	15.0	32.5
Near	38.85	94.6	-2.8	6.6	12.7	15.1	37.7
Far	38.85	100.7	20.1	10.3	15.0	17.8	25.6
Allowable [12]		< 1530	> -50	> 3.926	> 0	> 0	< 67.5
Acceptable		☑	☑	☑	☑	☑	☑

Table 7. KH Flexible Riser Hang Off and Subsea Interface Loads (Static).

Position	Offset [m]	Maximum Hang Off Tension [kN]	Maximum PLEM Connector Tension [kN]	Maximum Hang Off Bending Moment [kN.m]	Maximum PLEM Connector Bending moment [kN.m]
Nominal	0	96.7	7.6	7.0	54.1
Near	38.85	94.6	-0.4	9.3	19.7
Far	38.85	100.7	20.6	22.7	53.2
Allowable [12]	-	< 300	< 300	< 150	< 150
Acceptable	-	☑	☑	☑	☑

Table 8. KG Flexible Riser Results (Static).

Position	**Offset** [m]	**Maximum Effective Tension** [kN]	**Minimum Effective Tension** [kN]	**Minimum Bending Radius** [m]	**Touch Down Length** [m]	**Sag Bend Minimum Seabed Clearance** [m]	**Hog Bend Maximum Elevation from Seabed** [m]
Nominal	0	80.7	8.1	9.3	15.9	29.9	56.5
Near	38.85	78.1	0.3	6.7	16.0	25.7	60.62
Far	38.85	83.4	19.45	14.4	15.9	31.62	48.42
Allowable [12]		< 1530	> -50	> 3.926	> 0	> 0	< 67.5
Acceptable		☑	☑	☑	☑	☑	☑

Table 9. KG Flexible Riser Hang Off and Subsea Interface Loads (Static).

Position	**Offset [m]**	**Maximum Hang Off Tension [kN]**	**Maximum PLEM Connector Tension [kN]**	**Maximum Hang Off Bending Moment [kN.m]**	**Maximum PLEM Connector Bending moment [kN.m]**
Nominal	0	80.7	9.8	13.0	52.4
Near	38.85	78.1	3.4	3.5	53.1
Far	38.85	83.4	21.9	26.0	32.8
Allowable [12]	-	< 300	< 300	< 150	< 150
Acceptable	-	☑	☑	☑	☑

4.3. Strength Analysis

The vessel offsets for the strength analysis are applied in 5 directions for each riser, Near, Quartering Near, Transverse, Quartering Far and Far with respect to the plane of the flexible riser (Figure 6). This in turn examines each flexible riser for maximum vessel offsets conditions and a complete strength overview is analyzed. The offset used is both intact and damaged vessel excursions. The load case matrix for the strength assessment is shown in Table 10.

Table 10. Strength Assessment Load Case Matrix.

Operating Conditions	**Service**	**Riser Content Density [kg/m3]**	**Riser Content Pressure [barg]**	**Vessel Excursion**	**Environmental Loading**
Normal (Intact Mooring) Abnormal (Damaged Mooring)	Start of Life End of Life Marine Growth Sensitivity	Operational Mean (796) Operational Max (1000)	Operational Min.(0) Operational Max.(28.3)	Near Far Transverse Quartering Near Quartering Far	Directional 100yr RP Wave and Current

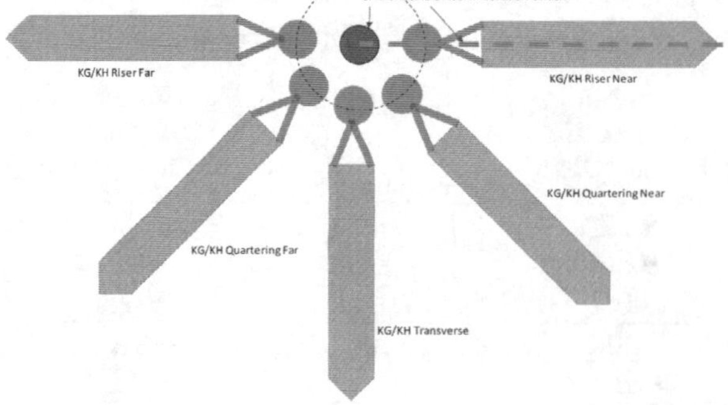

Figure 6. Vessel Offset and Environmental Directions (Strength Analysis)

IOP Conf. Series: Earth and Environmental Science 1081 (2022) 012034 doi:10.1088/1755-1315/1081/1/012034

4.3.1. KH and KG Flexible Riser Strength Analysis Results. The structural integrity of the KH and KG flexible risers are maintained and acceptable throughout the analysis. Results pertaining to the MBR, tension and interface loads are presented in the following three (3) sections, presented below.

4.3.1.1. KH and KG Flexible Riser Minimum Bending Radii. The overall MBRs of the KH and KG flexible risers are presented in Table 11 This MBR represents the minimum bending radius that the flexible riser experienced throughout the analysis, at any length along the risers. All MBRs presented are above the stipulated minimum MBR as per the acceptance criteria stated in the Supplier Specification. The MBR occurs in the sag bend of the flexible riser, during "Near" vessel offset conditions.

Table 11. KH and KG Flexible Riser MBR.

Operating Condition	Mooring Condition	FSO Condition	Overall Minimum Bend Radius	
			KH [m]	KG [m]
Normal	Intact	Fully Loaded	4.98	4.63
		Ballasted	4.63	4.12
Abnormal	Damaged	Fully Loaded	5.17	4.36
		Ballasted	4.94	4.38
Allowable [12]			**3.926**	

4.3.1.2. KH and KG Flexible Riser Tension. Overall maximum and minimum tensions of the KH and KG flexible riser are presented in Table 12. The overall minimum tension within the flexible pipe is negative, indicating compression. This compression is within allowable limitations. Likewise, the overall maximum tension is within the structural allowable limitations of the flexible.

Table 12. KH and KG Flexible Riser Tensions.

Operating Condition	Mooring Condition	FSO Condition	Overall Min. Tension		Overall Max. Tension	
			KH [kN]	KG [kN]	KH [kN]	KG [kN]
Normal	Intact	Fully Loaded	-9.03	-7.39	185.73	161.44
		Ballasted	-15.57	-11.82	178.28	154.04
Abnormal	Damaged	Fully Loaded	-7.83	-9.02	182.84	155.60
		Ballasted	-14.70	-11.33	169.08	146.49
Allowable [12]			**-50**		**1530**	

4.3.1.3. KH and KG Flexible Risers PLEM & SPM Hang-Off Interface Loads. The maximum KH and KG flexible pipe interface loads at the SPM hang-off and PLEM tie-in are presented in Table 13 and Table 14, respectively. All interface loads are within the allowable limitations.

Table 13. KH Flexible Riser Interface Loads.

Operating Condition	Mooring Condition	FSO Condition	SPM Hang Off Interface		PLEM Tie-In Interface	
			Max. Tension [kN]	Max. Bending Moment [kN.m]	Max. Tension [kN]	Max. Bending Moment [kN.m]
Normal	Intact	Fully Loaded	185.73	27.45	26.43	65.36
		Ballasted	178.28	39.24	34.54	75.06
Abnormal	Damaged	Fully Loaded	182.84	26.71	32.50	74.43
		Ballasted	169.08	37.94	37.94	68.02
Allowable [12]			**300**	**150**	**300**	**150**

Table 14. KG Flexible Riser Interface Loads.

Operating Condition	Mooring Condition	FSO Condition	SPM Hang Off Interface Max. Tension [kN]	SPM Hang Off Interface Max. Bending Moment [kN.m]	PLEM Tie-In Interface Max. Tension [kN]	PLEM Tie-In Interface Max. Bending Moment [kN.m]
Normal	Intact	Fully Loaded	161.44	24.76	24.76	82.47
		Ballasted	152.89	31.76	31.76	93.10
Abnormal	Damaged	Fully Loaded	155.60	29.21	29.21	87.14
		Ballasted	146.49	26.29	26.29	87.27
Allowable [12]			**300**	**150**	**300**	**150**

4.3.2. *Marine Growth Sensitivity.* The worst load case for MBR of the KG flexible riser was selected and the full Marine Growth profile applied to the DBM and flexible riser. The results of this load case are presented in Table 15. alongside the sensitivity assessment utilizing Marine Growth.

Table 15. KG Flexible Riser Marine Growth Sensitivity – MBR.

Marine Growth Included	Operating Condition	Mooring Condition	FSO Condition	Vessel Offset and Environmental Loading Direction	Overall Minimum Bend Radius
No	Normal	Intact	Ballasted	KH Riser	4.12 m
Yes	Normal	Intact	Ballasted	Quartering Near	5.25 m
Allowable [12]					**3.926**

As can be seen by Figure 2, Table 15 and Table 16, the minimum bend radius and maximum effective tension increase because the marine growth adds the weight of risers. It leads to hog bend elevation decrease and sag bend elevation increase. With the assumption of SPM and PLEM are the fixed points of riser, the additional weight results bigger tension on the riser to seabed. In this simulation, the marine growth is assumed to have 30 mm thickness and 1325 kg/sq.m for its density.

Table 16. KG Flexible Riser Marine Growth Sensitivity – Tension.

Marine Growth Included	Operating Condition	Mooring Condition	FSO Condition	Vessel Offset and Environmental Loading Direction	Overall Max. Effective Tension
No	Normal	Intact	Fully Loaded	KG Quartering	185.73 kN
Yes	Normal	Intact	Fully Loaded	Far	198.28 kN
Allowable [12]					**1530**

4.3.3 *Flexible Riser Design Pressure Sensitivity.* A sensitivity study utilizing the maximum design pressure of the flexible pipe has been performed the results shown in Table 17 for KG Riser and KH Riser. The study was conducted under fully loaded FSO at KG Quartering Far Position as well as normal mooring and operating condition because the maximum pressure could be reached in this condition. It is to be noted that the design pressure cases of 206 barg are meant to be performed using a 1yr RP environment. For conservatism, the same 100yr RP environment as used for the remainder of the study reported in this document was utilized. The worst MBR load case for the KG riser was selected to be presented as a comparison to the 206 barg internal pressure case.

Table 17. KG Flexible Riser Design Maximum Pressure Sensitivity.

Riser Internal Pressure	Overall Minimum Tension [kN]		Overall Maximum Tension [kN]		Minimum Bending Radius [m]	
	KG	KH	KG	KH	KG	KH
28.3	-8.30	4.55	139.09	162.08	4.66	8.58
206	-8.31	4.57	139.07	162.13	4.66	8.60
Allowable [12]	**-50**		**1530**		**3.926**	

As can be seen in the results tables, differences between the normal operating pressure and maximum design pressure on the riser response characteristics are numerically insignificant

4.4. Interference Analysis.

Interference analysis has a significant role to make sure clashing energy among the subsea structure below the limit. The excess clashing energy could produce the riser broken then cause hydrocarbon spill which is harm for the environment. The load case matrix for interference analysis is shown in Table 18.

Table 18. Interference Assessment Load Case Matrix.

Operating Conditions	Service	Riser Content Density [kg/m^3]	Riser Content Pressure [barg]	Vessel Excursion	Environmental Loading
Normal (Intact Mooring) Abnormal (Damaged Mooring)	Start of Life End of Life Marine Growth Sensitivity	Operational Mean (796) Operational Max (1000)	Operational Minimum (0) Operational Maximum (28.3)	Near Far North North East South East South South West North West	Directional 100yr RP Wave and Current

The interference analysis results summarizing the general findings of the KH & KG flexible riser clashing with adjacent mooring lines is presented in Table below. The results shown in this table are the minimum clearance distances that have been obtained from the FEA analysis for all considered load cases. The results are divided into two analysis categories, quasi-static and dynamic analysis.

- For **normal operating conditions,** no clashing is observed for all considered load cases. The minimum clearance observed between a flexible riser and a mooring line after the exposure of an extreme environment to the riser system is 0.10m. It is to be reiterated that this result is conservative.
- For **abnormal operating conditions** and with a fully loaded FSO, contact is observed. This contact is already observed in the quasi-static analysis, as such the dynamic analysis does not differ as is anticipated. For a ballasted FSO, no contact between the flexible risers and mooring lines is observed.

For all cases considered and as shown in Table 19 below, the quasi-static analysis shows larger clearances than the dynamic analysis, which is expected.

Table 19. Analysis Results Summary.

Operating Condition	Mooring Condition	FSO Loading Condition	Minimum Clearance between Flexible Risers (KH & KG) to all six (6) Mooring Lines	
			Quasi-Static Analysis	Dynamic Analysis
Normal	Intact	Fully Loaded	1.62 m	0.10 m
		Ballasted	2.86 m	0.41 m
Abnormal	Damaged	Fully Loaded	Contact	Contact
		Ballasted	3.83 m	0.59 m

Figure 7(a) on the following page present the quasi-static minimum clearances between each flexible riser and the adjacent mooring lines for each load case (720 total load cases). A separation line between normal and abnormal operation conditions is shown to ease the interpretation of this graphic. Points scattered within this plot are representative of the minimum clearances between both KH & KG risers and their nearest adjacent mooring lines. A plot point above the zero (0) clearance distance (shown as a red dotted line) indicates that no clashing has been observed. It is clear from the graphic that for abnormal operating conditions and certain load cases, the flexible risers are in contact with the mooring lines. No contact is observed for normal operating conditions.

Figure 7(b) shows the dynamic analysis, following the quasi-static analysis in the preceding image. As can be observed in the figure, most clearance values have decreased (i.e. flexible risers are closer to the mooring lines) during the dynamic analysis. This is to be expected, as wave loading will contribute to motion of the flexible risers & mooring lines. As with the quasi-static analysis, no clashing is observed for normal operating conditions and only a limited amount of load cases show clashing (plot points below zero (0) clearance) for abnormal operating conditions.

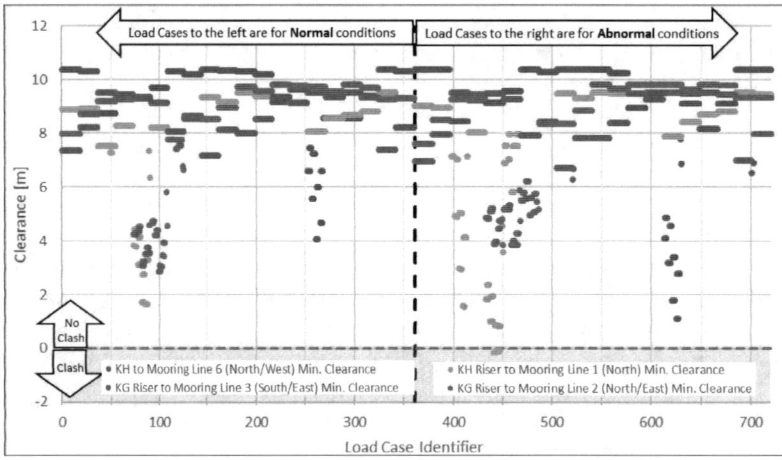

Figure 7(a). Quasi-Static Analysis Minimum Clearance Plot

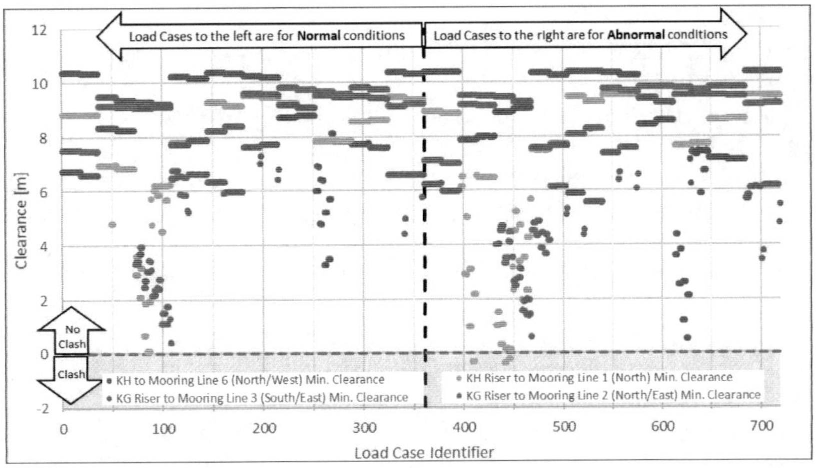

Figure 7(b). Dynamic Analysis Minimum Clearance Plot

4.4.1 *KH Flexible Riser Clashing Energy Analysis.* Due to the KH flexible riser exhibiting contact with mooring line #1 in some of the analyses, a clashing energy assessment is herein performed and reported. Clashing assessments are very computational intensive, as such only load cases which exhibit clashing have been analyzed. A total of 6 load cases out of 720 (i.e.: 0.8%) exhibited contact. The load cases to which clashing is isolated to, are the fully loaded FSO, single line mooring failure and end of life system configurations. The maximum clashing energy between the KH flexible riser and mooring line #1 are reported in Table 20. The maximum clashing energies exhibited by the analysis are small, indicating that the flexible riser contact impact with the mooring line is very light. In general engineering practice, a clashing energy value of less than 2kJ of a flexible pipe is deemed to be acceptable [12].

Table 20. KH Flexible Riser Clashing Energy.

Operating Condition	**Mooring Condition**	**FSO Condition**	**Riser System Life Cycle**	**Vessel Offset and Environmental Loading Direction**	**Maximum Clashing Energy** [kJ]
Abnormal	Damaged	Fully Loaded	EOL	KH Riser Near North	0.083 0.032

5. Conclusion

This paper presents an engineering design of flexible risers with distributed buoyancy modules configuration. After conducting several analysis namely mooring analysis, static analysis and strength analysis, the designed flexible riser could meet allowable values. In theory, distributed buoyancy module could be consider as the one of alternative to replace conventional subsea structure in the harsh environment (Natuna Sea) due to its advantage in installation process. The interference analysis results in several cases, KH riser will clash with mooring line 1 but the energy is insignificant. It means that the hydrocarbon spills that are produced by clashing could be avoided.

References

[1] Ogbeifun A M, Oterkus S, Race J, Naik H, Moorthy D, Bhowmik S, Ingram J 2021 *IOP Conf. Ser.: Mater. Sci. Eng.* **1052** 012022.
[2] Zhimin Tan, Cobie Loper, Yucheng Hou, Terry Sheldrake 2010 *Proc. Int. Conf. Offshore Mech. Arct. Eng. – OMAE* **79476** 373
[3] Chen W, Guo S, Lid Y, Gaie Y, Shen Y 2021 Int. J. Nav. Archit. Ocean Eng. **13** 650
[4] Papathanasiou T K, Markolefas S, Khazaeinejad P and Bahai H 2017 *IOP Conf. Ser.: Mater. Sci. Eng.* **276** 012023
[5] DNV-RP-C205. *Environmental Conditions and Environmental Loads*. Norway: BDet Norske Veritas; 2010.
[6] DNV-OS-F201. *Dynamic Risers*. Norway: BDet Norske Veritas; 2010.
[7] DNV-OS-S302. *Offshore Riser Systems*. Norway: BDet Norske Veritas; 2012.
[8] DNV-OS-E301. *Position Mooring*. Norway: BDet Norske Veritas; 2012.
[9] DNV-RP-F203. *Riser Interference*. Norway: BDet Norske Veritas; 2009.
[10]API 17L. *Specification for Flexible Pipe Ancillary Equipment*. USA: American Petroleum Institute; 2013.
[11] API-RP-2SK. *Design and Analysis of Station Keeping System for Floating Structures*. USA: American Petroleum Institute; 2005.
[12] NOV 6933. *Design Premise of Dynamic Analysis*. USA: National Oilwell Varco; 2017.

Economic feasibility study due to implementation of dual fuel engine of mini-LNG carrier in LNG distribution to power plants in Bali and Lombok

Emmy Pratiwi[1,2], I Made Ariana [1,2], Dhimas Widhi Handani[1,2], A A B Dinariyana[1,2], Ketut Buda Artana[1,2] and Hayy Nur Abdillah[1]
[1] Department of Marine Engineering, Faculty of Marine Technology, Institut Teknologi Sepuluh Nopember, Surabaya, Indonesia
[2] Center of Excellences Maritime Safety and Marine Installation, Institut Teknologi Sepuluh Nopember, Surabaya, Indonesia

emmy.p@its.ac.id

Abstract. The fact that the maritime transport contributes 2.8% of global greenhouse gas (GHG) emissions challenges the ship to reduce the CO2 emissions. International Maritime Organization (IMO) requires limiting the sulphur to 0.5% m/m in the fuel oil used on board ships operating outside designated emission control. One of strategies to comply the IMO regulations is by using the alternative fuels which may contain low or zero sulphur for ship main engine and that is liquefied natural gas (LNG). Instead of retrofitting the diesel engine which economically requires higher cost, engine conversion to dual fuel diesel engine is preferable in this study. Before the conversion of dual fuel diesel engine is implemented, the feasibility study should be performed, both technically and economically. This study focused on the economic analysis due to conversion of dual fuel engine using LNG fuel. The study case in this research is the plan of dual fuel engine conversion of Landing Craft Tank (LCT) as mini-LNG carrier for distributing the LNG ISO tank to power plants in Bali and Lombok. The economic analysis in this study was carried out to adjust several components in Capital Expenditure (CAPEX) and Operational Expenditure (OPEX) that occurred due to dual fuel engine conversion. Finally, the result of this papers shows that the dual fuel engine conversion for mini-LNG carrier for gas distribution is economically feasible since it gives the smaller cost compared to diesel engine.

1. Introduction

The issue of greenhouse gas (GHG) in recent years is getting more serious. Unfortunately, the international shipping of maritime activities contributes relatively huge and responsible for about 2.8% of global GHG emissions in 2007 [1]. In accordance with the Initial IMO Strategy on reduction of GHG emissions from ships, the total annual GHG emissions from international shipping should be reduced by at least 40% by 2050 compared to 2008 [2]. In response the GHG issue and to improve the air quality, IMO set a new limit the sulphur content in the fuel oil used on board ships on 1 January 2020. The rule limits, known as IMO 2020, requires the onboard ship use fuel oil with the sulphur content less then 0.50% m/m (mass by mass) for those ships which operating outside designated emission controls area and the limits become more stricter by 0.10% m/m for ships operating within the emission control areas [3]. However, most ships were using heavy fuel oil which contains much higher sulphur, consequently the combustion of engine produces the ships' emissions. Several strategies to comply the IMO regulations are provided and suggested. One of the strategies is for ship using the engine running on the alternative fuels which may contain low or zero sulphur. Liquefied natural gas (LNG) is considered as environmentally alternative fuel to oil based shipping fuel because LNG provides the reduction of sulphur oxide (SOx) and nitrogen oxide (NOx) emissions [4].

The study of implementation of LNG as marine fuel for the ship has already been proved. However, as the consequence of the implementation of LNG on board vessels, the technical aspect including the development of LNG-fuelled engines, LNG fuel tanks, gas supply systems and LNG infrastructure should be conducted. From the technical point of view, the utilization of LNG as marine fuel is feasible [5]. Nevertheless, the economical aspect is also important to be considered since there need some improvements and provide infrastructures in supporting gas fuel system. The economic aspect of the implementation gas fuelled engine conversion will be presented in this paper. This study is actually the continuation of previous study and part of authors' research maps about the small scale LNG distribution for power plant in Bali and Lombok Island, Indonesia, by utilizing LCT as mini-LNG carrier which carrying LNG stored in the ISO tank [6].

Based on the previous study, the utilization of mini-LNG carriers is feasible and requires relatively small cost to supply the LNG demand in Lombok and Bali, but the main engine of the LCT still use the diesel engine. Thus, the study aims to analyse the engine conversion in mini-LNG carrier whether will be still economically feasible and how dual fuel engine may provide lower distribution cost of LNG form LNG hub terminal to several power plant in Bali and Lombok. This study will describe changes in several aspects in terms of CAPEX and OPEX due to the conversion of gas-fuelled engines. However, the CAPEX and OPEX in this study not only assessed for the implementation of engine conversion, but also include for all investments form jetty, storage facility, regasification unit until the system distribution to the power plant and cost of operational and maintenance in receiving terminal. Economic analysis will also be developed for two scenarios of the cost of capital or CAPEX. The first scenario is using a new building of non-conventional LCT, and second scenario is using a non-conventional LCT conversion. Finally, the cost comparison between diesel fuel engine and conversion of dual fuel engine will be presented for both scenarios.

2. An Overview of the LNG Distribution Design in Bali and Lombok, Indonesia

The economic study of dual fuel engine conversion is performed for mini-LNG carrier which operate in the case of LNG distribution to power plants Bali and Lombok. Therefore, a brief overview of the LNG distribution design in Bali and Lombok will be explained first in this chapter. The detail of the LNG distribution design can be found in the previous study [6].

Instead of using conventional small-scale LNG (SSLNG) carriers, due to the characteristic of the location of power plant, this study proposes the utilization of Landing Craft Tank (LCT) as the non-conventional LNG carriers because it is considered to more suitable to overcome the low demand gas in power plant and the location which are scattered with narrow waterway. LNG is filled to ISO tank in the LNG terminal as the origin then the ISO tank is stored in the LCT. Unloading the LNG ISO tank from the ship is conducted in in the receiving terminal and then the truck will transport them to stacking yard. Finally, LNG stored in ISO tanks can then be supplied to fuel power plants after being regasified [6].

In the LNG distribution design, the LNG supply terminal is from Floating storage and regasification unit (FSRU) in Benoa Terminal, Bali. While the demand of power plant consists of Lombok Peaker, Gas Power Plant of Lombok, Gilimanuk and Pemaron located in Bali and Lombok Island.

Based on the mathematical model for optimizing the capacity-based design route, the route for each power plants are represented in the Figure 1. X_1, X_2, X_3 and X_4 represents the location of the demand, which is Lombok Peaker, MPP Lombok, Gilimanuk and Pemaron power plants. While the number of mini-LNG required to meet the gas demand based on the capacity of power plants, roundtrip and sailing frequency, can be summarized in Table 1. For example, Route A will be designed for supplying the ISO LNG tank from Terminal Benoa (X_0) to power plant in Lombok Peaker (X_1) then back to the origin carrying the empty LNG ISO tank.

IOP Conf. Series: Earth and Environmental Science 1081 (2022) 012035 doi:10.1088/1755-1315/1081/1/012035

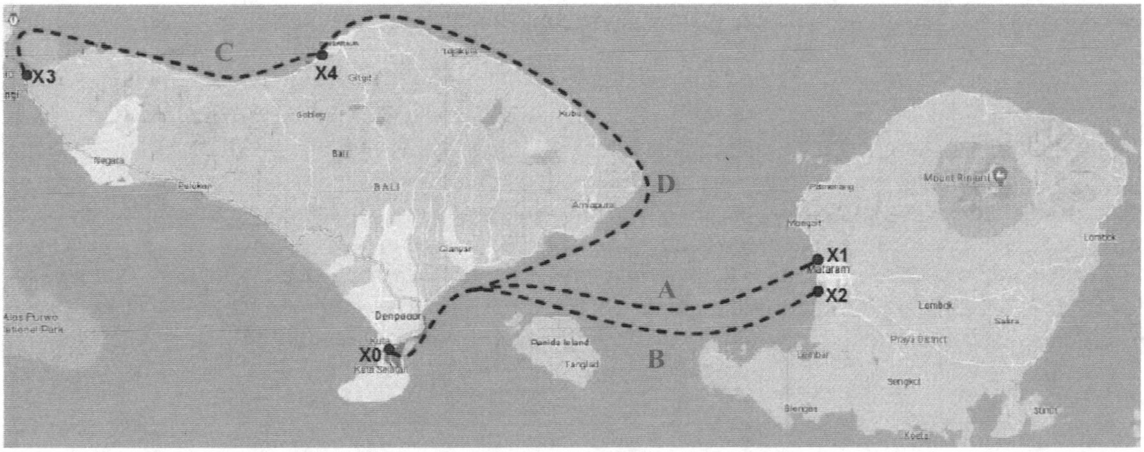

Figure 1. Design of LNG distribution route in Bali and Lombok Island, Indonesia [6].

Table 1. Summary of Number of *mini-LNG* carrier and total trip per year [6].

Route	Route	Number of Mini LNG Carrier	Total Trip/Year	Information
A	X0-X1-X0	1	180	Continuous
		1	180	Intermittent
B	X0-X2-X0	1	120	Continuous
		1	120	Intermittent
C	X0-X3-X0	2	53	Continuous
D	X0-X4-X0	1	80	Continuous

3. Design of Dual Fuel Diesel Engine in Mini-LNG Carrier.

In the technical calculation, the first step is to calculate the power of main engine to ensure that the specification of dual fuel engine match. Based on the power calculation, engine power required for one engine is 1096 kW, thus Wärtsila 6L20DF is a suitable main engine to be installed in the ship in terms of power requirements. The detail of selected dual fuel engine specification can be seen in Table 2. The engine is usually installed for dual fuel operation, it means that the engine can run in either gas or diesel operating mode.

Table 2. Specification of Dual Fuel Engine.

Specification	Wärtsila 6L20DF
Power	1110 kW
RPM	1200 rpm
Min Length	3244mm
Min Height	2424mm
Min Width	1690mm
Configuration	Inline 6, 4 Stroke DF Diesel

In addition to dual fuel engine installation, the consequence of dual fuelling is the need for additional LNG tanks for storage and a fuel supply system to the engine. In this system, the LNG is stored in ISO tank and placed on deck at the stern of the ship which was previously used to load cargo.

4. Economic Analysis

The economic analysis is used to evaluate whether the project investment is feasible and to provide the cost margin of gas sales from this distribution design. Two parameters are considered in the economic analysis, which are capital expenditure (CAPEX) and operational expenditure (OPEX). While the proceeds in the cash flow is derived from the revenue according to the amount of processed gas in the power plant. CAPEX is defined as the payment to the initial investment of assets. Component cost of CAPEX includes the investment for the transportation (LCT) and receiving terminal construction. OPEX is the expenses that occurred during the operation of LNG distribution, as well as the maintenance cost. The operational and maintenance cost consider arise at the receiving terminal as well as transportation costs using LCT and trucks. This study will focus more on changes in several aspects in terms of CAPEX and OPEX due to the conversion of dual fuel engines.

The economic study will calculate the several parameters including Payback Period (PP), Net Present Value (NPV) and internal rate of return (IRR) assuming a contract duration of 10 years. The payback period determines how long it would take the cash flow can recover the initial investments. The PP is written as:

$$PDB = \frac{Total\ Cost\ of\ Investement}{Project\ Cash\ Flow} \tag{1}$$

NPV and IRR are two discounted cash flow methods used to evaluate capital investments or projects. The NPV is the calculation of the difference between the present value of cash inflows and present value of cash outflows at the required rate of return of project over a period of time. The calculation of NPV use the following formula.

$$NPV = \sum_{t=o}^{n} \frac{R_t}{(1+i)^t} \tag{2}$$

R_t = net cash flow on year t
i = discount rate
t = the year whose cash flow is being discounted
n = project duration (years)

The internal rate of return (IRR) is a rate of return on an investment. The IRR in the investment is the interest rate that gives the net present value become zero, or where the sum of discounted cash flow is equal to the investment [7]. The feasibility of project can be determined by comparing the IRR with the Weighted Average Cost of Capital (WACC). If the IRR is greater than the WACC, then the project's rate of return is greater than the capital cost that was invested and should be accepted.

5. Results and Analysis

Before calculating the economic parameters, namely PP, NPV and IRR, the cost components for both CAPEX and OPEX will be presented first. There are some cost components which have changed because of the adjustment due to the conversion of diesel-fuel engine to dual fuel engines with gas-diesel fuel. Cost adjustments are explained in more detail in the following sub-chapters.

5.1. Adjustment in the Cost Components on CAPEX and OPEX

The economic analysis in this study was carried out to adjust several cost components for CAPEX and OPEX due to the main engine conversion. According to the technical analysis about calculation of main engine requirement, the dual fuel engine that designed to be installed in mini-LNG carrier is Wärtsila 6L20DF. Some of the cost components that are recalculated in this economic analysis include:

a) OPEX : Gas fuel cost and engine operating cost.
b) CAPEX : Capital cost for purchasing the dual fuel main engine and its supporting system (fuel system).

Basic assumptions and specifications of the Wärtsila 6L20DF engine related to the calculation of the economic study can be seen in Table 3 as follows:

IOP Conf. Series: Earth and Environmental Science 1081 (2022) 012035 doi:10.1088/1755-1315/1081/1/012035

Table 3. Basic Assumptions for Economic Calculations and Main Engine Specifications.

Exchange Rate of 1 USD	IDR 14.500
LNG Price	14 USD/MMBtu
MFO Price	13897 IDR/Lt
BHP Wärtsila 6L20DF	1.110 kW
Heat rate at 100% load (GAS)	8.062 kJ/kWh
Pilot fuel consumption at 100% load, gas mode	5 g/kWh

The significant different due to the implementation of dual fuel engine is the gas fuel cost. Hence, the fuel gas consumption for each route and total number of operating ships annually should be calculated. By using the following formulas for fuel gas consumption (FGC) and fuel oil consumption (FOC) as the pilot fuel consumption, dual fuel consumption can be estimated and the results of the calculation of fuel consumption are shown in Table 4.6. 2 below.

$$FOC = BHP \times SFOC \times t \quad (3)$$

$$FGC = BHP \times SFGC \times t \quad (4)$$

BHP denotes engine brake horsepower (kW), SFOC is specific fuel oil consumption during time t (hour) (gr/kWh), SFGC is specific fuel gas consumption (kJ/ kWh) and t is endurance (hour).

Table 4. Calculation of Dual Fuel Consumption.

Fuel Consumption of Dual Fuel Engine		Design route				
		Benoa-Lombok Peaker-Benoa	Benoa-MPP Lombok-Benoa	Benoa-Gilimanuk-Benoa	Benoa-Pemaron-Benoa	Standby Ship
BHP Wärtsila 6L20DF	kW	1.110	1.110	1.110	1.110	1.110
Annual Gas Fuel Consumption per Route (FGC/year)	kJ	90,4x10^9	86,5 x10^9	88,7x10^9	45,6x10^9	42,3x10^9
Annual Gas Fuel Consumption per Route	Btu	85,7 x10^9	81,97 x10^9	84,09 x10^9	43,3 x10^9	40,1 x10^9
Annual LNG Consumption	MMBtu	85.666	81.971	84.093	43.273	40.135
Gas Fuel Consumption per Trip	Ton	6	5	8	6	8
Annual Pilot Fuel Consumption (FOC/year)	MT	56	56	37	28	18
Fuel Cost per year	IDR (Billion)	17,39	16,64	17,07	8,78	8,15
	USD	**252.170**	**241.292**	**247.536**	**127.380**	**118.142**

Table 4 shows the calculation of dual fuel costs per year including fuel cost for gas in operating mode and diesel fuel as the pilot fuel for each route. The fuel cost for dual fuel engine will be included in the OPEX for mini-LNG carrier. To find out the feasibility economy in terms of fuel cost, then the fuel

IOP Conf. Series: Earth and Environmental Science 1081 (2022) 012035 doi:10.1088/1755-1315/1081/1/012035

consumption cost of the dual fuel engine will be compared with the existing fuel cost using diesel fuel only.

Table 5. Profit from the Difference Between Diesel and Dual Fuel Cost.

Cost Component	Unit	Benoa-Lombok Peaker-Benoa	Benoa-MPP Lombok-Benoa	Benoa-Gilimanuk -Benoa	Benoa-Pemaron-Benoa	Standby Ship
1. Existing Fuel Consumption (Diesel Fuel)						
Diesel Fuel Consumption/Trip/Ship	Litre	6022	5806	8381	6073	8381
Amount of trip per Year	Trip	165	165	110	165	105
Number of Ships	Unit	2	2	2	1	1
Diesel Fuel Consumption per Year	Litre	1.987.167	1.916.081	1.843.816	1.002.046	880.003
Diesel Fuel Consumption per Year	MT	1.969.282	1.898.836	1.827.222	993.028	872.083
Fuel Cost per year	IDR (Billion)	**27,6**	**26,6**	**25,6**	**13,9**	**12,2**
2. Dual Fuel Consumption						
Dual Fuel Cost per Year	IDR (Billion)	17,39	16,64	17,07	8,78	8,15
	USD	**252.170**	**241.292**	**247.536**	**127.380**	**118.142**
Profit on fuel cost per year	**IDR (Billion)**	**10,2**	**10,0**	**8,6**	**5,1**	**4,1**
	USD	**$ 705.147**	**$ 688.753**	**$ 589.799**	**$ 354.525**	**$ 281.495**

The calculation of existing fuel cost is shown in number 1 of Table 5. This result gathered from our previous calculation and represented here. Compared to the existing fuel cost, fuel cost for dual fuel engine is lower than the existing one. Therefore, there is a total profit per year occurred from the difference costs of approximately Rp. 38 billion per year for all route distribution or about $2.619.720. Profit on fuel cost per year in every route is presented in the las row of Table 5.

In addition to the cost of fuel consumption as an annual operating cost, the costs that arise because of engine conversion are the capital costs for purchasing the engine. The cost of purchasing the engine including its fuel system, LNG ISO tank as a gas fuel storage tank and shipyard costs for engine repowering conversion can be seen in Table 6.

Table 6. CAPEX of Dual Fuel Engine

CAPEX	Unit	Price
LNG Conversion (LNG Pac)	USD/Engine	1.753.800
LNG ISO Tank and Freight cost	USD/ISO Tank	136.000
Shipyard Cost for Repowering Engine Conversion	USD	138.889
CAPEX Total	USD	**$ 2.028.688,89**
	IDR (Billion)	**29,4**

The price of a dual fuel engine (LNG Pac) is assumed based on the results of a cost estimate from one of the engine makers (Wärtsila) which submitted costs for ships with a power of 2,800 kW and 15,492 kW. Then the price is calculated as per kW and the LNG Pac price is obtained as Table 7 in numbers 1 and 2. Apart from engine makers, there were previous researchers who stated that the estimated cost of purchasing a dual fuel engine was 639 USD/kW [8]. In this study, the first estimate is used which is

1580 USD/kW to calculate the LNG Pac cost. This price includes all components along with engineering fees, class approval, commissioning, and sea trial fees for 65 working days.

Table 7. Some References about the Estimated Cost of Dual Fuel Engine Conversion.

No	Cost References	Price	Power	Price	Unit	Selected
1	Engine maker	USD 4.424.000	2800 kW	**1.580**	**USD/KW**	v
2	Engine maker	USD 8.836.000	15492 kW	**570**	**USD/KW**	x
3	Previous study by Prastyasari et al., (2019) [8]			**693**	**USD/kW**	x

The difference in costs due to the use of the dual fuel main engine on this mini-LNG ship will then be accommodated in the calculation of the economic study with the parameters of payback period, NPV and IRR.

5.2. Capital Expenditure (CAPEX)

CAPEX is a capital modal for the investment in distributing LNG from terminal LNG hub to the user in power plant by utilizing mini-LNG carrier and truck as land transport. Basically, CAPEX consists of two main components, investment costs or capital for transportation including ISO LNG tank (LCT) and investment costs for the construction of receiving terminals, including jetty construction, equipment in LNG storage, regasification unit, piping system to power plant as well as the office building.

The capital cost for the LCT carrying LNG ISO Tank is developed for two investment scenarios, the scenario 1 is the LCT is a new building, and scenario 2 is the conversion of the existing LCT ship. According to the calculation of the estimated capital cost for LCT, it is obtained that the cost of non-conventional LCT Vessels for new building and conversion scenario are $5.601.342 and $4.726.001 respectively. The estimated cost includes the materials, construction, and procurement of 20 ft ISO Tank LNG tanks, as well as the cost of converting and modifying the engine to a dual fuel engine such as deck construction reinforcement, dual fuel engine purchasing and shipyard cost. This fee includes 10% VAT. The cost of LNG ISO Tank 20 ft is approximately USD 68,000 per tank, including shipping from China to Indonesia.

Furthermore, the investment cost of infrastructure and equipment in the LNG terminal hub and power plant will be calculated. Infrastructure components and equipment need to be adjusted to fulfil the requirements and specifications at each power plant in accordance with the LNG distribution design on each route. Unit prices for each cost component are obtained from information from several vendors and reports in previous studies. Investment costs for ships and in the receiving terminal as shown in Table 8.

Table 8. Component of CAPEX.

CAPEX	**Item**	**Price**	
Ship building	New Building (Scenario 1)	$	5.601.342
	Conversion (Scenario 2)	$	4.726.001
Receiving Terminal and Power Plant	Lombok Peaker	$	5.670.812
	MPP Lombok	$	6.132.211
	Pemaron	$	5.475.295
	Gilimanuk	$	5.766.254

Finally, the sum of CAPEX for each route can be derived and the summary of CAPEX for both scenario in each distribution route can be seen in the Table 9. This total CAPEX has been included the tax, permit and miscellaneous. There is slightly difference between these scenarios and shows that the CAPEX for first scenario is higher than the second scenario. It happens due to the component cost for building the ship is surely higher than converting the existing ships, while the CAPEX in terminal and powerplant for both scenario is same.

Table 9. Summary of CAPEX for both scenarios.

CAPEX	**Unit**	**Price**			
		Scenario 1		**Scenario 2**	
LNG Distribution to Lombok Peaker	set	$	21.935.545	$	19.659.659
LNG Distribution to Gas Power Plant in Lombok	set	$	22.535.363	$	20.259.477
LNG Distribution to Pemaron	set	$	14.399.628	$	13.261.685
LNG Distribution to Gilimanuk	set	$	22.059.619	$	19.783.733
Standby Ship	unit	$	5.601.342	$	4.726.001
TOTAL CAPEX (Include Tax, Permit, etc. 25% and Miscellaneous 5%)		$	86.531.496	$	77.690.556

5.3. *Operational Expenditure (OPEX)*

Operational expenditure or OPEX covers all costs incurred during the operation of LCT ships and LNG receiving terminals annually, including maintenance costs. Operational cost of LNG distribution using LCT consist of several item with cost components as listed in Table 10.

Table 10. Cost Components of OPEX for LCT ships.

O&M Costs of LCT as Mini LNG Carrier	Benoa-Lombok Peaker-Benoa	Benoa-MPP Lombok-Benoa	Benoa-Gilimanuk-Benoa	Benoa-Pemaron-Benoa
Dual Fuel Cost	$ 252.169,84	$ 241.292,42	$ 247.535,68	$ 127.379,85
Lubricating Cost Fresh Water Cost Consumable Cost Port Handling Cost Maintenance Cost Crew Cost Assurance & Administration Cost	$ 851.178,76	$ 849.351,88	$ 752.074,47	$ 426.696,03

The calculation OPEX in this study is an adjustment from previous study [6], especially those related to the operation of the main engine, such as fuel costs, lubrication costs and maintenance costs for dual fuel engines. The cost for gas and pilot fuel has been calculated in Table 4. Maintenance costs of ships with dual fuel engines after conversion are assumed to be approximately 20% higher [8].

Table 11 shows a summary of OPEX costs for each distribution route. The following OPEX calculation has added a 30% margin to accommodate possible differences in the price estimation process. The OPEX applies to both scenarios because it is assumed that there is no difference in operating and maintenance costs for both new building ship and conversion ship.

Table 11. Summary of Operation and Maintenance Costs for LCT.

O&M Cost of Mini LNG Carrier	Unit	Value
Benoa-Lombok Peaker-Benoa	US$/year	1.103.349
Benoa-MPP Lombok-Benoa	US$/year	1.090.644
Benoa-Pemaron-Benoa	US$/year	554.076
Benoa-Gilimanuk-Benoa	US$/year	999.610

Another component of OPEX is operating and maintenance costs at the LNG receiving terminal. In general, the operating and maintenance costs of the LNG terminal consist of annual maintenance costs, logistics costs, electricity needs at the plant and labour cost. OPEX for each LNG receiving terminal can be seen in Table 12. The operational cost consists of logistic cost, land rent and annual electricity cost at terminal.

Finally, a summary of the estimated operational costs (OPEX) for each scenario of LNG distribution to power plants is represented in Table 13. The total OPEX is the summation of operating and maintenance cost for LCT and LNG receiving terminal in each power plant.

Table 12. Operational and Maintenance Costs of LNG Receiving Terminal in Each Power Plant

O&M Cost of Receiving LNG Terminal	Maintenance Cost	Operational Cost	Labour Cost	TOTAL
		(US$/Year)		(US$/Year)
PLTMG Lombok Peaker	300.000,00	3.501.103	514.759	4.315.862
MPP Lombok	300.000,00	1.552.759	514.759	2.367.517
Pemaron	300.000,00	2.182.414	514.759	2.997.172
Gilimanuk	300.000,00	3.365.172	514.759	4.179.931

Table 13. Estimated Total OPEX for LNG Distribution to Power Plant.

OPEX	US$
Benoa-Lombok Peaker-Benoa	5.419.211
Benoa-MPP Lombok-Benoa	3.458.162
Benoa-Pemaron-Benoa	3.551.248
Benoa-Gilimanuk-Benoa	5.179.541

5.4. Calculation of PP, NPV and IRR

Economic analysis is carried out using investment feasibility criteria such as Net Present Value (NPV), Internal Rate Return (IRR) and Payback Period (PP) with the assumption that the project contract duration is 10 years. To be able to calculate the three criteria, CAPEX and OPEX have been obtained in the previous calculation. In addition, the estimated revenue from the sale of gas also needs to be calculated.

Table 14 shows a summary of total CAPEX, OPEX and the amount of gas processed to derive the revenue from all power plant. Revenue is calculated based on the selling price margin or costs for transportation from the LNG source to the gas demand in power plant. The revenue derived from the LNG distribution depend on the amount of the gas processed in the user. The larger the gas that utilized, as long as there is no need additional number of operating ships, then the revenue will be bigger. LNG In this study, margin variation is used to determine the sensitivity of margin in LNG market which is range from $2.00 to $4.00 with $0.5 interval.

Table 14. Summary of CAPEX, OPEX, Processed Gas for Both Scenarios.

Distribution Route	CAPEX		OPEX	Processed Gas (MMBTU)
	Scenario 1	Scenario 2		
LNG Distribution to Lombok Peaker	$ 21.935.545	$ 19.659.659	$ 5.419.211	9.067
LNG Distribution to MPP Lombok	$ 22.535.363	$ 20.259.477	$ 3.458.162	10.000
LNG Distribution to Pemaron	$ 14.399.628	$ 13.261.685	$ 3.551.248	5.333
LNG Distribution to Gilimanuk	$ 22.059.619	$ 19.783.733	$ 5.179.541	8.667
Standby Ship	$ 5.601.342	$ 4.726.001	$ 487.985	-
TOTAL	$ 86.531.496	$ 77.690.556	$ 18.096.146	33.067

Table 15. Revenue and Variation of Margin Rate.

	Margin	Total
Revenue (USD/MMBtu)	$ 2,00	$ 24.138.667
	$ 2,50	$ 30.173.333
	$ 3,00	$ 36.208.000
	$ 3,50	$ 42.242.667
	$ 4,00	$ 48.277.333

Based on the CAPEX, OPEX, and revenue presented in Table 14 and Table 15, then the PP, NPV and IRR are calculated. The first parameter is PP which represent how long it takes to recover the cost of an investment or to hit breakeven. In analysis of payback period, the proceeds in every year obtained from the nett revenue (earning after tax) and add with the depreciation cost. The value of proceeds will decrease by 0.05% annually due to the assumption that there is an increase in operating costs. Finally, the payback period has been determined based on the cumulative proceeds and initial or investment capital cost. Figure 2 and

Figure 3 show the payback period on the investment estimate for both scenario with difference revenue margins.

Figure 2 shows that with a project duration of 10 years, a margin of USD 4 gives breakeven or can recover the capital cost in year 4 while the lowest margin of USD 2 dollars still cannot provide a payback period of less than 10 years of project duration. Therefore, in case of scenario 1, the project will be worth investing if the margin is at least USD 2.5 per MMBtu.
For the scenario 2, as shown in

Figure 3, the quick payback period can be achieved in approximately 3.5 year if the margin is at USD 4, while at margin USD 2 still does not provide a payback period before the project duration ends. Therefore, the project will be feasible, in the term of payback period, if the margin is set at least USD 2.5.

After the parameter PP has been calculated, then other analysis investment eligibility criteria in economic will be calculated in the form of NPV and IRR. This value will give more broadly financial parameters about LNG investment feasibility.

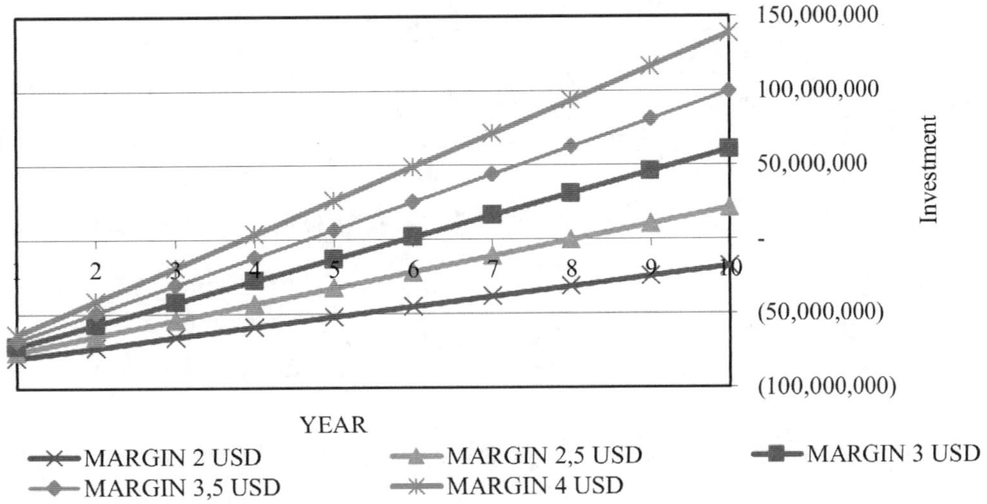

Figure 2. Payback Period of LNG distribution to Power Plant with Margin Variation (Scenario 1).

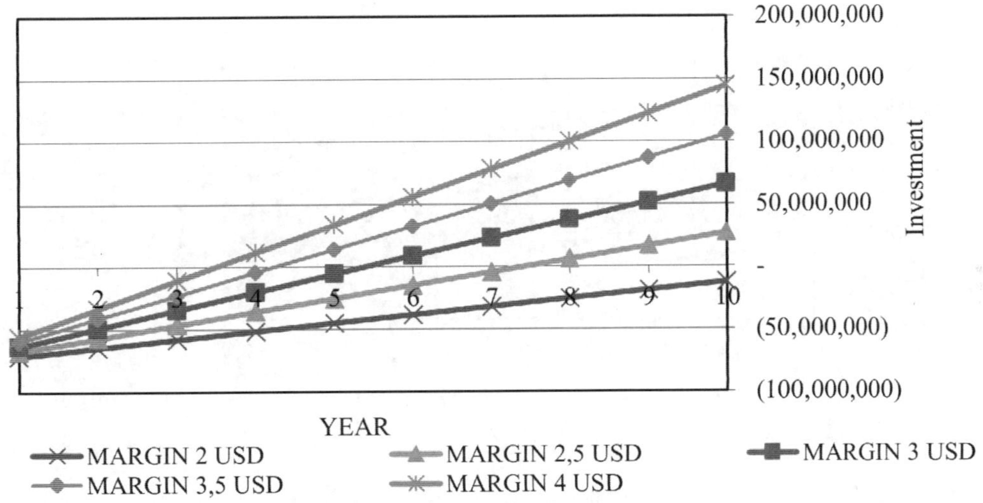

Figure 3. Payback Period of LNG distribution to Power Plant with Margin Variation (Scenario 2).

Figure 4 and Figure 5 display the financial parameters in the term of NPV and IRR for two scenarios with difference margin rate. The graph between year and NPV shows that the LNG sales margin is directly proportional to the NPV, the greater the LNG sales margin, the greater the NPV value. The investment will give profit if the NPV has positive value in the end of project duration where the IRR value must be bigger than the WACC of the company or the government's discount rate. The IRR below the required rate of return will cause the project loss. Therefore, the LNG distribution will be worth investing if the margin rate can give the IRR above the WACC or government's discount rate. This study assumed that the government's discount rate is 10% and WACC is in 11%, so the project will be feasible economically if the IRR is higher than 11%. Hence, scenario will be worth investing when the margin rate is above USD as well as the scenario 2. Both scenario of LNG distribution will be worth investing when the margin rate is above 3 USD with slightly difference NPV.

IOP Conf. Series: Earth and Environmental Science 1081 (2022) 012035 doi:10.1088/1755-1315/1081/1/012035

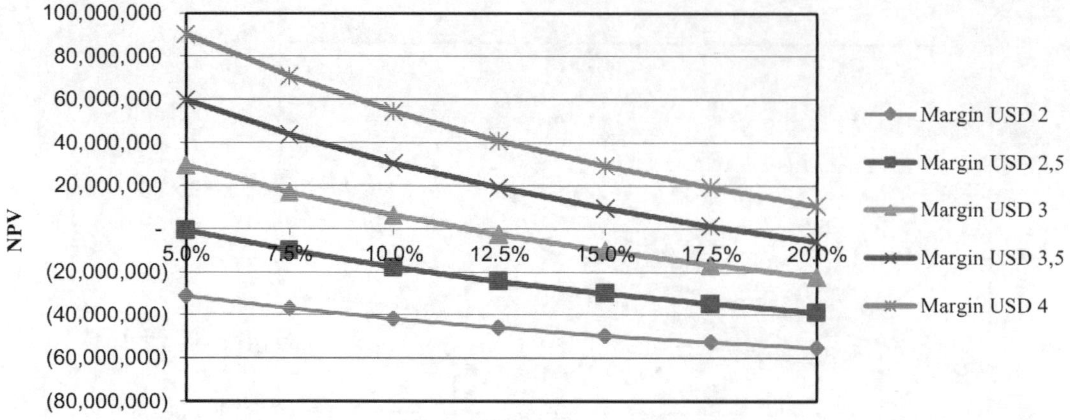

Figure 4. Internal Rate of Return (IRR) for LNG distribution to Power Plant with Margin Variation (Scenario 1).

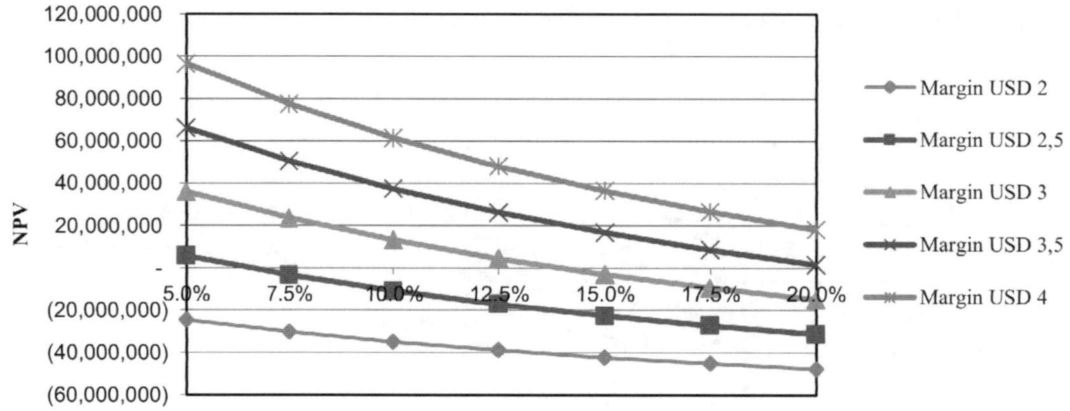

Figure 5. Internal Rate of Return (IRR) for LNG distribution to Power Plant with Margin Variation (Scenario 2).

Due to some margin that give the low IRR and PP is more than the project duration, so in this study a sensitivity analysis was also carried out to determine at what margin the required IRR, which is at 11%, can be obtained. For the first scenario, to get an IRR of 11%, the value of the margin rate for LNG distribution to the power plant should be equal to or greater than USD 2.94 and USD 2,79 for the scenario 2.

The results of the sensitivity analysis with variations in the margins rate show different IRRs for the distribution of LNG to power plants can be seen in the Table 16. Based on the economic analysis for the two scenarios, the second scenario, which is using a non-conventional LCT conversion ship, provides better feasibility than building a new ship in the first scenario. The scenario 2 provide better profitability of investments. It is caused by the lower capital cost in the LCT construction.

Table 16. IRR with Different Margin and Minimum Tolling Fee for Both Scenarios.

	Margin	IRR (Scenario 1)	IRR (Scenario 2)
Tolling Fee (USD/MMBtu)	**$ 2,00**	-3,6%	-2,5%
	$ 2,50	4,8%	6,5%
	$ 3,00	11,7%	14,0%
	$ 3,50	17,9%	20,6%
	$ 4,00	23,6%	26,9%
Minimum Tolling Fee (IRR 11%) for Scenario 1	**$ 2,94**	11,0%	11,0%
Minimum Tolling Fee (IRR 11%) for Scenario 2	**$ 2,79**	11,0%	11,0%

The main objective of this study is to prove whether the LNG distribution in Bali and Lombok will still be feasible and worth investing the LCT carried out the LNG ISO tank use the dual fuel engine. This problem will be answered if we compare the economic feasibility of the existing LCT with the LCT using dual fuel engine.

The result of the previous study by utilizing LCT using diesel engine provided that both scenarios is feasible for investment if the margin rate is above USD 3/ MMBTU with the IRR 8,5 and 14,0% respectively. However, in this study, dual fuel engine in the LCT can provide the higher IRR for the first scenario with the same margin rate, USD 3. Thus, the implementation of dual fuel diesel engine for LCT in this case study can provide better economical aspect.

6. Conclusion

This study presents the implementation of LCT as the mini-LNG carrier to supply gas demand in several gas-fuel power plant located in Bali and Lombok. The LNG ISO tank stored in the LCT has been proved that it gives lower distribution cost and suitable to be implemented specially to fulfil the LNG needs for scattered locations with narrow waterway. To comply the IMO regulations about emission reduction of the ship, the LCT use the engine running on gas fuel with low Sulphur. Not only the technical aspect, but the implementation of dual fuel engine should also be analyzed in the economical aspect. According to the economic analysis in this study, the LNG distribution will be worth investing if the margin rate is above the USD 3/MMBtu for both scenarios and can provide higher IRR than existing ships.

Finally, it can be concluded that the conversion of a mini-LNG ship with the use of a dual fuel main engine provides a lower distribution cost compared to the use of diesel engine ship. Although it requires large investment capital for dual fuel engines and higher maintenance costs, the lower operating costs due to fuel profit make the ships on board the dual fuel diesel engines for LNG distribution to the Bali and Lombok provide better economic feasibility. The profit obtained from the lower gas fuel consumption cost is the main factors that cause the better economic feasibility. This is very interesting finding that LNG as the marine fuel is feasible to be implemented, not only in the environmental perspective but also economically. However, further detail study needs to be performed to accommodate some limitations in this paper, such as gas price fluctuation, the availability of gas bunkering facilities, and possible loss due to reduced cargo transported because the LNG or reduced payload.

Acknowledgement

The authors would like to express our gratitude to the Ministry of Research and Technology/National Research and Innovation Agency (Kemenristek/ BRIN) of Republic Indonesia for the support in research funding by 2399/PKS/ITS/2021, the Center of Excellence for Maritime Safety and Marine Installation, Institut Teknologi Sepuluh Nopember, Surabaya and our research partner Agency for the Assessment and Application of Technology (BPPT).

References

[1] IMO, "Third IMO GHG Study 2014, Executive Summary and Final Report," International Maritime Organization (IMO), 2015. Accessed: Feb. 16, 2022. [Online]. Available: https://wwwcdn.imo.org/localresources/en/OurWork/Environment/Documents/Third%20Greenhouse%20Gas%20Study/GHG3%20Executive%20Summary%20and%20Report.pdf

[2] IMO, "Initial IMO GHG Strategy," International Maritime Organization (IMO), 2018. [Online]. Available: https://www.imo.org/en/MediaCentre/HotTopics/Pages/Reducing-greenhouse-gas-emissions-from-ships.aspx

[3] IMO, "IMO 2020 – cutting sulphur oxide emissions," *International Maritime Organization (IMO)*, 2020. Accessed: Feb. 11, 2022. [Online]. Available: https://www.imo.org/en/MediaCentre/HotTopics/Pages/Sulphur-2020.aspx

[4] S. Laribi and E. Guy, "Promoting LNG as A Marine Fuel in Norway: Reflections on the Role of Global Regulations on Local Transition Niches," *Sustainability*, vol. 12, no. 22, p. 9476, Nov. 2020, doi: 10.3390/su12229476.

[5] L. Sastre Buades, "Implementation of lng as marine fuel in current vessels: Perspectives and improvements on their environmental efficiency," 2017.

[6] E. Pratiwi, D. W. Handani, G. B. D. S. Antara, A. Dinariyana, and H. N. Abdillah, "Economic analysis on the LNG Distribution to power plants in Bali and Lombok by utilizing mini-LNG carriers," 2021, vol. 1052, no. 1, p. 012053.

[7] C. B. Smith and K. E. Parmenter, "Chapter 13 - The Economics of Efficient Energy Use," in *Energy Management Principles (Second Edition)*, C. B. Smith and K. E. Parmenter, Eds. Oxford: Elsevier, 2016, pp. 331–358. doi: 10.1016/B978-0-12-802506-2.00013-6.

[8] F. I. Prastyasari, D. W. Handani, A. A. B. Dinariyana, and M. A. Prakoso, "Feasibility study of diesel engine replacement on passenger ships: a system dynamics approach," *WMU J Marit Affairs*, vol. 18, no. 2, pp. 281–296, Jun. 2019, doi: 10.1007/s13437-019-00169-9.

IOP Conf. Series: Earth and Environmental Science 1081 (2022) 012036 doi:10.1088/1755-1315/1081/1/012036

Direct Measurement of Motion on Fast Ferries

Jamal[1,2], A Sulisetyono[1], W D Aryawan[1]

[1]Naval Architecture Department, Faculty of Marine Technology, Institut Teknologi Sepuluh Nopember (ITS), Kampus ITS Sukolilo, Surabaya 60111, Indonesia

E-mail: [2]jamal@polbeng.ac.id

Abstract. This paper describes a new identification method for the fast ferry motions by using the direct measurements on the ship's board. The inertial sensor type of the Analog Devices (ADIS) 16364 is selected as a as suitable instrument to measure accelerations and velocity of the ship motions namely heave, pitch, and roll in time domain. Furthermore, the ship motion responses are transferred as the energy spectral density in frequency domain which are identified using the Fast Fourier Transform (FFT) method. For the case study, the measurement equipment was placed on the navigation room of MV. Dumai Line 12 recording in 15 minutes within ship voyage from Bengkalis to Batam in September 2021. Finally, the natural frequencies of heave, pitch, and roll motions are identified about 0.568 rad/s, 0.117 rad/s, and 0.227 rad/s, respectively.

1. Introduction

Fast ferries or marine buildings of various sizes are only objects that are relatively tiny and sensitive to the influence of waves if they are in a very large ocean. The sea waves can affect the ship, which causes a threat to the ship's structure [1], affects the human factor [2], besides that it can also disturb the comfort of the crew, which can have an impact on the safety of the ship [3,4]. For this reason, ship building design cannot be separated from the calculation of ship responses that refer to international seakeeping criteria [2], such as two references that are often be used as standards, namely the North Atlantic Treaty Organization (NATO) [5] in the Standardization Agreement (STANAG) 4154 and the Nordic Co-operative Organization for Applied Research (NORDFOSK 1987) [6].

Several ship motion measurements have been made in the past, including one by NATO for the STANAG 4154 determination. The "UNO National Bio Dynamics Laboratory Ship Motion Simulator (NBDL)" simulation [25] and using Bayesian theory [26] were used to measure ship motions. There is also a direct measurement of the ship's motion, such as monitoring the characteristics of the state of the sea using the smartphone "Xiaomi Mi 8" [27]. Rumawas [12] has also looked at the comfortability of human factors on an offshore supply vessel (OSV) in Norwegian waters, employing an inertial sensor from Analog Devices (ADC) to directly measure the motions of the ship on board.

Other measurement tools related to direct measurements using the gyroscope sensor include this tool, which is suitable for measuring various high vibration movements [21, 23], can measure roll motions on ship stability in real time [24], and can even measure various angles of motion, such as hand and robotic arm movements [22].

Basically, the ship's response is defined as the cause of the motions by multiplying an input wave spectrum and the characteristics of the ship's response in regular waves which are converted to Response Amplitude Operators (RAO) [1]. However, the ship's response has never been studied and evaluated during ship's voyage in wave. For this reason, the direct ship motion measurements are carried out to evaluate the actual response of ship in seaway. Furthermore, the direct measurements can be used as a monitoring of the ship's response since the data logger has a suitable capacity to record. The direct measurement device, which is placed on ship's board is a suitable tool to make an analysis of the ship performance consider to safety and comfortability for the crew and passengers [3, 4, 10]. In addition, the data recorded may be used as a decision support system for the captain to act in avoiding potentially dangerous phenomena that may be experienced by ships in following and navigating the ocean [3].

This paper aims to evaluate the ship's response to oscillatory motion on fast ferries in the waters of the Malacca strait with a direct measurement method on board, which can measure the ship's motion validly according to actual environmental conditions. The measuring instrument used is the ADIS 16364, with the advantage that it can measure 6 degrees of freedom of the ship's motion, because this tool uses an accelerometer sensor and a gyroscope sensor, each of which can measure 3 translational motions and 3 angular motions. The method used to process the data is Fast Fourier Transforms (FFT), this method is used to process the time domain data into a frequency domain that is used to measure the ship's response validly.

2. Theoretical of the ship motion measurement

2.1 Ship motion in irregular wave

The seakeeping study of ships is closely related to the behavior of ships when subjected to waves and the resulting effects on people, systems, and mission capabilities [7, 8]. There are six motions of ship known as the degree of freedom (DOF) [9] that occur due to the ship's external forces as shown in Figure 1. The first three motions are defined as translational movements towards X, Y and Z axes which are namely surge, sway and heave, respectively. The other three motions are called as rotational movements around X, Y and Z axes, namely roll, pitch and yaw, respectively.

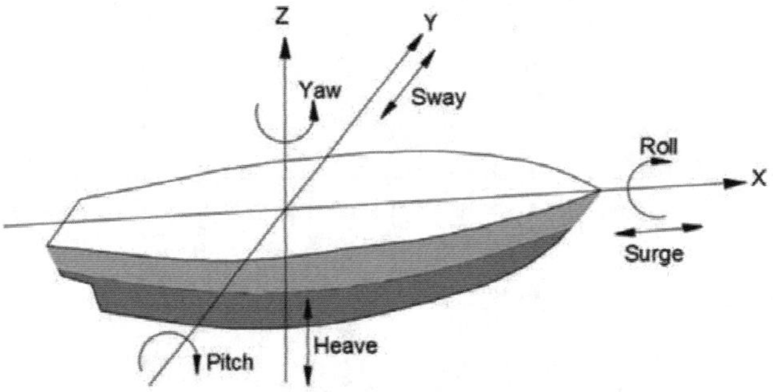

Figure 1. The six degrees of freedom (6 DOF)

A floating body on the surface of the water that is induced by wave exciting force, causes motion responses. The ship motion evaluation can be predicted using the spectral analysis approach which is basically a spectrum energy of ship response. The spectrum response of ship's motion in irregular or random wave, (S_R), is determined by the multiplication of the Response Amplitude Operator (RAO $_{(\omega)}$) and the wave spectrum ($_{(\omega)}$), as shown in Equation 1. RAO$_{(\omega)}$ is basically motion responses of ship under excitation of sinusoidal or regular wave that is measured on a ship model test in a tow tank or calculated numerically. The wave spectrum is total energy of wave for all possible range of wave frequencies that is obtained by measurement on site for long period of time or it is easily to determine with the standard formula of wave spectrum.

$$S_R = (RAO_{(\omega)})^2 \times S_{(\omega)} \tag{1}$$

Where, S_R is the response spectrum of ship motions (m^2/s); $S_{(\omega)}$ is wave spectrum (m^2/s); $(RAO_{(\omega)})^2$ is transfer function or ship characteristics, and ω is wave frequency (rad/sec). In the direct measurement approach, the wave spectrum and transfer function are not needed because the ship's motion response is defined on board directly.

2.2 ITTC standard measurement

The description of the ship's motion response obtained from the measurements device shows the form of superside components identic with an irregular wave, as illustrated in Figure 2. An irregular or random wave is defined as a superposition of many regular waves with the same direction but different wave frequencies or wavelengths [1]. In practice, the International Towing Tank Conference (I.T.T.C.) gives a limitation number of cycles (N) in ship motion measurements, which is very important to note. The cycle number of motion measurement must be recorded within the minimum number N = 50, or more N = 100 as the usual standard, while N = 200 and above is considered a very good measurement [11]. In addition, I.T.T.C. also provides as reference in the measurement or recording time of wave data or ship's motion data, which is between 15 to 20 minutes [11].

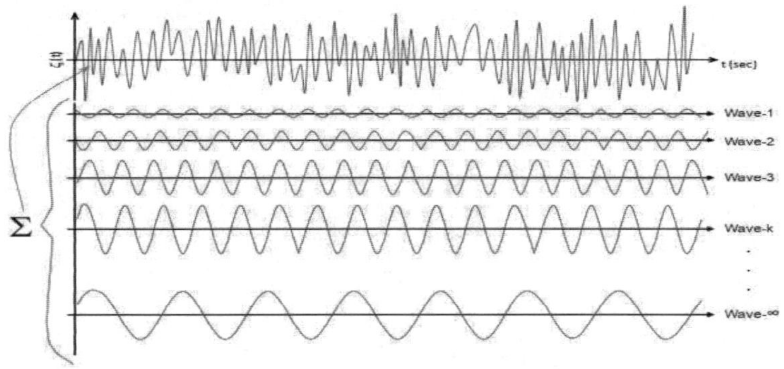

Figure 2. An infinite number of superpositioned random waves [1]

2.3. Fast Fourier Transforms (FFT)

Ocean waves can be described in general terms as random signals in electronics or periodic vibrations in mechanical systems [1]. The wave condition, which is generally referred to as a random process can be represented by the Fourier series containing periodic components with frequencies. According to some references, the random wave elevation can be written in the form of an equation (2) [1]:

$$\zeta(t) = \bar{\zeta} + \sum_{n=1}^{N} A_n \cos(\omega_n t) + B_n \sin(\omega_n t) \tag{2}$$

Where ζ is amplitudo, n is number of data, t is time per n data, ω_n is natural frequency, and the constant of A_n and B_n are defined in equation (3).

$$A_n = \frac{2\pi}{T_H} \int_0^H \zeta(t) \cos(\omega_n t)\, dt \tag{3}$$

$$B_n = \frac{2\pi}{T_H} \int_0^H \zeta(t) \cos(\omega_n t)\, dt \tag{4}$$

Where T_H is the time span of the whole process

Since the measurement instrument records the ship's motion in the time domain data, so that the Fast Fourier Transform (FFT) method as presented in Equation 3 is applied to process the ship's response in the frequency domain [12]. In practice, the discrete Fourier Transform (DFT) algorithm is suitable approach to deal with discrete data for calculating quickly and efficiently. The Discrete Fourier Transform equation in (3) is applied in a wide variety of fields, such as digital signal processing, solving partial differential equations, and to algorithms for multiplying large integers.

$$X(\mathrm{k}) = \sum_{j=1}^{N} x(\mathrm{j}) {\omega_N}^{(j-1)(k-1)} \tag{5}$$

IOP Conf. Series: Earth and Environmental Science 1081 (2022) 012036 doi:10.1088/1755-1315/1081/1/012036

Where *X*(k) is the result of the discrete Fourier transform of *X*(k) containing the signal magnitude values for k = 0, 1, ... *N*; *x*(k) expressed in *x*(j) is a discrete time signal; and *N* is the signal length of *x*(k). The frequency components, ω_n, are calculated by equation (5) [1].

$$\omega_n = \frac{2\pi n}{T_H} \text{ [rad/sec] for } n = 1,2,3 \ldots\ldots \infty \tag{6}$$

Equation (3) shows that a random wave is composed of a superposition of a large number of regular waves, which theoretically reaches an infinite number. The application of equation (3) in the field of marine and marine engineering has been introduced by Weinblum & St. Denis (1950) and St. Denis & Pierson (1953) [1], which shown as illustration in Figure 2. The equation to get the ship's response should be done smoothing and filtering the results of FFT [1,14,15]. This is intended to eliminate noise that occurs in the result data and to get more accurate results.

2.4 Filters using shape preserving interpolation

The data filtering is used to remove noise response based on the rational cubic spline interpolant method. This spline interpolation method has been widely used by various fields of science such as in the design of cars and aircraft fuselages [13] and as a result, it can also be used in the shipping industry. The method as shown in equation (5) is suitable to estimate or interpolate a given finite data set with second-order parametrics.

$$\begin{aligned}(x) &\equiv s_i(x)\\ &= \frac{A_{i0}(1-\theta)^3 + A_{i1}\theta(1-\theta)^2 + A_{i2}\theta^2(1-\theta) + A_{i3}\theta^3}{(1-\theta)^2\alpha_i + \theta(1-\theta)(2\alpha_i\beta_i + \gamma_i) + \theta^2\beta_i}\end{aligned} \tag{7}$$

The following conditions will assure that the rational cubic spline interpolant in (1) has C^2 continuity:

$$\begin{aligned}&s(x_i) = f_i,\\ &s(x_{i+1}) = f_{i+1},\\ &s^{(1)}(x_i) = d_i,\\ &s^{(1)}(x_{i+1}) = d_{i+1},\\ &s^{(2)}(x_{i+}) = {}^{(2)}(x_{i-})\end{aligned} \tag{8}$$

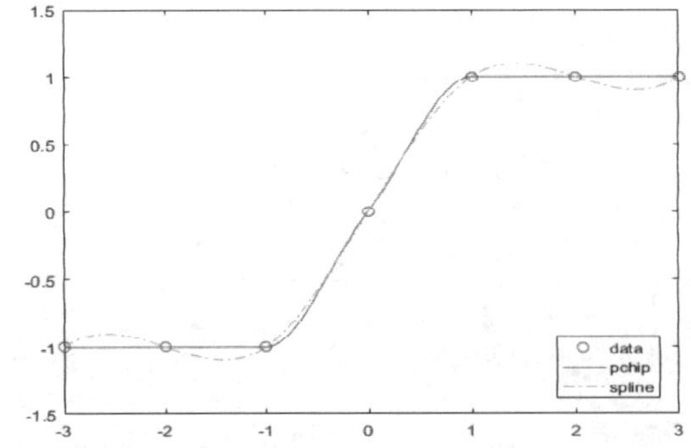

Figure 3. Difference between spline and Rational Cubic Spline Interpolant (pchip) [12]

Shape-preserving interpolation in Matlab refers to the Piecewise Cubic Hermite Interpolation (pchip) [12]. This method maintains the monotonicity and shape of the data. For surfaces, the Interpolant fit type uses the scattered Interpolant and nearest methods functions, and the Matlab grid data function for cubic and bi harmonic methods. The thin-plate spline method uses the tpaps function.

The type of interpolant to use depends on the characteristics of the data being fit, the required smoothness of the curve, speed considerations, post-fit analysis requirements, and so on. The linear and nearest neighbor methods are fast, however the resulting curves are not very smooth. The cubic spline and shape-preserving methods are slower, but the resulting curves are very smooth. In the picture of figure 3, there are 2 curves, namely a spline curve (dushed line curve) with a smooth and wide curve, while the second curve is a shape preserving interpolation curve using the chip formula (full line curve) with a narrower or filtered approach. Figure 3 show the filter method can produce the better results [12].

3. Experimental design

In direct measurement, three important factors must be considered to give the appropriate results including (i) the accuracy of the instrument device, (ii) the accuracy of the measurement process, and (iii) the processing of the measurement data recorded. In the previous paper [14], an analysis of the selection of tools and determination of the measurement location was introduced. Meanwhile, the identification method of fast ship motion using direct measurements technique is presented in this paper. (a case study of the MV Dumai Line 12) is proposed.

3.1. Accuracy and precision of instruments

The instrument used is the ADIS16364 sensor which is a high-precision tri-axis inertial sensor product from analog devices and uses an Analog to Digital Converter (ADC) [15]. The ADIS16364 sensor is equipped with 2 sensors, namely (i) Gyroscope sensor to measure the rotation motions such as roll, pitch and yaw in XYZ coordinate as shown in Figure 4. The data recorded is in the form of angular velocity in units of deg/s (^{0}/s). (ii) Accelerometer sensor is used to measure the acceleration of the translation motion of ship namely surge, sway and heave in XYZ coordinates as shown in Figure 4. The measurement data is recorded in the form of acceleration in units of m/s^2. All sensors are built in the IMU-Evaluation board microcontroller, which can be installed and connected to a personal computer via a universal serial bus (USB) cable, as seen in Figure 5. Table 1 discusses the specification, accuracy, and precision of ADIS16364.

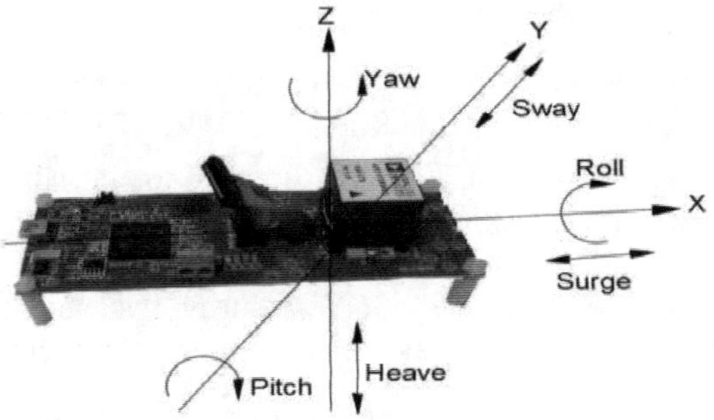

Figure 4. Direction of motion on the measuring instrument

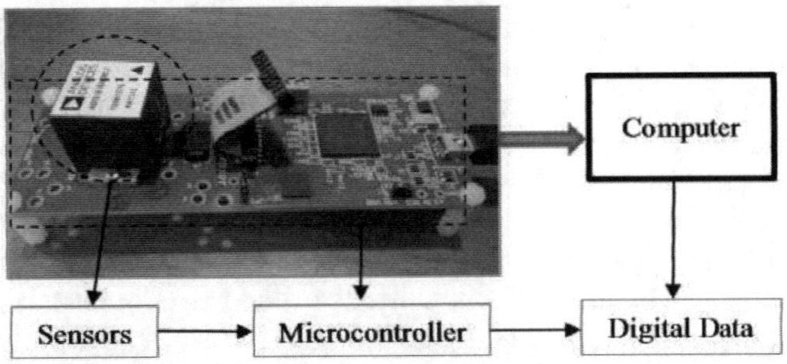

Figure 5. The operating principle of the ADIS16364 sensor

In addition, the ADIS16364 sensor's specifications of the instrument are designed with great accuracy and precision particularly for the motion measurement tool. Precision and accuracy are two ways scientists think of mistake. Accuracy refers to how close a measurement to the correct or accepted value. Precision refers to how close the measurement of the same item to each other, and precision does not depend on accuracy. Herewith it is possible to be very precise, but do not very accurate, and it is also possible to be accurate without being precise. Measurements in scientific works are of the best quality if the measurements used have measuring instruments that have the highest precision and accuracy. To ensure the traceability of the tools used and correctness shown by the motion measuring instrument used refers to specifications of the tools that have been guaranteed correct during production, the specifications include the accuracy and precision of the tools listed in the data sheet book that the precision of the gyroscope sensor is 0.05 ^{0}/sec/LBS, the accelerometer sensor precision is 1.00 mg/LBS. In the meantime, the accuracy of the gyroscope sensor is +350 ^{0}/s and the accelerometer sensor accuracy is +5.25 g.m/s^2, see Table 1.

Table 1. Specifications, accuracy and precision of the ADIS16364 sensor [15]

ADIS16364 spesification:
TA = 25 ^{0}C, VCC = 5,0 V, angular rate = 0 ^{0}/sec, dynamic range = +300 ^{0}/sec + 1g
Aplication : Medical Intrumentation, Robotics, Platform controls, Navigation
Calibration temperatur range: -20 ^{0}C to +70 ^{0}C

No	Item	Min	Typ	Max	Unit
1	Gyroscopes sensors				
a.	Gyroscope accuracy	+300	+350		^{0}/sec
b.	Gyroscope precision (Dynamic range = +300^0/sec)	0,0495	0,05	0,0505	^{0}/sec/LBS
2	Accelerometer sensors				
a.	Accelerometer accuracy	+5,00	+5,25		g
b.	Accelerometer precision	0,99	1,00	1,01	mg/LBS

The sensors are mounted in the forward-facing cab parallel to the ship's longitudinal axis and are connected to a personal computer via a universal serial bus (USB) cable. Setting the retrieval time is done by setting the "record length" on the monitor screen for the measuring instrument operation. In accordance with the specifications of the ADIS16364 measuring instrument, that the data retrieval speed are 833 samples per second. to get 15 minutes of measurement, the record length setting are 750.300 data points.

3.2 Selection of measurement locations
Data retrieval of measurement are carried out several time in different location during ship voyage. The measurement location was designed in advance before the ship operates, which is considered to have the potential for-excessive motion to occur. The four locations of measurement points are selected as presented in Figure 6. This paper presents only the data collected at the coordinates of 1°05'31.5"N 103°37'39.7"E (4th measurement) which is water area between the island of Tanjung Balai and the island of Batam [16]. At this location, the highest wave is occured about 0.4 meters according to Windy (Wind map & weather forecast; Windy.com) [17].

The MV Dumai line 12 fast ferries are passenger ship types operate from Bengkalis to Batam (the waters of the Malacca Strait). The principle dimensions of the MV Dumai line 12 are a length overall (L) 38.30 meters, a breadth (B) 7.40 meters, a depth (H) 2.70 meters, a draft (T) 1.30 meters, service speed (Vs) 32 knots, and the vessel can carry 379 passengers and 11 crews [18]. The general arrangement of ship is shown in Figure 6. The data collected is classified into the data before and during measurement. The data collected before the measurement activity includes (i) the main dimension of the ship, (ii) the general arrangement of the ship, (iii) bow and stern draft data, (iii) the wave height and the wind speed, and (iv) the sensor position placement on the ship. In the meantime, the data must be measured or recorded during the time of measurement include (i) the velocity and acceleration of the ship motions, (ii) the direction and speed of the ship, (iii) the direction and speed of wind, and (iv) the location point and the trajectory route of the ship which are read by the Global Positioning System (GPS) device on the ship.

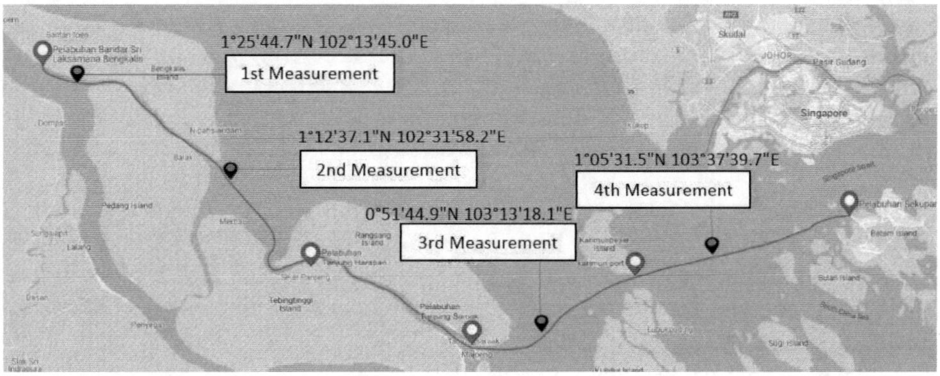

Figure 6. The coordinates of the ship motion measurement location

3.3 Placement of the measuring sensor

The placement of the motion sensor (ADIS16364) is determined based on the objectives and the existing standards of measurements. In addition, the placement of the sensor instrument in the navigation room follows the existing seakeeping criteria standards as stated in STANAG 2000 [5] or NORDFORSK 1987 [6]. Figure 7 shows the placement of the sensor instrument in the navigation room of MV Dumai line 12. In more detail, the sensor is placed on a table with coordinates of 1.08 meters of starboard side, 10.59 meters from FP and 3.87 meters above draft, or 5.17 meters above baseline, as seen in Figure 8.

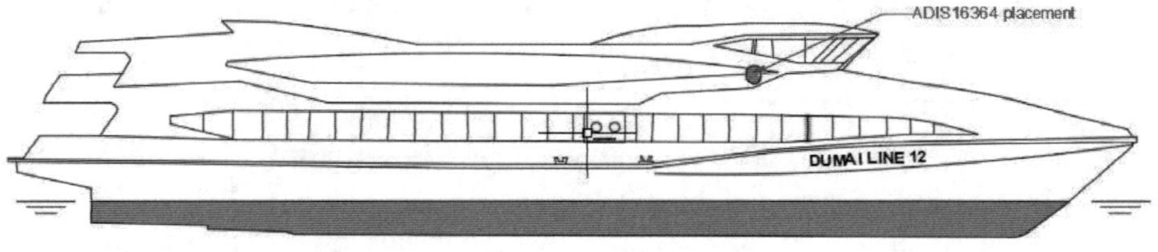

Figure 7. ADIS16364 placement, side view

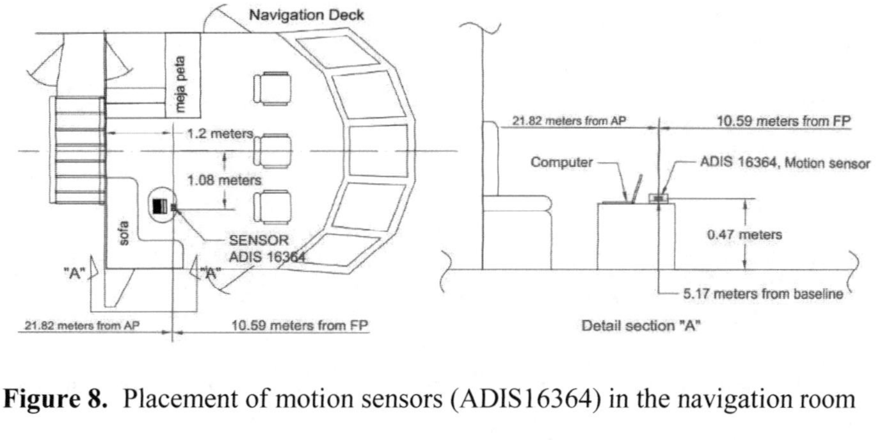

Figure 8. Placement of motion sensors (ADIS16364) in the navigation room

4. Results and discussion

The measurement data is taken within 15 minutes with a record length of 750,300 LBS, which means that every recorded data has 750,300 lines of data. The acceleration of heave motion is measured by the accelerometer of ADIS16364 and is plotted in the time domain as shown in Figure 9a. Through the integrating of acceleration data, the velocity of heave motion is defined as presented in Figure 10a. And the elevation of heave motion is obtained by differential of the velocity motion, see Figure 11a. And the angular velocity of roll and pitch motions are recorded by the gyroscope sensor of ADIS16364 and is presented in Figures 12a and 15a, respectively. The angular acceleration of roll and pitch motions is obtained by the integrating of the angular velocity of both motion that are presented in Figures 13a and 16a, respectively. The elevation of roll and pith motion which are shown in Figures 14a and 17a, respectively, are obtained by the differential of the angular velocity of roll and pitch motion, respectively.

Figures 9a – 17a show the original data recorded of acceleration, velocity and elevation marking as the grey color line. Since the signal data have noise, the data recorded is filtered using the shape-preserving interpolation approach that the results are presented in blue color line all spectrums of acceleration, velocity, and elevation in frequency domain as shown in Figures 9b – 17b are transformed from the time domain data by using the Fast Fourier Transform algorithm in MATLAB software.
Heave acceleration of the MV Dumai line is measured by the accelerometer within 15 minutes recording give 750,300 data. Figures 9a, 10a, and 11a show that the maximum amplitude of acceleration, velocity and elevation of heave motion are defined 7.60 m/s^2, 0.009 m/s, and 1.1×10^{-5} m, respectively. Figures 9b, 10b, and 11b show the spectrum of acceleration, velocity, and elevation of heave motions in which the natural frequency appeared about 0.569 rad/s.

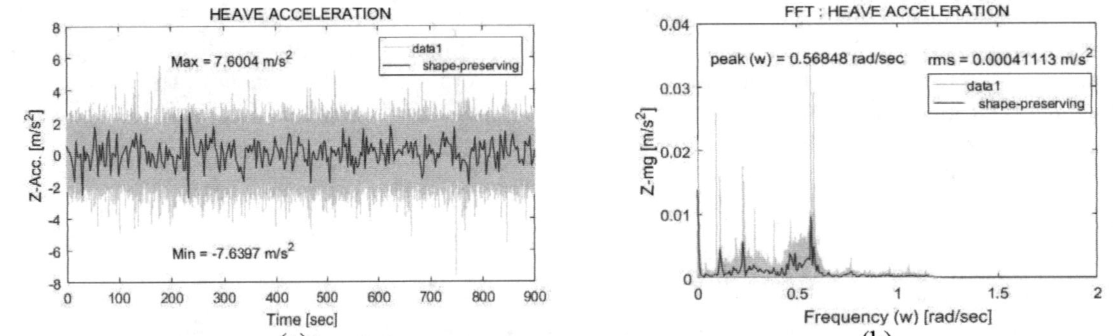

(a) (b)

Figure 9. (a) Heave acceleration in time domain, (b) Heave acceleration in frequency domain

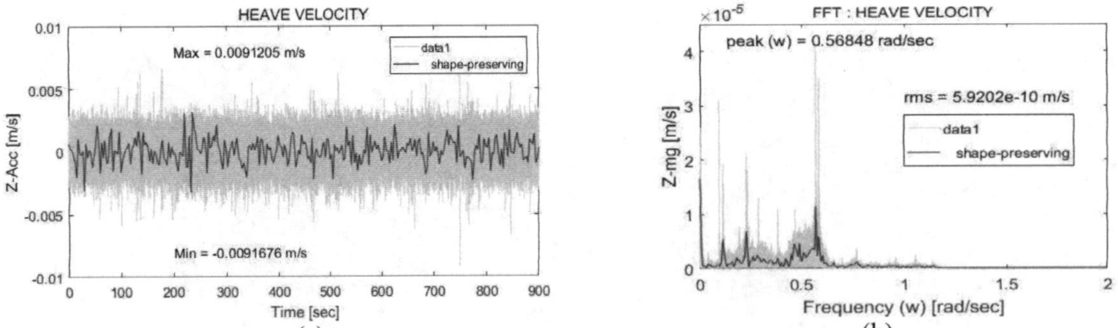

(a) (b)

Figure 10. (a) Heave velocity in time domain, (b) Heave velocity in frequency domain

IOP Conf. Series: Earth and Environmental Science 1081 (2022) 012036 doi:10.1088/1755-1315/1081/1/012036

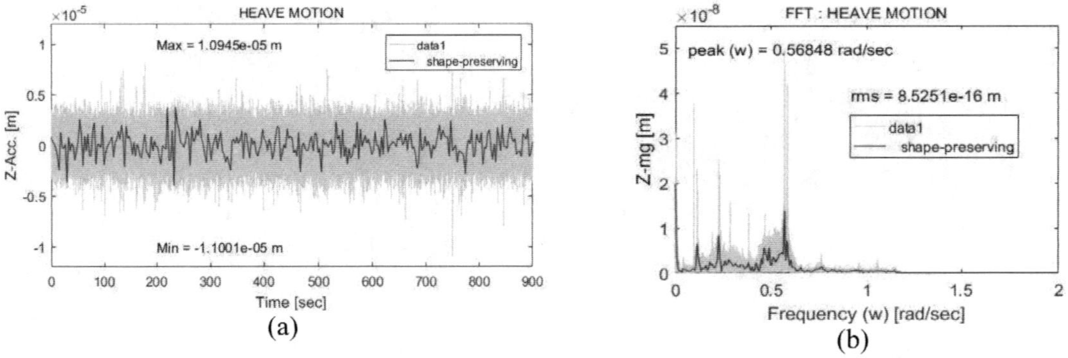

Figure 11. (a) Heave motion in time domain, (b) Heave motion in frequency domain

The maximum amplitude of acceleration, velocity and elevation of roll motions are about 1.0x10^4 deg/s^2, 12.9 deg/s, and 0.016 deg, as presented in Figures 12a, 13a, and 14a, respectively. The spectrum of acceleration, velocity, and elevation of roll motions in which the natural frequency occurred at about 0.117 rad/s, see Figures 12b, 13b, and 14b. The maximum amplitude of acceleration, velocity and elevation of pitch motion are about 2.98x10^4 deg/s^2, 35.8 deg/s, and 0.043 deg, as presented in Figures 15a, 16a, and 17a, respectively. The spectrums of acceleration, velocity, and elevation of pitch motions in which the natural frequency occurred at about 0.227 rad/s, see Figures 12b, 13b, and 14b.

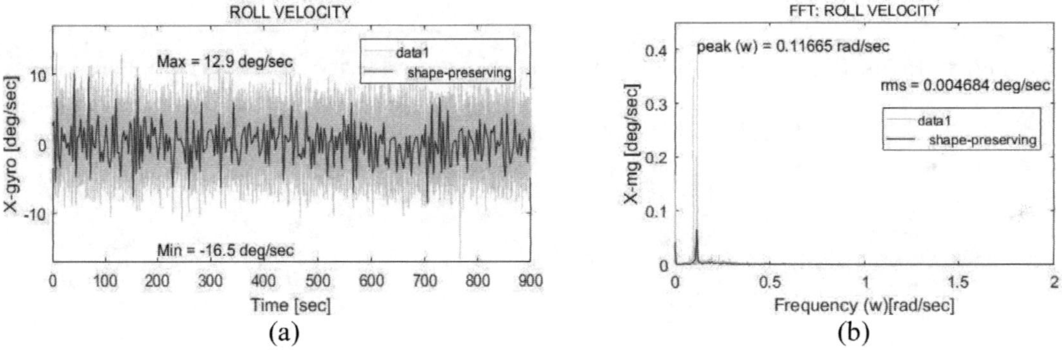

Figure 12. (a) Roll velocity in time domain, (b) Roll velocity in frequency domain

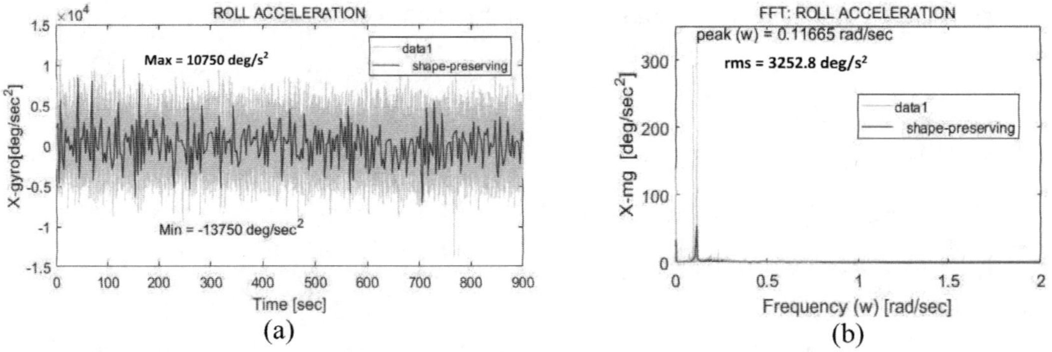

Figure 13. (a) Roll acceleration in time domain, (b) Roll acceleration in frequency domain

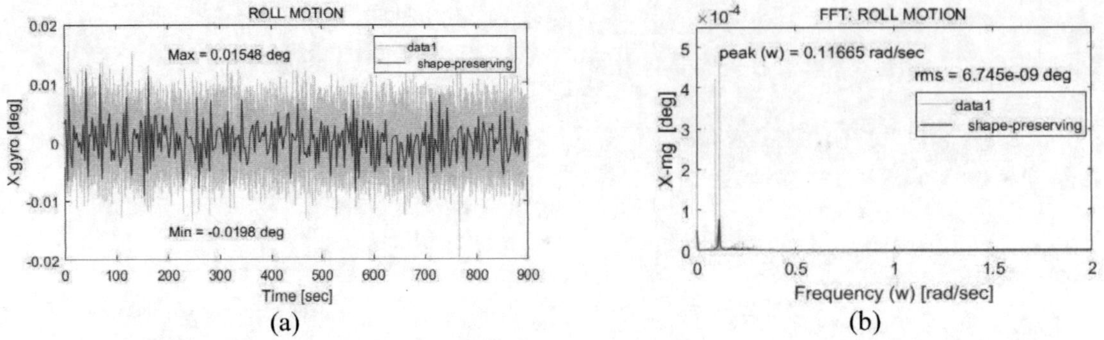

Figure 14. Roll motion in time domain, (b) Roll motion in frequency domain.

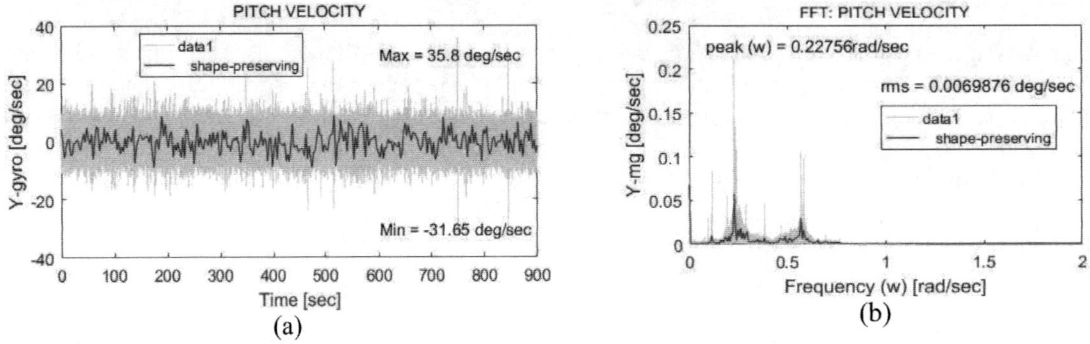

Figure 15. (a) Pitch velocity in time domain, (b) Pitch velocity in frequency domain

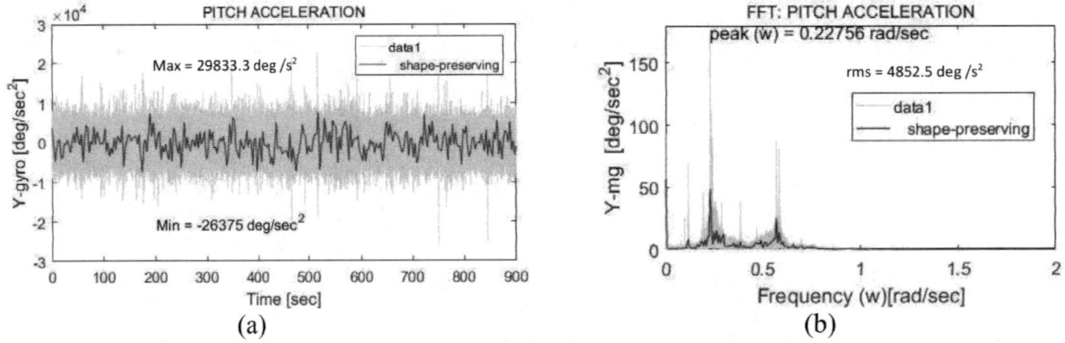

Figure 16. (a) Pitch acceleration in time domain, (b) Pitch acceleration in frequency domain

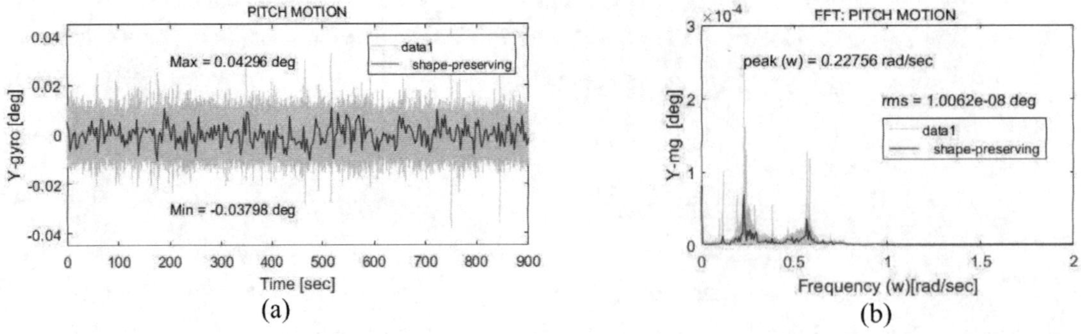

Figure 17. (a) Pitch motion in time domain, (b) Pitch motion in frequency domain

5. Conclusion

Direct measurements of the heave, roll, and pitch motions of the MV Dumai line 12 passenger ship sailing in the waters of the Malacca Strait is performed. The data recorded in term of time domain are filtered using the shape preserving interpolation . Furthermore, the response spectrum of motions in frequency domain is obtained using the FFT method in which the natural frequency of heave, roll, and pitch are defined about 0.568 rad/s, 0.117 rad/s and 0.227 rad/s, respectively.

Acknowledgments

We would like to express our gratitude to RISTEK-BRIN Indonesia for providing financial support under Penelitian Disertasi Doktor (PDD), as well as the Director of the Dumai line Bengkalis branch and the captain of the MV Dumai line 12, for allowing us to conduct data collection on board.

Reference

[1] Djatmiko E B 2012 *Prilaku Dan Operabilitas Bangunan Laut di Atas Gelombang Acak* (ITS Press Surabaya)

[2] Rumawas V, Asbjørnslett B E 2014 A *Content Analysis Of Human Factors In Ships Design* (Trans RINA, Vol 156, Part A3, Intl J Maritime Eng) Jul-Sep 2014. DOI No: 10.3940/rina.ijme.2014.a3.299

[3] Pascoal R, Soares G, Sørensen J C 2017 *Ocean wave spectral estimation using vessel wave frequency motions* (J Offshore Mech Arct. Eng) 2007, 129, 90–96

[4] Gaglione S, Piscopo V, Scamardella A 2016 *The Overall Motion Induced Interruptions as Operability Criterion for Fishing Vessels* (J Mar Sci Technol): DOI 10.1007/s00773-016-0373-2

[5] NATO 2000 *Standardization Agreement (STANAG): Subject: Common Procedures for Seakeeping in the Ship Design Process* (NATO) Military Agency for Standardization)

[6] NORDFORSK 1987 *Assessment of Ship Performance in a Seaway* (Copenhagen, Nordforsk)

[7] Carolyne Q J 2019 Seakeeping and Maneuvering (USNA EN455: 2019) Course Notes.

[8] Sulisetyono A 2012 *Seakeeping Analysis Of The Trimaran Ferry Ship In Short, Crested Sea For A Case Of East Java Water Condition* (RINA, Royal Institution of Naval Architects - International Conference on Ship and Offshore Technology) ICSOT Indonesia 2012: Developments in Ship Design and Construction 137-143

[9] Bhattacharya R. 1978 *Dynamics of Marine Vehicles* (ISBN-13: 9780471072065)

[10] Rumawas V, Asbjørnslett B E, Klöckner C A 2009 *Human Factors Evaluation In Ship Design : A Case Study On Offshore Supply Vessels In The Norwegian Sea , Part I : Theoretical Background.* (Naval Engineers Journal) 128(4):87

[11] Journée J M J, Massie W W 2001 *Offshore Hydromechanics First Edition* (Delft University of Technology)

[12] MATLAB Function Reference. http://matlab.izmiran.ru/help/techdoc/ref/fft.html

[13] Karim S A A, and Pang K V 2016 *Shape Preserving Interpolation Using C2 Rational Cubic Spline* (Journal of Applied Mathematic) Volume 2016, Article ID 4875358, 14 page http://dx.doi.org/10.1155/2016/4875358

[14] Jamal, Sulisetyono A, Aryawan W D 2020 *Review of the Seakeeping Criteria for the Study of a Passenger Ship Criteria in Indonesian Water* (ICETIR 2020: Engineering Faculty Universitas Jenderal Soedirman) IOP Conf. Series: Materials Science and Engineering 982 (2020) 012041. doi:10.1088/1757-899X/982/1/012041

[15] Analog Devices 2011 *Six Degrees of Freedom Inertial Sensor.* (ADIS 16364 Data sheet) https://www.analog.com/media/en/technical-documentation/data-sheets/adis16364.pdf

[16] Googlemap.com *Coordinat point at 1°05'31.5"N 103°37'39.7"E* https://www.google.com/maps/place/1%C2%B005'31.5%22N+103%C2%B037'39.7%22E/@1.149115,103.5535742,10.83z/data=!4m6!3m5!1s0x0:0x147fffafbe6340a5!7e2!8m2!3d1.0920794!4d103.6276998

[17] https://www.windy.com/?0.959,103.383,10,m:dGVai1U

[18] Dumai line 2022 *General Arrangement Drawing Of Dumai Line 12* (MV Dumai line 12-Bengkalis branch office)

[19] Takami T, Nielsen U D, Jensen J J 2021 *Real-Time Deterministic Prediction Of Wave-Induced Ship Responses Based On Short-Time Measurements.* (Ocean engineering 221 (2021)) 108503.https://doi.org/10.1016/j.oceaneng.2020.108503

[20] Nielsen U D, Brodtkorb A H, Jensen J J 2018 *Response Predictions Using The Observed Autocorrolation Function* (Marine structure Jurnal) Marine struktur 58 (2018) 31-52. https://doi.org/10.1016/j.marstruc.2017.10.012

[21] Zeyuan X, Danwei W, Guoxing Y, Zhijian H 2022 *A Novel Tracking Control Approach Of Amplitude Signals For Vibratory Gyroscopes Suppressing High-Frequency Disturbance* (Measurement: Vol. 195, 31 May 2022,) https://doi.org/10.1016/j.measurement.2022.110981.

[22] Cengiz T, Muhammed E 2022 *Classification Of Surface Electromyography And Gyroscopic Signals Of Finger Gestures Acquired By Myo Armband Using Machine Learning Methods* (Biomedical Signal Processing and Control: Volume 75, May 2022, 103588) https://doi.org/10.1016/j.bspc.2022.103588

[23] Shaohua L, Guanci Y, Junyang L, Hassen M O 2022 *Dynamic Analysis, Circuit Realization and Accelerated Adaptive Backstepping Control Of The FO MEMS Gyroscope* (Chaos, Solitons & Fractals: Volume 155, February 2022, 111735) https://doi.org/10.1016/j.chaos.2021.111735

[24] Garcia R F, Antonio C, Ameal F 2001 *Ship Stability Monitoring by Motion Frequency Analysis* (IFAC Proceedings Volumes: Volume 34, Issue 7, July 2001, Pages 39-43) https://doi.org/10.1016/S1474-6670(17)35056-5

[25] Pattison J H, Sheridan, D J 2004 *Human Performance Factors and Measures in Hull Form* (Selection. RTO AVT Symposium on "Habitability of Combat and Transport Vehicles: Noise, Vibration and Motion) October 2004, 4–7.

[26] Tristan P 2015 *Ship Seakeeping Operability, Motion Control, and Autonomy - A Bayesian Perspective* (IFAC-Papers On Line 48-16 (2015) 217–222; Science Direct)

[27] Silvia P, Antonio A, Vincenzo D C, Giampaolo F, Salvatore G, Anna I, Elena M, Pasquale P, Vincenzo P, Alessandra R 2021 *Sea State Monitoring by Ship Motion Measurements Onboard a Research Ship in the Antartic Waters* (Journal of Marine Science and Engineering) J. Mar. Sci. Eng. 2021, 9, 64. https://doi.org/10.3390/jmse9010064.

IOP Conf. Series: Earth and Environmental Science 1081 (2022) 012037 doi:10.1088/1755-1315/1081/1/012037

Propeller selection optimization on mini LNG ship based on engine propeller matching approach

I E Sandjaja[1,2], I M Ariana[1], Erwandi[2] and M Indiaryanto[2]

[1] Marine Engineering Department, Institut Teknologi Sepuluh Nopember, Surabaya, Indonesia.
[2] Kantor Balai Hidrodinamika, Badan Riset Inovasi Nasional, Surabaya, Indonesia.

Abstract. Mini LNG Carrier Ship is specially designed for the purpose of distributing Liquefied Natural Gas (LNG) throughout Indonesia and supporting the national program of generating 35.00 MegaWatt (MW). Indonesia's vast territory with long distances requires an efficient transportation, including Mini LNG Carrier Ship to reduce costs, especially fuel and emission levels that meet the EEDI criteria set by the International Maritime Organization (IMO). One of the efficiency factors of Mini LNG Ship Carrier is obtained by optimizing the design, especially to the propulsion system. The propulsion system consists of the main engine and propeller. This study focuses on the selection of ship propulsion propellers, especially the type and number of propeller blades. The good propeller will have an impact on the level of energy efficiency. The research method used is a computational simulation using the Engine Propeller Matching (EPM) method. The study tested the B-Series propeller with an alternative number of propeller blades are (z) = 3, 4, 5, 6, and. 7. Calculating simulation results show that the analysis of the EPM computational results at 10 knots service speed, shows that the B6.40 propeller has a better efficiency value than the other propeller types (z = 3, 4, 5, and 7). Propeller type B6.40 produces a good level of efficiency according to the planned 2 x 250 KW engine performance diagram.

1. Introduction

Indonesia is a country with the third largest natural gas reserves in the world and known as tenth largest natural gas producing countries in 2015 [1]. Utilization of natural gas called Liquefied Natural Gas (LNG) in various fields, especially the household consumption sector, the industrial sector, the power generation sector and the transportation sector. The Indonesian government has launched a 35,000 MW electricity program, where LNG is used to power gas turbine power plants [2]. LNG is also used in the transportation sector, as a fuel for large vehicles such as trucks and buses, as well as fuel for ships.

The use of LNG, which is a fuel with the advantage of being cheaper and environmentally friendly, and whose emission levels can be reduced by up to 85% when compared to petroleum fuels, so that the demand for LNG needs is expected to increase and further develop to reach around 40% of the total demand for LNG. 2016 to 2040 [3]. The increasing demand for LNG requires a means for distributing LNG properly, where in general LNG production is located in remote areas so that one of the alternative means of distribution is ideal by using ships [4]. The character of LNG has the advantage of being efficient in cargo specifications, which requires less space in transportation compared to other types of transportation (petroleum fuel), so that the cost of operating the ship is cheaper.

The Indonesian Hydrodynamic Laboratory (ILH), in 2020 has conducted research and has designed a special LNG carrier ship called the Mini LNG Carrier Ship, with its purpose is for distributing LNG throughout Indonesia as an answer to the challenges of future LNG needs [5]. The wide distribution and long distances in the LNG distribution operational chain in Indonesia make Mini LNG Carrier Ships to be efficient in their operations in order to reduce the level of fuel consumption and the allowable level of CO2 exhaust emissions when the ship operates according to the Energy Efficiency Design Index (EEDI) standard which has been established by the International Maritime Organization (IMO) in 2011 [3].

One of the efforts to improve the operational efficiency of the Mini LNG Carrier is through optimization of the propulsion system design. The propulsion system consists of the hull, main engine and propeller [6]. The good propulsion system will affect the level of efficiency so that it will reduce the level of fuel consumption on the ship [6].

This study aims to find the efficiency level of the mini LNG carrier ship through optimization of the propulsion system design which is more focused on choosing the type of B-Series propeller with the number of blades (z) which has the highest efficiency level. The alternative number of blades (z) on the type B-Series propeller is the number of blades z = 3, 4, 5, 6, and 7 [7]. The EPM numerical test was carried out on each variant of the number of blades (z) on the B-Series propeller with the provisions of the hull type previously determined and the main engine to be planned. This analysis and calculation aims to obtain the type of propeller with the number of blades (z) which has the best level of propeller efficiency for Mini LNG Carrier Ships.

2. Literature Review

Based on ILH research in 2021, the Mini LNG Ship was designed using a propeller type B4.40 [8], while for the development of this research, a study was carried out on the B-Series propeller with variations in the number of blades 3, 4 , 5, 6 and 7.

The initial stage in the research on selecting the propeller to be used, it is necessary to know in advance the amount of Effective Horse Power (EHP) used to move the ship at the desired speed, EHP can be obtained by using numerical calculations, software or by testing model tests. The formula for calculating the amount of ship resistance, is as follows [9], [10]:

$$\text{Rts} = \tfrac{1}{2}.\, \rho.Cts.S.V_s^2 \tag{1}$$

$$\propto = \tfrac{1}{2}.\, \rho.Ct_S.S \tag{2}$$

Where :
Rts = Ship total resistance of the ship..
ρ = Density of seawater.
Ct = Ship total resistance coefficient.
S = Wet surface area of the ship.
Vs = Ship speed.

Ship resistance amount in this study will be obtained by testing the model in the towing tank [2] from the amount of ship resistance obtained, the amount of effective horse power (PE) is needed, as follows [11]:

$$P_E = Rt_S.\, V_s \tag{3}$$

Effective Horse Power is required for a ship without a propulsion system. To move a ship that has equipped with a propeller will require thrust as follows [12]:

$$T = \frac{Rts}{(1-t)} \tag{4}$$

$$THP = T.V_a \tag{5}$$

Where:
T = Thrust.
THP = Thrust horsepower.
t = Thrust deduction factor.
V_a = Advanced velocity.

The ship can run at the planned speed if the magnitude of the resistance force is smaller or equal to the ship's thrust [13] [14] [15] [16]

$$T_{ship} = T_{Propeller} \tag{6}$$

$$\frac{R}{1-t} = K_T \rho n^2 D^4 \tag{7}$$

$$\frac{\propto V_s^2}{1-t} = K_T \rho n^2 D^4 \tag{8}$$

$$K_T = \frac{\propto V_s^2}{\rho n^2 D^4 (1-t)} \tag{9}$$

$$V_S = \left(\frac{Va}{1-w}\right)^2 \tag{10}$$

$$K_T = \frac{\propto\left(\frac{Va}{1-w}\right)^2}{\rho n^2 D^4 (1-t)} \tag{11}$$

$$J = \frac{Va}{n.D} \tag{12}$$

$$K_T = \frac{\propto\left(\frac{JnD}{1-w}\right)^2}{\rho n^2 D^4 (1-t)} \tag{13}$$

$$K_T = \frac{\propto J^2}{(1-t)(1-w)^2 \rho D^2} \tag{14}$$

$$\beta = \frac{\propto}{(1-t)(1-w)^2 \rho D^2} \tag{15}$$

$$K_T = \beta J^2 \tag{16}$$

Where:
K_T = Thrust coefficient
n = Propeller Rotation..
D = Propeller Diameter.
w = wake fraction.
J = Advance coefficient

The results of the KT value (Equation 14) are synchronized with the KT value from the Open Water Test so that the KQ value and J value are obtained to maintain the propeller efficiency [17]

$$K_Q = \frac{Q}{\rho n^2 D^5} \tag{17}$$

$$\eta_{prop.} = \frac{J\,.\,Kt}{2\pi\ x\ KQ} \tag{18}$$

Where:
K_Q = Torque propeller coeffisient.
ηo = Propeller efficiency
Q = Torqeu propeller.

Except for KT, KQ, and J values, other technical data can be used in planning the main engine to drive the propeller. The technical specifications of the main engine are needed to determine whether these engines are suitable or not to be used to drive the planned propeller. For the ship to move according to the desired speed, the planned propeller must be avoided from cavitating or it will affect the propeller performance.

Comparing TC calculations with TC Burril diagrams [18] to determine the occurrence of Cavitation. Calculating AP value [18]:

$$A_p = A_d \times \left(1.067 - \left(0{,}229\ x\ \frac{P}{D}\right)\right) \tag{19}$$

Where: $A_d = A_e$ (20)

Calculating $(Vr)^2$ [18]:

$$(Vr)^2 = V_a^2 + (0{,}7 \times \times \pi \times n \times D)^2 \tag{21}$$

Where :
Va = Speed advance (m/s)
n = propeller rotation (rps)
D = diameter behind the ship (m)

Calculating T_c value [18]:

$$T = \frac{EPH}{(1-t)\ x\,Vs} \tag{22}$$

Where :
EHP = Effective Horse Power
Vs = Service speed
T = Thrust deduction factor

Calculating Tc value [18]:

$$Tc = \frac{T}{Ap \text{ x } 0{,}5 \text{ x px } (Vr)^2} \tag{23}$$

Calculating the value of $\sigma_{0,7\text{ m}}$

$$\sigma_{0,7\text{ m}} = \frac{188{,}2+19{,}63.H}{Va^2+3{,}826n^2D^2} \tag{24}$$

Where :
H = Height of Axis from Base Line (m)
Va = Speed Advance (m/s)
n = Propeller Rotation (RPS)
D = Propeller Diameter (m)

3. Methodology

The optimization of the propulsion system in this study was carried out through a numerical calculation simulation method approach using the Engine Propeller Matching (EPM) method. The EPM method is a ship character modeling method designed based on the correlation between the hull, propeller and main engine.

The hull design of the Mini LNG Carrier has been determined based on the ILH study on activities in 2020 [2] and the main engine (250 x 2 KW) with the use of B-Series propeller type B.4.40 through Self Propulsion testing in ILH activities in 2021 [2].

Propeller B-Series has various types of variants, characteristics of the number of blades and shape. This study limits the character of the B-Series propeller on the number of leaf variants, namely (z) with the number of 2, 3, 4, 5, 6, and 7 propeller blades.

Optimization is done by selecting the type of propeller with the number of blades (z) which has a high level of efficiency through a numerical calculation test simulation based on the EPM rules for each variant of the number of available propeller blades, based on the following steps:

- Literature study is used to obtain data that supports research on optimization of the selecting process of propeller type used.
- Data analysis, the data obtained from the literature study is used to carry out the propeller optimization process.
- The propeller design calculates K_T, K_Q, and η_{Prop} using model resistance test data, calculations for blades variations 3, 4, 5, 6, and 7.
- Main engine planning based on step 3.
- Selection of propellers that have high efficiency.
- Considering the level of cavitation propeller.
- Selection of the type of propeller

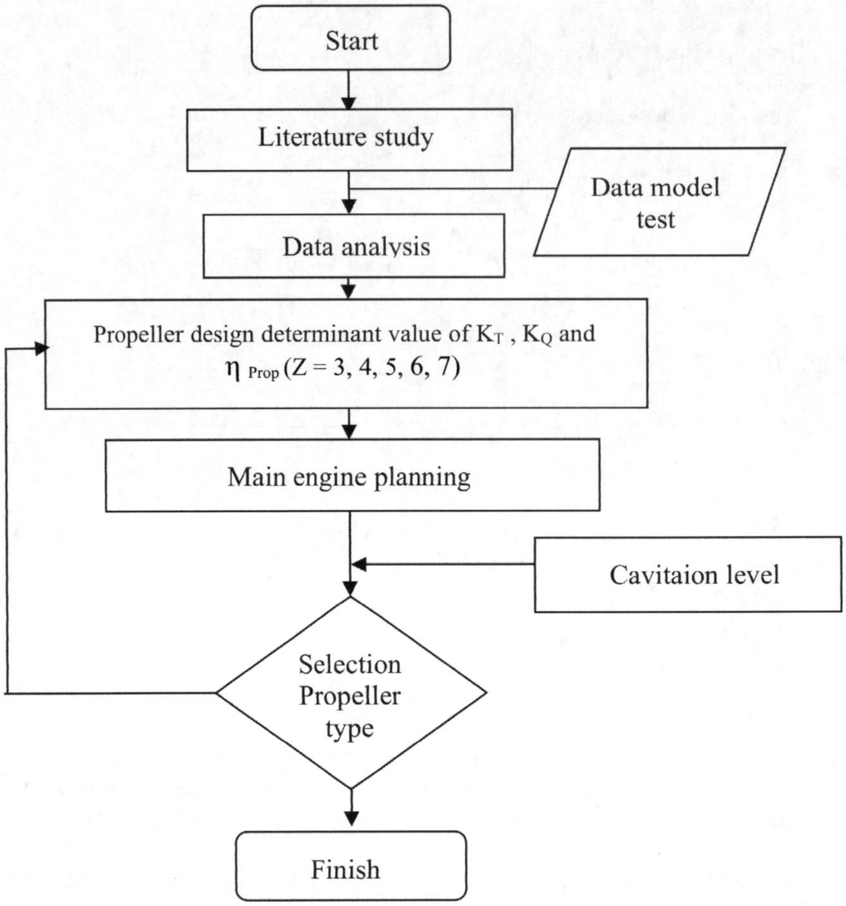

Figure 1. Research methodology flowchart

4. Result and Discussion

Propulsion system (propeller) planning, first it is known the amount of thrust required from the ship. Propeller planning in this research thrust is obtained from the calculation of the ship's resistance where the measurement of the ship's resistance is obtained from testing the ship model in the test pool. The ship data is as follows:.

Ship length (LWL)	= 46,44 m
Wide (Breadth Molded)	= 11,40 m
High (H)	= 3,50 m
Depth (T)	= 2,50 m
Propeller diameter (D)	= 1,5 m
Model scale (λ)	= 11,4282 m
Service speed	= 10 knots

IOP Conf. Series: Earth and Environmental Science 1081 (2022) 012037 doi:10.1088/1755-1315/1081/1/012037

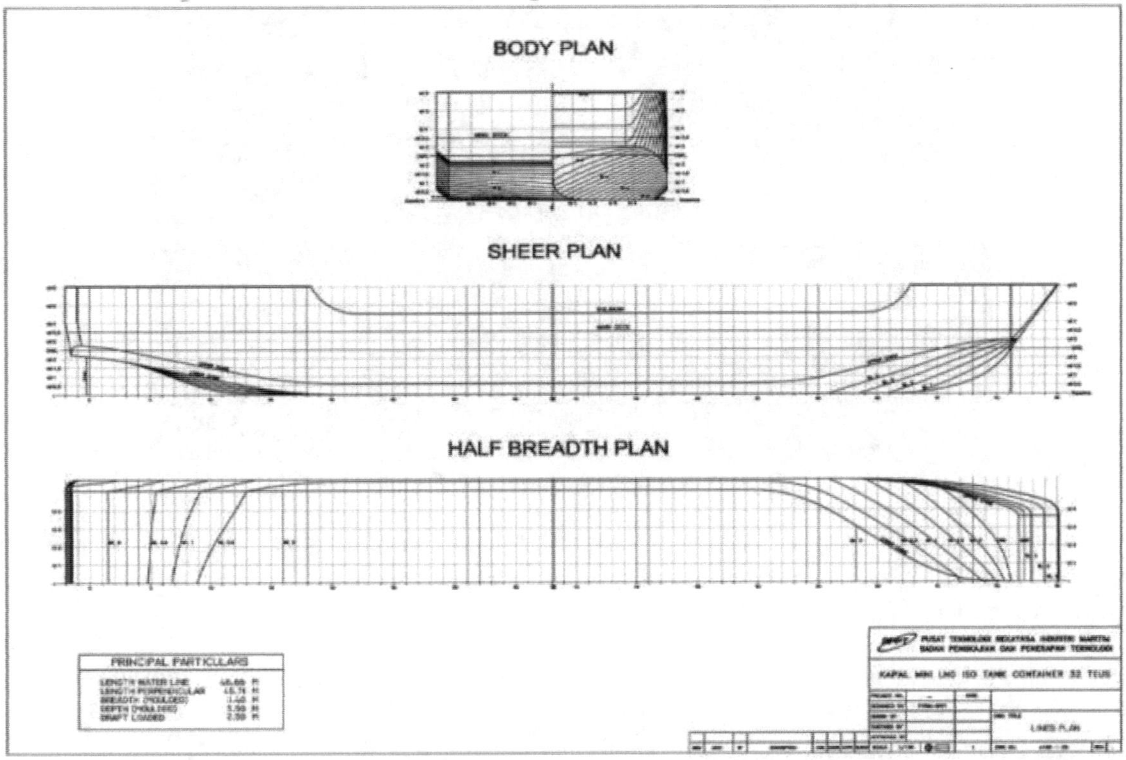

Figure 2. Ship lines plane

The ship model was tested at several speeds to obtain the data needed in the propeller design process, these data include:

Ship resistance (R) = 50,2 KN
Thrust (T) = 64,8 KN
Thrust deduction (t) = 0,225
Wake fractio (w) = 0,249

Table 1. Full loaded ship performance

VS KNOTS	N RPM	N HZ	PS KW	PE KW	ETA-D PE/PD	TH KN	R KN	THDF 1-R/TH t
5	120,0	1,999	35,7	21,7	0,640	10,9	8,4	0,229
6	145,2	2,420	63,7	38,4	0,636	16,1	12,5	0,228
7	171,2	2,853	105	62,9	0,630	22,6	17,5	0,228
8	201,4	3,357	175	102	0,615	32,1	24,8	0,227
9	238,3	3,972	297	167	0,591	46,6	36,1	0,226
10	**276,9**	**4,614**	**475**	**258**	**0,572**	**64,8**	**50,2**	**0,225**
11	318,2	5,303	732	387	0,556	88	68,3	0,223
12	366,3	6,106	1133	579	0,538	120	93,7	0,221

Figure 3. Resistance testing model test [2]

The ship can move according to the desired speed if the propeller's thrust is greater than or equal to the ship's resistance force (equation 6). Thrust propeller load characteristic (KTdes) calculated in equation 14 with a variation of J from 0.1 to 1 with a speed variation of 5 to 12 knots. KTdes rate depicted on the J, K_T, K_Q open water test curves of the propeller type. KTdes curve intersection to K_T is the synchronization condition point. Figure 4 depicts the Thrust generated to the ship movement at the planned speed shown by the K_T value calculated with intersection of K_T and KTdes curves.

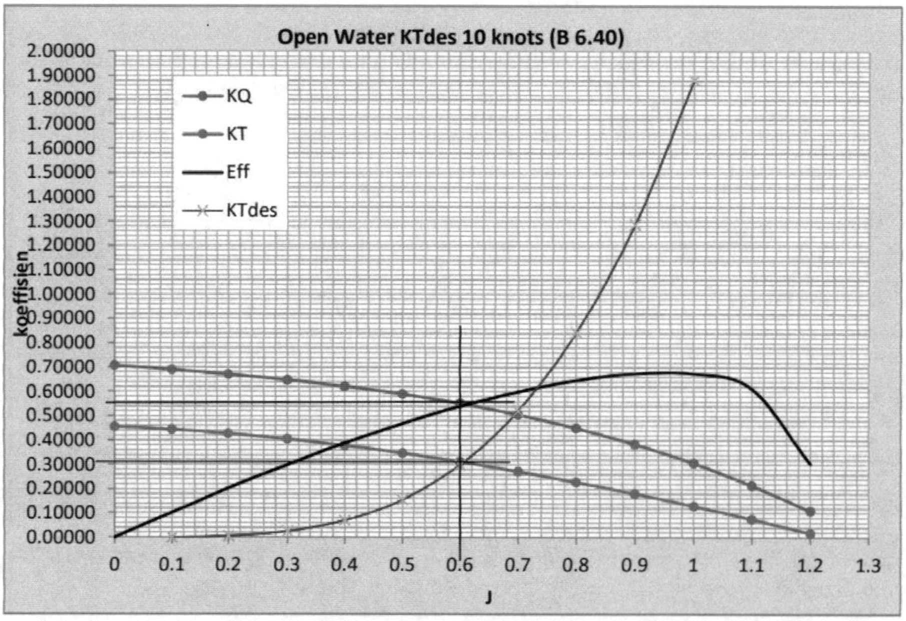

Figure 4. Synchronization of K_T and KTdes

Figure 4. J shows the curve value of K_T value, K_Q value is obtained and used to calculate the first engine planning in Table 1 below. Table 2 summarizes that service speed of 10 knots obtained the amount of power used for the propeller B.6.40. Appropriate performance of the propeller with the main engine is the synchronization process (shown in Figure 5).

The Mini LNG ship uses two main engines, planned with a power @ 250 KW and an engine speed of 1,800 Rpm equipped with a gearbox according to the market (1: 5,15). The feasibility of propeller performance determine, one of them is by avoiding cavitation events, table 3 shows the cavitation conditions that occur in this propeller variation

Table 2. Synchronization of KT and KT_{des} B 6.40

VS	Va	J	Nprop	N engine	K_Q	Q	K_T	T	DHP	SHP	η_{prop}
KNOTS			Rpm	Rpm		KN_m		KN	KW	KW	
5	1,92	0,660	117	602	0,0520	1,53	0,290	5,68	19	19,6	0,58611
6	2,31	0,650	142	733	0,0520	2,27	0,280	8,15	34	35,5	0,55732
7	2,69	0,645	167	862	0,0540	3,26	0,280	11,26	57	60,0	0,53255
8	3,08	0,640	193	994	0,0530	4,25	0,300	16,05	86	90,3	0,57685
9	3,47	0,620	224	1156	0,0540	5,86	0,300	21,70	137	144,6	0,54848
10	3,86	0,610	253	1307	0,0550	7,63	0,320	29,60	202	213,0	0,56514
11	4,24	0,590	288	1485	0,0540	9,67	0,310	37,00	291	306,4	0,53934
12	4,62	0,570	324	1674	0,0580	13,20	0,340	51,60	448	472,0	0,53207

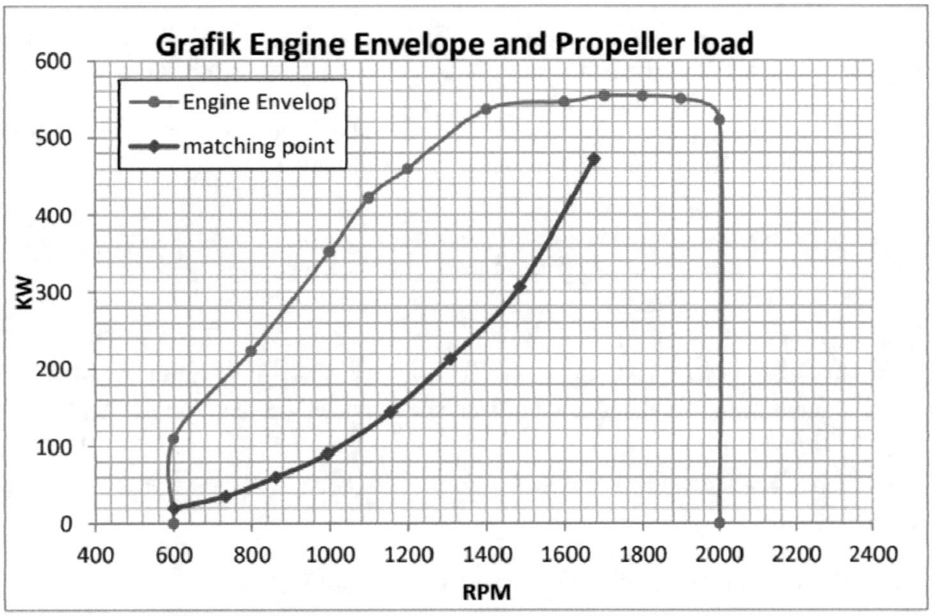

Figure 5. Engine Envelope and Propeller load Graphic

Table 3. Propeller selection calculation results

Type	Ap	n (RPM)	n (rps)	Vr^2	T	TC	$\tau 0{,}7R$	TC Burriil	η_{prop}
B.3.40	0,51217	258	4,300	215,9140	24,86	0,43864	1,01455	Kavitasi	0,5520
B.4.40	0,51217	262	4,367	222,1946	27,69	0,47477	0,98587	Kavitasi	0,5261
B.5.40	0,51217	258	4,300	215,9140	27,73	0,48928	1,01455	Kavitasi	0,5433
B.6.40	0,51217	253	4,217	208,1992	29,60	0,54163	1,05215	Kavitasi	0,5651
B.7.40	0,51217	253	4,217	208,1992	28,68	0,52479	1,05215	Kavitasi	0,5192

Table 3, The selected type of propeller to be used, in addition to the propeller having good efficiency, it is also necessary to consider whether or not cavitation occurs, because cavitation has an impact on the decrease in propeller efficiency, the propeller experiences erosion, and the occurrence of vibration [18]. Table 3 is the result of calculations for predicting cavitation according to Burrill, et al [19].

5. Conclusion

This study uses the optimization method of variation of the number of B-Series propeller blades with the results of the initial propulsion test for the number of blades 4, then the study was developed for the same propeller character with variations in the number of blades (z) = 3, 4, 5, 6, and 7 The research uses the EPM approach to get the optimization of the number of blades that will be used.

The cavitation process approach on the B-Series propeller can be done using a Burrill diagram. Based on the results of cavitation calculations in table 2, the propeller blades 3, 4, 5, 6, and 7 all experienced cavitation including leaf 4 which had undergone a propulsion test. The results of the B-Series B4.40 EPM propeller in table 3 have a propeller efficiency of 0.5261 with a cavitation number (0.7R) of 0.9859, while the results of model testing show the propeller efficiency of 0.517 [nurcholis 2021.14].

The B6.40 propeller pad test is obtained from the EPM results in table 3, the propeller efficiency is 0.5651 with a cavitation number (0.7R) of 1.0516.

The analysis of the EPM calculation results at a service speed of 10 knots shows that the B6.40 propeller has a better efficiency value than other types of propellers. Propeller type B6.40 has good efficiency in accordance with the planned 2 x 250 KW engine performance diagram.

Suggestion: Future research is expected to test models B4.40 and B6.40 in Cavitation tunnels and develop B-Series propeller development by reducing cavitation levels.

6. References

[1] Indonesia-investments, *"Indonesia-investments.com,"* 4 Juli 2016. [Online]. Available: https://www.indonesia-investments.com/id/bisnis/komoditas/gas-alam/item184. [Accessed 20 Februari 2022].

[2] Nurcholis, *"Uji Hidrodinamika Model Kapal (Inovasi Pengembangan Teknologi Maritim),"* Surabaya, 2021.

[3] F. Abbasov, Statistical Analysis of The Energy Efficiency Performance (EEDI) of New Ships, Transport & Environment Based on Official EEDI, 2017.

[4] Muhammad Riduwan, Dika V. Devintasari, "Perencanaan Pengiriman Pasokan Bahan Bakar Liquid Natural Gas (LNG) untuk Pemenuhan Pembangkit Listrik di Wilayah Indonesia Timur," *Jurnal Penelitian Transportasi Laut,* vol. 22, no. 2, pp. **111-124, 2020**.

[5] BRIN, "brin.go.id/desain-kapal-mini-lng," brin.go.id, 21 September 21 SEP 2021. [Online]. Available: https://www.brin.go.id/desain-kapal-mini-lng-brin-siap-menyuplai-kebutuhan-gas-di-bali-lombok-dan-ntt/. [Accessed 15 Februari 2022].

[6] S. Adji, Engine-Propeller Matching, Surabaya: Teknik Sistem Perkapalan, 2005.

[7] S. A. Harvald, Resistance Propulsion of Ship, Lyngby: The Technical University of Denmark, 1983.

[8] Baharudin, 2021, pp. **6-10**.

[9] E. V. Lewis, Principles of Naval Architecture Second Edition, Jersey City: The Society of Naval Architects and Marine Engineers, 1988, p. **61**.

[10] S. Adji, Engine-Propeller Matching, Surabaya: Teknik Sistem Perkapalan, 2005, p. **7**.

IOP Conf. Series: Earth and Environmental Science 1081 (2022) 012037 doi:10.1088/1755-1315/1081/1/012037

[11] S. A. Harvald, Resistance Propulsion of Ship, Lyngby: The Technical University of Denmark, 1983, p. **135**.

[12] E. V. Lewis, Principles of Naval Architecture Second Edition, Jersey City: The Society of Naval Architects and Marine Engineers, 1988, p. **152.**

[13] S. W. Adji, "Engine-Propeller Matching," *ACADEMIA,* p. **8**, 2005.

[14] S. Adji, Engine-Propeller Matching, Surabaya: Teknik Sistem Perkapalan, 2005, p. **10.**

[15] S. Leksono, "Sinkronisasi Propeller Dengan Mesin Induk Pada Kapal Ikan Untuk Meningkatkan Efisiensi dan Kinerja," *Wave Volume 10,* pp. **19-24**, 2016.

[16] S. W. Adji, Engine-Propeller Matching, ACADEMIA, 2005, p. **10**.

[17] E. V. Lewis, Principles of Naval Architecture Second Edition, Jersey City: The Society of Naval Architects and Marine Engineers, 1988, p. **145**.

[18] E. V. Lewis, Principles of Naval Architecture Second Edition, The Society of Naval Architects and Marine Engineers, 1988 Jersey City, p. **182**.

[19] S. W. Adji, "Engine propeller matching," *ACADEMIA,* p. **8**, 2005.

Study on Natural Gas/Diesel Dual-fuel Engine Energy Ratio: Effect of Natural Gas Injection Parameters

Frengki Mohamad Felayati[1], Dwisetiono[1], Hadi Prasutiyon[1], Betty Ariani[2]

[1]Department of Marine Engineering, Faculty of Engineering and Marine Science, Hang Tuah University, Surabaya, Indonesia 60111
[2]Department of Naval Architecture, Faculty of Engineering, Muhammadiyah Surabaya University, Surabaya, Indonesia 60113

Corresponding: frengki@hangtuah.ac.id

Abstract. Natural gas has been a promising demand for several years in Indonesia as a fuel for a diesel engine by converted into a natural gas/diesel dual-fuel engine. However, determining the energy ratio of the diesel and natural gas fuel is important due to the engine performance and emissions which affect the engine safety operation. This study presents the method to determine the natural gas and diesel fuel energy ratio on intake port natural gas injection mode through experiment. A direct injection diesel engine is converted to a natural gas/diesel dual-fuel engine by injecting natural gas into the intake port. The diesel injection parameters are unmodified for the experiment; besides the natural gas injection variations are studied to determine the energy ratio. Moreover, the engine is tested for low to high load conditions. However, natural gas injection duration, pressure, and injection timing variation affect the fuel energy ratio and indicated thermal efficiency (ITE). At low load, the optimum fuel energy ratio and ITE are achieved at a long injection duration (10 ms) and with advanced injection timing. Moreover, at high load, the optimum fuel energy ratio and ITE is achieved at high natural gas injection duration (12 ms), high injection pressure (3 bar), and advancing the injection timing.

1. Introduction

The abundant natural gas resource in Indonesia is projected to market interest over years [1]. Natural gas is mostly used for factories and power plants; it is slightly used for others such as transportation. However, natural gas is applicable to use for transportation internal combustion engines. Besides, it is less popular in Indonesia, especially for land-based transportation. Meanwhile, marine transportation keeps in progress for optimizing the natural gas as ship fuel to reduce the contribution to greenhouse gas (GHG) emissions [2,3].

Most of the ships are powered by a diesel engine which is mainly fueled by diesel fuel. A diesel engine is suitable for a heavy-duty industry that needs a high power output with high efficiency. Combining diesel with natural gas fuel (dual-fuel) is an improvement to keep the diesel engine running with nearly equivalent performance and lower GHG emissions [4]. Natural gas/diesel dual-fuel engines can be used to utilize natural gas. In dual-fuel mode, natural gas fuel dominates the energy fraction of the combustion process more than diesel fuel [5]. The low energy fraction of diesel fuel is the ignition source for the natural gas able to ignite in a compressed ignition engine.

Natural gas can be utilized in the diesel engine in dual-fuel engine mode by being injected into the intake port at low pressure [6]. However, the energy ratio of natural gas fuel should be determined due to the effects on the engine performance and emissions as reported in several studies [7–10]. The proper energy ratio of natural gas and diesel fuel enhances the combustion process and emissions [11]. Moreover, engine load conditions may vary the fuel energy ratio of natural gas/diesel to result in high combustion efficiency [12–14]. In contrast, it may affect knock at high loads or deteriorate the combustion quality which leads to safe operation and uncontrolled emissions, respectively [11,15].

For that reason, determining the energy ratio of the natural gas and diesel fuel is important due to the engine performance and emission. However, most of the research identifies the effect of natural gas/diesel fuel energy ratio on engine performance and emissions [10,14,16]. Nevertheless, it does less explain the method to determine the energy ratio of natural gas/diesel dual-fuel engine, especially for a manual controlled diesel engine or without an electronic control unit (ECU) to control the fuel injection. This study presents the method to determine the natural gas/diesel fuel energy ratio using natural gas injection into the intake port and direct injection of diesel by experiment. Several methods are observed to regulate the fuel energy ratio of natural gas/diesel, such as varying natural gas injection timing, pressure, and duration. All the variations are identified in low and high load conditions. The method's effects on the performance are briefly discussed; meanwhile, the effects on emissions are discussed in another study.

2. Methodology

In this section, the methodology for determining the natural gas/diesel fuel energy ratio is described. There are several sub-sections that are discussed, such as experimental setup, fuel properties, and injection strategies. The strategy of natural gas injection in detail is discussed as the focus of studying the energy ratio of natural gas/diesel fuel.

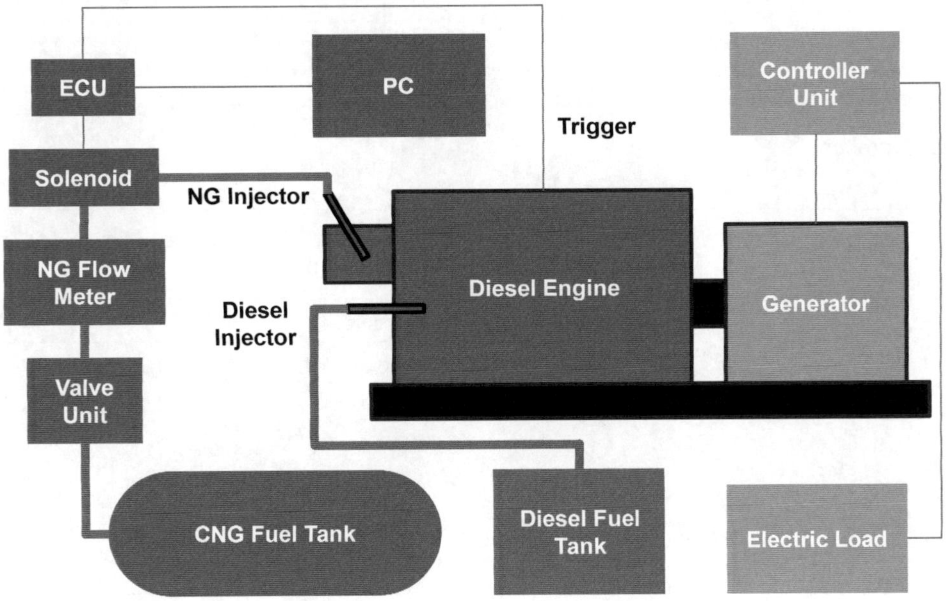

Figure 1. Experimental setup.

2.1. Experimental Setup

In this section, the experimental setup is briefly explained; a detailed explanation is reported in another study [17]. The experimental study used a dedicated diesel engine that is fueled by diesel. The engine is converted to a dual-fuel (natural gas and diesel) engine by injecting natural gas into the intake port and is controlled by the ECU. The engine specification detail is shown in another work [18].

Moreover, diesel fuel is injected directly into the combustion chamber. Moreover, the engine is used an electric load unit by an electric heater and is controlled by a controller unit. The illustration of experimental setup is depicted in Figure 1.

2.2. Fuel Properties

The experiment on the natural gas/diesel dual-fuel engine used compressed natural gas (CNG) and diesel fuels. Furthermore, the CNG fuel pressure is regulated to low pressure by a regulator. The fuel properties details are shown in Table 1.

2.3. Injection Strategies

The focus of this experiment is natural gas fuel injection rather than diesel fuel. Besides, the diesel fuel injection strategies are also described sufficiently as follows. The diesel injection in this experiment is unmodified from the original setting from the manufacturer. Meanwhile, the fuel mass injection is regulated by the manual handle attached to the engine which adjusts the flow rates of natural gas fuel substitution. However, the natural gas substitution is using the percentage of energy substitution (PES) calculation which is described in another study [17]. In brief, the PES is the natural gas fuel energy percentage compared to the total fuel energy.

Thus, the engine is fully operated in diesel fuel mode at first to the engine speed and load target. After that, the natural gas is injected at the target mass flow rates and leads to an increase in the engine speed. In that condition, the diesel fuel mass injection is decreased until the engine speed decreases to the normal target. The injection factors that affect the natural gas flow are regulated (e.g., pressure, timing, and duration). The natural gas injection parameters for the experiment are detailed in Table 2.

Table 1. Fuel properties.

	Natural Gas	Diesel
Density at 15°C (kg/m^3)	0.656	820
Cetane Number	-	53
Octane Number	120	-
LHV (MJ/kg)	50	43.2

Table 2. Natural gas injection parameters.

	Duration (ms)	Pressure (bar)	Timing (°CA BTDC)
Low Load (1 kW)	9	-	100
	10	-	130
	11	-	160
High Load (4 kW)	11	2	100
	12	3	130
	13	4	160

3. Result and Discussion

The experiment was conducted to analyze the proper method for determining the natural gas/diesel fuel energy ratio in manual diesel engines. However, there are some results on the engine performance in every natural gas injection parameter in low and high loads conditions, 1 kW (25%) and 4 kW (100%), respectively. The effect of natural gas injection parameters on the PES and ITE is described. The effect on PES determines the injection parameters that affect the PES; meanwhile, the effect on ITE determines the effectiveness of the combustion inside the chamber.

IOP Conf. Series: Earth and Environmental Science 1081 (2022) 012038 doi:10.1088/1755-1315/1081/1/012038

3.1. *Effect of Injection Strategies on Low Load*

Figure 2 depicts the effect of natural gas injection parameters applied in low load conditions. There are two parameters are used in this case, injection duration and injection timings at the same injection pressure (3 bar). The injection pressure is avoided by vary due to the rich fuel mixture condition which leads to power loss, very low ITE, and deteriorated emissions. Thus, the injection pressure remained to be low enough.

Figure 2 shows that by prolonging the injection, the duration increases the PES (10 ms) and slightly lowers the PES at the very long injection duration (11 ms). However, in this case, the injection timing is controlled at 160°CA BTDC. The injection duration of natural gas affects the natural gas mass flow to the combustion chamber. The longer injection duration increases the natural gas mass and energy. Besides, the data shows that there are other factors that affect the natural gas mass trapped in the combustion chamber due to non-linear curves made. It is possibly due to the part of natural gas fuel being trapped in the intake port as described by Felayati et al. [19]. Thus, the PES slightly decreased at a very long injection duration.

In this case, a long injection duration (10 ms) leads to an increase in the ITE and the longest injection duration deteriorates the ITE. It indicates that sufficient natural gas trapped in the combustion chamber with a long injection duration may increase the combustion quality [20]. However, the longer injection duration may prevent the air from entering the combustion chamber thus rich natural gas/air mixture occurs thus decreasing the ITE.

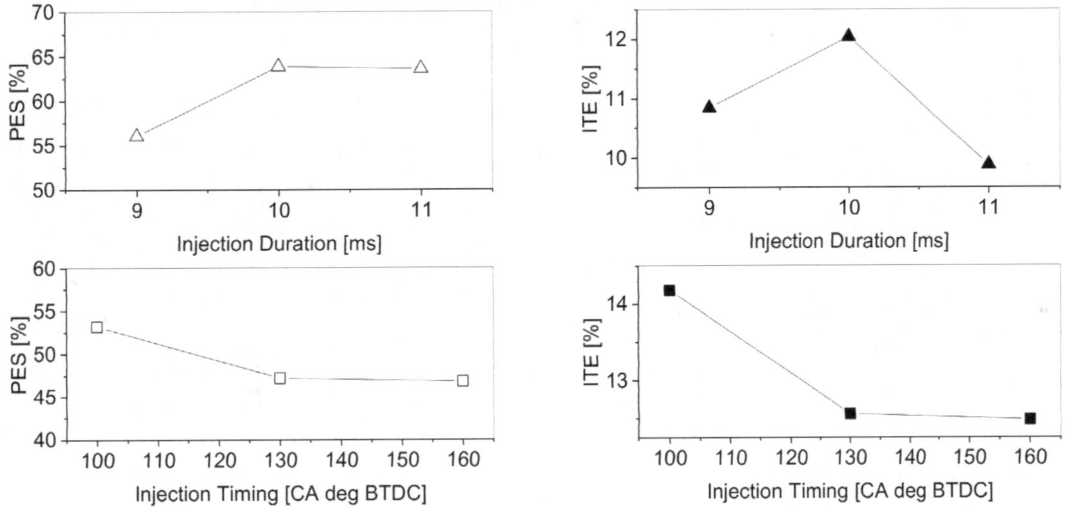

Figure 2. Effect of injection timing and duration on PES and ITE at low load.

Moreover, the injection timing also affects the PES of natural gas. Figure 2 also shows the effect of natural gas injection timing on the PES with a controlled injection duration of 8 ms. It shows that advancing the injection timing of natural gas (160°CA BTDC) lowers the PES. However, advancing the injection timing of natural gas may cause the natural gas trapped in the cylinder when the intake valve opens (IVO) to be more sufficient for combustion.

Besides, in this case, most of the injection timings are at intake valve close (IVC), and only at 160°CA BTDC is slightly at IVO. Thus, most of the natural gas entered into the chamber is in the next cycle. It means that advancing the injection timing of natural gas prolongs the natural gas/air mixture duration in the intake port [21]. The well-mixed natural gas/air leads to lower mixture stratification thus decreasing the combustion quality [19]. As a result, the ITE decreases as advancing the natural gas injection timing.

However, in this experiment results (Figure 2) clearly indicate that the natural gas injection duration and timing can be used as a method to determine the diesel and natural gas fuel energy ratio at low load conditions. As stated earlier, the injection pressure of natural gas is avoided use as the method to vary the fuel's energy ratio for some reasons mentioned. Thus, only natural gas injection duration and timing are able to vary in this case to determine the natural gas and diesel fuel energy ratio.

3.2. Effect of Injection Strategies on High Load

Figure 3 shows the natural gas injection parameter's effect on the PES and ITE with high load conditions (4 kW). In this case, the effects of injection duration, injection timing, and injection pressure are identified. However, the injection duration is regulated at 1800 rpm and 3 bar for the engine speed and injection pressure, respectively. Meanwhile, the injection pressure variations are set at 12 ms, 2000 rpm, and 4 lpm for the injection duration, engine speed, and natural gas flow rate, respectively. Both injection duration and injection pressure parameters are regulated at 160°CA BTDC for the injection timing. Moreover, the injection timing parameters are set at 2000 rpm, 12 ms, and 3 bar for the engine speed, injection duration, and injection pressure, respectively.

Figure 3 depicts that at high load conditions, a longer injection duration increases the PES. A longer duration of the injection means a higher concentration of the natural gas in the intake port [8]. Thus, it is reasonable that the PES increases as the injection duration increases. However, the ITE increases in a linear pattern as the injection duration increases. Besides, a longer injection duration at 13 ms lowers the ITE. It indicates that very high PES deteriorates the combustion in the cylinder thus lowering the ITE.

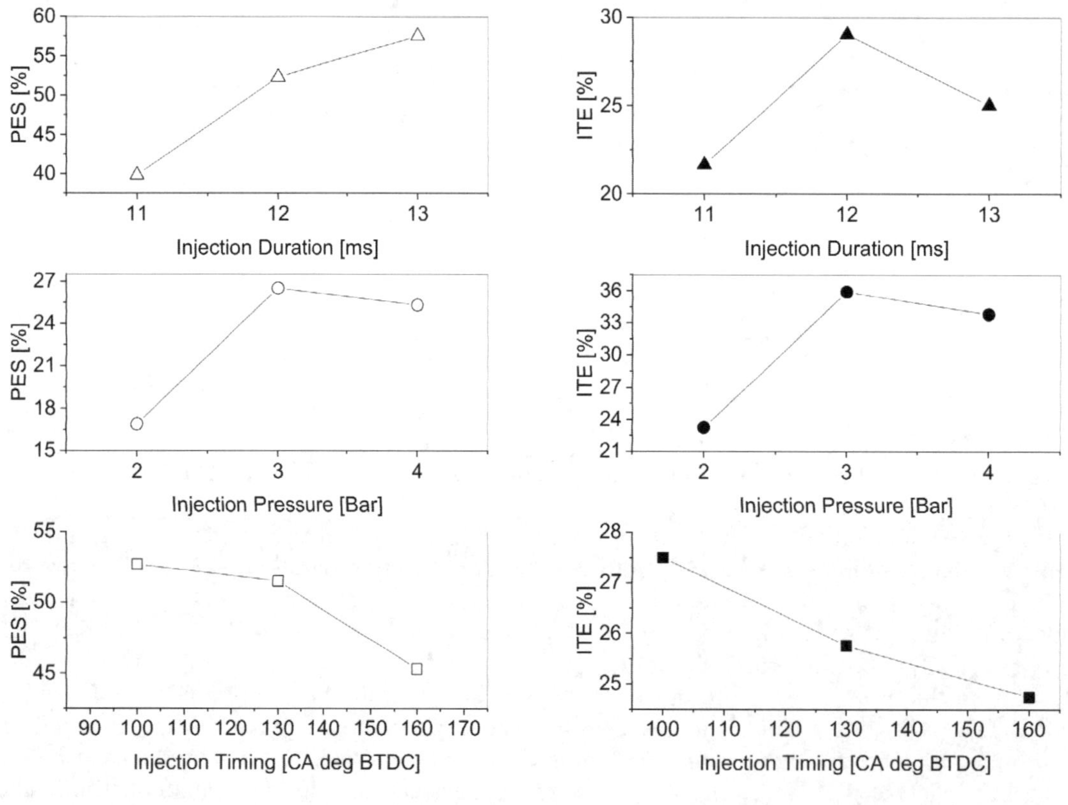

Figure 3. Effect of injection timing, pressure, and duration on PES and ITE at high load.

Furthermore, injection pressure variation affected the PES and ITE in the same pattern (Figure 3). The PES and ITE increase with the higher injection pressure but slightly decrease at the higher injection pressure (4 bar). It also shows that the injection pressure variations less affect the PES at a high value. It indicates that high injection pressure tends to vary the kinetic energy besides the natural gas mass trapped in the cylinder [21]. It leads to an increase in the combustion characteristics as increases the injection pressure (3 bar) then decreases in the high injection pressure (4 bar) as the ITE lowers.

Figure 3 also shows that in high load conditions, advancing the injection timing also lowers the PES and ITE in low load condition cases. As discussed in low load condition cases, the injection timings are conducted when IVC thus the natural gas fraction is trapped in the intake port and flows to the combustion chamber when IVO is in the afterward cycle. Advancing the injection timing means the longer time natural gas has to enter the combustion chamber. Thus, the natural gas/air mixed more homogeneously with advancing injection timing [19]. It caused the natural gas concentration to lower with advancing the injection timing. However, advancing the injection timing lowers the stratification of the natural gas/air mixture. Thus, it caused the ITE to be lower than advancing the injection timing due to lower combustion characteristics.

However, the results (Figure 3) indicate that natural gas injection duration, pressure, and timing can be used for determining the diesel and natural gas fuel energy ratios at high load conditions. Compared to the low load condition, the natural gas injection pressure is able to vary in high load conditions. Thus, the engine load condition should be considered in order to determine the fuel energy ratio of natural gas and diesel.

4. Conclusion

The method for determining the fuel energy ratio on natural gas/diesel dual-fuel engines is studied at low and high load conditions by an experiment using natural gas injection strategies. There are several conclusions can be mentioned from the experiment:

a. At low load conditions, the fuel energy ratio can be determined with natural gas injection timing and duration variation. The long natural gas injection duration (10 ms) achieved higher PES and ITE. Moreover, advancing natural gas injection timing increases the PES and ITE.
b. At high load conditions, natural gas injection pressure, duration, and timing variation are able to regulate the fuel energy ratio. The optimum PES with high ITE is achieved at a long injection duration (12 ms). High natural gas injection pressure (3 bar) leads to higher PES and ITE. Moreover, advancing natural gas injection timing leads to an increase in the PES and ITE.

However, this research is a brief study about the method of determining the natural gas injection variation effect on the diesel/natural fuel energy ratio. A parametric study is suggested that combines all of the natural gas injection strategies to achieve high performance and lower exhaust emissions.

5. References

[1] Felayati F M, Semin, Cahyono B, Prayogi U and Winarno A 2022 Future perspective of the ship alternative fuels in Indonesia *IOP Conf. Ser. Earth Environ. Sci.* **972** 012024

[2] Ariani B, Ariana I M and Fathallah M A 2019 Effect of natural gas injection timing on combustion performance & methane slip emission of diesel – NG dual fuel engine: An experimental study *AIP Conf. Proc.* **2187** 020003

[3] Pan P, Sun Y, Yuan C, Yan X and Tang X 2021 Research progress on ship power systems integrated with new energy sources: A review *Renew. Sustain. Energy Rev.* **144** 111048

[4] Karim G A 2015 *Dual-fuel diesel engines* (CRC Press)

[5] Felayati F M, Semin, Zaman M B and Gusti A P 2019 Consideration of Split Injection Strategy for Marine Diesel Engine Combustion Process *AIP Conference Proceedings* vol 030021

[6] Kakaee A-H, Rahnama P and Paykani A 2015 Influence of Fuel Composition on Combustion and Emissions Characteristics of Natural Gas/Diesel RCCI Engine *J. Nat. Gas Sci. Eng.* **25** 58–65

IOP Conf. Series: Earth and Environmental Science 1081 (2022) 012038 doi:10.1088/1755-1315/1081/1/012038

[7] Huang H, Lv D, Zhu J, Zhu Z, Chen Y, Pan Y and Pan M 2019 Development of A New Reduced Diesel/Natural Gas Mechanism for Dual-Fuel Engine Combustion and Emission Prediction *Fuel* **236** 30–42

[8] Zaman M B, Rahmatullah D T, Semin and Felayati F M 2019 Effect of natural gas injection pulse width to diesel dual fuel performance *AIP Conference Proceedings* vol 2187 (American Institute of Physics Inc.) p 020023

[9] Zhang Y, Fu J, Shu J, Xie M, Zhou F, Liu J and Zeng D 2019 A Numerical Investigation of The Effect of Natural Gas Substitution Ratio (NGSR) on The In-Cylinder Chemical Reaction and Emissions formation Process in Natural Gas (NG)-Diesel Dual Fuel Engine *J. Taiwan Inst. Chem. Eng.* **105** 85–95

[10] Liu J, Guo Q, Guo J and Wang F 2021 Optimization of a diesel/natural gas dual fuel engine under different diesel substitution ratios *Fuel* **305**

[11] Wang Z, Fu X, Wang D, Xu Y, Du G and You J 2021 A multilevel study on the influence of natural gas substitution rate on combustion mode and cyclic variation in a diesel/natural gas dual fuel engine *Fuel* **294**

[12] Felayati F M, Semin, Cahyono B and Bakar R A 2021 Numerical investigation of dual-fuel engine improvements using split injection natural gas coupled with diesel injection timings at low load condition *Int. J. Eng. Appl.* **9** 31–8

[13] Ariani B, Ariana I M and Fathallah A Z 2020 Experimental Investigation on Natural Gas Injection to Minimize Abnormal Combustion and Methane Slip in the Diesel - Natural Gas Dual Fuel Engine at Low Load *Int. Rev. Mech. Eng.* **14** 599–606

[14] Wang Z, Du G, Wang D, Xu Y and Shao M 2018 Combustion Process Decoupling of A Diesel/Natural Gas Dual-Fuel Engine at Low Loads *Fuel* **232** 550–61

[15] García Valladolid P, Tunestål P, Monsalve-Serrano J, García A and Hyvönen J 2017 Impact of Diesel Pilot Distribution on The Ignition Process of A Dual Fuel Medium Speed Marine Engine *Energy Convers. Manag.* **149** 192–205

[16] Wang Z, Fu X, Wang D, Xu Y, Du G and You J 2021 A multilevel study on the influence of natural gas substitution rate on combustion mode and cyclic variation in a diesel/natural gas dual fuel engine *Fuel* **294** 120499

[17] Felayati F M, Semin, Cahyono B, Bakar R A and Birouk M 2021 Performance and emissions of natural gas/diesel dual-fuel engine at low load conditions: Effect of natural gas split injection strategy *Fuel* **300** 121012

[18] Felayati F M, Semin, Cahyono B and Bakar R A 2019 Investigation of the effect of natural gas injection timing on dual-fuel engine emissions using split injection strategies *Int. Rev. Mech. Eng.* **13** 645–54

[19] Felayati F M, Semin and Cahyono B 2021 Methane Emissions Evaluation on Natural Gas/Diesel Dual-Fuel Engine during Scavenging Process *IOP Conference Series: Earth and Environmental Science* vol 698 (IOP Publishing Ltd) p 012036

[20] Song E, Liu Z, Yang L, Yao C, Sun J and Dong Q 2017 Effects of Nozzle Structure on The Gas Mixture Uniformity of Marine Gas Engine *Ocean Eng.* **142** 507–20

[21] Baratta M, Misul D and Xu J 2021 Development and application of a method for characterizing mixture formation in a port-injection natural gas engine *Energy Convers. Manag.* **227** 113595

The Application of Emergency Safety Device for The Improvement of Marine Operation of a Gas-Powered Ship

Agoes Santoso, Muhammad Badrus Zaman, Muhammad Irsyad Saihilmi
Department o Marine Engineering, ITS, Surabaya 60111, Indonesia

Abstract. A new Emergency Safety Device (ESD) based on a PLC module that can start and stop a modified dual-fuel diesel engine in automatically has been created. When fitted to a modified dual fuel engine, the laboratory scaled PLC module has been tested and found to perform well in any programmable emergency circumstance. Because the modified engine's compression ratio has been increased, the risk of hazards has increased as well. When any pre-defined dangers are recognized, the ESD logical code can automatically control the functioning of the fuel system to provide uninterrupted fuel delivery, but it can also halt the fuel supply, including gas and diesel fuel. Over-pressure, over-temperature, flow rate, rpm, and gas leakages are all potential hazards. To identify those potential hazards, then scoring method is used to model and simulate them quantitatively. The simulation data is used as input data for Arduino Source Code, which is used to process logical acts in the marine engine operation. Finally, the safety level is assessed using a safety standard approach such as the SHEL Method. The results of the works show that ESD eligible to control the operation of gas fuel system and it is considerable to be applied for gas-powered ship.

Keywords:
Emergency Safety Device, PLC Module, Integrated Automation, Dual Fuel Diesel Engine, Gas powered vessel, SHEL Method

1. Introduction

A modern ship is mainly developed by the existence of electronic based control system that specifically called as automatic control system. There many advantages may offered by the automation system and the most important benefit is the improvement of safety level as the highest objective of the modern ship operation [16]. Moreover, integrating the onboard automation system may contributes better performance of ship operation and its system [9]. This paper is implements of a new module called as *Emergency Safety Device* (ESD) to support research in modify the use of gas as fuel [12]. Gas facing of potential risks related to the safety of the engines, furthermore, when bigger power is expected from the modified engine [15]. ESD module expected can detect any initial danger and capable of react automatically to shut down the engine operation. The main objectives of this research [12] in investigating the maximum capability of the piston material should be accommodated without any risk for potential hazard then the potential growing of combustion power and heat always in safe conditions. It is a challenge in optimizing the control accuracy of the EDS without any risk of danger.

2. Review of the Gas-Powered Ship

Regulation 13 of TIER III MARPOL has been come into forces [2]. Optimism to the progress of the emissions control since the *Technical Code* of *Tier I* and *Tier II* and the implementation of the *Tier III* brings the related technology forward to *Zero Emission level* [26; and 28]. Term of *Near-Zero Emission Level* had been claimed [27; 29; and 30]. Using gas as engine's fuel is one method to deal with the TIER III. For complying IGF Regulation [24] in term of fuel redundancy then the best alternative is dual-fuel technology [3].

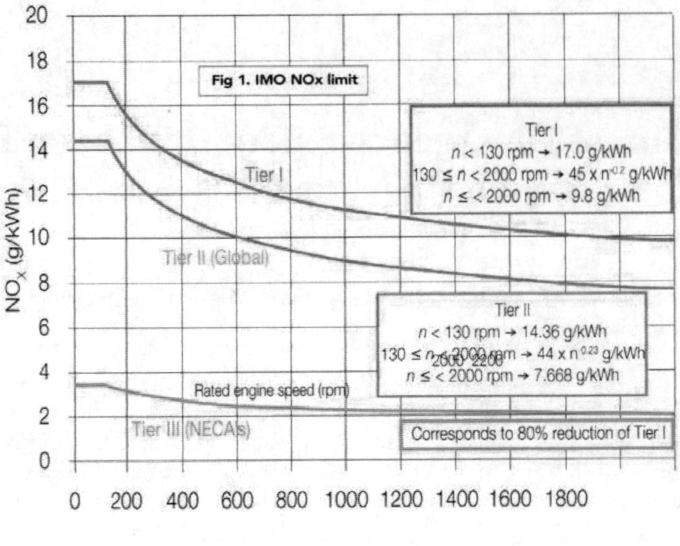

Figure 1. Permitted NOx Emission Level based on the MARPOL 73/78 [25]

Dual-fuel system has been widely developed in marine areas whether as established factory product or modification from conventional one [15]. Fuel gas in the form of LNG or CNG cannot be ignited directly in diesel cycle combustion process [12]. There are two methods in burning gas inside diesel engine. First is delivery the gas fuel in same time with the incoming air intake supply then at the cycle time of compression stroke the diesel fuel is injected to perform initial burning. In respectively, the flowing of both air intake and CNG will perform combustion cycle in continuously [4]. The second method called as direct injection gas burning when pilot diesel fuel injected to perform initial firing then following by regularly gas injection as dominated fuel [3].

3. The Idea Why a New ESD PLC Module Highly in Demand

Gas has lower energy content compared with diesel fuel. To achieve similar power output, in this research the gas quantity has been increased in proportional ways by improving the volume of the combustion chamber. The works cause higher compression ratio when the volume of the piston bowl is increased [12]. Those modification may cause potential damages in the in-cylinder combustion chamber when the higher power is resulted [4] that may cause higher combustion energy. Many papers have already resumed that modify diesel engine may contribute any certain potential damages in the engine parts [8]. It is the basic idea why the ESD is developed here. This new module will be a complement for the existing PLC module that mainly used as fuel gas control system (GC-PLC).

The new PLC module (named as ESD module) that generated in this work can be arranged as complemented (stand-alone) or event supplemented system (integrated). ESD Module proposed to protect of the safety level by setting it to prevent any possible risks from certain technical problems such as over-temperature, pressure drop, over-pressure, gas leakages, un-normal vibration level, noise, over-speed, overload, engine knocking, etc. ESD will shut down the diesel engine [6] for any fails.

There are several sensors that installed in the experimental setup of the modified dual fuel diesel engine. The PLC system also been setup to control the injection timing, both of gas fuel and diesel fuel. So far, the established system has already run well. But, due to the improvement of the combustion efficiency by modify the piston geometry to get higher power and lower emission [12], then the experiment work may cause some risks to several possible damages. For example, caused by higher engine compression ratio, thinner piston crown geometry, overheating and overpressure, etc. The common controller diagram that was developed in the modification of dual fuel diesel engine system is shown in Figure 2.

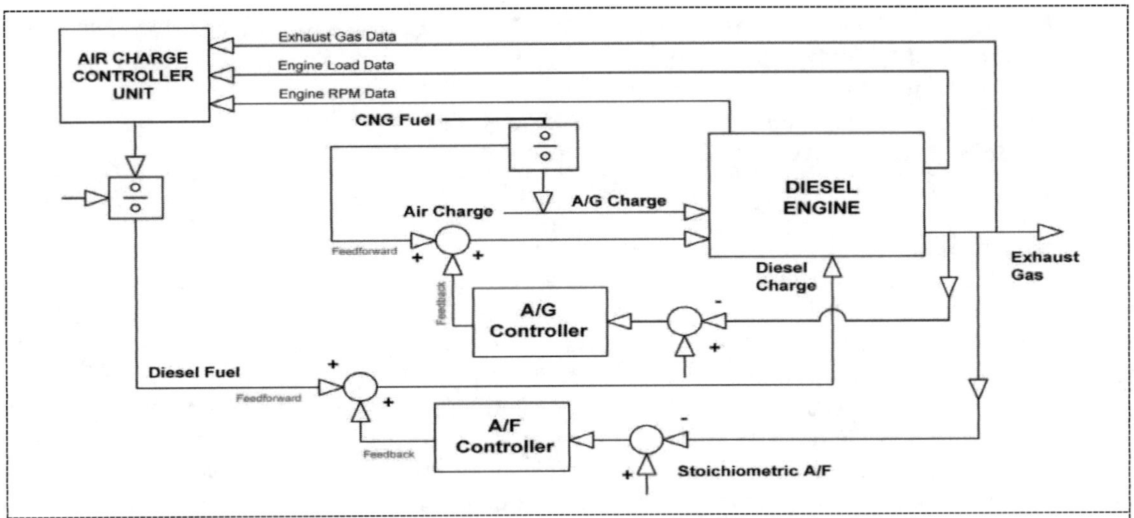

Figure 2. Schematic controller system for Dual Fuel Diesel Engine

4. Review of SHEL Model for Safety Analysis.

There several models known that can be used to be a tool in system theory approach, especially in aspect of safety and accident. Some of the analytical models are [32]:

- Functional Resonance Accident Methods (FRAM).
- Systems-*Theoritic* Accident Model and Processes (STAMP)
- The Management Oversight and Risk Tree (MORT)
- Man Technology Organization Method (MTO or HTO)
- SHEL Method
- *AcciMap* Accident Analysis Technique

SHEL or SHELL method will be chosen here to analyze the performance of the developed new ESD module. The method quite easy but powerful enough to help decision in the evaluation of the safety and potential accidents that may be caused by an engineering structure or model [23]. The model is named after the initial letters of its components (software, hardware, environment, liveware) and places emphasis on the human being and human interfaces. The SHELL model adopts a systems perspective that suggests the human is a rarely, if ever, the sole cause of an accident. Then the application of the SHELL model spread up to the mostly all engineering aspects includes maritime safety [33; 34; and 35]. This model analysis will be used in this research works to evaluate and test the safety performance that may possibly conducted by the new generated ESD module. Figure 3 shows the general model of SHELL Method [23]. The model will be implemented in this study to help in solving safety problems and making better decision.

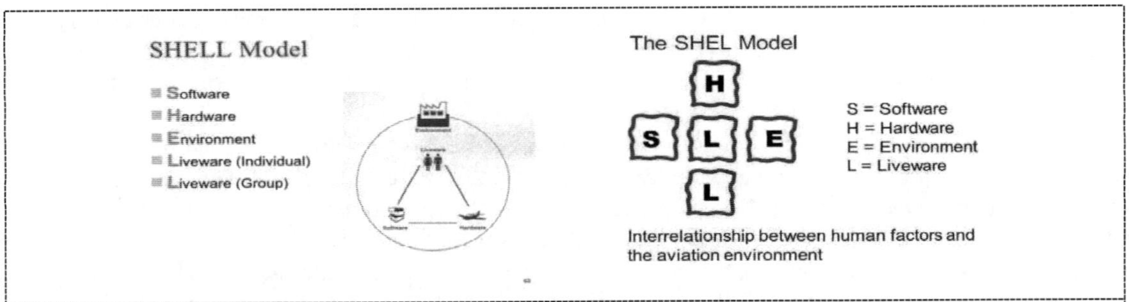

Figure 3. SHELL Model [34]

For prevention and safety analysis use tool called as SHEL Model to solve problem where the model can be defined by some important points as follow:

- Software = Regulation, Operation Management, Crew, etc.
- Hardware = Navigation Equipment, Electricity, Pipe, Machinery
- Environment = Weather, Wind, Current, Wave
- Double L = Crew

5. ESD System Setup

Today's technology of electronic automatic control system is not complicated compared with a previous decade because of the computer tools available in easy and fast [7]. The new ESD module planned also to be developed based on the existing VIAS (*property of Praxis Automation* [5]) which has also specific menu for real-time online system called as an *Internet Ship View* (ISV) that it makes possible for any shipping office to monitor its ship operation, recording, and even interupting several classified duties such as start/stop the engine for any high-level security conditions [6].

It is common that diesel engine has been provided by a safety device unit. Engine experimental test bed in this works is a modified engine from conventional YANMAR TF85MH [13] as shown in Figure 4. The new ESD module is installed in an integrated PLC module to control the operation of the modified dual fuel engine and conduct the data acquisition [12]. Previous success work has been done in the development of DCFS Module under VIAS [6]. The fundamental works in this research activity is similar with the steps and procedures when the industrial PLC module such as VIAS is implemented.

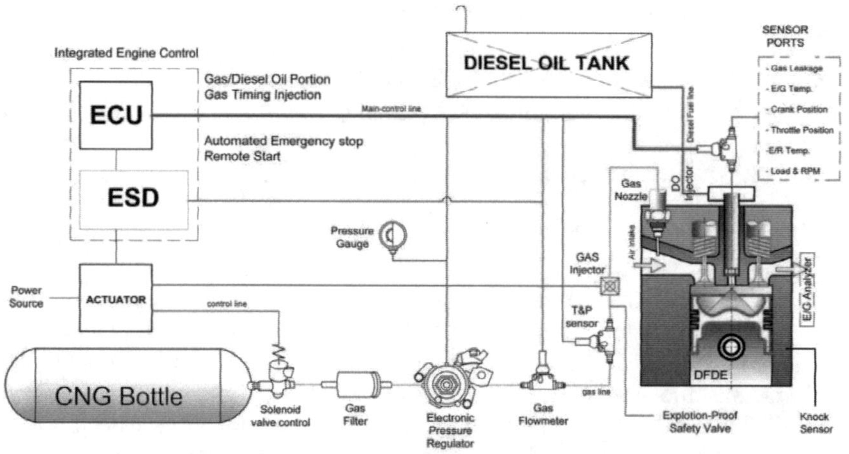

Figure 4. Arrangement of Emergency Safety Device (ESD)

5.1. Data and System Analysis

The highest priority of experimental works is safety operation. The system must be capable of predict any sign of accident in early [16]. Hazard situation can be recognized and controlled by ESD without disturb to the objectives of the research goals. The accuracy of ESD is a key point in investigating as much maximum strength as possible of the new piston modifications in step by step adjustments without danger the experimental works.

In this continuing work, the common brand *Arduino PLC* [17] is still used to control gas portion and gas injection timing. The new generated ESD module that also developed under Arduino PLC will works independently but it can be improved in a larger scheme such as the application of VIAS-ESD onboard a ship, especially in wider scheme such as gas-powered ship.

Using the existing engine test bench setup in the Marine Power Laboratory in ITS that is already installed with an Arduino PLC module to control the gas injection timing and portion then the module will be improved to equipped with the ESD concept. PLC set will consist of two elements. First is existing gas controlled (GC) PLC module and the second one is the new ESD PLC module.

Five technical parameters will be evaluated as input data under the ESD module, there are:
- In cylinder pressure (over-pressure problems)
- In cylinder temperature (over-temperature problems)
- Engine rotation (RPM) (over-speed problems)
- Gas flow rate (over and under-flow problems)
- Gas leakage inside the experimental room

6. Analysis And Discussions

Main idea in the analysis of ESD is determine working procedure and its limitation when the module installed in the available modified dual fuel diesel engine. Any potential damages that are detected by each sensor will be quantified then the level of the risk evaluated based on the well-known safety analytical method. SHEL or SHELL Method will be implemented here to evaluate and analyse the safety level of the developed new ESD module.

6.1. Analysis of the Experimental result

All function in term of hardware and software has been proven work properly under stand-alone tests. So far it can be concluded that the experimental results using the new ESD Module has been run well. The input data can be set related with the real value of the developed diesel engine, the data has been processed under logical safety standard and the output data precisely can be used to trigger any desire actions depend on the pre-setting hazard level. The flexibility in determining Safety Index and Hazard Level make the ESD Module system become upgradable in analysing any scale of hazards when it is used to accompany a prosperous works such as laboratory experimental research. Experimental works can be done in grow up conditions without risk for endangering conditions. The safety limits can be reached higher and higher without any real disaster can be realized.

6.2. Implementation of a Safety Method to evaluate the performance of ESD Module

The Scoring Method has been structured in logical programming of the ESD Module code. Safety Index are also accounted based on the consequences category as adopted in the knowledge of SHEL Method. The SHEL Method not only to be used to evaluate the safety but also to generate specific leverage of the influence of human error in any accidents. It is one of the important ways to test the effectivity of automation system.

Standard condition data based on the conventional diesel engine YANMAR TF-85MH that will be used as baseline input data can be stated as follow:
- Maximum in-cylinder temperature : 2000^0C (at TDC)
- Maximum in-cylinder pressure : 65 bar (at TDC)
- In-cylinder Mass flow : 35 m/s (at TDC)
- Engine revolution : 2200 RPM (maximum)
- Gas leakages : as detected by any room's sensor

By using the results of computer simulation works as previous research [29] then the increments values can be listed in the Table 1.

Table 1. Scoring Values of Excessive Five (5) Technical Parameters

No	Parameters Evaluated	Determined Scoring Values				
		0	1	2	3	4
1	Max. In-cyl. Temperature (^{0}C)	2000	2100	2200	2300	2400
2	Max In-cyl. Pressure (bar)	65	70	75	80	85
3	Max. In-cyl. Mass flow (m/s)	35	37.5	40	42.5	45
4	Engine Revolution (RPM)	2200	2250	2300	2350	2400
5	Gas Leakage on Room	NO	YES			

The scoring values is a pre-determined set based on the beginning level of the experimental works. The increment values are really may happens and has been predicted under previous simulation works [29]. After the earlier scenarios proven successful, then the values can be raised to get higher performances when the modified diesel engine burn gas as fuel.

From Table 1 can be described that the maximum in-cylinder temperature of the diesel engine has value of 2000^{0}C. The previous computer modelling results that using modified piston crown can increase the temperature up to 2400^{0}C. For example, when the input sensor predicts 2300^{0}C then ESD processing unit will identify that the scoring value is 3. Another example, when the maximum in-cylinder mass flow predicted at 45 m/s, then the ESD logic will put scoring value of 4. If three other parameters still under the baseline data, then ESD logic will summarize the total value of 7. In the other word, it can be stated that the consequences category is 7.

Slight difference for gas leakages scoring level. There are only binary conditions, so it is described as NO and YES condition. NO conditions get 0 value and YES condition get value of 10. Therefore, when the gas leakages occurred in whatever the value of the other parameters, then the safety index become 10 in total value. Gas leakages condition will override other four technical parameters in acting to stop engine. Table 2 and Table 3 show the example of scoring values for entire possibly data. There are totally 626 simulations on the list to be evaluated.

Table 2. Simulation number 1 (first data)

1	Temperature	0	1	2	3	4
2	Pressure	0	1	2	3	4
3	Mass flow	0	1	2	3	4
4	RPM	0	1	2	3	4
5	Gas Leakage	0	10			

Value = 0 (SAFETY INDEX)

Table 3. Simulation number 626 (last data)

1	Temperature	0	1	2	3	4
2	Pressure	0	1	2	3	4
3	Mass flow	0	1	2	3	4
4	RPM	0	1	2	3	4
5	Gas Leakage	0	10			

Value = 10 (SAFETY INDEX)

Hazard scoring value was determined refer to the successive data and summarized in Table 4. The simulation results data can be used to set up baseline data to state the system is in safe or failure condition. For example, total score of 8 or more (in any contribution from the four parameters) will trigger ESD to stop the engine running. Even the score less than 8 but the gas leakage occurred then ESD will stop the engine. It is a correct strategic when deal with high pressure gas operation.

Table 4. Consequences Assessment

CONSEQUENCES CATEGORIES	DESCRIPTION	ACTIONS
0	Baseline	
1	Very Very Safe	
2	Very Safe	
3	Safe	
4	Less Safe	Light Flashing
5	Toward Unsafe	
6	Slight Unsafe	Sound Buzzing
7	Almost Unsafe	
8	Unsafe	Engine Shut Down
9	Very Unsafe	
10	Totally Unsafe	

The values of the consequences categories in Table 4. are determined based on the total values of the five parameters randomly as shown in Table 2 and Table 3. Whatever the total value will give consequences action from ESD Module. The ESD Module is programmed to do three level actions:

- When the total value come to 4 or 5 then the light of ESD Module will flashing. It is indicating an early warning for the modified diesel engine operation.
- When the total value come to 6 or 7 then the sound alarm will be buzzing with flashing alarm light as above. Both alarms are indicating more serious potential hazard.
- When the total value come to 8 and over then ESD Module will stop both gas and diesel supply valves then the engine will shut down.
- From the 626 listed simulation data, the maximum possible value is 16 but the system already setup the value 8 to do shut down the engine. From the list then can be summarized as follow:
  - Total value of 0 to 3 is 35 data or 5.6%
  - Total value of 4 and 5 is 87 data or 13,9%
  - Total value of 6 and 7 is 148 data or 23.63%
  - Total value of 8 and over is 356 data or 56.87%

From those perspectives, the potential hazard already been predicted at 23.63% chances when the alarm will noisy, then 56.87% data are probability to stop the engine operation. The figure also gives information how big the potential hazard can affect the running of the modified dual fuel diesel engine. The value of 7 to 10 are dominated the potential hazard by contributing the number of total data of 314 or 50.16%. The biggest possibility scoring value is 8 or more. It is one of the reasons to put value of 8 to stop the engine operation. By setting the minimum value of 8 to trigger the engine stop then, whatever the value of four technical parameters, the engine will be stopped when the gas leakages happened.

Further experimental works can occupy the new ESD module in flexible pre-setting consequences categories. When the above categories are already proven in the matter of safety without any potential hazard then the action can be set in higher value. For example, flashing light set in categories 5 or 6, sound alarm will noise at categories 7 or 8, and finally when the category 9 or 10 reached then the ESD Module will trigger the valves to close and the engine will stop running. Etc.

6.3. Advanced analysis

Based on the advanced safety analysis method such as SHEL Method then the new developed ESD Module can be implemented confidentially in the level of commercial. The model of human error is a very important data to study the level of safety. The application of SHEL model is adopted based on Scoring Method and Safety level as described above data is shown in Table 5. Results of analysis are used for determining hazard avoidance, especially to investigate how human error may affect the safety of this kind automation system. All possible applications are listed here.

Table 5. SHEL Model criterion Based on the Scoring Data

SHEL Model Application	Explanation for Applications
S	Regulation, Operation Procedure, Arduino IDE Code
H	Engine Unit, Arduino Uno PLC, parts sensor, alarm, stop-valves, emergency manual stop, CNG bottle, Gas sensor, GC-PLC, Modified pistons
E	Ambient, Noise, Crowd people, Laboratory lay-out, Engine test-bench
L	Researcher, Technician, Supervisor

More analysis by using SHEL model is shown in Table 6. Each of components listed in Table 5. will described one by one in Table 6. Main consideration as basically also regulated by SHEL Method is how the interaction between human as liveware to the other components.

Table 6. Analysis of Human Interference Using SHEL Model

SHEL Model Application	Explanation for Applications
L – S	- How researcher follow Regulation, safety conduction, safety equipment - How researcher follow SOP, deal with gas - How researcher built the Code under Arduino PLC
L – H	- How researcher prepare the engine - How researcher prepare PLC module - How researcher fit sensor properly - How researcher fit alarm system properly - How researcher can access the emergency stop button - How researcher conduct safe CNG bottle - How researcher fit gas sensor properly - How researcher prepare Gas Control PLC properly - How researcher modified piston perfectly
L – E	- How researcher adapt with the laboratory ambient - How researcher adapt with the laboratory noise - How researcher adapt with the people crowded - How researcher familiar with laboratory lay-out, emergency escape, safety equipment - How researcher prepare engine test-bench properly
L – L	- How researcher communication with technician and supervisor - How researcher follows technician and supervisor guidance and instructions

Based on the human interferences with the component software, hardware, environment, and liveware then the possible hazard caused by human error can be listed in Table 7. below.

Table 7. Analysis of Human Error Based on SHEL Model

SHEL Model Application	Explanation for Applications	Score
L – S	- Researcher does not follow safety Regulation	1
	- Researcher does not follow experimental SOP	1
	- Code error	0
L – H	- Engine partly or completely broken in suddenly	0
	- PLC Module does not work in suddenly	0
	- Sensor units does not work in suddenly	0
	- Alarm units does not work in suddenly	0
	- Researcher cannot push emergency stop button	1
	- CNG bottle leakage in safety valve	0
	- Gas sensor cannot detect gas leakage	0
	- Gas Control PLC error	0
	- Certain modified piston damaged during operation	0
L – E	- High temperature due to blower broken	0
	- High noise due to many engines running together	0
	- Very crowded due to class practical works	0
	- Laboratory lay-out is bad, un-clear emergency escape	0
	- Less fire & safety equipment	0
	- Researcher lack in detect engine test-bench condition	1
L – L	- Communication with technician/supervisor are not proper	1
	- Researcher neglect technician/supervisor guidance	1

From twenty possible hazard data as listed in Table 7 there are six hazards directly caused by human error. It is 30% from the whole possible hazard data. Human error dominated in software and liveware model in contributing potential accident. Then, the new developed ESD module can be concluded does not much interference by human error. The possible human error of 25% as listed in the two reasons of L-S model can be fixed easily. By detailing each of potential source of the hazard that has been developed based on observation and experience during previous experimental works, then SHEL analysis may be implemented to minimize the risks of accident especially caused by human error. All human error listed in Table 7 are commonly avoidable easily. High risk in using CNG as fuel in the experimental works especially when the genuine part such as engine piston are modified mechanically and dimensionally, then the biggest portion of hazards, can easily be suspected that they come to the hardware model.

7. Conclusion

New Emergency Safety Device (ESD) Module can protect the modified dual fuel diesel engine from more than 80% cases of the potential hazard before the engine operation stop. In this initial setting already show how the powerful of the system to guard the safety of this experimental engine. Next experimental works can be continued with stringent level when known that this initial simulation has been proven in secure condition. Any potential accidents caused by higher working loads in the modified piston crown when using gas as fuel can be eliminated by this new ESD module. SHEL Method has been tested and show that the possibly hazard caused by human error can be happened in value up to 30%. The remaining 70% of potential hazards that majority may be caused by hardware and system can be handled by the existence of the new ESD Module. As an automation system, the new ESD Module has been proven success in eliminating human error.

The new ESD module has been proven running well in stand-alone arrangement and could not face any difficulties to advance it into an integrated system with the existing Engine Control Unit. Furthermore, the new ESD module effectively can be developed as a standard module that accompany any gas-powered vessel, especially when the ship is installed with any modified diesel engine type.

8. References

1. Aravind C, Suji Prasad S.J, and Ponni Bala M 2020 *Int. J. Scientist & Technology Research* **9** (01), Jan2020.

2. Bakas I 2015 Module 5 Propulsion and Power Generation of LNG driven Vessels (23rd to 27th November 2015- University of Piraeus) Greece.

3. Elgohary MM, Seddiek IS, and Salem AM, 2014 *Proc IMechE Part M: J Engineering for the Maritime Environment* p1–11.

4. Herdzik J, 2011 *Journal of KONES Powertrain and Transport* **18** (2).

5. Lin J, 2017, PRAXIS Ship Automation, *Guest Lecturer slide presentation* for Department of Marine Engineering ITS, 22 November 2017, Surabaya Indonesia.

6. Saihilmi MI, Santoso A, and Kusuma IR, 2020 The 5th Int. Conf. SENTA. *IOP Material Science and Engineering*, Faculty of Marine Technology, Institut Teknologi Sepuluh Nopember (ITS), Surabaya Indonesia.

7. Saihilmi MI, Zaman MB, Santoso A, and Kusuma IR, 2020 The 8th *Int. Conf. on Ocean and Coastal Engineering*, Environmental and Natural Disaster Management (ISOCEEN) 2020", SCOPUS Index, Department of Ocean Engineering, ITS Surabaya Indonesia.

8. Santoso A 2017 *Technical Report of the design of Gas Powered RoPax in part of Feasibility Study of the Project Development Dual Fuel Roro Ferry in the Route Merak Bakahueni*, PT. Gajah, Bandung.

9. Santoso A, Kusuma IR and Lin J, 2018, *1st Maritime Safety International Conference* MASTIC, p32 Bali Indonesia.

10. Santoso A, Lin J, Kusuma IR and Koenhardono ES, 2017 *Seminar Nasional Maritim, Sains, Teknologi Terapan 2017*, 21 November 2017, p.227, Surabaya Indonesia

11. Santoso A, Saihilmi MI, 2019, *Developing Integrated Automation for Emergency Safety Devices to Control Power On/Off Main Diesel Engines that using CNG as Fuel*. Department of Marine Engineering ITS Surabaya Indonesia.

12. Santoso A, Semin, Cahyono B, and Sampurno B, 2020 *Int. J. on Engineering Application (IREA)*, **1** (8), pp.1-9, Rome.

13. Semin, Zaman MB, and Santoso A, 2019 *Int. Review of Mechanical Engineering (IREME)*, **13** (3), Rome

14. Sheena P.A, Suneetha M.N, Prasanna K.H, 2016, *Int. J. of Electrical and Electronics Research ISSN 2348-6988 (online)* **4** (3), pp: (74-81), Month: July - September 2016.

15. Tiwari A 2015 *Int. J. of Energy Science and Engineering* **1**(5) p163-169

16. Zaman MB, Santoso A, Kobayashi E, Wakabayashi N, and Maimum A, 2015 *Int. J. on Marine Navigation and Safety of Sea Transportation*, **9** (1) March 2015 Poland

17. Mir A and Swarnalatha R, 2018 *J. of Engineering Science and Technology*, **13** (12), pp.4131-4144 Dubai.

18. Ziaul Haque Munim 2019 *Supply Chain Forum: An Int. J.* **20** (4), 266-279.

19. Markovic, Dusan & Koprivica, Ranko & Pešović, Uroš & Ranđić, Siniša. 2015 *Acta agriculturae Serbica*. 20. 145-153. 10.5937/AASer1540145M.

20. Chae, Chong & Kim, Mingyu & Kim, Hyung-Ju. 2020 *Applied Sciences*. 10. 4564. 10.3390/app10134564.

21. Kooij, Carmen & Colling, Alina & Benson, C. 2018 *When will autonomous ships arrive? A technological forecasting perspective*. 10.24868/issn.2515-818X.2018.016.

22. Feng, Dongying & Xiao, Jinchao & Wei, Sanxi & Zhang, Lei. 2015. *Marine Engineering Frontiers*. **3** (10).14355/mef.2015.03.002.

23. Zaman MB, 2015 *Ship Safety Navigation System*, ITS Press Publisher, Surabaya

24. International Maritime Organization (IMO) 2016 *The International Code of Safety for Ships Using Gases or Other Low-Flashpoint Fuels (IGF Code)*, IMO Publishing Co, London.

25. International Maritime Organization (IMO) 2019 *MARPOL Annex VI and NTC 2008 with Guidelines for Implementation* MARPOL 2017 Edition with Supplement February 2019, Sales number IC664E, London.

26. Mestemaker, B.T.W., Castro, M.B.G., Blom, E.C. and Cornege, H.J. 2019 Zero Emission Vessels from a Shipbuilders Perspective, *Research Gate*, Publication no. 334480164.

27. Walter, B. and Gatellier, B. 2003 *Oil & Gas Science and Technology – Rev. IFP*, **58** (1), pp. 101-114.

28. Castro, M.B.G. 2019 *Proc. of MOSES2019, 2nd Int. Conf. on Modelling and Optimization of Ship Energy Systems*, 8-10 May 2019, Glasgow, Scotland, UK.

29. Matysek, A. et. all. 2005 Near Zero Emission Technologies, *abare eReport 05.1. January 2005, Prepared for the Australian Government*, Dept. of Industry, Tourism and Resources, Canberra.

30. Dijkstra, J.W. and Marel, J. 2006 Near Zero Emission Technology for CO2 Capture from Power Plants", *8th Int. Conf. on Greenhouse Gas Control Technologies*, Trondheim, Norway.

31. Anonime 2016 Great Programmable Logic Controller Uses, https://program-plc.blogspot.com/2016/09/great-programmable-logic-controller-uses.html.

32. Open electronic 2014 Arduino as a programmable Logic Controller, https://www.open-electronics.org/arduino-as-a-programmable-logic-controller-plc/.

33. Cassama F 2015 A Study on Marine Accident Causation Models Employed by Marine Casualty Investigators, *WMU Dissertation*, World Maritime University, Sweden.

34. Lee Y.C., 2016 *Journal of the Korean Society of Marine Engineering*, **40** (8), pp.721-725.

35. Itoh H., Matsuoka T., Murohara Y., (2004), "An Extension of m-SHEL Model for Analysis of Human Factors at Ship Operation", Inproc. of the 3rd *Int. Conf. on Collision and Groundings of Ships (ICCGS 2004)*, pp.118-122, October 25-27, Japan.

IOP Conf. Series: Earth and Environmental Science 1081 (2022) 012040 doi:10.1088/1755-1315/1081/1/012040

Barrel Distortion Correction for Digital Optic-Based Motion Tracking in Ship Hydrodynamic Model Experiments

Z Irawanto[1,2], I M Ariana[1] and Erwandi[2]

[1] Department of Marine Engineering, Institut Teknologi Sepuluh Nopember, Indonesia
[2] The National Agency for Research and Innovation, Indonesia

Email: ariana@its.ac.id

Abstract. In hydrodynamic ship-model experiments, motion tracking of ship model trajectory is employed to predict the maneuverability of the ship when operating in its actual environment. There are some systems for motion tracking, such as inertial measurement unit, radar, laser, and infrared sensors. The system has proven reliable, however, the disadvantage is that they are all relatively expensive. This paper proposes a low-cost digital optic-based motion tracking system using consumer-grade video camera. In order to examine the accuracy of the system, some experiments to apply the method were carried out in maneuvering offshore engineering basin at the Indonesian Hydrodynamics Laboratory, The National Agency for Research and Innovation. To cover a broader area of the trajectory, a wide-angle camera lens was utilized, consequently, the image recorded by the camera experienced a barrel distortion. As a reference for the measurements, gridlines spaced 2 meters apart were drawn on the floor of the basin. Using the grid pattern, the distortion was corrected by finding the relationship between the grid pattern in the object space and in the distorted image. A fourth-order polynomial regression was applied to solve the problem. The results showed that the method was effective for correcting distorted images with a relative standard error of 0.42%.

1. Introduction

In order to predict the maneuverability of ship when operating at sea, hydrodynamic experiments were carried out on the ship model. The maneuverability of a ship in both restricted and open seas is an important factor for ensuring the safety of the ship. In the experiments, motion tracking is applied to track the trajectory of the model. Some of the motion tracking systems used in the experiment are the inertial measurement unit (IMU) sensor [1], radar [2], laser [3], and infrared sensor [4].

Recently, research on digital optics, as well as its application to motion tracking, is growing along with the development of digital optical technology. Several researches on optical-based motion tracking have been carried out in various marine applications. Using an active camera, tracking in real time was applied to the maritime surveillance system on ships in harbours, measuring the distance and speed of ships [5]. This method is also used for a warning system to prevent accidents [6]. Another method is tracking ships at night using light spot detection [7]. Other studies are motion tracking based on horizon detection, which are applied to measure ship attitude [8], to track ships in the open sea [9], and to develop navigation systems on autonomous ships [10]. Erwandi and Suzuki [11-13] proposed a low cost technique to measure the wave height caused by ship model motion in towing tank using digital video camera. The technique called as projected light distribution method to measure the unsteady wave generated by ship forced to oscillate in heave motion while the ship model moves forward in constant speed. A simple digital camera is employed to record the series of images of the unsteady wave in radiation condition and diffraction condition. The pixel brightness of the images is proportional to the second derivative of surface wave equation.

Several studies on the application of low-cost digital optics for motion tracking have been carried out on hydrodynamic experiments on ship models. A single camera can applied to track the planar motion of the ship on a 2-dimensional plane. Motions on a horizontal plane can be tracked using a camera above the basin to detect yaw and sway [14], while motions on a lateral plane for surge, heave and pitch can be tracked by a camera from the side of the ship model [15]. Using multiple cameras and applying videogrammetry techniques in hydrodynamic experiments, the 6DOF motion of the model ship can be measured [16].

A digital optical-based commercial motion tracking system has been applied in many fields and has been widely used in experiments in several hydrodynamic laboratories. Some of the systems are Qualisys [17], OptiTrack [18] and Krypton [19]. The system has proven reliable, however, the disadvantage is that they are all relatively expensive and such a system does not allow users to modify or access raw data.

This paper describes a technique of a low-cost digital optical-based motion tracking system for the trajectory measurement of the ship-model maneuvering test, using consumer-grade video cameras. A method is applied to correct the distorted image due to the use of a wide-angle camera lens. In this paper, it is assumed that the motion is purely planar motion in the horizontal plane. Rotation and translation in the vertical direction such as roll, pitch, and heave are neglected. It is assumed that the image distortion is circularly symmetrical and the center of the distortion is in the center of the image. The radial distortion is assumed to be dominant so that the tangential correction can be neglected.

2. Model Experiment

Experiments of turning maneuver on ship model were conducted in maneuvering offshore engineering basin (MOB) at the Indonesian Hydrodynamics Laboratory, The National Agency for Research and Innovation. The dimensions of the basin are 60 m long, 35 m wide and 2.5 m deep.

2.1. Ship model

Maneuver experiments were carried out on a 1:84.6 model scale of LNG carrier. The main dimensions of the model are presented in table 1.

Table 1. Main dimensions of ship model.

Length between perpendicular	3.144	m
Beam	0.492	m
Draft	0.132	m
Displacement volume	0.152	m^3

2.2. Measurement equipment

A low-cost digital video camera motion tracking system was developed for the trajectory measurement of the ship-model maneuvering test. The camera is installed on the ceiling of the MOB at a height of 32 meters from the basin floor and is connected to a computer. A wide-angle camera lens is utilized in the experiments to cover a broader area of the ship-model trajectory.

2.3. Test program

The test was carried out using the free running model method at the initial ship speed of 6.5 knots, at the model draft of 0.132 m, and at the model displacement of 0.152 m^3. The turning maneuver test was carried out on the starboard and port side direction at a steering angle of 35 degrees.

2.4. Test procedure

The ship's trajectory is determined based on the position of 2 LED targets mounted on the deck of the ship. The first target is mounted on the ship's center of gravity (CG) in horizontal plane, while the second target is mounted on the forepart (FP) of the ship model.

Applying the free running method, model experiments are carried out using a controllable drive system so that the model equipped with propellers and rudders moves within six degrees of freedom. Propeller rotation rate is adjusted until it reaches the expected model speed. The rudders are turned at the expected angle via the control system. The model speed and rudder angle are kept constant until the model rotates at least two revolutions.

In the experiment, a camera is mounted above the basin and connected to a computer to record the motion of the ship model during the test. The experiments are performed under low light conditions to ease the targets detection. Measurement data in the form of videos are processed offline after the experiment.

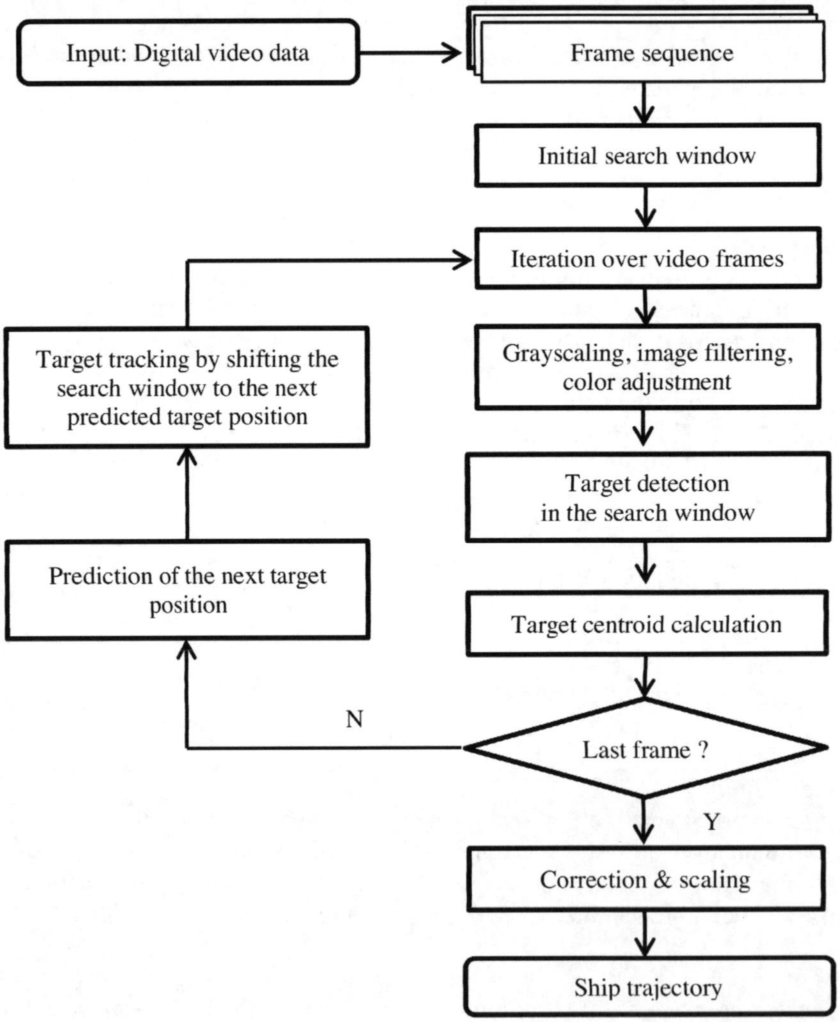

Figure 1. Flowchart of ship model trajectory tracking.

IOP Conf. Series: Earth and Environmental Science 1081 (2022) 012040 doi:10.1088/1755-1315/1081/1/012040

2.5. Data processing
The measurement video data are processed after the experiment (post-processing) applying a computer program created using Microsoft Visual Basic 6.0. Videos are recorded at a resolution of 720 x 480 pixels and a frame rate of 30 frames per second. Videos are saved in uncompressed AVI format to keep the image quality. Video data are processed into numerical data of motion by applying motion tracking techniques.

3. Motion Tracking
Motion tracking, also known as motion capture, is a technique for detecting and following moving objects. A low-cost digital optic-based motion tracking system was developed for the trajectory measurement of the ship-model maneuvering test. The motion of the ship model is recorded by a consumer-grade video camera mounted above the basin. Flow chart of motion tracking is presented in figure 1. A technique for detecting and tracking targets is applied to the measurement system and will be described below.

A video contains a collection of images called frames, and each frame consists of a number of *rows* x *columns* of pixels. Each pixel has three numerical RGB components (Red, Green, Blue) which represent the color of the pixel area, and each color component has a values between 0 and 255. The color value can be used to detect the target position in the video image.

The data obtained from the experiments are some video recordings of the motion of the ship model. Each frame of the video contains information of the position and heading angle of the ship model at a certain time t, which can be identified based on the position of the 2 LED targets captured by the camera. Figure 2 shows several frames of the maneuver experiments performed under low light conditions. For each frame, there are 2 visible LED targets as bright dots on a dark background. Using computer programs, the targets can be detected easily based on the color value for each pixel in the frame image.

Figure 2. Several video frames of maneuver experiment at low light condition.

Target localization is applied to reduce the search area to speed up the search process. To reduce noise, a filter is applied to the video image. For each video frame, 3 measurement variables can be obtained, namely measurement time t, target-1 position (x_1, y_1), and target-2 position (x_2, y_2). The measurement time t can be derived from the video frame number divided by the frame rate (30 fps). Target detection is performed for each video image from the beginning to the end of the video frames.

A correction method is applied for barrel distortion which will be described in section 4. Applying the correction method, the position of the targets in the distorted image are corrected and transformed into the model space, so that we get (x_1', y_1') and (x_2', y_2') for target positions in the model space. The position of the ship model (x, y) is measured at the center of gravity of the ship which is represented by the position of target-1 (x_1', y_1') . The ship model heading angle is calculated from the angle formed by target-1 and target-2. Furthermore, the scale factor is used to get the value at full-scale.

3.1. Target detection

The motion of the ship model is measured based on the position of the target mounted on the deck of the model. The model test is carried out under low light conditions, and two LED lights are used as targets, so the targets can be identified easily. The target position is detected based on the color parameter value. In this paper, the detection is based on the grayscale value, and the maximum grayscale value is the criterion for finding the target position. Grayscale has a value range from 0 to 255, where black is 0 and white is 255. The gray value for a given RGB value is calculated according to the following approach [20]:

$$gray\ scale = 0.3\ R + 0.59\ G + 0.11\ B \tag{1}$$

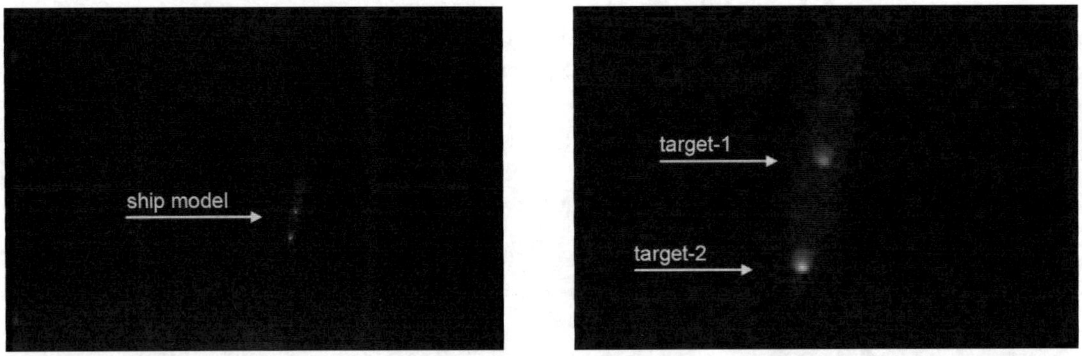

(a) (b)

Figure 3. Position of ship model (a) and the targets (b) in a video sequence.

(a) (b)

Figure 4. Target-2 at 64x magnification. (a) Target in RGB color. (b) target in grayscale. Red crosshair indicates pixel with maximum gray value.

IOP Conf. Series: Earth and Environmental Science 1081 (2022) 012040 doi:10.1088/1755-1315/1081/1/012040

Target detection is performed on each frame of the video by iterating over the pixels in the search window. The target position is determined based on the pixel position that has the maximum grayscale value. Figure 3a shows the position of the ship model in a video sequence, while the target positions are shown in figure 3b. Target-1 is in the middle of the ship model and uses a green LED, while target-2 is in forepart of the ship model and uses a red LED. The 64x magnification of the target is shown in figure 4a for RGB color mode, while for grayscale mode it is shown in figure 4b. Pixel with maximum gray value is indicated by red crosshair. For each frame of the video, the positions of target-1 and target-2 in the image coordinate system are obtained.

3.2. Target localization

To accelerate the target detection process, a search window is applied to localize the target. Search window is the area around the target to be detected as shown in figure 5. The figure shows the search windows for target-1 and target-2 with dimensions of 16 x 16 pixels around the target. During the tracking process, the search window moves following the target position. By applying target localization, target detection is carried out only in the search window, not in the entire image area.

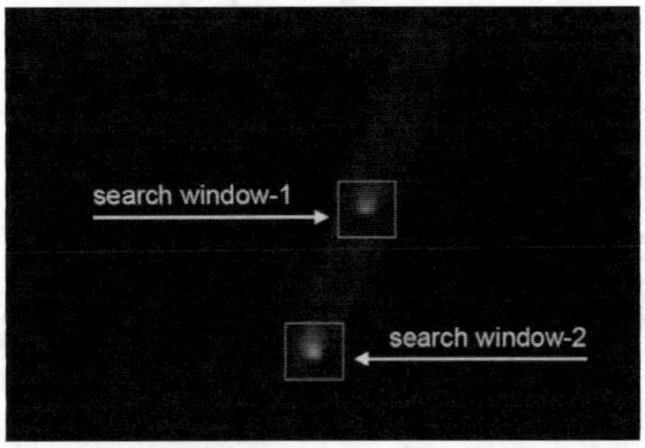

Figure 5. Target localization using search windows.

3.3. Centroid of target

In the target detection stage, the target position is determined based on the maximum gray value. Occasionally the point with the maximum gray intensity does not represent the center of the target accurately. In order to enhance the measurement accuracy in subpixel, a centroid calculation is performed to the target. The centroid is calculated on the area covering the target.

There are a number of methods can be used to calculate the centroid of a target image, such as binary centroid, gray weighted centroid and squared gray weighted centroid [21]. The binary centroid is a method which the gray image in the target window is converted to the binary image according to the selected threshold, then the centroid is calculated by

$$(\bar{x}, \bar{y}) = \left[\sum_{j=1}^{m} \sum_{i=1}^{n} i.I_{ij}, \sum_{j=1}^{m} \sum_{i=1}^{n} j.I_{ij}\right] / \sum_{j=1}^{m} \sum_{i=1}^{n} I_{ij} \quad (2)$$

Where $\bar{x}$ and $\bar{y}$ are the target centroid coordinates; m and n are the size of the windows where the centroid is being calculated; I_{ij} is the intensity at the ijth pixel location. In the binary centroid method, the value of I_{ij} is one or zero depending on the threshold.

A gray weighted centroid is calculated by the same equation, but no threshold applied to the intensity values. The equation for a squared gray weighted centroid is similar to the gray weighted centroid but the intensity values are squared, so it gives more weight to the higher intensity value. The gray weighted centroid method is then used in this paper. The centroid is calculated on an area of 7x7 pixels around the target.

3.4. Image filtering

The model test was carried out in low light conditions, and the distance between the target and the camera was relatively far. As a consequence, a lot of noise appears in the video images. Usually the video images have an unevenly distributed gray value, and lie in a small range, so the images have poor contrast. A number of methods can be used to reduce noise, such as mean, median and Gaussian filters [22]. In order to enhance the image quality, a histogram equalization method is used to redistribute the gray intensity evenly [22]. The method modifies the contrast by stretching the distribution of gray value to the expected range. Adopting the method, the following equation is used to discard gray intensity below the lower limit I_l and above the upper limit I_u and to scale the intensity I_i to a new intensity value I_i'.

$$I_i' = \begin{cases} 255 & I_u < I_i \\ 255\,x\left(\frac{I_i - I_l}{I_u - I_l}\right) & I_l \le I_i \le I_u \\ 0 & I_i < I_l \end{cases} \tag{3}$$

Before the detection process, the image quality of the object in the video is analyzed to determine the appropriate filter and its parameters.

3.5. Target tracking

Various algorithms have been developed to predict the target position, such as Kalman Filter [23] and Particle Filter [24]. In this paper, a simple technique is applied for target tracking. The motion of the target can be tracked by assuming the target position of the current frame *Xt* is the center point of the search window *Xs* for tracking in the next frame as the following equation:

$$Xs_{i+1} = Xt_i \tag{4}$$

For better results, a small improvement was applied to the equation (4). The center point of the search window in the next frame is determined by predicting the target position for the next frame. While the target position in the next frame is predicted by the target speed in the current frame as follows:

$$\hat{X}t_{i+1} = Xt_i + (Xt_i - Xt_{i-1}) = 2Xt_i - Xt_{i-1} \tag{5}$$

Then the center position of the search window is

$$Xs_{i+1} = 2Xt_i - Xt_{i-1} \tag{6}$$

4. Barrel Distortion Correction

Utilizing a wide-angle camera lens in the ship-model experiments, a broader area of ship-model trajectory could be covered. However, the video image recorded by the camera is seriously distorted. In the distorted image, a straight line in real life appears to be curved convexly away from the center of the image. This type of distortion is called barrel distortion or fish eye effect.

Several methods have been proposed for barrel distortion correction. Generally, the distortion correction method should calculate center of distortion and correction both radial and tangential component [25]. In a barrel distorted images, it is usually assumed that the radial distortion is dominant so that the tangential correction can be neglected. Some methods use the calibration pattern and then compare it to the pattern on the distorted image to find an equation for distortion correction.

Smith et al. [26] proposed a method based on a mathematic model for the distortion correction of endoscope images, by calculating the set of expansion coefficients in orthogonal Chebyshev polynomials. Tsai [27] purposed a second-order distortion correction that is radially symmetrical from the image center. Haneishi at al. [28] proposed a polynomial model to correct endoscope images based on the assumption that the distortion is circularly symmetrical. A method proposed by Bukhari and Dailey [29] does not require a calibration pattern. The distortion parameters are estimated from the circular arcs in the distorted image.

4.1. Mathematical modelling for barrel distortion

Adopting Haneishi's method, this paper uses a circular symmetrical model. In order to simplify the problem, it is assumed that the center of the distortion is in the center of the image. Consider (x, y) is the coordinate of a point in a distorted image and (x_c , y_c) is the center of distortion in the image. The radial distance r of the point from the center of distortion is calculated by

$$r = \sqrt{(x - x_c)^2 + (y - y_c)^2} \tag{7}$$

Using polynomial approximation, the corrected value of radial distance is expressed as

$$r' = a_1 r + a_2 r^2 + a_3 r^3 + \ldots + a_n r^n \tag{8}$$

where n is the degree of the polynomial and a is a set of coefficients. Then, the corresponding point in the corrected image is given by

$$\begin{aligned} x' &= x_c' + r' cos\theta \\ y' &= y_c' + r' sin\theta \end{aligned} \tag{9}$$

where (x_c' , y_c') is the center of corrected image and

$$\theta = tan^{-1}\left(\frac{y - y_c}{x - x_c}\right) \tag{10}$$

4.2. Calibration

As a reference for the measurements, gridlines spaced 2 meters apart were drawn on the floor of the basin as shown in figure 6. The distance of the longitudinal and transverse grid lines to the center of the tank are presented in table 2. The intersection of the longitudinal and transverse grid lines arranges an 11x11 grid points. Figure 7a shows the pattern of the grid points in object space. The grid pattern is then used as a reference to correct the distortion of the video image.

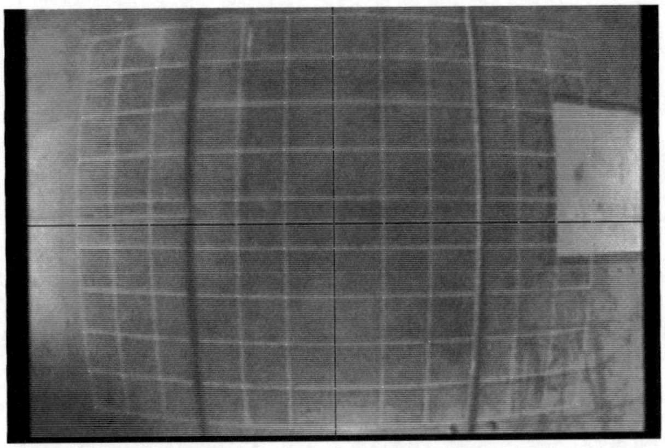

Figure 6. Grid lines and barrel distortion effect on the image.

In a high distorted image, the straight grid lines are distorted into curved grid lines, and the grid points are distributed in a curved pattern. Using an image editor, the position of the grid points in the distorted image are measured and the results are shown in figure 7b. Using the equation (8) for barrel distortion correction, the relationship between the grid points in the object space and the grid points in the distorted image can be found. Polynomial regression is used to solve the problem, and the results are described in the discussion.

Table 2. Grid line position.

	Distance to center of tank (m)										
Vertical lines	-12	-10	-8	-4	-2	0	2	4	8	10	12
Horizontal lines	-9	-7	-5	-3	-1	0	1	3	5	7	9

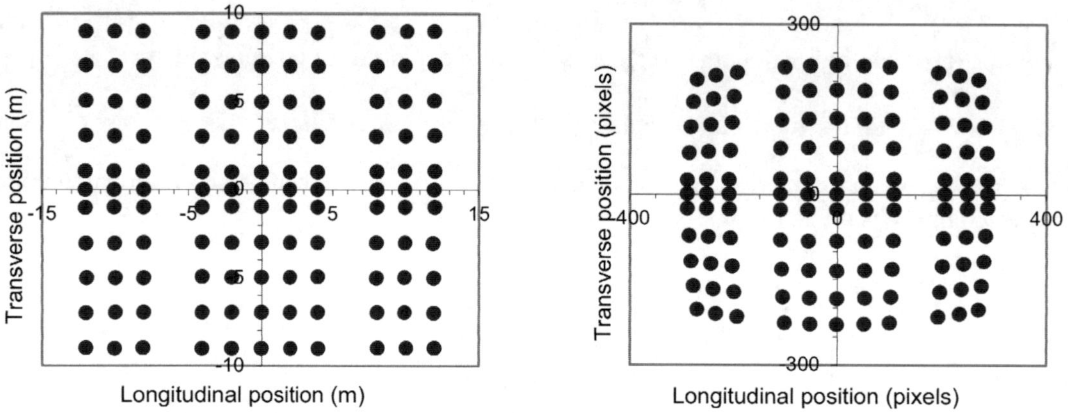

(a) (b)

Figure 7. Grid points in object space (a) and in distorted image (b).

5. Results and Discussion

5.1. Correction parameters

Using polynomial regression, the grid points in the distorted image are corrected and transformed into object space. Polynomial order is one of the parameters that must be determined to solve the regression. To determine the reasonable order of polynomial regression, residual standard error is used as indicator of the goodness of fit. The residual standard error or also called standard error is the standard deviation of the residual values, or the difference between a set of observed and predicted values, as the following equation

$$SE = \sqrt{\frac{1}{n-m} \sum_{i=1}^{n} \left(r'_i - \hat{r}'_i\right)^2} \tag{11}$$

where r' and $\hat{r}'$ are the true values and the predicted value of the radial distance of grid points, respectively; n is the number of data points and m is the number of variables in the regression analysis. Equation (11) can be expressed in relative terms, which is called the relative standard error, by dividing the standard error by the mean, as follows:

$$RSE = SE/\bar{r}' \tag{12}$$

The relationship between the order of polynomial and the standard error is investigated, and the result is shown in figure 8. The curve shows that the standard error drops dramatically as the order increases from 2 to 3, then above order 4 the standard error does not change significantly. Therefore, order 4 is chosen for the polynomial regression.

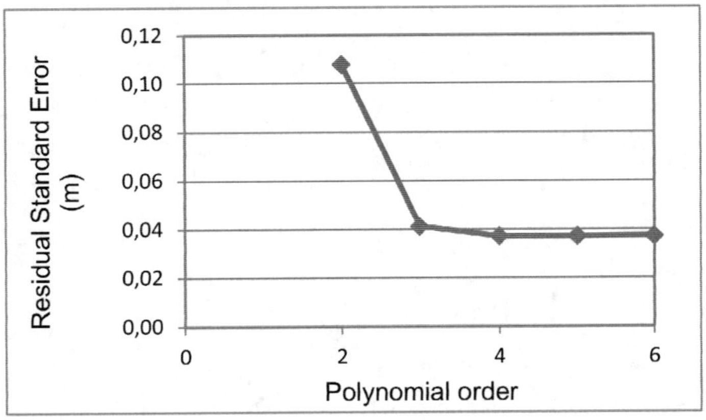

Figure 8. Relationship between the polynomial order and the standard error.

Figure 9 shows the relationship between the radial distance of the grid points in the distorted image and the related grid points in the object space. The least square method is applied to get the best-fitting curve to set of points by minimizing the sum of the squares of the residuals of the points from the curve. The polynomial approximation of the relationship is represented by a solid curve. The coefficients for polynomial equation that best fit the data are presented in table 3. The curve of best fit has a standard error of 0.037 meters and a coefficient of determination R^2 of 0.99999.

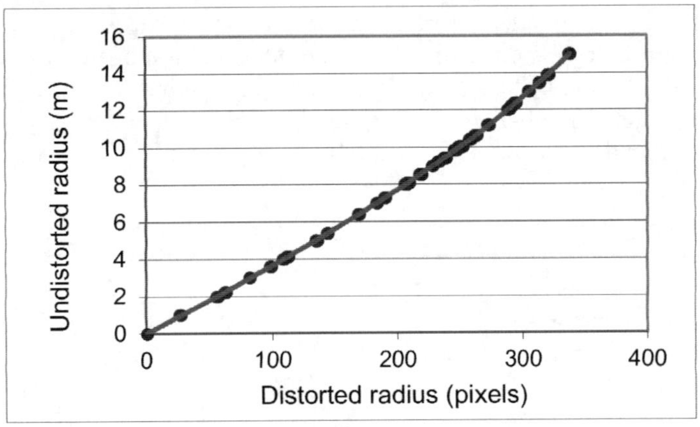

Figure 9. Relationship between the radial distance in the distorted image and the related radial distance in the object space.

Table 3. Polynomial coefficients.

a1	a2	a3	a4
3.488E-02	2.131E-05	-6.734E-08	2.599E-10

Using the coefficients in table 3 for the polynomial equation, grid points in a distorted image can be transformed to corrected grid points in object space. Comparing the difference between the corrected radius and the actual radius of grid points, the average value of the absolute difference is 0.42 percent.

Figure 10 shows the polynomial approximation for corrected grid points compared to their actual position in model space. The figure shows that the correction method is effective for correcting a distorted image. The correction method is then used to correct the position of the ship model in the video images.

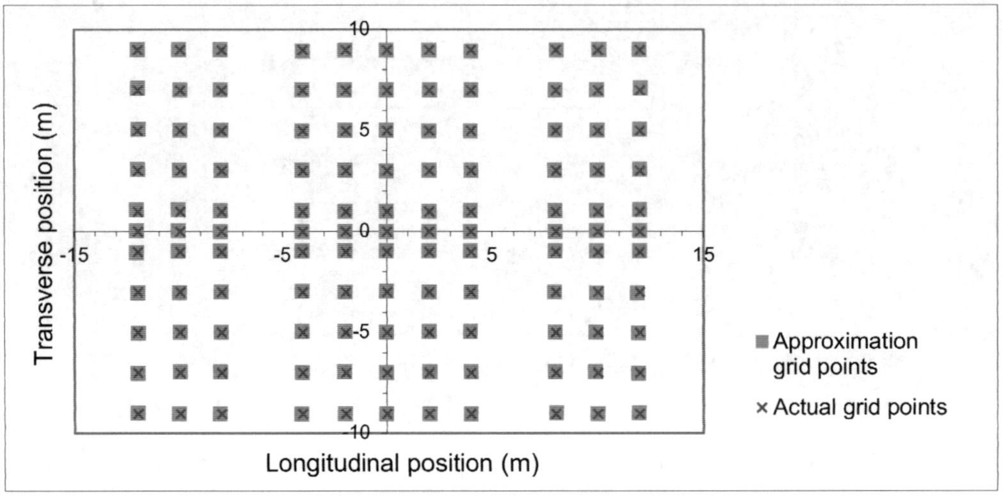

Figure 10. Comparison between the approximation grid points and the related actual grid points in model space.

5.2. Model test results

In this experiment, the data is stored in digital video format with a resolution of 720 x 480 pixels and a frame rate of 30 fps. Using a computer program, the video data is processed to get the trajectory of the ship model. The ship's trajectory is determined based on the position of the first target at the center of gravity of the ship model. while the ship's heading angle is measured from the angle between the first target and the second target. The number of images processed for one turning maneuver experiment is 10050 images or 335 seconds, and it gets 2.5 revolutions.

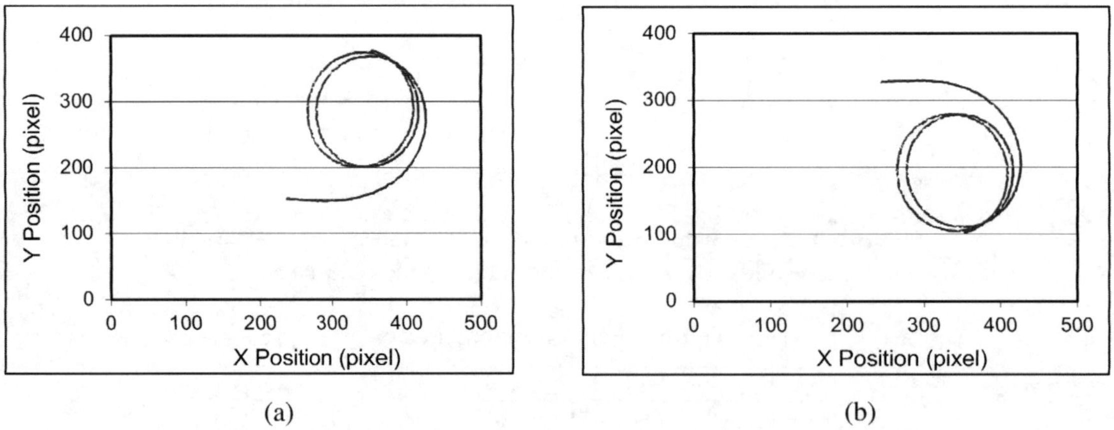

Figure 11. Tracking turning circle maneuver in image coordinate system. (a) Port side direction. (b) Starboard direction.

Figure 11 shows the trajectory of the ship model on the turning maneuver test in image coordinate system. The result has distorted values due to barrel distortion in the image. A polynomial approximation to barrel distortion is applied to correct the results. After correcting for distortion and multiplying by the scale factor, the ship's trajectory for the full-scale is obtained. The full-scale ship's trajectory on the turning maneuver test for portside and starboard directions are shown in figure 12, while the parameter values for turning circle are presented in table 4 in both dimensional and non-dimensional terms.

The experiment results are then compared with the International Maritime Organization (IMO) criteria for ship maneuverability [30]. IMO requires the advance should not exceed 4.5 ship lengths and the tactical diameter should not exceed 5 ship lengths in the turning circle maneuver at maximum rudder angle of 35°. Generally the experiment results are acceptable based on IMO regulations. The maximum advance and tactical diameter do not exceed the requirements of IMO maneuvering standards.

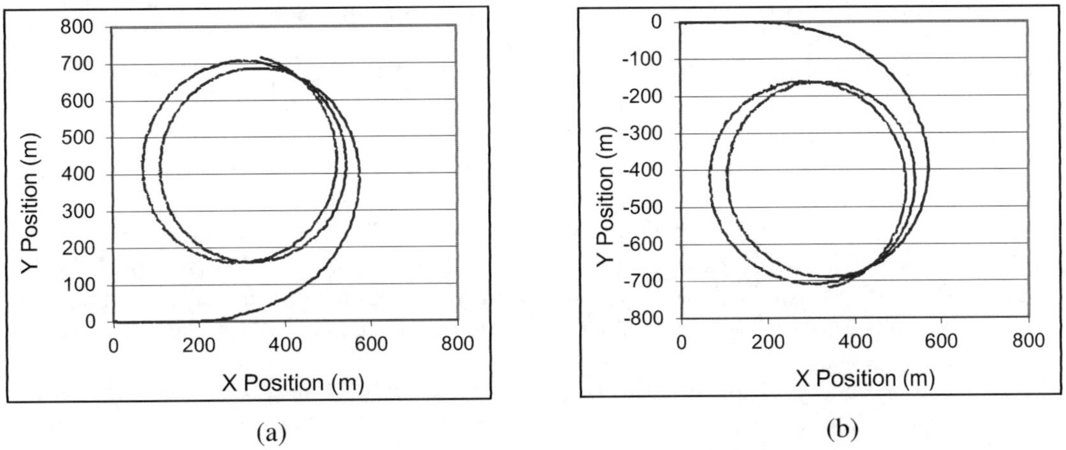

Figure 12. Tracking turning circle maneuver in full-scale.
(a) Port side direction. (b) Starboard direction.

Table 4. Turning circle results.

	Port side direction		Starboard direction		IMO
	(m)	Non-dimensional	(m)	Non dimensional	
Advance	544.8	2.05 Lpp	542.0	2.04 Lpp	4.50 Lpp
Transfer	240.4	0.90 Lpp	244.1	0.92 Lpp	
Tactical diameter	659.8	2.48 Lpp	656.0	2.47 Lpp	5.00 Lpp
Turning diameter	495.7	1.86 Lpp	495.7	1.86 Lpp	

6. Conclusion

This paper presents a series of model experiments on the turning circle ability of a 1:84.6 scale LNG carrier at calm water. A low-cost digital optic-based motion tracking system was developed for the measurement. The motion of the ship model is recorded by a consumer-grade video camera mounted above the basin. Applying a computer program, the video data can be transformed into numerical data on the motion of the ship model.

Utilizing a wide-angle camera lens in the ship-model experiments, a broader area of ship-model trajectory could be covered. However, the image recorded by the camera could be seriously barrel distorted. The polynomial approach is used for mathematical modelling for the distortion.

Furthermore, a gridlines spaced 2 meters apart on the floor of the basin is applied for calibration. A fourth-order polynomial regression was selected to find the relationship between the grid pattern in the object space and in the distorted image. The analysis was carried out on the corrected and the actual grid points. The polynomial equation for distortion correction has a relative standard error of 0.42%. The results show that the correction method is effective for correcting a distorted image.

In the purposed motion measurement system, the digital video data is processed offline. A Color-based tracking method is applied to detect objects and their positions. The results are then corrected by the polynomial approximation for the barrel distortion. In general, the results of the model experiments show that the turning ability of the ship meets the IMO criteria.

References

[1] Zhu R & Zhou Z 2004 A real-time articulated human motion tracking using tri-axis inertial/magnetic sensors package *IEEE Transactions on Neural systems and rehabilitation eng*. **12** 295-302.

[2] Chiani M, Giorgetti A & Paolini E 2018 Sensor radar for object tracking *Proc.of the IEEE* **106** 1022-41.

[3] Mertz C, Navarro-Serment L E, MacLachlan R, Rybski P, Steinfeld A, Suppe A, Urmson C, Vandapel N, Hebert M & Gowdy J 2013 Moving object detection with laser scanners *J. Field Robotics* **30** 17-43.

[4] Colantonio S, Benvenuti M, Di Bono M G, Pieri G & Salvetti O 2007 Object tracking in a stereo and infrared vision system *Infrared physics & technology* **49** 266-271.

[5] Loomans M J, de With P H & Wijnhoven R G September 2013 Robust automatic ship tracking in harbours using active cameras *2013 IEEE Int. Conf. on Image Processing* pp 4117-21.

[6] Ye X W, Jin T & Ang P P 2020 Computer vision-based monitoring of ship navigation for bridge collision risk assessment *Machine vision and navigation* (Springer, Cham) pp 787-807.

[7] Liu L, Liu G, Chu X M, Jiang Z L, Zhang M Y & Ye J 2019 Ship Detection and Tracking in Nighttime Video Images Based on the Method of LSDT *J. Phys.: Conf. Ser*. vol 1187 (IOP Publishing) no 4 p 042074.

[8] Del Pizzo S, Gaglione S, Angrisano A, Salvi G & Troisi S 2018 Reliable vessel attitude estimation by wide angle camera *Measurement* **127** 314-324.

[9] Fefilatyev S, Goldgof D, Shreve M & Lembke C 2012 Detection and tracking of ships in open sea with rapidly moving buoy-mounted camera system *Ocean Eng*. **54** 1-12.

[10] Prasad D K, Rajan D, Rachmawati L, Rajabally E & Quek C 2017 Video processing from electro-optical sensors for object detection and tracking in a maritime environment: a survey *IEEE Transactions on Intelligent Transportation Systems* **18** 1993-2016.

[11] Erwandi & Suzuki T 1999 Analysis of waves generated by a ship oscillating on calm water with forward speed using projected light distribution method *J. The Society of Nav. Archit. of Japan* **186** 7-15.

[12] Erwandi & Suzuki T 2001 Analysis of diffraction waves around a ship using projected light distribution method *J. The Society of Nav. Archit. of Japan* **189** 1-12.

[13] ITTC 2005 The Seakeeping Committee *Proc.of the 24th ITTC* vol 1 p 202.

[14] Hollyhead C J, Townsend N C & Blake J I 2017 Experimental investigations into the current-induced motion of a lifeboat at a single point mooring *Ocean Eng*. **146** 192-201.

[15] Figuero A, Rodriguez A, Sande J, Peña E & Rabuñal J R 2018 Dynamical study of a moored vessel using computer vision *J.Marine Sci. and Technol*. **26** 11.

[16] Nocerino E, Menna F, & Troisi S 2013 High accuracy low-cost videogrammetric system: an application to 6DOF estimation of ship models *Videometrics, Range Imaging, and Applications XII; and Automated Visual Inspection* vol 8791 (SPIE) pp 143-157.

[17] Malas B, Banks J, Cappelletto J & Thornton B 2019 Applications of motion capture technology in a towing tank *Proc.of the Advanced Model Measurement Technology for the Maritime*

Industry, Rome, Italy, 9-11.

[18] Nguyen T T D, Nguyen V M, Yoon H K & Kim Y H 2020 An Experimental Study on the Motion of the Floater Moored near Port in Waves Generated by a Ship *J. Navigation and Port Research* **44** 363-374.

[19] Ghamari I, Greco M, Faltinsen O M & Lugni C 2020 Numerical and experimental study on the parametric roll resonance for a fishing vessel with and without forward speed *Applied Ocean Research* **101** 102272.

[20] Adobe Systems Incorporated 1999 *PostScript Language Reference Manual* 2nd edn (Addison-Wesley Publishing Company) chapter 6 p 304.

[21] Shortis M R, Clarke T A & Short T 1994 Comparison of some techniques for the subpixel location of discrete target images *Videometrics III* vol 2350 International Society for Optics and Photonics pp 239-250.

[22] Jain R, Kasturi R & Schunck B G 1995 Image filtering *Machine vision* vol 5 (New York: McGraw-hill) chapter 4 pp 112-139.

[23] Kalman R E 1961 A New Approach to Liner Filtering and Prediction Problems, Transaction of ASME *J. Basic Eng.* **83** 95-108.

[24] Moral P D & Jacod J 2001 Interacting particle filtering with discrete observations *Sequential Monte Carlo methods in practice* (New York: Springer) pp 43-75.

[25] Shah S & Aggarwal J K 1996 Intrinsic parameter calibration procedure for a (high-distortion) fish-eye lens camera with distortion model and accuracy estimation *Pattern Recognition* **29** 1775-88.

[26] Smith W E, Vakil N & Maislin S A 1992 Correction of distortion in endoscope images *IEEE Transactions on Medical Imaging* **11** 117-122.

[27] Tsai R Y June 1986 An efficient and accurate camera calibration technique for 3D machine vision *Proc. IEEE Computer Vision Pattern Recognition* Miami, FL pp 364–374.

[28] Haneishi H, Yagihashi Y & Miyake Y 1995 A new method for distortion correction of electronic endoscope images *IEEE Transactions on Medical Imaging* **14** 548-555.

[29] Bukhari F & Dailey M N 2013 Automatic radial distortion estimation from a single image *J. mathematical imaging and vision* **45** 31-45.

[30] MSC, IMO 76/23 2002 Resolution MSC 137 (76), standards for ship manoeuvrability. Report of the maritime safety committee on its seventy-sixth session, annex 6 *International Maritime Organization*.

Structural Reliability Analysis for the Construction Design of the High-Speed Ship with CFRP Material

M A Hafiz[1] and A Sulisetyono[1]

[1]Department of Naval Architecture, Institut Teknologi Sepuluh Nopember, Surabaya, Indonesia

Abstract. A high-speed ship mostly experiences the excessive motions in seaway that influence wave loads act to ship's hull extremely. For this reason, the structure of a high-speed ship must be designed more strength than a low-speed vessel. In this paper, the construction design of the high-speed ship with material of CFRP are evaluated consider to the vertical bending moment under various wave excitation conditions. The eight design variations of the midship section are proposed to be evaluated against the uncertainty load stress due to the vertical motions of ship in random wave. The diffraction theory is adopted to solve the random wave loads due to various wave condition of heading angles, heights, and periods in which the maximum vertical bending moments of the rigid ship are obtained. Structural reliability analysis is applied for assessing structural safety and reliability of the ship's structure. The structural strength of the construction design which are required for the reliability analysis have been developed based on permissible stress of BKI regulations. Safety levels associated with each mode of failure of all construction designs are determined and compared. Finally, the construction design is selected with the safety index 1.07.

1. Introduction

In general, a high-speed ship able to travel at a speed of more than 30 knots [5]. It is also stated in the 2000 HSC Code which classified a high-speed ship according to the maximum speed of exceeds $3.7\nabla^{0.16667}$ (m/s), where ∇ is the displacement volume of the ship (m^3) [13]. In a high-speed ship design, there are four challenges consider to the ship's performance including manoeuvrability, suitability of the main engine capabilities, high fuel consumption cost, and ability cargo transporter [5]. However, the strength of a high-speed ship's construction must be highly considered because the nature of the ship's operation at high-speed influences the extreme wave loads that affect to the ship's structure [19].

Naturally, a ship in wave is a free floating under six degrees of freedom motion namely surge, sway, heave, roll, pitch, and yaw. The ship motions are occurred due to the exciting force of the waves that also induces the wave loads act to the ship's structure [18]. The vertical motion of a high-speed ship such as heave and pitch are dominant motions contribute to the vertical bending moment of ship's structure longitudinally. Investigation of the distribution of wave loads along the ship's length informs that the largest bending moment occurs in the midship area whose value is calculated on ship operating in calm and wavy water conditions [17]. And the strength of the ship structure needs to be considered in design stage of the ship's construction especially at a critical area of the midship.

Strength prediction of ship's construction generally use two approaches, namely the deterministic method which analyses the strength of the ship's construction due to the constant load, and the probabilistic method which considers the random nature of the waves to the uncertainty of the bending moment and the stress of the ship's structure [12]. In the safety assessment related to the strength of a ship's construction design, the approach of structural reliability is widely adopted. However, structural reliability is a suitable method for estimating the margin between the load acting on the structure due to waves and the strength of the design construction which are both modelled randomly. The strength of ship construction can be defined as the ultimate limit conditions related to the maximum load carrying capacity of the structure, where this limit status includes several modes, for example, excessive yielding (plastic mechanisms) and instability (buckling failure) [12]. The practical method for predicting ship's motions and loads is commonly use panel method as suitable numerical method applying the diffraction theory, the green's function theory, the Rankine source theory, and others [10, 11, 19]. And the ultimate limit condition can be determined based on the classification regulation, the material test, and the finite element analysis [1, 4].

In general, the strength of the ship's construction is basically dependent to a combination of the profile arrangements or scantling and the material used [12]. In the last decade, the carbon composite material is commonly used for structures of a hight speed boat, because having advantages such as lightweight, sufficient strength, non-corrosion, good flexural, and quite resistant to heat [22]. Composite material consists of two or more constituents namely the reinforcement and the matrix, where the reinforcement is classified into fibers and particles, and the matrix is made up of polymeric materials, iron, aluminium, and others [15]. Carbon Fiber Reinforced Polymer (CFRP) is a type of composite material whose main content is carbon fiber with polymer adhesive [6]. The CFRP has been widely used as the main material for patrol boats, corvettes [8], military assets [7], tubes for submarines [16], and others.

This paper intends to perform the strength evaluation of a high-speed ship with CFRP material due to the load of vertical bending moments on the longitudinal ship's hull considering the random wave characteristics of the frequency, height, and relative direction to the ship heading. The eight designs of the midship section construction are proposed for the CFRP combat ship in which each construction design has different in inertia of the transverse body. The strength of various construction designs is evaluated against loads of vertical bending moment and furthermore a stress of the ship's structure is obtained by dividing of the bending moment with the inertia of the section construction [2]. Finally, the reliability analysis is adopted to assess the strength of ship's structure under excitation of random wave.

2. Methodology

2.1. Construction design

Combat fast boats have the main dimensions as shown in table 1, and the shape of the hull is V hull which is presented in terms of the lines plan as shown in figure 1. Eight variations of ship's construction design are introduced which are called the midship designs of A, B, C, D, E, F, G, and H, as presented in figures 2 – 9. Briefly, the construction designs of A and E are defined as the longitudinal construction types, B and F designs are the transverse construction types, and C, D, G, and H designs are the combined construction types. All constructions have a support plate, senta, centre girder, and fenders. Besides that, designs A, C, E, and G are the single-sided constructions with variations on the bottom construction. And designs B, D, F, and H have double sides with variations of the distance between sides as well as the bottom construction. The size and type of the construction profiles for all designs can be seen in table 2. In this study, the construction material used is CFRP whose properties refer to the test results of the unidirectional carbon fiber material which has been discussed in [4], where the mechanical properties of the material are shown in table 3.

Table 1. Ship's main dimensions

Parameter	Dimension	Units
Length Overall	11.995	Meter
Length of Waterline	10.467	Meter
Breadth	3.1	Meter
Depth	2.172	Meter
Draught	0.55	Meter
Maximum Speed	10	Knots
Cruising Speed	9	Knots
Displacement	9.7	Tons
Length of Center Bouyancy	4.399	Meter

IOP Conf. Series: Earth and Environmental Science 1081 (2022) 012041 doi:10.1088/1755-1315/1081/1/012041

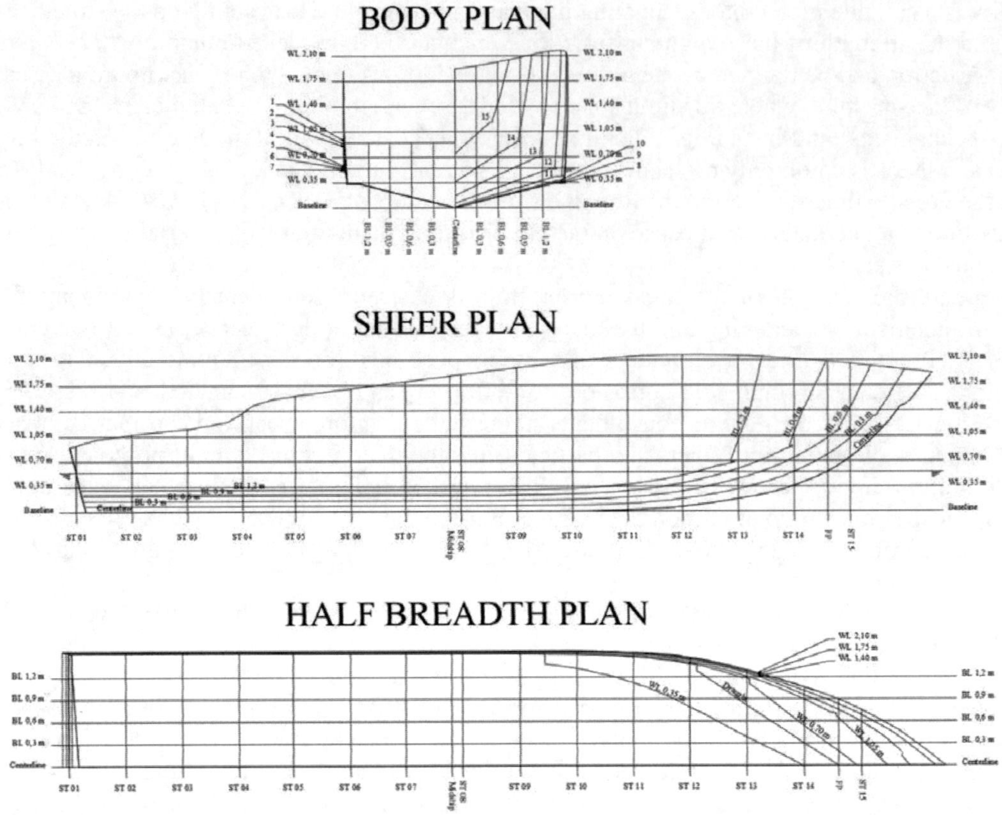

Figure 1. Lines plan

Table 2. Dimension of construction profile for each construction design.

Construction Design	Construction Profile (mm)					
	Bottom Plate	Side Plate	Fender S. Plate	D. Bottom Plate	Centre Girder	Side Girder
A	20	12	12	12	30	15
B	20	12	12	12	30	-
C	20	12	12	12	30	-
D	20	12	12	12	30	15
E	40	24	24	24	60	30
F	40	24	24	24	60	-
G	40	24	24	24	60	-
H	40	24	24	24	60	30

Construction Design	Construction Profile (mm)				
	Bottom Transverse	Senta	Deck Girder	Side Long.	Web Transverse
A	-	200x100x30	200x100x30	75x15	-
B	100x30	-	-	-	200x100x30
C	100x30	-	-	-	200x100x30
D	-	200x100x30	200x100x30	75x15	-
E	-	400x200x60	400x200x60	150x30	-
F	200x60	-	-	-	400x200x60
G	200x60	-	-	-	400x200x60
H	-	400x200x60	400x200x60	150x30	-

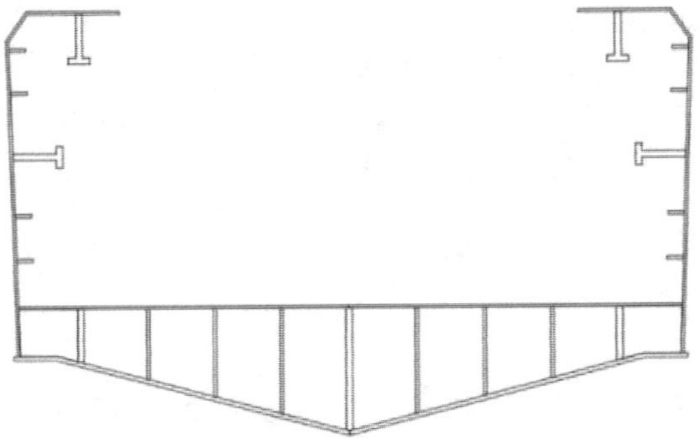

Figure 2. Midship section of design A.

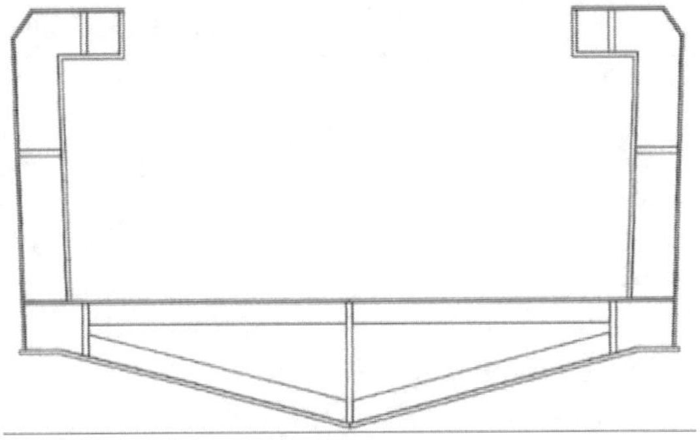

Figure 3. Midship section of design B.

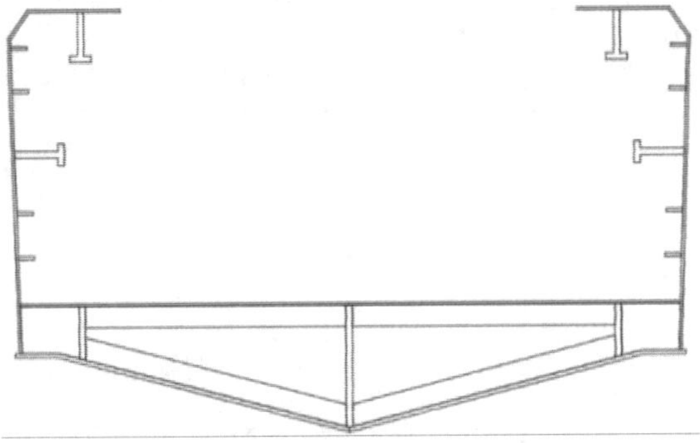

Figure 4. Midship section of design C.

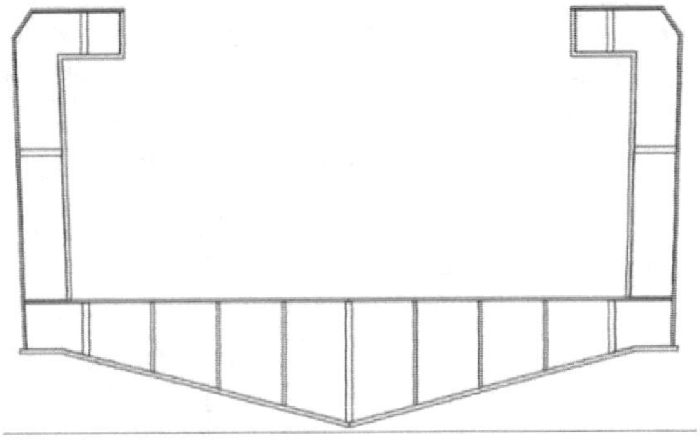

Figure 5. Midship section of design D.

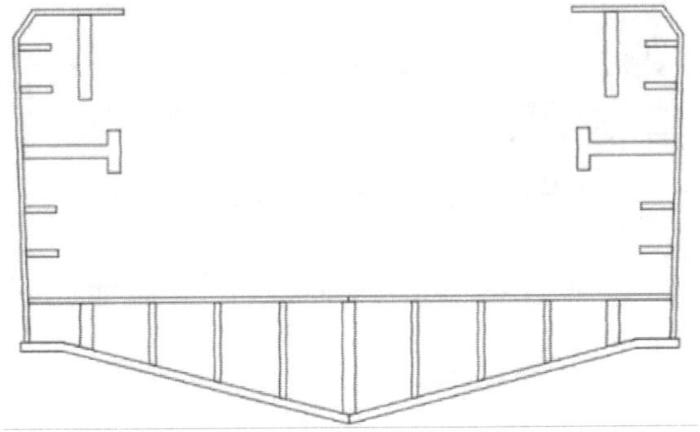

Figure 6. Midship section of design E.

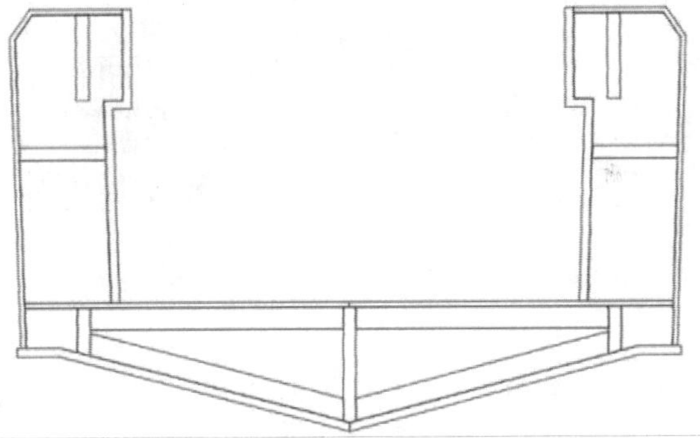

Figure 7. Midship section of design F.

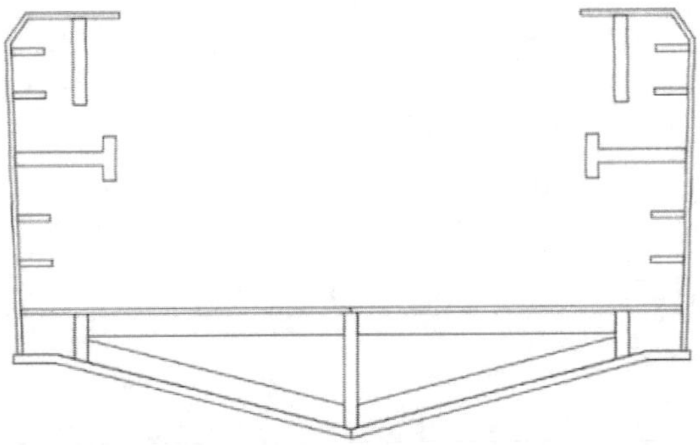

Figure 8. Midship section of design G.

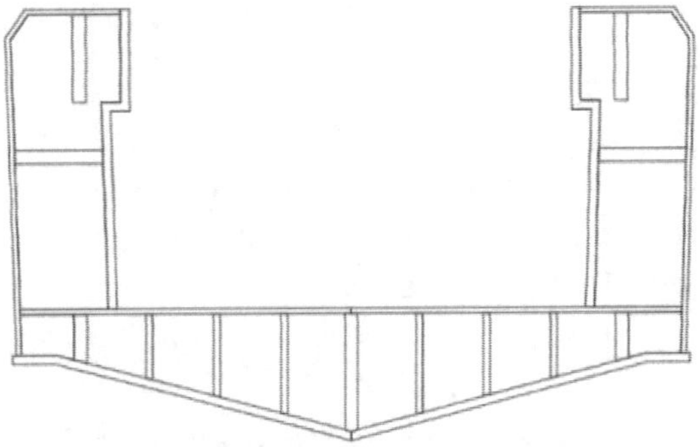

Figure 9. Midship section of design H.

IOP Conf. Series: Earth and Environmental Science 1081 (2022) 012041 doi:10.1088/1755-1315/1081/1/012041

Table 3. Material properties [4]

	Density (g/cm^3)	Elasticity Module (Mpa)	Shear Module (Mpa)	Poisson Ratio	Tensile Strength (MPa)	Shear Strength (Mpa)	Shear Strength (Mpa)
X Direction		78,715	2,195	0.4	826	20	118
Y Direction	1.31	4,930	2,195	0.25	37	20	27
Z Direction		4,930	2,195	0.25	37	20	27

2.2. Ship Operational Condition

The design of the construction is developed by considering the strength of the structure in dealing with wave loads that work to the ship's hull while it is operating at sea. In this case, the ship owner requires the operational limits in moderate wave conditions at the maximum speed of ship. Based on classification of wave malignancy as shown in table 4, the wave characteristics are obtained including the wave heights of 0.1 m, 0.5 m, 1 m, and 2 m, and the zero-up crossing period of 0.59 s, 0.77 s, 1.11 s, and 1.44, respectively. The incoming waves relative to the ship's direction are determined as shown in figure 10 including head sea, beam sea and following sea. The wave loads calculation are performed in 12 cases of ship operational conditions.

Table 4. Wave malignancy classification [21]

Sea State	Description of Sea	Significant Wave Height (m)	Wind Speed (Knots)	Beaufort Scale
0	Calm (Glassy)	0	0.0	0.0
1	Calm (Rippled)	0.01 – 0.10	0.01 – 6	1 – 2
2	Smooth (Wavelets)	0.11 – 0.50	7 – 10	3
3	Slight	0.51 – 1.25	11 – 16	4
4	Moderate	1.26 – 2.50	17 – 21	5
5	Rough	2.51 – 4.00	22 – 27	6
6	Very Rough	4.01 – 6.00	28 – 47	7 – 9
7	High	6.01 - 9.00	48 – 55	10
8	Very High	9.01 - 14.00	56 – 63	11
9	Phenomenal	> 14.00	> 63	12

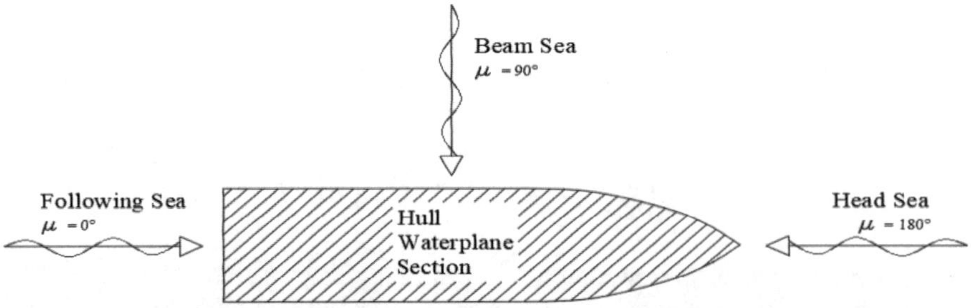

Figure 10. Description of the type of wave direction.

For the ship load analysis, the random waves are suitable to be expressed in terms of a wave energy spectrum which is developed using the standard formula of the Pierson-Moskowitz [1] as described in equation (1).

$$S_{(PM)} = \frac{H_S{}^2}{4\pi}\left(\frac{2\pi}{T_Z}\right)\omega^{-4}\exp[-\frac{1}{\pi}\left(\frac{2\pi}{T_Z}\right)\omega^{-4}] \quad (1)$$

Where $S_{(PM)}$ is the Pierson-Moskowitz wave spectrum in $m^2/(rad/s)$, H_S is the significant wave height in meters, T_Z is the zero-up crossing period in seconds, and ω is the wave frequency in rad/s.

2.3. Load Stress Calculation

The study only considers vertical wave loads, because the dominant motion of a small ship when speeding are vertical motions, namely heave and pitch. And the most load that influences to strength of the ship's structure longitudinally is the vertical bending moment on the midship area. The vertical bending moment (M_B) is calculated in two free surface conditions of the calm water (M_S) and the wavy water (M_W) [17] as described in equations (2) and (3).

$$M_B = M_S + M_W \tag{2}$$

$$M_W = M_\zeta + M_Z + M_\theta \tag{3}$$

However, the wave bending moment (M_w) is summations of the bending moments due to the wave (M_ζ), the heave motion (M_z), and the pitch motion (M_Θ).

The wave bending moment of the high-speed ship in irregular wave are calculated by using 3D diffraction theory or Panel method. The full scale of hull surface is modelled and discretised into number of panels as shown in figure 11. The grid or panel independent study is conducted to ensure the numerical results accuracy on the computation. The optimized nodes and elements number are about 16,485 and 16,250, respectively.

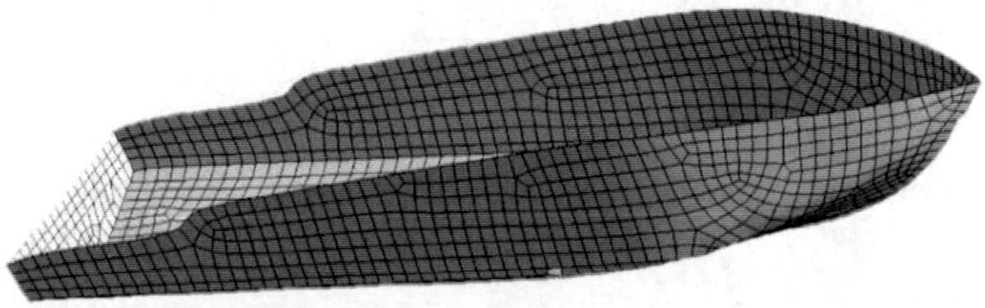

Figure 11. 3D ship model applied for motion analysis.

As in the case of a corvette [20], the wave load of the combat ship is predicted using 3D diffraction theory in which the results are presented in the form of an energy spectrum of bending moment. It is different with the previous study [20] that the response of bending moment due to irregular waves are evaluated in the time domain approach. In this case, the ship motions are simulated for a long period of 30 second under excitations of the various wave conditions. The maximum response of the bending moment is taken for computing the maximum stress that is occurred in the construction design. The maximum stress (σ) is commonly occurred at the bottom of the midship construction that is calculated by multiplication of the vertical bending moment and the distance between the centre of gravity and the bottom of the ship (Z_B), then the results is divided by the inertia moment of the midship construction area (I_{yB}), as presented in equation (4).

$$\sigma = \frac{M_B \, x \, Z_B}{I_{yB}} \tag{4}$$

Both the moment of inertia and the length distance of the centre of gravity to the bottom point (Z_B) for all construction design is calculated and presented in table 5. However, the stress bending moment due to vertical motions is dependent on the design construction that give various inertia moments.

Table 5. Moment of inertia and Z_B data for each variation of construction design.

Construction Design	A	B	C	D	E	F	G	H
Moment of Inertia (m^4)	0.5391	2.1016	0.7510	1.8891	0.8792	3.1194	1.0928	2.9058
Z_B (m)	0.6581	0.9005	0.5094	1.1142	0.7714	0.9775	0.6253	1.1079

2.4. Safety Index Assessment

Uncertainty of the bending moment and the stress of the ship structure due to random wave loads are considered. The structural reliability analysis is adopted to accommodate the two uncertainties in the loads and the strength of the construction. In this case, all possible stress values in the construction due to variations in wave loads are expressed in the form of a probability density function. Likewise with the strength of the ship's construction, where the uncertainty of the construction quality due to the fabrication process is also represented in the probability density function model. The strength distribution is developed based on the variation of the manufacturing processes as well as the various fatigue type in the equation (5) which is written in BKI (Badan Klasifikasi Indonesia) regulation as the permissible stress equation.

$$\sigma_{perm} = \frac{R_m.R_1.R_2.R_3.R_4.R_5.R_6.R_7.R_8}{S} \quad (5)$$

Where R_m is the tensile strength of the laminate used is about 826 MPa, see table 3; S is the factor value for the type of load occurred that has been classified includes the static loads with factor value of 1.1, the extreme loads with a factor value of 1.1, and the statistical loads (sea loads) with a factor value of 2, for other reduction factor can be seen in table 6.

Table 6. Reduction factor for permissible stress [3]

No.	Reduction Factors for Laminates	
1	Statistical long-term loading (sea loads)	$R_1 < 0.75$
2	Static long-term loading (tank walls)	$R_2 < 0.5$
3	Raised temperature	$R_3 < 0.9$
4	Ageing	$R_4 < 0.8$
5	Manufacturing i.e.	
	Hand lay-up	$R_5 = 0.9$
	Prepreg	$R_5 = 1.0$
6	Fatigue	
	$N < 10^2$	$R_6 = 1.0$
	$N < 10^3$	$R_6 = 0.8$
	$N < 10^6$	$R_6 = 0.4$
7	Different material properties	$R_7 < 0.9$
8	Moisture	$R_8 < 0.8$

Two distributions which are a function of stress are combined to become an intersection of the two distribution graphs called the joint probability density function. The safety index of each construction design is calculated using equation (6) which is a comparison between the difference between the mean values and the standard deviations taken from both load and strength distributions.

$$\beta = \frac{\mu_S - \mu_Z}{\sqrt{\sigma_S{}^2 + \sigma_Z{}^2}} \quad (6)$$

where the safety index (β) is an indicator of quality, μ_S is the mean strength, μ_Z is the mean load, σ_S is the standard deviation of the strength, and σ_Z is the standard deviation of the load.

3. Results and discussions

The calculation results of the vertical bending moment in calm water conditions are shown in figure 12, where the distribution of static bending moments along the ship's construction is shown. The values for all sections or stations of the construction are presented in which the largest vertical bending moment is about 90,436.93 Nm, generally occurred in the middle area of the longitudinal ship [9]. The static bending moment is summed with the dynamic bending moment to obtain the total bending moment.

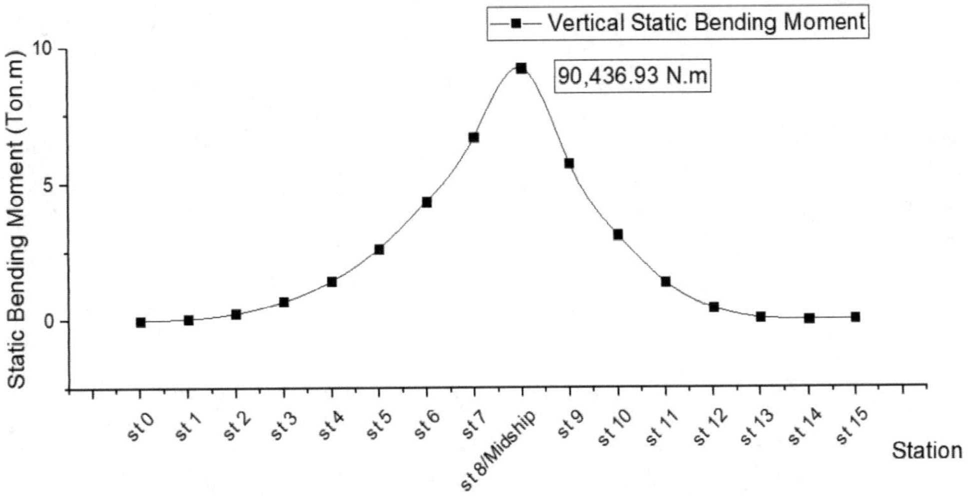

Figure 12. Graph of vertical bending moment in calm water conditions.

The wave bending moment of the ship's structure is calculated consider to the wave excitation force with varying heights of 0.1 m, 0.5 m, 1 m, and 2 m which are presented in the wave energy spectrum in figure 13. The wave spectrum is generated by equation (1) for the frequency range of 0.3 rad/ sec to 2 rad/sec. Furthermore, the vertical bending moment is numerically predicted by the panel method for simulating the ship's motion in the time domain. Figure 14 shows the simulation results of the vertical bending moment for a case of wave height of 0.1 m and heading angle of following sea, beam sea, and head sea. The other incoming wave cases of wave height of 0.5 – 2 m are presented in figures 15 - 17, respectively. It is seen in figures 14, 15, 16, and 17 that the response of bending moment in heading angle of head sea are occurred highest compared to the others heading angle of beam and following sea. For load analysis purposes, the maximum vertical bending moment that occurs during 30 seconds of simulation is taken for all study cases. table 7 describe the results of static, dynamic, and total vertical bending moments due to the 12 (twelve) cases of ship operational condition.

In table 7, the total vertical bending moment has the smallest value about 91.04 kNm while the incoming wave from the side of ship (or beam sea), and the largest is about 323.83 kNm when the incoming wave from the head of ship (head sea). This phenomenon shows the heading angle of ship's direction contribute to the vertical bending moment response as well as the wave height. In general, the static bending moment give the vertical bending moment higher than dynamic bending moment except in case of 2 m wave height at heading angle of following and head sea.

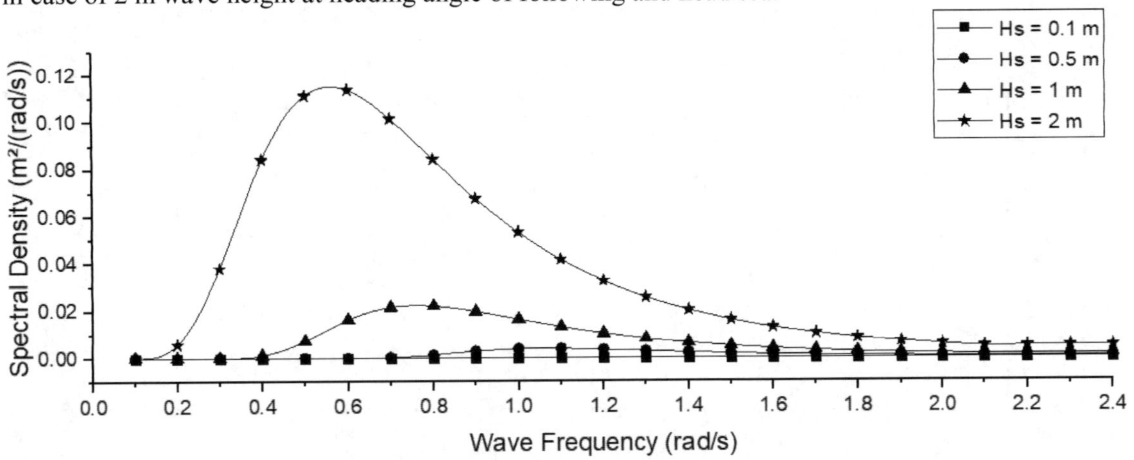

Figure 13. The Pierson-Moskowitz wave spectra at variations in wave height of 0.1 m to 2 m.

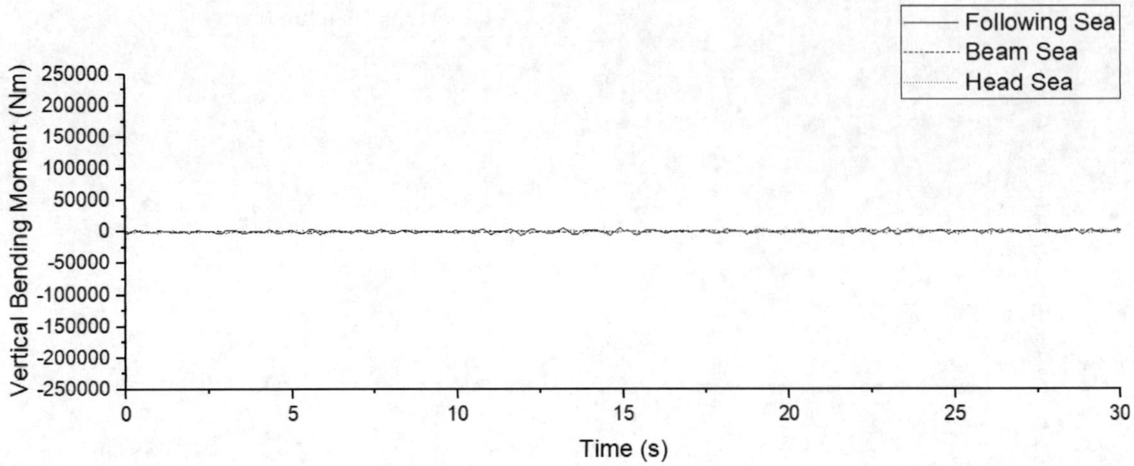

Figure 14. Vertical bending moment for a case of significant wave height of 0.1 m and heading angle of 0°, 90°, and 180°

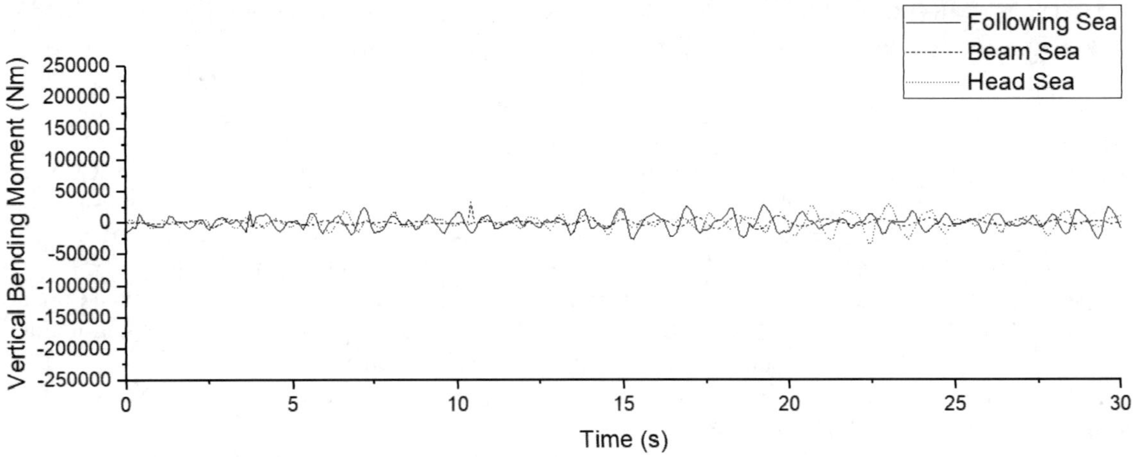

Figure 15. Vertical bending moment for a case of significant wave height of 0.5 m and heading angle of 0°, 90°, and 180°

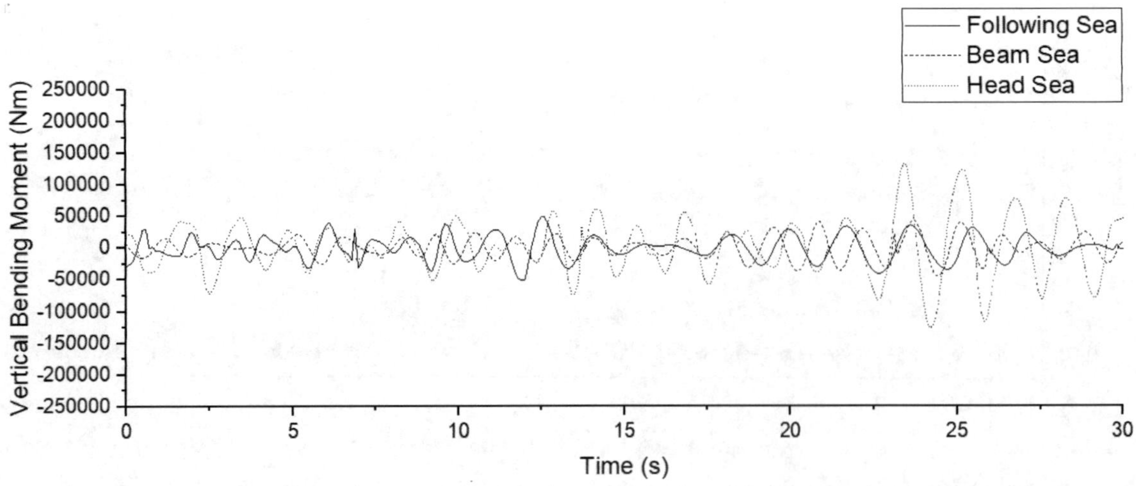

Figure 16. Vertical bending moment for a case of significant wave height of 1 m and heading angle of 0°, 90°, and 180°

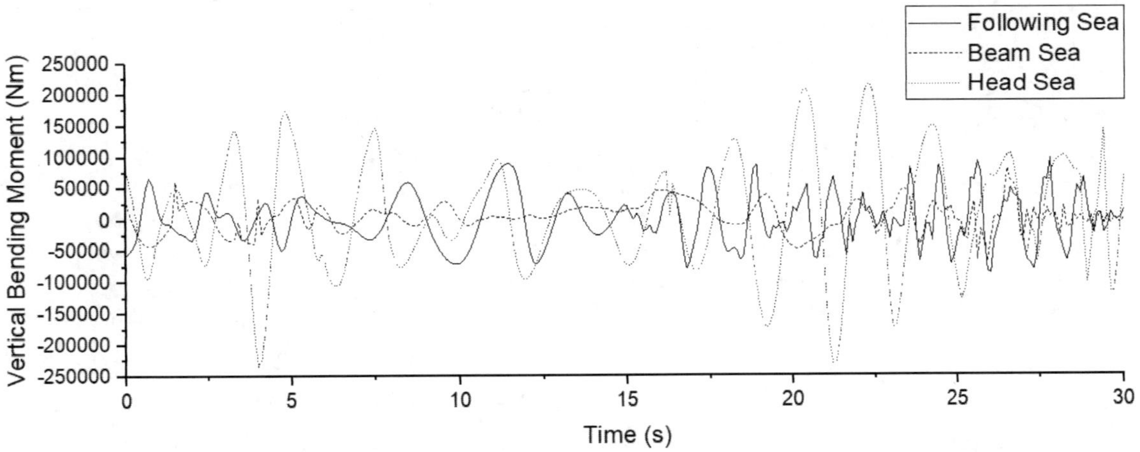

Figure 17. Vertical bending moment for a case of significant wave height of 2 m and heading angle of 0°, 90°, and 180°

Table 7. Vertical bending moment maximum

Cases	Wave Direction	Wave Height (m)	M_S (Nm)	M_w (Nm)	M_B (Nm)
Case 1	Following Sea	0.1	90,436.93	5,748.52	96,185.44
Case 2		0.5		28,347.86	118,784.79
Case 3		1		50,121.51	140,558.44
Case 4		2		95,981.25	186,418.18
Case 5	Beam Sea	0.1		605.34	91,042.26
Case 6		0.5		11,231.65	101,668.58
Case 7		1		41,813.82	132,250.74
Case 8		2		80,244.88	170,681.80
Case 9	Head Sea	0.1		4,319.56	94,756.48
Case 10		0.5		34,798.52	125,235.45
Case 11		1		134,084.61	224,521.54
Case 12		2		233,396.09	323,833.02

The stress due to wave loads is calculated using equation (6), and the results are described in table 8. Table 8 shows the maximum stress occurred in 8 (eight) variations of construction design under consideration of 12 (twelve) ship operational cases. In the same case of ship operational, the stress appeared in each construction design, namely A - H shows the various value since the inertia moment is varied from each construction design, see table 5. The highest value of the maximum stress shown in table 8 is about 395.32 MPa which occurs in the construction design A in the case number 12. While the lowest value is seen in the construction design F, which is about 28.53 MPa in the case number 5.

The stress value of strength due to uncertainty in the fabrication process is calculated using equation (5), where the variables that affect the stress value of the ship construction including Hand lay-up, Prepreg, and Fatigue are assumed to vary in value that describe uncertainty of human error due to fatigue, lack of concentration, etc. during the production process. In this study, the manufacturing coefficients which is presented in table 6 are reduced about 85% of the best condition. Likewise, the reduction of other coefficient factors including R_1, R_2, R_3, R_4, R_7, and R_8 are about 0.74, 0.49, 0.89, 0.79, 0.89, and 0.79, respectively. Table 9 explains the calculation results of the construction stress under uncertainty condition of manufacturing process.

Table 8. Maximum load stress of the construction designs due to the various cases of the ship operation.

Construction Design	Stress on various vertical bending moment (MPa)					
	Case 1	Case 2	Case 3	Case 4	Case 5	Case 6
A	117.42	145.01	171.59	227.57	111.14	124.11
B	41.21	50.90	60.23	79.88	39.01	43.56
C	65.24	80.57	95.34	126.45	61.75	68.96
D	56.73	70.06	82.90	109.95	53.70	59.96
E	84.39	104.22	123.32	163.56	79.88	89.20
F	30.14	37.22	44.05	58.42	28.53	31.86
G	55.04	67.97	80.43	106.67	52.09	58.17
H	36.67	45.29	53.59	71.08	34.71	38.76

Construction Design	Stress on various vertical bending moment (MPa)					
	Case 7	Case 8	Case 9	Case 10	Case 11	Case 12
A	161.44	208.36	115.67	152.88	274.08	395.32
B	56.67	73.13	40.60	53.66	96.20	138.76
C	89.71	115.77	64.27	84.95	152.29	219.65
D	78.00	100.67	55.89	73.86	132.42	191.00
E	116.04	149.75	83.14	109.88	196.99	284.13
F	41.44	53.49	29.69	39.24	70.36	101.48
G	75.67	97.66	54.22	71.66	128.47	185.30
H	50.42	65.08	36.13	47.75	85.60	123.47

Table 9. Stress for strength based on manufacturing variation and type of fatigue.

No.	Manufacturing	Fatigue	Strength (MPa)
1	Hand Lay-up ($R_5 = 0.9$)	$N < 10^2$ ($R_6 = 1.0$)	134.60
2		$N < 10^3$ ($R_6 = 0.8$)	107.68
3		$N < 10^6$ ($R_6 = 0.4$)	53.84
4	Prepreg ($R_5 = 1.0$)	$N < 10^2$ ($R_6 = 1.0$)	121.14
5		$N < 10^3$ ($R_6 = 0.8$)	96.91
6		$N < 10^6$ ($R_6 = 0.4$)	48.46
7	Hand Lay-up (Human Error). ($R_5 = 0.765$)	$N < 10^2$ ($R_6 = 1.0$)	114.41
8		$N < 10^3$ ($R_6 = 0.8$)	91.53
9		$N < 10^6$ ($R_6 = 0.4$)	45.76
10	Prepreg (Human Error). ($R_5 = 0.85$)	$N < 10^2$ ($R_6 = 1.0$)	102.97
11		$N < 10^3$ ($R_6 = 0.8$)	82.38
12		$N < 10^6$ ($R_6 = 0.4$)	41.19

The strength of the construction and the load acting on the structure which are presented in tables 8 and 9 can be described respectively in the form of a probability density function which is a normal distribution. Then the combination of both probability distribution of the strengths and the loads for all variations of the construction design is shown in the graph of the joint probability density function as described in figures 18-25. In assessing the strength of the fast boat construction design with CFRP material, the theory of structural reliability level 3 is adopted by solving equation (6), where the quality of the structure strength is determined based on the safety index. For the safety index calculation, the mean (μ_S) and the standard deviation (σ_S) of the construction strength are obtained using the graph distribution of probability density function about 86.74 and 30.81, respectively. Table 10 shows the calculation results of the safety index for all construction designs that are declared unfit for designs A,

C, D, and E because of the loads received by the construction exceeds the strength of the construction so that the safety index is negative. However, the B, F, G, and H construction designs produce a positive safety index so that these designs are considered reliable for applying in CFRP fast boat construction under conditions of wave heights up to 2 m and wave frequency ranges of 0.3 - 2 rad/sec. The good construction has a safety index of more than 0 [12, 14], which means designs B, F, and H are reliable construction designs. The strongest construction is shown in design F with safety index about 1.07.

Table 10. Safety indexes calculation.

no.	Parameter	Mean (μ_Z)	Standard Deviation (σ_Z)	$\mu_S - \mu_Z$	$\sqrt{\sigma_S^2 + \sigma_Z^2}$	β
1	Load-Construction Design A	183.72	79.80	-96.98	85.54	-1.13
2	Load-Construction Design B	64.48	28.01	22.25	41.64	0.54
3	Load-Construction Design C	102.08	44.34	-15.34	53.99	-0.28
4	Load-Construction Design D	88.76	38.55	-2.02	49.35	-0.04
5	Load-Construction Design E	132.04	57.35	-45.30	65.10	-0.70
6	Load-Construction Design F	47.16	20.48	39.58	37.00	1.07
7	Load-Construction Design G	86.11	37.40	0.63	48.46	0.01
8	Load-Construction Design H	57.38	24.92	29.36	39.63	0.74

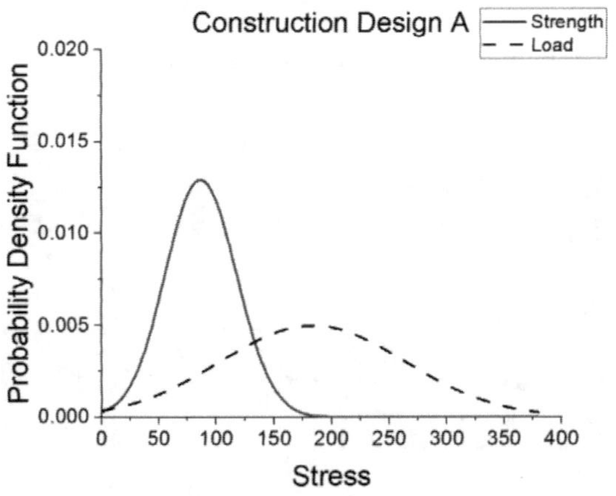

Figure 18. Joint probability density function of design A.

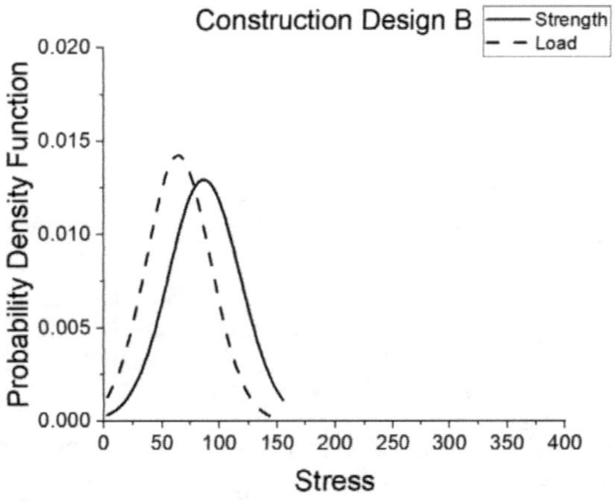

Figure 19. Joint probability density function of design B.

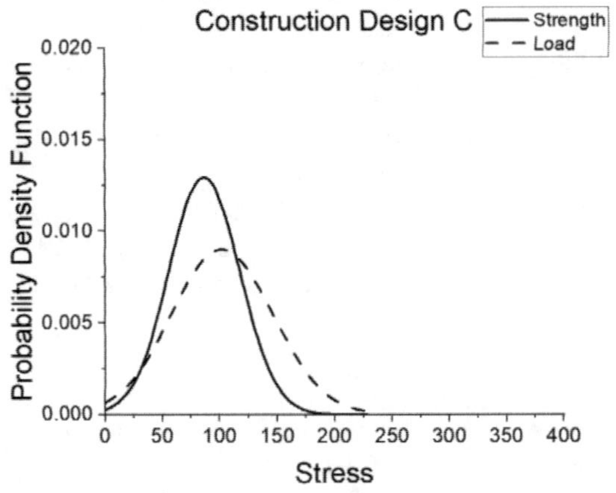

Figure 20. Joint probability density function of design C.

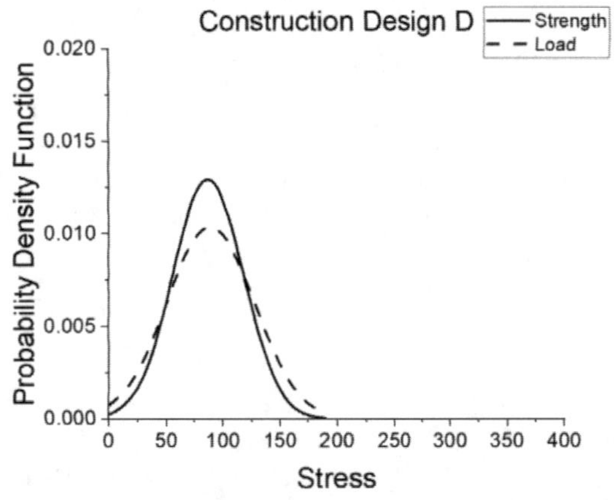

Figure 21. Joint probability density function of design D.

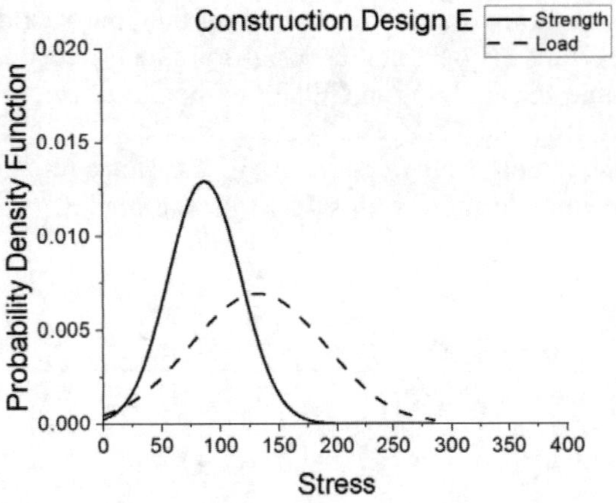

Figure 22. Joint probability density function of design E.

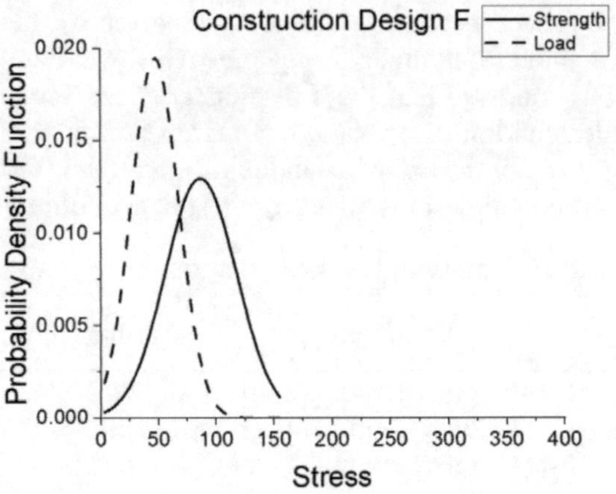

Figure 23. Joint probability density function of design F.

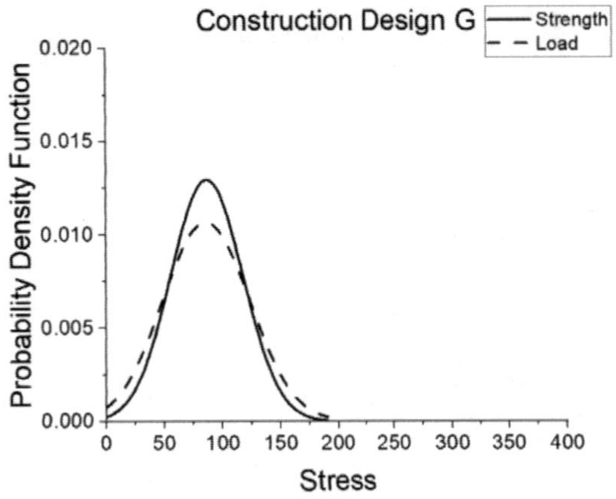

Figure 24. Joint probability density function of design G.

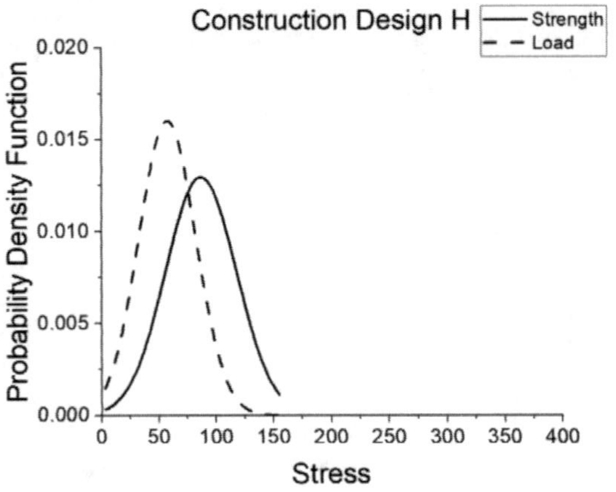

Figure 25. Joint probability density function of design H.

4. Conclusions

Eight variations of the construction design of the high-speed ship with CFRP materials are proposed to evaluate their reliability in receiving wave loads due to the vertical motion of ships in random waves. Diffraction theory is adopted to predict the vertical bending moment due to the random wave loading in 12 different conditions of heading angle, height, and period. Structural reliability analysis is applied to assess the safety and reliability of the ship structure. The probability density function of load stress and structural strength of the construction design are developed to obtain the safety index of all construction designs. The calculation results concluded that the construction designs of A, C, D and E are declared unfeasible, and the B, F, G, and H construction designs are reliable applying on the 12 m CFRP ship with varying degrees of strength. Finally, the construction design F is selected with a safety index of 1.07.

5. References

[1] American Bureau of Shipping (ABS) Head Office 2016 *Spectral-Based Fatigue Analysis for Vessel* (Houston: American Bureau of Shipping).

[2] Arianto P Y, Sulisetyono A, and Putranto T 2017 *Analysis of Stress Due to Wave Load on the Corvette Warship* (Surabaya: SENTA the International Conference on Marine Technology 2017) vol 1 pp 91-98.

[3] Badan Klasifikasi Indonesia Head Office 2021 *Rules for Small Vessels up to 24 m* Consolidated Edition 2021 (Jakarta: Badan Klasifikasi Indonesia).
[4] Ekşı S and Genel K 2017 *Comparison of Mechanical Properties of Unidirectional and Woven Carbon, Glass and Aramid Fiber Reinforced Epoxy Composites* (Antalya: ICCESEN International Conference on Computational and Experimental Science and Engineering) vol. 3 pp 879–882
[5] Faltinsen O M 2005 *Hydrodynamics of High-Speed Marine Vehicles* (Cambridge: Cambridge University Press).
[6] Gibson R F 1994 *Principles of Composite Material Mechanics* (Michigan: McGraw-Hill, Inc).
[7] Gargano A, Das R, and Mouritz AP 2019 *Finite Element Modelling of The Explosive Blast Response of Carbon Fibre-Polymer Laminates* Compos. B: Eng. **177**.
[8] Gargano A, Pingkarawat K, Blacklock M, Pickerd V, and Mouritz A P 2017 *Comparative Assessment of The Explosive Blast Performance of Carbon and Glass Fibre-Polymer Composites Used in Naval Ship Structures* Compos. Struct. **171** pp 306-316.
[9] Jialong Jiao, Songxing Huang, Shan Wang, and Soares C. Guedes 2021 *A CFD–FEA Two-Way Coupling Method for Predicting Ship Wave Loads and Hydroelastic Responses* Appl. Ocean Res. **117**.
[10] Jiayi He, Huiyu Wu, Ren-Chuan Zhu, Chen-Jun Yang, and Francis Noblesse 2020 *Practical Flow-Representations for Arbitrary Singularity-Distributions in Ship and Offshore Hydrodynamics, With Applications to Steady Ship Waves and Wave Diffraction-Radiation by Offshore Structures* Eur. J. Mech. B Fluids **83** pp 24-41.
[11] Jiayi He, Huiyu Wu, Ren-Chuan Zhu, Chen-Jun Yang, and Francis Noblesse 2021 *Practical evaluation of flows due to arbitrary singularity distributions in the 3D theory of ship motions in regular waves at* $0{,}3\leq\tau$ Eur. J. Mech. B Fluids **85** pp 8-20.
[12] Mansour A E 1990 *An Introduction to Structural Reliability Theory* (Washington: Ship Structure Committee).
[13] Maritime Safety Committee 2000 *The International Code of Safety for High-Speed Craft, 2000.* The Maritime and Coastguard Agency.
[14] Motok M, Nikola M, and Stefan R Reliability based structural design of river–sea tankers: Still water loading effects Mar. Struct. **83**.
[15] Mukhammad A F H 2014 *Mechanical Feasibility Study of Hemp Fiber-Polyester Randomized Composites as SNI Standard Helmet Materials* (Semarang: TRAKSI Mechanical Engineering Scientific Magazine) vol. 14(2) 11.
[16] Runbo Zhang, Zihan Li, Qiuyu Sun, Guocai Yu, Xintao Wang, and Linzhi Wu 2022 *Design and characterization of the carbon fiber tube reinforced polymer composite for full ocean depth submersibles* Compos. Sci. Technol. **217**.
[17] Santosa B 2013 *Ship Strength handbook* (Surabaya: ITS-Department of Naval Architecture and Shipbuilding Engineering).
[18] Sulisetyono A 2012 *Seakeeping analysis of the trimaran ferry ship in short crested sea for a case of east java water condition* (Ambon: RINA, International Conference on Ship and Offshore Technology (ICSOT) Indonesia).
[19] Sulisetyono A and Kadhafi M 2017 *Fatigue Life Prediction for Warship Operation in Indonesian Water* Appl. Mech. Mater. **874** pp 140-146.
[20] Sulisetyono A and Putranto T 2016 *Wave Load Analysis of the Corvette Ship in the Sea Water Indonesia* Appl. Mech. Mater. **862** pp 291-295.
[21] World Meteorological Organization 2011 *Manual on codes.* (Geneva: WMO Publications).
[22] Zhang K, Gu Y, li M, and Zhang Z 2014 *Effect of Rapid Curing Process on The Properties of Carbon Fiber/Epoxy Composite Fabricated Using Vacuum Assisted Resin Infusion Molding* Mater. Des. **54** pp 624–631.

IOP Conf. Series: Earth and Environmental Science 1081 (2022) 012042 doi:10.1088/1755-1315/1081/1/012042

Robust Laboratory Scale Seakeeping Test Wave Measurement Method Use Ultrasonic Sensor

Hardi Zen[1,2], Indra Ranu Kusuma[1] and Endang Widjiati[2]

[1]Department of Marine Engineering, Faculty of Marine Technology, Sepuluh Nopember Institute of Technology (ITS)
[2]National Research and Innovation Agency (BRIN), Indonesia

Abstract. The seakeeping test is one of the tests done in the Indonesia Hydrodynamic Laboratory (IHL) test tank on a model ship or floating unit. In the test tank, a wave can be generated by using electric and mechanic wave flaps that are pre-programmed on a certain condition of the wave, i.e. different frequencies and amplitude. To measure waves, produced by a wave generator, a wave measuring instrument is needed that can measure changes in water level in the test tank. The current wave measuring instrument is in the form of a 600 mm twin wire wave probe. This instrument is can be used to measure waves in the test tank but has limitations because it is only placed in one location without being moved. Another method is developed to be used in making alternative measuring instruments using ultrasonic sensors. In this paper, this alternative wave measuring instrument shows that measuring waves even when moving the measurement point to a new point location or following a towing carriage can be done and gives better measurement results. Based on these results, this new measuring instrument can be used in a more complex configuration test in the tank.

1. Introduction

The ship's seakeeping test carried out by the test tank laboratory uses waves that are scaled to follow the scale of the ship model but have some weaknesses from the generated waves, namely the waves are very directed and there is no influence from natural conditions or the surrounding climate [1]. Seakeeping measurements are realistically carried out at sea with truly realistic conditions due to climatic effects, but the weakness of practical seakeeping tests is that it costs a lot if the real conditions of the ship are not met, such as speed and reliability in receiving ocean waves [2].

According to the International Maritime Organization (IMO) regulations, the performance or reliability of a ship is affected by the conditions under which wind and waves occur [3]. Various studies on random waves have emerged several deterministic methods that can predict wave focusing numerically and experimentally [4]. In general, random waves are special events that occur in the path of a wave. As long as the theory is used, ocean conditions are considered independent components with specific direction, amplitude, frequency, and phase [5]. a method is needed to measure air altitude in real-time and continuously, namely ultrasonic. Ultrasonic measuring instruments have many advantages in measuring waves when conducting seakeeping tests in the Maneuvering and Ocean Engineering Basin test pool, such as being easy to move to any position above the water surface and ultrasonic measuring instruments being low cost and easy to obtain.

2. Literature Review

In this section, the measurement components of the seakeeping test will be described.

2.1. Water Level Based on Reflection on Light

The use of portable sensors can be easily installed in most types of ships mounted on the bottom of the bow. The first sensor can measure a surface distance of 0.2 -8 m with a deviation of 5cm, while the second sensor has a more extended range [6]. Meanwhile, the use of wavelength division multiplexing (WDM) measures is based on the reflection of the light source through an optical fiber tip. (OFT) as in picture 1 [7]

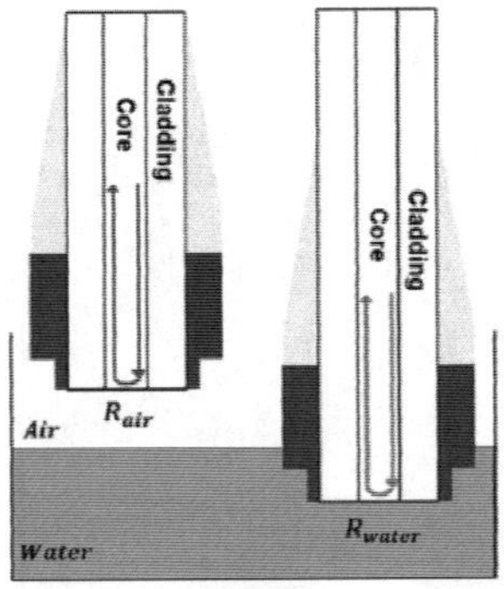

(a)

Figure 1 (a). Schematic diagram of the OFT

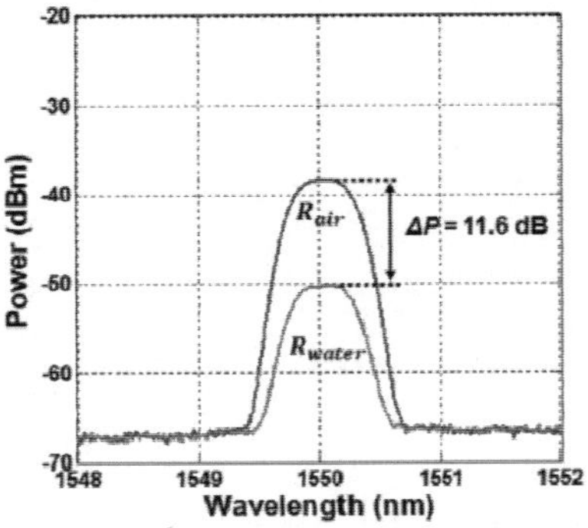

(b)

Figure 1 (b). The measured spectrum is according to the refractive index of the material (R_{Air}: reflection in air, and R_{water}: reflection in the water).

Using the arm processor microcontroller through the sensor also functions as a water characteristic analog signal changer [8].

2.2. Water Level Based on Pressure

The sensor-based silica graphene diaphragm Fiber Bragg Grating (FBG) was developed to measure water level and pressure using a uniform grating BG sensor made of SMF-28 fiber with a central wavelength of 1550 nm, bandwidth 0.5 nm, length 4 mm, and reflectivity. 90%. The sensor's response to the water level is monitored in terms of the Bragg wavelength shift, which is proportional to the strain. The diaphragm is deformed when inserted into the water column under hydrostatic pressure, as shown in Figure 2 [9].

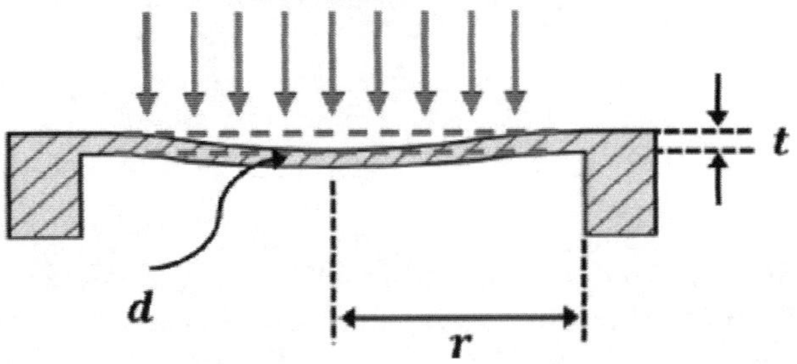

Figure 2. Deformation of the diaphragm due to pressure

2.3. Water Level Based on Twin Probe

Wave probes operate by measuring the current that flows between two stainless steel wires that are immersed in water. This current is converted to an output voltage that is directly proportional to the immersed depth [10]. Each wave probe comprises two parallel stainless steel rods with a plastic head and foot. The head is fixed to the calibration stem and a mounting block is supplied that allows the calibration stem to be fixed to any vertical surface.

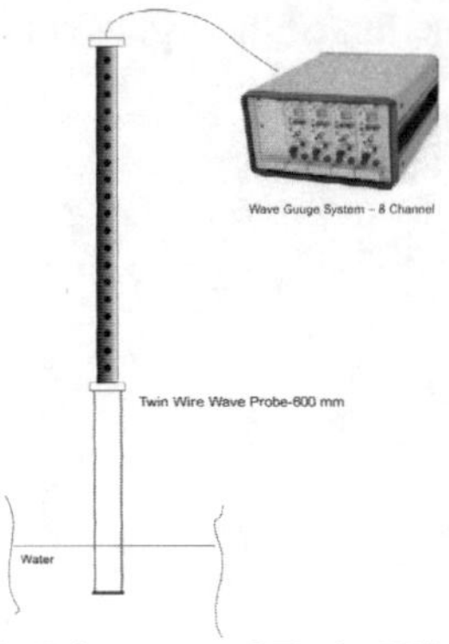

Figure 3. Diagram of Twin Wire Probe

2.4. Water Level Base Ultrasonic

The use of advanced technologies such as biomedical applications and various nondestructive inspection modalities in recent decades has enormous potential in ultrasonic sensors. The applications of waves in robotics, entertainment, security, and industrial uses have been highlighted by ultrasonic because of the tremendous benefits of this technology. Design and build a sea wave height measuring instrument with Arduino Uno ATmega 328 to function as a simple microcontroller-based digital tool that can measure ocean wave height in real-time with offline or online data storage systems, and data measurements can be used as a basis for planning the construction of ports, jetties, cages beach floating nets, floating buildings, fishing and shipping activities. Overall, the measuring instrument has been able to operate well and store data offline. The data read is unstable by moving water surfaces or waves, which are always different in each experiment.

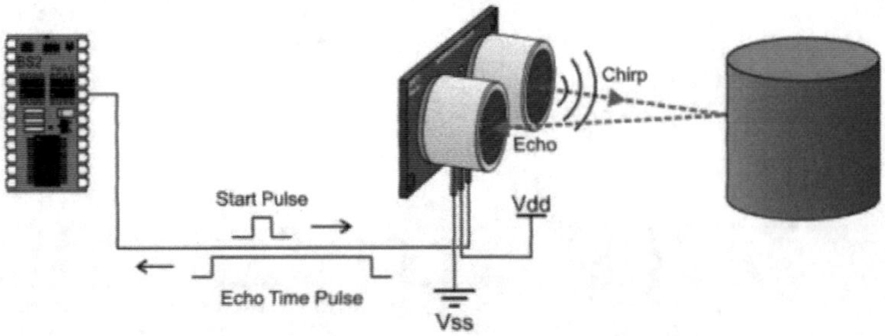

Figure 4. How Ultrasonic Sensors Work

Ultrasonic sensors can convert sound into electrical quantities and vice versa. The way this sensor works is in a method somewhere in the sound wave that can see the existence or distance of an object with a specific frequency. It is called an ultrasonic sensor because it uses sound waves to detect an object. This sensor consists of an ultrasonic transmitter and receiver.

3. Methodology

Steps taken in this measurement method are setting up the hardware components, installing the necessary software, calibration process, and testing the previous steps by doing the measurement. Every step will be described in the following subsections.

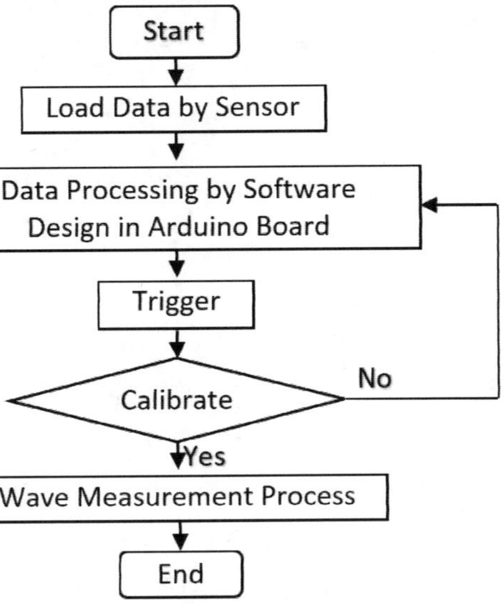

Figure 5. Wave Measurement Process

3.1. Hardware Design

The hardware design of the water wave height measuring device uses several components, such as the JSN-SR04T ultrasonic sensor and the Arduino Uno board, as shown in Figure 5. The ultrasonic sensor is connected to the Arduino board via the trigger and then played to the PC via a USB connection.

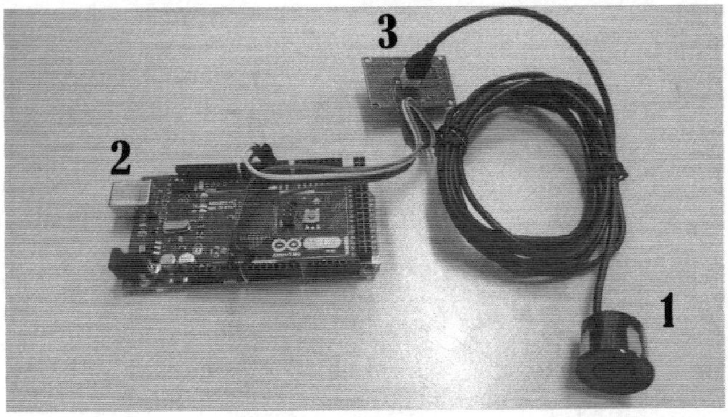

1. JSN-SR04T ultrasonic sensor
2. Arduino Board
3. HC-SR04 trigger,

Figure 6. JSN-SR04T Waterproof Ultrasonic Sensor

3.2. Software Design

At this stage, the introduction of the sensor device is done by giving commands to the Arduino board through the Arduino IDE 1.8.16 software. The controls on the Arduino board include many parameters, including device recognition, sensor reading duration, data rate, and output. The use of open-source python can provide the final results of the measurements needed.

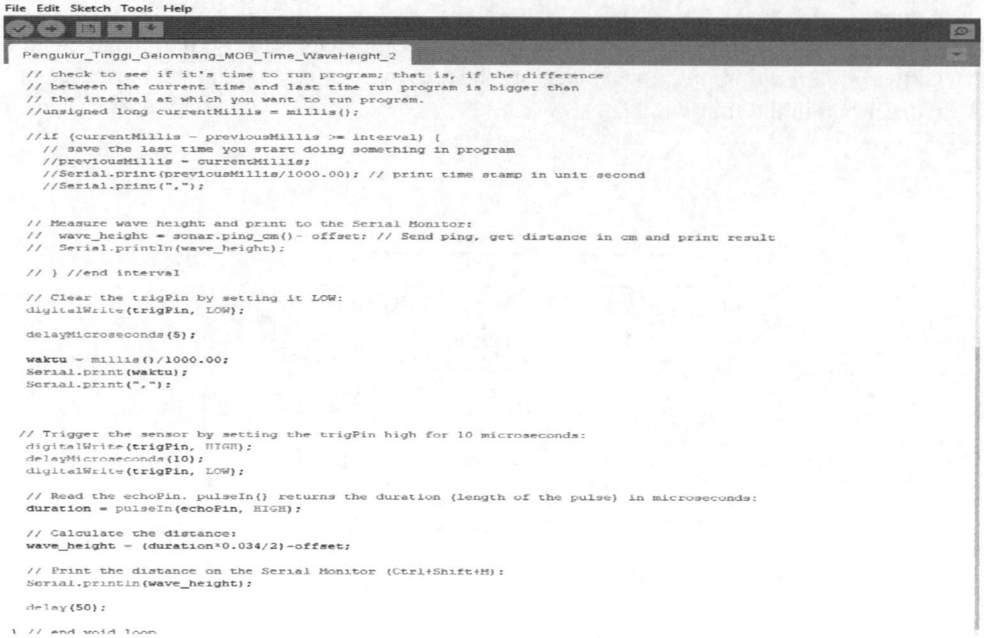

Figure 7. Arduino Program

to measure the distance can be made the following formula ;

wave_height = (duration*0.034/2) - offset;

where :

wave_height = wave height (mm)

offset = sensor distance to the water surface/reflection area

duration*0.034/2 = the speed of sound

3.3. Calibration

In the calibration process, a flat object is used as the ultrasonic reflection medium attached to the arm of the marking table as a reference for changes in height. The ultrasonic sensor is placed on the arm of the 2nd marking table. Calibration is carried out starting from the minimum reading distance capability of the 200 mm ultrasonic sensor according to the sensor's specifications, namely the measurement area of 200 mm – 8000 mm.

Figure 8. Sensor Calibration Process on the Marking Table

The data read by the Arduino sensor is calibrated with the data displayed by the marking table in mm using a scatter graph in Microsoft Excel to then look for the difference in readings that occurs with the regression equation. The equations obtained will be reprogrammed in the Arduino software to match the marking table measurements.

3.4. Wave Measurement
At this stage, all measuring instruments are functioning perfectly and are ready to perform their functions. Wave measurements were carried out in the MOB test pool using regular waves generated by an electric wave generator with varying heights and wave periods. The ultrasonic measuring instrument is positioned in line and not far from the wave probe, which is permanently installed in the wave area. The reading results from the sensor are displayed in the form of tables and graphs.

Figure 9. Regular Wave Data Collection Process

4. Results and Discussions
Refer to the calibration process described in the previous section, the following result is obtained,

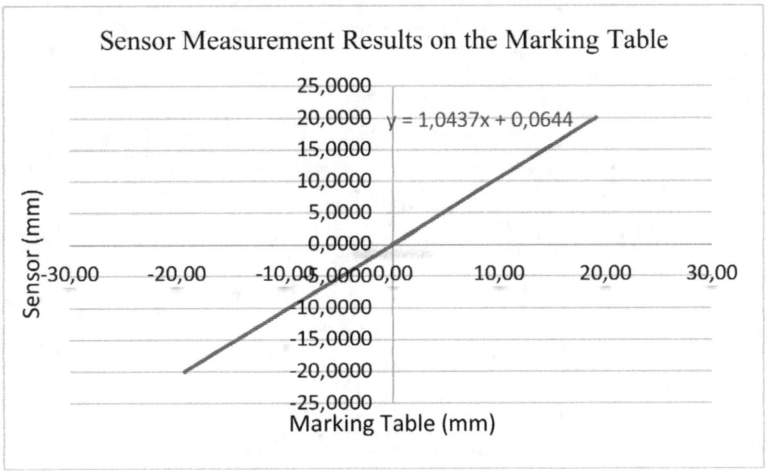

Figure 10. Calibration Result

The results of the sensor readings on the marking table show that there is a linear difference between the results shown in the marking table. From these data, a linear equation can be obtained Y=1.0437x + 0.0644.

From the formula obtained, we can enter it into the Arduino program so that the following measurement results follow the actual measurement results.

wave_height = (((duration*0.034/2)-offset)*1.0437)+0.0644;

IOP Conf. Series: Earth and Environmental Science 1081 (2022) 012042 doi:10.1088/1755-1315/1081/1/012042

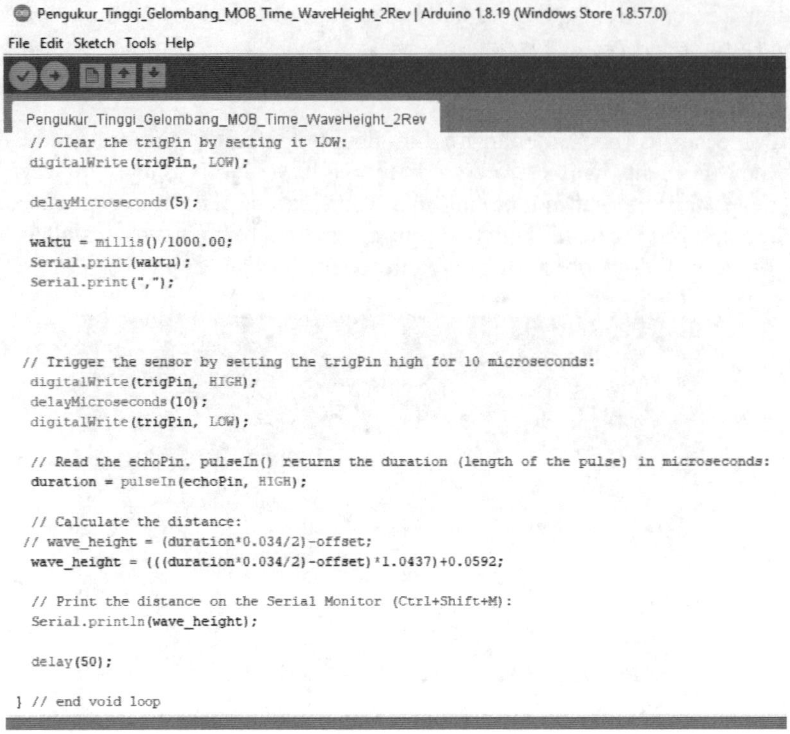


```
 // Clear the trigPin by setting it LOW:
 digitalWrite(trigPin, LOW);

 delayMicroseconds(5);

 waktu = millis()/1000.00;
 Serial.print(waktu);
 Serial.print(",");


// Trigger the sensor by setting the trigPin high for 10 microseconds:
 digitalWrite(trigPin, HIGH);
 delayMicroseconds(10);
 digitalWrite(trigPin, LOW);

 // Read the echoPin. pulseIn() returns the duration (length of the pulse) in microseconds:
 duration = pulseIn(echoPin, HIGH);

 // Calculate the distance:
// wave_height = (duration*0.034/2)-offset;
 wave_height = (((duration*0.034/2)-offset)*1.0437)+0.0592;

 // Print the distance on the Serial Monitor (Ctrl+Shift+M):
 Serial.println(wave_height);

 delay(50);

} // end void loop
```

Figure 11. Program After Calibration

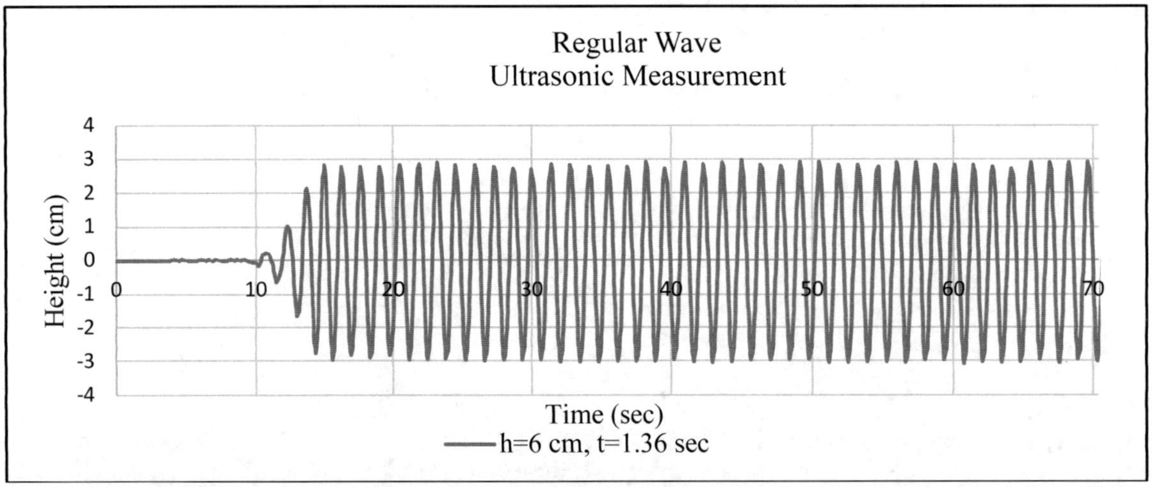

Figure 12. Regular Wave Measurement Use Ultrasonic, h=6 cm, t=1,36 sec

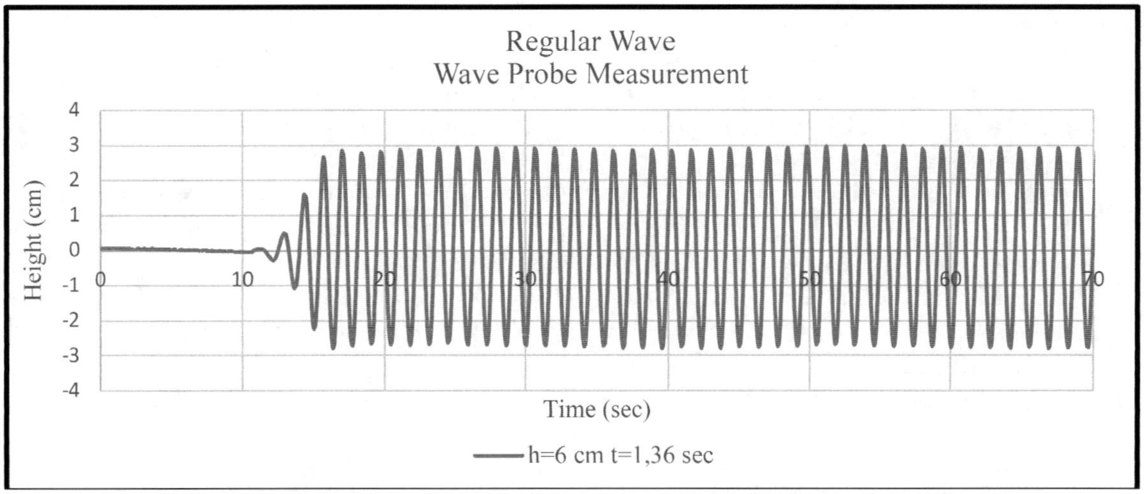

Figure 13. Regular Wave Measurement Use Wave Probe, h=6 cm t=1,36 sec

From the measurement results using an ultrasonic sensor, the wave height is 6.09 cm, and the standard deviation is 1.97, while for the wave probe measurement, the wave height is 5.92 cm, and the standard deviation is 1.89.

Based on experiences during the measurement process, and also comparing with the literature mentioned in previous sections, Table 1 below gives the advantages and disadvantages of applying different methods of measuring wave heights.

Table 1. Advantages and Disadvantages of Different Wave Measurement Methods

	Advantages	Disadvantages
Light	• cover longer distances • cover distance = 20 – 800 cm	• The circuit is too complicated • Requires high maintenance since it uses fiber • Cost around US$ 100
Pressure	• measure water level and temperature in real-time • cover distance = 0 – 100 cm	• Need pressurized media • Not suitable for locations that move frequently • Cost around US$ 226
Twin Wire Probe	• Resistant to corrosion • Cover distance = 0 – 60 cm	• Cost around US$10,000 • Spending time moving the location of the measuring point
Ultrasonic	• measure in different places • Cost around US$ 10 • cover distance = 20 – 800 cm	Measurement range is only in the area of 20 - 800 cm

5. Conclusions

From the results of the literature, the use of ultrasonic sensors in measuring water levels is very suitable for small measurement areas and more precisely for laboratory scales, in addition to being cheap, easy to obtain, and easy to assemble. They are not recommended for measurements with high sea level waves. For continuation, it is necessary to validate the ultrasonic sensor measurement data with a calibrated wave probe. There is a very small difference in the results of ultrasonic wave measurements with twin wire probes, only 0.17 cm for wave height

References

[1] Jiau, J. l., Ren, H. l., Sun, S. z., & Adenya, C. A 2016 *Investigation Of A Ship's Hydroelasticity And Seakeeping Performance By Means Of Large-Scale Segmented Self-Propelling Model Sea Trials*. *Journal of Zhejiang University-SCIENCE A (Applied Physics & Engineering)* Vol 17 (6) pp 468-484

[2] Jiao, J., Chen, C., Sun, S., Adenya, C. A., & Ren, H 2018 *Reproduction Of Ocean Waves For Large-Scale Model Seakeeping Measurement: The Case Of Coastal Waves In Puerto Rico & Virgin Islands And Gulf Of Marine (Ocean Engineering)* Vol 153 pp 71-87

[3] Lee, J.-H., Kim, Y., Kim, B.-S., & Gerhardt, F 2021 *Comparative Study On Analysis Methods For Added Resistance Of Four Ships In Head And Oblique Waves* (*head and oblique waves)* Vol 236 p 109552

[4] Lu, W., Li, J., Li, X., Tian, X., Wu, X., & Zhang, X. 2019 *Experimental investigation on the statistics of rogue waves under a random wave background (Ocean Engineering)* Vol 186 106075

[5] Clauss, G., & M.Klein 2011 *The New Year Wave in a seakeeping basin, Generation, propagation, kinematics and dynamics (Ocean Engineering)* Vol 38 pp 1624–1639

[6] Christensen, K. H., Röhrs, J., Ward, B., Fer, I., Broström, G., Saetra, Ø., & Breivik, Ø. 2013 *Surface Wave Measurements Using A Ship-Mounted Ultrasonic Altimeter (Methods in Oceanography)* Vol 6, pp 1-15

[7] Lee, H.-K., Choo, J., Shin, G., & Kim, S.-M. 2020 *On-Site Water Level Measurement Method Based On Wavelength Division Multiplexing For Harsh Environments In Nuclear Power Plants* (*Nuclear Engineering and Technology)* Vol 52 pp 2847-2851

[8] Krishnamoorthy, S., Karthikeyan, M., & Nishanth, B. 2021 *Design Of Water Level Regulator Using Arm Processor Microcontroller By Means Of Sensors* (*Materials Today: Proceedings)*

[9] Ameen, O. F., Younus, M. H., Aziz, M., Azmi, A. I., Ibrahim, R. R., & Ghoshalda, S. 2016 *Graphene Diaphragm Integrated FBG Sensors For Simultaneous Measurement Of Water Level And Temperature* (*Sensors and Actuators)* pp 225–232.

[10] Peter, L. 2015 *A Wire Probe for Measuring High Frequency Sea Waves* (*IOP Science)*

IOP Conf. Series: Earth and Environmental Science 1081 (2022) 012043 doi:10.1088/1755-1315/1081/1/012043

Analysis of a Lower Deck Rigging Configuration and Reusable Padeye Design based on Finite Element Model for Improving Stacking Operation Efficiency

Robbi Yahya Giri Kautsar[1], Rudi Walujo Prastianto[1], Mohammad Jazil Rifqy[2] and Roficho Filzah Farhana[2]

1) Department of Ocean Engineering, Faculty of Marine Technology, Institut Teknologi Sepuluh Nopember, Surabaya, Indonesia
2) Structural Engineer, PT Gunanusa Utama Fabricators, Cilegon, Indonesia

rudiwp@oe.its.ac.id

Abstract. In the topside fabrication process, the decks will be fabricated first simultaneously and then they will be arranged sequentially to form the topside through a stacking operation. The integrity of the deck structure must be maintained during and after the stacking process. The integrity of the deck structure and rigging system is strongly influenced by the design of the rigging system. Meanwhile, the design of the rigging system is limited to the equipment available to the company. So, it is necessary to analyze a proper rigging configuration with the existing limitations. In this case study, the rigging design and stacking analysis of the lower deck structure were carried out. From the analysis results, the 4 crane rigging configuration is used, with the highest crane utilization occurring on the crane is 89%. In the rigging system, the smallest safety factor on the sling is 1.17, and the shackle is 1.12, while on the deck structure the maximum UC is 0.32 and the maximum deflection that occurs is 2.17 cm. The reusable padeye UC is 0.53 in the numerical method analysis with a maximum von misses stress is 191 MPa which was concentrated at the lower end of the stiffener plate.

1. Introduction

In the future, oil and gas energy will continue to be produced, oil and gas will remain the world primary energy supply, around 44% in 2050 [1]. To cover this demand, oil and gas energy companies will try to increase their production by continuing to explore and exploit new oil and gas fields. Around 27% of oil and gas fields are in offshore areas so offshore facilities play an important role [2]. Technically, offshore facilities are divided into two types, fixed structures for shallow water areas and floating structures for deepwater areas. Before operating, offshore facilities were constructed at the fabrication yard by the fabrication company. The fabrication challenges facing the construction of such platforms may be reverted to the satisfaction of certain levels of personnel safety, fabrication quality and cost, facilities and equipment operability, and pre-planned load-out schedule [3].

One component of the fixed and floating offshore structure is the topside. Topside is a structure located above the water level to avoid waves action, with specific operational functions such as accommodation, drilling, or production facilities. The topside consists of several level decks with tubular columns. Then, the deck is framed of beam girders, either built up or Welded Plate Girder (WPG), and welded plates. Offshore structures are usually first fabricated in small units. After fabrication, these will be moved to an open area for assembly. Offshore contractors tend to do as much work as possible on the ground to minimize work in the air. This method is productivity driven. In fabrication, one can do a much better and faster job on the ground and in a weather protected workshop [4].

To fabricate a topside structure, the most common method is to sub-assembly each deck panel on ground level, and stack them one level by another.

A typical deck panel fabrication sequence is as below:

1. Lay the main beams on temporary support/stool,
2. Secondary beams join to the main beams,
3. Weld flat steel deck plate onto the beam frame and
4. Fit-up vertical column and braces.

Before all of deck are assembled, these may need to go through many stacking operations. Stacking operations are classified as heavy lift operations and milestone activities in the fabrication phase. The main problem in stacking operations is to maintain the integrity of the deck structure during stacking operations and to design the optimum rigging system. Meanwhile, the design of the rigging system is limited to the equipment available to the company. The integrity of the deck structure and rigging system is affected by the design of the rigging system. So the study of optimum rigging system design is necessary. If the deck is a truss deck, the obvious problem is that during fabrication, we do not have truss action in the deck, so the deflection of the deck may be very large such that final fit-up could pose major problem. For opened deck, the deflection will not pose a problem, but we have to make sure the deck leg work points do not shift during assembly [4].

Lifting equipment consists of crawler cranes, slings, shackles, and padeyes. The designed rigging system must meet safety standards without overdesign. Without lifting equipment, offshore platforms cannot be built effectively. Heavy lifts in major offshore projects are required to be conducted safely and cost-effectively. It is always a challenge for a structural engineer to produce an optimized design for both the lifted structure and lifting rigging system for use with the selected crawler crane that will lead to cost savings.

Commonly, padeye is designed for one times lifting operation. There are many lifting operations in a topside fabrication, the use of disposable padeye designs is inefficient and requires a lot of steel [5]. For fabrication efficiency, this study will analyze the design of a reusable padeye. Finite Element Analysis (FEA) is used in this study. Especially in a structural analysis, a finite element model can represent and analyse a detailed structural component with greater precision than conventional simplified hand calculations. This is because the actual shape, load and constraints, as well as material property can be specified with much greater accuracy than that used in hand calculations [6], [7], [8], [9],[10].

These constraints need to be resolved efficiently in order to arrive at a cost-effective solution. So in this study, the three objectives of the research are:

- Evaluate rigging configuration with the minimum number of cranes and selected rigging equipment to avoid overdesign
- Investigate lifting schemes which can provide cost-effective solutions and safe operations for stacking operation
- Design of reusable padeye by Finite Element Analysis (FEA) to improve stacking operation for one topside fabrication

2. Research Methods

In this section, the method is described from the early stage until the research was completed. We describe the stages using a flowchart. The research methods are described in Figure 1.The stacking operation analysis was carried out on the lower deck of topside wellhead platform. The topside weight is 733.29 tons, without equipment (piping, instrument, mechanical, and appurtenances). The program used in this research is SACS (Structural Analysis Computer System) which is based on offshore structural analysis, the phase of modeling and lifting analysis using static analysis [11]. This software has the ability to offer a wide array of analysis and design to its user and generally could solve the finite element analysis from a simple two-dimensional frame to complex three-dimensional frame [12].

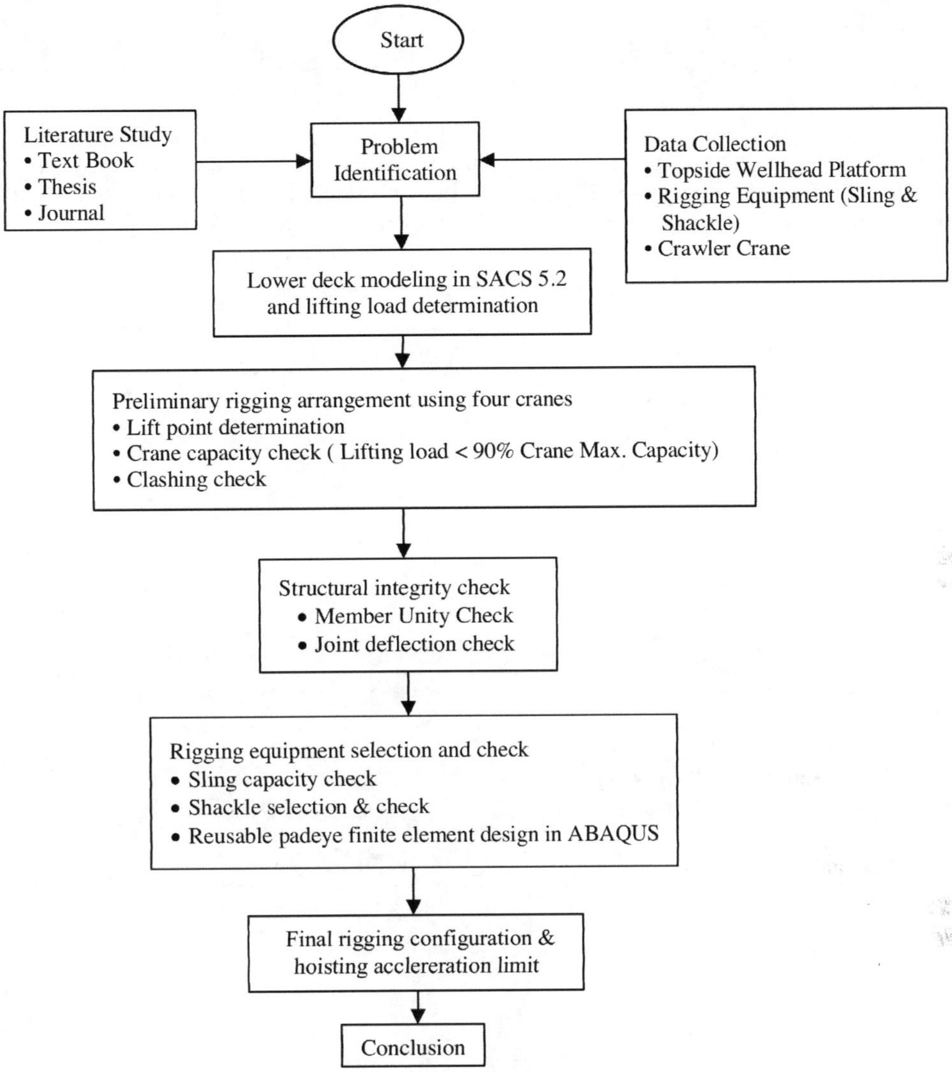

Figure 1. Research methods.

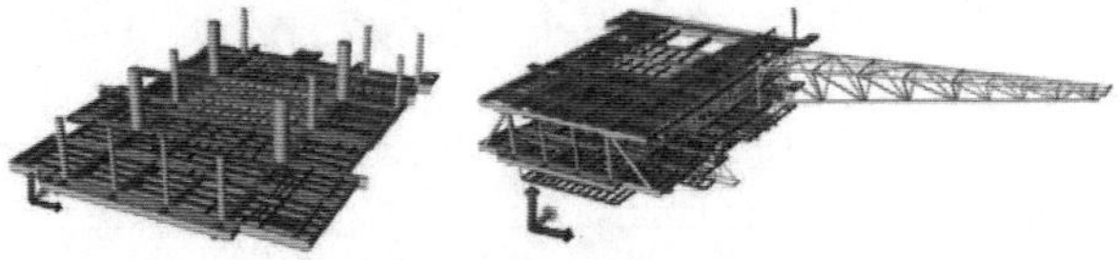

Figure 2. Lower Deck and Topside Model.

The topside is arranged of 4 deck levels, sequentially from bottom to top, namely the sump deck, lower deck, mezzanine deck, and upper deck. Then, to maximize the lifting load and optimize the crane capacity, the selected lower deck weight for stacking operation is 308.7 tons (with plates and tubulars) are shown in Figure 2 above. The weight and center of gravity of the lower deck structure are shown in Table 1 below as follows.

IOP Conf. Series: Earth and Environmental Science 1081 (2022) 012043 doi:10.1088/1755-1315/1081/1/012043

Table 1. Deck Properties.

Deck Level	Mass (ton)	Centre of Gravity (m)	
		X	Y
Lower	308.7	0.238	0.025

The following are lifting equipment selected and available at the company, such as crawler cranes, slings, and shackles. Safety analysis will be carried out on selected rigging equipment. Equipment data is shown in Table 2 until Table 4 and the shackle dimension is shown in Figure 3.

Table 2. Crawler Crane Capacity.

Type	Max. Capacity (ton)
#1 CKE-2500	250
#2 K-7250	250
#3 K-7250	250
#4 CCH-2800	280

Table 3. Sling Properties.

Properties	Quantity	Unit
Diameter	3.25	inch
MBL/Minimum Breaking Load	492	ton
SWL/Safety Working Load	88	ton
Mass	0.662	ton
Length	16	m

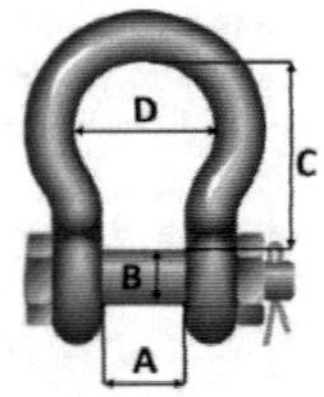

Figure 3. Shackle Dimension.

Table 4. Shackle Properties.

Properties	Quantity	Unit
SWL/Safety Working Load	85	ton
Mass	69.85	kg
A	127	mm
B	83.82	mm
C	330.2	mm
D	200.15	mm

3. Results and Discussion

3.1. Rigging Analysis

Rigging analysis is carried out to determine the magnitude of sling tension and the crane reaction. The magnitude of the crane reaction is determined by the lift point attachment on the deck. The distribution of deck loads on each crane must be evenly distributed to optimize crane capacity. The rigging system design is shown in Figure 4. In this study, the stacking operation was carried out at an elevation of 11.24 m. Stacking elevation is obtained by the height of the lower deck leg above the skid shoe.The crane boom is adjusted to the hook point, the lifting capacity of the crane will decrease as the crane boom length increases and the decreasing of lifting angle. Clashing checks are carried out to ensure that the edge of the deck does not clash with the crane boom. The results of the analysis are shown in Table 5. This analysis is still in the preliminary rigging stage, in the crane capacity analysis, the crane load analyzed does not include the weight of rigging equipment such as slings, shackles, and padeyes.

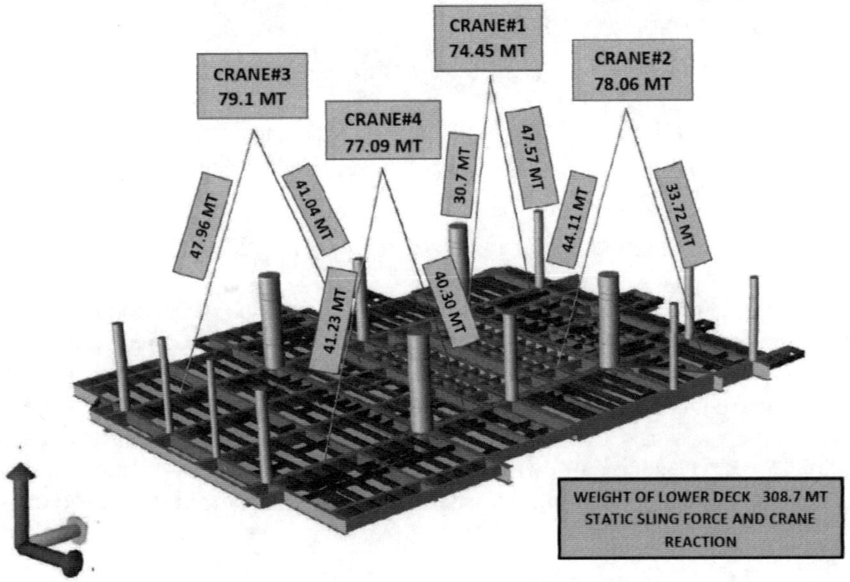

Figure 4. Static Rigging Analysis.

Table 5. Preliminary rigging analysis.

Crane Type	Load (ton)	Capacity (ton)	Utilization (%)	Boom (m)	Radius (m)	Clearence (m)
#1 CKE-2500	74.45	107.4	69	39.6	10	1.34
#2 K-7250	78.06	103.3	76	39.6	10	1.13
#3 K-7250	79.1	103.3	77	39.6	10	1.43
#4 CCH-2800	77.09	90.9	85	42	12	0.94

The deck structure integrity analysis is then carried out. This analysis includes the UC (Unity Check) value that occurs in the deck and rigging system. In lifting conditions, there will be a tension force on the slings due to the dead load of the deck. From the results of the analysis, it was found that the stress that occurs in the slings and decks is still below the allowable stress, with a maximum UC occurring in members 6013-6025 of 0.32 as shown in Figure 5.

IOP Conf. Series: Earth and Environmental Science 1081 (2022) 012043 doi:10.1088/1755-1315/1081/1/012043

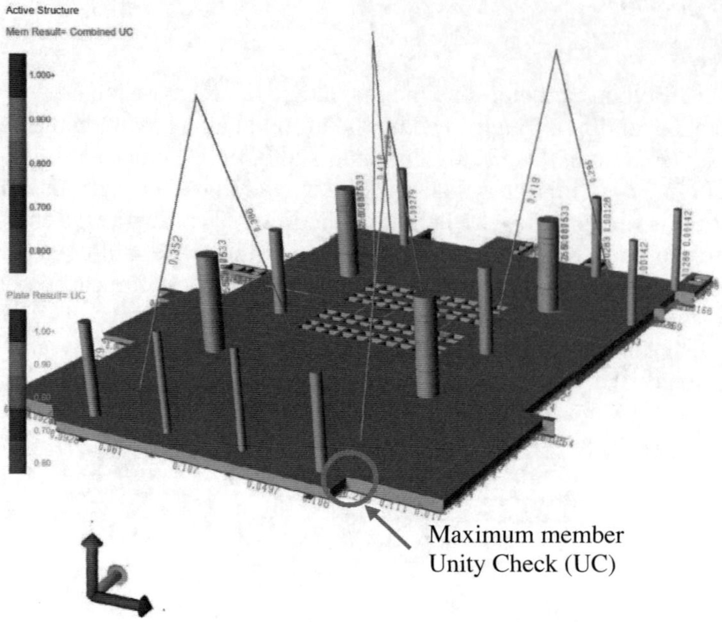

Figure 5. Rigging System Unity Check (UC).

The deck is a flexible type structure, so it is easy to deform. The deflection occurs due to the dead load of the structure due to the span created by the rigging configuration. There are two types of spans, back span and cantilever span (overhanging), to clarify the shape of the deflection, the magnitude of the deflection can be factored, as shown in Figures 6 and 7. To ensure that the deflection that occurs on the deck is still elastic, it is necessary to do a deflection check on the joint of interest for each type of span. Elastic deflections are also a source of difficulty in maintaining tolerances in the location of nodes [6]. From the analysis results, the deflection that occurs on the deck is still in the elastic phase as shown in Table 6.

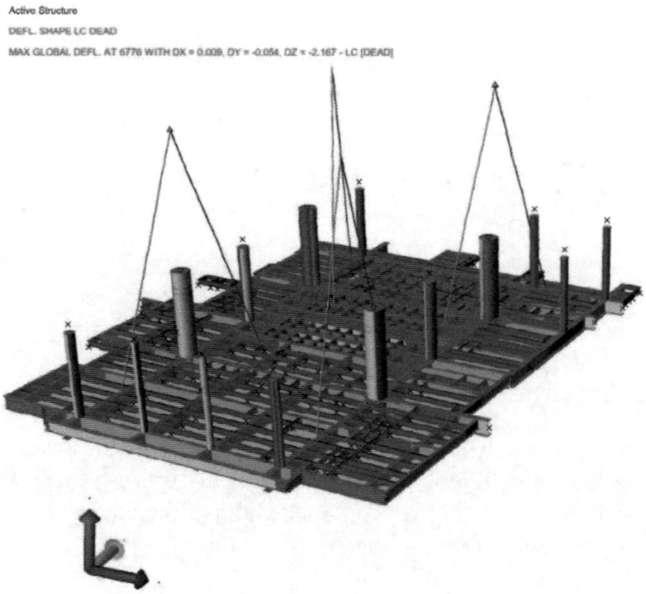

Figure 6. Deflection Shape Factor = 1.

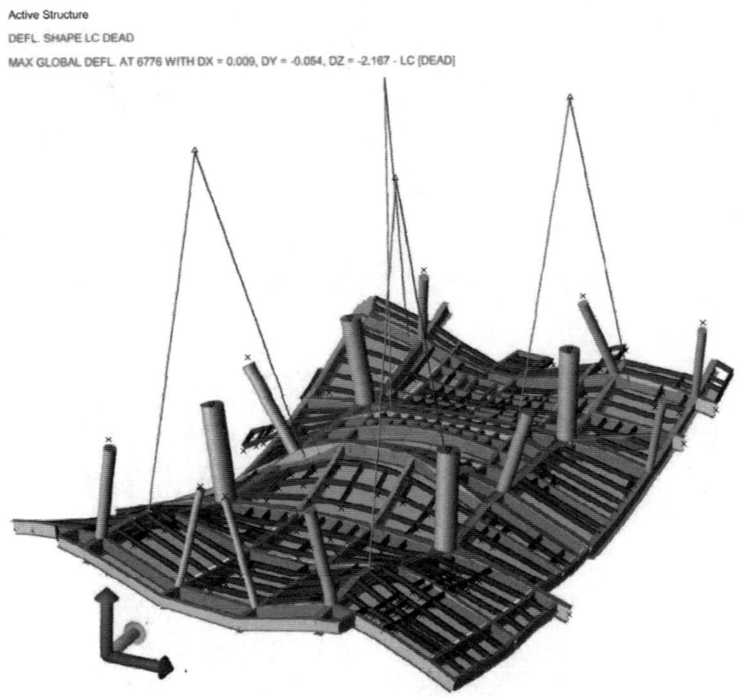

Figure 7. Deflection Shape Factor = 10.

Table 6. Deflection Check.

Joint ID	Span		Deflection Limit		Deflection Value (cm)	REMARKS
	Length (m)	Type	Limit	Limit Value (cm)		
6642	13.2	Back	L/360	3.67	1.09	PASS
6526	11.6	Back	L/360	3.22	0.85	PASS
6279	9.5	Back	L/360	5.28	0.88	PASS
6776	8.9	Cantilever	2L/360	4.94	2.17	PASS
6293	5.05	Cantilever	2L/360	2.8	0.85	PASS
6273	7.41	Cantilever	2L/360	4.12	1.28	PASS
6020	4.69	Cantilever	2L/360	2.6	0.95	PASS

After analyzing the integrity of the deck structure, the next step is to analyze the integrity of the rigging system. The safety factor of the rigging system must be below 2 to avoid overdesign and troublesome rigger work. The factored tension value on the sling is obtained by multiplying the contingency factor value is 1.03, DAF (Dynamic Amplification Factor) value is 1.15, and SKL (Skew Load Factor) value is 1.33 which refers to the DNVGL-ST-N001 standard for lifting operation to ensure the safety of the rigging system design due to dynamic loads and uncertainty of external factors [13].The selection of shackle dimensions is based on the maximum tension that occurs in the sling and the diameter of the shackle must be able to be entered by the sling. From the results of the analysis, the value of the safety factor has met the criteria as shown in Tables 7 and 8.

Table 7. Sling Check.

Type	Member Sling	Tension (ton)	Factored Tension (ton)	SWL (ton)	Safety Factor
CRANE #1	SL 1.1	47.57	74.94	88	1.17
	SL 1.2	30.70	48.37	88	1.82
CRANE #2	SL 2.1	33.72	53.12	88	1.66
	SL 2.2	44.11	69.49	88	1.27
CRANE #3	SL 3.1	41.04	64.65	88	1.36
	SL 3.2	47.96	75.56	88	1.16
CRANE #4	SL 4.1	40.30	63.49	88	1.39
	SL 4.2	41.23	64.95	88	1.35

Table 8. Shackle Check.

Type	Member Sling	Factored Tension (ton)	SWL (ton)	Safety Factor
CRANE #1	SL 1.1	74.94	85	1.13
	SL 1.2	48.37	85	1.76
CRANE #2	SL 2.1	53.12	85	1.6
	SL 2.2	69.49	85	1.22
CRANE #3	SL 3.1	64.65	85	1.31
	SL 3.2	75.56	85	1.12
CRANE #4	SL 4.1	63.49	85	1.34
	SL 4.2	64.95	85	1.31

3.2. Reusable Padeye Design

Padeyes are important lift components, which link module structure and shackles. In lift arrangement, a shackle locks up a padeye by inserting shackle pin through padeye hole, while the shackle bow connects to a sling [4].These areas are to be studied by a suitable finite element analysis for the adequacy of the structural design. Generally, the padeye is only composed of the main plate, to increase the strength of the padeye, additional plates such as cheek plates and stiffener plates can be added, as shown in the Figure 8. The check plate functions to increase the shear, axial, and bearing strength of the padeye, then the stiffener plate functions to increase the bending strength of the padeye in the in-plane and out-plane directions.

Designing padeye on each deck to be lifted, is very inefficient, will require a lot of plate material and time to fabricate. So it is necessary to design a padeye that is re-usable, that it can be reused on the condition that the padeye's strength capacity still meets the required design load. Once the padeye has been used, it can be cut and reused in the next stacking operation. In this topside fabrication case study, there are 8 stacking processes with the following details.

- Sump Deck (3 x Stacking)
- Lower Deck (1 x Stacking) – Highest Lifting Load (Reference Padeye Design)
- Mezzanine Deck (3 x Stacking)
- Upper Deck (1 x Stacking)

With 8 stacking processes, it means that padeye must be used 8 times in the topside fabrication process. The lower deck is the deck with the heaviest lifting load, so the tension that occurs in the padeye can be used as a design load reference in the padeye design. The initial height of the padeye to the pinhole is 280 mm, it is assumed that in the cutting process the padeye height is reduced by 10 mm so that after 8 times cutting process the padeye height becomes 210 mm.

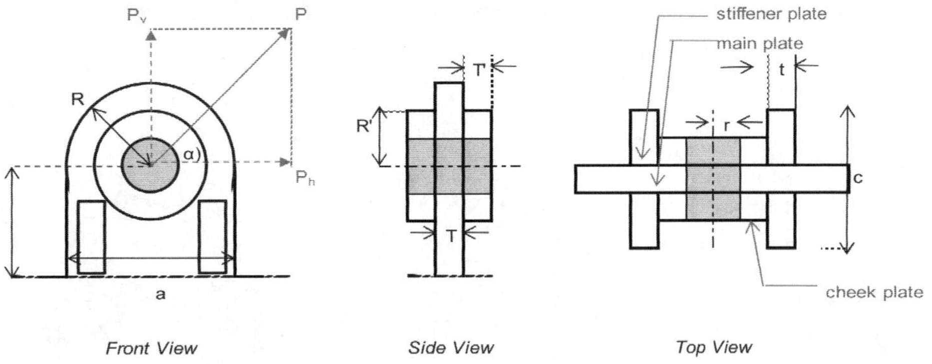

Figure 8. Reusable Padeye Dimension.

In this analysis, the dimensions of the padeye model are shown in Table 9. Determination of padeye hole diameter is based on the size of the pin shackle diameter, which must pass to avoid excessive contact stress. The specification of the material used is API 2H Grade 50 steel, with a yield stress of 360 MPa. The bottom of the padeye attached to the deck is constrained fixed. The load on the padeye modeled on the pinhole is 74.94 tons with a loading angle is 61.6 °. The load on the padeye is obtained from the highest factored tension that occurs in the shackle due to the tension force of the sling, then the loading angle is obtained from the critical angle of the sling in the rigging system. Finite element analysis was performed on the padeye at the first of the stacking operation up to 8 times, to ensure the strength of the padeye under the same loading conditions. Meshing analysis was carried out on this model, to ensure that the analysis results obtained were convergen [14]. From the modeling, results obtained weight of padeye is 111.85 kg.

Table 9. Reusable Padeye Dimension.

Main Plate		Cheek Plate		Stiffener Plate	
Properties	Value (mm)	Properties	Value (mm)	Properties	Value (mm)
a	380	R	160	c	200
b	280	T	25	t	25
R	190				
r	45				
T	50				

The output of this analysis is to obtain the maximum von misses stress that occurs in the padeye, to determine the level of local padeye strength. Von Misses stress is calculated with the following equation.

$$V_{misses} = \sqrt{(\sigma_x^2 - (\sigma_x \sigma_y) + \sigma_y^2 + (3t_{xy}{}^2))} \qquad (1)$$

where σ_x is the normal stress x component, σ_y is the normal stress y component and t_{xy} is the shear stress.

From the results of the analysis on the padeye at the first of stacking operation, the maximum von misses stress is 191MPa which is concentrated at the lower end of the stiffener, as shown in Figure 9. The distribution of high von misses stresses also occurs in the padeye hole area that is in contact with the shackle. The tension force on the deck due to the padeye at the lower end of the stiffener can make the padeye detach from the deck, therefore the thickness of the weld also needs to be taken into account. In practice, the padeyes are to be examined by NDT (Non-Destructive Test) [15]. The UC padeye at the first of the stacking operation is 0.53.

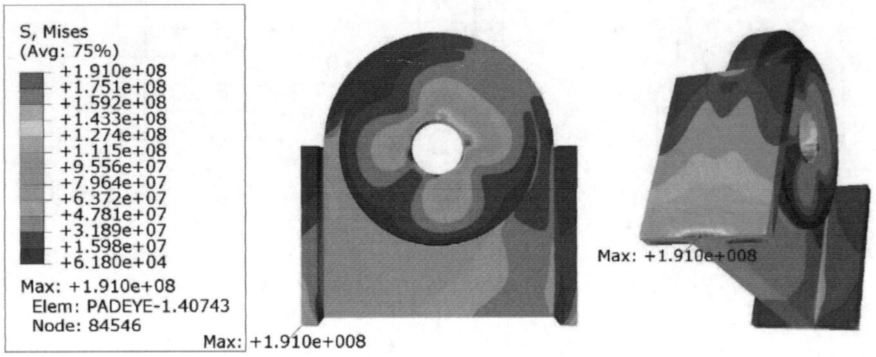

Figure 9. Reusable padeye at first stacking (height = 280 mm).

After 8 stacking operations, the padeye height decreased to 210 mm. With the same loading conditions as the padeye at the first of the stacking operation, maximum von misses stress of 188 MPa is obtained which is concentrated at the padeyehole, as shown in Figure 10. So, the obtained UC is 052. The results of this stress value do not show a significant difference with the padeye stress at the first of the stacking operation. From this analysis, the reusable padeye was designed safely for 8 stacking operations.

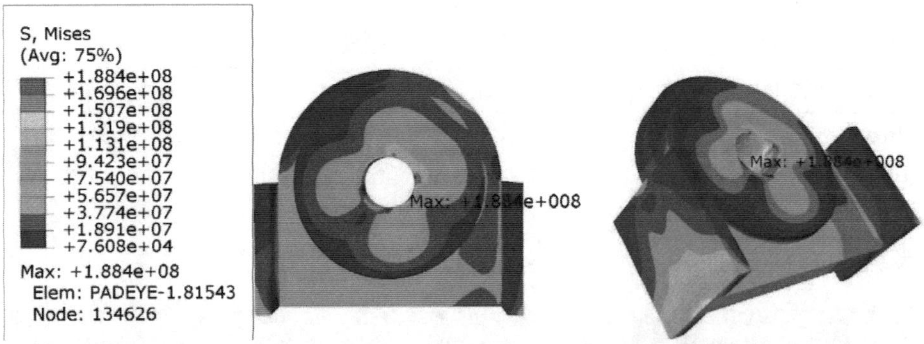

Figure 10. Reusable padeye after 8 x stacking (height = 210 mm).

3.3. Dynamic Analysis

After the weight of the rigging equipment is obtained, it can be added to the deck load to become the total load on each crane reaction for crane capacity analysis, as shown in Table 10. The maximum crane capacity must be below 90%, while the remaining 10% capacity is used as crane reserve capacity due to dynamic load that occurs on the deck during a stacking operation. The results showed that the crane capacity used was still adequate, with a critical crane capacity is 89% on the #4 CCH-2800 crane, as shown in Table 10 below.

Table 10. Final Crane Capacity.

Type	Load (ton)	Rigging (ton)	Total Load (ton)	Capacity (ton)	Utilization (%)	Failure Load Factor
#1 CKE-2500	74.45	3.49	77.94	107.4	73	1.378
#2 K-7250	78.06	3.49	81.55	103.3	79	1.267
#3 K-7250	79.1	3.49	82.59	103.3	80	1.251
#4 CCH-2800	77.09	3.49	80.58	90.9	89	1.128

Static analysis was carried out on the rigging system, to cover dynamic effects, the load on the sling was factored with DAF (Dynamic Amplification Factor). Crane reactions can increase due to deck movement, one of the most critical phases is lifting the deck from the temporary support/stool. The deck that was originally stationary will be hoisted until it reaches a certain lifting speed. Changes in speed will cause hoist acceleration which is a dynamic effect. The hoisting acceleration must be limited, to avoid crane collapsing. The hoisting acceleration limit is obtained by multiplying the reserve load factor on the most critical crane, which is multiplied by the acceleration due to gravity. From this analysis result, the hoisting acceleration limit is 1.25 m/s^2. During traveling, the road as the crane route must be flat and the traveling crane speed must be slow and uniform to avoid dynamic loads, the lower deck lifting plan is shown in Figure 11.

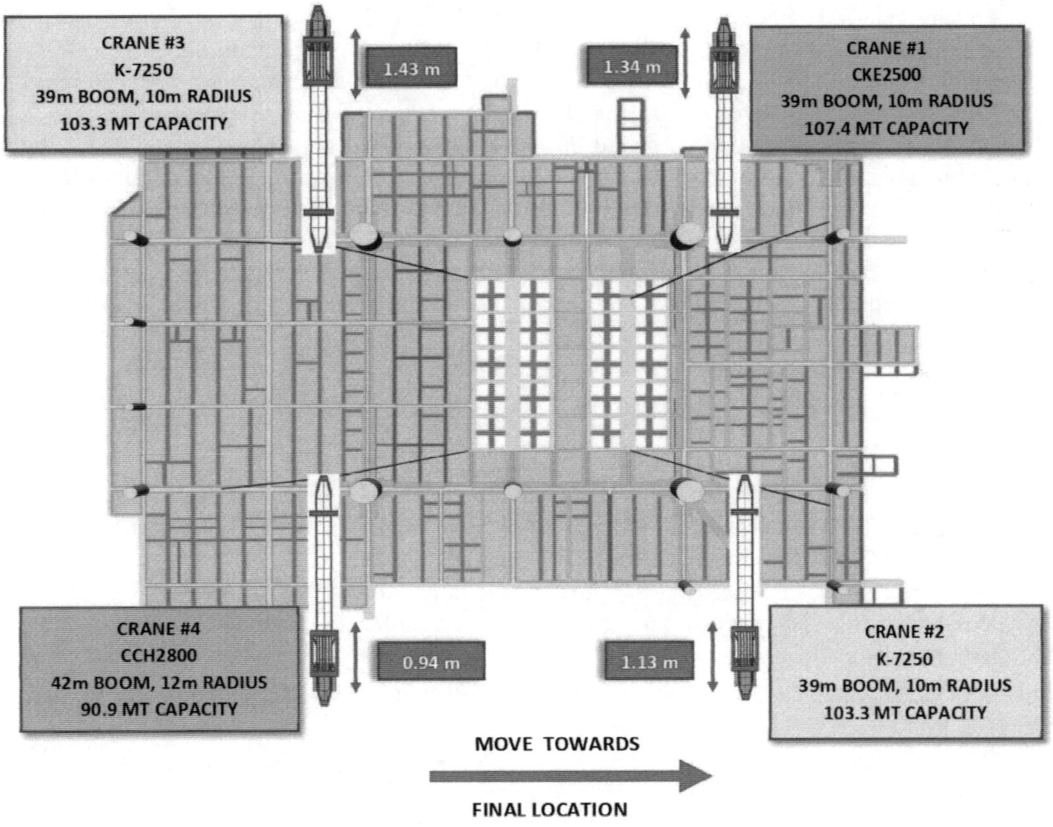

Figure 11. Lower deck stacking procedure (top view).

4. Conclusion

This study confirms that the design of the rigging system using four cranes has met the safety criteria with efficient methods. The critical crane capacity is 89% occurred in the CCH2800 crane. There are 2 slings on each crane, with a minimum sling safety factor of 1.17 and a shackle is 1.13. The reusable padeye has been designed for 8 stacking processes, with a maximum UC is 0.53. The integrity of the deck structure has met the criteria, with a maximum UC is 0.32 and a deflection is 2.17 cm. The crane hoisting acceleration limit is $1.25 m/s^2$ to avoid crane collapse. Further research can be carried out on optimization of padeye design and rigging system design based on system strength and cost.

References

[1] DNV GL, 2017 DNV GL Energy Transition Outlook – Oil and Gas.
[2] Prasad Singh S S Agarwal J R and Mani N, 2019 *Offshore Operations and Engineering* (Florida: CRC Press)
[3] Hafez K A and Ismael M M, 2013 Practical investigation of a monopod fabrication method and the numerical investigation of its up-righting process *Int. J. Nav. Archit. Ocean Eng.* **5**, 3 p. 431–453.
[4] Liang L, 2004 Heavy Lift Installation Study of Offshore Structures (Singapore: National University of Singapore)
[5] Liu Z C Zhou B and Tan S K, 2013 Finite element analysis and structure optimum design of lifting padeye *Adv. Mater. Res.* **658** p. 399–403.
[6] Reddy D V and Swamidas A S J, 2016 *Essentials of Offshore Structures* (Florida: CRC Press)
[7] Prastianto R W Hadiwidodo Y S and Widhestomo S W, 2020 Stress Distribution along the Weld Toes of Tubular KT and KDT Joints under Balance Axial Loads and In-Plane-Bending Moments in *Proceedings ofthe 6th International Seminar on Ocean and Coastal Engineering, Environmental and Natural Disaster Management (ISOCEEN 2018)* p. 170–175.
[8] Prastianto R W Hadiwidodo Y S and Fuadi I F, 2018 Stress concentration factor distribution of inclined brace in multiplanar offshore tubular double kt joints Stress concentration factor distribution of inclined brace in multiplanar offshore tubular double kt joints *IOP Conf. Ser. Earth Environ. Sci.* **162**.
[9] Prastianto R W Hadiwidodo Y S and Fuadi I F, 2018 Stress Concentration Factor of A Two-Planar Double KT Tubular Joint due to In-Plane Bending Loading in Steel Offshore Structures in *ISOCEEN 2017* **177**.
[10] Prastianto R W, 2020 Mooring Analysis of SPAR Type Floating Offshore Wind Turbine in Operation Condition due to Heave , Roll , and Pitch Motions *IOP Conf. Ser. Earth Environ. Sci.* **618**.
[11] Bentley, 2018 *SACS Manual* .
[12] Hadiwidodo Y S and Prastianto R W, 2018 STRUCTURAL CORROSION MODELING FOR RESERVE STRENGTH ASSESSMENT OF FIXED JACKET PLATFORM *Int. J. Civ. Eng. Technol.* **9**, 8 p. 1006–1012.
[13] DNV GL, 2016 *DNVGL-ST-N001 Marine operation and marine warranty* .
[14] Dassault, 2011 *Abaqus/CAE User's Manual* .
[15] Roy K D Dev A K and Aksu S, 2018 Installation Engineering of Topside Modules on Ship Shaped Offshore Floating Structures in *Proceedings of 4th PAAMES and AMEC2010*.

Development of Mini SART-DMOM in Finding Location of Fishing Vessel

Trika Pitana*[1], Dwi Priyanta[1], Salman Nur Wido Susilo[1],Dian Prasetyawati[2]

[1]Departement of Marine Engineering, Institut Teknologi Sepuluh Nopember Surabaya, Indonesia
[2]Naval Architecture, University of Muhammadiyah Surabaya

*trika@its.ac.id

Abstract. The high number of fishing vessels accidents in the Java Sea is due to the large number of fishing vessels that do not have safety support equipment in sailing, such as SART (Search and Rescue Transponder). So fatalities occur because it is difficult to find the location of fishers when an accident occurs. Also, the high cost of SART equipment on the market makes fishers unable to afford it. Research related to Mini SART-DMOM (Search and Rescue Transponder-DMOM) is an innovation tool with a working system almost the same as SART on the market, and this equipment has a lower price compared to SART sold on the market. In assembling the prototype, the main components used are GPS Neo 6m, Arduino Uno, LoRa and Wemos (ESP 8266). The prototype assembly is divided into Mini SART-DMOM and Mini SARR-DMOM (Search and Rescue Receiver-DMOM). In the prototype test that was carried out, it ran smoothly where the Mini SART-DMOM could get location via GPS. Lora sent the location data from the LoRa Transponder to the LoRa Receiver. The data received by the Receiver was sent by the wifi Board (Wemos ESP 8266) to be monitored live remotely via Web Blynk. However, in this trial, the maximum recorded range was required to be improved.

Keywords: Fishing vessel accidents, SART, LoRa, Transponder, Receiver.

1. Introduction

Water transportation in Indonesia has quite severe risks. It is revealed that most ships used in Indonesia are canoes or boats with outboard motors [1]. There is the fact that Indonesia is indeed an archipelago country with an area mainly made up of oceans. One-third of Indonesia's territory is land area, and two-thirds of Indonesia's region is the ocean. Therefore, the sea plays a vital role in the life of the Indonesian people. Sea transportation plays a crucial part for Indonesia as an archipelagic country. This aims to realize inter-island connectivity in Indonesia in various fields such as tourism, trade, and others. The increasing importance of sea transportation in Indonesia is also balanced by the growing number of accidents in Indonesia's marine areas. From the data obtained from the results of ship accident investigations that have been carried out by the National Transportation Safety Committee (NTSC) in the period 2018 to 2020, it is stated that many ship accidents occur in Indonesian waters with a total of 1152 cases of ship accidents. After analyzing all of these cases, NTSC noted that the most ship accidents happened in the Java Sea waters as many as 219 ship accidents [1]. The type of fishing vessel almost dominates the ship accident in the waters of the Java Sea. A total of 104 cases of fishing vessel accidents occurred in the Java Sea [1] [2].

Many ship accidents occurred in the waters of the Java Sea and are mainly dominated by fishing vessels, indicating that improvements are needed in terms of the safety of marine transportation. Many types of accidents occur on fishing vessels sailing in the Java Sea. In addition, there are many types of causes of ship accident fishing vessel accidents that occur in the Java Sea are dominated by the fact that the fishing vessel does not install navigation equipment due to the economic reason. Many fishers cannot afford to purchase a navigational equipment and other maritime support equipment. Communication equipment is crucial to communicate with each other, especially in the shipping. When the fishing vessels are lost, many of them are found 1 x 24 hours or even more due to the absence of incoming information regarding the difficulty of finding the location of the fishing vessel due to the absence of a complete navigation system [3] [4] [5].

There is a communication tool called Automatic Identification System (AIS), a navigation tool that can send and receive ship data information to either other ships or VTS (Vessel Traffic Service) stations [3] [5] [6] [7] [8]. There is a ship safety support tool under the auspices of the AIS standard, namely AIS-SART (Search and Rescue Transponder). This tool only serves to transmit radio signals used to locate the ship when an accident occurs [5]. This device cannot be used as a data receiver. AIS-SART uses the AIS Class A standard. The AIS Class A standard equipment has a high standard compared to the AIS Class B standard. Therefore, the AIS-SART equipment sold in the market today is quite expensive. However, fishing vessels use AIS equipment but must have a minimum (Gross Tonnage) size of GT 60. Currently, many fishing vessels have a (Gross Tonnage) GT below 60 [3]. Therefore, that ships cannot use equipment such as AIS Class B, and many fishers cannot afford the AIS-SART (Search and Rescue Transponder) device as it exists today due to the high cost of the equipment. Therefore, Mini SART (Search And Rescue Transponder)-DMOM is more economical than AIS-SART on the market. Because this Mini SART uses the LoRa (Long Range) system as well as an efficient solution to solve the problem [7] [9] [10].

2. Methodology

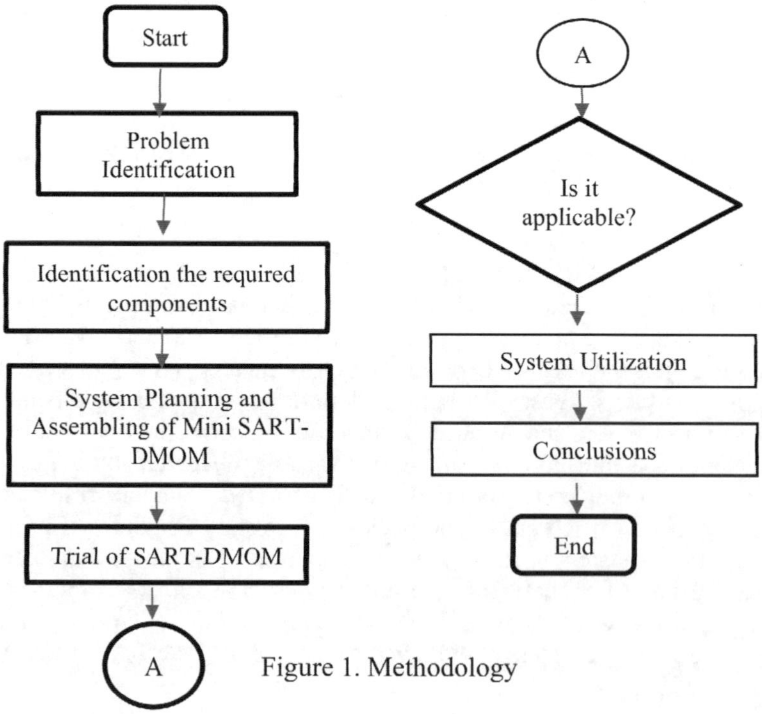

Figure 1. Methodology

IOP Conf. Series: Earth and Environmental Science 1081 (2022) 012044 doi:10.1088/1755-1315/1081/1/012044

Figure 1 is a research flow chart or stage diagram in the process of developing and designing a Mini SART (Search And Rescue Transponder)-DMOM. The methodology started from problem identification, identification of required components, system planning and assembling of mini SART-DMOM, trial component and finally summary of research.

2.1. Problem Identification
There are two main problems during conducting a research: to develop Mini SART- DMOM dan Mini SARR-DMOM and to check the operational of those equipment in finding the location of fishing vessel.

2.2. Identification of the required components
There are two prototypes to be used. Those are SART-DMOM and Mini SARR-DMOM. The components used in the Mini SART-DMOM are GPS, Arduino, LoRa, Emergency Button, LCD, Antenna, and Power Supply. However, the components for SARR-DMOM are Arduino, Wifi Board, LoRa, Power supply, and Antenna.

2.3. System Planning and Assembling
During the planning and assembly stages, the Mini SART prototype will be combined with LoRa for data transmission via radio signals. The programming is also required to detect the location of fishing vessel when the fishing vessel lost is occurred.

Programming is divided into two programs on the Mini SART-DMOM Transponder and Mini SARR-DMOM. Programming is done on Arduino components through the Arduino Application. After determining the components and assembling the tool, proceed with a trial on the device.

2.4. Trial of SART-DMOM
The objective of this step is to determine whether the SART-DMOM run either very well or not. SART-DMOM is checked whether the equipment is possible transmitting to SARR-DMOM. In addition, it is checked whether SARR-DMOM is possible to receive data of location from SART-DMOM.

3. Results And Discussion

3.1. System Planning

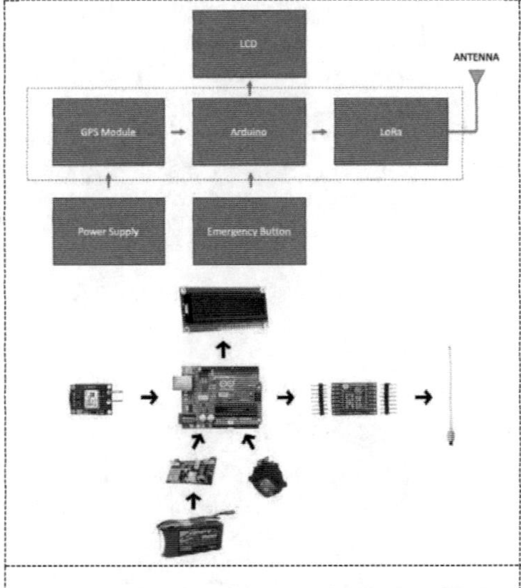

Figure 2. Components of MINI SART-DMOM

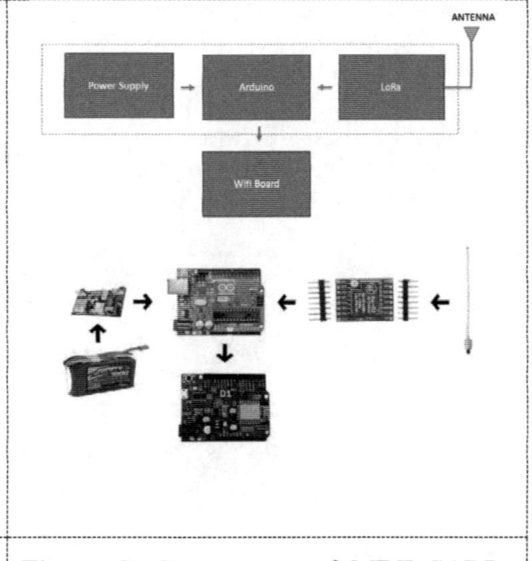

Figure 3. Components of MINI SARR-DMOM

IOP Conf. Series: Earth and Environmental Science 1081 (2022) 012044 doi:10.1088/1755-1315/1081/1/012044

Table 1. Specification of Mini SART-DMOM Transponder

Component	Type
Arduino	Arduino Uno
GPS	GPS Neo 6M Module
LCD	LCD 16x02 + 12C
LoRa	LorRa Ra-02 SX1278
Wifi Board	Wemos D1 WIFI ESP8266
Antenna	433 MHz, 5dBi
Battery	1000 mAh

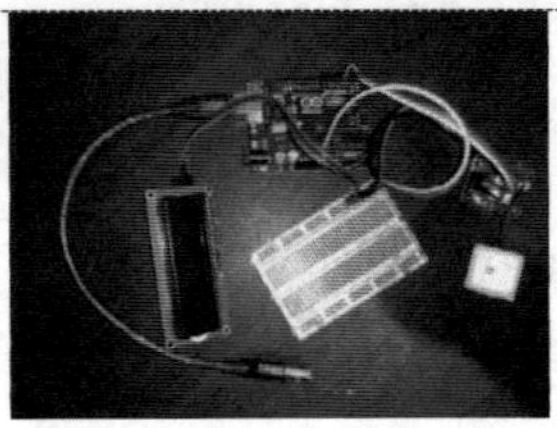

Figure 4. MINI SART-DMOM Assembly

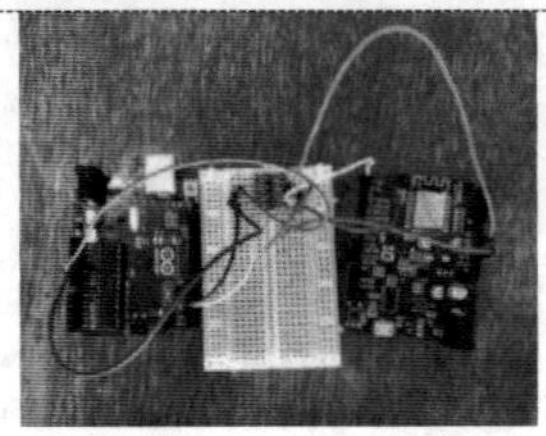

Figure 5. MINI SARR-DMOM Assembly

Figures 2 and 3 are system planning for the Mini SART-DMOM and Mini SARR-DMOM. Selection of selected components based on several indications. These indications include price, spare parts are easy to be obtained in Indonesia, and components can be compatible with other components. Selection of selected components based on several indications. These indications include price, spare parts are easy to get in Indonesia, and components can be compatible with other components. The selected component types can be seen in Table 1.

3.2. Assembly

In the Mini SART-DMOM assembly, we assembled two prototypes. The first assembly is done with a link between the GPS and Arduino. After that, the Mini SART-DMOM added an LCD. Then added LoRa into the two prototypes whose function was to transmit data via a 433 MHz radio signal between the Mini SART-DMOM Transponder and Mini SART Receiver. Then the Mini SART-DMOM Receiver is added with a Wifi Board, namely WEMOS ESP 8266. The Wifi Board functions to send data to be monitored via WEB Blynk. Figures 4 and 5 are an assembly of several Mini SART-DMOM components. The Mini SART-DMOM Transponder and Mini SART-DMOM Receiver results can be seen in Figures 6 and 7.

Figure 6. MINI SART-DMOM Transponder

Figure 7. MINI SART-DMOM Receiver

IOP Conf. Series: Earth and Environmental Science 1081 (2022) 012044 doi:10.1088/1755-1315/1081/1/012044

3.3. Programming

In the Mini SART-DMOM programming, some programming is done. For the Mini SART-DMOM Transponder, GPS programming is carried out via Arduino, which functions to get the Longitude and Latitude Coordinate Locations. Then do the LCD programming so that the location data can appear in the LCD. Then do Lora programming, which goes into Arduino programming, which serves to transmit data on the Mini SART-DMOM Transponder and receiver data on the Mini SART-DMOM Receiver that uses radio signals at a frequency of 433 MHz. Besides that, do the Wi-Fi Board program to monitor data into the Blynk WEB. Figures 8 and 9 below are some examples of programming carried out in this study.

```
void displayInfo()
{
  if (gps.location.isValid())
  {
    Serial.print("Latitude: ");
    Serial.println(gps.location.lat(), 6);
    Serial.print("Longitude: ");
    Serial.println(gps.location.lng(), 6);
    Serial.print("Altitude: ");
    Serial.println(gps.altitude.meters());
  }
    lcd.setCursor(0,0);
  lcd.print("LAT :");
  lcd.print(gps.location.lat(),6);
  lcd.setCursor(0,1);
  lcd.print("LONG :");
  lcd.print(gps.location.lng(),6);
  }
```

Figure 8. MINI SART-DMOM TransponderProgramming

```
void loop() {

  // try to parse packet
  int packetSize = LoRa.parsePacket();
  if (packetSize) {
    // read packet
    while (LoRa.available())
    {
      int inChar = LoRa.read();
      inString += (char)inChar;
      val = inString.toInt();
      gpsLat = inString;
    }
    inString = "";
    LoRa.packetRssi();
  }
```

Figure 9. MINI SART-DMOM Receiver Programming

3.4. Mini SART-DMOM Trial.

In the trials carried out, the tool can run smoothly. Mini SART-DMOM can get location coordinates (Longitude and Latitude) via the built-in GPS. As in figure 10 below. Then LoRa, which is on the Mini SART-DMOM Transponder, sends the location data to LoRa in the Mini SART-DMOM Receiver. Furthermore, in the Receiver, the data is channeled to the wifi Board. Namely, Wemos ESP 8266 so that the location data can be monitored via Web Blynk. In the trials on land that have been carried out, it can be concluded that the Mini SART-DMOM Transponder and Receiver prototype can run smoothly. Still, this prototype is deemed unable to be tested and used at sea because the prototype cannot detect it remotely. The results of the trial data can be seen in Table 2.

Figure 10. MINI SART-DMOM Transponder Detect Location

Table 2. Mini SART-DMOM Trial Data

Trial	Weather	Location	Antenna	Time to Get Location	Range
1.	Sunny	Populational Settlement	The Transponder and receiver use a small antenna	5 – 7 minutes	34.68 m
2.	Sunny	Small Populational Settlement	The Transponder and receiver use a small antenna	4 – 5 minutes	42.92 m
3.	Cloudy	Small Populational Settlement	The Transponder and receiver use a large antenna	3 – 5 minutes	5 m
4.	Cloudy	Small Populational Settlement	The Transponder and receiver use a large antenna	3 – 5 minutes	30.72 m
5.	Sunny	Field without Obstacles	The Transponder uses a large antenna, and the receiver uses a small antenna	Less than 1 minute	31.50 m
6.	Sunny	ITS Soccer Field	The Transponder and receiver use a large antenna + booster	2 – 3 minutes	50.33 m

3.5. Cost

The cost of making the Mini SART-DMOM is divided into two because, in this study, two prototypes were made, namely the Mini SART-DMOM and the Mini SARR-DMOM. The list of costs incurred can be seen in the explanation below. The prices listed below are much cheaper than the SART prices sold on the market today.

Table 3. Cost Mini SART-DMOM

No	Component	Price
1.	Arduino Uni and Data Cable	Rp140.000
2.	GPS Neo 6M	Rp90.000
3.	Switch	Rp3.000
4.	Battery Lipo 1300 mAh	Rp260.000
5.	Battery Jack xt30	Rp8.000
6.	Module LoRa Ra-02 SX1728 SPI includes Antenna	Rp150.000
7.	LCD Modues + 12 C	Rp42.000
8.	Jumper Cable	Rp15.000
9.	Black Boc Case	Rp30.000
	TOTAL PRICE	Rp738.00

Table 4. Cost Mini SARR-DMOM

No	Component	Price
1.	Arduino Uni and Data Cable	Rp140.000
2.	Module LoRa Ra-02 SX1728 SPI includes Antenna	Rp150.000
3.	Wemos Wifi Board and Data Cable	Rp200.000
4.	Logic Level Converter 5v – 3v	Rp20.000
5.	Perforated PCB	Rp9.000
6.	Jumper Cable	Rp15.000
7.	White Box Case	Rp30.000
	TOTAL PRICE	Rp564.00

3.6. Discussion

In the trials on land that have been carried out, it can be concluded that the Mini SART-DMOM and Mini SARR-DMOM prototype can run smoothly. Still, this prototype is deemed unable to be tested and used at sea because the coverage of prototype is very short, whereas the maximum distance is only 50 m. Several indications might cause the range not to be far. That comes from the antenna. It has the required signal amplifier for the antenna. This device needed a special antenna. Power can also indicate the cause of the device's failure to reach long distances. Power is too small later will also affect the antenna. When compared to the AIS Board DE70321T, the AIS Board DE70321T has a greater power of 7.2 V. Therefore, a prototype with AIS Board can reach long distances. The Mini SART-DMOM is much cheaper than the SART price sold in the market today. In the selection of components, comparisons are made in terms of price. SART sold in the market today is around 6 million rupiahs more.

4. Conclusion

In research on the problem of high fishing vessel accidents and the high cost of safety support equipment, a prototype innovation was created with Mini SART-DMOM. Mini SART-DMOM has an affordable price and can detect fisher's coordinates. This prototype can be an alternative and useful for stakeholders to develop SART for fishers wever, what has been done in research related to the Mini SART-DMOM Coverage of prototype is still very short to detect the location (about 50 m). Normally, coverage area of LoRa is about 40 km. Therefore, the recommendation for Mini SART-DMOM is to add an antenna signal booster to detect locations at a long distance.

Acknowledgments

The authors would like to thank Digital Marine Operation and Maintenance Laboratory, and Marine Engineering Department, ITS, which has supported working on this Mini SART-DMOM Prototype research.

References

[1] KNKT, "REKAPITULASI KECELAKAAN KAPAL," [Online]. Available: http://knkt.dephub.go.id/knkt/ntsc_maritime/maritime.htm.

[2] A. I. S. W. ,. M. R. ,. A. Sereati Hasugian, "Mapping of Vessel Accident Characteristics in Indonesia Based on Investigation of KNKT," *Warta Penelitian Perhubungan,* pp. 229-240, 2017.

[3] S. W. M. S. W. S. M. Agus, "AUTOMATIC IDENTIFICATION SYSTEM (AIS) BERBASIS MIKROKONTROLER UNTUK PENGAWASAN NELAYAN DI WAKATOBI," *Jurnal UMJ Samnastek,* pp. 1-7, 2018.

[4] A. Maulidi, "Desain Sistem Automatic Identification System (AIS) berbasis mini PC," *Journal PPNS,* pp. 105-108, 2019.

[5] D. G. T. S. Deandra Nurul Fadilah, "STUDY OF AUTOMATIC IDENTIFICATION SYSTEM (AIS) DATA USAGE FOR INDONESIAN MARITIME ZONE SURVEILLANCE," *Jurnal Teknologi Penginderaan,* pp. 229-250, 2019.

[6] Arlintas, "Cara Kerja dan Manfaat AIS Kapal," 3 April 2020. [Online]. Available: https://arlintas.com/.

[7] M. Gucma, "LOW COST AIS SYSTEM FOR SAFE NAVIGATION," *KONBin,* 2008.

[8] K. P. R. Indonesia, "Automatic Identification System (AIS)," 25 Maret 2019. [Online]. Available: http://dephub.go.id/org/disnavtanjungpinang/post/read/automatic-identification-system-%28ais%29.

IOP Conf. Series: Earth and Environmental Science 1081 (2022) 012044 doi:10.1088/1755-1315/1081/1/012044

[9] N. Ducrot, LoRa Developer Guide, Orange IoT Solutions, 2016.

[10] P. M. A. H. K. Ayub Repa Batong, "Analisis Kelayakan LoRa Untuk Jaringan Komunikasi Sistem Monitoring Listrik Di Politeknik Negeri Samarinda," *PoliGrid,* pp. 55-64, 2020.

IOP Conf. Series: Earth and Environmental Science 1081 (2022) 012045 doi:10.1088/1755-1315/1081/1/012045

Structural Strength Study for Vertical Axis Hydrokinetic Turbine Frame

N Ikhwan[1], T Y Yuwono[1], A Wikarta[1], A P Harendsa[1], D A Pratama[1], D A Sekarrini[1], D A Fatjrin[1], G A Nurbawono[1]

[1]Department of Mechanical Engineering, Sepuluh Nopember Institute of Technology, Surabaya, Indonesia

Abstract. One of the renewable energy sources is ocean currents which can be extracted with the Darrieus vertical axis hydrokinetic turbine (VAHT). This research aims to analyze the deformation and structural stresses that occur in the Darrieus VAHT support frame due to the loads. In the VAHT installation, it is necessary to have a support structure in the form of a solid frame, so that the turbine remains in a stable condition when operating. The VAHT frame must be able to withstand the load from a rotating turbine and have a small drag force. This study conducted a numerical simulation using the finite element analysis method. The effect of the cross-sectional shape of the structural beams on the drag due to flow was also investigated. The result is the frame designed with the ASTM A36 Grade B pipe structure has maximum stress of 63,116 MPa and it is relatively smaller and remains within the safe limits of the material. In addition, based on CFD simulation, the drag caused by the flow across the structure with this design is also smaller. Therefore, this study can be used as a consideration in making decisions to design the Darrieus VAHT frame.

1. Introduction

Digitalization nowadays increases the need for electricity many times compared to the last few decades. According to the International Energy Agency, worldwide electricity consumption in 2019 reached 22,848 TWh, or almost double the electricity consumption in 2010. This massive consumption rate is followed by higher production levels too. In 2019, non-OECD countries produced a total of 1,850 TWh, while OECD countries produced a total of 11,194 TWh. A total of 63% of the world's electricity production is generated from non-renewable sources, such as coal, oil, and natural gas. Meanwhile, the share of renewable energy is dominated by hydropower, which accounts for 16% of total electricity production [1].

Table 1. Sources of electricity generation worldwide

Source	
Coal	36,7 %
Renewables	26,7 %
Natural Gas	23,5 %
Nuclear	10,3 %
Oil	2,8 %

One of the renewable energy sources with great potential is hydropower. Given the large water surface on earth, it is estimated that in 2014 there will be an electrical capacity of 500 to 2,750 MW that can be extracted from the ocean [2]. The increase in production capacity cannot be separated from the increasing number of research on hydropower technology, some of which are horizontal axis turbines, vertical axis turbines, and oscillating hydropower. Among these various technologies, 76% of research and development on marine technology is focused on Horizontal Axis Marine Turbines

IOP Conf. Series: Earth and Environmental Science 1081 (2022) 012045 doi:10.1088/1755-1315/1081/1/012045

(HAMTs). Meanwhile, based on Elsevier's abstract database on Scopus, there is only about 7% research on VAMTs compared to research on VAWTs, although VAMTs can be used to extract energy from ocean currents on the same principle as VAWTs [3].

One of the important elements in VAMTs technology is the support and frame that supports the turbine during operation. There are many types of VAMTs supports, such as floating supports and supports embedded in the seabed. Meanwhile, the design of the frame used must be able to withstand operational loads without disturbing the work of the rotor. The operational loads such as the drag force due to the turbine, the weight of the system, and the drag force from the interaction of the flow with the submerged frame. The cross-sectional shape of the turbine frame structure affects the resulting drag. By comparison, a rectangular cross-section with sharp edges has a large drag coefficient while making the front end of the square rounded, reducing drag by up to 45%. By streamlining the rear end of the square into a sharp trailing edge, drag can be reduced by up to 85%. This shows that the cross-sectional shape of the frame structure will greatly affect the resulting drag. Surface roughness also affects the wake and the separation point from the flow to the body being passed. So, there are several ways to reduce the drag force that arises in the submerged frame.

In this study, the strength of the frame structure of the Darrieus turbine is simulated using the finite element analysis method. The frame is designed with ASTM A36 Steel material in the form of a cylindrical pipe. The reason for using a cylindrical pipe for the frame geometry refers to the research that has been done by Gautier [4]. It is expected that the stresses and total deformation information will be generated, and it can be used as a reference to determine the safety of the frame design. In addition, fluid flow simulations were carried out to determine the effect of the cross-sectional shape of the frame structure on the drag force caused by the interaction between sea currents and the frame.

2. Methodology

2.1 Frame Design

The design of the Darrieus turbine frame is shown in Figure 1. The Darrieus turbine frame used is a modified turbine from a previous study. The difference with the previous Darrieus turbine frame is that the frame is composed of an ASTM A36 circular pipe. The simulated Darrieus Turbine Frame is only 3.8 meters high from the bottom frame shown in figure 2. It is adapted to the Darrieus Turbine System which is submerged in water. The dimensions of the Darrieus turbine frame used are shown in Table 2, whereas the loads on turbine frame are shown in Table 3.

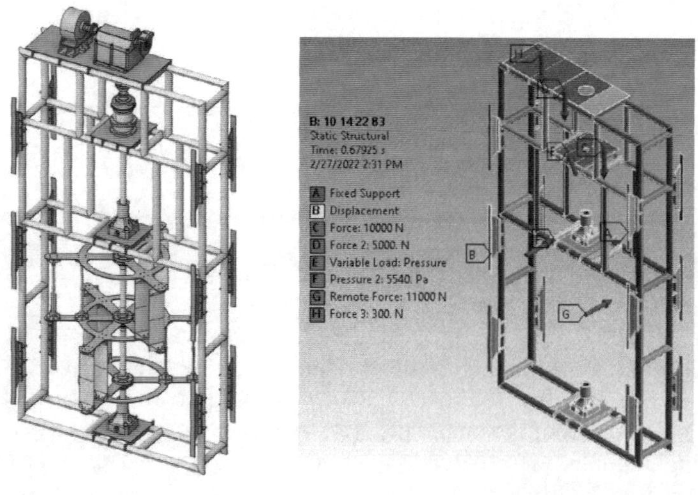

(A) (B)

Figure 1. (A) 3D Design of System Darrieus Turbine (B) Loads on Turbine Frame

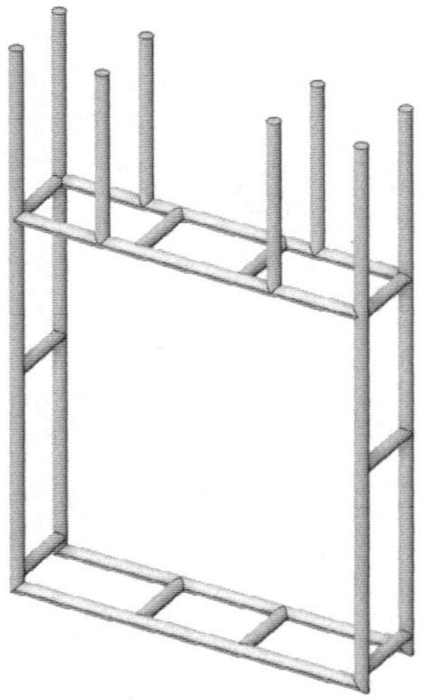

Figure 2. Simulated 3D Design of Darrieus Turbine Frame

Table 2. The geometry of Darrieus Turbine Frame

Parameter	Unit (mm)
Diameter of Circular Cylinder pipe	88.9
Height of Frame	4900
Height of submerged Frame	3800
Length of Frame	2463.9
Wide of Frame	610

Table 3. The Loads on Darrieus Turbine Frame

Load	Value
Frame Weight (C)	10 kN
Gearbox (D)	5 kN
Turbine Drag (G)	11 kN
Generator (H)	300 N

2.2 CFD Model

The computational fluid dynamic (CFD) method was carried out to extract the data from the vertical axis hydrokinetic turbine frame. Using ANSYS Fluent to simulate the turbine frame gives some advantages such as being cheaper and safer than the realistic model. The numerical study was done in the three-dimensional steady-state condition which includes three main steps pre-processing, solving, and post-processing.

The pre-processing step consists of making the geometry, creating the flow domain, mesh generation, and setting up a solver. Based on Bruno [5], the flow domain is 25620 mm long, 23227.8 mm width, and 25900 mm height. The geometry was made in SpaceClaim. The next step is meshing the flow domain which is shown in figure 2. The mesh uses the unstructured tetrahedral element with a prism layer is applied to the turbine frame wall to maintain the accuracy of the result.

The boundary condition is shown in figure 2. The velocity inlet boundary condition was set in the inlet where the flow enter the domain. The pressure outlet boundary condition was set in downstream of the flow with the pressure equal to atmospheric pressure. The other side of the domain was set as a symmetry boundary condition so that the normal velocity is equal to zero.

The boundary condition is shown in figure 2. The velocity inlet boundary condition was set in the inlet where the flow enter the domain. The pressure outlet boundary condition was set in the downstream of the flow with the pressure equal to atmospheric pressure. The others side of the domain was set as symmetry boundary condition so that the normal velocity is equal to zero. The SST k-ω turbulence model is used because it gives accurate prediction of the amount flow separation under adverse pressure gradient [6].

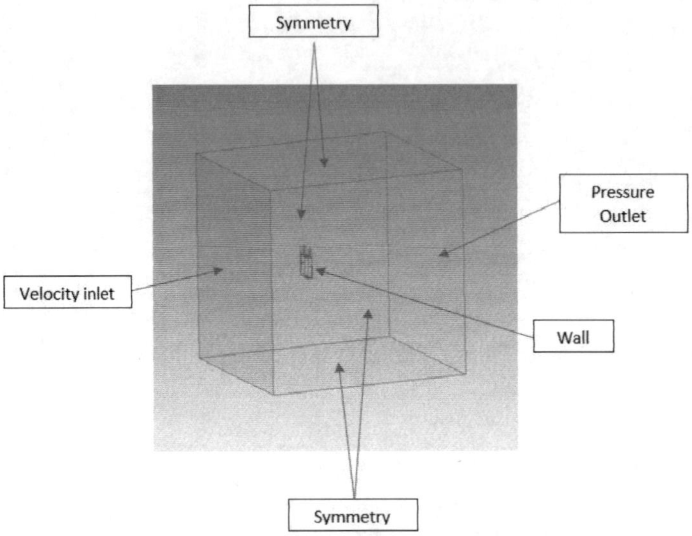

Figure 3. Flow Domain and Boundary Condition

(a)

(b)

Figure 4. Computational grid (a) all domain.; (b) cylinder

The solving or processing step is the step where the governing equations are solved using iterative method until the convergence criteria is achieved.

The post-processing step involves extracting the desired flow properties from the computed flow field. The extracted data are drag force, moment, pressure and velocity contours, and turbulent viscosity contour.

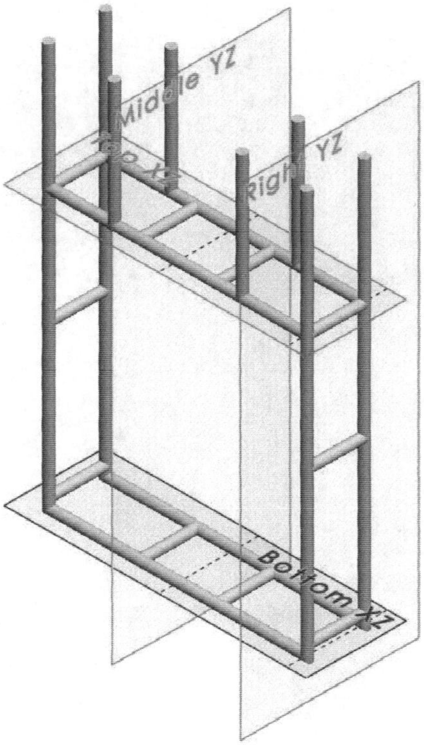

Figure 5. Frame Geometry for Simulation

IOP Conf. Series: Earth and Environmental Science 1081 (2022) 012045 doi:10.1088/1755-1315/1081/1/012045

3. Result and Discussion

3.1 flow Analysis Results of the Turbine Frame

Analysis of fluid flow in the turbine frame using a flow with a speed of 2 m/s in the direction of the z axis. The simulation is carried out by dividing the turbine frame into four intersecting areas, namely the middle (middle yz plane), right (right yz plane), top (top xz plane), and bottom (bottom xz plane) intersection areas. Figure 6 shows the simulation results in the form of pressure contours and velocity contours shown by Figure 7 for each intersection area. The large pressure at the stagnation point causes a fairly large drag force with a value of 2.77 kN and a drag coefficient (C_d) of 4.523.28. wake that arises behind the turbine frame will cause a large drag force, so that the total moment of the resulting force will also be large with a value of 5,09 kN.m. Wake is a fluid flow phenomenon where the flowing fluid will form a vortex due to the difference in pressure. The emergence of wake on the turbine frame can be seen from the value of the turbulent viscous ratio which is quite large at each intersection area, namely 5185.51.

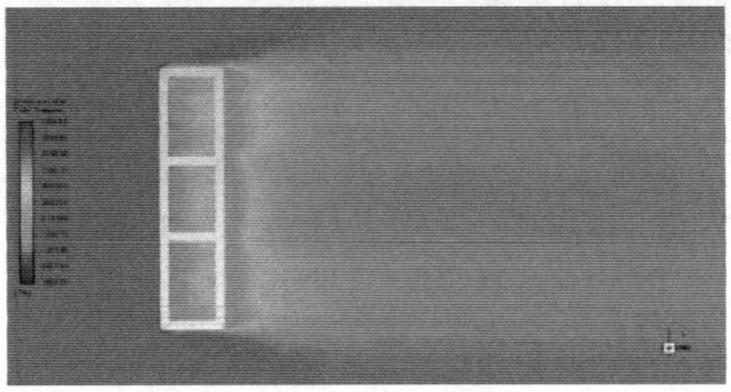

(a) top - xz

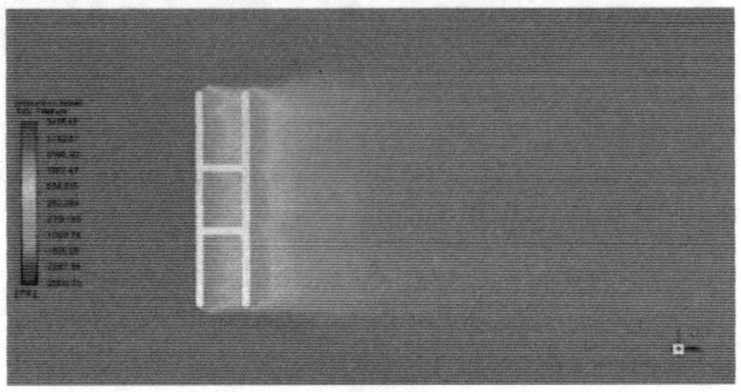

(b) bottom - xz

(c) right - xz

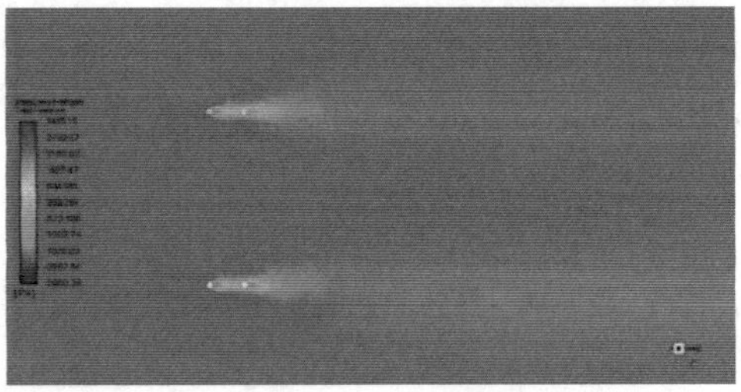

(d) middle - xz

Figure 6. Turbine frame simulation pressure contour with a fluid flow of 2 m/s

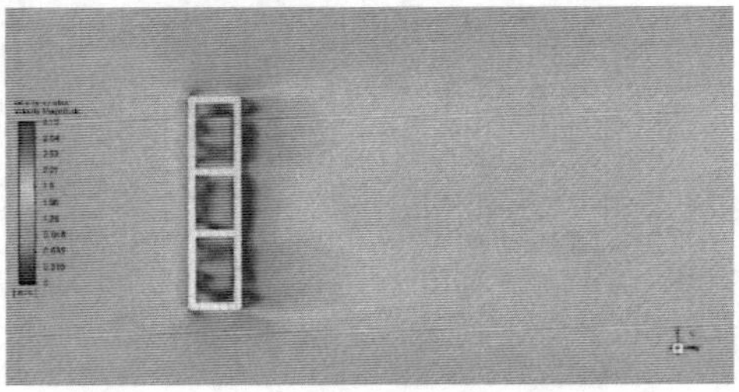

(a) top - xz

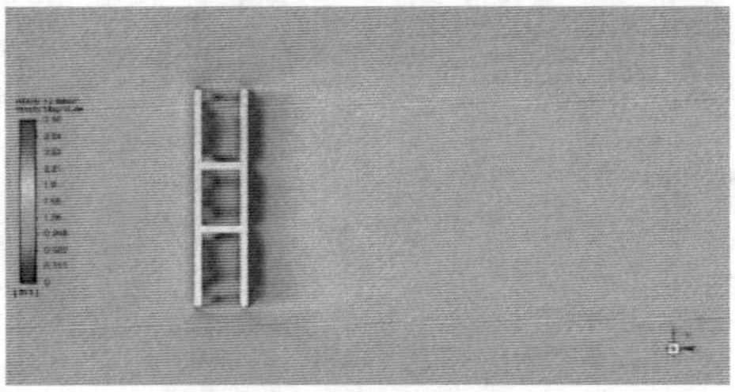

(b) bottom - xz

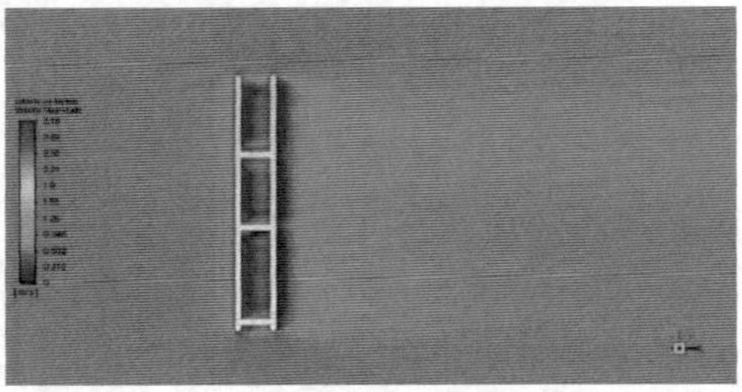

(c) right - xz

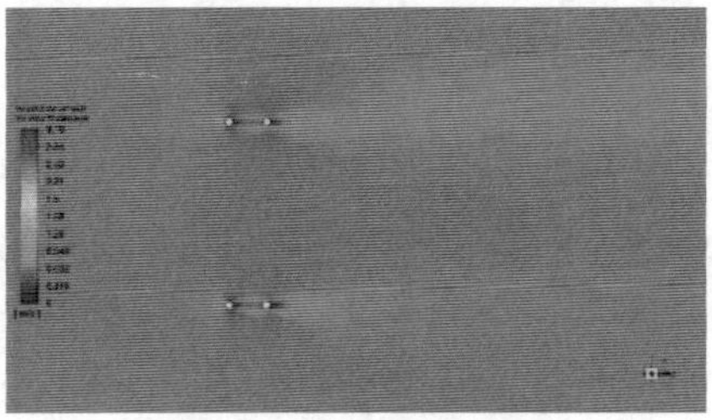

(d) middle - xz

Figure 7. Turbine frame simulation velocity contour with a fluid flow of 2 m/s

The next fluid flow analysis is at a speed of 2.5 m/s in the direction of the z axis. The simulation is carried out by dividing the turbine frame into four intersecting areas, namely the middle (middle yz plane), right (right yz plane), top (top xz plane), and bottom (bottom xz plane) intersection areas. Figure 8 shows the simulation results in the form of pressure contours and velocity contours shown by Figure 9 for each intersection area. The large pressure at the stagnation point causes a fairly large drag force with a value of 4.26 kN and a drag coefficient (C_d) of 6,988.83. wake that arises behind the turbine frame will cause a large drag force, so that the total moment of the resulting force will also be large with a value of 7.87 kNm. Wake is a fluid flow phenomenon where the flowing fluid will form a vortex due to the difference in pressure. The emergence of wake on the turbine frame can be seen from the value of the turbulent viscous ratio, which is quite large at each intersection area, namely 6351.33.

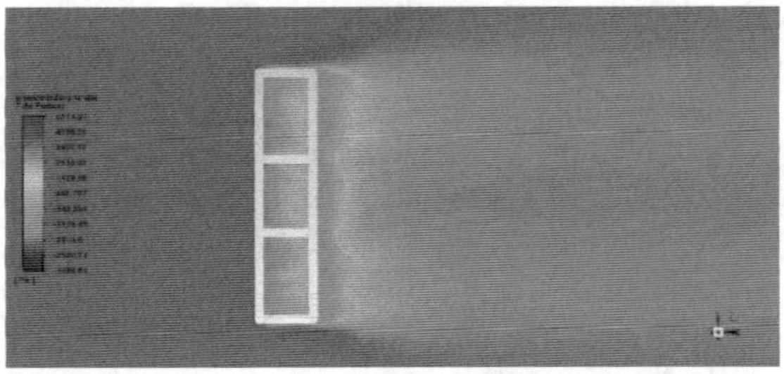

(a) top - xz

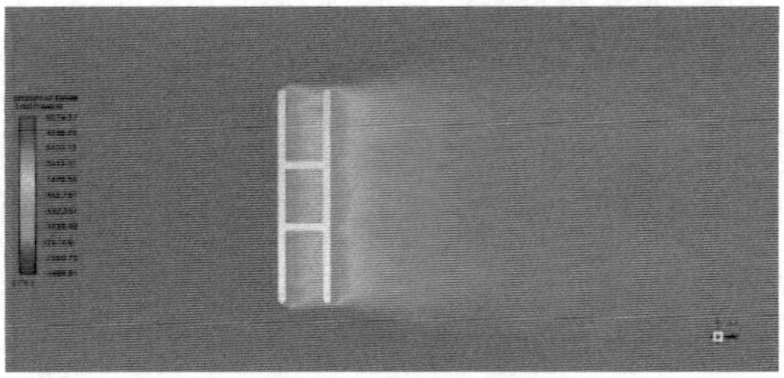

(b) bottom - xz

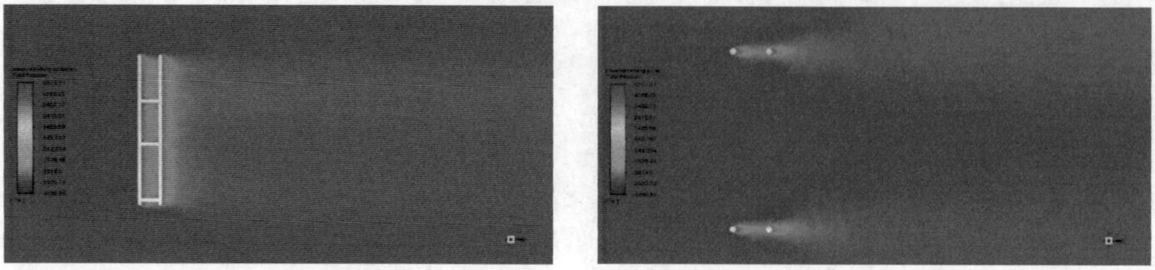

(c) right - xz (d) middle - xz

Figure 8. Turbine frame simulation pressure contour with a fluid flow of 2.5 m/s

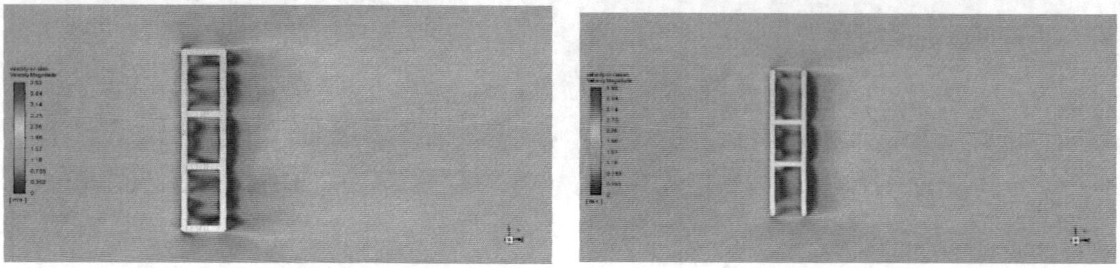

(b) top - xz (a) bottom - xz

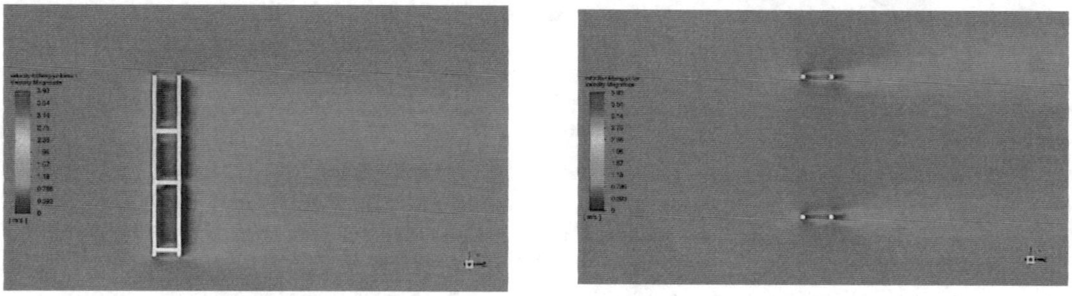

(c) right - xz (d) middle - xz

Figure 9. Turbine frame simulation velocity contour with a fluid flow of 2.5 m/s

Another flow velocity variation value used in this study is 3 m/s with the intersection area as in the previous flow velocity simulation. In this variation, a pressure contour as shown in Figure 10 and the velocity contour will also be generated as shown in Figure 11 due to the large pressure at the stagnation point of the frame, a drag force of 6.09 kN arises with a drag coefficient of 9.945.19. In addition, a large drag force can also be generated due to the emergence of a wake that occurs at the back of the frame. The greater the drag force, the greater the resulting moment. With this speed variation, the moment in the frame has a value of 11.19 kN.m. The occurrence of a wake on the turbine frame can be shown by the value of the turbulent viscous ratio, where in this simulation the turbulent viscous ratio value is 7418.52.

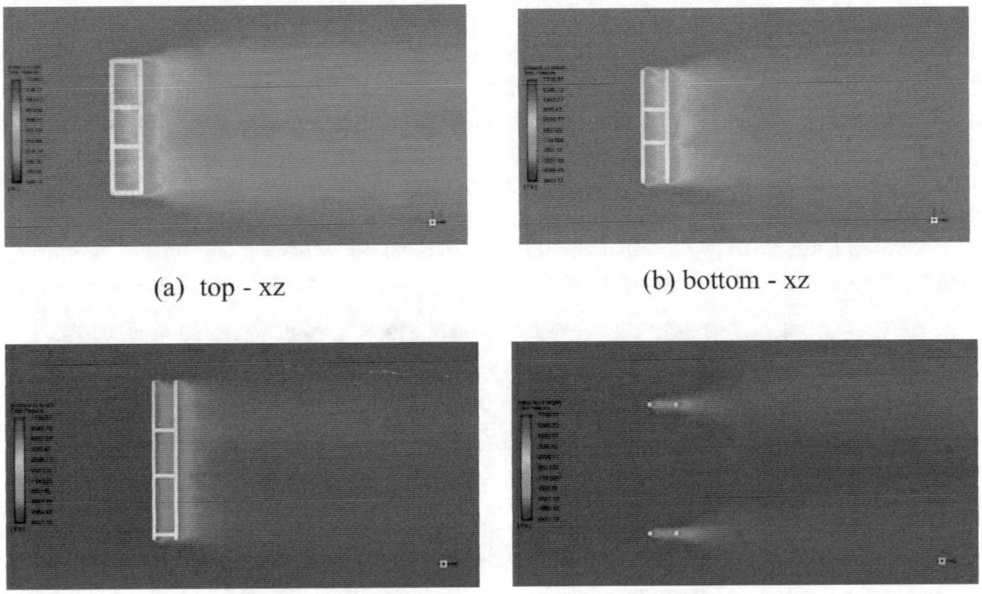

(a) top - xz (b) bottom - xz

(c) right - xz (d) middle - xz

Figure 10. Turbine frame simulation pressure contour with a fluid flow of 3 m/s

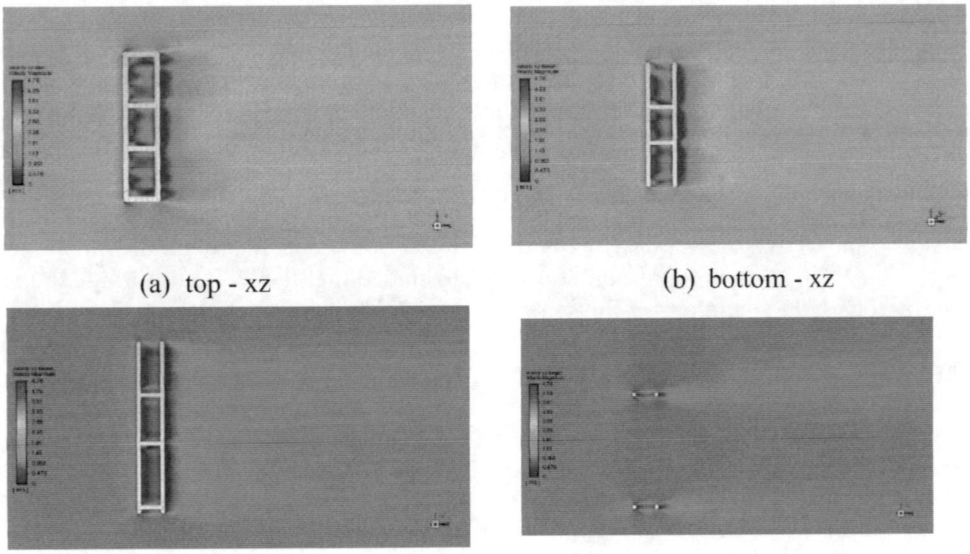

(a) top - xz (b) bottom - xz

(c) right - xz (d) middle - xz

Figure 11. Turbine frame simulation velocity contour with a fluid flow of 3 m/s

3.2 Stress Analysis Results of the Turbine Frame

Stress analysis result of the turbine frame is used Boundary Conditions Fixed Support's that is on the top sliding foot, displacement that is on the middle sliding, weight of the system is approximately 10 kN, the weight of the gearbox is approximately 5 kN, the stress due to current drag force water on the vertical and horizontal components, the remote force assume approximately 11 kN due to the turbine drag force, and the last is the generator weight assume approximately 300 N. The maximum stress results is generated from the simulation that occurs at points connection with a value of 33,32 MPa. Because ASTM A36 has 400 MPa of tensile strength value and 250 MPa of yield strength value, the turbine frame is acceptable.

Figure 12. Maximum stress in turbine frame simulation

4. Conclusion

Changes in the speed of sea water currents hitting the darrieus turbine frame cause changes in the drag force generated. As the current velocity increases, the pressure at the stagnation points in the frame increases and the wake and turbulence increase. In this study, the largest drag force occurs at a speed variation of 3 m/s, which is 6.09 kN. The drag force produces a moment of 11.19 kN.m. High turbulence at this speed variation is indicated by the turbulent viscous ratio of 7,418.52. While the frame analysis shows that the greatest stress value in the frame occurs around the middle joint with a value of 82.539 MPa. With these results, it can be concluded that the Darrieus turbine frame structure can be used properly in conditions of flow velocity with variations of 2-3 m/s.

5. References

[1] IEA (2021), Electricity Information: Overview, IEA, Paris https://www.iea.org/reports/electricity-information-overview

[2] Melikoglu, M. 2018. *Current status and future of ocean energy sources: A global review*. In Ocean Engineering (Vol. 148, pp. 563–573). Elsevier Ltd.

[3] Nguyen, M. T., Balduzzi, F., & Goude, A. 2021. *Effect of pitch angle on power and hydrodynamics of a vertical axis turbine*. Ocean Engineering, 238.

[4] Gautier, R., Biau D., Lamballais E. 2013. *A reference solution of the flow over a circular cylinder at Re = 40.* Computers & Fluids, pp. 103 – 111. Elsevier Ltd.

[5] Bruno, L., Fransos, D., Coste, N., Bosco, A. 2009. *3D flow around a rectangular cylinder: A computational study*. Journal of Wind Engineering and Industrial Aerodynamics, pp. 263–276. Elsevier Ltd.

[6] Zhang, D. 2017. *Comparison of Various Turbulence Models for Unsteady Flow around a Finite Circular Cylinder at Re=20000.* Journal of Physics: Conference Series.

Monohull or Twin hull Shape Selection Based on the Performance of Hull Planning on Fixed Displacements

Aang Wahidin, Bambang Teguh Setiawan, Priyambodo Nur Ardi Nugroho

Politeknik Perkapalan Negeri Surabaya

Abstract. The selection of the hull shape below the waterline becomes very important to improve performance and minimize emission during the design stage of the fast boat hull. To understand the performance comparison between the monohull and twin-hull of the same length and width, an analysis was carried out on an 8 m fiberglass ship with Maxsurf. Performance analysis was carried out with Hydromax for hydrostatic performance and ship stability and Hullspeed to analyze the performance of ship resistance and power. Hydrostatic Wetted Surface Area and transverse metacenter radius for the monohull were 13.883 m^2 and 1.486 m, while a catamaran shows 12% to 24% higher WSA and 2% to 3% higher BMt compared to a monohull. The resistance and power of the ship at 40 knots for monohull were 24267.07 N and 1116.09 HP. In contrast, the catamaran shows up to 74% increase. Stability arm and radius of gyration show as follow: 0.297 m and 0.631 m for a monohull, while a catamaran shows up to 140% higher max GZ and 222% GMt. The results of this study can be used as a reference for determining the use of a monohull or catamaran hull in the early stages of a fast boat design.

Keywords: monohull, twin-hull, planning

1. Introduction

Currently, more attention is being paid to eco-friendly ships that minimize emissions and reduce the impact of global warming (Husain, 2014). The selection of the hull type becomes an important issue at the initial design stage. In the fast ship hull (planning hull) aspect, the hull's shape below the waterline becomes very important to improve performance and reduce fuel consumption. To improve the performance of the planning hull, an uppendig is added to the hull below the waterline, such as chine, spray rails, bilge strakes, and keel strakes (David Stimson, 2002). In ship design planning, it is expected that the ship's hull will also provide additional lift so that the ship will quickly return to the upright position when the ship is at high speed or also called stability at high speed (Bertram V, 2002).

When selecting the ship's shape below the waterline, there were rarely studies regarding the selection of monohull or twin hull forms. In the Displacement, LCG/L, KG/H, the angle of the rise of the floor is the same between the monohull and twin hull. This paper discussion regarding the advantages and disadvantages of a monohull or twin hull hulls of the same length and width. As the object of the study, a performance analysis was carried out on an 8 m fiberglass ship built at the Politeknik Perkapalan Negeri Surabaya Non-Metal Workshop by varying it into several forms of ship hull models. The results of this study can be used as a basis (reference) to determine the use of a monohull or catamaran hull in early-stage fast boat design (planning hull).

2. Literature Review

2.1. *Hull form planning*

In the planning hull ship, the cross-sectional shape below the water line is planned to have a constant shape starting from the middle of the ship (midship) to the rear (stern). The V bottom hull has two popular shapes, namely the moderate V and deep V. Both of these shapes have a different V shape on the back (transom), whereas for the moderate V shape, the V angle on the transom is between 15 to 19 degrees while the deep V shape has an angle V on the transom between 20 to 24 degrees. The inclination angle of V is also called deadrise (*David Stimson, 2002*). Speed/length ratio parameters that distinguish a ship as a displacement hull, semi-planing hull, or planning hull are as follows:

- Displacement hull: has a speed/length ratio of less than 1.4
- Semi-planing hull: has a speed/length ratio range between 1.4 to 2.0
- Planing hull: has a speed/length ratio range of more than 2.0

2.2. *Prisoner of the ship*

Resistance in the monohull and twin-hull can be divided into skin resistance or viscosity resistance, wave-making resistance, and form resistance, including pressure drag and water spray. In hydrodynamics, the residual resistance is defined as the difference between the total resistance and the viscous resistance. For a high-speed planning hull, the total resistance is generally divided into the following components:

- Viscosity resistance, which occurs in the "wetted surface area", is shown by the formula:

$$R_f = \frac{\rho \times V^2{}_m \times \lambda \times b^2 \times C_F}{2\cos\beta \times \cos\tau} \quad (1)$$

- The pressure resistance is the horizontal component of the hydrodynamic pressure that occurs at the bottom of the ship. The amount is:

$$R_p = \Delta \times tg\lambda \quad (2)$$

So the total resistance "high speed planning hull": $R = R_f + R_p$ (3)

The wave resistance represents the energy lost due to the waves formed when the ship sails. In comparison, eddy resistance states the energy lost due to the separate flow, especially behind the ship. The residual resistance for low-speed ships is around (8 to 25)% of total resistance, while it is up to (40 to 60)%.

2.3. *Ship resistance and performance*

In ship design, designers always aim that the primary size of the ship and the shape of the ship can minimize the resistance that will occur in the movement of the ship later. The size of the ship's resistance that will occur, in addition to depending on the size of the ship's speed, also depends on the parameters of the ship's line plan, for example, the calculation of the shape of the ship and the relationship between the wet surface elements of the ship and the resistance.

One of the main size parameters of the ship that affects the size of the ship's resistance is: the length of the ship (L), where the length of this ship can give the size of the Froude number $Fn = \frac{V_O}{\sqrt{gL}}$, which means it can also determine the size of the ship's wave resistance (Rw)

2.4. *Calculating effective thrust*

The method that is widely used in calculating the effective motor power (EHP) of ships is the Pamelia method, which is for ordinary ships with parts of the ship that usually protrude.

IOP Conf. Series: Earth and Environmental Science 1081 (2022) 012046 doi:10.1088/1755-1315/1081/1/012046

$$EHP = \frac{V \times V_{O^3}}{L \times C} \times \frac{x}{\lambda} \times \sqrt{\psi} \tag{4}$$

Where: V = Ship displacement in m^3, L = Lwl= length of the ship submerged in water in meters, Vo = ship speed in knots, x = coefficient that depends on the protruding part of the ship, and = multiplication number

2.5. *Fast ship stability (planning hull)*

Furthermore, to calculate the stability of a small ship, it must meet the general criteria (Renilson, M and Hardon, P, "International Journal of Small Craft Technology" Qintq, UK.), which are listed in Table 1 below:

Table 1. Ship Stability Criteria for small craft

Criteria	Pass criteria
GM(m)	Greater than 0.35
The area under the GZ curve from 0 to 300 (m rad)	Greater than 0.055
The area under the GZ curve from 0 to 400, or the angle of down flooding if it occurs at a lower angle (m rad)	Greater than 0.09
The area under the GZ curve from 300 to 400, or the angle of down flooding if it occurs at a lower angle (m rad)	Greater than 0.03
GZ above 300 (m)	Greater than 0.2
GZmax angle (deg)	Greater than 0.25

3. Methodology

This research begins by analyzing the performance of an 8 m fiberglass ship built at the Surabaya State Shipping Polytechnic Non-Metal Workshop by varying the shape of the hull model with Maxsurf software. The hull's shape is varied into one hull (monohull) and three hull forms (twin-hull). The shape of the hull (twin-hull) it is varied based on the distance between the three hulls, namely once (1x) the width of the twin-hull hull, one and a half times (1.5x) the width of the twin-hull hull, and twice (2x) the width of the twin-hull hull. Furthermore, from the variation of the model, performance analysis was carried out with Hydromax software for hydrostatic performance and ship stability and hull speed software to analyze the performance of ship resistance and power. To compare performance on different ship lengths in the same way as in the 8-meter long model, it is modeled for variations in ship length, including 12m, 16m, 20m, 24m, 28m, 32m, and 36m.

4. Results and Discussion

4.1. *Identification of hull planning*

Small ships with high speed (hull planning) are grouped into three types: the hull is held up by the buoyancy force, the hull is held in place by the dynamic lift rather than the buoyancy, and the hull lifted from the water surface is held back. The dynamic lifting force on the wing (foil) submerged in water (Digul Siswanto, 1986). Based on the identification of several planning ships that have been built at PPNS Non-Metal Workshop, we are interested in ships with dynamic lift type specifications, namely: Patrol Vessels, with main sizes: Loa: 8.0 m, B: 2.2 m, H: 1.0 m, Draft: 0.4 m, Engine: 2 x 85 HP (170 HP) and Service speed (Vs): 25 – 30 Knots.

This is because the ship is equipped with technical data and complete supporting drawings, so as a sample, the primary size of the fast ship used in the discussion in this study is the ship. Furthermore, a hull planning design is carried out from the ship's data starting from one hull (monohull) and two hulls (twin-hull). The form of two hulls (twin-hull) it is varied based on the distance between the two hulls, namely once (1x) the width of the twin-hull hull, one and a half times (1.5x) the width of the twin-hull hull, and twice (2x) the width of the twin-hull hull.

4.2. Monohull hull design and catamaran (twin-hull)

In a planing hull ship, the cross-sectional shape below the water line is a constant shape starting from the middle of the ship (midship) to the back (stern). These two forms have a different V shape on the back (transom), where the moderate V shape has a V angle on the transom between 15 to 19 degrees, while the deep V shape has a V angle on the transom between 20 to 24 degrees. The slope angle of V is also called deadrise

From the model data above, a single hull (monohull) and superstructure design process was carried out using the Maxsurf Pro Version 11.12 software with one of the models, as shown in the following Figure 1 to Figure 4 below.

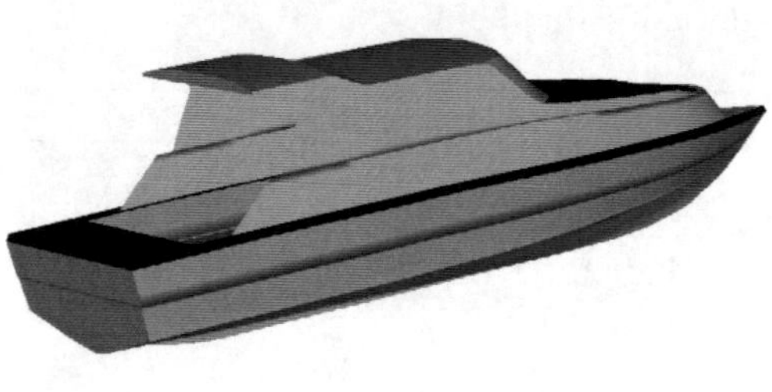

Figure 1. Single Hull Surface (monohull)

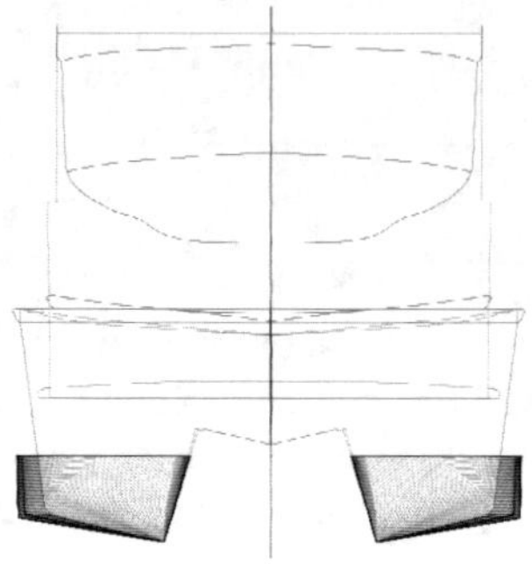

Figure 2. Hull Spacing 1x Hull Width twin-hull

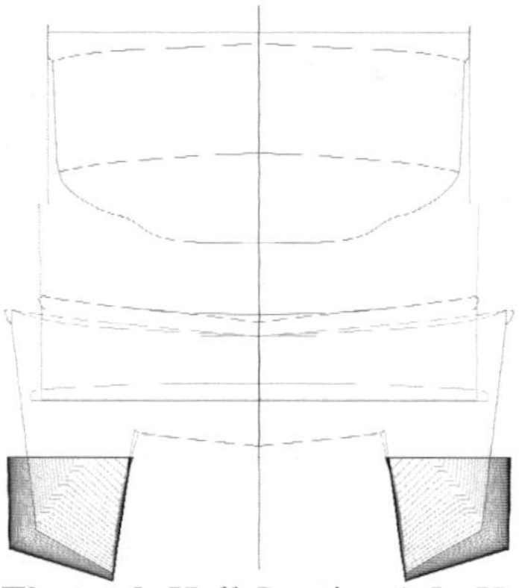

Figure 3. Hull Spacing 1.5x Hull Width twin-hull

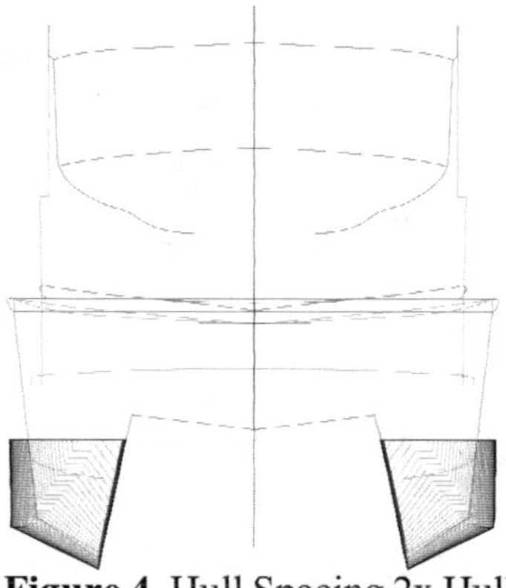

Figure 4. Hull Spacing 2x Hull Width twin-hull

Prior to the analysis, to ensure that the model has the same displacement ratio and the location of the LCB, the model is corrected for the difference in Displacement and the position of the longitudinal center of buoyancy (LCB) with the following equation.
Ship Displacement Correction:

$$\text{Displacement correction} = \frac{D_m - D_C}{D_m} \times 100\% \times 0.5\% \tag{5}$$

where: Dm = Displacement of monohull in tons, Dc = Displacement of catamaran in tons
And correct for the location of the Longitudinal Center of Buoyancy (LCB) with the following equation:

$$\text{LCB correction} = \frac{LCB_m - LCB_C}{L_{wl}} \times 100\% \times 0.1\% \tag{6}$$

where: LCBm = LCB location of monohull ship in tons, LCBc = LCB location of catamaran in tons, Lwl = length of the Load water line (Lwl) in meters

4.3. Hydrostatic Performance

To analyze the performance characteristics of the ship (hydrostatic) as a step to determine the optimal choice of hull shape characteristics. The shape of the hull surface produces a parametric table as shown in Table 2:

Table 2. The shape of the hull surface

Characteristics		Monohull	Catamaran 1x	Catamaran 1.5x	Catamaran 2x
Displacement	Tonne	2.498	2,488	2.459	2.486
Volume	m^3	2.437	2.427	2,399	2.425
Draft to Baseline	m	0.452	0.452	0.487	0.487
Immersed depth	m	0.452	0.382	0.538	0.584
Lwl	m	7.154	6.622	6.585	6.649
Beam wl	m	2,097	2.194	2.197	2.197
WSA	m^2	13,883	15,562	16,473	17,186
Max cross sect area	m^2	0.477	0.46	0.465	0.456
Airplane area	m^2	11,414	8,643	6.12	5,982
Cp		0.713	0.797	0.783	0.799
Cb		0.36	0.64	0.632	0.599
cm		0.509	0.803	0.808	0.749
Cwp		0.761	0.871	0.868	0.863
LCB from zero pt		-0.784	-0.851	-0.878	-0.789
LCF from zero pt		-0.753	-0.657	-0.659	-0.595
KB	m	0.316	0.297	0.269	0.258
BMt	m	1.486	1,882	1,721	1,698
BMl	m	12.95	10,723	7.73	7.513
GMt	m	1,802	2.179	1955	1,921
GMl	m	13,266	11.02	7,964	7.736
KMt	m	1,802	2.179	1.99	1,956
KMl	m	13,266	11.02	7.999	7,771
Immersion (TPc)	tonne/cm	0.117	0.089	0.063	0.061
MTc	tonne.m	0.047	0.035	0.025	0.024
RM at 1deg	tonne.m	0.079	0.095	0.084	0.083

By using the Hydromax software with an analysis process using the number of sections 200 and height parameters, the following characteristic curve is produced as seen in Figure 5 to Figure 8 below:

Monohull Model

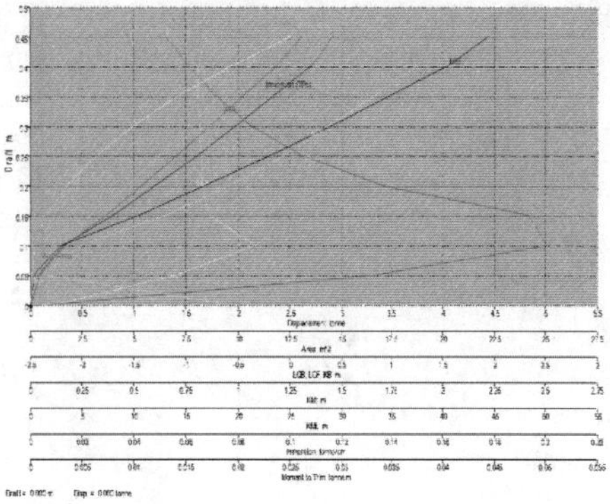

Figure 5. Monhull model hydrostatic curve

Twinhull 1.0x. models

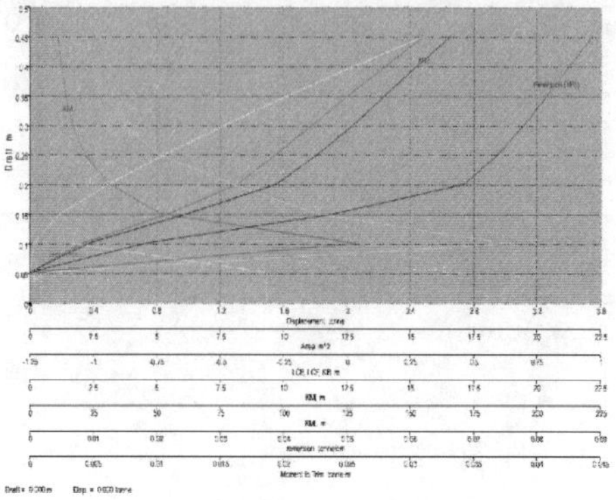

Figure 6. Figure 1.0x. twin-hull hydrostatic curve

Model Twinhull 1.5x

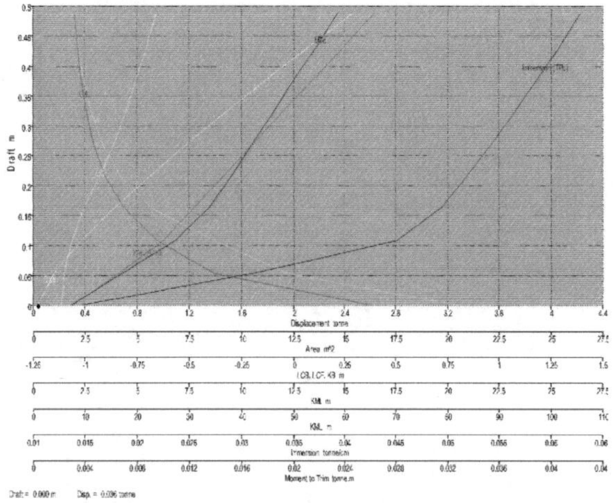

Figure 7. Figure 1.5x twin-hull hydrostatic curve

Twinhull 2.0x. models

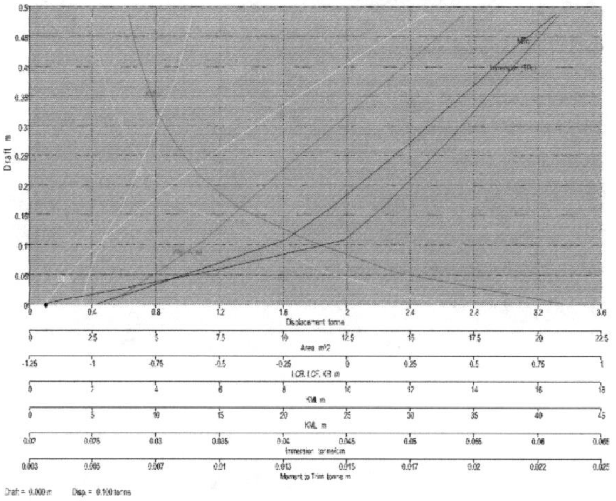

Figure 8. Figure 1.5x twin-hull hydrostatic curve

4.4. Performance of ship's resistance and motor power

To analyze the ship's performance related to the ship's power and resistance capabilities from several design samples, an analysis of the ship's resistance and power is carried out using the Hullspeed software tool. By varying the speed of the ship from 0 knots to 40 knots, the results are depicted in Figures 9 to Figure 12 as follows:

IOP Conf. Series: Earth and Environmental Science 1081 (2022) 012046 doi:10.1088/1755-1315/1081/1/012046

Monohull Model

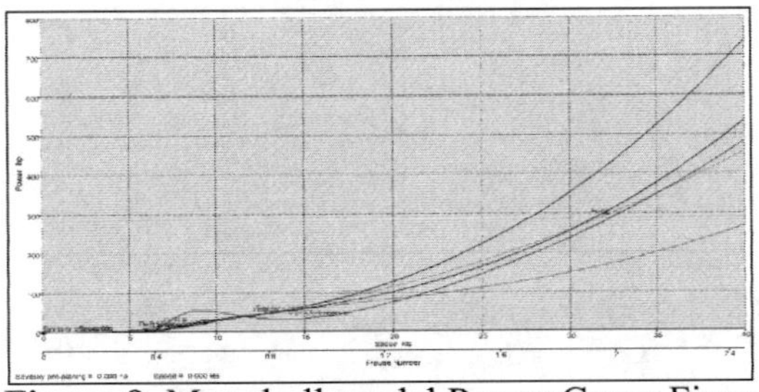

Figure 9. Monohull model Power Curve Figure Van Oortmeersen Power method. (hp) at speed Vs = 40 knots = 482.74 hp

Twinhull 1.0x. models

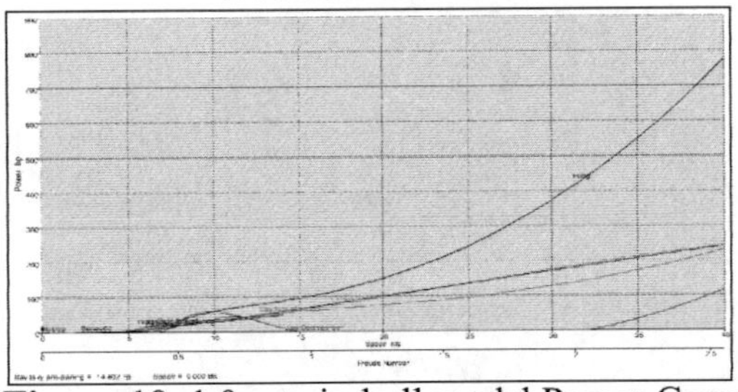

Figure 10. 1.0x. twin-hull model Power Curve Van Oortmeersen Power method. (hp) at speed Vs = 40 knots = 231.16mobile phone

Model Twinhull 1.5x

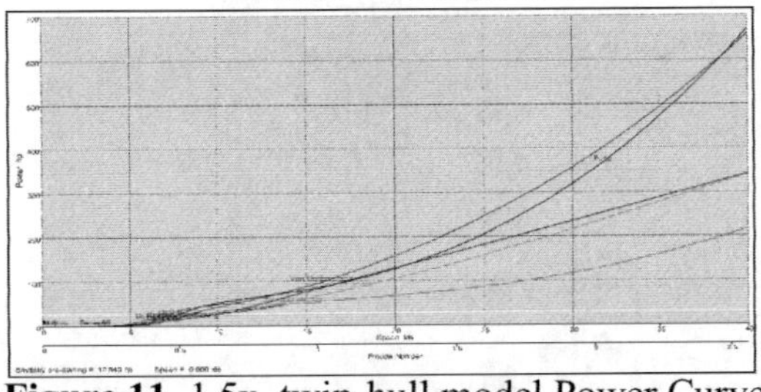

Figure 11. 1.5x. twin-hull model Power Curve Van Oortmeersen Power method. (hp) at speed Vs = 40 knots = 217.04mobile phone

Twinhull 2.0x. models

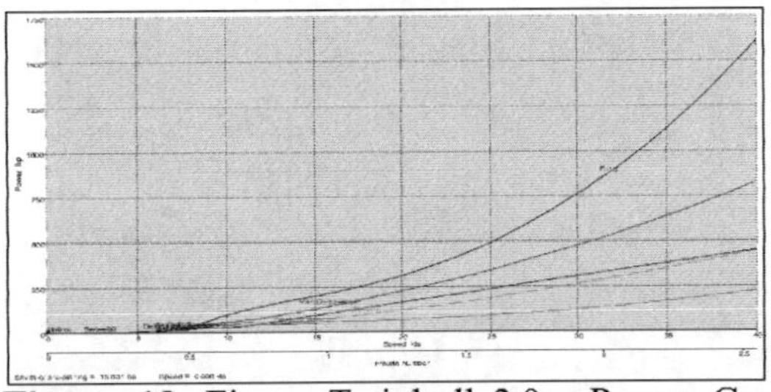

Figure 12. Figure Twinhull 2.0x. Power Curve Model
Van Oortmeersen Power method. (hp) at speed Vs = 40 knots = 224.95mobile phone

Based on the separation of the waveforms generated by the model can be seen in the following Figures 13 to Figure 20 below:

Monohull Model

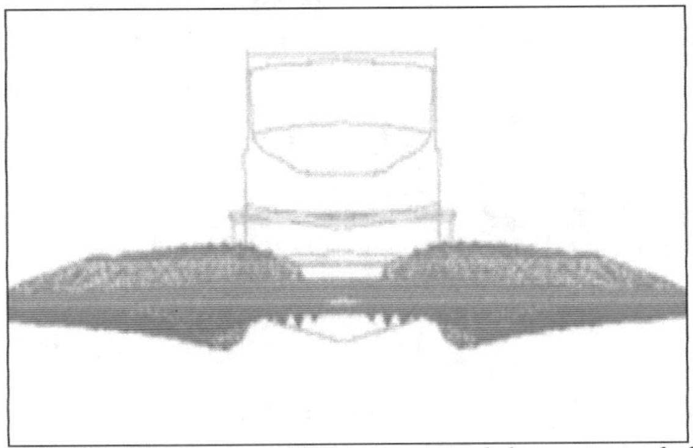

Figure 13. Image of wave model generated by monohull ship

Twinhull 1.0x. models

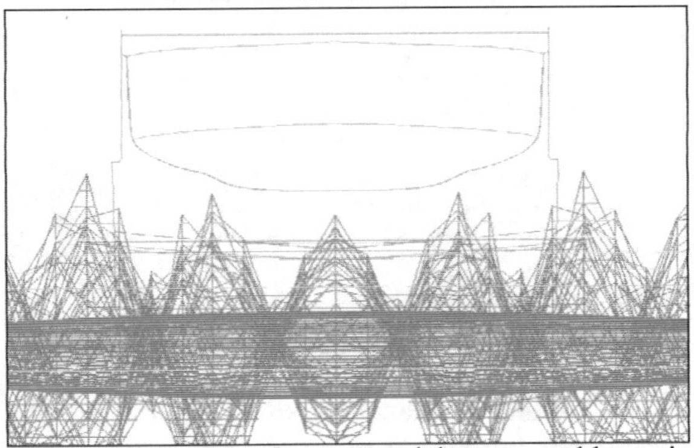

Figure 14. Image of wave model generated by twin-hull ship 1.0x

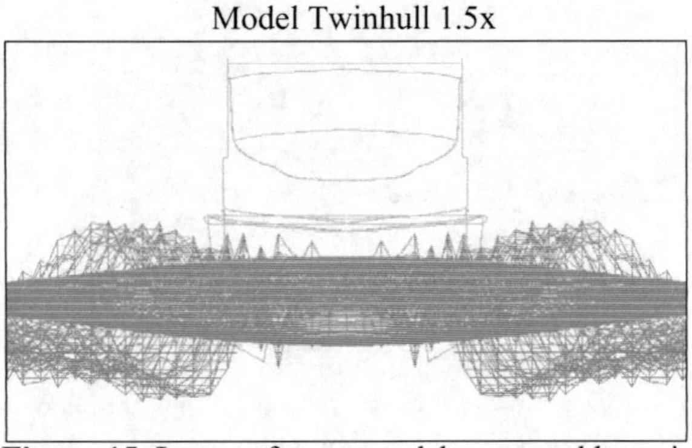

Figure 15. Image of wave model generated by twin-hull ship 1.0x

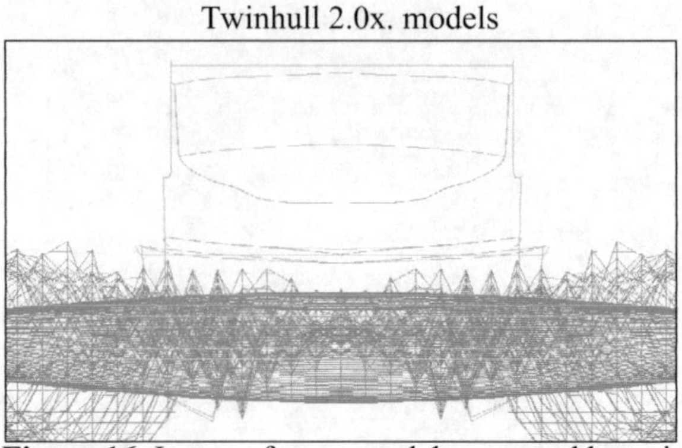

Figure 16. Image of wave model generated by twin-hull ship 1.0x

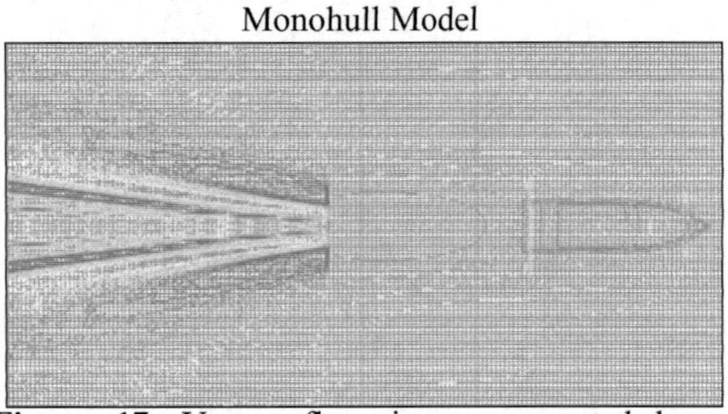

Figure 17. Vortex flow image generated by a monohull ship model

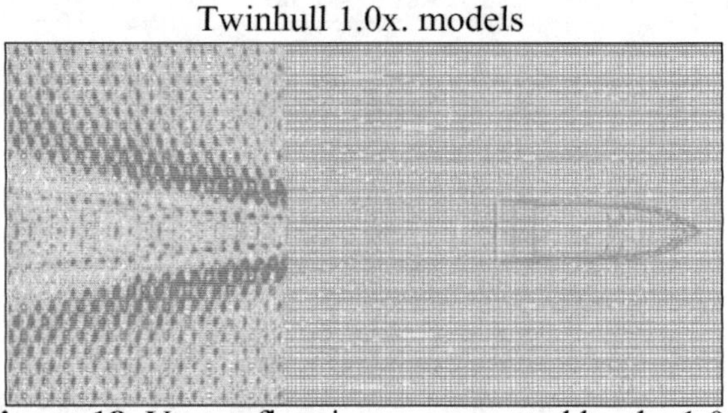

Figure 18. Vortex flow image generated by the 1.0x twin-hull ship model

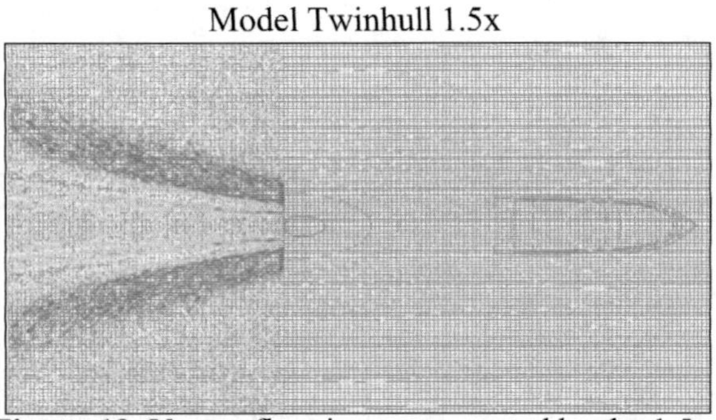

Figure 19. Vortex flow image generated by the 1.5x twin-hull ship model

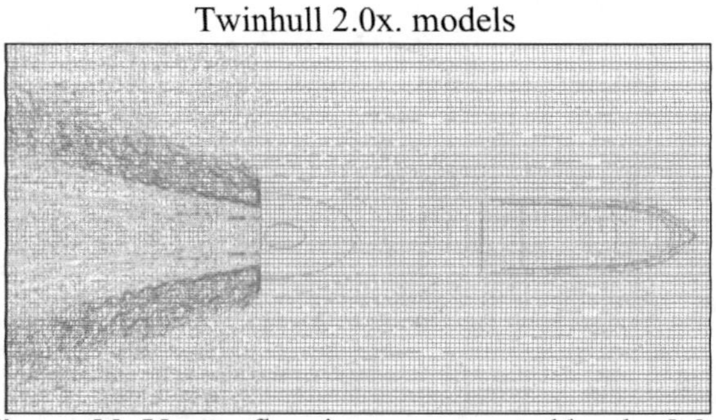

Figure 20. Vortex flow image generated by the 2.0x twin-hull ship model

4.5. Performance stability

The results of the analysis of the stability of the model at full load conditions based on IMO criteria using Hydromax on the monohull model resulted in the following ship conditions as illustrated in Figures 21 to 24.

IOP Conf. Series: Earth and Environmental Science 1081 (2022) 012046 doi:10.1088/1755-1315/1081/1/012046

Monohull Model

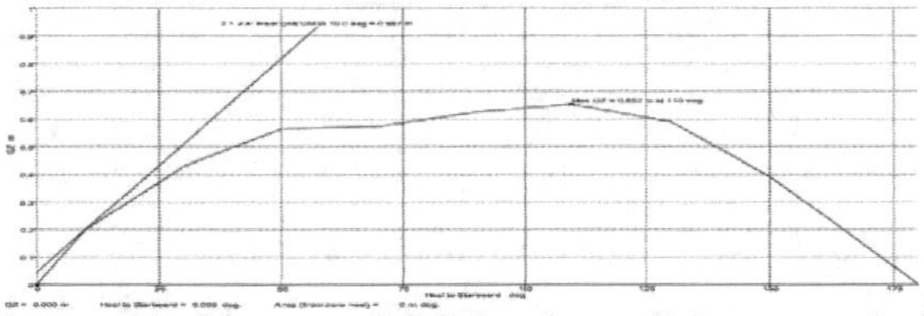

Figure 21. Picture of full load condition monohull model stability curve
Angle of maximum GZ is:110.0 deg
Initial GMt : 0887 m
IMO criteria minimum GZ 30 deg and Initial GMt is 0.150 m

Twinhull 1.0x. models

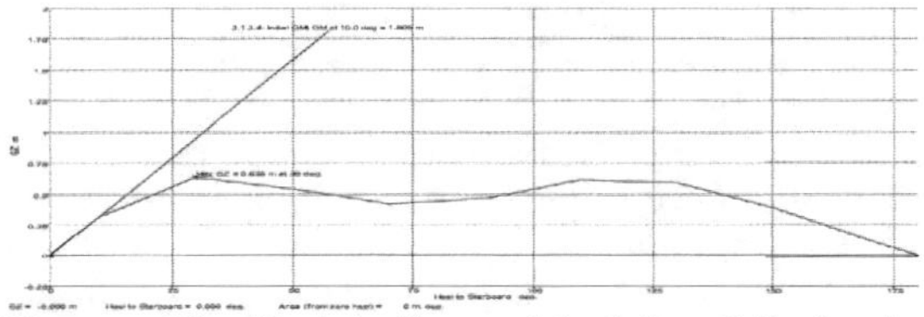

Figure 22. Twinhull model 1.0x full load condition stability curve
Angle of maximum GZ is: 30 deg
Initial GMt : 1,805 m
IMO criteria minimum GZ 30 deg and Initial GMt is 0.150 m

Model Twinhull 1.5x

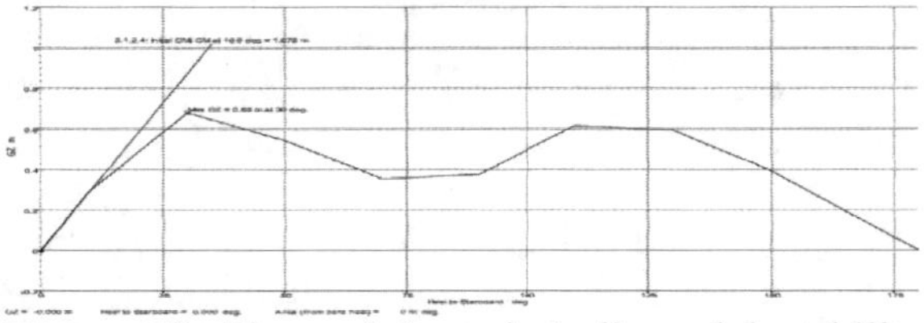

Figure 23. Figure 1.5x twin-hull model stability curve full load condition
Angle of maximum GZ is: 30 deg
Initial GMt : 1,678 m
IMO criteria minimum GZ 30 deg and Initial GMt is 0.150 m

Twinhull 2.0x. models

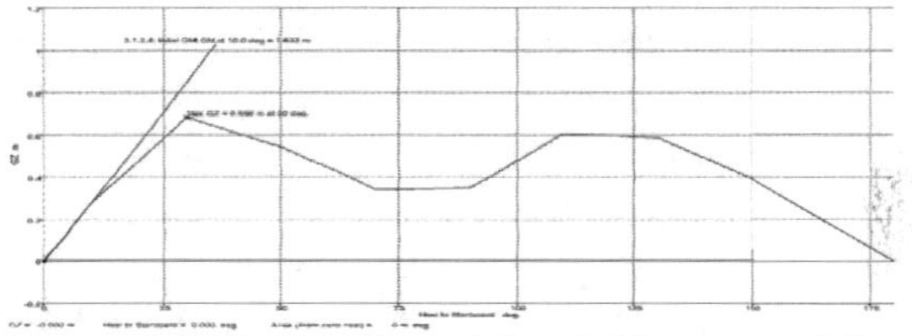

Figure 24. Twinhull model stability curve 2.0x full load condition
Angle of maximum GZ is: 30 deg
Initial GMt : 1,678 m
IMO criteria minimum GZ 30 deg and Initial GMt is 0.150 m

4.6. Difference between catamaran hull and monohull in fixed Displacement

To find out the differences in characteristics between the monohull hull and the twin-hull hull at a fixed displacement (D) and a fixed length (L) and a fixed-width (B), then the results of the comparison of the characteristics of the monohull and Catamaran hull shapes with 3 comparisons of the distance of the catamaran hull on the model length. 8 m ship is as in Table 3:

Table 3. The results of comparing the characteristics of the monohull and Catamaran hull shapes with 3 comparisons of the distance of the catamaran hull on the model length. 8 m ship

Criteria		Monohull	Hull Distance Catamaran		
			1x	1.5x	2x
Loa(m)		8.0	8.0	8.0	8.0
Bm(m)		2.2	2.2	2.2	2.2
Lwl(m)		7,154	6,622	6.585	6,649
Bwl(m)		2,097	2,194	2,197	2,197
T(m)		0.452	0.452	0.487	0.487
(tons)		2,498	2,488	2,485	2,486
LCB/Lwl	(m)	-0.11	-0.15	-0.12	-0.12
LCB	(m)	-0.784	-0.971	-0.788	-0.789
Cp		0.713	0.797	0.783	0.799
Resistance/RT (N) at Vs=40 kn		10496.05	5026.13	4719,16	4891,15
Power (Hp) at Vs=40 kn		482.74	231.16	217.04	224.95

Criteria		Monohull	Hull Distance Catamaran		
			1x	1.5x	2x
Max GZ	(m)	0.652	0.638	0.68	0.686
GZ max angle	(level)	1100	300	300	300
Initial GMt at full load condition		0.887	1,805	1,678	1,678
Draft at FP	(m)	0.591	0.646	0.728	0.713
Draft at AP	(m)	0.347	0.29	0.268	0.28
Trim	(m)	0.244	0.356	0.46	0.433

Meanwhile, to compare the characteristics of the 8-meter model, it can be compared with the monohull and twin-hull models with the same hull distance ratio as in the 8 meter model for ship lengths ranging from 12m, 16m, 20m, 24m, 28m, 32m, and 36m.

5. Conclusions

The most prominent finding to emerge from this study is that monohull catamaran hull shape has a significant comparison of ship characteristics in calm water conditions, namely:

a. On the monohull hull, the hydrostatic Wetted Surface Area (WSA) characteristic is 13,883 m2, while the catamaran with 1x hull width is 15,562 m2, 1.5x hull width is 16,473 m2, and 2x hull width is 17,186 m2. The characteristics of the transverse metacenter radius (BMt) on the monohull hull is 1,486 m, while the catamaran with a hull distance of 1 x hull width is 1,882 m, a hull distance of 1.5 x hull width is 1,721 m, and a hull distance of 2 x hull width is 1,698 m.

b. The resistance and power of the ship show that it is lower than the monohull at high speed (40 knots). Namely, on the hull of the monohull, the resistance (RT) is 24267.07 N with Power (Hp) is 1116.09 hp, while the catamaran with a hull distance of 1x the width hull resistance (RT) is 27590.17 N with Power (Hp) is 1268.93 hp, hull distance 1.5x width of hull resistance (RT) is 34164.13 N with Power (Hp) is 1571.28 hp and hull distance 2x width of the ship's hull (RT) is 42281.97 N with Power (Hp) is 1944.64 hp

c. parametric study of the shape of the hull was also carried out, and the results showed some similarities and differences in the behavior of the two forms of the hull. On a catamaran, the selection of the distance between the two hulls shows a significant difference in wave behavior.

d. The assumption that the center of gravity is following the shape of the hull based on the location of the center of gravity per unit area and the ship's load at the maximum condition (full load condition) or full Displacement produces a stability arm (max GZ) and radius of gyration (GMt), on the ship's hull. monohull stability arm (max GZ) 0.297m and radius of gyration (GMt) 0.631 m, while the catamaran with a hull distance of 1x the hull width of the stability arm (max GZ) 0.60m and radius of gyration (GMt) 1.601m, hull distance 1.5x the width of the hull of the stability arm (max GZ) 0.699m and the radius of gyration (GMt) 1.931 and the hull distance is 2x the width of the hull of the stability arm (max GZ) 0.713m and the radius of gyration (GMt) 2.033 m

The analysis results on the 8m ship model can be used to select the hull type in the initial design stage. However, to increase confidence in determining the hull type, it can also be seen recommendations on ship models with lengths of 12m, 16m, 20m, 24m, 28m, 32m, and 36m. Of course, the choice of hull type must be based on a more detailed study that must consider other parameters such as production costs, operating costs, seakeeping, etc. It is suggested that the association of the actual model is investigated in future studies to validate the performance.

References

[1] Bhattacharyya, Rameswar., (2006) Dynamics of Marine Vehicles. First edition. John Wiley & Sons, Inc.
[2] David Stimson, 2002, How to Build the Ocean Pointer: A Strip-Built Outboard Skiff

[3] Husain, Arastu Sajjad, 2014, Improving Hull Design for Better Efficiency and Environmental Protection, International Society of Ocean, Mechanical and Aerospace Scientists and Engineers.
[4] IntelliStab, Marine Service, Research Project 557 Stability of Criteria Applied to Small Fishing Vessels and Associated Survivability (2006), School of Marine Science & Technology Newcastle upon Tyne United Kingdom.
[5] International Maritime Organization (IMO)(2004) Sub-Committee on Stability and Load Lines and Fishing Vessel safety, Review of Intact Stability Code.
[6] Seif, MS, E. Amini., Performance Comparison Between Planning Monohull and Catamaran at High Froude Number. Iranian Journal of Science & Technology, Transaction B, Vol.28, No.B4. 2004.
[7] Bertram, V. (2002). Practical ship hydrodynamics. Butterworth Heinemann.
[8] Renilson, M and Hardon, P, "International Journal of Small Craft Technology" Qintq, UK.)

IOP Conf. Series: Earth and Environmental Science 1081 (2022) 012047 doi:10.1088/1755-1315/1081/1/012047

A Preliminary Web-Based Intermediary Application Design For Ship Repair Planning Services

Nurhadi Siswantoro[1], Hari Prastowo[1], Hafiz Rahadyan Nugroho[1,2], Muhammad Badrus Zaman[1], Trika Pitana[1], Dwi Priyanta[1], Hari Prastowo[1]

[1]Department of Marine Engineering, Institut Teknologi Sepuluh Nopember (ITS), Surabaya, Indonesia

[2]Department of Maritime Studies, Systems Engineering and Logistics, Hochschule Wismar, Germany

corresponding author's e-mail: nurhadi@ne.its.ac.id

Abstract. According to data from the 2019 Ministry of Transportation Statistics Book, the number of national shipping fleets or ships in the country from 2015-2019 has doubled. However, from the potential of the current operating fleet, it shows that more than 75% of the fleet are over 20 years old. With such conditions, it is necessary to plan for ship maintenance and repairs to be as good as possible and to spend as little time as possible. This research aims to design a web-based intermediary application for ship repair planning services that can be accessed by using a website browser. The initial stage is to observe the condition of ship repair and planning from shipyard and shipowner perspectives through the literature and field study. From the result of the observation, then analyze what problems occur in the field and then arrange what functions or features are needed to be included in the design of the application to be able to solve these problems. Application design is done by making mock-ups, designing interfaces, creating databases and coding of the application system. After that, the application is tested and verified to find out that the application can run as intended and know the strengths and weaknesses of the application. The features in this application are designed to intermediary shipyard and shipowner in the ship repair planning process. The result of this preliminary web-based application design is by using this application, ship owners can search for docking space in available shipyards according to the schedule and place desired by the shipowner. In addition, by using this application, the shipowner can find out when the ship's certificate expires so that they can prepare a schedule for conducting a survey. Then, both the shipyard and the shipowner can use this application to monitor and report on the progress of ship repair work.

1. Introduction

According to data from the 2019 Ministry of Transportation Statistics Book, the number of national shipping fleets or ships in the country has doubled. In 2015, the original fleet of 16,142 vessels increased to 32,587 in 2019. However, according to Hasbullah (2016), from the potential of the current operating fleet, it shows that more than 75% of the fleet are over 20 years old. The age of this ship is a long-standing problem for most national shipping companies. The consequence of operating ships over 20 years old causes the cost of repairing and maintaining ships to increase relatively [1]. The increasing number of ships and the age factor of ships that are no longer young will certainly

IOP Conf. Series: Earth and Environmental Science 1081 (2022) 012047 doi:10.1088/1755-1315/1081/1/012047

require shipyards for repair and maintenance of these ships. Don't forget also to make sure the ship's certificates are still in their validity period. Ship repair and certification is a must to guarantee a certificate of seaworthiness and safety. The survey period that is required or used as a reference is the docking survey because it can maintain the ship as a whole, especially the hull below the waterline, besides that in docking the ship can maintain ship performance standards and can guarantee the operation of the ship until the next survey period.

But, in some cases, the shipowner finds it difficult to adjust the ship's docking time to the survey period that needs to be carried out. This can happen due to several factors faced by the shipowner. The first, according to Firstdhitama (2018), is due to the tight operational schedule of the ship operational time and area, and also the length of time it takes for repair work. The hectic schedule or work contract is also coupled with the ship's operational area which can be quite far from the shipyard that is usually selected for its repairs [2]. Second, according to Samuel (2014), ship owners tend to carry out ship repairs close to the maximum due date for surveys, as well as approaching high season schedules, for example approaching the Eid homecoming season. This is also worsened by the fact that many managements still do not have integrated data regarding the survey schedule for their fleets [3].

On the other hand, from the shipyard side, the problems that occur include the scheduling process for the activities of each ship repair work in the shipyards in general, which is still done manually, so this method is slow and ineffective. Although some work has been done using computers, the parts are still separate from each other and not yet integrated. This results in inefficiency and often causes errors that can hamper the performance and competitiveness of a shipyard in the ship repair service process [4]. For example, it is in terms of scheduling information and procurement of repair materials. Scheduling information systems are often not integrated into other departments, such as the repair department which directly handles repair work in the field, making it difficult to determine the planning method and list of repair project work and materials needed. Sometimes when there is a material delay, from one production department to another, the production department that requires materials from another production department still has to submit a written request for goods to the responsible production department. Furthermore, the responsible production division still has to re-apply to the material procurement department.

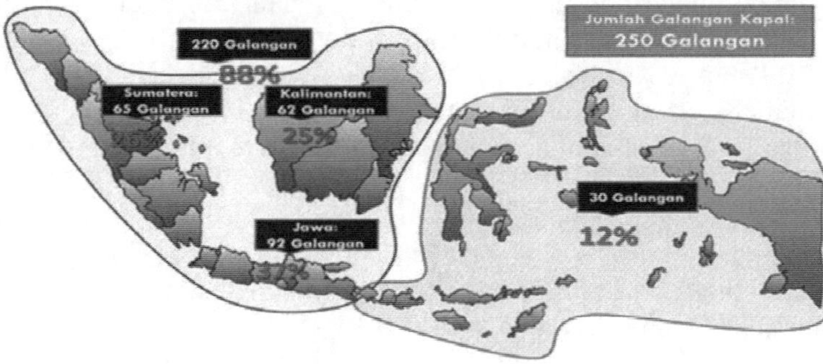

Figure 1. The Distribution of Shipyards in Indonesia

The next problem from the shipyard side is the uneven distribution of shipyards. According to data obtained from the Association of Indonesian Ship and Offshore Industry Companies (IPERINDO) in 2013 shown in Figure 1, the number of shipyards in Indonesia is around 250 shipyards. From that number, about 88% of shipyards are located in western Indonesia. The remaining 12% are in eastern Indonesia. Most of the major shipyards are concentrated in Batam, Tanjung Priok (Jakarta), and Surabaya. Shipowners tend to choose to do repairs or dock at their subscribed shipyards. However, because it is far from the port where their ship operates, ships that want to dock must sail quite far from the shipyard. This results in a high-cost economy for ship operators who want to repair their ships. In addition, for shipyards that are not spread out in areas that are the center of shipyards in Indonesia or are not shipyards where the previous shipowner subscribed, then these shipyards will be more difficult to get ship repair projects even though in fact these shipyards are also able to provide facilities and services needed. The other reasons for the delay in ship repair works are multi-faceted.

One of the identified reasons is project management and planning. Project scheduling including manpower, materials, reusable resources and facilities is a very complex task. Poor allocation of resources, keeping workers waiting for prerequisite activities, resulting in fluctuating workloads leading to high overtime costs and possible delays in delivery [5].

In making this research, the author was inspired by several previous studies. The inspiration can be in the form of ideas or topics that will be discussed that have similarities with the topics that the author will take. And it can also be used as inspiration for methods in designing this web-based application. Dave Hansel on 2016 make a research with the title of *"Perancangan Aplikasi Komputer Berbasis Android Untuk Estimasi Biaya Reparasi Kapal Interaktif"*. From this paper, the author sees an opportunity for a process or field that can be further refined in the ship repair process. This research discusses of making an android-based computer application that functions to estimate the cost of ship repairs [6]. The author also inspired by research by Renny Rahmadi Putra on 2016 entitled *"Perancangan Aplikasi Berbasis Komputer Untuk Proses Manajemen Mutu Pada Pembangunan Kapal Baru"*. This research talks about quality management in shipyards is currently not implemented systematically. The results of the inspection process for the construction of new ships are still stored in separate computer cabinets and folders [7]. The author also read a research by Rani Nurwanti on 2016 with title of *"Analisa Peningkatan Kualitas Layanan Jasa Reparasi Kapal di Galangan Kapal Jawa Timur"*. This research talks about the quality of repair services and the level of user satisfaction significantly affect the sustainability of the shipyard. The purpose of this research is to improve the quality of ship repair services based on the assessment of ship owners/shipping companies who have done ship repairs at shipyards in the East Java area [8]. And the last but not least, the author inspired by the research from Marlen Jenri Hutapea on 2017 with title of *"Perancangan Aplikasi Berbasis Android untuk Manajemen Proyek Reparasi Kapal"*. This research have similarities in the form of management project on ship repair services [9]. But from this research, the author can develop an application with better function and also cope the downside of the previous research. The main objective of this study is to design an android-based computer application for ship repair project management.

Based on the problems that have been described previously, in this research, the author designs a web-based intermediary application that can connect the shipyard with the shipowner in planning ship repair projects. This application will help ship owners to schedule the time, place and shipyard that will be selected for ship repair, which services are selected, as well as reminders of the docking period or repair schedule based on the due date period from ship certification to ship owners. Then, for the shipyard it will also be facilitated in scheduling docking so that it can maximize so that the repair schedule is always filled, and the ship repair process continues. With this application service, it will support ship repair scheduling activities and improve services in providing information and administrative convenience both from the shipowner and shipyard side. This application is expected to overcome the problems that occur and can produce a system that can collect, process, store, and distribute information quickly, precisely, and accurately

2. Methods

This research is aimed at designing a web-based intermediary application for ship repair planning services. Please, see Figure 2 (two) to understand the research flow chart or stage diagram of this research

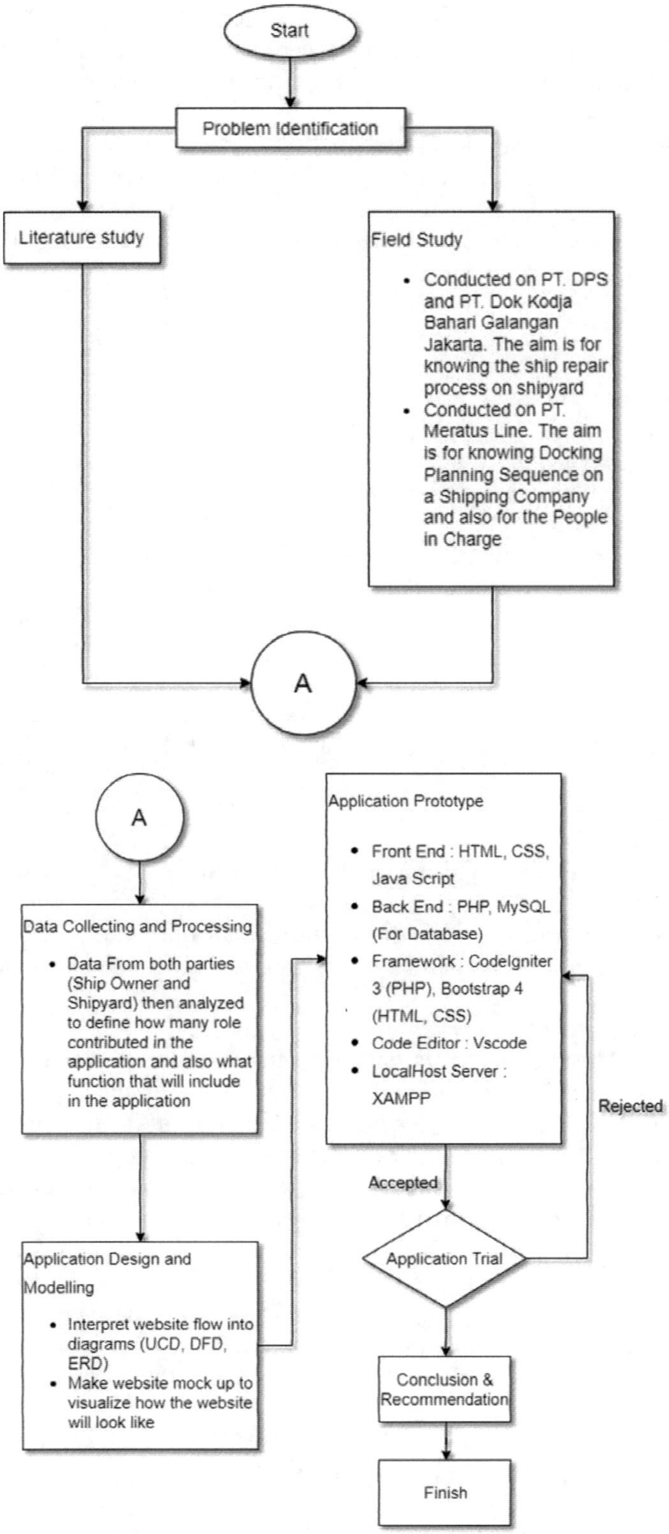

Figure 2. Research Methodology Flowchart

IOP Conf. Series: Earth and Environmental Science 1081 (2022) 012047 doi:10.1088/1755-1315/1081/1/012047

2.1. Problem Identification
The first step taken in doing this research is to identify the background that forms the basis of this research. The background can certainly be in the form of problems that will be faced, and solutions must be found for these problems.

Problem identification / problem statement is an attempt to state explicitly what research questions need to be answered or find a way out. This research contains two questions that are expected to be answered in order to find a solution to the problem. In other words, this problem formulation is a complete and detailed question regarding the scope of the problem to be studied based on problem identification and problem limitations. In addition, the formulation of the problem must be related to the purpose of the research

2.2. Literature Study and Field Study
The next step in doing this research is to examine the theories that support this research in the form of literature studies and also go directly to the field in the form of field studies. Literature study is a theory that will be used to support the work of this research in order to have a basis stating that the theory is in accordance with what will be done. Through literature study, you will also get a better understanding of what will be done to complete this research. The most appropriate media as an assistant in this literature study is through books, scientific journals, completed final assignments related to what is being done, papers, and also through the internet.

Field studies must also be carried out to make it easier to do this research. When conducting a field study, it will be seen whether the theory that has been studied has a relationship or is relevant to the conditions in the field. With the relationship between literature studies and field studies based on what actually happened, it will be a very good helper in doing this research. In the field study, it will be seen what kind of work is done in project management and also the ship repair process itself in a company.

2.3. Data Collecting and Processing
After the data that has been collected is obtained, the next step is processing the data that has been obtained. The data obtained is then analysed to obtain a model in the application to be designed. Each data obtained is then sorted, and grouped based on the plan from the design that has been made. From this data processing, parameters and indicators are generated that will be included in the application.

2.4. Application Design and Modelling
After the data that has been collected is obtained, the next step is processing the data that has been obtained. The data obtained is then analysed to obtain a model in the application to be designed. Each data obtained is then sorted, and grouped based on the plan from the design that has been made. Based on the results of this data processing, it is expected to be able to produce a mock up which is a real picture of a design concept, how the concept will look later when the application has been designed and the results of the mock up will look very much like the original form.

The design of the mock up that has been approved will be designed into a planned application. The form of the application that is designed in essence must resemble the mock up that has been made because it becomes a guide in designing the application. With the mock up that has been made, the programmer will find it easier to design the application. This is because the application designer will make a design according to what has been planned without thinking about what application will be designed. The point is that the mock up that has been prepared is made as complete and clear as possible so that the designer does not have to imagine again what the shape of the application will look like from each step in the operation of the application.

2.5. Application Prototype
After the mock up of the application is formed, then Database Design is carried out, as the database is used as a data storage medium and information is entered into the application. Then the application coding is done, coding is the main process in application programming so that the application can run according to the designed application modelling.

2.6. Application Trial
The next stage is testing the program and function of the application that has been designed. Applications that have been designed and approved by the supervisor will then be tested. When testing of this application is being carried out, what is the hope of the author is that the function of this application can be as expected so that through this application it can help provide services and simplify the ship repair planning process.

2.7. Conclusions and Suggestions
After all stages have been carried out in the research, meaning that the application has been tested and improvements have been made to the application, conclusions will be drawn from working on this application. The conclusions are essentially in line with the objectives of the research, which also answers the problem formulation that has been determined in the previous chapter. Do not forget that there are recommendations in the suggestions section that can be given to improve or develop this research for the future.

3. Result and Discussion

3.1. Project Management
Project management can be defined as a process of planning, organizing, leading, and controlling a project by its members by utilizing resources as optimally as possible to achieve predetermined goals [10]. The basic functions of project management consist of the management of scope of work, time, cost, and quality. In the world of management in general, including in project management, the objectives of the project must be met, namely by completing the project despite environmental problems, pressure on completion time [11]. In fact, the lack of project management processes for ship repair in Indonesia is still very high, this has an adverse effect on the ship repair process. Based on the results of the literature study that I got, according to Zaman (2018) one of the main causes of delays in the ship repair process in the study case at PT. Adiluhung Saranasegara is a poor scheduling so that the plan and realization of the work are not achieved [12]. According to Padaga (2018) in a case study at PT. PAL Indonesia (Persero) the project management process was disrupted due to several things, among others, due to supervision when the work was not going well, evaluation of work results was not carried out because work notes were not conveyed to other departments, and lack of coordination between the shipyard and the shipowner [13]. From some these problems, it was also decided to provide a feature to oversee ship repair project activities as well as a feature to report results and job evaluations that can be accessed by internal and external parties at the shipyard and shipowner who has responsibility for the jobdesk.

3.2. Ship Repair Process
A repair project at a shipyard is a ship repair project managed by the shipyard in accordance with an approved contract. The nature of the project is a job order, and the completion schedule depends on the agreement of both parties. However, the length of time a project runs in a shipyard can change from the time agreed in the contract. This leads to the fact that one day the project can run smoothly and be completed on time, and there are also projects that experience delays. Projects can be completed on time or late, it is greatly influenced by the shipyard management in working on the project. If the shipyard management has been managed properly, the results will be good and can be on time and vice versa. So to realize this, a good system in shipyard project management is needed so that the work on the project is on time and the results are good.

The work activities in the shipyard or ship repair business generally consist of docking activities, ship cleaning below the waterline, plate replacement, propulsion and rudder system repairs, machinery and electrical repairs. From the overall ship repair activities described in general, some of the work is carried out by non-permanent workers or commonly referred to as sub-contractors [6].

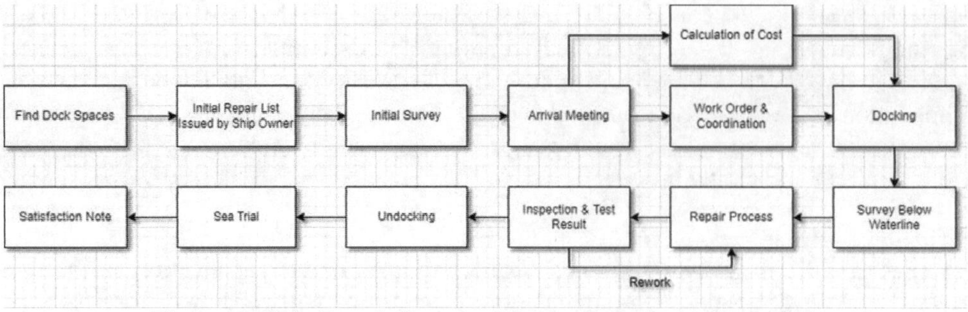

Figure 3. Ship Repair Process / Business Process Flow [13]

According to Padaga (2018), ship repair work / business process has a flow that must be carried out before the ship begins to be repaired, see Figure 3. The sequence of ship repair work / business process that usually occurs in Indonesia can be explained as follows :

1. The existence of Owner Requirements issued by ship owners or shipping companies
2. Calculation of the estimated cost by the marketing department from the shipyard based on the repair list data that has been submitted to the shipyard, after that submitting the cost calculation back to the Owner or shipping company.
3. If the estimation calculation that has been submitted to the owner or shipping company is accepted, a repair list will be made for review
4. Repair list for review that has been made will be distributed to parties involved in the project so that they can be studied first
5. An Arrival Meeting (AM) was held which was attended by crew members, Owner Surveyors, workshop managers, parties from the relevant departments, and Project Managers. The repair list for review was discussed at this meeting and checked whether there was work that needed to be removed or added. From the results of this meeting, a repair list of working firs will be generated which will later be used in the repair project
6. If in the process of work there is a workshop or the party working on the project sees something that needs additional improvement, an additional order list can be made, but the order must be approved by the Owner.
7. If the additional order is approved by the Owner, the PM administration will make a list of additional orders that have been approved, with a note that the additional cost is not more than 10% of the estimated repair cost.
8. After the repair work is completed, including additional work, a Satisfaction Note will be made by the PM
9. Satisfaction Note will be submitted to the marketing party for recalculation of the actual costs used during the project
10. Calculations made by the marketing party will then be given back to the Owner
11. Based on the calculations made by marketing, the Owner can negotiate the price with the marketing party
12. The owner must pay off the payment according to the results of negotiations with the marketing division before the ship sails again.

3.3. Ship Repair Process in PT. Dok & Perkapalan Surabaya and PT. Dok Kodja Bahari Jakarta

The ship repair process that takes place at DPS is carried out based on the Standard Operation Procedure (SOP) that has been made by the shipyard. Stages of the ship repair process at PT. DKB Jakarta is broadly the same as in PT. DPS Surabaya. Based on observations at the shipyard, the processes that occur in the shipyard are :

1. Pre-Docking
   a) Submission of Repair List

      Here, the owners who work are represented by the owner surveyor (OS). At this stage, it starts from the preparation of a list of repairs which is compiled based on input from the crew

IOP Conf. Series: Earth and Environmental Science 1081 (2022) 012047 doi:10.1088/1755-1315/1081/1/012047

and owner surveyors and from the list of repairs in the previous survey/year. At this stage the condition of the ship is still sailing to the shipyard.

b) Initial Survey and Arrival Meeting

The initial survey was carried out to ensure the condition of the ship before entering the dock and matched it to the repair list they submitted. This is done to avoid unequal views on the condition of the ship which results in losses on the one hand.

Furthermore, an Arrival Meeting will be held, namely a meeting between the ship owner and the shipyard to discuss the list of repairs, repair schedules, repair times and costs. At this stage it is possible to reduce or add work as agreed by both parties. And there is a fee agreement that is adjusted to the amount of work that has been agreed upon.

c) Material Calculation & Price

Material calculations are carried out to obtain the initial price for the construction contract. The material itself is the result of calculations from a repair list that has been approved by the shipyard. This work is usually carried out by the shipyard

d) Work order

Here the shipyard issues a work order based on the findings from the repair list at the shipyard, later this work order is distributed to all persons in charge of ship repair work.

e) Coordination

Coordination is carried out to equalize perceptions and briefings of all parties involved and responsible for the project. This stage is carried out by the project leader. Where the project leader coordinates with related parties before the start of work.

2. Docking

a) Preparation

The preparation stage is the stage before the ship enters the floating dock or docks at the floating dock. At this preparatory stage, the shipyard begins to prepare electricity, fresh water, lighting, security, and other things needed during docking work.

b) Docking

At this stage the ship is put into a dock or moored to a floating dock with the direction of the dock master and the help of tugboats.

c) Survey before work

This survey is carried out to see the condition of the ship which is below the waterline. Because the part of the ship below the waterline has not been surveyed, the survey is carried out when the ship is already above the floating dock.

d) Repair Process

At this stage, the ship has entered the dock and preliminary work will be carried out in the form of cleaning the hull, checking the thickness of the hull plate and others. At this stage the class and the surveyor owner will carry out inspections so that it is possible for additional work to be carried out by the shipyard if additional damage is found. But try to stick to the schedule that has been planned. And if there is a change, it is immediately notified to the Owner with a note that the amount of additional work does not exceed the number of times the ship has to board the dock or if there will be a change in schedule it will be discussed together between the owner and the shipyard. The process of carrying out ship repairs at the shipyard is carried out as usual.

e) Ship Repair Inspection and Test Results

This activity is carried out to check the quality of ship repairs. Furthermore, the ship is in a position ready to launch into the sea so it is necessary to do a final check on all parts of the underwater. So that the class and the surveyor owner will conduct another inspection of the work that has been done until no problems are found in the underwater hull.

3. Undocking

The ship is removed from the dock then the last and final finishing work is carried out before being prepared for the sea trial.

4. Sea Trial

At this stage the ship is already above the water so that the underwater work is done. Then a sea trial was conducted to determine the performance of the ship and other systems. At this stage the documents and reports will be finalized and the ship will be ready to sail. After this process is complete, a satisfaction note will be made by the Project Leader to report the overall repair and realization work along with the results that have been carried out by the shipyard.

3.4. Docking Planning Conducted by Meratus Ship Management (MSM)

This Procedure defines the activities to be carried out by the Ship Management shore staff and Vessel's staff in order to prepare the vessels for the repair works and drydock at a Shipyard and the management of the relevant works while being carried out, in compliance with the MSM. This include criteria for planning, carrying out, or receiving internal/external inspections, both ashore and on board ships to verify compliance with the policy, classification regulations, requirements. There are also PIC for the sequences that later will determine the distribution of user roles in the application. Here are the list of person in charge and their responsibility in the docking planning sequence :

a) **General Manager (GM)**
- To instruct and support the Ship Manager and Superintendents for preparation, evaluation, selection and execution of repair work specification, reporting the works, schedule and budget updates to Company Management.
- To fix the budget approved/ review by Management and communicate it to technical staff involved.
- To update the Company Management about the works progress and significant works scope required changes affecting budget and time at Shipyard in order to get Management approval.
- To provide the final approval of Shipyard invoice.

b) **Ship Manager (SM)**
- To provide the necessary technical information and support to the Superintendent for the repair works, especially repair works having a large impact on the budget/works duration.
- To assist the GM in the selection of Shipyards invited to bid and in the evaluation of the Shipyards' bids received.
- To review the vessels repair work specification prepared by the Superintendent.
- To evaluate and compare the repair offers received from the Shipyards to comment and submit to the GM for his evaluation.
- To attend the vessel one week before vessel completed docking to determine certainty when the vessel is ready to operate.

c) **Superintendent (SUP)**
- To prepare the Ship repair/ dry-dock works specification with the assistance of the management on board for engine and propulsion section.
- To coordinate with Class.
- To provide a regular daily update to the Ship Manager about the works progress, repair schedule and budget of the vessel at Shipyard.
- To analyze, discuss and review the final works list and relevant Shipyard invoice.
- To submit results of the above to the Ship Manager and GM for their evaluation and confirmation.

d) **Dock Coordinator / Dock Monitoring**
- Request for offers from several Shipyards and make comparisons.
- Monitor daily reports from SUP and HSP.
- Review invoice from Shipyard.

e) **Hull Superintendent (HSP)**
- To prepare the Ship repair/ dry-dock works specification with the assistance of the management on board, exclude engine and propulsion section.
- To monitor the day-to-day in the shipyard
- To discuss with Shipyard non-compliances versus the contractual commitments.
- To organize and coordinate all activities related to the repair works.

f) Master, Chief Officer and Chief Engineer

- To maintain all safety precautions, as required by Company SMS, by Vessel Flag Administration and by Shipyard internal procedure on board during repairs.
- To provide the necessary instructions and to provide follow up and control to any subcontractor carrying out a work which is not included into Shipyard's contract responsibilities on board of vessel, or by the ship staff, according to the instructions and orders received by the Superintendent.

PT. Meratus Line already has a Docking Planning Sequence that they used to as we can see in the Figure 4. The docking sequences starts a year before the planned schedule and will continue until 2 weeks after the undocking.

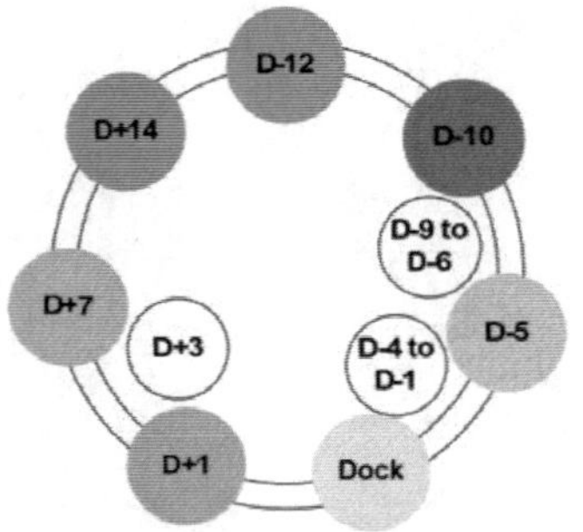

Figure 4. Docking Planning Sequences (Meratus Ship Management)

3.5. Application Design Basic Framework

Application modeling is the initial stage in designing applications. The application modeling aims to make it easier to communicate the intent and purpose of the application to programmers or users. In the application modeling will be known how the application will work. The application modeling is made in the form of a diagram. Application modeling will be divided into 3 forms:

1. Use Case Diagram
2. Entity Relationship Diagram
3. Data Flow Diagram.

3.5.1. Use Case Diagram

Use Case Diagram supports ERD and DFD so that it can be run in the application. UCD describes the authority and duties of the User to run the application. Use cases are developed by system analysis together with users. Use cases explain simply the system functions from the user's point of view. UCD in the application can be described in the following figure. Please see Figure 5 to understand the Use Case Diagram for shipowner and shipyard.

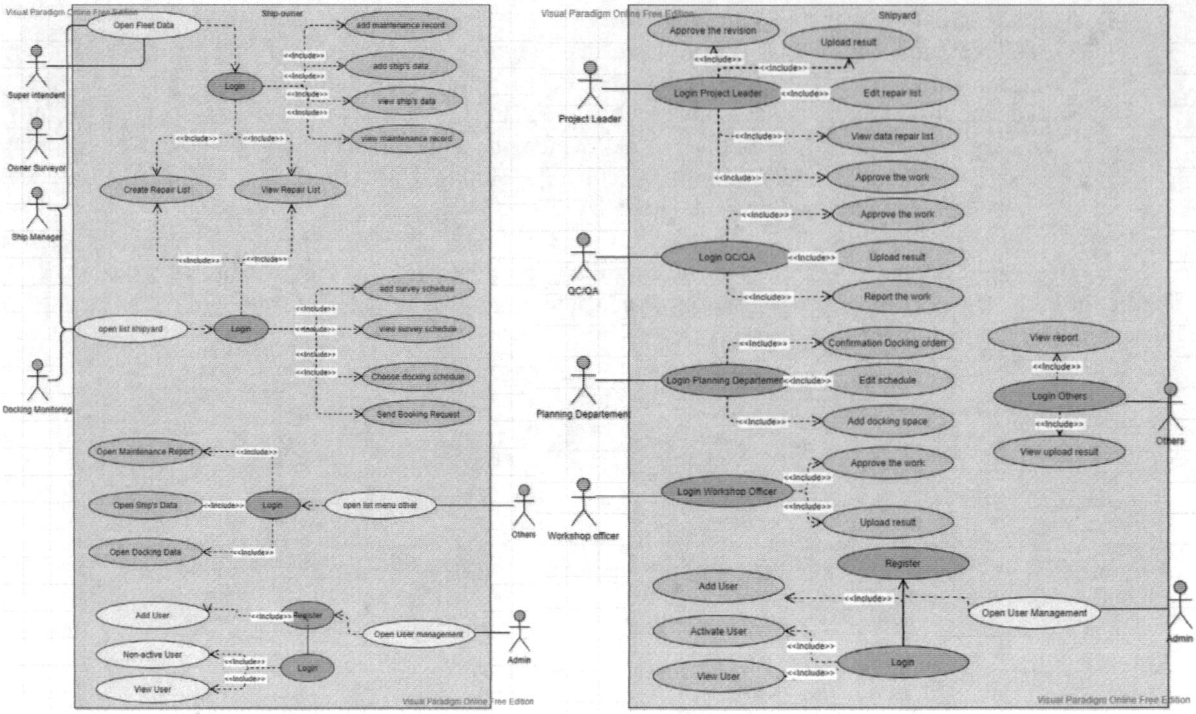

Figure 5. UCD for Shipowner and Shipyard

3.5.2. Data Flow Diagram (DFD)

From UCD it will be known how the data flow is run by the application. Then after that made a Data Flow Diagram (DFD). To describe the logic of the system requirements, namely what processes are needed by the system and how the entry and exit of information in the system is used Data Flow Diagrams (DFD). Flowcharts or data flow diagrams can be depicted in the following figure.

In the Figure 6, shown the DFD for Ship Owner or Shipping Company from the very beginning of the data flow process (login) until it finished. can also be seen in the picture, when a user logs in, they must have a role that has been assigned to that user. with this role sharing will automatically divide users according to the authority they can do in the application. From the login page, each user will be directed to the dashboard, then they can choose what menu they will work on. Even in each menu, the data or actions directed by the user will affect the next process until the sequence of processes is complete.

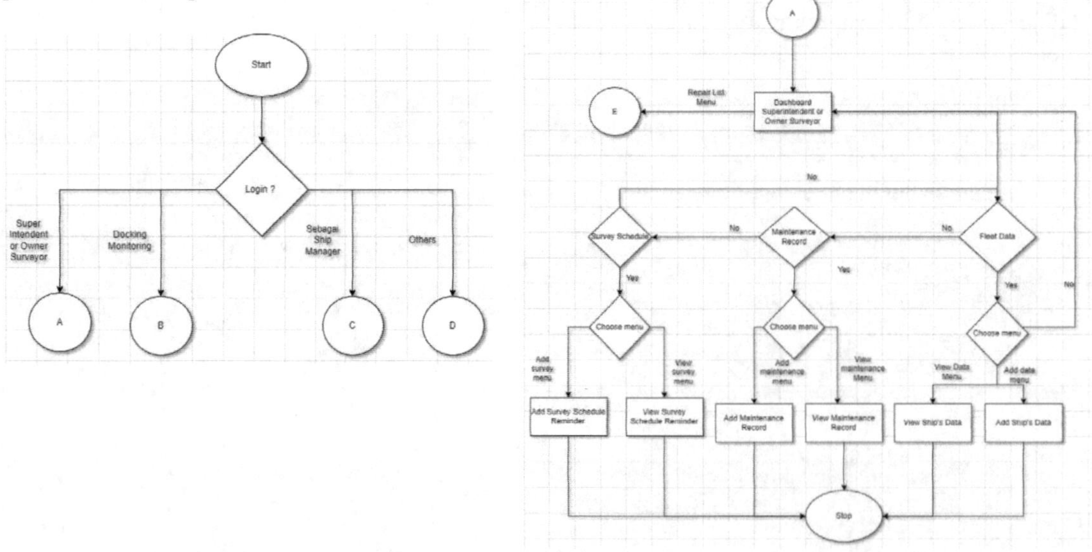

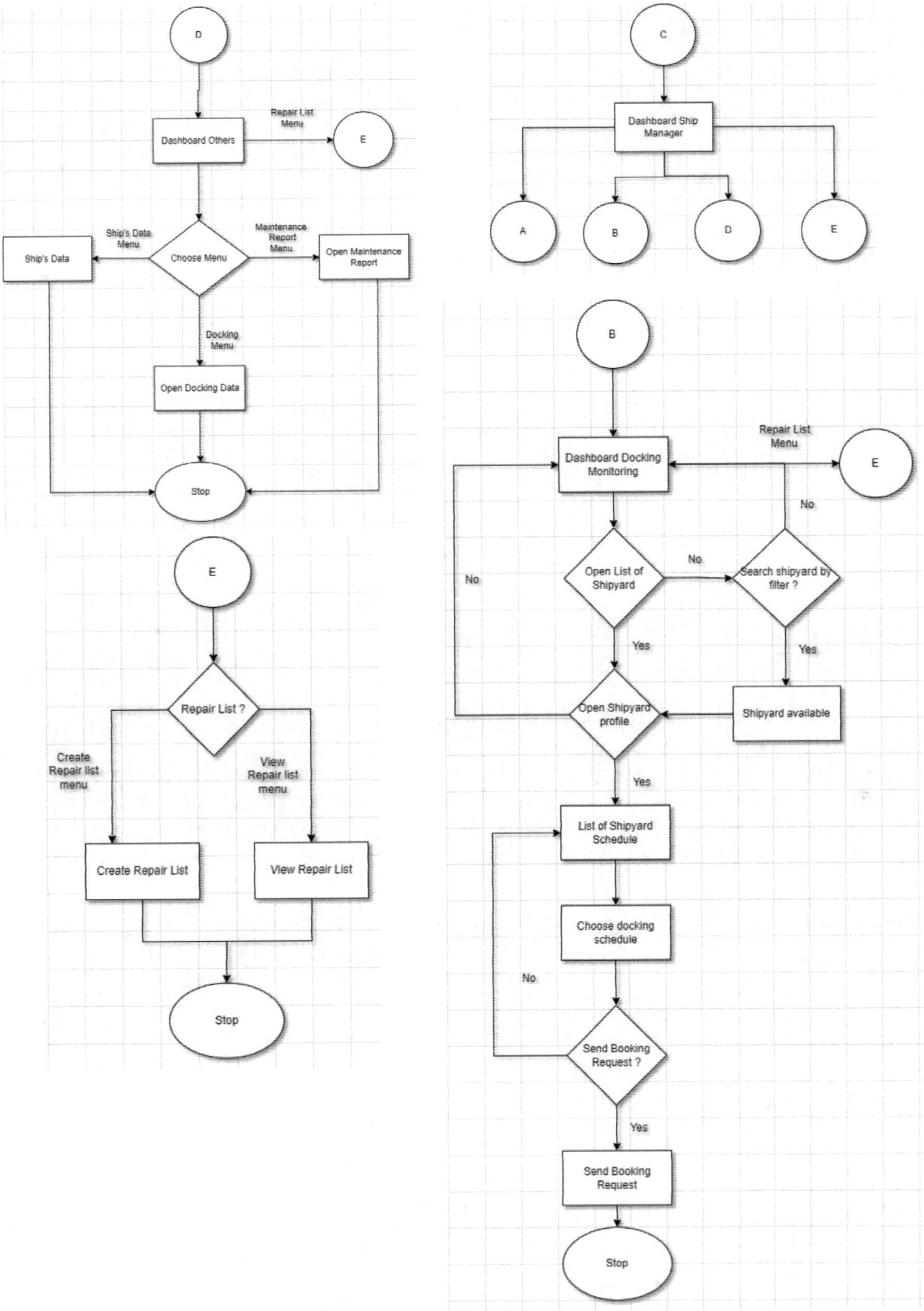

Figure 6. Data Flow Diagram

3.5.3. Entity Relationship Diagram (DFD)

After creating a data flow diagram along with the authority of each user, then a relationship diagram for each entity is designed (Entity Relationship Diagram). ERD is the first step in determining the flow and describing the relationships that occur from the application made. The relationship between these entities is used to design data structures and relationships between data. With ERD, it can be seen clearly what relationships exist in the system and how they relate to one another, and also the attributes in the entity table.

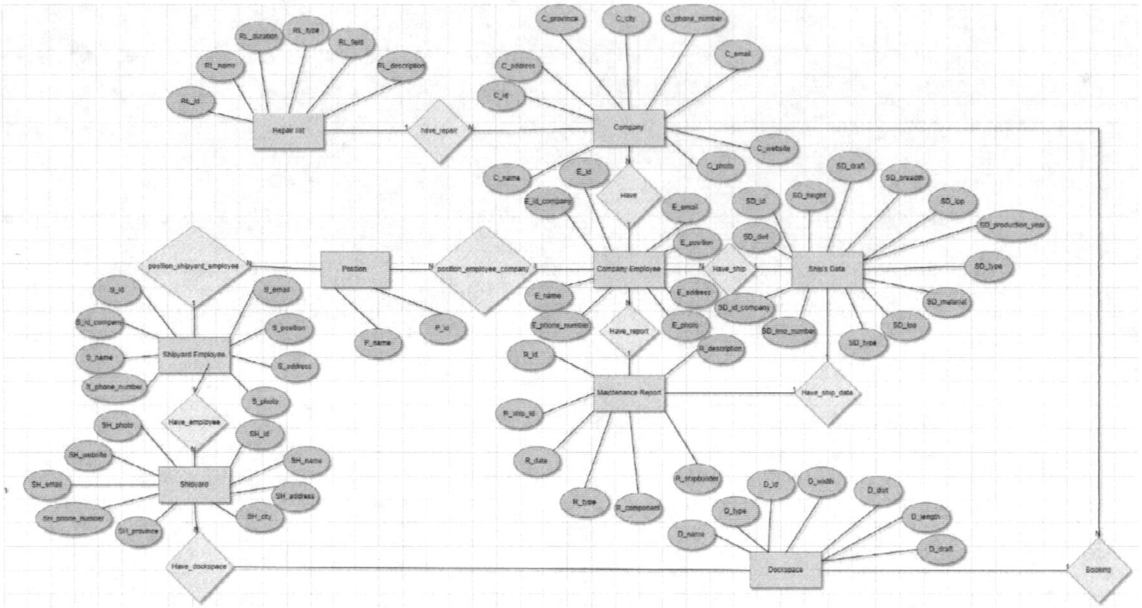

Figure 7. Entity Relationship Diagram

In Figure 7, the ERD shown the entity that contain the data of certain things. Then, inside the entity the are some atributes or we can know it as the specific data measured. Lets take an example. The company is one of the entity, inside the company there are many specific data like the company name, addres, phone number, city, email, etc. That specific data is called atributes.

3.6. Application Prototype

For the current application design, the author use the XAMPP virtual server, Virtual Studio Code as a source code editor and also use Bootstrap for the front-end framework, MySQL database, and CodeIgniter for developing web framework (back-end) using the PHP programming language.

Website Structure

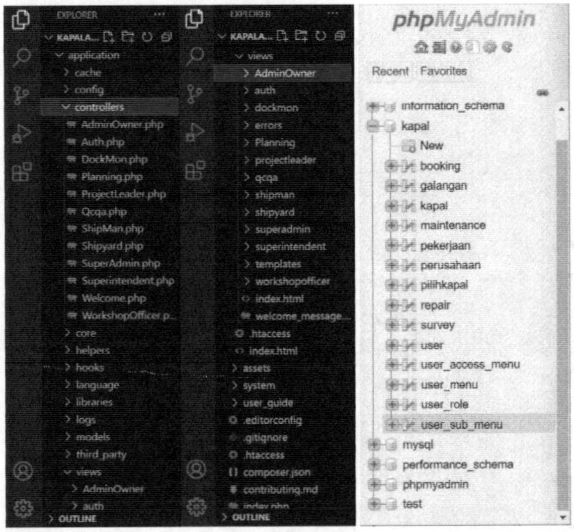

Figure 8. Website Structure (Visual Studio Code and phpMyAdmin)

The coding structure of this website is mainly divided into 2 types, namely the controllers and views sections, see Figure 8. The Controller section is in charge of giving orders on the website, it can be in the form of buttons or like filling out forms. Meanwhile, the coding function for views is mainly to help shape or display such as in the form of a table or shaped like a photo frame for uploading images. Then for the the MySQL database management is also divided as in the phpMyAdmin software.

Website Application Interface

Figure 9. Home Page

The following is the interface display of this website-based application. This website has the domain shiprepair-dmom.com and can be accessed via a browser. When you arrive at the website page, you will be greeted by a home page as shown in Figure 9. This website has the main function as a tool to search for available shipyards (see Figure 10), order and book docking space (see Figure 11), and can also help with reporting. progress of work on the ship being repaired as shown in Figure 12.

IOP Conf. Series: Earth and Environmental Science 1081 (2022) 012047 doi:10.1088/1755-1315/1081/1/012047

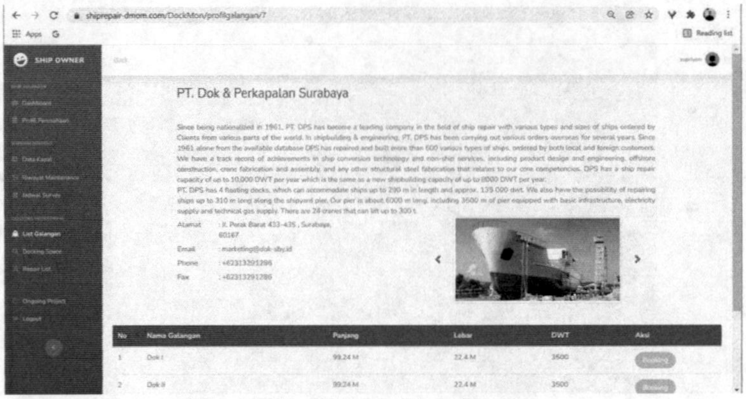

Figure 10. Shipyard Profile Page

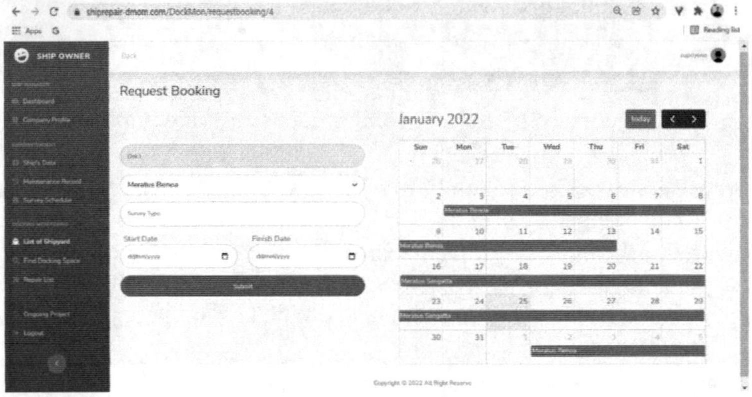

Figure 11. Request Booking Page

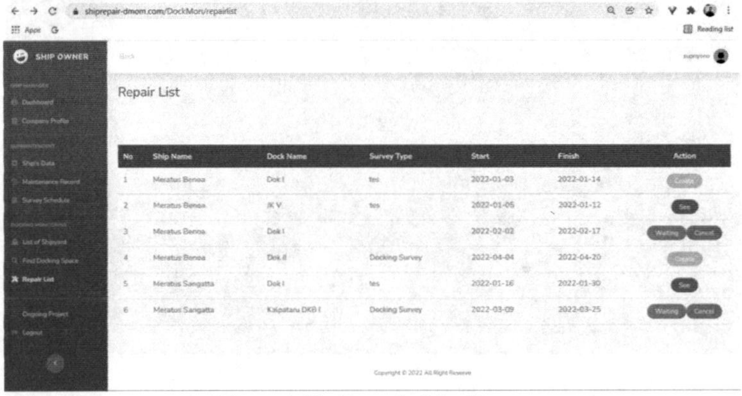

Figure 12. Repair List Page

4. Conclusion

The first objectives of this study is to identify the process of ship repairs planning from the ship owner perspective and shipyard perspective at this time. Based on the field study and from literature review, the process of ship repair and docking planning has some flaws. That flaws makes this process risky for errors to occur because the system used is still manual and not integrated. In addition, the relationship that is lacking from each party can cause this process to be late or not on schedule.

The second objectives of this study is to to design a web-based intermediary application for planning ship repair services. Based on the result of Application Prototype, it can facilitate for both

Shipyard and Ship Owner in Ship Repair and Docking Process mainly for finding and booking available dock spaces and project management during it's repair progress.

5. References

[1] M. Hasbullah 2016 Strategi Penguatan Galangan Kapal Nasional Dalam Rangka Memperkuat Efektifitas Dan Efisiensi Armada Pelayaran Domestik Nasional 2030 *Jurnal Riset dan Teknologi Kelautan*, Vol. **14**, No. 1, 103-112

[2] W. Firstdhitama 2018 Perencanaan Penjadwalan Reparasi Kapal Ferry Dengan Menggunakan Metode Flash *Wave : Jurnal Ilmiah Teknologi Maritim*, Vol. **12**, No. 1, 1-12

[3] U. B. Samuel 2014 Perancangan Aplikasi sistem Repair Schedule Pada Kapal Penyeberangan Merak-Bakauheni Berbasis Web Menurut Aturan Biro Klasifikasi Indonesian. *KAPAL : Jurnal Ilmu Pengetahuan & Teknologi Kelautan*, Vol. **11**, No.3, 144-153

[4] E. S. Hadi & P. Manik 2008 Pengembangan Sistem Informasi Berbasis Komputer Dalam Menunjang Kegiatan Penjadwalan Reparasi Kapal di Galangan PT. DOK dan Perkapalan Kodja Bahari (Persero) Unit Produksi Jakarta II *KAPAL : Jurnal Ilmu Pengetahuan & Teknologi Kelautan*, Vol. **5**, No. 3, 166-172

[5] K. Salimifard, G. Jamali, and S. Behbahaninezhad 2012 Resolving Resource Conflicts in A Ship Repair Project *International Journal of Modeling and Optimization*, Vol **2**, No. 5, October 2012, 618-621

[6] D. Hansel & W. Triwilaswandio 2016 Perancangan Aplikasi Komputer Berbasis Android untuk Estimasi Biaya Reparasi Kapal Interaktif *JURNAL TEKNIK ITS* Vol. **5**, No. 2, (2016) ISSN: 2337-3539, G47-G51

[7] R. R. Putra & W. Triwilaswandio, 2016 Perancangan Aplikasi Berbasis Komputer untuk Proses Manajemen Mutu pada Pembangunan Kapal Baru *JURNAL TEKNIK ITS* Vol. **5**, No. 2, (2016) ISSN: 2337-3539, G326-G331

[8] R. Nurwanti & W. Triwilaswandio 2016 Analisa Peningkatan Kualitas Layanan Jasa Reparasi Kapal Di Galangan Kapal Jawa Timur *JURNAL TEKNIK ITS* Vol. **5**, No. 1, (2016) ISSN: 2337-3539, G41-G46

[9] M. J. Hutapea, T. W. Pribadi, & I. Baihaqi 2017 Perancangan Aplikasi Berbasis Android untuk Manajemen Proyek Reparasi Kapal *JURNAL TEKNIK ITS* Vol. **6**, No. 2 (2017), 2337-3520, G158-G163

[10] Project Management Institute 2013 *A Guide to The Project Management Body Of Knowledge (PMBOK Guide) – Fifth Edition* (Newtown Square, Pennsylvania USA: Project Management Institut, Inc.)

[11] Dinsmore, P. C 1990 Ideas, guidelines and techniques for applying project management solutions in the general business arena: Lessons for executives. *International Journal of Project Management*, **8**(1), 33-38. https://doi.org/10.1016/0263-7863(90)90006-W

[12] Zaman, M. B et al 2019 Risk Assessment in Ship Repair Scheduling *International Journal of Mechanical Engineering and Technology (IJMET)* Vol **10**, Issue 03, 1-8

[13] L. K. Padaga, I. Rochani, & Y. Mulyadi 2018 Penjadwalan Berdasarkan Analisis Faktor-Faktor Penyebab Keterlambatan Proyek Reparasi Kapal : Studi Kasus MV. Blossom. *JURNAL TEKNIK ITS*, Vol. **7**, no. 1, 1–6

Acknowledgments

The authors thank to Digital Marine Operation and Maintenance Laboratory, Department of Marine Engineering, and Institut Teknologi Sepuluh Nopember for supporting facilities this research.

IOP Conf. Series: Earth and Environmental Science 1081 (2022) 012048 doi:10.1088/1755-1315/1081/1/012048

Development of Decision Support System for Prevention Measures and Fire Accident Handling on Container Ship

Muhammad Badrus Zaman[1], Nurhadi Siswantoro[1], Ratu Balqis Fernando[1], Trika Pitana[1], Dwi Priyanta[1], Hari Prastowo[1]

[1]Department of Marine Engineering, Institut Teknologi Sepuluh Nopember (ITS), Surabaya, Indonesia

corresponding author's e-mail: drus_zaman@yahoo.com

Abstract. According to data 2018-2021 on the status of the National Transportation Safety Committee, the largest percentage of ship accidents (29%) in Indonesia are caused by fire or exploded. The main purpose of this research is to design a website-based application that can be accessed by phone to support ship crews in preventing and handling fires on board using a decision support system method, where the website-based application itself is an application that can be run using web and browser technology bases of internet or intranet. In the preparation of the research, first, observations were made on the cargo of dangerous goods transported by the container ship (MV. Sinar Penida). A website-based application for fire prevention and handling system is designed after making a mock-up. Application design is done by making mock-ups, designing interfaces, creating databases, and coding the application system. The features have been equipped with an emergency muster list, fire handling based on the class of cause, and a cargo separation feature based on the ship's bay plan using the Decision Support System (DSS) method or decision support system that can increase the effectiveness in decision making.

1. Introduction

The purpose of the decision support system itself is to increase the level of effectiveness of decision making by managers and overcome cognitive processing and storage [1]. Transportation carried dangerous goods must comply with applicable regulations. The International Maritime Dangerous goods Code (IMDG Code) is a regulation used to prevent losses due to accidents caused by cargo [2]. According to the National Transportation Safety Committee (KNKT) 2021 shipping safety recommendation status data in Indonesian waters from 2018 to 2021, there are various types of ship accidents, such as sinking, burning, running aground and colliding. The data shows that the most common types of accidents are fires or ship explosions with a percentage of 29% [3]. This research refers to the ISM Code (International Safety Management Code), which is an international standard for a safety management system or management for safe ship operations. The purpose of implementing the ISM code itself is to ensure maritime safety so that incidents do not occur that cause fatality and severe ship damage, causing marine pollution [4]. Accidents are a threat to shipping or commerce because even though safety standards are constantly being improved, shipping is always fraught with risks [5]. Shipping accidents can be caused by human error, collisions until explosions. In order to prevent the consequences of all these possible incidents, the factors causes human error need to be addressed [6]. Knowledge of ship crews about handling fires is very important. Because of the crew has less knowledge it will cause fire handling to be slow and if there is mishandling during firefighting it will make the fire burn bigger and spread to other parts. In addition, the factors causing the high number of marine accidents, especially ship fires, are caused by the lack of awareness of the parties concerned. Crews should be provided with appropriate training in regulations, precautions and packaging procedures for the handling and transport of dangerous goods and fire management in the event of an explosion or fire caused by the cargo being carried.

Table 1. KNKT Investigation Results

No	Year	Number of Accident	Type of Accident					
			Sinking	Listing/ Capsizing	Explosion	Collision	Grounding	Others
1	2021	2	1	0	1	0	0	0
2	2019	16	2	0	4	6	1	3
3	2020	0	0	0	0	0	0	0
4	2018	30	6	2	9	3	5	5
Total		48	9	2	14	9	6	8

Source: Data status of recommendation for shipping safety KNKT 2021

Dangerous cargo is defined as goods whose nature, if in handling, processing, hoarding/deviation does not follow the standards that have been set, it can cause harm to humans, materials, and the surrounding environment [7]. With the development of this software, it can prevent fires due to dangerous cargoes that have leaks or damage to packaging and mismatches in the placement of cargo and make it easier for ship crews to handle fires and prevent mishandling due to lack of crew knowledge. The decision support system method is used in the design of applications on features aimed at preventing incorrect placement of cargo that causes fires in order to filter and analyze data about the cargo manifest, collect comprehensive information that can be used to solve problems and make effective decisions, namely in the form of cargo separation according to the IMGD Code.

2. Methods

Decision support system have constituent components consisting of data management, model management, communication (dialogue subsystem), and knowledge management [8].

a. Data Management:
Contains a database consisting of relevant data managed by software or a database management system (DMS)

b. Model Management:
Provide management and analytical capabilities on system

c. Communication:
In the form of an interface so the user can give orders to the system

d. Knowledge:
Optional it can be used to support other subsystems

2.1. Data Collection

The data collection used is divided into 2 categories, primary data, and secondary data. The secondary data were taken from the literature studies which also include government publications, websites, and others. The primary data were taken from muster list emergency, ship general data, crew list, from the PT. Samudera Indonesia.

2.2. Data Processing

The data used during the research was obtained through literature and PT Samudera Indonesia which was processed into a basic framework and basic business application, data flow diagrams, entity diagrams and a mock-up of the application display. The data processing of ASAP (As Soon As Possible) website-application mention below:

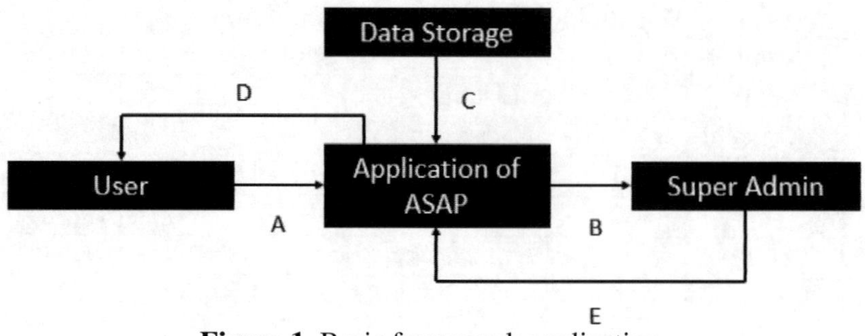

Figure 1. Basic framework application

A. Process A shows the data input process in the form of generalship data and data cargo manifest made by the User.
B. Process B shows the data that has been inputted by the next User sent to Super Admin for account approval.
C. Process C is inputting data that has been received by the system will be processed to become output.
D. Process D obtained output in the form of data, such as ship bay plan layout data, cargo being transported, and cargo segregation in the cargo hold.
E. Process E occurs if there is a user who wants to do account approval, Super Admin gets a notification from ASAP and agrees to account can be activated. Super admin can see the data that has been inputted by the user.

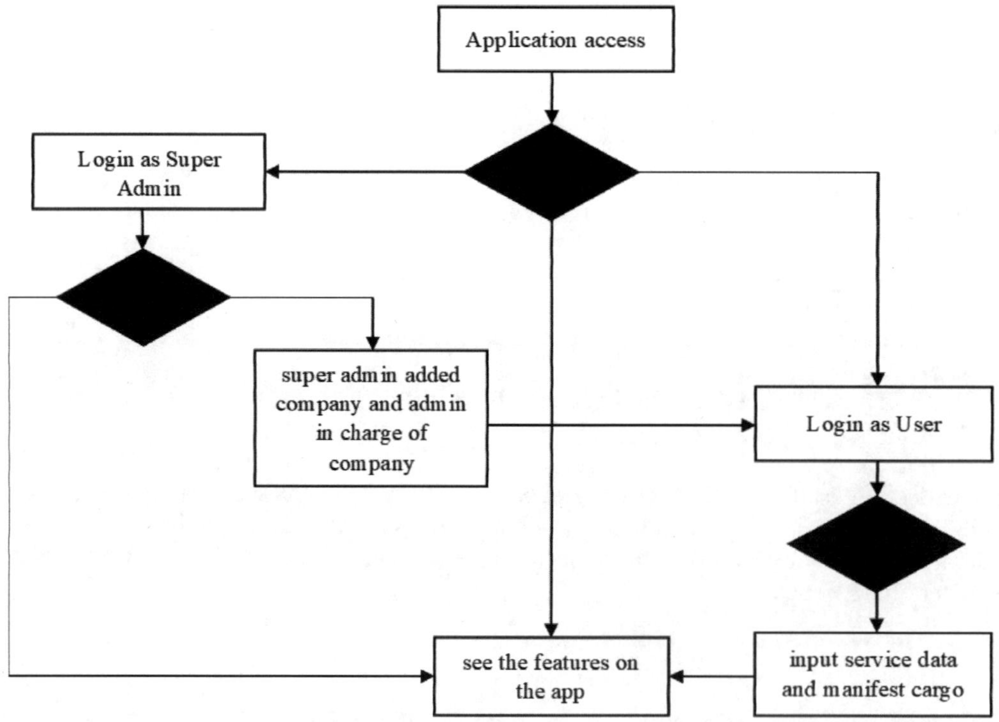

Figure 2. Basic business application

A data Flow Diagram (DFD) is a diagram that describes a form of data transfer from one process to another. In other words, DFD describes the flow of data transfer planned in this application. This DFD can help to understand a system logically, structured, and clear.

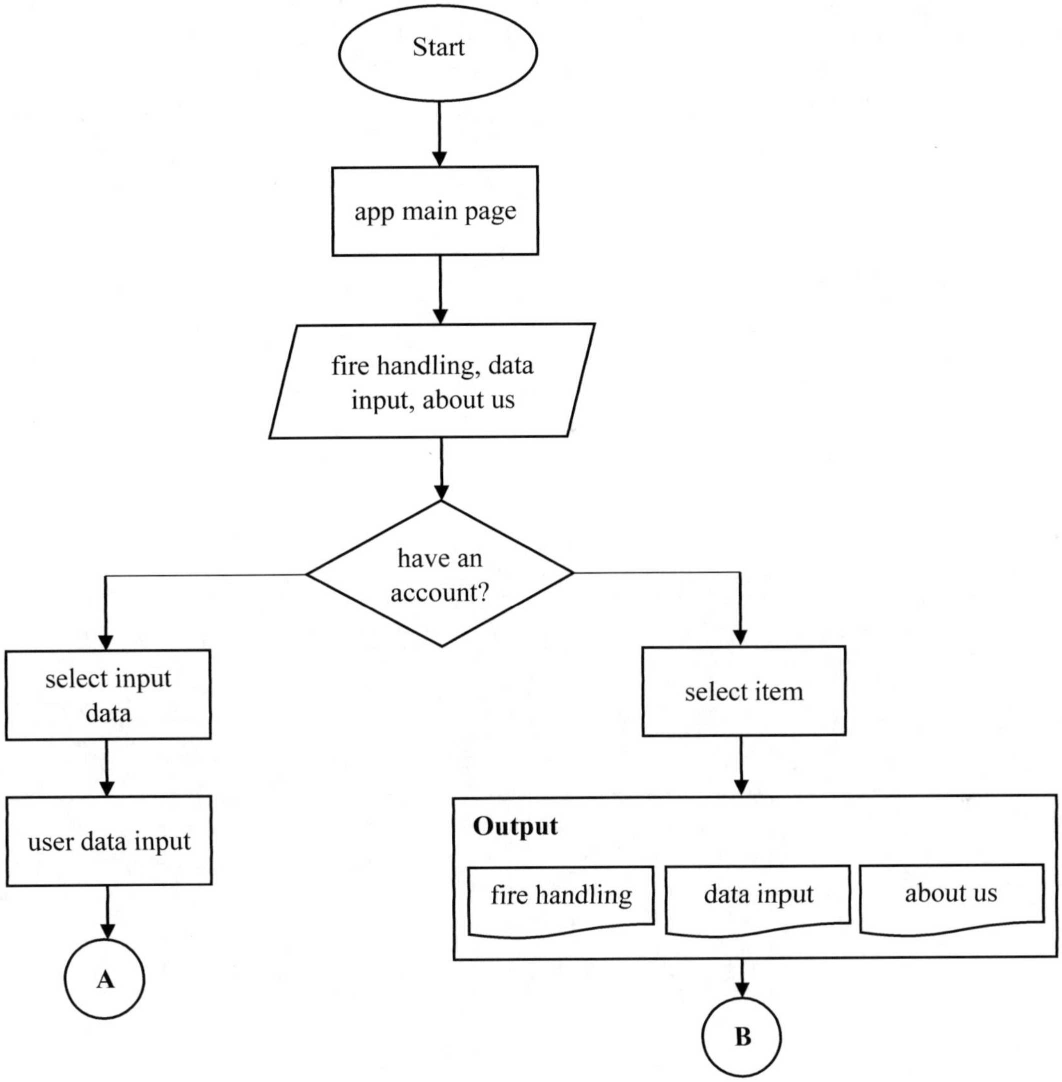

Figure 3. Data Flow Diagram (1/2)

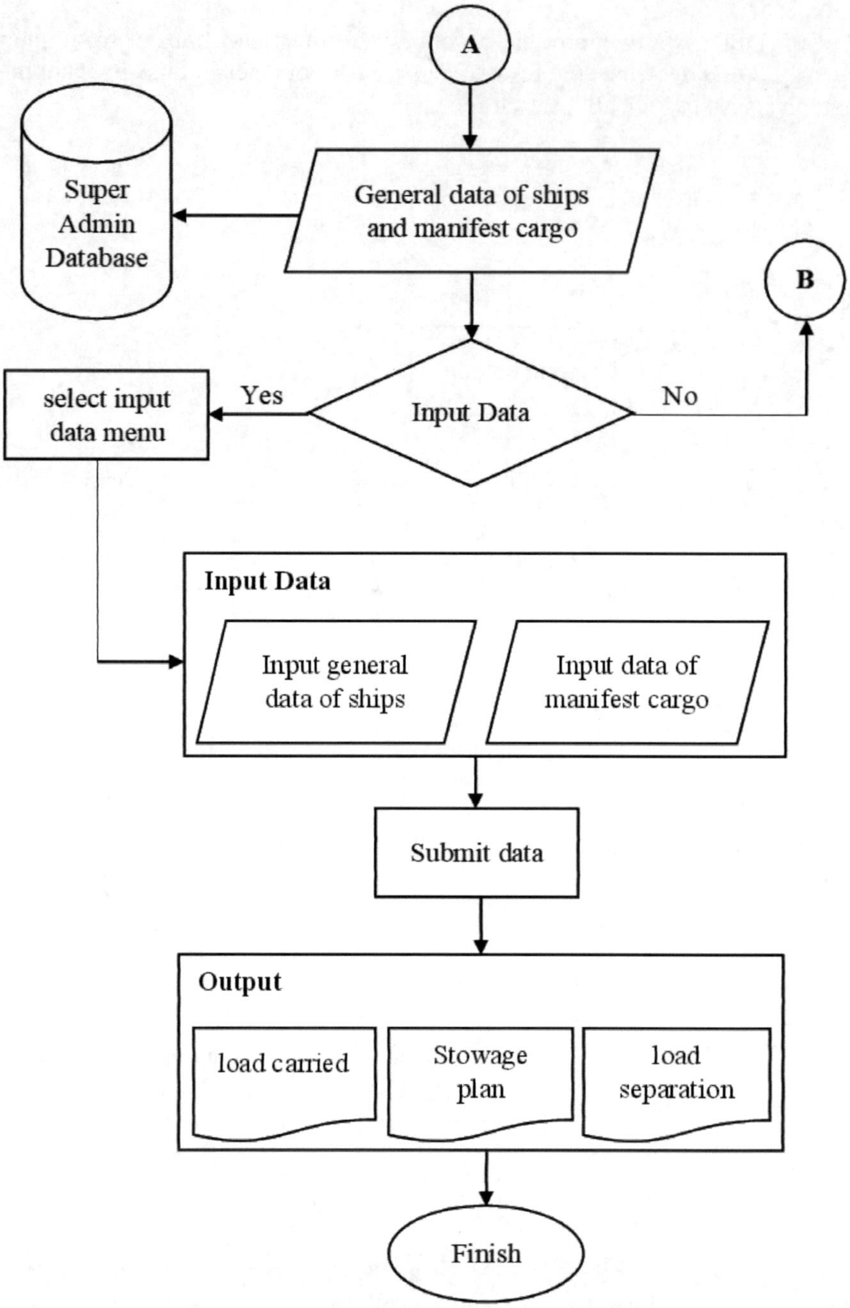

Figure 4. Data Flow Diagram (2/2)

The preparation of ERD (Entity Relationship Diagram) is carried out to find out the relationship between the concepts that exist in the compiled system, design ideas from the database to be compiled. Each entity has a relationship with the attributes used, where the relationship is described with a symbol of a connected line. The user entity has a relationship with the attribute email address and password, while the muster list entity has attributes of choice from the consequences of fires which will later show the role of each crew member. The hazard muster list entity has a relationship with the user entity. The user entity also has a relationship with the data input entity, where the data input has attributes of general data of the ship, cargo manifest and hazard identification of the cargo being transported, from the data set that has been inputted, will display the ship's stowage bay plan containers. This section shows the stowage bay plan to prevent mistakes in laying cargo that can cause accidents, one of which is: explosion/fire. On firefighting entities connected to the user, entity has output on how to handle fires based on their cause.

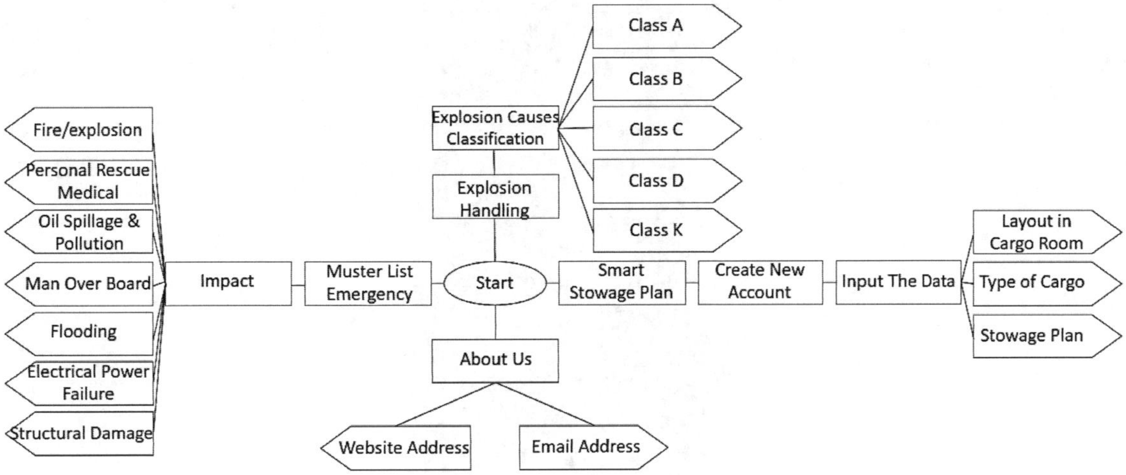

Figure 5. Entity ddiagrams

In this prototype design used JavaScript as a programing language focus on developing an application uses a web browser base due to the nature of JavaScript which is easy to learn, dynamic, multiplatform and has add ons for certain features. JavaScript is a pure object programming language because all program code is wrapped in classes [9]. For the database management used mySQL to make easier for users to access data containing information either personally or publicly on a website, where the SQL is based on a relational algebra [10].

3. Result and Discussion

3.1 App View

The starter Page is the first page displayed by the application. The starter display on the ASAP application itself has several elements such as at the top there is a logo, the name of the application is ASAP, the slogan of the application, and there is a red box at the bottom that shows the start of the application. When the start button is pressed, the application will start and show the next screen, namely the main or home screen.

Figure 6. Starter page view from desktop view

Figure 7. Starter page from mobile view

This page contains the application logo, the user welcome text, there are three red lines in the upper right corner of the display that shows the menu, where the menu options will display HOME, SOLUTION and there is also a LOGIN display. In the solution display, there are any services or features offered by the ASAP application, namely emergency muster lists, fire handling by class, data input and about us. In the input data feature, the user must enter the general data of the ship and the manifest data of the cargo first so that the system can process it into a display of the ship's stowage bay plan.

Figure 8. Menu options on the application display

3.2 Prototype Design

In the system design process, the PHP programming language is used which can be displayed via an HTML browser using the CodeIgniter framework. As an example, in the development of coding so that the system can run, the following is one of the codes used in the preparation of the prototype.

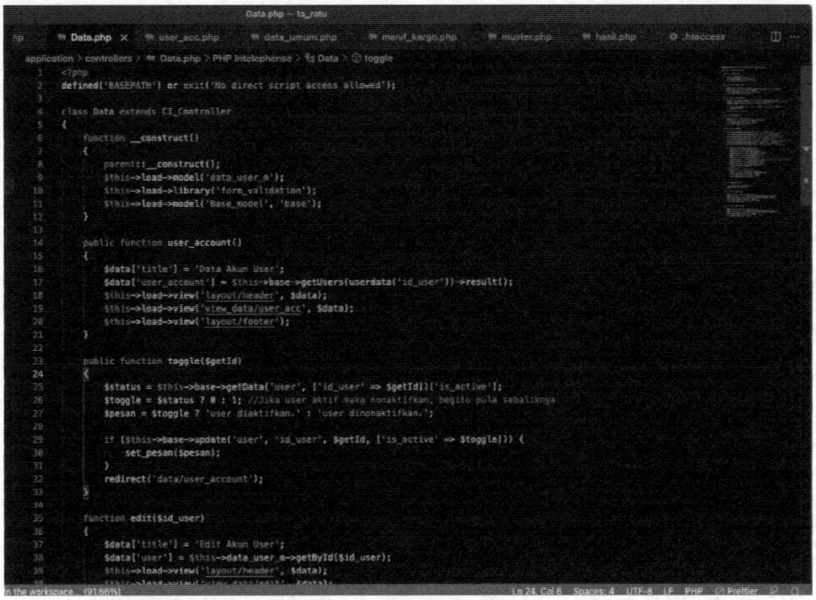

Figure 9. FrontPage prototype design process view

When the prototype has successfully to operate and don't have any trouble when it run, so others of ship emergency muster list and ship's cargo can be added, and the website-application becomes more complex.

4. Conclusion

Based on the results of the discussion, analysis, and validation of "Development of a Decision Support System for Prevention and Handling of Fire Type Accidents on MV Sinar Penida Container Ships", the following conclusions were obtained.

1. In the follow-up to the handling of fires that may occur onboard the ship, a website-based application prototype was designed, named ASAP. In the preparation of the prototype, data was obtained from PT Samudera Indonesia, namely data in the form of an emergency certificate or emergency muster list which presents the actions and procedures that must be carried out by each crew member when danger occurs, especially those caused by fire. In the developed software, the emergency certificate is displayed on the muster list selection menu. In avoiding the occurrence of errors in fire handling, the software is also equipped with fire handling features based on the cause class which is displayed on the fire handling options menu.
2. In the software, a decision support system is developed for users to get the preparation of cargo, especially dangerous cargo, or stowage plans on container ships in accordance with the IMGD Code. To determine the level of danger of the hazardous cargo that will be transported by the ship, a smart stowage plan feature is made in the software by implementing a decision support. After the system shows the level of dangerous cargo, the system will process it into a ship layout that displays a bay plan with the cargo being transported.
3. Hazardous cargoes are classified into several classes according to the IMGD Code. The grouping is based on the nature of the material, starting from being explosive, sensitive to water and so on. The distribution of hazardous cargo is divided into 9 classes, some of which are further subdivided into sub-divisions, namely class 1 which is an explosive charge category that consists of several sub-divisions, namely class 1.1, 1.2, 1.3, 1.4, 1.5, 1.6, class 2 indicates a gas category charge. hazardous substances which are divided into subdivisions 2.1, 2.2, 2.3, class 3 is categorized as flammable liquids, class 4 indicates the cargo category of goods which are divided into subdivisions 4.1, 4.2, 4.3, cargo class 5 indicates loads that are

oxidizing agents that subdivided into subdivisions 5.1, 5.2, class 6 indicating cargo is a cargo categorized as toxic goods which are divided into subdivisions 6.1, 6.2, class 7 cargo which is a substance with corrosive properties, class 8 is a charge that has radioactive properties, and class charge 9 which can cause irritation if exposed to the body or inhaled.

5. References

[1] Zulita, L. N. 2013 Sistem Pendukung Keputusan menggunakan metode SAW untuk menilai dosen berprestasi (Studi kasus di Universitas Dehasen Bengkulu).
[2] Joanne Ellis 2010 Undeclarated Dangerous Goods – Risk Implicstions for Maritime Transport.
[3] Komite Nasional Keselamatan Transportasi 2021 Data status rekomendasi keselamatan pelayaran KNKT 2021.
[4] IMO 2002 ISM Code and Resived Guidelines on Implementation of ISM Code. London.
[5] IMO 2011 MARPOL Annex 3 73/'78.
[6] Butt, N., Johnson. D., Pike, K., Pryce – Roberts, N., Vigar, N 2013 15 Years of Shipping Accidents: a review for WWF. Southamptom.
[7] Ridwan 1995 Diktat Keselamatan Kerja dan Pencegahan Kecelakaan. Jakarta.
[8] Lita Asyiriati Latif et al 2018 Sistem Pendukung Keputusan Teori dan Implementasi. Yogyakarta.
[9] Rosa & M. Shalahuddin 2018 Rekayasa Perangkat Lunak Edisi Revisi. Bandung.
[10] Fotache, M., Strimbei, C. 2015 SQL and data analysis. Some implications for data analysits and higher education. Romania.

Acknowledgments

The authors thank to Digital Marine Operation and Maintenance Laboratory, Department of Marine Engineering, and Institut Teknologi Sepuluh Nopember for supporting facilities this research.

Impact Assessment of Breaking Waves Criteria Subjected to Wave Energy Converter

Rudi Walujo Prastianto[1], Fuad Mahfud Assidiq[2], and Mukhtasor[1]

[1] Department of Ocean Engineering, Institut Teknologi Sepuluh Nopember, Sukolilo, Surabaya, Indonesia
[2] Department of Ocean Engineering, Universitas Hasanuddin, Bontomarannu, Gowa, Indonesia

E-mail: assidiqfm@unhas.ac.id

Abstract. Breaking wave effect on Wave Energy Converter (WEC) represents an important issue of site selection analysis especially fringing reefs area. To learn a more acceptable sense of the influence of this effect, an introspection on the estimate of initial breaker points is proposed in the present study. The paper describes an investigation of the impact assessment of wave-breaking criteria subjected to WEC recently desired as a benchmark by the previous three wave-breaking methods. The numerical simulation is carried out by the Computational Fluid Dynamics (CFD) solver. The solver is based on the Finite Volume Method (FVM) to create the discretization of the governing equations. Large Eddy Simulation (LES) is solved in a global approach together with Volume of Fluid (VoF) for capturing free surface. Reported justifications are compared with the experimental data to validate the accuracy of the numerical approach and then applied to generate wave-breaker point location on a fringing reef in Numerical Wave Tank-based (NWT) of 16 scenarios. The results are discussed with respect to relative wave-breaker locations, relative wave-breaker height range, and wave breaking probability under various wave steepness, water depth, and slope fringing reef. It is given regarding the discrepancy between different methods and recommendations for important guidances are outlined.

1. Introduction

Breaking waves are one of the most complicated hydrodynamic models in a coastal area [1]. In terms of nearshore region such as reef dimensions, reef roughness environmental loads, wave run-up, and sediment transport are governed by local hydrodynamic phenomena specifications. Understanding breaking wave generation across the reef is critical not only for extremely low water depths, but also for the impact of wave-reef interaction. Breaking waves in some theoretical schemes are applied to providing a specific description of shallow-water wave nonlinearity, the eddy viscosity, and the macroscopic dissipation [2]. It is also useful for modeling the propagation of long-period waves in shallow water depth [3].

Describing in nonlinear effects [2], breaking wave characteristics together with wave run-up in the reef zone [4], wave-structure interaction and reef current celerity distribution [5], and effects of reef morphology configurations [6] are some of the studies for modeling wave dynamics over reefs. Following that, a numerical wave tank (NWT) was developed, capable of simulating breaking wave and, in many cases, replacing a physical wave tank. Another study focuses on using this computationally wave breaking method, which includes air entrainment [7], wave breaking on a sloping beach [8], solitary wave run-up process [9], and a hybrid computational scheme of breaking wave dynamics model at specific locations [10].

Furthermore, complex surf zone geometries, surf zone energy dissipations, and nearshore breaking wave have all been supported by smoothed particle hydrodynamics (SPH) [11]. Despite the fact that SPH has become a common technique for ocean engineering problems, its performance in terms of boundary conditions and model convergence is still being improved [12]. Newly developed Boussinesq equations, on the other hand, are better suited to simulating breaking waves on complex reef morphology variations [13]. Even Boussinesq equations are the universal tool due to their computational efficiency, there are still a few drawbacks, such as wave breaking not being well captured and the inability to resolve breaking wave vertical flow due to polynomial approximation. The study generates NWT based on Large Eddy Simulation (LES) to improve the issues of simulating the solitary breaking wave over the reefs using Boussinesq. The LES model may be able to explain significant large-scale unsteadiness flow in greater detail [14]. Because the solitary wave transformation over a reef has been extensively studied in the past, the wave-reef interaction analysis is expected to encourage ocean engineering problems [2].

Yasuda et al. [5] provided experimental and numerical knowledge of breaking solitary waves over flat reef. Particle image velocimetry (PIV) was used to measure free-surface capturing and velocity distribution in various locations. It also used the fully nonlinear boundary integral method (BIM) to numerically model the wave model. Regardless of the fact that the aforementioned studies described intriguing findings, this paper was only given a preliminary examination. The Navier-Stokes equations collaboration between the Volume of Fluid (VoF) terms for free-surface monitoring [15] and LES for turbulence approach [14] is computed by the 3-D NWT. In contrast to a previous study [16], this one focuses on the evaluation of breaking wave criteria when subjected to a wave energy converter under varying tidal conditions.

2. Numerical Modelling

The high-level schemes for spatial and temporal discretization approach of the three-dimensional wave breaking criteria subject to wave energy converter installation is carried out using computational fluid dynamic (CFD). The model solves the two-phase fluid turbulent flow problem using large eddy simulation (LES) in a finite volume framework [14,17,18]. The volume of fluid (VoF) is purposed to realize the interface air and water accurately based on split lagrangian scheme. Futhermore, pressure modelling solution is obtained from the generalized minimal residual method (GMRES) which clicks on the convergence rapidly, efficient, and recommended for all flow modelling types [18,19].

The numerical wave tank (NWT) is 22.95 long, 0.55 m wide, and 0.60 high to emulate the wave tank used in the experiments. See previous study [16] for detailed laboratory settings. The dynamic boundary method (DBM) for wave generation is used, with a TMA spectrum scheme [20,21,22,23]. The breaking waves are absorbed at the outlet of NWT using a numerical beach method [24]. The detailed implementation of boundary condition can be followed in the previous work [25,26,14] and Figure 3(c) in the section 3.1. Furthermore, courant frederick lewis (CFL), automatically adjusted on 0.50, is adopted to resolve and emphasize the time steps of 0.01 automatically during 200 wave propagation simulations. Structured mesh is applied to discretize the computational domain, For the whole directions, the mesh size is initially set to be $\Delta x = \Delta y = \Delta z =$ 0.03 m across the domain as illustrated in Figure 1(a). Meanwhile, mesh refinement is constructed circumstantially in the free surface, slope reef, and flat reef zone which are presented in Figure 1(b–c-d), respectively. The total computational mesh consists of 275400 cells. The simulation duration is not appointed to be 40 initially to guarantee the significant transient effect of breaking wave height and mean water depth. Over 8 processors data processing, it takes approximately 10 to 36 hours per simulation on a bundle server with Intel Core I7 7700HQ CPU - 2.80GHz. To assess the performance of the numerical simulation, the common coefficient of determination R^2, the model Skill value, and mean absolute

IOP Conf. Series: Earth and Environmental Science 1081 (2022) 012049 doi:10.1088/1755-1315/1081/1/012049

percentage error (MAPE) are adopted and calculated by [27,28] as mentioned in equation (1) and (2), respectively.

For LES modelling breaking waves over fringing reef, it is crucial to examine the breaking wave criteria methods at the incipient of breaker point where strong turbulence is created. It could be calculated by equation (3) and (4) concerning locations of relative breaking wave height, equation (5) as the estimation ranges of relative breaking wave heights, and equation (6) and (7) for probability of breaking waves. The detailed description of those can be established in [29,30,31]. Ultimately, the three criteria are considered here to cover most of the existing breaking wave detection methods possibly and related to each other.

$$skill = 1 - \frac{\sum \left| Y_{num} - Y_{\exp} \right|^2}{\sum \left(\left| Y_{num} - \overline{Y_{\exp}} \right| + \left| Y_{\exp} - \overline{Y_{\exp}} \right| \right)^2} \tag{1}$$

$$MAPE = \frac{\sum \left| \frac{Y_{\exp} - Y_{num}}{Y_{\exp}} \right|}{n} \times 100 \tag{2}$$

$$\left(\frac{H}{h} \right)_b = \frac{A\left\{ 1 - \exp\left[-1.5\pi \, {h_b}/{L_0} \left(1 + B \tan^{4/3} \theta \right) \right] \right\}}{{h_b}/{L_0}} \tag{3}$$

$$\left(\frac{H}{h} \right)_b = \left(\frac{H}{h} \right)_{cr} + \left(\frac{h_b}{L_0} \right)^{-1.023} \times \left\{ \frac{H_0}{L_0} - \left[0.410 \times \frac{h_b}{L_0} + 0.950 \times \left(\frac{h_b}{L_0} \right) \right]^2 \right\} \tag{4}$$

$$\left(\frac{H}{h} \right)_b = C \frac{L_0}{h_b} \left\{ 1 - \exp\left[-1.5\pi \frac{h_b}{L_0} \left(1 + 11 s^{4/3} \right) \right] \right\} \tag{5}$$

$$F_{br} = \exp\left[-9.1733 \times 10^{-4} \left(\frac{\alpha}{\varepsilon_s} \right)^2 \right] \tag{6}$$

$$\varepsilon_s = \frac{H_s}{g T_p^2} \tag{7}$$

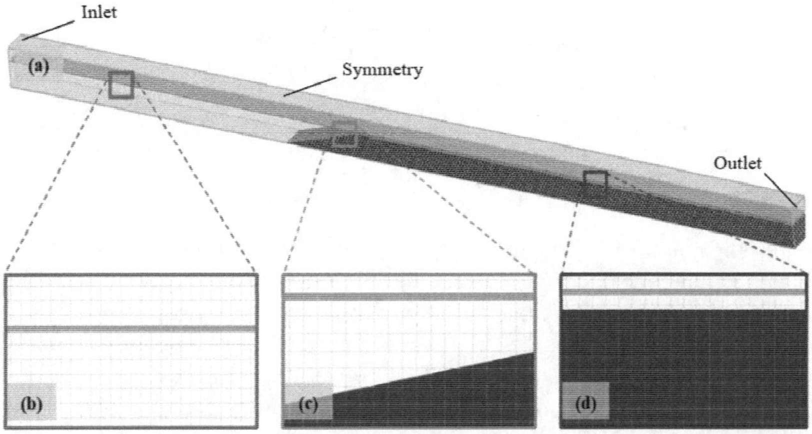

Figure 1. The structured meshes and boundary conditions of numerical wave tank (NWT)

3. Results and Discussion

The results and discussion of the CFD study are presented in this section. In order to the breaking wave modelling is implemented in real conditions, Spermonde islands, one of Indonesian role model located in South Sulawesi, are adopted in this study which correspond to the fringing reef configuration of [16] as illustrated in Figure 2. The recording for 30 years of sea states of Spermonde, the details are listed in Table 1, is carried out to simulate it by varying water depth h which represented tidal conditions as shown in Table 2 [32]. To compose easier several scenarios, the code practice is applied such as *A-1* which means the wave parameter H_s= 0.321 m, T_p= 5.791 s establish in h= 0.364 m, etc. Therefore, 16 scenarios of code practice are being simulated that is to say *A-1*, *B-1*, *C-1*, *D-1*, *A-2*, *B-2*, *C-2*, *D-2*, *A-3*, *B-3*, *C-3*, *D-3*, *A-4*, *B-4*, *C-4*, and *D-4*. Besides that, the Froude number and Reynold number scaling are set up based on [33,34].

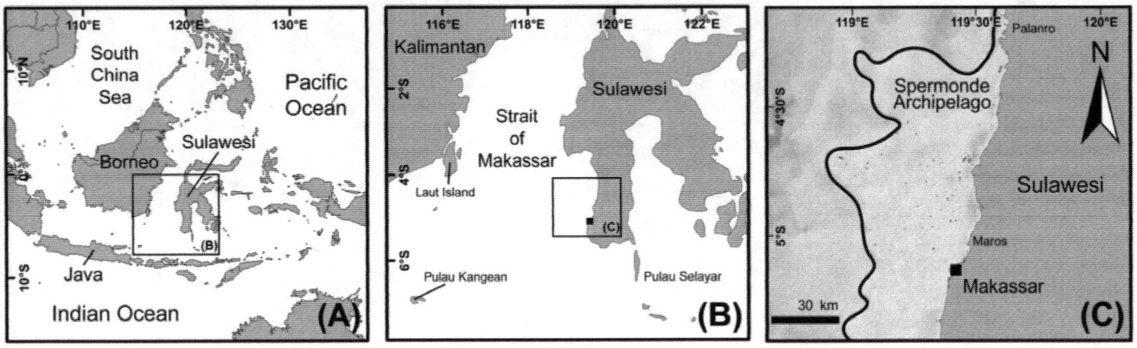

Figure 2. Spermonde islands, the role model of this assessment study

Table 1. Sea states parameter of Spermonde islands

Codes	Existing		Conversion			
	H_s [m]	T_p [s]	Scale Factor	H_s [m]	Scale Factor	T_p [s]
A	0.321	5.791	λ	0.013	$\lambda^{1/2}$	1.158
B	0.607	4.784	λ	0.024	$\lambda^{1/2}$	0.957
C	0.750	5.036	λ	0.030	$\lambda^{1/2}$	1.007
D	1.000	4.532	λ	0.040	$\lambda^{1/2}$	0.906

Table 2. Parameter of various water depths h

Codes	h_r [m]	h [m]
1	0.014	0.364
2	0.028	0.378
3	0.042	0.392
4	0.056	0.406

3.1. Validation of Numerical Model

Figure 3 compares the numerical and the experimental cross-fringing reef distribution of the free-surface elevations on, where cross-fringing reef distribution, on the parallel view, is correlated with significant wave heights H_s, mean water level depth h_{MWL}, and wave gauge positions. Figure 3(a), on the whole wave gauges, appears that the present numerical H_s generally agree well with the experimental results with R^2, Skill, and MAPE values equal to 0.991, 0.998, and 3.402%. When comparing the predictions to other models at flat reef surface, such discrepancies may be primarily due to a complexity flow, shoaling phenomenon as the initial of stopping propagation waves, and wave energy dissipation effects significantly. In addition, it could also be observed that H_s decreases rapidly in the flat reef caused by the breaking wave process.

Figure 3(b) depicts the present numerical h_{MWL} satisfactorily provide prediction results with R^2, Skill, and MAPE values equal to 0.980, 0.993, and 5.392%, respectively. Decreasing water depth could be fallen out in view of cross-fringing reef distribution between the slope reef area and the flat reef tip induced by shoaling before waves break. This suggests that unstable wave symmetry and mass conservation in NWT must be balanced by decreasing water depth in the reef edge. On the other hand, it revealed that wave propagation numerically along flat reef exposes the highest water depth. Therefore, the large magnitude of wave energy of flat reef has been reduced by breaking waves and bottom friction from itself. Overall, it seems that our prediction better captures the breaking wave height and breaking wave depth incident.

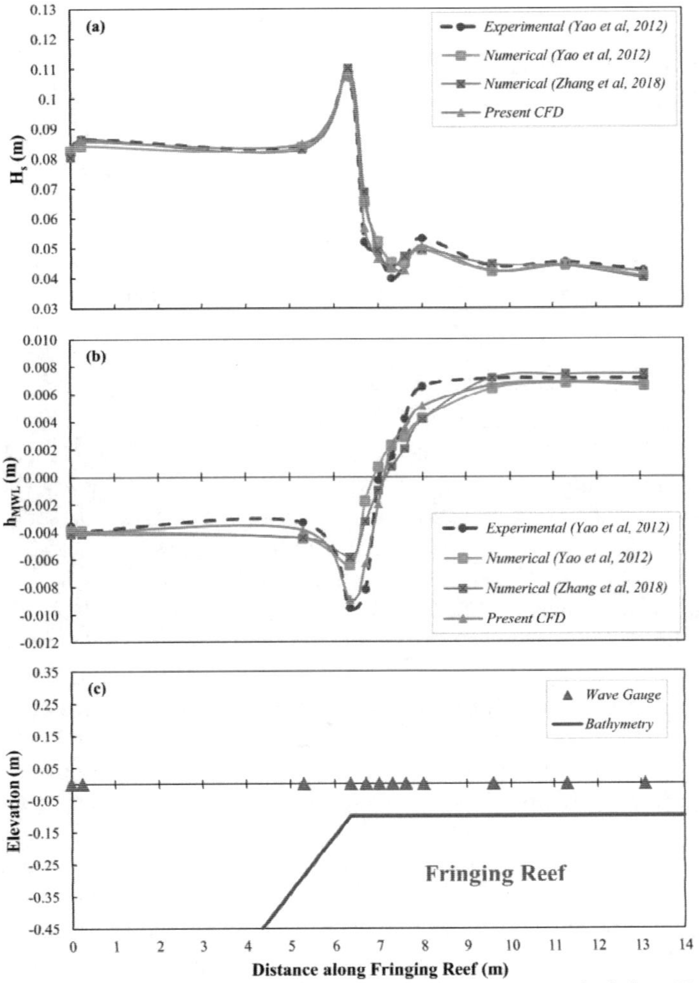

Figure 3. The numerical modelling validation, (a) Significant wave heights H_s, (b) Mean water level depth h_{MWL}, and (c) Wave gauge positions

3.2. Implementation of Breaking Wave Criteria Methods

3.2.1. Locations of Relative Breaking Wave Height

Many criteria have been proposed to represent breaking wave on the fringing reef. However, most of these criteria only apply to constant water depth and gentle slopes. Consequently, [29] developed a novel criterion method based on cases of breaking waves occurring on a steep slope, in both the slope reef and the flat reef are mentioned in equation (3) and (4), respectively.

The results using this method are illustrated in Figure 4, where the coefficient of determination R^2 are 0.9018 and 0.9109, which could be said to be emperically and numerically related to the relative breaking wave height $(H/h)_b$. Based on these results, it is found that increasing the water depth at breaking wave location causes by decreasing $(H/h)_b$. If $(H/h)_b$ decreases, the dependence of slope reef angle and the critical breaking wave height is relatively more dominant. As a complement to the final results of this method, it could be concluded that in the cases *B-1*, *C-1*, *D-1*, *B-2*, *C-2*, *D-2*, *B-3*, *C-3*, *D-3* included in the slope reef area. While the cases of *A-1*, *A-2*, *A-3*, *A-4*, *B-4*, *C-4*, *D-4* are in flat reef area.

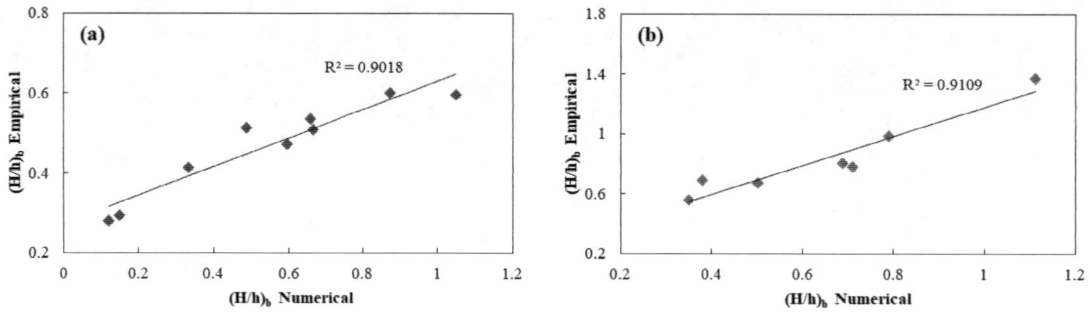

Figure 4. The coefficient of determination R^2 between the numerical results and the empirical formula from Xu et al (2019), (a) Slope reef and (b) Flat reef

3.2.2. Estimation Ranges of Relative Breaking Wave Height

This second criterion method provides empirical constants with minimum and maximum values representing the lower and upper thresholds of breaking wave scatter data, respectively. In equation (5) proposed by [30] estimates the local maximum ratio of the relative breaking wave height parameter $(H/h)_b$ for a certain wave period. The estimation range of theses criterion is based on the grouping of water depth parameter *h*, so that it could be written as Group I (*A-1*, *B-1*, *C-1*, *D-1*), Group II (*A-2*, *B-2*, *C-2*, *D-2*), Group III (*A-3*, *B-3*, *C-3*, *D-3*), and Group IV (*A-4*, *B-4*, *C-4*, *D-4*).

In the Figure 5, there is an increasing of relative breaking wave height $(H/h)_b$ causes by decreasing water depth of relative breaking wave h_b/gT_p^2. This is due to the higher $(H/h)_b$ then shoaling effect is more visible and it could be occurred breaking wave quickly when minimum – maximum slope is to be gentler. Therefore, the energy dissipation increases drastically as the waves propagates along the flat reef. These criteria are divided into three categories, comprising qualify, overestimate, and underestimate. According to Figure 5(a) to 5(d), it could be concluded that the results obtained are quite well. In the qualifying category, there is only maximum two cases per group that match or in other words around 43.75% of breaking waves are in the qualify category. It ts different with the overestimate category, the simulation of maximum estimated breaking waves is only 25% due to the tracking of wave propagation in NWT which is not stable in constant water depth. Most likely the equation (6) only applies to the cases of a gentle slope generally. On the other hand, the underestimate category obtained 31.25% of the minimum estimated breaking waves. It could be interpreted that the breaking wave height is relatively unbroken, and the whole wave asymmetry is stable well. From the three previous categories explanation, it could be assumed that several cases that meet the qualifying category are consist of *D-1*, *A-2*, *B-2*, *B-3*, *C-3*, *B-4*, *C-4*. Then, the overestimated category are in *A-1*, *C-2*, *D-2*, *D-4*. Finally, *B-1*, *C-1*, *A-3*, *D-3*, *A-4* are in the underestimated category.

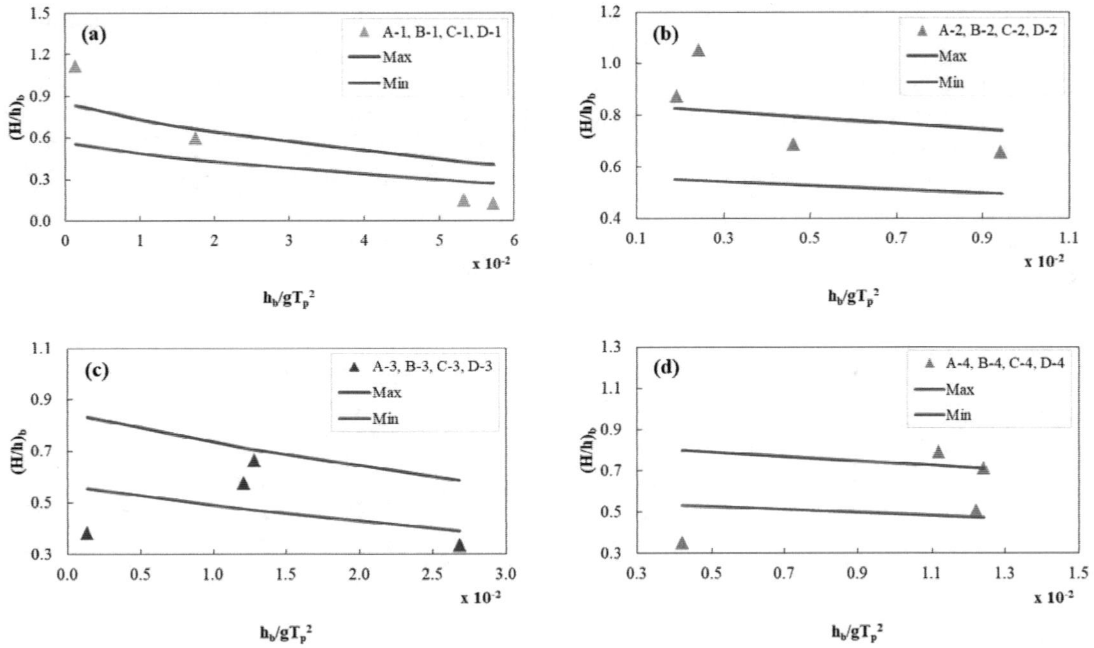

Figure 5. The breaking wave criteria according to Goda (2010) are in varying water depths h, (a) 0.364 m, (b) 0.378 m, (c) 0.392 m, and (d) 0.406 m

3.2.3. Probability of Breaking Wave

When the waves will break through the breaker point location periodically with varying velocity at temporal time, the possibility of breaking waves will change their position or unbreak. Therefore, [31] proposed the linear wave theory converting concept from temporal to spatial aspects in order to some precise estimates of breaking waves as indicated by equation (6) and (7), respectively. According this method, the breaking wave occurs when the wave crest acceleration is faster than the wave body acceleration itself. Nevertheless, this criterion focuses on the probability of breaking wave with respect to the significant wave steepness ε_s.

According to Figure 6, it is known as the ε_s increases significantly causes by increasing in the probability of breaking wave occurrence F_{br}. The parameters of the probability of breaking wave from total propagation waves are determined based on the intersection line between F_{br} and $\varepsilon_{s\ minimum}$ as the smallest probability of occurrence. Then, the intersection line between F_{br} and $\varepsilon_{s\ maximum}$ is as the largest probability of occurrence. Judging from the explanation, there are three possible events resulting from these parameters, comprising the probability obtained intersection lines from $\varepsilon_{s\ minimum}$ to $\varepsilon_{s\ maximum}$ is the first event, the second event is obtained from one of the $\varepsilon_{s\ minimum}$ or $\varepsilon_{s\ maximum}$ only, and the last event does not occur intersection line from $\varepsilon_{s\ minimum}$ or $\varepsilon_{s\ maximum}$ at all.

After identifying the three probability events from previous statement, the sequential case sets listed in Figure 6(a) to 6(p) could be verified in terms of the F_{br}. Some cases that fall into the first event consist of *C-1*, *D-1*, *B-2*, *C-3*, *D-3*, *C-4* are attached in Figure 6(c-d-f-k-l-o). The results of intersection line of the probability from each wave propagation cases are converted into the number of breaking waves, comprising case of *C-1*= 1.080 – 8.820 times event, *D-1*= 1.030 – 8.840 times, *B-2*= 0.833 – 9.016 times, *C-3*= 0.990 – 8.145 times, *D-3*= 0.061 – 5.760 times, and *C-4*= 0.917 – 8.460 times. In summary, the wave slope, relative wave height, and relative water depth have a high impact on the first probability event. Meanwhile, the cases that fall into the second probability events are case of *B-1*= 1,380 times, *B-3*= 1,170 times, *B-4*= 1,092 times, and *D-4*= 8,460 times which are illustrated in Figure 6(b-j-n-p). The wave slope parameter is the main factor in this event. This result could be seen carefully in each case there is only an intersection line, both minimum and maximum probabilities.

On the other hand, Figure 6(a-e-g-h-i-m), the third probability event includes cases of *A-1*, *A-2*, *C-2*, *D-2*, *A-3*, and *A-4*. It is no intersection line at all because the wave parameter simulation results have the wave slope, relative wave height, and relative water depth tends to be of small, so that the adoption of the two previous methods to the six cases of this third probability does not generate in breaking wave criteria.

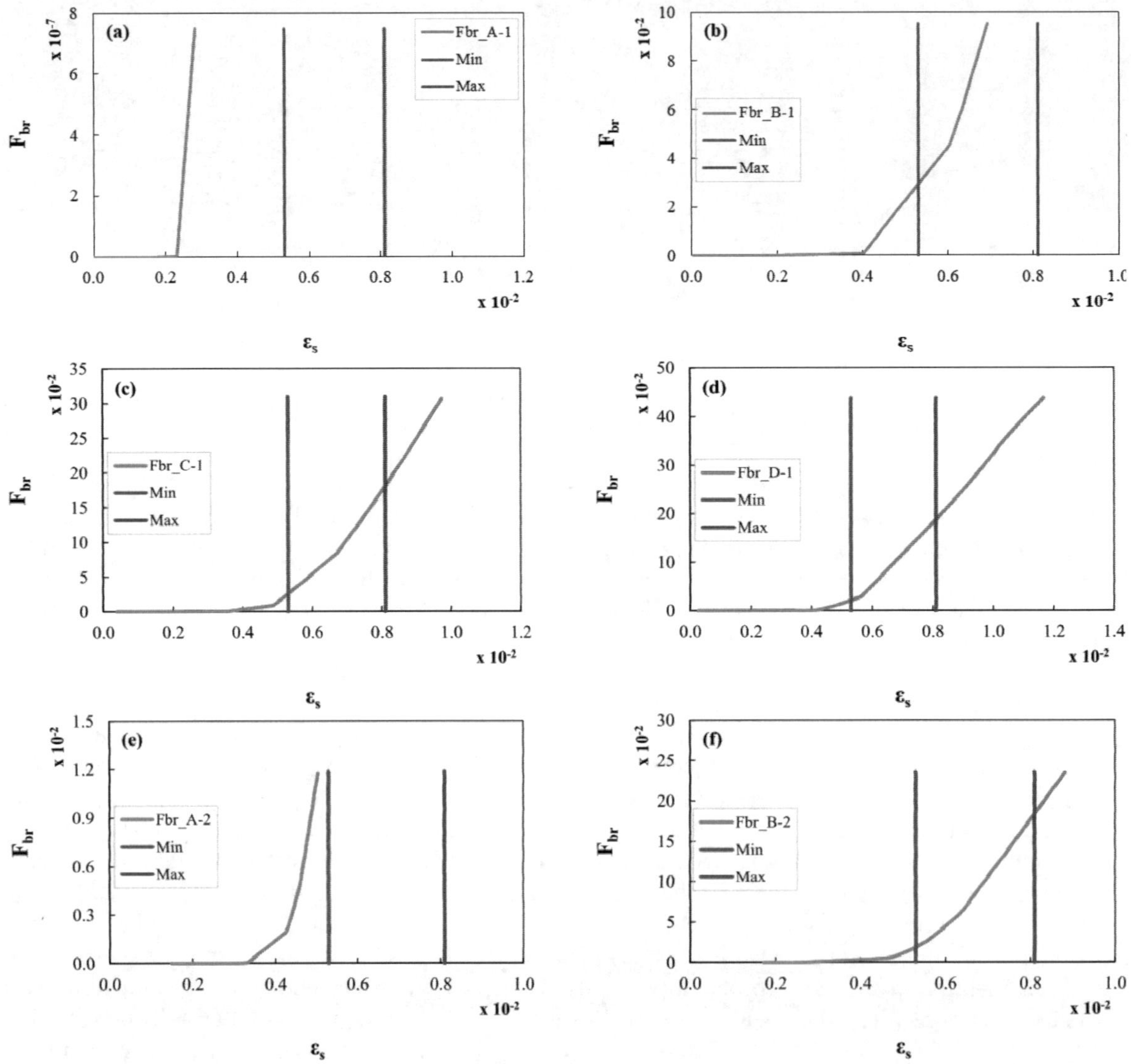

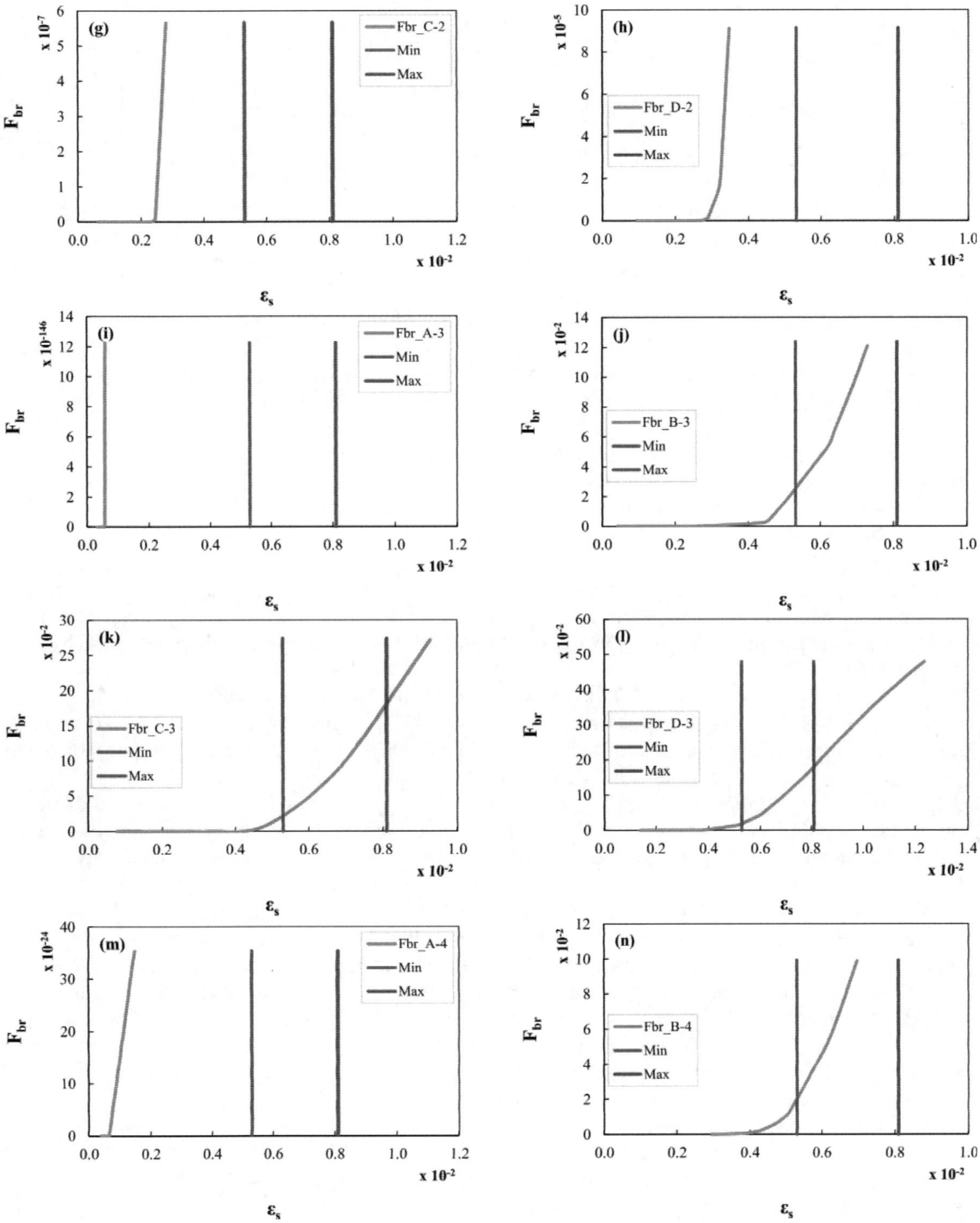
(g)
Fbr_C-2
Min
Max
x 10-7
Fbr
x 10-2
εs
(h)
Fbr_D-2
Min
Max
x 10-5
Fbr
x 10-2
εs
(i)
Fbr_A-3
Min
Max
x 10-146
Fbr
x 10-2
εs
(j)
Fbr_B-3
Min
Max
x 10-2
Fbr
x 10-2
εs
(k)
Fbr_C-3
Min
Max
x 10-2
Fbr
x 10-2
εs
(l)
Fbr_D-3
Min
Max
x 10-2
Fbr
x 10-2
εs
(m)
Fbr_A-4
Min
Max
x 10-24
Fbr
x 10-2
εs
(n)
Fbr_B-4
Min
Max
x 10-2
Fbr
x 10-2
εs

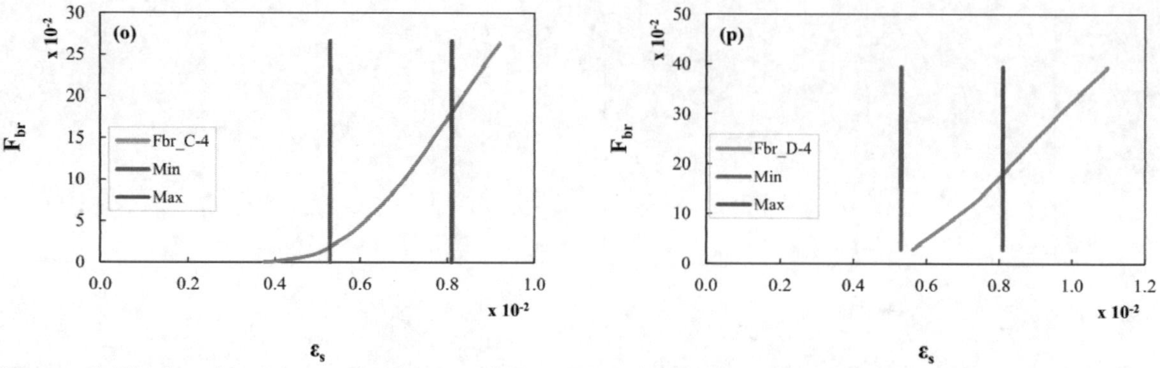

Figure 6. The breaking wave criteria according to Massel (2007) under various scenarios, (a) *A-1*, (b) *B-1*, (c) *C-1*, (d) *D-1*, (e) *A-2*, (f) *B-2*, (g) *C-2*, (h) *D-2*, (i) *A-3*, (j) *B-3*, (k) *C-3*, (l) *D-3*, (m) *A-4*, (n) *B-4*, (o) *C-4*, and (p) *D-4*. Green Lines - probability of breaking wave occurrence F_{br}; blue lines – minimum significant wave steepness; red lines - maximum significant wave steepness

3.3. Impact Assessment for Positioning Wave Energy Converter

In this subsection, we apply the well-validated breaking wave criteria to examine the scenarios of the relative breaking wave height, the relative water depth, wave steepness, and probability of breaking wave occurrence that may affect the positioning wave energy converter (WEC). The lists of scenario that justify to be assessed in more impact detail are *D-1*, *B-2*, *C-3*, and *C-4*. The characteristics between the propagating wave height and wave nonlinearity from each wave gauge will be studied further as presented in Figure 7 and Figure 8, respectively. In order to understand the characteristics easily, the propagation of significant wave height H_{gauge}/H_i is compared with distance of each wave gauge on the NWT X_{gauge}/L_{NWT}.

The first well-validated scenario, *D-1*, illustrated in Figure 7(a), shows that the discrepancy of pre-breaking wave height is the smallest among others of 0.07. It induced by the slope angle of breaker point while the breaking waves are taking place in the slope reef area. The occurrence of small breaking wave height is due to the low water depth that means shoaling is quite small. Meanwhile, the degradation of propagating wave height of 0.64 taken place in post-breaking due to energy dissipation over water depth transformation. Second, Figure 7(b) is presented the breaker point *B-2* in the slope reef. When the wave propagates towards the fringing reef, the incident wave height stands apart post-breaking wave height of 0.21. In addition, it was found that decreasing wave height of 0.78 after the waves is in post-breaking. Moreover, the breaking wave *C-3*, illustrated in Figure 7(c), is similar to previous scenarios too. It was found that the pre-breaking wave height runs counter to 0.345 of the largest of all post-breaking. This is produced by sloping area of breaker point. Similar to the difference in pre-breaking, the degradation of the post-breaking wave height was the largest of all scenarios at 0.78 due to the significant energy dissipation. At the las moment, the breaking wave *C-4* occurs in flat reef area, shown in Figure 7(d). When the wave propagates towards the fringing reef, pre-breaking wave height differ 0.197 than average breaking wave height. It may be seen that decreasing post-breaking wave height is about 0.40.

In observing the fluid flow, the Froude number F_r is applied and to investigate the wave nonlinearity, Ursell number U_r is adopted. Based on Figure 8, it can be said that the Froude number of all selected scenarios generally falls into the subcritical flow category ($F_r < 1$). All simulation results of each investigated scenario tend to be similar in terms of the increasing F_r from pre-breaking to breaking conditions. Therefore, the breaking wave celerity is very important to harnessing wave energy. When breaking condition leads to post-breaking, the F_r of *D-1*, *B-2*, and *C-3* scenario is steeper degradation than *C-4*. Another phenomenon that needs to be known is that the increasing celerity between pre-breaking and breaking was not included in the maximum value. In other words, the highest celerity taken place right after the post-breaking incident while the F_r converts to 0.52,

0.17, and 0.19. Only scenario *C-4* crops up the same celerity before and after the breaking event of 0.09. This is because the water depth effect on fringing reef h_r that greatly affects the breaking wave celerity. On the other hand, the U_r results from all scenarios satisfy in the shallow water ($U_r > 15$) referring to Dawson (1981). Further, the U_r value of incident wave was less than thirty. It should be noted that scenario *D-1*, and *B-2* are obtained U_r greater than 100. This contradicts the U_r results of *C-3*, and *C-4* less than 100. This abnormality is highly depended on breaking wave height and breaking wave length, respectively. If the U_r is greater, the wave steepness and inconsistency wave geometry will also increase.

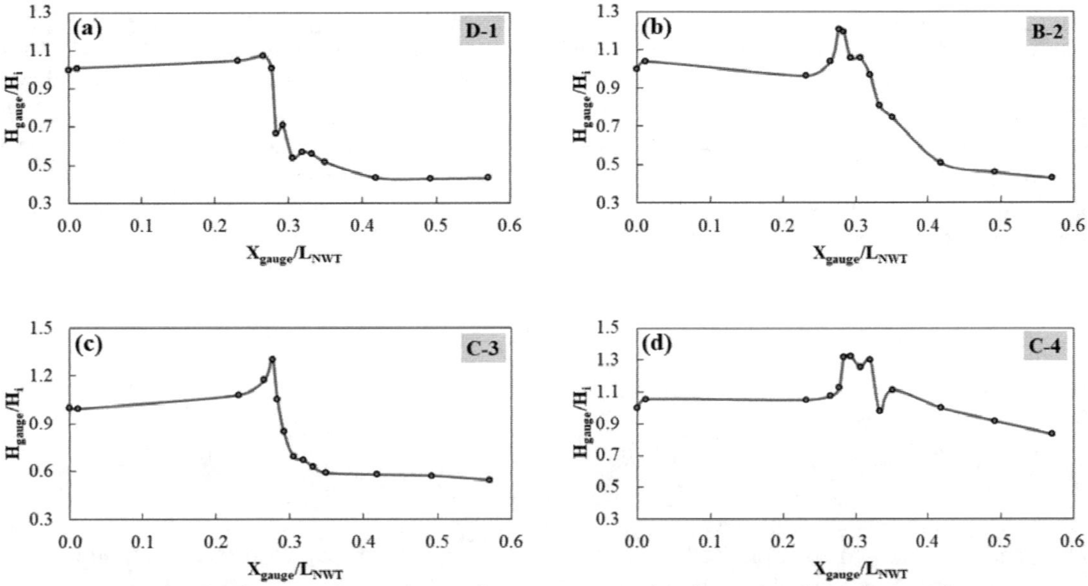

Figure 7. The transformations of wave propagation in each selected scenarios

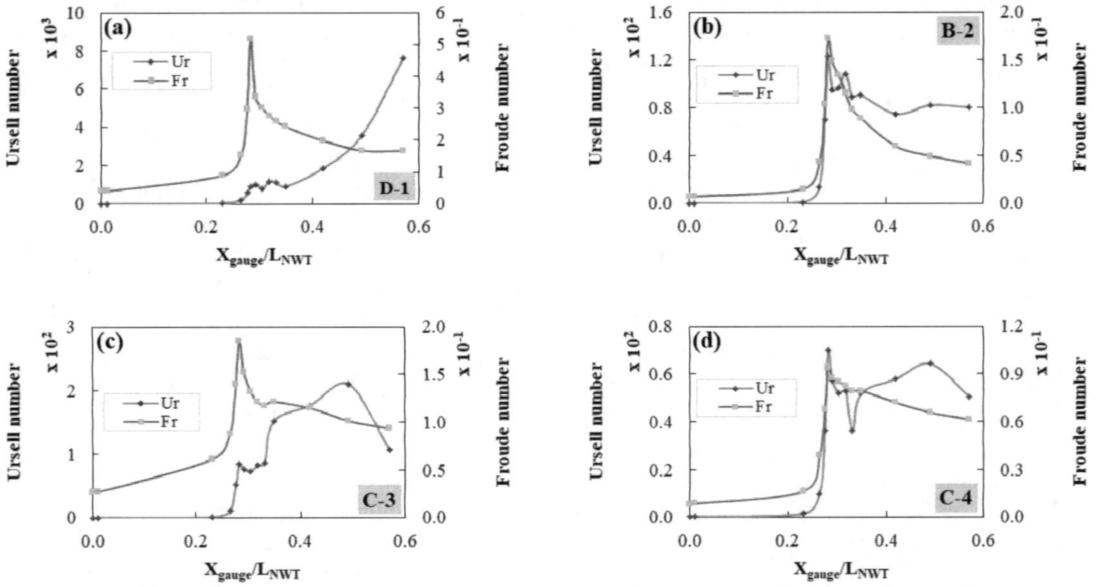

Figure 8. The Froude number F_r and Ursell number U_r in each selected scenarios

4. Conclusion

The propagation of breaking waves over a fringing reef in NWT is investigated through LES modelling. The free surface is tracked by the VoF method. Three existing numerical and experimental studies with varying wave height, wave period, and water depth based on different breaking wave parameters are employed to validate the numerical model. The breaking waves are classified based on three common criteria available in literature. The investigations show that present CFD model outperforms the previously used numerical models in view of its capability to better emulate the breaking waves on the slope or flat reef else. The present model is then implemented to examine the impact assessments for positioning WEC. All common criteria are reasonably applicable to predict breaking wave incident. The results show that the four selected scenarios are most sensitive to the whole scenarios. These findings indicate that the energy dissipation occurs significantly during post-breaking wave height degradation such as *C-3*, *B-2*, *D-1*, and *C-4*, respectively. From pre-breaking to breaking conditions, celerity of breaking waves is expected to experience larger harnessing wave energy. Moreover, relative wave-breaker locations, relative wave-breaker height range, breaking wave probability, wave steepness, water depth, and slope fringing reef are induced by fluid flow and wave nonlinearity for generating violent breaking wave. Further studies could shed more light on the relationship between breaking wave force and WEC.

References

[1] Roeber V and Cheung K F 2012 Boussinesq-type model for energetic breaking waves in fringing reef environments *Coast. Eng.* **70** 1–20

[2] Lin P 2004 A numerical study of solitary wave interaction with rectangular obstacles *Coast.Eng.* **51** pp 35-51

[3] Chella M A, Bihs H, Myrhaug D *et al* 2017 Breaking solitary waves and breaking wave forces on a vertically mounted slender cylinder over an impermeable sloping seabed *J. Ocean Eng.*Mar. Energy **3** pp 1–19

[4] Zhang S, Zhu L, Li J 2018 Numerical simulation of wave propagation, breaking, and setup on steep fringing reefs *Water* **10** p 1147

[5] Yao Y, Huang Z H, He W R and Monismith S G 2018 Wave-induced setup and wave-driven current over Quasi-2DH reef-lagoon-channel systems *Coast. Eng.* **138** pp 113–125

[6] Yao Y, Zhang Q, Chen S and Tang Z 2019 Effects of reef morphology variations on wave processes over fringing reefs *Applied Ocean Research* **82** pp 52-62

[7] Kamath A, Chella M A, Bihs H and Arntsen Ø A 2017 Energy transfer due to shoaling and decomposition of breaking and non-breaking waves over a submerged bar *Eng. Appl.* Comput. Fluid Mech. **11** pp 450–466

[8] Lin P and Liu P L F 1998b Turbulence transport, vorticity dynamics, and solute mixing under plunging breaking waves in surf zone J. Geophys. Res. 103 pp 15677–15694

[9] Lin P, Chang K A and Liu P L F 1999 Runup and rundown of solitary waves on sloping beaches ASCE J. Waterw. Port Coast. Ocean Eng. 125 pp 247–255

[10] Fang K, Zou Z, Dong P, Liu Z, Gui Q and Yin J 2013 An efficient shock capturing algorithm to the extended boussinesq wave equations Appl. Ocean Res. 43 pp 11–20

[11] Wei Z and Dalrymple R A 2018 Surf zone wave heating by energy dissipation of breaking waves Coastal Engineering Proceedings 1 p 2

[12] Lowe R J, Buckley M, Altomare C, Rijnsdorp D *et al*. 2019 Numerical simulations of surf zone wave dynamics using Smoothed Particle Hydrodynamics *Ocean Modelling* **144** p 101481

[13] Darvishi E, Fenton J D and Kouchakzadeh S 2016 Boussinesq equations for flows over steep slopes and structures J. Hydraul. Res. 55 pp 324–337

[14] Yao Y, He T, Deng Z, Chen L and Guo H 2019 Large eddy simulation modeling of tsunami-like solitary wave processes over fringing reefs Nat. Hazards Earth Syst. Sci. pp 1281–1295.

[15] Lubin P, Vincent S, Caltagirone J and Abadie S 2003 Fully three-dimensional numerical simulation of a plunging breaker C. R. M´ecanique 331 pp 495–501

[16] Yao Y, Huang Z, Stephen G, Monismith, Edmond Y M 2012 1DH Boussinesq modeling of wave transformation over fringing reefs *Ocean Engineering* pp 30-42.

[17] Davidson, J. dan Costello, R. (2020), "Efficient Nonlinear Hydrodynamic Models for Wave Energy Converter Design—A Scoping Study", Journal of Marine Science and Engineering, Vol. 8, hal. 35.

[18] Zhiyin, Y. (2015), "Large Eddy Simulation: Past, Present and the Future", Chinese Journal of Aeronautics, Vol. 28, hal. 11-24.

[19] Melville, W.K. (1982), "The Instability and Breaking of Deepwater Waves", Journal of Fluid Mechanics, Vol. 722, hal. 165–185.

[20] Windt, C., Davidson, J., Schmitt, P. dan Ringwood, J.V. (2019), "On the Assessment of Numerical Wave Makers in CFD Simulations", Journal of Marine Science and Engineering, Vol. 7, hal. 47.

[21] Fernandes, A. M. dan Fonseca, N. (2013), "Finite Depth Effects on the Wave Energy Resource and The Energy Captured by a Point Absorber", Ocean Engineering, Vol. 67, hal. 13-26.

[22] DNV-RP-C205 (2010), Recommended Practice: Environmental Conditions and Environmental Loads, Det Norske Veritas, Norway.

[23] CEM EM-1110-2-1100 (2015), Coastal Engineering Manual part II, US Army Corps of Engineers, Washington DC.

[24] Schmitt, P. dan Elsaesser, B. (2015), "A Review of Wave Makers for 3D Numerical Simulations", Proceedings of the MARINE 2015—Computational Methods in Marine Engineering VI, Rome, hal. 437–446.

[25] Aggarwal, A., Alagan Chella, M., Bihs, H., dan Myrhaug, D. (2020), "Properties of Breaking Irregular Waves over Slopes". Ocean Engineering, hal 216.

[26] Gomes, A., Pinho, J.L.S., Valente, T., Antunes do Carmo, J.S.V. dan Hegde, A. (2020), "Performance Assessment of a Semi-Circular Breakwater through CFD Modelling", Journal of Marine Science and Engineering, Vol. 8, hal. 226.

[27] Willmott, C. J. (1981), "On the Validation of Models", Physics and Geography, Vol. 2, hal. 184–194.

[28] Myttenaere, A., Golden, B., Grand, B. Le. dan Rossi, F. (2016), "Mean Absolute Percentage Error for Regression Models", Neurocomputing, Vol. 192, hal. 38–48.

[29] Xu, J., Liu, S., dan Li, J. (2019), "Experimental Study of Wave Propagation Characteristics on a Simplified Coral Reef", Journal of Hydrodynamics, Vol. 32, hal. 385–397.

[30] Goda, Y. (2010), "Reanalysis of Regular and Random Breaking Wave Statistics", Coastal Engineering Journal, Vol. 52, hal. 71–106.

[31] Massel, S.R. (2007), "Ocean Waves Breaking and Marine Aerosol Fluxes: Wave Breaking Criteria and Probability of Breaking", Atmospheric and Oceanographic Sciences Library, Springer, New York.

[32] Kench, P. S. dan Mann, T. (2017), "Reef Island Evolution and Dynamics: Insights from the Indian and Pacific Oceans and Perspectives for the Spermonde Archipelago", Frontier on Marine Science, Vol. 4, hal. 145.

[33] EMEC (2009), Marine Renewable Energy Guides : Tank Testing of Wave Energy Conversion Systems, The European Marine Energy Centre, London.

[34] Payne, G. (2008), Guidance for the Experimental Tank Testing of Wave Energy Converters: version 01b, Super Gen Marine, University of Edinburgh, Edinburgh.

IOP Conf. Series: Earth and Environmental Science 1081 (2022) 012050 doi:10.1088/1755-1315/1081/1/012050

Performance analysis of dual-stage vertical axis hydrokinetic turbine (VAHT) with azimuth variation at open canal hydrokinetic power plant

R Hantoro, E Septyaningrum, G Nugroho, M F A Alam, M D Fawwaz, F D Alfarisyi, M I Alfarizi and R A Prasetyo

Department of Engineering Physics, Institut Teknologi Sepuluh Nopember, Surabaya, Indonesia

Abstract. For several decades, the use of hydropower generation using hydrokinetic turbines has grown rapidly, aiming to get energy alternatives and reduce the dependency on fossil energy. Despite it has low efficiency, the vertical axis hydrokinetic turbine (VAHT) has been used to utilize hydro energy and has a promising result. A previous study had stated that the utilization of cascaded blade enhances the performance of VAHT. However, this technology produces fluctuating torque and power during its rotation. Hence the dual-stage VAHT is proposed to overcome this issue to improve VAHT's efficiency. This research analyzes the dual-stage VAHT's performances by variating its inter-stage azimuth of 45°, 60°, 75°, and 90°. For each stage, there is six blades that attach at three arms. The computational fluid dynamics (CFD) simulation is employed as its capability to provide in-depth information regarding flow phenomenon, force, and torque. This simulation has proven that the dual-stage configuration has an impact on reducing torque fluctuation of VAHT and show that different of inter-stage azimuth angle brings to different torque fluctuation pattern. The average torque value produced by the new turbine with variations in shift between stages of 45°, 60°, 75° and 90° is 10,119.21 Nm, 10,656.13 Nm, 10,490.31 Nm, and 10,457.3 Nm, respectively.

1. Introduction

Nowadays, hydrokinetic energy has become an alternative to reduce dependence on fossil energy as its overwhelming resources. International Hydropower Association (2020) reports that total energy production from hydropower is 1,330 Gigawatt, an increase of 1.6% or 22 Gigawatt from 2019. According to the theory of global marine, Indonesia itself has potential hydro energy up to 17,400 TWh/year that comes from its 3.5 million km2 water area [1-2]. That potential can be realized in the form of tidal waves, oceans, marine currents, rivers, and canal energy conversion although hydrokinetic energy has a lot of room for improvement especially in the matter of its efficiency.

The best option to utilize hydrokinetic energy is using turbines that have a similar working principle as wind turbine system. The turbines that are used for power generation can be classified into 2 categories depending on the orientation of its axis: vertical axis hydrokinetic turbine (VAHT) and horizontal axis hydrokinetic turbine (HAHT) [3]. As most rivers or canals have low current, many researches have proven that VAHT is more suitable since it can generate power with such speed, it is independent of the current direction and easy to maintain [4-7]. Since its first development in 1991, VAHT researches have been conducted to improve its performance such as blade geometries, foil configuration, and variable pitch control mechanism [8-16]. A Previous study of VAHT's performance analysis conducted by Hantoro, et al. proves that VAHT has promising results to be implemented but one of its problems is it has a large torque fluctuation that affects its efficiency as well as its vibration [17].

The use of dual-stage configuration is aiming to reduce that problem. The result of its implementation shows that VAHT efficiency increased and lead to lower torque fluctuation [18-19]. Moreover, Baruah, et al. get an average result of 27.9% in power coefficient increase by implementing dual-stage configuration in vertical axis wind turbine system and a similar result is expected in VAHT [20]. Based on those promising results, this research tries to conduct a performance analysis of dual-stage VAHT configuration with azimuth between stages' variation. The variations include 45°, 60°, 75°, and 90°. Data that are expected to be obtained are forces caused by the water flow of 3 m/s by using computational fluid dynamics simulation that can be used to calculate the torque of the dual-stage VAHT.

2. Methodology

2.1. Turbine Design

The turbine design used in this study is shown in Figure 1. The turbine used is a modified turbine from a previous study. The dimensions of the turbine used are shown in Table 1. The turbine has 12 blades which are divided into two stages, namely the upper stage and the lower stage which are 45°, 60°, 75°, 90°. Each stage consists of 3 arms with 2 blades per arm. The configuration of the shift between turbine stages is shown in Figure 2.

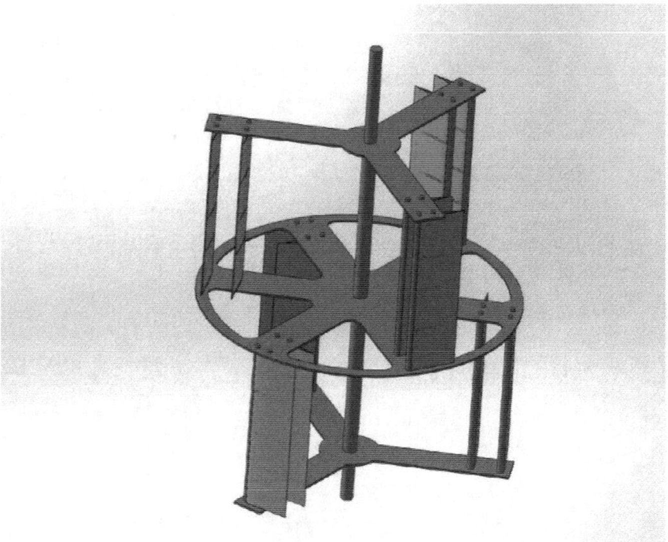

Figure 1. Model example of 60° dual-stage VAHT.

Table 1. Geometry of turbines.

Parameter	Unit (mm)
Cord Length	251.39
Span	855
Diameter	1250
Turbine Aspect Ratio	1380
Blade Aspect Ratio	3.4
Hydrofoil	NACA 0018

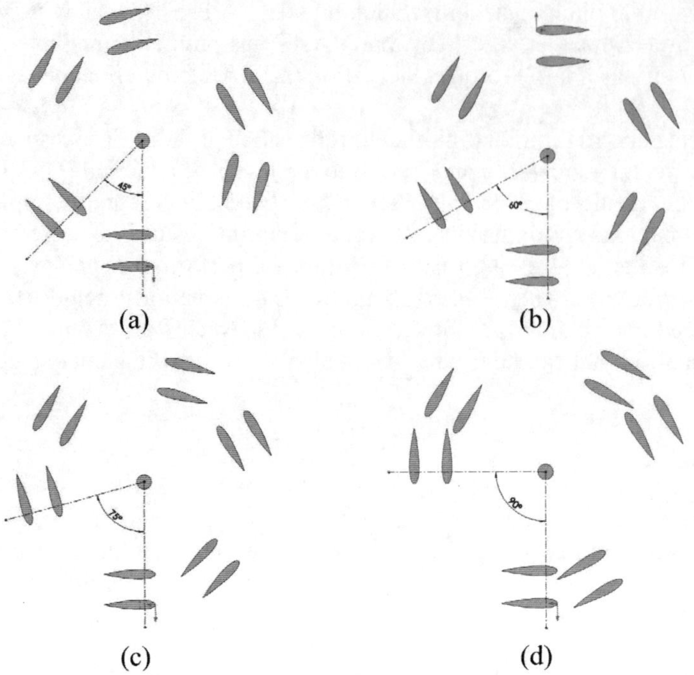

Figure 2. Variation of azimuth between stages (a) 45°, (b) 60°, (c) 75°, (d) 90°.

2.2. Meshing

After the geometry making process is complete, then the next step is the meshing process. In the meshing process, domain solving is done into small parts, in each of these parts is considered a volume control. In each volume control will be calculated by the Energy Conversion, Mass, Momentum, using CFD software. Small size meshing, making calculations more accurate but requires severe computing.

Table 2. Grid Independent Test

No.	Element size	Elements	Torque (Nm)
1	21	4340846	19,912.09
2	20.5	4505317	21,519.33
3	20	4641441	23,443.48
4	19.5	4870268	23,592.47

In getting the best meshing results, several experiments are needed with variations in the size of the mesh element. The meshing process, using the Face Sizing method with Element Sizing is 20 mm. From the results of the experiment, the number of elements is obtained in the table below:

Table 3. Total Elements and Nodes

	45	60	75	90
Nodes	824611	830779	825743	820657
Elements	4605332	4641441	4603719	4581818

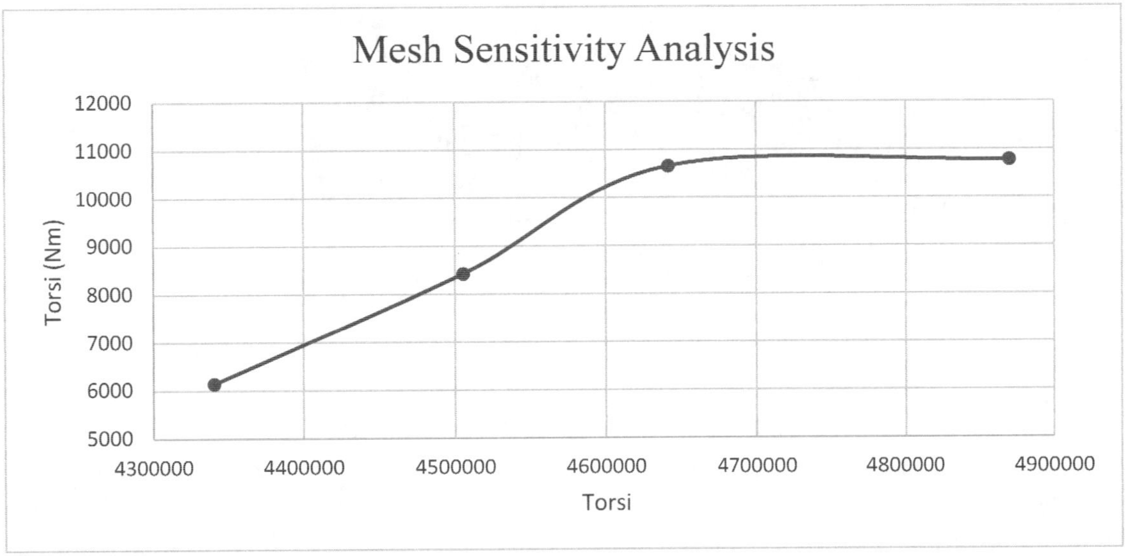

Figure 3. Mesh Sensitivity Analysis

Based on the graph above, for a good meshing it is in the 4641441 element for a 60 degree shift variation, because when the number of elements is added to 4870268 the torque value has a small error.

Table 4. Mesh Sensitivity

Torque (Nm)	6,135.73	8,419.59	10,656.12	10,781.32
Error	16%	7%	8%	1%

The mesh parameter used in the simulation are shown in the table below.

Table 5. Mesh Parameter

Parameter	Detail	Value
Method	Geometry: Rotating Domain, Tetrahedrons	Default
Face Sizing Rotating Domain	Element Size	20 mm
Face Sizing Interface	Element Size	50 mm
Skewness	Maximum Skewness	0.62391

2.3. Numerical Setup

To get the data about what you are looking for, a numerical simulation is carried out using the computational fluid dynamics (CFD) method. Due to the limited data and the high cost for experimental studies, numerical simulation is considered a promising method to perform. The numerical simulation consists of three main stages, namely pre-processing (geometry, mesh generation, and setting), processing (solver), post-processing (result).

In the pre-processing stage, the first thing to do is make the geometry. The simulated geometry consists of rotational and stationary domains. The stationary domain is 8D long, 4.8D wide. The boundaries of the stationary domain are defined as inlets and outlets. The rotational domain consists of a turbine blade and a shaft. The turbine arm is neglected due to considerations of simplification and the economy of computational time. The modeling carried out in this study is shown in Figure 3.

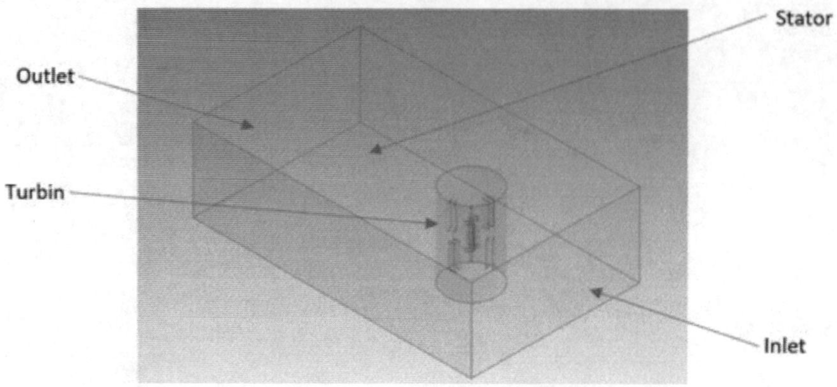

Figure 4. Geometry design and boundary condition.

After the geometry stage is complete, then proceed with the meshing stage. The grid is fully tetrahedral unstructured in both external and rotating domains as shown in Figure 4. The unstructured tetrahedral grids are chosen due to the capabilities to discretize complex geometries with fast and minimum user intervention. Then after the geometry and meshing stages are carried out the next stage is to determine the boundary conditions and turbulence models from the simulation. The turbulence model used is the SST (Shear Stress Transport) k-ω turbulence. The k-ϵ turbulence model has limitations in boundary layer flows with adverse pressure gradients [21]. On the other hand, the SST k-ω turbulence model gives accurate predictions of the amount of flow separation under adverse pressure gradients and has been used in several cases of CFD simulations on wind and water turbines.[22] Then in a study of dynamic stall on Shear Stress Transport (SST) k−ω turbulence model showed the results which is better than the turbulence − model [23 - 24]. While the boundary conditions consist of inlet, outlet, blade. Boundary conditions and simulation set up are shown in Table 2.

Table 6. Boundary condition setup

Parameter	**Condition**
Turbulence Model	*K-Omega SST*
Rotating Domain	*Mesh Motion* dan *Frame Motion* of 35 RPM
Boundary Condition	
Inlet	*Velocity Inlet*
Outlet	*Pressure Outlet*
Blade	*Wall*
	Moving (rotating wall)
Initialization Method	*Hybrid Initialization*

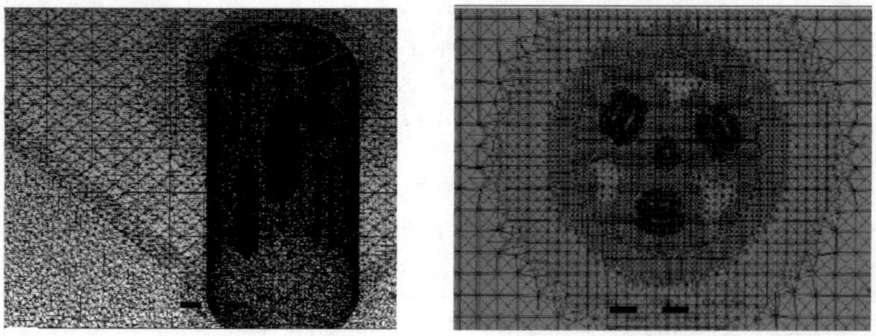

(a) (b)

Figure 5. Grid generation in (a) all domains, (b) blade (top view).

The processing stage is carried out by inputting all the parameters and boundary conditions that have been determined in the previous stage. The results of the simulation can be in the form of curves, contours, 2D and 3D animations, vectors, etc. The final stage of the simulation is the solution or result (post-processing) of the simulation in the form of torque fluctuations and speed contours.

3. Results and Discussion

3.1. Torque Fluctuation

In this simulation, each model is simulated using a speed of 3 m/s and a turbine rotation of 35 RPM. In this case it is assumed that water flows from the positive z-axis to the negative z-axis, with the turbine rotating counterclockwise. The position of the arm when rotating is represented in azimuth values. As long as the turbine rotates, it will produce hydrodynamic forces (lift and drag forces) which are then projected into tangential and radial forces. The force data is then processed to obtain turbine torque. The magnitude of this force is influenced by the azimuth angle of the turbine. Force and torque data collection is carried out at every 15° angle, so that one full rotation requires 24 data collection points. In this study, 24 simulations were carried out with variations in shift between stages of 45°, 60°, 75°, and 90°. Turbine torque simulation results are shown in Figure 5.

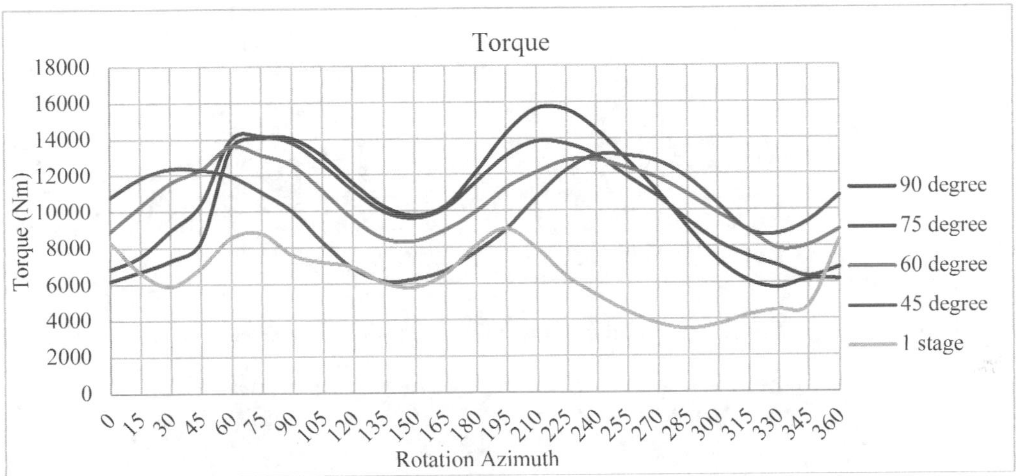

Figure 6. Dual-stage VAHTs and single-stage VAHT torque comparison.

The simulation results show that the torque value has a fluctuating shape during the rotating turbine (360°). However, due to the different configurations of each model, the fluctuation pattern will also be different. This is caused by changes in the angle of attack on each blade of each turbine arm. In the model with a shift of 90°, the maximum torque occurs at an azimuth angle of 225° and the minimum torque value occurs at an azimuth angle of 0° and 360°. While in the 75° shift model, the maximum torque value occurs at 60° azimuth and the minimum value at 345° azimuth angle. The model with a shift of 60°, the highest torque value occurs at an azimuth angle of 60° and the minimum torque at an azimuth of 150°. In the model with a shift of 45°, the highest torque value occurs at an azimuth angle of 240° and the minimum torque at an azimuth of 135°. The simulation also shows that the average torque value produced by the new turbine with variations in shifts between stages is 45°, 60°, 75°, and 90°, which are 10,119.21 Nm, 10,656.13 Nm, 10,490.31 Nm, and 10,457.3 Nm, respectively.

The difference between the maximum and minimum torque values at 45° shift is 6,942.18 Nm, at 60° shift is 5,752.17, at 75° shift is 7,832.53 Nm, and at 90° shift is 9,896.01 Nm. From the difference in the maximum and minimum torque values, it can be seen that the smallest difference occurs in the model with a shift of 60°. This shows that the configuration has the most stable torque fluctuation. The occurrence of torque fluctuations is due to differences in the hydrodynamic forces that hit the turbine when the turbine rotates. This force is proportional to the pressure experienced by the turbine blades. The smaller the pressure, the smaller the force obtained. In a turbine with a shift of 60° the turbine blade position has an even distribution on each azimuth so that the force obtained from the turbine for each azimuth is also more evenly distributed, this causes fluctuations in the 60° shift model to have better fluctuations.

3.2. Speed Contour

The characteristics of the flow through the turbine can be analyzed through the velocity contours of the flow. The velocity contour is displayed on the four variations of the shift between stages so that the different effects of the four variations can be seen. The distribution and direction of the velocity of the water flow in the turbine can be seen through the velocity contours shown in Figure 7, Figure 8, Figure 9, and Figure 10. The flow velocity pattern is taken at a water flow velocity of 3 m/s.

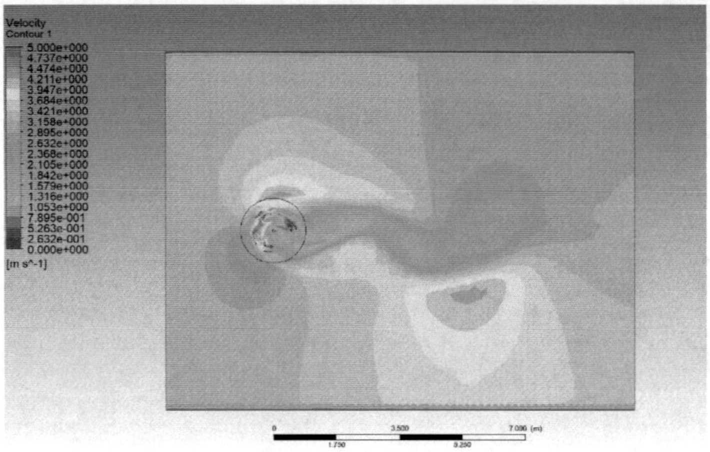

Figure 7. Speed contour of 45° azimuth variation.

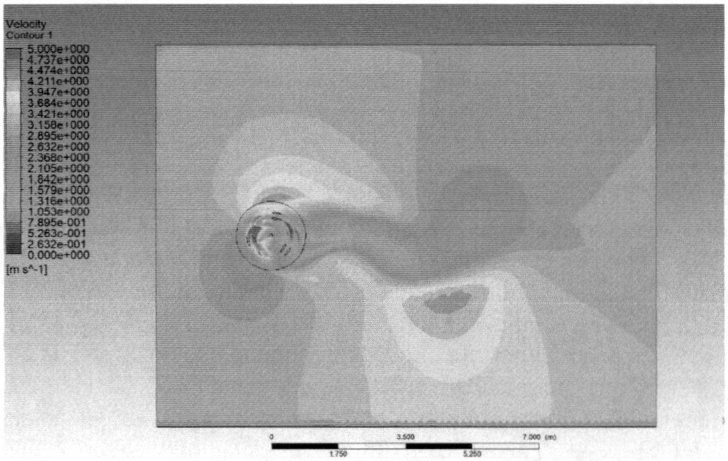

Figure 8. Speed contour of 60° azimuth variation.

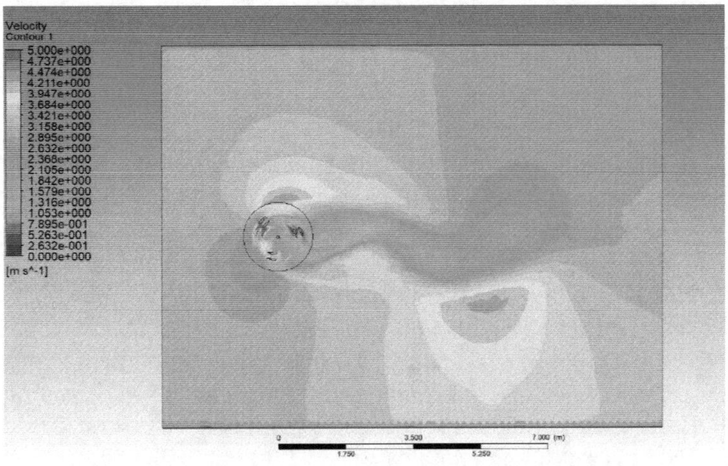

Figure 9. Speed contour of 75° azimuth variation.

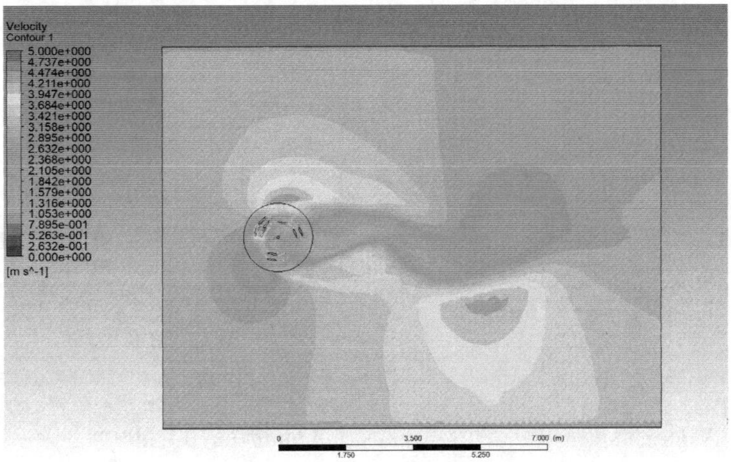

Figure 10. Speed contour of 90° azimuth variation.

The increased flow velocity can increase the Reynolds number, if the Reynolds number is greater, the drag force will also be greater. The decrease in flow velocity behind the turbine which is lower than the speed at the inlet due to energy extraction at the turbine, this is in accordance with the total work equation which is a change in kinetic energy where high kinetic energy at the inlet and low kinetic energy at the outlet done by the turbine is work. The decrease in kinetic energy at the outlet is indicated by a decrease in the flow velocity.

From Figure 7, Figure 8, Figure 9, and Figure 10, variations in shift between stages will affect the distribution of speed that passes through one blade and has an effect on the other blades. It can be seen that the shift between 60° stages has a speed distribution between one blade and another that is more evenly distributed in the upstream than the variation of shift between 45°, 75° and 90° stages in the upstream. From observing the velocity contour, it can be seen that the flow velocity decreases after passing through the turbine due to energy extraction by the turbine so that there is a decrease in speed.

4. Conclusion

In this research, 4 types of turbines have been evaluated, namely turbines with shifts between stages of 45°, 60°, 75° and 90°. From the variation of shift between stages, the second-generation turbine with a shift between stages of 60° has a torque ripple that is more stable than the variation of shift between other stages, namely 45°, 75° and 90°. The average torque value produced by the new turbine with variations in shift between stages of 45°, 60°, 75° and 90° is 10,119.21 Nm, 10,656.13 Nm, 10,490.31 Nm, and 10,457.3 Nm, respectively.

5. Acknowledgement

The authors thank to PT. Pembangkitan Jawa Bali (PJB) UP Paiton and the Ministry of education for making this research possible.

6. Reference

[1] I. R. H. Irfan, "Respon to Pontoon and Pendulum Motion at Wave Energy Converter Based on Pendulum System," *E3S Web of Conferences 43,01022 (2018),* 2017.

[2] J. K. a. G. Bhuyan, "Ocean Energy: Global Technology Development Status," *Report prepared by Powertech Labs for the IEA-OES (Implementing Agreement on Ocean Energy Systems - International Energy Agency,* 2009.

[3] T. Wenlong, S. Baowi & M. Zhaoyong, "Conceptual Design and Numerical Simulations of A Vertical Axis Water Turbine Used for Underwater Mooring Platforms," *International Journal of Naval Architecture and Ocean Engineering,* **vol. 5**, 625-634, 2013.

[4] A. Posa, C. M. Parker, M. C. Leftwich & E. Balaras, "Wake Structure of a Single Vertical Axis Wind Turbine," *International Journal of Heat Fluid Flow*, **61**, pp. 75-86, 2016.

[5] I. Ross & A. Altman, "Wind Tunnel Blockage Corrections: Review and Application to Savonius Vertical-Axis Wind Turbines," *Journal of Wind Engineering and Industrial Aerodynamic,* **99**(5), pp. 523-538, 2011.

[6] G. Tescione, C. J. S. Ferreira & G. J. W. van Bussel, "Analysis of a Free Vortex Wake Model for the Study of the Rotor and Near Wake Flow of a Vertical Axis WInd Turbine," *Renewable Energy, Elsvier,* **vol. 87**(P1), pp 552-563, 2016.

[7] H. Beri & Y. Yao, "Double Multiple Streamtube Model and Numerical Analysis of Vertical Axis Wind Turbine," *Energy Power Engineering,* **3**(3), pp. 262-270, 2011.

[8] Y. Takamatsu, A. Furukawa, K. Okuma & K. Takenaouchi, "Experimental Studies on a Preferable Blade Profile of High Efficiency and the Blade Characteristics of Darrieus-Type Cross-Flow Water Turbine," *JSME Internasional Journal,* **34**, 1991.

[9] F. Scheurich, T. M. Fletcher & R. E. Brown, "The Influence of Blade Curvature and Helical Blade Twiston the Performance of a Vertical-Axis Wind Turbine," *48th AIAA Aerospace Science Meeting Including the New Horizons Forum and Aerospace Exposition,* pp. 1-16, 2010.

[10] M. R. Castelli & E. Benini, "Effect of Blade Inclination Angle on a Darrieus Wind Turbine," *ASME. J. Turbomach,* **134** (3), 031016, 2011.

[11] R. Gupta & A. Biswas, "Computational Fluid Dynamics Analysis of a Twisted Three-Bladed H-Darrieus Rotor," *Journal of Renewable and Sustainable Energy*, **2**(4), pp.1-15, 2010.

[12] M. C. Claessens, "The Design and Testing of Airfoils for Application in Small Vertical Axis Wind Turbines," *Master of Science Thesis, Delf University*, 2004.

[13] W. Roynarin, "Optimisation of Vertical Axis Wind Turbines," *Northumbria University,* 2004.

[14] R. Balaka & A. Rachman, "Pitch Angle Effect for Horizontal Axis River Current Turbine," *Procedia Engineering*, **50**, pp. 343-353, 2012.

[15] S. Hwang, Y. Lee and S. Kim, "Optimization of Cycloidal Water Turbine and The Performance Improvement by Individual Blade Control," *Applied Energy*, **86**(9), pp.1532-1540, 2009.

[16] A. Schonbore & M. Chantzidakis, "Development of a Hydraulic Control Mechanism for Cyclic Pitch Marine Current Turbines," *Renewable Energy*, **32**(4), pp.662-679, 2007.

[17] R. Hantoro & E. Septyaningrum, "Novel Design of a Vertical Axis Hydrokinetic Turbine – Straight-Blade Cascaded (VAHT-SBC): Experimental and Numerical Simulation," *J. Eng. Technol. Sci.,* **vol. 50**, no. 1, pp. 73-86, 2018.

[18] E. A. Attia, H. Saber & H. E. Gamal, "Performance and Dynamic Characteristics of a Multi Stages Vertical Axis Wind Turbine," *Journal of Vibroengineering*, **vol. 18**, issue 6, pp. 4015-4032, 2016.

[19] D. W. Wekasa, C. Wang, Y. Wei & W. Zhu, "Experimental And Numerical Study Of Turbulence Effect On Aerodynamic Performance Of A Small-Scale Vertical Axis Wind Turbine," *J. Wind Eng. Ind. Aerodyn.*, **vol. 157**, pp. 1 14, 2016, doi:10.1016/j.jweia.2016.07.018.

[20] M. Baruah, M. Swarnakar & Nripen Das, "Computational Investigation of Double Stage Vertical Axis Wind Turbine," *AIP Conference Proceedings*, **2091**, 2019.

[21] Yang, Bo & Lawn, Chris,"Fluid dynamic performance of a vertical axis turbine for tidal currents," Renewable Energy, Elsevier, vol. 36(12), pages 3355-3366, 2011.

[22] Ye Li, Sander M. Calisal, "Modeling of twin-turbine systems with vertical axis tidal current turbine: Part II—torque fluctuation".,Ocean Engineering, Volume 38, Issue 4,2011

[23] Ekaterinaris, J.A.; Platzer, M.F."Computational prediction of airfoil dynamic stall". Progress Aerosp. Sci. , 33, 759–846. 1998

[24] Cheng, Q.; Liu, X.; Ji, H.S.; Kim, K.C.; Yang, B. Aerodynamic Analysis of a Helical Vertical Axis Wind Turbine. *Energies* , *10*, 575. 2017.

IOP Conf. Series: Earth and Environmental Science 1081 (2022) 012051 doi:10.1088/1755-1315/1081/1/012051

Comparison Analysis of Options to Comply with IMO 2020 Sulphur Cap Regarding Environmental and Economic Aspect

Danuja Wijayanto[1], Gede Bagus Dwi Suasti Antara[2]

[1] Center of Excellences Maritime Safety and Marine Installation, Institut Teknologi Sepuluh Nopember, Surabaya, Indonesia
[2] Department of Marine Engineering, Faculty of Marine Technology, Institut Teknologi Sepuluh Nopember, Surabaya, Indonesia

danujawijayanto2@gmail.com, dwi.suasti@gmail.com

Abstract. IMO enforce new regulation known as IMO 2020 Sulphur Cap which limits the sulphur content in the fuel oil used onboard ships. This new global sulphur cap is a significant challenge for the shipping industry. The compliance options may be classified into three main options: using alternative fuels, low-sulphur fuel oil, and scrubbers. Each option comes with some advantages and disadvantages which creates a trade-off between the options. Some options may require a higher initial investment in terms of capital expenses, and some are more expensive during operations than others. A comparison analysis will be conducted with respect to the environmental and economic aspects.

1. IMO 2020 Sulphur Cap Compliance

In recent years, environmentally driven regulation is most likely affecting shipping. Since most ship still relies on fossil fuels, it has become a priority on the global policy agenda [1]. One of the recent restrictions introduced by the International Maritime Organization (IMO) is the 2020 Sulphur Cap Compliance. The regulation limits the sulphur content of fuel oil from 3.5% m/m (mass by mass) to 0.5% for ships operating outside the Emission Control Area (ECA). It includes fuel used in main, auxiliary engines, and boilers must have no more than 0.5% m/m sulphur content [2].

The sulphur content regulation comes as part of the International Convention for the Prevention of Pollution from Ships (MARPOL) Annex VI which was adopted by the IMO in 1997. This regulation urges limits on the air pollutant carried in the ship's exhaust gas. Those gas are sulphur oxides (SOx), nitrous oxides (NOx) and particulate matters. Before SOx, the first global limit is 4.5% set in 2005. Then it was reduced to 3.5% in 2012 and 0.5% in 2020. The lower limits imposed in (ECAs) are much stricter with 0.1% and came into force at the beginning of 2015 [3].

SOx are produced through the oxidation of sulphur contained in the fuel. It has a negative effect both on human health and the environment. For humans, SOx may be linked to stroke, asthma, cardiovascular disease, lung cancer, and pulmonary disease. In terms of the environment, SOx may cause acid rain and may harm crops, forests, and aquatic species. Furthermore, SOx contamination may cause ocean acidification [4]. A study on the effects of the IMO 2020 found that if shipping industries comply with the regulation, it may lead to a 77% drop in total sulphur oxide emissions from the ship. The reduction is equivalent to 8.5 million metric tonnes of SOx. Delaying the implementation of regulation from 2020 to 2025 may contribute to more than 570,00 additional premature deaths compared to the implementation from 2020. Reducing this emission will also have a considerable impact on vulnerable areas and may improve the condition of coastal communities [5].

For shipping industries, the regulation creates new challenges. Particularly for container shipping, the 2015 SOx requirements alone affect the shipping cost by an amount of USD 500 million. The cost for shipping for Asia-North Europe trade lanes has increased by 1.2%-3.6%. The 2020 IMO Sulphur Cap believe to have a much more consequential effect on shipping costs. The calculation shows that it may give a 20% to 85% increase in shipping cost, depending on the assumptions regarding speed, fuel price, and ship size [6].

The stringent environmental regulations create a challenge for container shipping. It is believed that the trends will continue from 2020 onwards [1]. Therefore, the shipowner must decide to invest in complying with the regulation so that they may remain competitive in the market.

There are three main feasible options available to comply with the regulation. The first option is the direct option to change the fuel to low-sulphur residual fuel. The second option is using abatement technology such as installing a scrubber to clean up sulphur from the ship's exhaust system. The third option is applying a cleaner alternative fuel such as LNG or methanol [1]. There is no one-size-fits-all solution, and the best option may depend on ship type, ship size, routing, operational patterns, and the availability of fuel itself [7]. A case study of container vessels with three applied options will be assessed to scrutinize the effect of the IMO 2020 Sulphur Cap on shipping industries by looking at the economic and environmental aspects.

2. Options to Comply with IMO 2020 Sulphur Cap

2.1 Switching to a Low Sulphur Fuel

Switching fuel into low sulphur fuel is one direct option to comply with IMO Sulphur Cap 2020. Examples are low-sulphur residual fuel oil (LSFO), very-low-sulphur fuel oil (VLSFO), or low-sulphur distillates such as Marine Gas Oil [1]. The main advantage of using low sulphur fuel is it requires less modification since it is useable for most engine configurations [7]. The case study provided in this study is using MGO. This distillates fuel represents the easiest means of meeting the IMO 2020 Sulphur Cap. MGO is already widely used in main engines which normally run-on heavy fuel oil [3].

Switching to MGO which has low viscosity may create operational challenges. The instability in blended fuel may result in poor ignition, furthermore, it may cause fouling in cylinders, turbocharges, and exhaust systems. Therefore, there is an increased risk of machinery breakdown which may result in a risk to safe navigation [3]. The International Chamber of Shipping (ICS) guides shipping companies and crews in preparing IMO 2020 Sulphur Cap. Ship-owners and operators are recommended to ensure ISO 8217:2017 are described whenever ordering low sulphur fuel [8]. Using MGO may require more operational expenditure due to the higher cost of MGO compared to Heavy Fuel Oil (HFO). The modification and installation of the compliance options using MGO fuel and Fuel-switching are estimated to be 6 USD/kW and the maintenance cost of MGO is estimated to be 0.6 USD/kW [9]. While the price of the MGO will be set on the price of Singapore Bunker for MGO at 698 USD/MT [10].

2.2 Installing Scrubber Technology

The second option to comply with the IMO Sulphur Cap is installing exhaust gas cleaning technology called SOx Scrubber. Scrubber technology is an equivalent compliance method introduced at Marine Environment Protection Committee (MEPC) 57 in 2008 [3]. The guideline for exhaust gas cleaning system (EGCS) was prescribed in Resolution MEPC.259 (68) which was adopted on 15 May 2015. The guideline provides the requirements for testing, survey, certification, and verification of EGCS [11].

There are two kinds of scrubber technology: dry scrubber and wet scrubber. The dry scrubber uses hydrated lime-treated granulates instead of water or any liquid to conduct the scrubbing process. While the wet scrubber uses liquid media bypassing the exhaust gas with the wash liquid. Common liquids are chemically treated freshwater or untreated seawater. Furthermore, the wet scrubber technology is classified into closed-loop and open-loop. The closed-loop system uses seawater and a scrubber water cleaning system. The closed loop produces a sludge that may be used in a cement kiln or concrete. The open-loop system operates with seawater and there is no scrubber water cleaning. It will discharge directly to the sea. There is also a hybrid system that combined the principle of closed-loop and open-loop systems [12]. The vessel which installed scrubber technology does not need any change to the engines or fuel treatment, however, it needs investment cost for installing the scrubber [7].

The cost of retrofitting the scrubber will be 327 USD/kW [13]. The price for retrofitting the scrubber is lower for the larger engine. For vessels less than 5,000 kW the scrubber cost may go up to 370 USD/kW and for the engine of 10,000 kW, the price may fall around 250 USD/kW. For a larger engine, the cost can be as low as 120 USD/kW [14]. Specifically, the scrubber will have capital expenditure for installation and more operational expenditure due to the additional fuel used to power the scrubber, maintenance cost, the cost for disposing of the waste, and financing cost [15].

Retrofitting with a scrubber will also increase the OPEX of the vessel, it is estimated the OPEX, particularly for operation and maintenance will raise about 34,000 USD per year [15]. Using a scrubber also raises the energy consumption which is calculated to raise fuel consumption by 3% [16].

By installing a scrubber, the number of TEUs on board the ship will also decrease due to scrubber's component which may be installed on the ship. It will decrease about 1% of the total TEU on board the ship [13]. For the case study, the price of the IFO/HFO is around 478.5 USD/MT [10].

2.3 Using Alternative Fuels

The third option to comply with the IMO Sulphur Cap is by using alternative fuels which will remove completely SOx, such as Liquefied Natural Gas (LNG) or methanol [1]. Compared to methanol, LNG may be more favourable as alternative fuel due to the technically proven solution and there is rapid development in the LNG bunkering infrastructure. Changing to LNG will not only remove SOx but also significantly reduce NOx as well as particulate matters (PM). Furthermore, using LNG will reduce up to 25% of CO2 emissions. It also has a positive impact on the Energy Efficiency Design Index (EEDI) of the vessel [7].

Retrofitting vessels with dual-fuel engines which consume LNG may require a high investment in CAPEX. The average conversion cost for converting to a dual-fuel engine is around 220-340 USD/kW. Specifically, the cost for retrofitting the vessel using LNG, including the modification of the main engines, natural gas supply system including natural gas storage spaces, supply piping and valves, gas detection units and alarms, exhaust ventilation system, and other components [17]. While for the OPEX it is estimated to increase about 10% compared to the existing vessel on the operation cost, such as crew, spare parts, and maintenance. Furthermore, in a container vessel, the vessel requires more physical space onboard which is estimated to take up almost 3% of a vessel's TEU slots [18].

To estimate LNG fuel consumption, a comparison between the heating value is used between the Diesel Oil and LNG. The diesel oil's lower heating value (LHV) which is used in the comparison is 42 MJ/kg while the LNG is 48 MJ/kg [19]. The LNG bunkering price in Singapore use 8.732 USD per mmbtu [20].

3. Case Study

The case study will give a clear example of the implementation of IMO 2020. A container vessel that serves the route from Port of Panjang in Lampung to Port of Singapore. The illustration map is shown in Figure 1 while the vessel picture is shown in Figure 2. The vessel will require 3 days to complete one trip from Port of Panjang to Port of Singapore. It offers a weekly schedule and direct and dedicated service between the two ports [21]. The name of the vessel used in this route is Meratus Samarinda. The ship data is then acquired by collected from various sources such as marinetraffic.com [22] and ships.jobmarineman.com [23]. The vessel data is shown in Table 1.

IOP Conf. Series: Earth and Environmental Science 1081 (2022) 012051 doi:10.1088/1755-1315/1081/1/012051

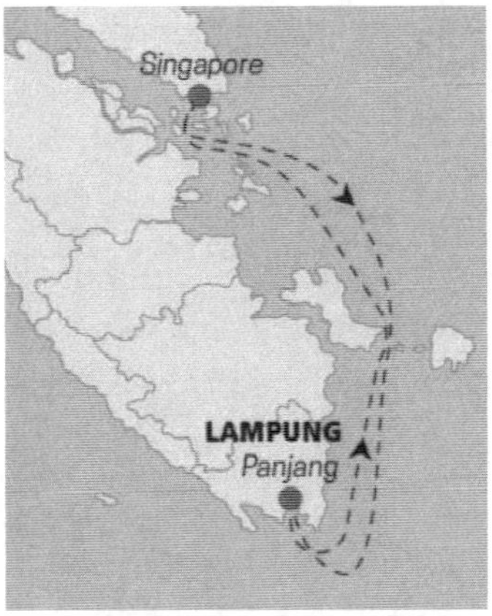

Figure 1. Ship Route [21]

Figure 2. Meratus Samarinda [22]

Table 1. Ship Information [22], [23]

Ship Information	
Ship Name	: Meratus Samarinda
IMO Number	: 9808481
Vessel Type	: Container Ship
MMSI	: 525125029
Call Sign	: YDIZ2
Flag	: Indonesia
GT	: 7542
DWT	: 8139 T
LOA	: 123 m
Breadth	: 24.4 m
Draught	: 5 m
Depth	: 7 m
TEU	: 526
Engine	: 2 x Yanmar 6EY26W
Total Power	: 3840 kW
RPM	: 750 rpm
Speed Max	: 17 knots
Former Name	: PAC Cekap

Based on the route and vessel data, the analysis will conduct considering the emissions produce which include SOx, NOx, and CO_2, and the economic point of view regarding Capital Expenditure, Operational Expenditure, and Fuel Cost, with respect for each compliance.

4. Analysis and Discussion

4.1 Emission Analysis

The emission analysis was carried out to estimate the number of emissions produced by the ship. The methodology to estimate the emission is following the studies which have been carried out by US Environmental Protection Agency (EPA) and Canadian inventory efforts [24]. Some examples of study which used in this study were conducted by Banawan, et. al. [17]. The equation for quantifying ship emission is shown as:

$$\Sigma E_{\text{total}} = \text{total } E_{\text{standby}} + \text{total } E_{\text{manouvering}} + \text{total } E_{\text{cruise}}$$

Where ΣE_{total} is the total emissions quantity during one trip. The emissions during standby, manoeuvring and cruising can be calculated with calculation below [24]:

$$\Sigma E = P \, \mathrm{LF} \, e_{fd} \, T$$

Where:

E	: Emission quantity (g)
P	: Engine power at maximum continuous rating (kW)
e_{fd}	: Emission factor (g/kWh)
LF	: Load Factor (per cent)
T	: Engine running time (h)

The engine power according to ship information in Table 1 is 3840 kW. The engine load factor is using 80% for cruising, 20% for manoeuvring, and 5% in standby mode. According to the information, the vessel trip is 72 hours, 0.5 hour for manoeuvring, and 24 hours in standby/loading-unloading mode [21]. While the emission factor is shown in Table 2 for cruising and Table 3 for manoeuvring and in port operation (standby/loading-unloading).

Table 2. Emission Factor Cruising Operation [24], [25], [26], [27]

	Cruising Operation					
	NOx	NOx (with Scrubber)	SOx	SOx (with Scrubber)	CO_2	CO_2 (with scrubber)
HFO	14	14.14	11.5	0.97	677	643.15
MGO	13.2		1		645	
LNG	2.16		0		548.25	

Table 3. Emission Factor Manoeuvring and In-Port Operation [24], [25], [26], [27]

	Maneuvering and In-Port Operation (Standby/loading-unloading)					
	NOx	NOx (with Scrubber)	SOx	SOx (with Scrubber)	CO_2	CO_2 (with scrubber)
HFO	11.2	11.312	12.7	0.95	745	707.75
MGO	10.6		1.1		710	
LNG	2.16		0		603.5	

In operation with LNG, the emission factor is considering LNG as primary fuel and MGO as pilot fuel. Therefore, the emission is 95% LNG and 5% MGO [17]. The result of the emission for 1 voyage could be seen in Figure 3 for SOx emission, Figure 4 for NOx emission, and Figure 5 for CO_2 emission. Although this paper focuses on the option to comply with the IMO 2020 Sulphur Cap, the effect to another air emissions such as NOx and Green House Gasses, in this case CO_2, is also examined to further understand the effect of each option in terms of environmental aspects.

As it can be seen in Figure 3, before applying the abatement technology, the number of SOx from the combustion in existing reference is reaching 2,607,014 grams. Applying the three option proves to be effective to reduce the number of SOx. The best reduction is achieved by using LNG with only 12,439 grams, follows by the installation of a Scrubber with 219,290 grams, while the application of MGO reduction is 226,675 grams.

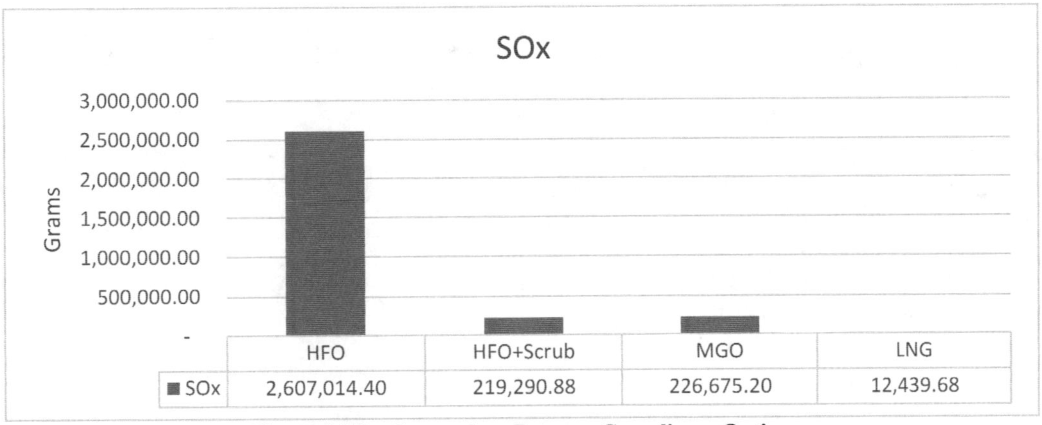

Figure 3 SOx Comparison Between Compliance Options

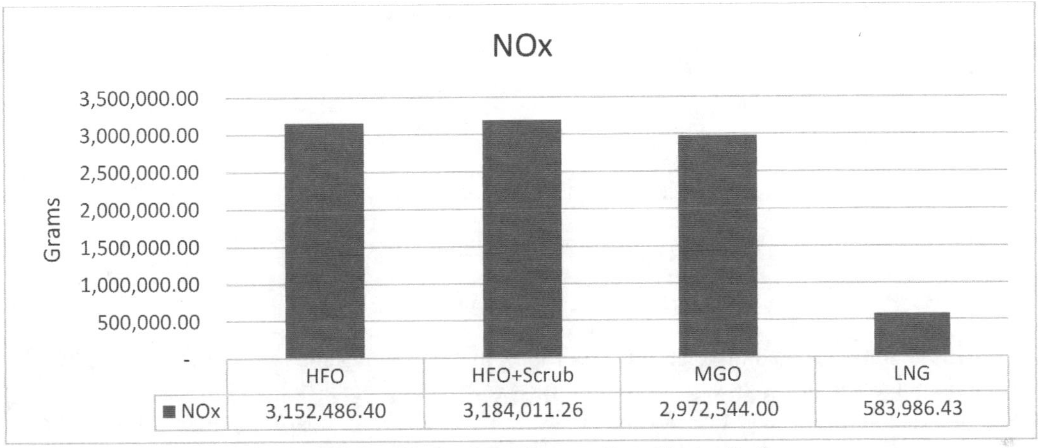

Figure 4 NOx Comparison Between Compliance Options

Another air emission, which may result from fuel oil combustion is NOx. From Figure 4, the application of the compliance options is not really affecting the reduction of the NOx. The best reduction is achieved by switching to a cleaner alternative fuel such as LNG. Switching to LNG produce only 583,986 grams from 3,152,486 gram of existing reference.

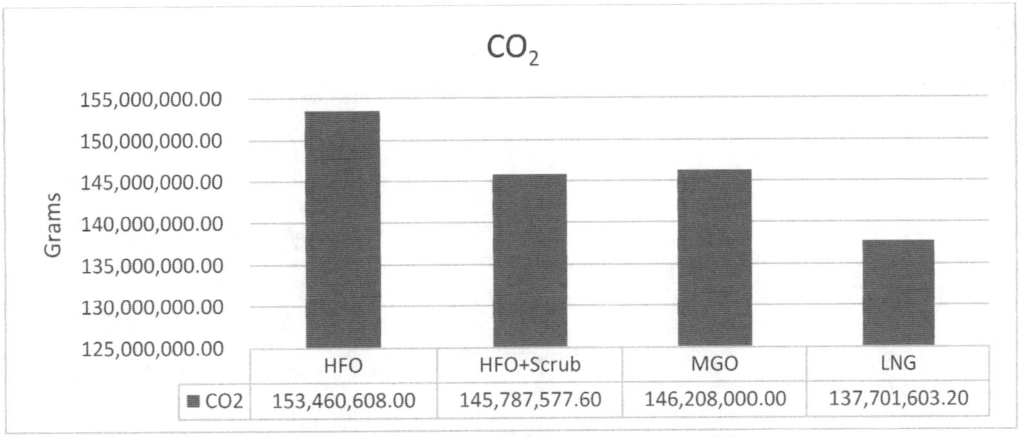

Figure 5 CO_2 Comparison Between Compliance Options

From CO_2 perspective in Figure 5, the options which still use conventional oil such as HFO and MGO only reduce relatively small amount of CO_2. Switching to LNG gives the best reduction of CO_2 from 153,460 grams CO_2 with HFO to 137,701 grams with LNG.

4.2 Economic Analysis

The economic analysis is conducted by analyse the three options by looking into the Capital Expenditure, Operation Expenditure, and the Fuel Cost. The operational data is obtained from the Drewry Ship Operating Costs Annual Review and Forecast [28] [29]. While the fuel cost is obtained from shipbunker.com [10]. The fuel cost of LNG is acquired from S&P Platts [20]. All price from the references is levelled to 2021 price.

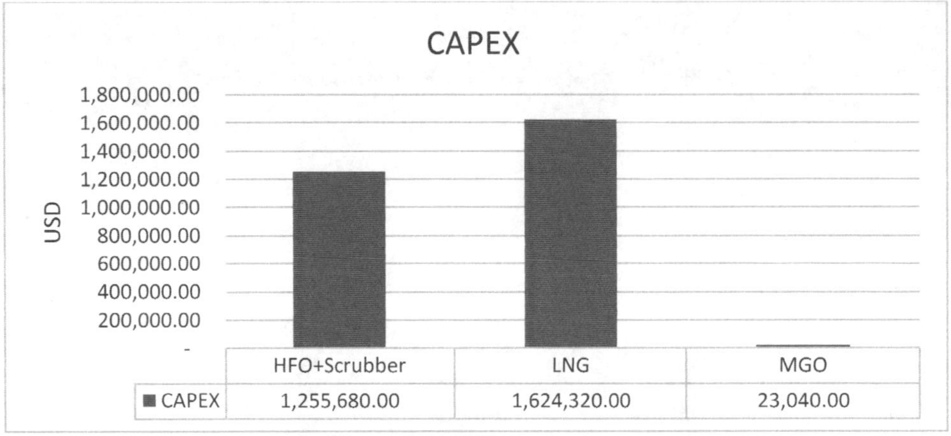

Figure 6 CAPEX Comparison

The CAPEX comparison between three compliance option is shown in Figure 6. The CAPEX comparison is considering the cost of component of retrofitting each option. For HFO + Scrubber the CAPEX is include scrubber equipment cost [13], the LNG CAPEX cost is include the modification of the main engines, natural gas supply system including natural gas storage spaces, supply piping and valves, gas detection units and alarms, exhaust ventilation system, and other components [17], and the MGO capex include the modification and installation in the engine and engine system [9]. Among all of the option the MGO offer the lowest CAPEX due to minor modifications to the ship with only 23,040 USD while the modification of LNG has the highest CAPEX with 1,624,320 USD.

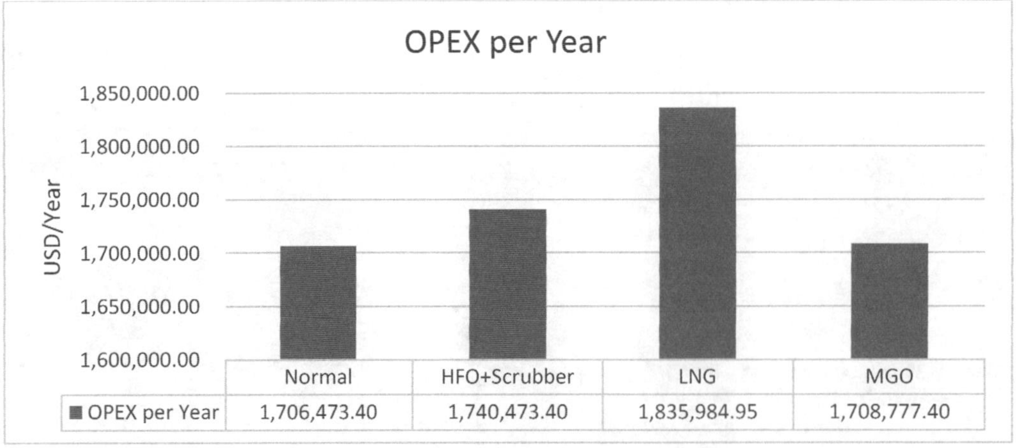

Figure 7 OPEX Comparison

IOP Conf. Series: Earth and Environmental Science 1081 (2022) 012051 doi:10.1088/1755-1315/1081/1/012051

The OPEX comparison between the three-compliance option and the existing reference is shown in Figure 7. The OPEX is following the Drewry OPEX categorization such as Manning Cost, Insurance, Stores, Spares, Lubricating Oils, Repair & Maintenance, and Management & Administration. The OPEX in HFO + Scrubber include 34,000 USD per year in Repair & Maintenance [15], for LNG it is 10% higher for Manning Cost, Spares, Lubricating Oils, Repair & Maintenance [18], while for MGO, the Repair & Maintenance Cost increase at the amount of 0.6 USD/kW [9].

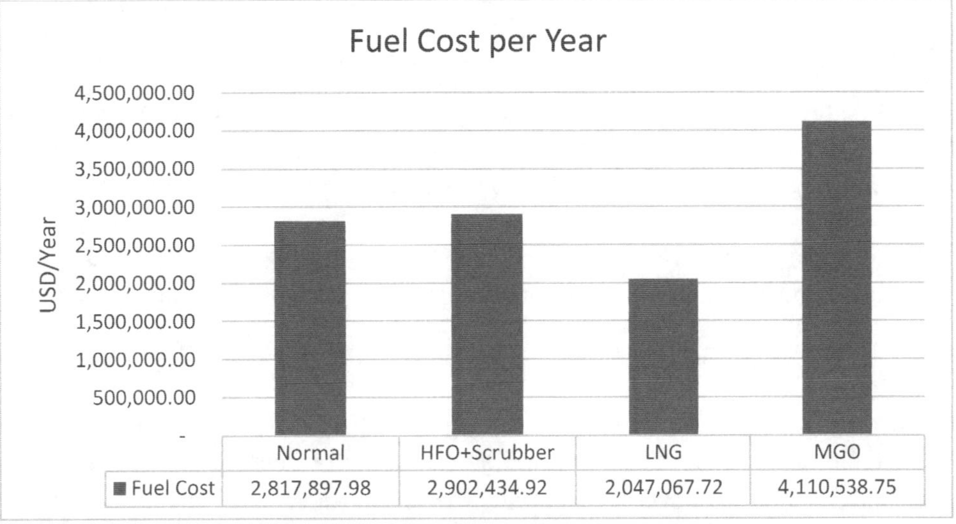

Figure 8 Fuel Cost Comparison

The fuel cost comparison per year is provided in Figure 8. The MGO has the highest fuel cost per year due to the high fuel cost compared to HFO. While the LNG have the lowest fuel cost per year among all the option with 2,047,067 USD/year.

5. Conclusion

Every compliance option has its own advantage and disadvantage in terms of emissions and economics. From an emissions aspect, regarding the IMO 2020 Sulphur Cap, HFO + Scrubber and MGO may have similar effects on SOx reduction. In terms of other emissions such as NOx and CO2, the two options also have similar effects. Further reduction in emissions may be achieved by changing to LNG Fuel. In terms of the economic aspect, MGO gives advantages of low CAPEX and OPEX rather than HFO + Scrubber. However, the decision needs further analysis since the MGO has the highest fuel cost per year due to its fuel price. It doubled the fuel cost per year compared to HFO.

Another option, if shipping owners seek to further reduce the emissions and target to reduce the CO_2, the option to use LNG may become an option since the LNG is also available in Singapore Port. In terms of the economic aspect, LNG may have higher CAPEX and OPEX, however, in the long run, depending on the fuel price, LNG may have the advantage of low fuel cost per year compared to other options.

Since this paper only compared the compliance option based on the emission and economic aspects, it also needs further study in terms of the technical aspect. Based on the vessel in the case study, it may be not fit to use HFO + Scrubber onboard the ship due to the vessel size being small. While for LNG, there are references that the existing engine may be converted to a dual-fuel engine [30]. However, it also needs to be more discussed in terms of the technical aspect.

Therefore, for the ship used in the case study, in the short term, MGO may be a viable option regarding compliance with IMO 2020 Sulphur Cap. While for the long term, LNG may become a better option considering the advantages of emissions and fuel cost.

IOP Conf. Series: Earth and Environmental Science 1081 (2022) 012051 doi:10.1088/1755-1315/1081/1/012051

References

[1] UNCTAD, Review of Maritime Transport, New York: United Nations Publications, 2019.

[2] IMO, "Sulphur 2020 - Cutting Sulphur Oxide Emissions," 2020. [Online]. Available: http://www.imo.org/en/MediaCentre/HotTopics/Pages/Sulphur-2020.aspx.

[3] R. Clark, "The 2020 Global Sulphur Cap: An Overview," Reed Smith, Hong Kong, 2018.

[4] EPA, "Sulfur Dioxide Basics," 28 January 2021. [Online]. Available: epa.gov/so2-pollution/sulfur-dioxide-basics.

[5] J. J. Corbett, J. J. Winebrake, E. W. Carr, J.-P. Jalkanen, L. Johansson, M. Prank, M. Sofiev, S. G. Winebrake and A. Karppinen, "Health Impacts Associated with Delay of MARPOL Global Sulphur Standards," Finnish Meteorological Institute, Helsinki, 2016.

[6] OECD, "Reducing Sulphur Emissions from Ships: The Impact of International Regulation," OECD/ITF, Paris, 2016.

[7] DNV-GL, "Global Sulphur Cap 2020," DNV-GL Maritime, Hamburg, 2016.

[8] ICS, "Guidance to Shipping Companies and Crews on Preparing for Compliance with the 2020 'Global
Sulphur Cap' for Ships' Fuel Oil in Accordance with MARPOL Annex VI," Marisec Publications, London, 2019.

[9] Z. Zhang and D. Liu, "A cost-benefit analysis of fuel-switching vs. use MGO: A CHINA-Europe container route case," in *IOP Conf. Series: Earth and Environmental Science*, Shanghai, 2021.

[10] Ship & Bunker, "Ship & Bunker for HFO and MGO," Ship & Bunker, 2022. [Online]. Available: https://shipandbunker.com/.

[11] MEPC, "2015 GUIDELINES FOR EXHAUST GAS CLEANING SYSTEMS," IMO, London, 2015.

[12] T. A. Tran, "Research of the Scrubber Systems to Clean Marine Diesel Engine Exhaust Gases on Ships," *Journal of Marine Science: Research & Development,* p. 7: 243, 2017.

[13] P.-C. Wu and C.-Y. Lin, "Cost-Benefit Evaluation on Promising Strategies in Compliance with Low Sulfur Policy of IMO," *Journal of Marine Science and Engineering,* vol. 9, no. 3, 2020.

[14] DNVGL, "The Fuel Trilemma: Next Generation of Marine Fuels," DNVGL, Høvik, 2015.

[15] ING, "How the 2020 Sulphur Cap Reshapes Global Shipping," ING, Amsterdam, 2019.

[16] S. A. Mohseni, E. v. Hassel, C. Sys and T. Vanelslander, "Economic evaluation of alternative technologies to mitigate Sulphur emissions in maritime container transport from both the vessel owner and shipper perspective," *Journal of Shipping and Trade,* 2019.

[17] A. Banawan, M. M. E. Gohary and I. S. Sadek, "Environmental and economical benefits of changing from marine diesel oil to natural-gas fuel for short-voyage high-power passenger ships," in *Proceedings
of the Institution of Mechanical Engineers, Part M: Journal of Engineering for the Maritime Environment*, Egypt, 2010.

[18] MAN, "Costs and Benefits of LNG as Ship Fuel for Container Vessels," MAN Diesel & Turbo, Denmark, 2011.

[19] ESSOM Co., Ltd., "Faculty of Engineering and Applied Science: Chemical Engineering," 2022. [Online].
Available: https://chemeng.queensu.ca/courses/CHEE332/files/ethanol_heating-values.pdf. [Accessed 21 April 2022].

[20] S&P Platts, "LNG BUNKER WEEKLY: Japan conducts first ship-to-ship bunkering; GAC, Pilot sign HOA," S&P Platts, 2022. [Online]. Available: https://www.spglobal.com/platts/en/market-insights/

latest-news/shipping/102220-lng-bunker-weekly-japan-conducts-first-ship-to-ship-bunkering-gac-pilot-sign-hoa.

[21] Indonesia Shipping Gazette, "Meratus Line Buka Layanan Internasional PING Service," Indonesia Shipping Gazette, 5 September 2021. [Online]. Available: https://indoshippinggazette.com/2021/meratus-line-buka-layanan-internasional-ping-service/. [Accessed February 2022].

[22] Marine Traffic, "Meratus Samarinda," Marine Traffic, 2022. [Online]. Available: https://www.marinetraffic.com/en/ais/details/ships/shipid:4984904/mmsi:525125029/imo:9808481/vessel:MERATUS_SAMARINDA. [Accessed February 2022].

[23] Marine MAN Ship Management, "Meratus Samarinda," Marine MAN Ship Management, February 2022. [Online]. Available: https://ships.jobmarineman.com/meratus-samarinda-9808481/. [Accessed February 2022].

[24] SENES Consultants Limited & Air Improvement Resources, Inc., "Review of Methods Used in Calculating Marine Vessel Emission Inventories," SENES, Vancouver, 2004.

[25] H. Winnes, E. Fridell and J. Moldanova, "Effects of Marine Exhaust Gas Scrubbers on Gas and Particle Emissions," *Journal of Marine Science and Engineering,* vol. 8, no. 299, 2020.

[26] J. Yang, T. Tang, Y. Jiang, G. Karavalakis, T. D. Durbin, J. W. Miller, D. R. Cocker III and K. C. Johnson, "Controlling emissions from an ocean-going container vessel with a wet scrubber system," *Fuel,* vol. 304, no. 121323, 2021.

[27] K. Johnson, W. Bill, E. Carlos, V. Castellucio, A. Asa-Awuku and W. Miller, "Black Carbon and Other Gaseous Emissions from an Ocean-Going Vessel Auxiliary Engine Equipped with a Scrubber," University of California , California, 2013.

[28] Drewry, "Ship Operating Costs Annual Review and Forecast 2011-2012," Drewry, London, 2011.

[29] Drewry, "Review and Forecast Ship Operating Costs 2012/2013," Drewry, London, 2012.

[30] YANMAR, "Marine Dual Fuel Engine," YANMAR, Hyogo, 2004.

Analysis of Systemic Safety Issues in Domestic RoRo Ferry Port Operation Using the Concept of Safety-II

A Nurwahyudy[1], T Pitana[1] and S Nugroho[1]
[1]Institut Teknologi Sepuluh Nopember, Faculty of Marine Technology, Surabaya, Indonesia

Aleik.207019@mhs.ne.its.ac.id, pitana@its.ac.id, snugroho@seatrans.its.ac.id

Abstract. Capsize of the Indonesian registered domestic RoRo ferry Yunicee in 2021 has brought public attention. The case was considered a repetition of the previous accident and presents drawbacks of ferry transport service. In the aspect of overall transport safety, a port is considered as part of safety defense at an early stage. Failure in risk mitigation could bring the ferry operation to a higher risk of an accident. Investigation into the case found a cluster of issues in some port functions was continuously neglected following several non-technical reasons since no sufficient approach has been made to present it. This paper adopts the concept of Safety-II to draw a detailed view of the coupling of port operation's functions and viewing its performance to support safe ferry operation. Categorization of functions developed from the organizational breakdown from the perspective of governance as described in the current port operation regulation as well as typical port operation. Relevant investigation reports are used to present potential functions which could be considered the risk agents. Hence, the function interconnectivity model would provide the complexity of port operation but presents a clear view of which function should be highlighted for improvement in supporting vessels' safe operation.

1. Introduction

On 29 June 2021, an Indonesian registered roll-on roll-off (RoRo) passenger vessel motor vessel (MV) Yunicee sank at Bali Strait [1]. The accident took place while the ship was waiting for berthing clearance. The ship was carrying 41 passengers and 25 units of various types of vehicles. The loading manifest indicated that the ship was in an overloaded state hence its maximum draft was overruled. Some other significant conditions indicated the presence of a significant amount of seawater flooding the cardeck. At about 5 minutes after the ship was in idle mode waiting for berthing instruction, the ship began to heel to the port side and progressed rapidly. A few minutes later, the ship continued listing and finally fully capsized at about 500 meters to the berthing wharf of Gilimanuk Ferry port. As a consequence, from a total of 75 people on board, 11 passengers were found deceased while the other 13 were declared missing. The ship itself sunk at about 70 meters deep with all vehicles carried [1].

This tragic accident brings back public memories of a similar occurrence that took place near the Yunicee capsize. Not just are the consequence of the accident relatively the same but also the process and factors that contributed to the accident are almost similar. Capsize of MV Rafelia in March 2016 has been investigated with significant findings and the findings have been issued to every involved party for their concern [2]. This was done with the expectation that many improvements should be made. The investigation by the National Transportation Safety Committee (NTSC) for the other similar cases involving domestic RoRo ferry operations identified that most of the ships were at an overloaded and overdraft state [14]. A significant amount of seawater flooded the car deck so reducing the ship's stability following the presence of the free surface moment [2][3][4][5]. This condition has raised the question of what improvements have been made so far to create a similar occurrence and result in such tragic loss for many sides.

Table 1. The natural frequency of casualties (accident and incident) and the fatality rate for domestic RoRo ferry operations every year from 2007 to 2020

Year	Productivity		Reported Occurrence			Natural Freq	Fatality Rate
	Trip	Passenger Carried	Accident (Acc)	Incident (Inc)	Fatality	$f = Acc/Trip$	Per year
2007	N/A	40,557,832	1	1	50	N/A	1.23E-06
2008	427,134	46,926,166	1	0	1	2.341E-06	2.13E-08
2009	433,226	61,011,280	1	0	1	2.308E-06	1.64E-08
2010	360,879	39,683,788	0	0	0	0	0
2011	383,052	52,589,081	4	1	81	1.044E-05	1.54E-06
2012	482,041	58,673,855	1	0	7	2.075E-06	1.19E-07
2013	479,104	62,036,587	0	0	0	0	0
2014	519,911	64,638,965	3	0	6	5.770E-06	9.28E-08
2015	N/A	65,220,965	1	0	0	N/A	-
2016	N/A	65,105,614	2	0	7	N/A	1.08E-07
2017	N/A	63,321,392	3	5	6	N/A	9.48E-08
2018	558,896	63,231,271	5	5	43	8.95E-06	6.80E-07
2019	77,073	64,571,653	2	7	5	2.59E-05	7.74E-08
2020	533,253	26,871,710	1	5	0	1.88E-06	0

The table above presents the yearly natural frequency of fail and safe operation which is indicated by comparing two levels of occurrences, accidents, and incidents that occurred from 2007 – to 2020. The data above is compiled from a database issued by the Directorate General of Land Transportation, Ministry of Transportation. Some data were not available due to changing policies in the database system [6]. Accident and incident data were acquired from the NTSC database [7]. It is critical that to understand why and how an accident took place, every aspect of the ferry operation should be reviewed and analyzed. From the shore-based operation side to the shipborne side with many perspectives and methodologies available [8].

The paper aimed to identify the reoccurrence of ferry casualty by focusing on the port function in supporting vessel safe operation. A ferry port is relatively considered a distant factor to be directly involved in an accident or incident. But it has a significant role in supporting and providing the ship operation as well as maintaining safety with all necessary needs [8]. A ship relied on the port support in many aspects such as providing supplies, access for the cargo carried, remedies for any emergency on board, and many other aspects [9]. To achieve the objective, the paper utilized the concept of Safety-II to understand the current state of the domestic ferry operation scheme. Safety-II concept states that accidents did not take place on daily basis as the development of the system was not meant to create accidents but to achieve the goal as designed [10]. To develop the most accurate picture of how a ferry port works, not only the general concept of the ferry port operation system is being studied (work as imagined) but also how it works based on normal daily basis operation (work as done) [10]. Relevant references including current regulations, local ferry port development and operation procedures, and any findings from the relevant investigation of domestic RoRo ferry ports were accessed. To understand the problem, the paper reviews relevant investigation reports and extracts related information about the port operation. The results from the report review compared with functions coupling in the domestic RoRo ferry port operation. The outcome of this comparison is the identification of key function/s which inherit significant safety issues. Discussion to relate the outcome with safety improvement is presented and a recommendation is proposed.

2. Domestic RoRo ferry port operation functions

As a country with more than 17,000 islands, Indonesia deems maritime transport necessary to support every aspect of the Nation's development [11]. For Indonesia, domestic RoRo ferries play a significant role in serving the nations' integrity [11][12]. The current domestic RoRo ferry system has been developed to connect its major islands and works as a transport hub for other transport modes [12]. In the general perspective, domestic RoRo ferry fleets connect islands and enhance possibilities for regional socio-technical development. In a more specific view, the transport system supports logistic distribution, and access to equalize economic development by providing low-cost transport across the nation [12].

Based on the statistical data issued by the Directorate General of Land Transportation, Ministry of transportation in 2021, there were 218 ferry ports established all over the country and serviced by 472 units of ferry [6]. Its successful operation is indicated by the number of passengers and vehicles transported as well as the total ports of call. Before the Covid-19 pandemic situation, five years of data show an increasing trend in total passenger and vehicle transport. On average about 64 million passengers and 17 million vehicles were transported. In 2020, the total trip produced by the domestic ferry fleets reached 176.635 calls. However, the pandemic reduced the number up to 40% for the passengers and 60% for the vehicles transported [6].

Under current regulations, a (ferry) port is defined as any place consisting of land and/or water with certain limits as a space for government and commercial users for berthing, embarking and disembarking passengers, loading and unloading cargo for ship, equipped by terminal facility and anchorage which is supported by safety and security enforcement as well other supporting activities and intermodal exchange area [11][12][13]. The development of the ferry ports in Indonesia is mainly categorized under two operational schemes i.e. commercial port and pioneer port [11][14]. Commercial ports are operated by the state-owned ferry company PT ASDP Indonesia Ferry, whereas the pioneer ports are operated by the local governments [11]. In addition, the development of ferry ports mainly focuses on the following: the hub of transport networks, economic activity main access, transport interchange activity, industrial or commerce support, distribution of public and commercial goods, and establishing national integrity and insight [11][18][19]. The port was also established as an extension of the government and commercial sector's roles [20][18]. Therefore, a domestic RoRo ferry port was constructed to connect land or railway transport network with ferry service so it connects seamlessly two distant modes [11]. A ferry port from the perspective of government involvement should accommodate the following aspect i.e. Regulate, guidance, control, oversight, safety, and security [24]. For implementing this role, Government assigns a harbormaster with the main functions: providing breakwater, maneuvering zone, waterway, navigational aids, protection to the environment, ensuring ease of access to the service users and cargoes, ensuring port safety and security [19][23][24].

The standard for ferry port service requires that a domestic RoRo ferry port needs to have three different zones to segregate the port from other unwanted visitors. The zones comprise Zone A for people, Zone B for the vehicles, and Zone C for vital installations such as fuel storage/bunkers, electrical installations, water supply, and a moveable bridge control room. Each zone is divided into several other sub-zones to accommodate the various port services to the users. For commercial ports, the functional service of the ferry port covers vessel support service (bunker, freshwater supply, mooring/unmooring, assist tug, access for passenger and vehicle), passenger service, vehicle service, and other services than previously mentioned [14]. The port shall have the following main facilities to support its function, the dedicated and secure passenger terminal, passenger access (gangway) from the passenger terminal to the ferry, a measurement gate and weighing facility, official space for ferry traffic management, safety, and security control to overall port activity, bunker facility, electric, freshwater, and communication access, firefighting installation, and parking spaces for awaiting

IOP Conf. Series: Earth and Environmental Science 1081 (2022) 012052 doi:10.1088/1755-1315/1081/1/012052

vehicles before boarding the ferry. The supporting facility may include commercial space, waste and sludge reception, space for port expansion/development, and other relevant activities.

The following chart is extracted from relevant regulations to illustrate general domestic ferry user flow, from the entrance point up to loading to the ferry [22][25].

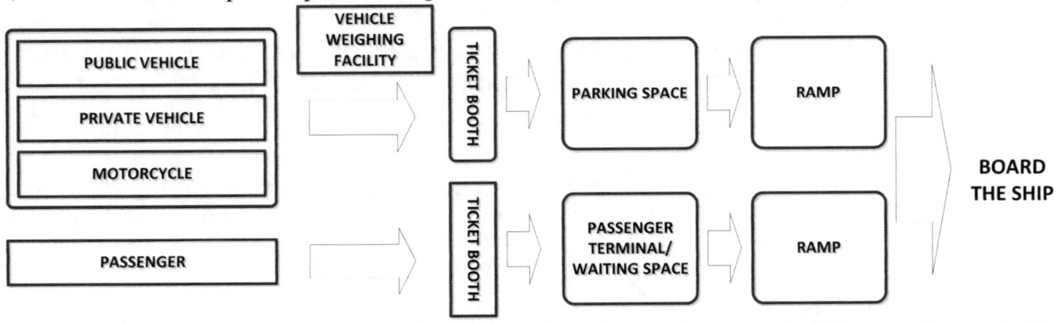

Figure 1. Flow pattern for passenger and vehicle to board ferry based on the Director-General of Land Transportation Regulation no SK.242/HK.104/DRJD/2010 [22]

3. The perspective of Safety-II in casualty analysis

3.1. Systemic safety issues

The safety issue is a factor that can reasonably be regarded as having the potential to adversely affect the safety of future operations. Systemic safety issues are indicated by any organizational problem rather than individual or technical issues. A safety factor itself is defined as an event or condition that increases the risk of safety [30]. It is something that if happens in the future would increase the likelihood of an occurrence as well as the severity of the adverse consequences [14][29]. A system is developed to achieve its desired output or objective that can be accepted by an agreed standard. Therefore, the safety issues within the system refer to the problem within an organization or a variety of internal and external organizational influences that impact the effectiveness of its risk controls, organization culture as well as higher regulatory supports [27].

3.2. Concept of Safety-II

Safety can be viewed from a different perspective. Hollnagel views safety as the absence of any harmful situation such as an Incident or accident [10]. Therefore, the focus of analysis under this perspective identifies the possible factors that contributed to the undesired event [28]. Hence the risk is identified as a result. This may be viewed in several aspects such as technology, human, and the organization. The involvement of humans in those areas is reviewed heavily as humans are fallible to be involved in any level of system operation. This concept is therefore considered Safety-I [10]. This perspective is widely used in many safety improvement programs.

The safety-I assumes that any mishaps can be described by identifying its supporting event in the past. Thus, Safety-I is unable to oversee that a system could expand and be adjusted to achieve the desired outcome. A system continues to develop as none in normal daily operation faces complex demand. The development is based on the ability of all systems to react and adapt to the changes. Each system's function variably interacts to compensate for the changes while maintaining its performance to achieve an agreed outcome. In consensus, a system itself was not designed to fail, but rather worked dynamically and adaptively to compensate for uncertain situations. Therefore, a system develops and is provided with many safety barriers such as working procedures, physical signs, safety tools, and many others. However, the fact shows that nothing is robust and readily available when a change of situation in complex form has to be dealt with. The system adapts to change dynamically as some steps might not be prescribed in any working procedure or prescribed working manuals [10].

The safety-II concept presents a view that functions in a system interact dynamically to achieve the objective by having adjustment and development [10]. The Safety-II sees that a daily basis operation hugely varied to respond to different situations. Under the concept of Safety-II, the safety management system is urged to see many things as going right rather than some factors that might be performing

wrongly. Humans will be viewed heavily. They are the only factor that could perform such adaptation rather than technical factors or environment. To accommodate growing needs and development of system complexity, it is necessary to see the system not just how it is performing under standard and creating undesired outcomes (Safety-I) but also view how the system adopts the changes by varying its performance so the objective of the system is achieved (Safety-II). Safety-II acknowledges that the system working right because every function in the system manages to adjust their work rather than following fully the procedure (work as imagined). Safety-II concept views failure as not a unique, individual event, but as a condition that varies in performance to compensate for the complexity and dynamic situation [10].
From those perspectives, Hollnagel proposed that viewing safety by looking at something that went wrong would not suffice and rather to see system must go right. To understand this concept, a novel model functional resonance analysis model is developed [10][33].

3.3. Model of functional resonance under Safety-II concept

To easily adapt the concept of Safety-II, a functional resonance analytical model (FRAM) is introduced. The model mainly aimed to overview functions variability based on achieving the system-designed performance. The basic conception under the FRAM model is the principle of equivalence of success and failures, the principle of approximate adjustments, the principle of emergence, and the principle of functional resonance [27]. The principle of the equivalent of success and failure describes that both outcomes have a similar process. Success is considered a favorable condition while failure is considered an unacceptable outcome [27].

In the case of emergence, the observed outcomes are of course also observable or real, but the same is not necessarily true for what brought them about. The outcomes may, for instance, be due to transient phenomena or conditions that only existed at a particular point in time and space. These conditions may, in turn, have emerged from other transient phenomena. The causes are thus reconstructed (or inferred) rather than found. They may therefore be impossible to eliminate or contain in the usual manner, but it may still be possible to control the conditions that brought them into existence, provided we understand how to work normally is done [27].

There are four steps to using the FRAM [27]. Step one delineates the functional entities that are important for the given scenarios or tasks. For this understanding, the system functions rather than system structure are of utmost importance as a basis of analysis. Every function is given characteristics based on the 6 areas namely Input, Output, Time, Control, Precondition, and resource. There are two aspects under the FRAM model to distinguish a function [27][29]. The foreground function defines as a function that becomes the main focus of the analysis whereas the background function defines as the support acts for the function to be fully operational based on the prescribed rules or regulations. Step 2, identify the variability of the function performance. It can be performed in many ways based on the nature of the function which is categorized under 3 distinct types humans, technology, and organization. Three types of variability characterized the function i.e. exogenous, endogenous, and upstream-downstream coupling. Step 3, analysis conducted to identify instantiation within the system. The instantiation is developed by system performance variability. This is critical to observe how far the function performance may vary and affect total system performance either to amplify, dampen, or even create unwanted outcomes. Step four under FRAM is monitoring, managing, and or terminating the performance variability. The variability may come in form of flexibility to adapt, or too stiff to react which later results in a brittle situation. Therefore, critical thinking should be used either to allow such flexibility or to modify the variability so then the goal of the system is acquired [31].

4. Analysis methodology

4.1. Analyze workflow

To identify the safety issues within the domestic RoRo ferry port operation, the paper utilizes the concept of functional resonance to present and describe the correlation among functions. A ferry port as mentioned above needs to successfully provide all necessary support for a ferry to operate safely.

- **Step one**, develop domestic RoRo ferry port work functions based on the relevant reference. Initially, the model was developed following a review of relevant references (work as imagined). To have more accurate and illustrate the daily operation of the port, a brief overview and interview with relevant parties were conducted so more functions were identified. Therefore, the models later improved, and the so-called "work as done" model. The result of this stage is a comprehensive and present real daily situation in ferry port operation.
- **Step two**, Review each function within the system by the variability of its performance. Each function is identified following the criteria as prescribed under the functional resonance model. Each function also is categorized based on the nature of its condition, type of operation as well as the purpose of the function. There are three types of function i.e. human, technology, and organization.
- **Step three**, collect relevant and potential findings on safety issues from reliable resources. This is mainly from the published relevant casualty investigation report. At this point, the paper overview the relevant investigation report to list the safety issues which is correlated to the operation of the ferry port. Each nature of the accident was examined to detect issues in the port operation. Each safety issue is categorized into four areas namely infrastructure, supervision/oversight, rules/regulation, and vessel/vehicle. The category of safety issues is based on the standard issued by the NTSC in presenting the findings.
- **Step four**, determine the function for each safety issue and rank it based on its proportion to the final event of nature of the accident. In this step, each safety issue is identified with its function. The selection of function refers to the information provided in the report and matches it to the nature of each function operation.
- **Step five**, compare the findings proportion with the states of function's performance variability. This final step is to conduct a result comparison between safety issues identified in the system of domestic RoRo ferry port with potential variability of its functions.

4.2. Developing a functional model for domestic RoRo and ferry port operation

A critical process in identifying safety issues is constructing the business process of the system. This is mainly to understand how the system was designed and focus on its objective. From relevant references and interviews with relevant the tables below present identified functions in the RoRo and ferry port. There are 23 functions in the port to accomplish its objectives as described in the previous sections. Of those functions 8 functions are Human type, 7 functions are organization type and 8 functions are technology type.

The systemic view highlight that there is no process working independently. Each function has background and foreground activity that is strongly connected with other functions. Each port function above identifies its entities based on 6 aspects namely Input, Time, Control, Precondition, Resource, and Output [31][32] [34].

The function model presents the characteristic of each function view by its frequency and amplitude.

Table 2. Characteristic of domestic RoRo ferry port function based on the frequency and amplitude

Name of Function	Function Type	Possible Source of Variability
Domestic Ferry Transport Users (H1)	Human	High Frequency, large amplitude
Port Gate and Security Check (T1)	Technology	low frequency, large amplitude
Weighing and measuring facility (T2)	Technology	low frequency, large amplitude
Ticket Booth (H2)	Human	High Frequency, large amplitude
Parking Space (T3)	Technology	Low frequency, large amplitude
Passenger Awaiting Lobby (T4)	Technology	Low frequency, large amplitude
Ticketed passenger access to ship (T5)	Technology	Low frequency, large amplitude

Vehicle Controller to the Moveable Bridge (H3)	Human	High Frequency, large amplitude
Ticket checkpoint (H4)	Human	High Frequency, large amplitude
Tally (H5)	Human	High Frequency, large amplitude
A moveable bridge (T6)	Technology	Low frequency, large amplitude
Vehicle stowing and securing (H6)	Human	High Frequency, large amplitude
Local Port Service and Navigation Support (O1)	Organization	Low frequency, large amplitude
Port Clearance (H7)	Human	High Frequency, large amplitude
Shipowner/ Shore Based Team (O2)	Organization	Low frequency, large amplitude
Shore berthing team (H8)	Human	High Frequency, large amplitude
weather information service (O3)	Organization	Low frequency, large amplitude
Port Security and Safety Control (O4)	Organization	Low frequency, large amplitude
Ship's Departure Preparation (O5)	Organization	Low frequency, large amplitude
Ship's consumable supply facility (T7)	Technology	Low frequency, large amplitude
Port-channel/Waterway (T8)	Technology	Low frequency, large amplitude
Harbourmaster and Port Authority office (O6)	Organization	Low frequency, large amplitude
Port Operator and Management (O7)	Organization	Low frequency, large amplitude

In summary, the domestic ferry port function model has a strong tendency to have a large amplitude with various frequencies of variability in function to take place.

4.3. References used for compiling safety issues in the domestic RoRo ferry port operation

The paper review 35 marine investigation reports involving domestic RoRo ferry operation issued by The NTSC from 2007 to 2020. The reports provide findings based on the nature of the occurrence. The nature of the accident is considered the outcome of the safety issues. There are 6 (six) different nature of accidents: Fire/explosion, capsize/sunk, collision/contact, grounding, fatality, and other types of accidents [7].

Any information in the reports was examined to identify issues in port operation which were considered a notable contributing factor in an accident or incident. It is reasonable that port operations have not directly caused incidents. The investigative body might relate to port operation whenever its performance is considered taking part in escalating the total operation of the ferry to a higher risk of casualty [29]. In some cases, there may not be any contribution from the port. For example, fully shipboard occurrences such as a collision in open seas or grounding far from the port jurisdiction are considered as no port contribution.

5. Results of models' analysis

5.1. Functional resonance model in domestic RoRo ferry port operation

As this model was developed for port operation only, the input for the model comes from several functions. The extent of the port operation function is not just limited to the servicing passenger, vehicles, and cargo from the port entry point up to boarding the vessel, but also considering the support to all departure preparation of the vessel as a port designed to do so [20][21][22]. Therefore, additional functions are added such as traffic controller, weather information service, and ship operator.

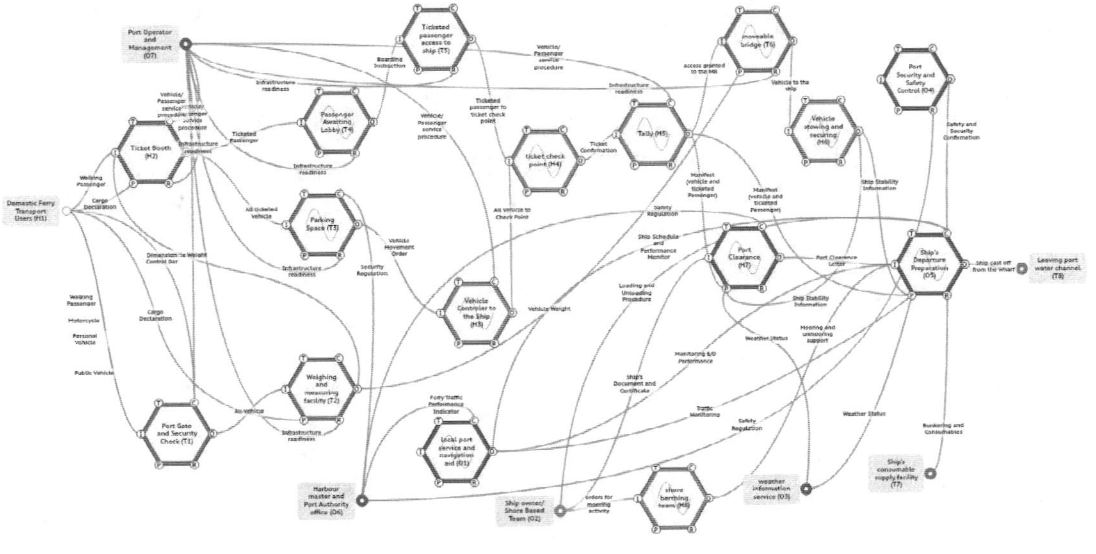

Figure 2. Domestic RoRo ferry port functions and their coupling

As the main input for the port operation, there are domestic ferry users, port operator, and management, harbormaster and port authority, shipowner/Shore-based team, weather information service, and ship consumable supply facility. The main output of the model is the ship able to safely leave the port. In total there are 45 upstream couplings and 42 downstream couplings from 23 functions within domestic ferry port operation.

The chart above presents a rather visible view of coupling in every function within domestic RoRo ferry port operation. Each function is coded with a certain combination identifier to present the type of function and its position within the model. Following the model, two functions have more upstream and downstream coupling i.e. Port Operator and Management (O7) and Ship's Departure Preparation (O5).

The port operator and management (O7) function has more foreground function as it plays a significant role in maintaining overall port operations such as Infrastructure readiness, Vehicle/Passenger service procedure. The output of service procedure from the port operator and management is also and control for other functions such as the Tally (H5) vehicle controller (H3) and port gate and security check (T1) function. In addition, the output of infrastructure readiness is a resource for other functions such as passenger awaiting lobby (T4), and access to the ship (T5) function

From the perspective of port operation, ship' departure preparation has more background functions than many of the other functions established to support the ship departure preparation. The ship departure requires Port Clearance Letter from the harbormaster, monitoring E/D Performance, and having Mooring and unmooring support for the input. As a precondition, the ship departure function receives support from the weather information service, clearance from the local port service, clearance from vehicle loading operation as well as and manifest of both passenger and vehicle carried from tally function. The function of port security and safety control provides a control for the prior departure as well as sailing procedures from the shipowner.

Port clearance (H7) function is also acquired more background function rather than its foreground. The only objective of the port clearance function is to issue a clearance letter for the vessel. However, the background function shows that many other functions are coupled to the port clearance process. As the input, the port clearance function requires a list document from the shipowner which normally comprises a request letter supported by a master sailing declaration letter, ship's safety certificate, and evidence payable service completion. As a resource, the port clearance (H7) process requires a manifest list from the Tally (H5), and weather information service (O3) as a precondition before the issuance of the clearance letter.

5.2. Identification of function's contribution to the casualty

Table 3. The proportion of safety issues in overall occurrence involving domestic ferry operation focusing on the function in the port function

	Infra-structure	Oversight/ Supervision	Rules/ Regulation	Vessel/ Vehicle
Domestic Ferry Transport Users		13	4	1
Harbour Master and Port Authority Office	1	24	3	
Local Port Service and Navigation Support	3	15	12	1
Parking space		1		
Port Channel/Waterway		5	1	
Port Gate	3	4		
Port Security and Safety Control		17	1	
Ship Operator/shore-based team		34	16	1
Ship's Departure Preparation		73	1	12
Tally		9	4	
ticket checkpoint		1		
vehicle controller to the ship			1	
Vehicle Stowing and Securing		17		
weather information service		6		
Weighing and Measuring Facility	1	7	2	
Port Clearance		6		

From the review to selected investigation reports, the paper identified 300 safety issues in the domestic RoRo ferry port operation. In general, the most notable issues fall under the category of oversight/supervision, where rules/regulation is the next notable category.

16 out of 23 functions in domestic ports were found to have contributed to the occurrence of accidents. The function of ship departure preparation was found to be associated more with safety issues. From the reports, there are 73 issues under the category of oversight/supervision, 12 issues under the category of vessel/vehicle, and 1 category related to rules/regulation. The next function with more safety issues is the ship operator/shore-based team coming with 34 issues under the category of oversight/supervision and 16 issues related to rules/regulation. The function of harbormaster and port authority inherit 24 issues under the category of oversight/supervision, 3 issues under the category of rules and regulation. The issues are more notable under capsize/sunk and Fire/explosion type which contributed by 13 functions (coming with 92 safety issues and 152 safety issues respectively), Grounding type contributed by 7 functions, Collision/Contact type contributed by 4 functions, structure failure contributed by 2 functions (coming with 3 safety issues), and fatality onboard contributed by 1 function (coming with 1 safety issues).

The fire/explosion occurrence type is identified as the most inherent notable safety issue. The issues mostly under category oversight/supervision with ship's departure preparation, ship operator/shore-based team, vehicle stowing and securing, and domestic RoRo ferry transport users as main functions contributed to such occurrence. It is interesting to point out that even if it is a relatively small contribution, the infrastructure of the port gate also plays a role to develop a situation that leads to the fire accident/ incident. Under sunk/capsize type of occurrence, the function of ship's departure preparation, ship operator/shore-based team, harbormaster, and port authority office play significant roles in such occurrence. Similar to the fire/explosion type, the issues came under the category of oversight/supervision followed by rules/regulations.

5.3. Correlation between function coupling with identified safety issues based on each nature of occurrence in domestic ferries

The functional resonance model result indicates the correlation among foreground, background activity, and cumulated safety issues identified in each function. The paper utilizes the Spearman rank correlation coefficient (R_s) to measure the value of a correlation between two pairs of data. Spearman's is widely used in many aspects such as social, geographic, and environmental variables. The basic assumption used under this statistic calculation is there is no correlation at all between foreground, background activity with safety issues in every function (H_0). Whereas, a strong correlation between functions and safety issues is considered as an alternative (H_1). The strength of a correlation denominates with a range of numbers where 0 is considered a very weak correlation and 1 is a strong correlation. Value of positive and negative for R_s translate there is correlation and no correlation respectively. Another indicator to accept or reject the assumption, P-value is utilized. P-value measures the probability that any watched relationship is due to chance. P-value range between zero and one where zero value presents a very strong suggestion to reject basic assumption (H_0) and 1 value presents weak or no suggestion to reject H_0.[35]

The coefficient (R_s) is calculated using the below equation:

$$R_s = 1 - \frac{6.\sum d^2}{n(n^2-1)}$$

To check the correlation between function's coupling and safety issue, Spearman's rank correlation was conducted to check the correlation for each Downstream Coupling (DC) and Upstream Coupling (UC) with identified cumulative safety issue (SI) for each function. The table below presents a compilation of foreground function, background function, and safety issues for each of the function types.

Table 4. Compilation number for function's downstream couple (DC), Upstream Couple (UC), and safety issues (SI) in the domestic RoRo ferry port

Function name	DC	UC	SI
Ship's Departure Preparation (O5)	1	9	86
Ship Operator/shore-based team (O2)	3	0	51
Local port service and navigation support (O1)	3	1	31
Harbour Master and Port Authority Office (O6)	3	0	28
Domestic Ferry Transport Users (H1)	3	0	18
Vehicle stowing and securing (H6)	3	1	17
Port Security and Safety Control (O4)	1	1	17
Tally (H5)	3	2	13
Weighing and Measuring facility (T2)	2	4	10
Port Gate and Security Check (T1)	2	2	8
Port Clearance (H7)	1	5	6
weather information service (O3)	2	0	6
Port channel/Waterway (T8)	0	1	6
Parking Space (T3)	1	2	1
vehicle controller to the ship (H3)	1	2	1
Ticket checkpoint (H4)	1	2	1

The result of the rank correlation calculation is shown below:

Table 5. Spearman's rank correlation coefficient for function in the domestic RoRo ferry port

Correlation	Rs	P	Conclusion
Downstream coupling and identified safety issues	0.6199	0.05	Rs is a moderate positive correlation with a 95% probability that H_1 is correct. To conclude there is a correlation between foreground function with safety issues
Upstream coupling and identified safety issues	-0.2537	0.5	Rs is a weak negative correlation with a 50% probability that H1 or H_0 is correct. To conclude, there is no correlation between background function with safety issues

6 Discussion

The results of the analysis present variation in safety issues in the domestic RoRo ferry ports based on each nature of the accident. While the number is quantifying the factors identified, it also indicated the tendency of issues to take place. It is obvious that not every occurrence is contributed by a similar number of factors but the presence of the factor is considered the matter. Taking example under occurrence type of capsize/sunk and fire/explosion, more functions were contributed compared to other occurrences such as grounding or fatality on board. This is mostly regarded by the nature of the accident itself. Two last mentioned are considered as shipboard operations rather than the capsize/sunk type. Domestic RoRo ferry port thus presents a significant contribution to such occurrence. In a narrower perspective, the function of shipborne operation as well as ship operator shore-based team play a significant role with the main aspect is under oversight/supervision category.

6.1. Frequency of fail and safe operation

Despite frequent notable accidents year by year, domestic ferry operations are considered still successful operations [9][37]. The frequency of failure and safe operation can be viewed as a natural frequency for an occurrence to take place [26]. Natural frequency is calculated by dividing the number of occurrences by its productivity numbers every year. **Table 1.** The natural frequency of casualties (accident and incident) and the fatality rate for domestic RoRo ferry operations every year from 2007 to 2020 shows that on average. The frequency of accidents every year is still considered very small. From the perspective of the casualty frequency, there was a sharp increment trend from 2007 to 2019, but later it declined significantly in the year 2020. The fatality rate per million passengers was found to be reduced from 2007 to 2020. The data also indicated that despite a growing number of occurrences but the outcome was considerably reduced.

Following natural frequency for occurrence, there should be a shift in focus for attention to improving safety by keeping all functions performed as stipulated in the safety standard rather than toward mitigating known and prominent safety issues. Hollnagel expresses that this model of investment in safety is rather reasonable to do than partially improving function with a higher tendential to mitigating notable safety issues [10].

6.2. Likelihood of function's variability in perspective of functional resonance

Following the concept of Safety-II, functions within the system performance are coupled to achieve the designed outcome [11]. Either the resonance of functions comes in a positive, damping situation or even negative resonance. Under positive resonance, every function within performs variably, not necessarily following the prescribed procedure but simply adapting the situation and modifying the performance sufficiently. The outcome is clear that the objective of the system is well achieved even exceeding the standard or target given [32]. Under damping resonance, not every function performs as designed and resulting in total operation reaching non-favorable conditions or events creating small incidents. In negative resonance, most of the functions perform beyond the safe standard. Each function did not manage to handle the situation as per design. The escalation of the situation so-called amplitude resonance could not be handled by other responder's functions so then the outcome is catastrophic multiple loss and severe property damage. The negative resonance is mostly stochastic and rather hard to predict as the indication of substandard performance is mostly unnoticeable and has small attention from the relevant authority in supervision [10][27][32].

In the domestic RoRo ferry operation, more notable safety issues observed under the category of fire/explosion occurrence with 13 functions had presumably resonated in negative conditions. This can be interpreted as most of the functions were unable to perform their prescribed duty following safety guidance or practices. In case of fire on the cardeck, control to the presence of dangerous goods onboard it is missing starting from the lorry stacking, entering the port zone, declaration of cargo, stowing, and securing the lorry onboard ferry. Following this information, every function was unable to identify the presence of dangerous goods and later put the ship at greater risk of fire [38][39][40].

In the case of ship capsize due to stability issues, the failure of function observed started from identification of vehicle weight, failure in stowing and securing the vehicle, inaccurate or absence of stability calculation during ship's departure preparation [2][3][4]. Some case presents failure of function while the ship was alongside the berth whereas most of the capsize cases took place while the ship was in under or full away condition [41][44][45].

Under the category of fire/explosion occurrence, 13 functions comprise 5 human types, 5 organization types, and 3 technology types. Both human and organization type functions are considered to have typical large amplitude. Related to the possibility of the resonance, human actions are more inherently high frequency while organization type is typically in low frequency. In the case of fire on board ferries, the variability of domestic transport users is high. This is considered reasonable as the nature of user transport is unpredictable. Passenger behavior and cargo carried on the vehicle are uncertain [36][46]. The absence of cargo check before entering the port and the vessel has played a significant negative resonance while the shipboard performance was sometimes overwhelmed with the fire outbreak condition thus resulting in total damage for the ship [39][40][42][43][46].

From those samples of negative resonance, a qualitative conclusion could be drawn. Human variability performance is considered high and to be influenced by many factors of internal and external. It should be understood that the effect of its variability is significant and might take place in a short period. On the other hand, organization-type functions tend to have low frequency and large amplitude. This can be interpreted by, any performance/behavior variability in an organization regardless is small could result in higher consequences. Technology type function can be regarded as a damping effect for any negative performance from a function. Technology types are considered low in variability and have a large amplitude of the result. This translates that technology type function nearly zero to vary unless intervention like maintenance or upgrade is made. The large consequence might occur when the variability of the technology type function took place [46]. The amplification of the consequence from technology consider significant when the operation is following the design limits. On the other hand, the consequence of operating the technology beyond design limits would also bring major consequences.

6.3. Comment on the statistical correlation between functions' coupling with the safety issue.

The statistical calculation to overview the correlation between function's coupling with safety issues presents intriguing findings. A strong correlation between downstream coupling and safety issues is considered reasonable. As the output of the function is considered to affect the other function's performance. Whereas background coupling has just a reference for the coupling itself. Take an example of the ship's departure preparation function which has the highest notable safety issue. The ship operation is mandated to perform safety levels at their best [28][13]. By doing this, port support plays a significant role. Before departure, any necessary safety check is performed. Any request for supplies of safety appliances and equipment is obtained during berthing. This translates as several functions coupled to the ship's activity such as bunkering, passenger and cargo operation as well local port service and navigational support. Any failure or lack of support for the ship would bring unfavorable consequences. This downstream and upstream coupling is also identified as affecting factors for variability

IOP Conf. Series: Earth and Environmental Science 1081 (2022) 012052 doi:10.1088/1755-1315/1081/1/012052

The utilization of a statistical model is considered critical to supporting the argument that functions' coupling is related to safety issues. However, this model analysis highly depends on the data available. Therefore, the accuracy of the information followed by its quantity will improve the result. Larger data will most likely represent the population in a much real situation. For this assumption, further research is recommended to examine not just one source of reference but also consider accessing other sources of information such as direct surveys, near-miss reporting, audit results, and social/public reports.

6.4. Limitation of the analysis

The domestic functional resonance model development is limited by the writer's perception of the working procedure which is supported by relevant references. In addition, the subjective perception is also made when reviewing the report to identify the issues within and correlated them with port operation as well as its functions. The subjectivity might lead to the disproportions of the results and the projection of different perspectives. To improve the results of the models, it is suggested that to obtain a more accurate picture of everyday operation in the domestic RoRo ferry port, more comments and suggestions are invited from relevant parties. This can be done by conducting direct current observation, focus group discussion, or in-depth interviews with other parties such as port companies from other places, ship operators, local regulators, and domestic RoRo ferry users. Following that activity, close monitoring of ferry performance is strongly suggested by creating mutual safety database so then any issues within an operation can be identified and analyzed.

7 Conclusion

Ferry ports play a significant role in supporting the safety of ferry operations by providing robust support for ship's supply, providing haven during an emergency, and as one of the safety barriers before departure. The port as required needs to set of functions to sufficiently operated and transfer cargo, vehicles as well as passengers safely to the ship. In addition, the port needs to be working seamlessly so support to the vessel is sustainable. Performance data indicated that the ratio between success and failed operation is still wide. Such performance achievement should be maintained and continue to improve.

However, notable safety issues in port operation remain and contributed to accidents mostly the capsize/sunk and fire accidents. In more detail, issues in oversight/supervision should have been considered more in the function of shipborne operation, Ship operator/Shore-based team, local port service and navigation support, harbormaster, and port authority. The factor of oversight/supervision mostly covers inconsistency of regulation implementation, absence of supervision, and lack of guidance in the regulation implementation are major issues in the concerned function. Under the Safety-II concept, functional resonance in a negative way had resulted from total safety performance into the undesired condition which escalate the situation to a higher risk of an accident. In contrast, positive functional resonance improves the safety of the operation by dynamically and intuitively changing function performance so the objective of the system is achieved. In the daily operation of the domestic RoRo ferry port, each function performs as required by the standard and agreed working procedure.

Further research is required by the expanding area of concern to have more view on safety issues and identify other issues in higher coupling functions. As the port is one of the functions in the overall ferry transport system, adding other functions and examining each of those in a much more detailed sub-function could later explore the situation and identify the notable issues which later could be mitigated and considered. The Safety-II model could be developed by adopting other quantitative risk assessment models which work similarly under the concept of Safety-I as suggested by Hollnagel.

8 References

[1] NTSC, 2021. Preliminary report and immediate recommendation investigation into capsize of the Indonesian registered MV Yunicee at Bali Strait on 29 June 2021, Jakarta: NTSC.

[2] NTSC, 2016. Investigation report into capsize of the Indonesian registered MV Rafelia at Bali Strait on 3 March 2016, Jakarta: NTSC.

[3] NTSC, 2012. Investigation report into capsize of Indonesia registered ropax ferry MV. Windu Karsa at Bone Bay on 27 August 2011, Jakarta, Indonesia: Author.
[4] NTSC, 2016. Investigation report into the grounding of the Indonesian registered MV Lestari Maju at Selayar Strait on 4 July 2018, Jakarta: NTSC.
[5] NTSC, 2021. Presentation on capsizing of the Indonesian registered MV Bili at Mentaya River on 20 February 2021, Jakarta: NTSC.
[6] DGLT, 2021. *Perhubungan darat dalam angka (Land transport in number),* Jakarta: Direktorat Jenderal Perhubungan Darat.
[7] KNKT, 2021. http:\\knkt.go.id. [Online].
[8] TC, I., Heinrich, J.-U. S. & Assheuer, S., 2019. *Domestic Ferry Safety in Indonesia*. London: International Maritime Organisation.
[9] IMO, 2014. Safety of ro-ro ferries. [Online]
[10] Hollnagel, E., Wears, R. L. & Braithwaite, J., 2015. From Safety-I to Safety-II: A white paper, *Resilient Health Care Net*: University of Southern Denmark, University of Florida, Macquarie University.
[11] DGLT, 2005. Master plan perhubungan darat (land transport master plan). Jakarta: Author.
[12] Haryo, B., 2013. Kebutuhan dan tantangan pengembangan angkutan sungai danau dan penyeberangan nasional (demand and challenge in the development of national inland waterway and ferry transport). Jakarta: GAPASDAP.
[13] Rahman, N., Nurwahyudy, A. & Weisbrod, R., 2019. *Improving Ferry safety for Lake Toba and the Republic of Indonesia As whole*, New York: Worldwide ferry safety association.
[14] MoT, 2004. Ministrial Decree no 52 year 2004 on Ferry Port Operation (Penyelenggaraan pelabuhan penyeberangan). Jakarta: Ministry of Transportation.
[15] Transit Cooperative Research Program, 2003. *TCRP: Report 100*, Washington DC: Transport Research Board, US.
[16] Isabelle Ryckbost, L. S. H. V., 2016. *ESPO Code of good practices for cruise and ferry ports*. Brussel: European Sea Port Organisation(ESPO).
[17] Ligteringen, H. & Velsink, H., 2017. *Port and Terminals*. Delft: Delft Academic Press / VSSD.
[18] Indonesia, P. R., 2008. Undang-undang Republik Indonesia Nomor 17 tahun 2008 tentang PELAYARAN (Shipping act). Jakarta: Kementerian Hukum dan Hak Asasi Manusia.
[19] Indonesia, P. R., 2009. Peraturan Pemerintah No 61 tahun 2009 tentang kepelabuhanan (Government Regulation on Sea Port). Jakarta: Menteri Hukum dan Hak Asasi Manusia.
[20] Indonesia, P. R., 2021. Perturan Pemerintah no 31 tahun 2021 tentang penyelenggaraan bidang pelayaran (Government regulation on Shipping Effectuation). Jakarta: Menteri Hukum dan Hak Asasi Manusia.
[21] MoT, 2016. PM 27 tahun 2016 tentang pengaturan dan pengendalian kendaraan yang menggunakan jasa angkutan penyeberangan (control and oversight for vehicle that using ferry transport). Jakarta: Ministry of Transportation.
[22] DGLT, 2021. PR DRJD no 4 tahun 2021 tentang pedoman pelaksanaan kegiatan local port service (LPS) di Pelabuhan Penyeberangan (Implementation Guidelines on local port service in a ferry port). Jakarta: Ministry of Transportation.
[23] OSHA, 2010. Roll-On roll-off (Ro-Ro) ship and dock safety. Washington DC: Author.
[24] MoT, 2015. PM Perhubungan nomor 39 tahun 2015 tentang Standar Pelayanan Pelabuhan Penyeberangan (Standard for Ferry Port Service). Jakarta: Kementerian Perhubungan.
[25] DGLT, 2010, Director General of Land Transportation regulation SK.242/HK.104/DRJD/2010 on technical guidelines on ferry port service, DGLT
[26] MoT, 2016. PM Perhubungan nomor 115 tahun 2016 tentang tata cara pengangkutan kendaraan di atas kapal (loading vehicle onboard vessel). Jakarta: Kementerian Perhubungan.
[27] Hollnagel, E. & Goteman, O. E., 2004. The Functional Resonance Accident Model. *Proceedings of Cognitive System Engineering in Process Plant*.
[28] Kristiansen, S., 2005. Maritime Transportation, Safety Management, and Risk Analysis. Burlington: Elsevier.

[29] ATSB, 2008. Analysis, causality, and proof in safety investigations, Canberra, Australia: Author.
[30] IMO, 2008. IMO Resolution MSC. 255 (84) adoption of the code of the international standards and recommended practices for a safety investigation into a marine casualty or marine incident (casualty investigation code), London: Author.
[31] Praetorius, G., Hollnagel, E. & Dahlman, J., 2015. Modeling Vessel Traffic Service to understand resilience in everyday operations, *Reliability Engineering & System Safety*, pp. Pages 10-21,
[32] Furniss, D., Curzon, P. & Blandford, A., 2016. Using FRAM beyond safety: a case study to explore how sociotechnical systems can flourish or stall. *Theoretical issues in ergonomic science*.
[33] Hollnagel, E. & Fujita, Y., 2013. The Fukushima disaster - systemic failures as the lack of resilience. *Nuclear engineering and technology*, 45(1), pp. 13-20.
[34] Lee, J. & Chung, H., 2018. A new methodology for accident analysis with human and system interaction based on FRAM: case studies in the maritime domain. *Safety Science*, Volume 109, pp. 57-66.
[35] Philip H. Ramsey, 1989. Critical Values for Spearman's Rank Order Correlation. *Journal of Educational Statistics*, Volume 14, pp. 245-253.
[36] NTSC, 2007. Investigation report into fire onboard Indonesia registered ropax ferry MV. Levina I at Tanjung Priok port on 22 February 2007, Jakarta, Indonesia: Author.
[37] Nurwahyudy, A., 2014. Contemporary issues in domestic RoRo ferry passenger operation in developing countries, Malmo: World Maritime University.
[38] NTSC, 2007. Investigation report into fire onboard Indonesia registered ropax ferry MV. Nusa Bhakti at Lombok Strait on 13 January 2007, Jakarta, Indonesia: Author.
[39] NTSC, 2008. Investigation report into fire onboard Indonesia registered ropax MV. Dharma Kencana I at Mentaya Hilir Selatan river on 18 May 2008, Jakarta, Indonesia: Author.
[40] NTSC, 2009. Investigation report into fire onboard Indonesia registered ropax ferry MV. Mandiri Nusantara at the Java Sea on 30 May 2009, Jakarta, Indonesia: Author.
[41] NTSC, 2012. Investigation report into the collision between Indonesia registered ropax ferry MV. Marina Nusantara with Indonesia registered coal carrier barge Bg. Pulau Tiga 330-22 at Barito River on 26 September 2011, Jakarta, Indonesia: Author.
[42] NTSC, 2012. Investigation report into fire onboard Indonesia registered ropax ferry MV. Laut Teduh-2 at Sunda strait on 28 January 2011, Jakarta, Indonesia: Author.
[43] NTSC, 2012. Investigation report into fire onboard Indonesia registered ropax MV. Musthika Kencana II at the Java Sea on 4 July 2012, Jakarta, Indonesia: Author.
[44] NTSC, 2013. Investigation report into the collision between Singapore registered gas carrier MV. Norgas Cathinka with Indonesia registered ropax ferry MV. Bahuga Jaya at Sunda Strait on 26 September 2012, Jakarta: Author.
[45] NTSC, 2016. Investigation report into capsize Indonesia registered ropax MV. Wihan Sejahtera at Madura Strait on 16 November 2015, Jakarta: Author.
[46] Nurwahyudy, A., 2019. *Fire onboard RoRo Ferry: Causation*. Bangkok, worldwide ferry safety association.
[47] NTSC, 2016. Investigation report into damage of ramp door on Indonesia registered ropax ferry MV. Nusa Putera at Madura Strait on 27 December 2018, Jakarta: Author.

Acknowledgments
The authors wish to acknowledge assistance from National Transportation Safety Committee for providing full access to the investigation files and documents, the Directorate of Inland Waterway and Ferry for the statistical data, and the Human Resource Training Body of the Ministry of Transportation for granting the funds for this research

IOP Conf. Series: Earth and Environmental Science 1081 (2022) 012053 doi:10.1088/1755-1315/1081/1/012053

Application of Hybrid Power Generation System on a Tanker Ship to Support the Development of Eco Ship Technology

ES Koenhardono[1,3] , DP Putri[2] and IR Kusuma[1]

[1]Department of Marine Engineering, Institut Teknologi Sepuluh Nopember
[2]PT (Persero) Pertamina

Abstract. The equator divides Indonesia into two parts, the areas located in North Latitude and South Latitude. This results in Indonesia having an abundant source of solar energy. Therefore, the use of hybrid power generation systems (solar cells and diesel generators) on ships can reduce generator fuel consumption, as well as reduce exhaust emissions from ships. This paper discusses the design of a hybrid power generation system on a tanker technically and economically. The total capacity of the solar panels installed on the ship can generate electrical energy of 885.2 kWh in a day. This electrical energy is used to supply electrical loads consisting of lighting, navigation equipment, radio communication, galley and laundry equipment, and air conditioner and refrigerator. The application of hybrid power plants on tankers can produce fuel consumption savings of 15.5% per year when compared to the use of conventional power generation systems. Meanwhile, the break event points due to the use of a hybrid generating system is less than 4 years or equivalent to the nominal total cost of Rp. 23,980,000,000.00 or $ 1,803,007.52.

1. Introduction

IMO is an international organization responsible for limiting GHG emissions from the international shipping industry. The IMO strategy in reducing GHG emissions is implemented in stages. IMO has targeted a reduction of carbon emissions from all international shipping by 40% by 2030, and a reduction of 70% by 2050, compared to conditions in 2008 [1,2,3]. Many strategies can be taken to reduce GHG emissions, such as hull cleaning to reduce drag [4,5]; speed optimization [6,7]; utilization of energy-saving lamps [8]; and use of solar power for accommodation services [9,10,11,12]. This paper discusses reducing GHG emissions by combining two methods, namely using energy-saving lamps and utilizing solar energy, so that this technology will produce environmentally friendly ships or Eco Ships. Another advantage besides reducing GHG emissions is the reduction of ship operating costs, so that the ship owner's profit is increasing

2. Methodology

In this research, technical and economic analysis was carried out to design a hybrid power generation system (solar cells and diesel generators) on ships. The technical analysis in this study begins with determining the ship used as the object of research, a tanker ship. The next step is to identify the area that can be used for laying solar panels on the ship. By identifying the area for laying solar panels, it can be seen the potential availability of solar energy on the tanker ship. Then an analysis of the distribution of the electrical load on the ship that can be supplied by solar panels is carried out by considering the availability of energy that can be produced by solar panels. The next steps in designing a hybrid power generation system (solar cells and diesel generators) on-board include changing the type of lamp used on-board from

[3]eddy-koen@its.ac.id

Fluorescent Lamp (FL) lamps to Light Emitting Diode (LED) lamps, calculating and selecting equipment specifications. solar panels, as well as modifying the electrical wiring diagram on the ship.
The economic analysis in this study uses the Break-Even Point (BEP) calculation method. The economic calculation includes the calculation of the investment and operational costs of the hybrid power generation system on board. The calculated operational costs are the cost of generator fuel, equipment replacement, and maintenance of the installation of the hybrid power plant system at the ship.

3. Result and Discussion

In this research, technical and economic analysis was carried out to design a hybrid power generation system (solar cells and diesel generators) on ships. The technical analysis in this study begins with determining the ship used as the object of research, a tanker ship. The next step is to identify the area that can be used for laying solar panels on the ship. By identifying the area for laying solar panels, it can be seen the potential availability of solar energy on the tanker ship. Then an analysis of the distribution of the electrical load on the ship that can be supplied by solar panels is carried out by considering the availability of energy that can be produced by solar panels. The next steps in designing a hybrid power generation system (solar cells and diesel generators) on board include changing the type of lamp used on board from Fluorescent Lamp (FL) lamps to Light Emitting Diode (LED) lamps, calculating and selecting equipment specifications. solar panels, as well as modifying the electrical wiring diagram on the

3.1. Analysis of the electrical load on the ship

Electrical loads on tankers are grouped into engine room machinery parts, deck machinery parts, air conditioners and refrigerator devices, galley and laundry equipment and electrical equipment. Electrical equipment is divided into navigation equipment, radio communication and lighting. Based on the one-line diagram on the tankers studied, the total electrical load in each group of electrical loads can be seen in Table 1.

Table 1. List of electrical loads on tankers.

Electric Load Group	Power (kW)
Engine room machinery dan deck part	2800,6
Galley and laundry equipment	116,8
Air conditioner and refrigerator device	179,1
Navigation equipment and radio communication	15,2
Lighting	28

Figure 1. 3D model of solar panel placement for power generation system hybrid on the tanker

IOP Conf. Series: Earth and Environmental Science 1081 (2022) 012053 doi:10.1088/1755-1315/1081/1/012053

3.2. Analysis of space available for the installation of solar panels on tankers

At this stage, the optimization of the area on the ship that can be a location for installing solar panels is carried out. From the optimization results, the areas that can be used as locations for installing solar panels are the wing deck with an area of 123.24 m2 and a cargo area with an area of 756.73 m2. The selection of solar panel installation locations considers the available area, the dimensions of the solar panels used, and the availability of maintenance space for solar panels. By using the SunPower E20-435-COM solar panel, the number of solar panels that can be placed on the wing deck is 57 panels and 350 panels in the cargo area. A 3-dimensional image of the placement of all solar panels can be seen in Figure 1.

With an average solar irradiation time in Indonesia of 12 hours, the maximum irradiation time is assumed to be 5 hours per day, so the total power that can be generated is 885.2 kW per day, as seen in Table 2.

Table 2. Maximum total energy produced per day.

	Number of panels	The maximum total power produced per panel (watt)	Maximum total energy produced per hour (watt-h)	Maximum total energy produced per day (kwh)
Wing deck	57	435	24795	123,95
Cargo area	350	435	152250	761,250
Total	407		177045	885,2

3.3. Analysis of lighting system load

Energy derived from solar panels is planned to meet the energy needs of the ship's accommodation space, which includes lighting, navigation equipment, radio communications, galley, and laundry equipment, as well as air conditioning and refrigerator device loads in seagoing operating conditions. However, due to the limited availability of environmentally friendly energy, it is necessary to replace the type of lighting on the ship that is more energy-efficient. At this time, the tanker uses FL-type lights and will be replaced with LED lights. The basis for consideration of changing the type of lamp is based on the value of the intensity of lighting in each room on the tanker which is adjusted to the BKI standard. The results of the calculation as a replacement for this type of lamp can be seen in Table 3.

Table 3. The results of the calculation of replacing TL lamps with LED lamps.

Junction	Location	Power with TL lamps	Power with LED lamps
JM	Navigation Light	0.82	0.82
L-1	Group Switch Panel & Nav Bri Dk Lts	6.65	0.252
L-3	Light Panel Board (C Deck Lts)	2.49	0.389
L-4	Light Panel Board (B Deck Lts)	1,695	0.427
L-5	Light Panel Board (A & Upper Deck Lts)	3,691	0.587
L-6	Light Panel Board (On Upper Deck Lts)	3,460	0.75
L-7	Light Panel Board (Engine Room Lts)	9,240	2,906
	Total	28,406	6,130

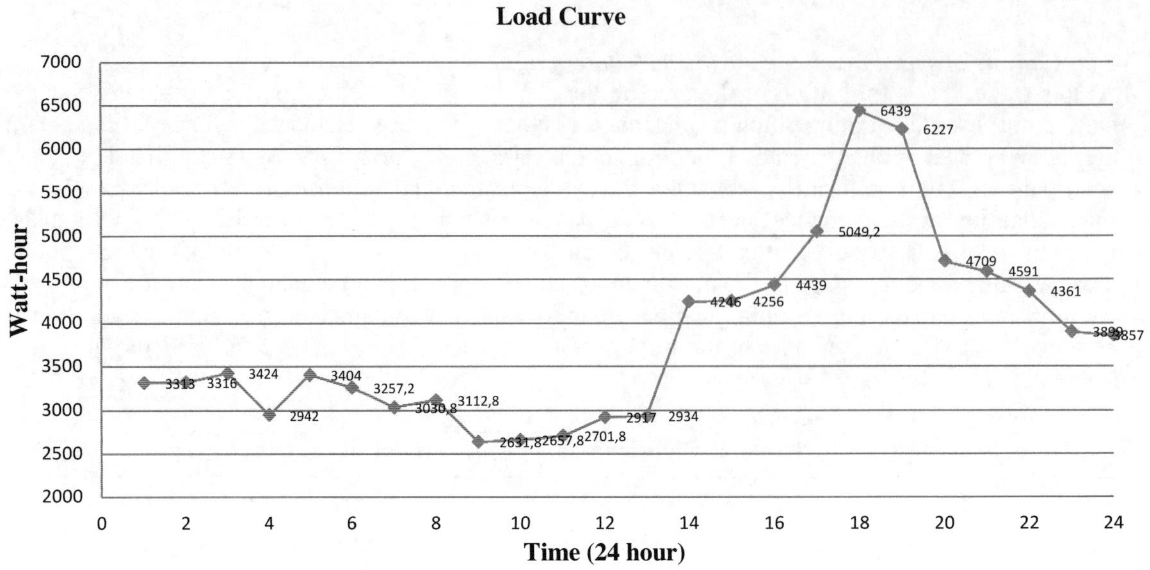

Figure 2. Lighting lamp peak load graph

After changing the type of TL lamp to LED lamp, the total lighting load on the ship was reduced to 6.1 kW. Based on the results of interviews, the average operating time of all lighting lamps on tankers is estimated to be 7.77 hours per hour day. Figure 2 shows fluctuations in electrical power requirements for the ship's lighting system in 24 hours, where the peak load that occurs at 18:00 is 6439 watts. And the total load of lighting lamps per day is 90.8 kW.

3.4. Distribution of the electrical load to be supplied by solar cells on tankers

Based on the capacity of solar energy that can be installed on tankers, electrical loads that can be supplied by solar panels include lighting, navigation equipment, radio communication, galley, and laundry equipment, as well as several components including an air conditioner and refrigerator device at seagoing operating conditions. as listed in Table 4.

Table 4. Electrical load to be supplied by solar cells on tankers

Load Type	Power required (kWh/day)
Lighting	90,8
Navcom Equipment, Galley Equipment, Air Conditioner System	566,7
Total	657,5

3.5. Solar Cell System Design

The total energy required to supply lighting, navigation equipment, radio communication, galley, and laundry equipment, as well as several air conditioners and refrigerator devices is 657.5 kWh/day. The solar energy that can be generated from the installed solar panels is 882,021 kWh/day, so there is a surplus of energy generated from the solar panels. Therefore, this excess energy can be stored in the battery.

The selection of solar panel system components is done by comparing several specifications of solar panels, batteries, solar chargers, and inverters, then selecting the most optimal, as listed in Table 5.

Table 5. Specifications for the installation of a hybrid power generation system

Type	Specification	Number of units
Solar Panel Sun Power E20-435-COM	435 w	407
Solar charge Controller Blue solar 150/100	12/24 VDC, 145 V/100 A	25
Battery 12V/500 Ah	500 Ah / 12 V	295
Inverter	10.000 W	3

3.6. Calculation of Investment and Operational Costs of Conventional Power Generation Systems
The calculation of investment costs for conventional power generation systems includes the cost of changing FL-type lighting. Operational costs in conventional systems include generator fuel consumption costs as well as maintenance and replacement costs. The total investment costs and operational costs for the first year of a conventional power generation system are listed in Table 6.

Table 6. Investment costs and first-year operational costs of conventional power generation systems

Investment Fee	Rp. 148.967.209,06
Operational Cost	Rp. 6.172.903.116,64
Total IDR	Rp. 6.321.870.325,70

The calculation in Table 4 is the total cash flow for the first year of a conventional power generation system. In this study, the economic calculation is carried out until the 25th year, as shown in Figure 3.

3.7. Calculation of Investment and Operational Costs of Hybrid Power Generation Systems
The cost of installing hybrid systems such as solar panels, batteries, inverters, and solar charge controllers includes investment costs for hybrid power generation systems. Meanwhile, the operational costs of the hybrid power generation system include the cost of generator fuel consumption as well as maintenance and replacement for the installation of the hybrid power generation system. The total investment costs and operational costs of the hybrid power generation system are shown in Table 7.

Table 7. Investment costs and first-year operational costs of hybrid power generation systems

Investment Fee	Rp. 3.956.868.470,80
Operational Cost	Rp. 5.218.806.436,51
Total IDR	Rp. 9.175.674.907,31

The cash flow for the hybrid power plant system for 25 years is shown in Figure 3.

3.8. Break Even Point (BEP) Analysis
The productive life of the ship is 25 years. Therefore, the break event point analysis for investment costs and operational costs between conventional systems and hybrid systems using the ship's productive life is 25 years. The graph of the intersection of investment and operational costs between a conventional system and a hybrid system over a period of 25 years is shown in Figure 3.
The point of intersection of investment and operational costs between the conventional system and the hybrid system is clearer in Figure 3 where the investment and operational costs between the conventional system and the hybrid system reach the same nominal amounting to Rp. 23,980,000,000.00 in the 3,69 th year.

3.9. Generator Fuel Consumption Savings Analysis

The comparison of generator fuel consumption when using a conventional system and after using a hybrid system is shown in Figure 4. Based on the analysis results from Figure 4, it can be concluded that using a hybrid system power plant (diesel generator and solar panels) on tankers can save generator fuel consumption. by 15.5% per year.

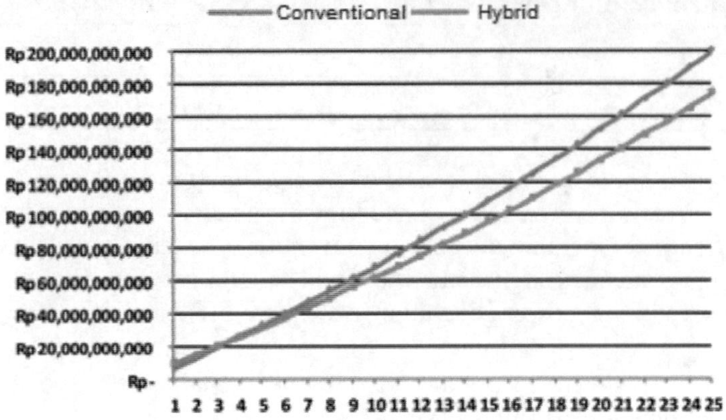

Figure 3. Investment and operational costs of conventional and hybrid power generation systems (25 years)

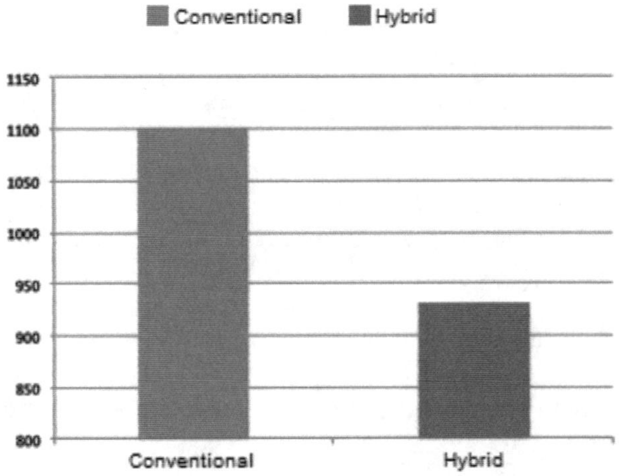

Figure 4. Comparison of generator fuel consumption between conventional and hybrid power generation systems

4. Conclusions

The results of the feasibility analysis, the use of a hybrid power generation system on tankers has good prospects in the future. This is based on the following results:

- Laying solar panels on tankers on the wing deck area of 123.24 m2 or 57 solar panels and in the cargo area of 756.73 m2 of 350 solar panels.
- The total amount of electrical energy produced is 882,021 kWh/day to supply electrical equipment of 566.7 kWh/day. Surplus electrical energy can be stored in the battery
- Fuel savings due to modification into a hybrid power generation system reaches 15.5% per year.
- The Break Even Point (BEP) on the use of a hybrid power generation system occurred in the 3.69th year with a nominal value of Rp. 23,980,000,000.00.

Based on the results of the technical and economic analysis, it can be concluded that the use of a hybrid power generation system is very prospective to be applied on ships to support Eco ship programs in the future.

5. References

[1] Aldo Chircop, "The IMO Initial Strategy for the Reduction of GHGs from International Shipping: A Commentary", The International Journal of Marine and Coastal Law vol. 34, no 3, pp 1-31, 2019.

[2] Tae-Hwan JoungSeong-Gil KangJong-Kap LeeJunkeon Ahn, 2020, "The IMO initial strategy for reducing Greenhouse Gas (GHG) emissions, and its follow-up actions towards 2050", Journal of International Maritime Safety Environmental Affairs and Shipping vol 4 no 1 pp 1-7, 2020.

[3] Patrizia Serra & Gianfranco Fancello, "Towards the IMO's GHG Goals: A Critical Overview of the Perspectives and Challenges of the Main Options for Decarbonizing International Shipping", Sustainability, 12, 3220, 2020

[4] Elidolu, G., Uyanık, T., Karabıyık, P. & Arslanoğlu, Y., "Hull cleaning effects on CO2 emissions estimation with machine learning methods", 5th International Mediterranean Science and Engineering Congress (IMSEC 2020), Turkey, 2020

[5] Oliveira, D.R., et al, "A novel tool for cost and emission reduction related to ship underwater hull maintenance", Journal of Cleaner Production, 356, 2022

[6] Norlund, E.K., & Gribkovskaia, I., "Reducing emissions through speed optimization in supply vessel operations", Transportation Research Part D Transport and Environment 23:105-113, 2013

[7] Psaraftis, H.N., "Speed Optimization vs Speed Reduction: the Choice between Speed Limits and a Bunker Levy", Sustainability 11, 2249, 2019.

[8] Su, C.L.et al, "Energy-efficient electronic light sources for marine vessels", Conference of Industry Applications Society Annual Meeting (IEEE IAS), 2012.

[9] Zhu, Y.Z. et al, "Influences of solar energy on the energy efficiency design index for new building ships", International Journal of Hydrogen Energy, 42,(30), p19389-19394, 2017.

[10] Aijjou, A., Bahatti, L., & Raihani, A., "Influence of Solar Energy on Ship Energy Efficiency: Feeder Container Vessel as Example", International Journal of Electrical Energy 7(1):21-25, 2019.

[11] Karatug, C. & Durmusoglu, Y., "Design of a solar photovoltaic system for a Ro-Ro ship and estimation of performance analysis: A case study", Solar Energy, vol 207, p 1259-1268, 2020.

[12] Kurniawan, A. 2016. "A Review of Solar- Powered Boat Development". IPTEK, The Journal for Technology and Science, 27(1).

Form Factor and Wake Fraction of Bulky Trimaran Hull Using CFD

ES Koenhardono[1], A Baidhowi[1] , and IS Arief[1]

[1]Departemen of Marine Engineering, Institut Teknologi Sepuluh Nopember

ahmadbai@gmail.com

Abstract. The value of form factor (1+k) and wake fraction significantly influence the ship propulsion design, including trimaran vessel. This paper presents the form factor and wake fraction of a single screw bulky trimaran using CFD. The model vessel is 1.92 m trimaran with scale to the actual vessel is 31.25 and evaluated in the range of Fn 0.1 to 0.2 based on ITTC recommendation. The results show that the form factor of the bulky trimaran is 1.433, which is close to the NPL series Trimaran value. The wake fraction was evaluated in two model speeds, 1.43 knot, which represent low speed and 6.26 knot which means high speed. The results show that the wake fraction pattern is similar to a single propeller vessel with the wake's peak at 12 o'clock with the maximum value of wake around is 0.1.

1. Introduction

Sea transportation that is fast, safe, comfortable, and environmentally friendly is necessary and cannot be avoided. One of them is the development of a trimaran ship that has three slender hulls. Compared to monohull ships, trimaran ships have lower resistance at high speeds, more stability, and a more expansive deck. Trimaran hulls have been applied for several vessels such as naval vessels, ferries and other types of applications [1][2][3]. The trimaran's slender hull shape resulted in the limited engine room available. Therefore, the bulky trimaran ships are expected to overcome this weakness. ITTC provide guidelines regarding resistance prediction [4]. One of the crucial factors in the resistance prediction is the form factor (1+k). This paper attempts to identify the form factor of a bulky trimaran hull shape using CFD.

The concept of resistance calculation was introduced in the last 18 centuries by William Froude. The resistance component consists of friction resistance (CF) and residual resistance (CR). This concept was refined in 1966 by adding wave resistance (CW) and resistance due to the form of the hull (CFO). The form factor is included in the hull form resistance.

Commonly, the position of the propeller is behind the hull. Therefore, the wake affects the flow shape that passes through the propeller blade. The magnitude of the wake is caused by the potential and viscous components of the flow over the hull. The interaction between hull and propeller around the propeller creates non-uniformity of inflow; thus, the wake simulation plays an essential role in hydrodynamic performance [5] [6].

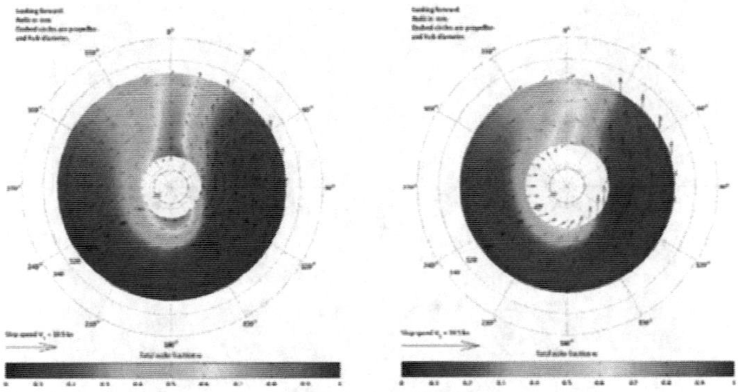

Figure 1. Typical Wake Fraction [7].

2. Research Methodology

2.1. Form Factor Simulation

This paper uses a trimaran vessel with principal particular as shown in Table 1.

Table 1. Specifications for the installation of a hybrid power generation system.

	Model	Real	Unit
LWL Main Hull	1.92	60	m
Main Hull Draft	0.11	3.44	m
Side Hull Draft	0.04	1.25	m
dist between CL	0.27	8.44	m
LCG	0.769	24.03	m
VCG	0.101	3.15	m
WSA	0.7451	23.2837	m^2
λ		31.25	

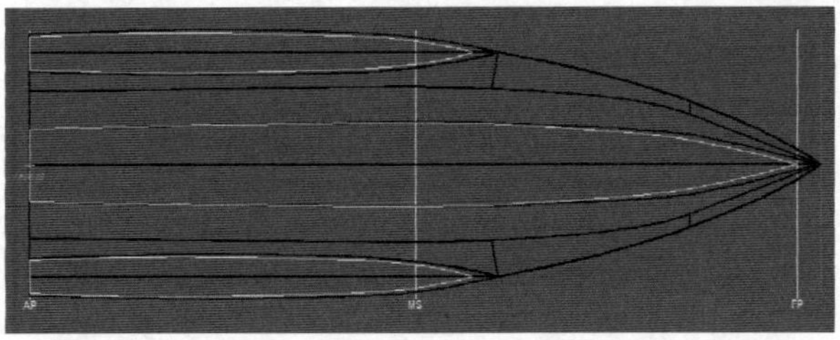

Figure 2. 2D drawing of the Bulky Trimaran

The model's characteristic is that the side hull is relatively tiny compared to the main hull, with the ratio of S/L being 0.1427. The other character of this model is the draft of the main hull is high compared to the side hull. The purpose of the design is to provide higher displacement to provide a high payload and expansive deck area.

Figure 3. 3D drawing of the Bulky Trimaran

Based on the ITTC recommendation, the resistance of a vessel can be written as:

$$C_{TM}(Re, Fr) = (1 + k)C_{FM}(Re) + C_w(Fr) \qquad (1)$$

Where:
C_{TM} : is the total coefficient resistance
$C_{TM} = \frac{R_{TM}}{\frac{1}{2}\rho_M S_M V_M^2}$
Re : Reynolds Number
$R_e = \frac{VL}{\sqrt{v}}$
Fr : Froude Number
$F_r = \frac{v}{\sqrt{gL}}$
C_{FM} : friction coefficient model
$C_{FM} = \frac{0.075}{(log_{10}Re-2)^2}$
λ : geometrical model scale
ρ_M : density water of model

For a low-speed region with $0.1<Fr<0.2$, the function will be Fr^4, the regression of the plot between C_{TM}/C_{FM} and Fr^4/C_{FM}. The form factor of the resistance can be obtained from the constant of the regression equation, as shown in Figure 4.

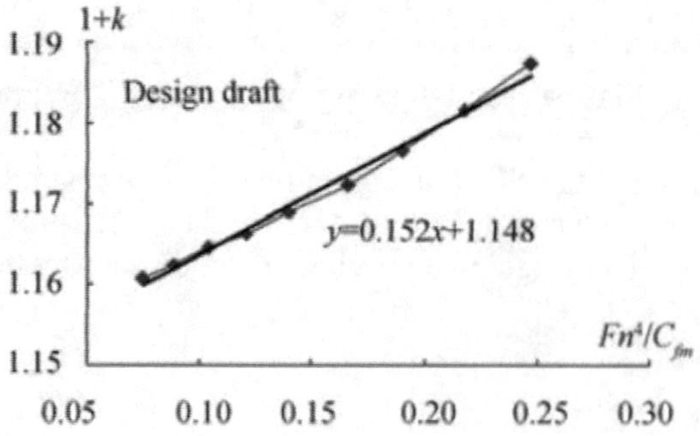

Figure 4. Example of Hughes-Prohaska plot [8].

Several tests were carried out to obtain the value of the form factor. The form factor value can be determined by using several tests' regression. A series of resistance tests in the towing tank was conducted using the similarity law for several speeds.

Table 2. Model simulation speed.

Model Speed (knot)	Actual speed (knot)	Fn
1	5.59	0.67
1.25	6.99	0.84
1.43	7.99	0.96
1.5	8.30	1.01
1.79	10.01	1.20
3	16.77	2.02
5	27.95	3.37
5.37	30.02	3.61
6.26	34.99	4.21
7	39.13	4.71

The process of the model test must comply with the similarity law. The model speed and ship speed use the same Froude number [9] [10]. The domains of the CFD model were configured as shown in Figure 5. The number of cells in this model is around 1,7million

2.2. Domain Setup

Numeca already employs C Wizards feature that provide convergence check for the meshing, Meshing basically is dicretization of the body surface into small quadrilaterals that led to the designation of panel methods [11]. Each element in CFD must be satisfies some parameters.The parameters are negative cell, concave and twisted cell. In a single cell as part of a complex system, mapping between the physical coordinate (x – y) and the natural coordinate (ξ – η) for a heavily volumetrically distorted elements leads to mapping of an area outside of the physical element into an interior area in the natural coordinates as shown by Fluid is volumetric, volumetric element distortion occurred in concave element. For concave elements, there are areas outside the elements which will be transformed into coordinate system, this concave element volume integration will results in negative value [12].

Table 3. Domain Setup

Upstream to the Hull	2Lpp
Downstream to the hull	5Lpp
Tank wall to the midsection	3Lpp

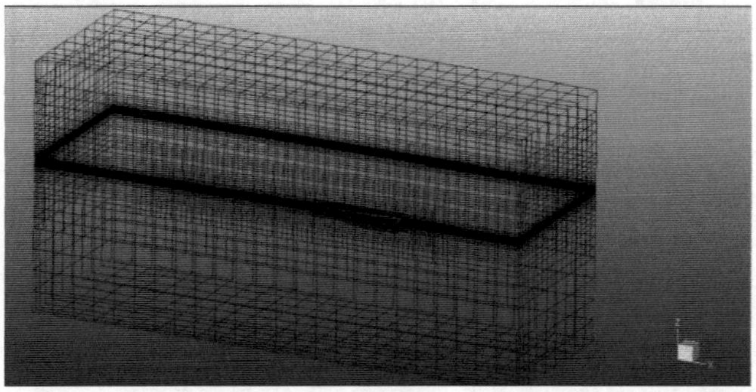

Figure 5. Domain setting

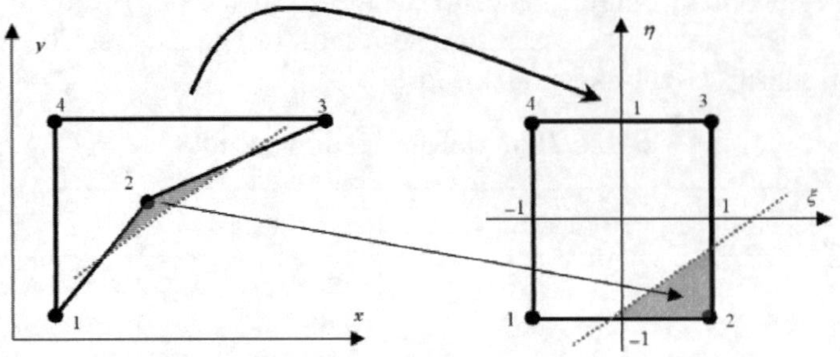

Figure 6. Cell Mapping

after several times of meshing iteration and refinements, the meshing quality has satisfied the zero concave, negative cell and twisted cells as shown by **Table 4.**

Table 4. Meshing Quality Check

Parameter	
Concave	0
Negative Cell	0
Twisted	0

2.3. Wake Fraction Simulation

The wake fraction (w) is the ratio of the difference of ship speed (v) and the propeller advance speed (v_A) to ship speed (v) [13] [14]. The formula can be written.

$$w = \frac{v - v_A}{v}$$

Where:

v : ship Speed (m/s)

v_A : propeller advance speed (m/s)

The model was simulated at two speeds to obtain the wake fraction value. The model speed for wake fraction evaluation was 1.43 knots and 6.26 knots.

3. Results and Discussion

3.1. Model Simulation Results

The test results can be shown in Table 5 and Table 6.

Table 5. Model simulation result for resistance.

Model Speed (m/s)	Model Speed (knot)	Fr	R_{TM} (N)	C_{TM}
0.514	1	0.1204	0.69	0.0070
0.643	1.25	0.1505	1.09	0.0071
0.736	1.43	0.1722	1.43	0.0071
0.772	1.5	0.1806	1.61	0.0073

Table 6. Model simulation result for C_{FM}, Fr^4/C_{FM}

Re	C_{FM}	Fr^4/C_{FM}	C_{TM}/C_{FM}
908053.4548	0.0048	0.0439	1.4640
1135066.8184	0.0046	0.1125	1.5535
1298516.4403	0.0044	0.1983	1.6024
1362080.1821	0.0044	0.2425	1.6563

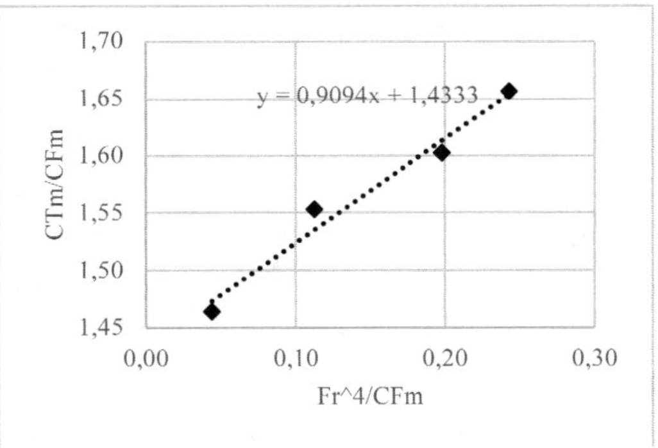

Figure 7. Linear regression based on data plot

Based on the model simulation, there are ten simulation results, but based on the ITTC, the recommended regression method is linear. Based on the data plot, a regression equation can be obtained

$$\frac{C_{TM}}{C_{FM}} = 0.9094\left(\frac{Fr^4}{C_{FM}}\right) + 1.4333$$

The form factor for the bulky trimaran is 1.433 for S/L 0.143 to analyze the form factor of the bulky trimaran with other trimarans.

3.2. Comparison to the similar trimaran

Several research concepts of trimaran have been studied. One of them is based on the S-NPL series with the variation of S/L [15]- [16], where the basic hull shape can be shown in **Figure 8**.

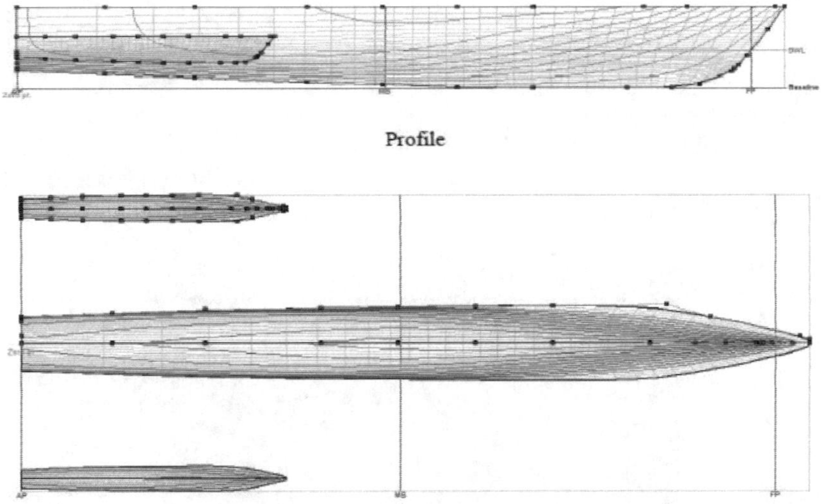

Figure 8. S-NPL Based Trimaran

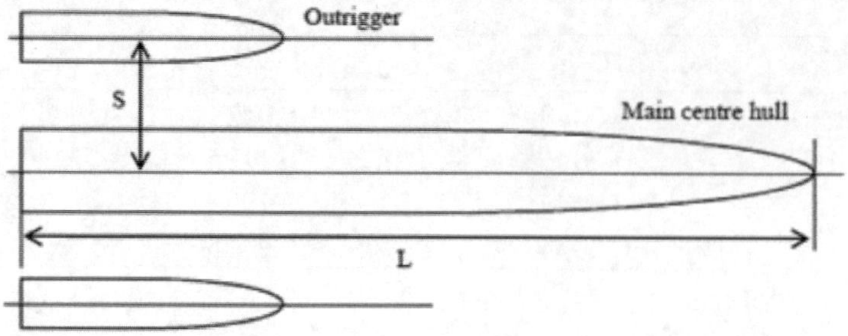

Figure 9. Prohaska Bulky Trimaran model plot

Table 7. Model simulation speed

Model	LWL (m)		Draft	B/T	S/L
	Main hull	Outrigger	(m)		
4a	1.6	0.53	0.103	1.5	0.2
4b	1.6	0.53	0.089	2	0.2
4c	1.6	0.53	0.08	2.5	0.2

The NPL trimaran hull form factor based on CFD simulations is presented in Table 8.

Table 8. Form Factor of NPL Trimaran.

Model	Form factor (1+k)
4a	1.21
4b	1.32
4c	1.44

Based on the comparison of the form factor, it can be concluded that the Bulky trimaran's form factor is close to NPL Trimaran with S/L 0.2 and B/T 2.5.

In order to reach time and resource optimization, the concept of actuator disk is used [15]. Actuator disk is a concept that injects the propeller performances into the actuator disk model to dynamically get the correct thrust and torque values for any inflow. This feature has an accuracy capable of competing with fully modelled propellers using sliding grids meshes but for the CPU time cost of traditional resistance and flow cases[17].

3.3. Wake fraction of bulky trimaran

The model simulation conducted for wake fraction used speed 1.43 knot and 6.26 knot, which represent low and high speed. Besides the wake fraction, the wave profile also can be observed where the results of the wave profile of 1.43 knot can be shown in Fig 10 and Fig 11 for 6.26 knot.

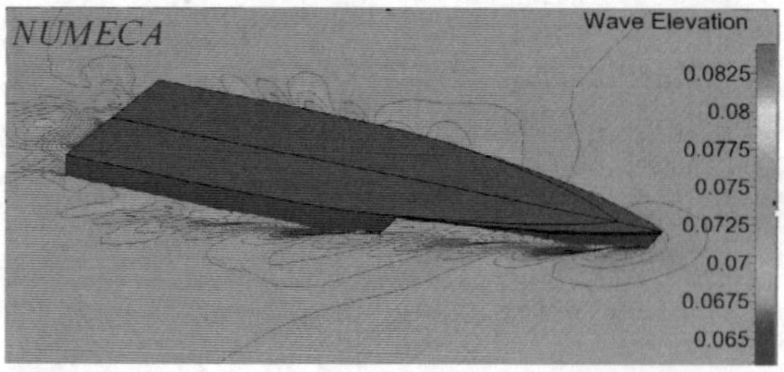

Figure 10. Model wave pofile speed 1.43 knot

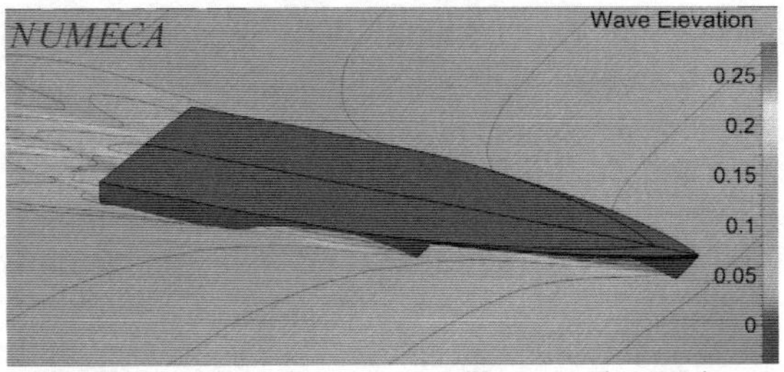

Figure 11. Model wave profile speed 6.26 knot

Based on the simulation, it can be concluded that the model's wave elevation is still relatively low while the wave elevation for 6.26 knot is relatively high.
The bulky trimaran is a single screw propeller located in the main hull shown by Figure 12. The diameter of the model propeller is 68.5mm.

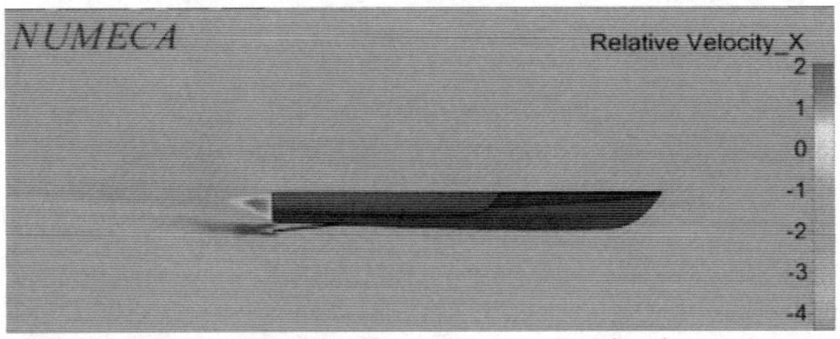

Figure 12. Position of the bulky trimaran's single screw propeller

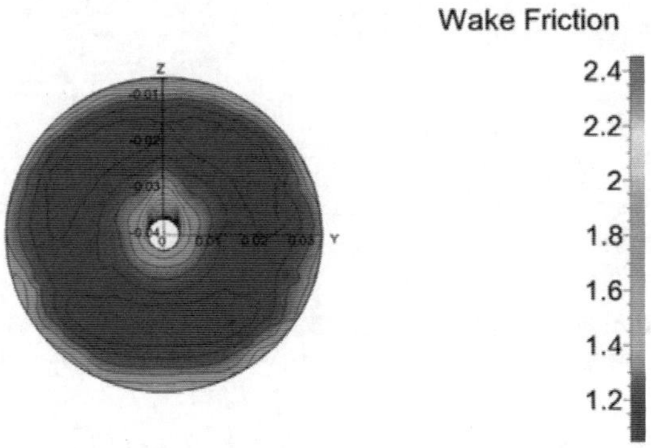

Figure 13. Wake fraction model at 1.43 knot

Based o Fig 13, the wake peak is at 12 oçlock position with value around 0.03. This phenomena indicates the non–uniform distribution of velocities at that position which similar to other single screw ship eventhough the model is trimaran.

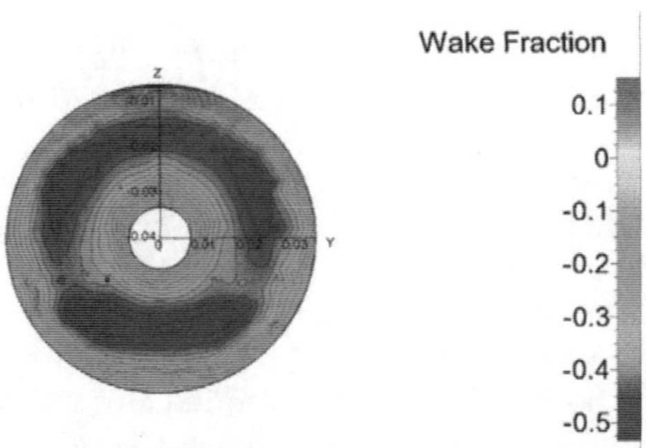

Figure 14. Wake fraction model at 6.26 knot

Figure 14 shows the wake peak is at 12 oçlock positions with a value around 0.1. This phenomenon indicates the non–uniform distribution of velocities at that position, similar to other single screw ships even though the model is trimaran. The other conclusion that can be drawn is that the wake is increased along with speed. The characteristics of the wake could be different to a single screw vessel if the vessel is using multiple propellers. The other configuration which can be evaluated is a triple screw with each propeller for each hull or twin-screw with a propeller installed in each side hull.

4. Conclusion

Based on the previous discussions, there are several conclusions can be drawn. The form factor value of bulky trimaran vessel is close to the form factor of NPL series trimaran. Form factor values of NPL series trimaran ranged from 1.21, 1.32 and 1.44, while the form factor of a bulky trimaran with S/L 0.143 is 1.433.

The wake fraction of a bulky trimaran is similar to a single propeller vessel. The peak of the wake is in the position of 12 o'clock, with the highest value 0.1. This characteristic might be changed due to the position and number of the propeller. The phenomena of wake fraction due to the change of the number of propellers and position of the propeller can be a very interesting research topic in the future.

Acknowledgements

This paper is the output of the Prototype Development Research. We want to thank the Institute for Research and Community Service, Sepuluh Nopember Institute of Technology, for funding this research. This simulation work was supported Marine Manufacturing and Design (MMD) Laboratory of Marine Engineering Department, Institut Teknologi Sepuluh Nopember, Surabaya.

References

[1] I. S. Silva *et al.*, "Development of a Trimaran ASV," *IFAC-Pap.*, vol. 51, no. 29, pp. 8–13, 2018.

[2] K.-J. Kang, C.-J. Lee, and D.-H. Kim, "Hull Form Development and Powering Performance Characteristics for a 2,500ton Class Trimaran," in *Practical Design of Ships and Other Floating Structures*, Elsevier, 2001, pp. 151–157.

[3] C. Bertorello, D. Bruzzone, P. Cassella, and I. Zotti, "Trimaran Model Test Results and Comparison with Different High-Speed Craft," in *Practical Design of Ships and Other Floating Structures*, Elsevier, 2001, pp. 143–149.

[4] ITTC, "ITTC – Recommended Procedures and Guidelines :Practical Guidelines for Ship Resistance CFD." ITTC, 2014. [Online]. Available: https://ittc.info/media/4198/75-03-02-04.pdf

[5] I. M. Kamal, K. Z. Iwan, N. Bose, J. Binns, S. Chai, and G. Davidson, "Powering for Medium Speed Wave-Piercing Catamarans comparing Waterjet and Screw Propeller Performance using Model Testing," in *Fourth International Symposium on Marine Propulsors. Austin, Texas, USA*, 2015.

[6] A. Amiadji, A. Santoso, and B. Sabastian, "Design of Catamaran Propulsion System with Demihull Distance Variation on Hospital Ship," *Int. J. Mar. Eng. Innov. Res.*, vol. 1, no. 3, 2017.

[7] K. Kim, F. Tillig, N. Bathfield, and H. Liljenberg, "Hydrodynamic optimization of twin-skeg LNG ships by CFD and model testing," *Int. J. Nav. Archit. Ocean Eng.*, vol. 6, no. 2, pp. 392–405, 2014.

[8] J. Wang, H. Yu, Y. Zhang, and X. Xiong, "CFD-based method of determining form factor k for different ship types and different drafts," *J. Mar. Sci. Appl.*, vol. 15, no. 3, pp. 236–241, 2016.

[9] J. S. Carlton, "Ship Resistance and Propulsion," in *Marine Propellers and Propulsion*, Elsevier, 2012, pp. 299–332.

[10] "volume-iii-resistance-committee.pdf.".

[11] J. L. Hess, "Panel Methods in Computational Fluid Dynamics," p. 20.

[12] G. R. Liu and S. S. Quek, "Modeling Techniques," in *The Finite Element Method*, Elsevier, 2014, pp. 301–345. doi: 10.1016/B978-0-08-098356-1.00011-4.

[13] A. F. Molland, S. R. Turnock, and D. A. Hudson, *Ship Resistance and Propulsion: Practical Estimation of Ship Propulsive Power*, 2nd ed. Cambridge: Cambridge University Press, 2017.

[14] M. Murdijanto, I. K. Utama, and A. Jamaluddin, "An Investigation into the Resistance/Powering and Seakeeping Characteristics of River Catamaran and Trimarana," *Makara J. Technol.*, vol. 15, no. 1, pp. 25–30, 2011.

[15] "C. Hwan Son," CFD Investigation of Resistance of High-Speed Trimaran Hull Forms", Florida Institute of Technology, bachelor Thesis, 2015.

[16] A. S. Bekhit, "Numerical simulation of the ship self-propulsion prediction using body force method and fully discretized propeller model," in *IOP Conference Series: Materials Science and Engineering*, 2018, vol. 400, p. 042004.

[17] R. Mikkelsen, Technical University of Denmark, F. M. Department of Mechanical Engineering, and MEK, "Actuator disc methods applied to wind turbines," Lyngby, 2003.

IOP Conf. Series: Earth and Environmental Science 1081 (2022) 012055 doi:10.1088/1755-1315/1081/1/012055

Analysis of Self-propulsion Test for Propeller KP-505 with and without Pre-Swirl Stator

Bayu Agastiya, I Made Ariana and Achmad Baidowi

Marine Engineering Departmen, Institut Teknologi Sepuluh Nopember, Surabaya

bayu.206019@mhs.its.ac.id, ariana@its.ac.id, ahmadbai@gmail.com

Abstract. In recent times, environmental issues have become a major concern. The installation of Energy Saving Device (ESD) on ships is an alternative method developed to reduce energy use. One type of ESD that is widely used is the installation of a Pre-Swirl Stator (PSS). PSS consists of stator blades mounted in front of the propeller to regulate flow before entering the propeller. This paper describes the effect of PSS installation on the hull of a KP-505 propeller. The asymmetrical PSS design was developed with three blades on the port side and one on the starboard side. The analysis was carried out to obtain the Propulsive Coefficient (PC). The research was conducted using Computational Fluid Dynamic (CFD)-based simulation. The simulation results show that the addition of PSS increases the ship's resistance up to 4%. However, the PC has also increased significantly up to 10.3, so it can reduce power requirements by up to 5.6%.

1. Introduction

Ships are the primary mode of transportation in distributing goods from one place to another. Data from UNCTAD, (2019) shows that 80% of the volume of world trade is currently carried out through sea transportation [1]. However, there is also a negative impact to this attributable to the exhaust gas emissions produced by the ship. A ship's activity is proportional to the exhaust gas emissions it produces [2]. In addition, the new regulation issued by IMO Tier III regarding limiting the amount of emissions produced by ships has forced the shipping industry to invest more effort in reducing ship exhaust emissions.

Installation of ESD on ships is one of the most widely used efforts to save energy. ESD technology was introduced in the mid-20th century and peaked in the late 1970s and early 1980s due to the fossil fuel crisis [3]. At the beginning of its development, the use of ESD was not well accepted because it was considered unsuitable for installation on ships. Some of the problems with ESD included construction failures on the ESD, damage to the ESD, and design limitations on the ESD that did not allow it to be installed on ships. However, with the increasing number of studies on ESD, the application of ESD can be accepted in almost all types of ships nowadays [4].

One type of simple ESD that can be a solution to save energy is the installation of PSS on the hull [5]. PSS installation in front of the propeller aims to produce a flow effect that rotates opposite to the propeller's direction. The propeller will experience this rotation as an additional blade to increase the resulting thrust force [6]. However, the increase in thrust force must be greater than the total additional resistance caused by the addition of stator blades. Energy-efficient devices are defined for use in three zones, namely zones I, II, and III, as shown in Figure 1 [7]. The three zones are: I) pre-devices, which are installed in front of the propeller, II) main device, which is mounted on the propeller, and III) post-propeller devices which are installed behind the propeller.

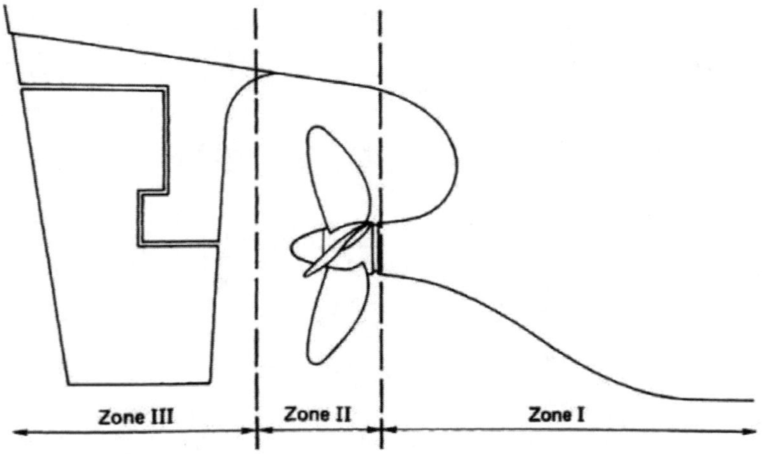

Figure 1. Zone clasification of energy saving device [7].

In several studies related to PSS installation, a wide variety of different results need to be reviewed. A proper configuration is required for the PSS design by adjusting the ship's characteristics. PSS is closely related to the flow before it enters the ship ergo the type of ship and the stern's shape greatly affects the design of PSS [8]. Proper PSS modeling will provide optimal energy savings on a ship. Kim et al., (2013) carried out a PSS design starting from the initial design stage, optimizing PSS selection, to the validation stage through a sea trial test. This resulted in PSS reducing power up to 4% [9]. Furcas and Gaggero, (2021) also carried out the design and optimization of PSS selection using 1, 2, and 3 stator blades. The number of 3 stator blades was found to save the most optimal energy with a power reduction of up to 8% [10]. However, the effect of PSS on the PC has not been widely researched, even though the propeller efficiency must also consider the interaction relationship between the ship's hull and the propeller, as well as the relationship between the use of open water torque and behind the ship torque. This research presents an analysis of the asymmetric PSS mounted behind the KP-505 propeller on the Kriso Container Ship (KCS) ship using CFD-based simulation. Asymmetric PSS is considered the most effective device and is widely applied to various ships without any problems [11]. For propellers with RH rotation, the asymmetric PSS is installed with the number of blades on the left side being more than that on the right side. This is done because the flow component that leads up is lower than the one that leads down. In addition, it also reduces production costs and the load on the stern by eliminating unnecessary blades [12].

Furthermore, self-propulsion tests were conducted to obtain the wake flow and thrust deduction factor. Calculation of the efficiency of the propulsion system due to the addition of PSS is done by calculating the value of the PC, which involves the calculation of three types of efficiency, namely open water efficiency, hull efficiency, and relative rotative efficiency.

2. Methods and Material

2.1. Methodology

Figure 2 shows the research methodology as a step-by-step process in the research. Six steps begin with problem identification, followed by model making, simulation, validation, result and discussion, and conclusion.

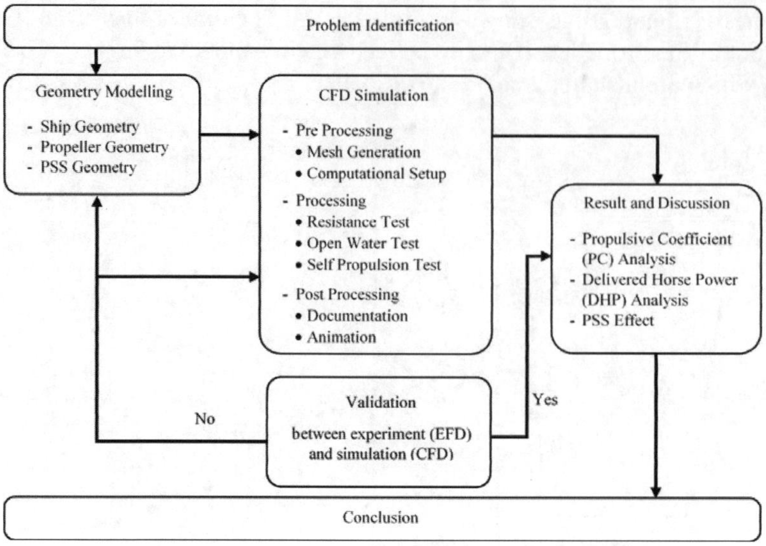

Figure 2. A framework of research methodology.

2.2. Geometrical Data

The ship used in this research is a 3,600 TEU KCS. The ship is a theoretical ship model created to be used for research purposes to provide data and fluid flow explanations and CFD validation for modern container ships with bulbous bows. The design of the KCS ship model was made in 2 (two) places, 1) Korea Research Institute for Ships and Ocean Engineering by conducting towing tank experiments to obtain resistance and wave fields, and 2) National Maritime Research Institute by conducting towing tank experiments to obtain surface pressure of the hull and downstream velocity field with and without propeller. Table 1 presents a detailed dimension regarding the KCS ship. Meanwhile, the propeller model uses the KP-505 propeller, which is installed on the KCS ship. The detailed dimensions of the KP-505 propeller are presented in Figure 3.

Table 1. Principal dimensions of KCS.

Main particular	Symbol	Ship	Model
Scale		1	31.6
Length	Lpp (m)	230	7.278
Maximum beam	B (m)	32.2	1.019
Depth	D (m)	19	0.601
Draft	T (m)	10.8	0.3418
Block coefficient	Cb	0.650	0.650
Wetted surface	S (m^2)	9424	9.437
Froude Number	Fn	0.26	0.26
Velocity	Vs (m/s)	12.345	2.196

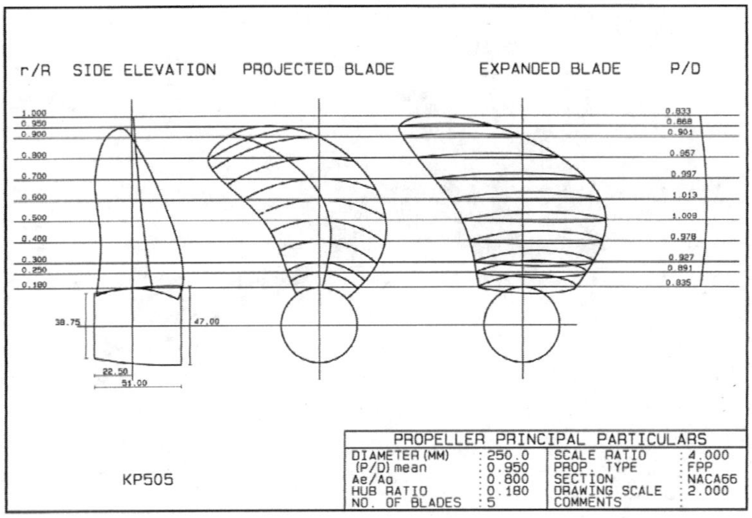

Figure 3. Main dimensions of the Propeller KP-505 [13].

3. Result and Discussion

3.1. Resistance Test

The resistance test simulation was carried out by pulling the ship without a propeller at a speed of Vs. The resistance test results were then used to determine the required thrust force to move the ship at a service speed (Vs). Table 2 shows the simulation result of the resistance test. Based on the simulation results, PSS can increase the ship's resistance up to 4%.

3.2. Open Water Test

The following simulation carried out was the open water test stage. An open water test was used to determine the characteristics of the KP-505 propeller. In the simulation process, the inputs required were propeller geometry, rotation (n), and advanced velocity (V_a). The geometry model was input in a cylindrical domain and rotation (n) domain, which was entered as a fixed parameter for the rotating propeller. The value of V_a was obtained by varying the advance ratio (J) value at each point between 0.0 – 1.0, calculated using the formula $V_a = n\, D\, J$.

The open water simulation was carried out using the dual-domain method. The first domain is the propeller and the second domain is the shaft. This method was chosen because only the propeller domain rotates while the shaft domain does not rotate, thereby reducing losses on the shaft.

Figure 4 presents the meshing grid for open water computing. Table 3 shows the number of meshing grids for the open water test.

Table 2. The result of the resistance test.

Type	CT x 10^{-3}		Err %
	Experiment	Simulation	
Without PSS	3.560	3.574	0.40%
With PSS	-	3.720	-
Gap	-	4.09%	-

Figure 4. Meshing grid for open water computing.

Table 3. The number of meshing grid for open water test.

	Background	Propeller	Total
Mesh Size	80,354	1,586,994	1,667,348

Furthermore, the Thrust and Torque values were obtained from the simulation results. The value of the thrust coefficient (*KT*) and torque coefficient (*KQ*), as well as the propeller efficiency, were calculated using the following equation:

$$KT_0 = T / (\rho\, n^2\, D^4) \quad (1)$$

$$KQ_0 = Q / (\rho\, n^2\, D^5) \quad (2)$$

$$\eta_0 = (KT/KQ)\,(J/2\pi) \quad (3)$$

The obtained values of KT, KQ, and J, were then inserted into the *KT- KQ-J* graph, as shown in Figure 5. Figure 5 shows the KT-KQ-J graph's comparison based on the EFD experiment and CFD simulation results.

3.3. Self-propulsion Test

A self-propulsion test is an experiment on a test tank using a model ship and a model propeller to determine the interaction results of the two. Several inputs are needed to perform self-propulsion testing: model geometry, propeller diameter (*D*), and propeller rotation (*n*).

There was a difference in the frictional resistance coefficient for the self-propulsion test case on the model ship. Therefore, adding an external force or SFC (skin friction correction) was necessary. The addition of SFC was calculated using the equation:

$$SFC = \{(1+k)(C_{FOM}-C_{FOS})-\Delta C_F\} x \frac{1}{2}\rho V_s^2 A_w \quad (4)$$

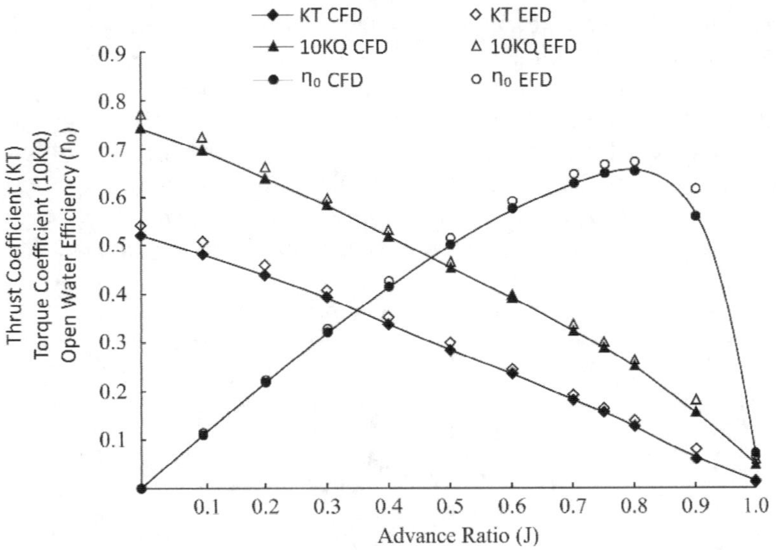

Figure 5. The result of the open water test.

In the analysis process, the results of self-propulsion were closely related to the results of the resistance test and open water tests.
Calculation of self-propulsion test analysis using the ITTC, (2002) method with the following equation [14]:

$$KT_B = T/\rho\, n^2 D^4 \quad (5)$$

$$KQ_B = Q/(\rho\, n^2\, D^5) \quad (6)$$

$$t = (T + SFC - R_T)/T \quad (7)$$

$$w = 1 - (n\, D\, J/V_s) \quad (8)$$

$$\eta_{RR} = KQ_0/KQ_B \quad (9)$$

$$\eta_H = (1-t)/(1-w) \quad (10)$$

$$PC = \eta_o\, \eta_{RR}\, \eta_H \quad (11)$$

The result for self-propulsion simulation is present in Table 4.

3.4. Effect of the Pre-Swirl Stator

Proper PSS design is crucial in contributing to the propeller performance. As of now, there is no definite model regarding the PSS design. The inflow of the PSS is highly dependent on the shape of the hull (especially the stern of the ship); therefore, the design of the PSS on different ships will have a different effect [15]. The design of the PSS is relatively simple. Fundamentally, it simply requires geometry that has an angle of attack that can optimally direct the flow to the propeller [10]. In this study, a PSS model of an asymmetrical type was developed with a total of 4 blades, with three blades on the port side and one blade on the starboard side. The foil type used was the NACA-0006 cross-section, chord length is

IOP Conf. Series: Earth and Environmental Science 1081 (2022) 012055 doi:10.1088/1755-1315/1081/1/012055

0.15D, and angle of attack by 10°. The stator is mounted in front of the propeller at a distance of 0.2D. Figure 6 presents a detailed model of the PSS.

Furthermore, in this analysis, the rotational speed of the propeller between ships without PSS and with PSS were given different values. The rotational speed of the propeller was given as zero condition. The analysis results in propeller efficiency, hull efficiency, relative rotative efficiency, and PC are shown in Table 5. It can be seen that the addition of PSS can increase PC value up to 10.3%. A relative velocity pattern is found in terms of flow, as shown in Figure 7. The difference can be seen clearly in that the stator works very well by causing the flow to be more evenly distributed.

Table 4. The result for simulation of the self-propulsion test.

Type		*n*	*KT*	*10KQ*	*J*	*(1-t)*	*(1-w)*
Without PSS	EFD	9.50	0.170	0.289	0.728	0.853	0.792
	CFD	9.328	0.172	0.311	0.735	0.873	0.781
	Err%	-1.81%	1.16%	7.59%	0.96%	2.38%	-2.28%
With PSS	CFD	8.91	2.664	0.347	0.676	0.865	0.686

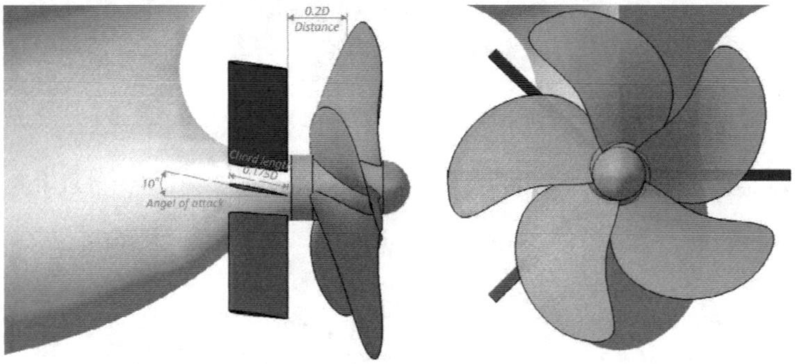

Figure 6. Detailed model of the PSS.

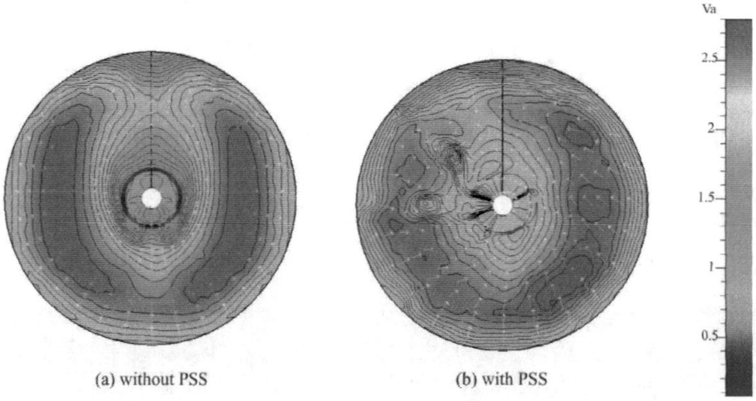

Figure 7. Relative velocity without and with PSS.

Table 5. Comparison result of PSS effect.

Type	n (rps)	CT x 10^{-3}	η_o	η_{RR}	η_H	*PC*	*DHP* (Watt)
Without PSS	9.328	3.574	0.649	0.993	1.122	0.7231	246.522
With PSS	8.91	3.720	0.617	1.034	1.262	0.7981	232.615
Gap	-4.59%	4.09%	-4.93%	3.26%	12.98%	10.3%	-5.6%

4. Conclusion
This paper presents a self-propulsion analysis to obtain the PC, which involves calculating of three types of efficiency, namely open water efficiency, hull efficiency, and relative rotative efficiency. Furthermore, an asymmetrical PSS model was installed on the KCS ship to determine the effect of the resulting flow before entering the propeller and its effect on the PC. The analysis results show that the addition of PSS can increase the ship's resistance by up to 4%. However, the PC also increased significantly up to 10.3, which can positively impact by reducing power requirements by up to 5.6%. The stator affects flow before it enters the propeller so that the flow is more evenly distributed, causing the reduction in torque requirement and increase in wake flow. However, additional resistance must be considered in the case of additional installation of PSS.

5. References
[1] United Nations Conference on Trade and Development (UNCTAD) 2019 *Review of Maritime Transport*
[2] Peng X, Wen Y, Wu L, Xiao C, Zhou C, and Han D 2020 A Sampling Method For Calculating Regional Ship Emission Inventories *Transportation Research Part D: Transport and Environment* **89** 102617
[3] Jong J H and Zondervan G J D 2006 *A framework for Energy Saving Device (ESD) Decision Making* (Netherlands: MARIN)
[4] Holtrop J and Mennen G G J 1982 An Approximate Power Prediction Method *International Shipbuilding Progress* **29**
[5] Nadery A and Ghassemi H 2020 Hydrodynamic Performance of the Ship Propeller under Oscillating Flow with and Without Stator *American Journal of Civil Engineering and Architecture* **8** 56–61
[6] Zondervan G J, Holtrop J, Windt J and Terwisga T V 2011 On the Design and Analysis of Pre-Swirl Stators for Single and Twin Screw Ships *2nd Int. Symp on Marine Propulsor* (Germany: Hamburg)
[7] Carlton J S 2007 *Marine propellers and propulsion* 2nd edition (Oxford: Butterworth-Heinemann)
[8] Nadery A, Ghassemi H and Chybowski L 2021 The effect of the PSS configuration on the hydrodynamic performance of the KP505 propeller behind the KCS *Ocean Engineering* **234** 109310
[9] Kim K, Leer-Andersen M, Werner S, Orych M and Choi Y 2013 Hydrodynamic optimization of pre-swirl stator by CFD and model testing *Int. Shipbuilding Progress* **60** 233-276
[10] Furcas F and Gaggero S 2021 Pre-swirl stators design using a coupled BEM-RANSE approach *Ocean Engineering* **222** 108579
[11] Shin Y J, Kim M C, Kang J G, Kim H U and Shin I R 2019 Validation of optimally designed stator-propeller system by EFD and CFD *brod shipbuilding* Vol 70 No 3 p. 133-151
[12] Kim M C, Chun H H and Kang Y D 2004 *Design and experimental study on a new concept of preswirl stator as an efficient energy-saving device for slow speed full body ship* Transactions of The Society of Naval Architects and Marine Engineers Vol. 112 T2004-1
[13] NMRI (National Maritime Research Institute) *KRISO Container Ship (KCS)*
[14] ITTC 2002 *Testing and Extrapolation Methods Propulsion, Performance Propulsion Test* 7.5-02-03–01.1
[15] Nowruzi H and Najafi A 2019 An experimental and CFD study on the effects of different pre-swirl ducts on propulsion performance of series 60 ship *Ocean Engineering* **173** 491–509

Numerical Study into the Resistance of a Trimaran Hull at Various Longitudinal Spacing

E Yuliora[1], I K A P Utama[1], I K Suastika[1]

[1]Department of Naval Architecture, Institut Teknologi Sepuluh Nopember, Surabaya, Indonesia

Abstract. Although the design of trimaran hull has gained popularity at the end of the 20th century, there are few databases available to numerical analysis of trimaran hull. This paper presents a numerical study of the resistance of trimaran hull at various length ratio (S/L) 0.2, 0.3, 0.4 and Froude numbers 0.2 to 0.6 carried out by CFD simulation using FINE™/MARINE by NUMECA software and experimental data. Tri-A has a lower resistance than the trimaran B and C at low Froude number, but at hight Froude number 0.5 and 0.6, the Tri-A has a low resistance than the others. These results are compared with towing tank test data and obtained a lowest difference in the resistance of 0.98 % at TRI-A3. The lowest resistance for the trimaran model is Tri-C5 (Fr=0.6; S/L=0.4) with a total resistance coefficient (C_T) of 65.89×10^{-4}, this is due to favorable wave interference with a minimum IF of 0.02. The Interference Factor (IF) follows the same pattern in each configuration, however at maximum Fr, the minimal IF produced on Tri-A3(Fr=0.4; S/L=0.2), Tri-C4 (Fr=0.5; S/L=0.4), and Tri-C5 is 0.22, 0.12, and 0.02, respectively. Where the lowest IF indicates a positive effect because the resistance of the interference hull is lower hull at high speed. The results of this study are expected to be able to strengthen the data base in presenting the effect of interference resistance of the trimaran, which can then be applied directly in determining the magnitude of the ship's propulsion at the preliminary design stage.

1. Introduction

Due to the interaction between the mainhull and sidehull, which creates interference caused by variations in flow velocity, pressure changes, and wave interactions induced by each hull, the multihull resistance component, notably trimaran, has a more complex phenomenon than monohull. The geometric shape of the hull, the arrangement of the sidehull positions relative to the mainhull, the separation distance between the hulls, fluctuations in the Wetted Surface Area (WSA), ship acceleration, turbulence models, form factors, and external factors of ocean waves all contribute to a more complex hydrodynamic phenomenon.

A trimaran is a ship with three hulls. Russian researcher coined the term trimaran to describe ships with one main hull and two small side hulls, yet this trimaran ship feature has been utilized in South Asia long before the scientific and technical revolution [1]. Extensively before Europeans arrived, outriggers were utilized for long ocean trips to explore all of the Pacific Ocean's islands. Bertorello et al [2] compared the trimaran's residual resistance and overall resistance to monohull and catamaran vessels in a study on the trimaran. Based on the results of his research, it was found that the trimaran always shows low resistance to the Froude Volume Displacement ($Fr\nabla$) 1.8 to 3.0, while monohull ships give their best performance at $Fr\nabla > 3$, moreover, interference effect more profitable on trimarans over monohulls and catamarans.

The resistance of the trimaran's hull configuration was optimized by varying the ratio of the distance between the hulls longitudinally S/L, which is constant at 0.078, and the sidehull is symmetrical and asymmetrical, where the position of the trimaran's stern sidehull is parallel with aft mainhull creates less resistance than a configuration with an aft sidehull that is not parallel to the aft mainhull, but the asymmetrical sidehull has a beneficial influence on the trimaran drag for the total resistance [3]. Luhulima et al [4] did the same thing, optimizing the trimaran configuration with constant S/L and varied R/L, namely with the sidehull parallel to the stern (A), the sidehull in the middle of the ship (B), and the sidehull parallel to the bow (C) (symmetrical sidehull), the results of CFD analysis with Fr 0.4 show configuration A has the smallest resistance, which is 1.7 % smaller than configuration B and configuration B is 3.9 % larger. The transom's hull design reduces wave and splash events behind the transom while restoring flow to the stern, resulting in a reduction in total drag [5].

Computational Fluid Dynamics (CFD) is a method of numerically calculating, predicting, and approaching fluid flow in the control of dimensions, area, and volume using computer aid to execute calculations on each of its dividing elements. Furthermore, CFD takes advantage of the advancement of high-speed computer technology, which allows for the analysis of a model with extremely high precision but requires a large amount of computer memory [6] and is reasonably inexpensive, accurate, and has a short duration [7]. In the process of estimating total resistance, both Experimental Fluid Dynamics (EFD) and CFD offer advantages and limitations that complement each other [8] [9][10].

The goal of this study is to compare the data from the towing tank test with the results of a CFD simulation utilizing NUMECA's FINETM/MARINE software. The trimaran model has been-evaluated against a semi-displacement hull shape with a transom-symmetrical hull shape and differences in the separation distance between the sidehull and mainhull. In the Froude number (Fr)= 0.2, 0.4, and 0.6, S/L = 0.2, 0.3, and 0.4. The findings of this research should help to improve the database for showing the influence of trimaran interference, which can then be used to determine the magnitude of the ship's propulsion at the preliminary design stage.

2. Computational Method

A commercial software FINE^{TM}/MARINE by NUMECA with unsteady RANSE (Reynolds-Averaged Navier-Stokes Equations) solver was used to perform the trimaran hydrodynamic analyses. The Navier-Stokes equation describes how fluids flow in a non-linear system of differential equations. The balance of the forces operating on the fluid, particularly the object and surface forces, is described by this equation. The Navier-Stokes equation assumes that the fluid is incompressible and a continuous state flows. The equation factor of the RANSE is Reynolds's decomposition, where the instantaneous quantity is decomposed into the sum of the mean times and their fluctuations, proposed first by Osborne Reynolds. The RANSE is mainly used to describe turbulent currents. This equation can be used with approximations based on knowledge of the turbulent flow properties to give an approximate time-average solution for the Navier-Stokes equation. This can be written in the following notation:

$$\rho\bar{u}_i\frac{\partial\bar{u}_i}{\partial x_j} = \rho\bar{f}_i + \frac{\partial}{\partial x_j}\left[-\bar{p}\delta_{ij} + \mu\left(\frac{\partial\bar{u}_i}{\partial x_j} + \frac{\partial\bar{u}_j}{\partial x_i}\right) - \rho\overline{u'_i u'_j}\right] \tag{1}$$

Continuity Equation and Momentum Equation (Navier-Stokes Equations, Reynolds Equations) can be represented to predict fluid calculations. Water is an incompressible fluid, meaning that its density (ρ) is constant. The mass density of water does not change with changes in pressure, but can still change with changes in temperature. The continuity and momentum equations for the flow of water when written in Cartesian coordinates are:

$$\frac{\partial u}{\partial x} + \frac{\partial v}{\partial y} + \frac{\partial w}{\partial z} = 0 \tag{2}$$

$$\frac{\partial u}{\partial t} + \frac{\partial uu}{\partial x} + \frac{\partial uv}{\partial y} + \frac{\partial uw}{\partial z} = -\frac{1}{\rho}\frac{\partial p}{\partial x} + \frac{1}{\rho}\frac{\partial \tau_{xx}}{\partial x} + \frac{1}{\rho}\frac{\partial \tau_{yx}}{\partial y} + \frac{1}{\rho}\frac{\partial \tau_{zx}}{\partial z} + g_x \tag{3}$$

$$\frac{\partial v}{\partial t} + \frac{\partial vu}{\partial x} + \frac{\partial vv}{\partial y} + \frac{\partial vw}{\partial z} = -\frac{1}{\rho}\frac{\partial p}{\partial y} + \frac{1}{\rho}\frac{\partial \tau_{xy}}{\partial x} + \frac{1}{\rho}\frac{\partial \tau_{yy}}{\partial y} + \frac{1}{\rho}\frac{\partial y}{\partial z} + g_y \tag{4}$$

$$\frac{\partial w}{\partial t} + \frac{\partial wu}{\partial x} + \frac{\partial wv}{\partial y} + \frac{\partial ww}{\partial z} = -\frac{1}{\rho}\frac{\partial p}{\partial z} + \frac{1}{\rho}\frac{\partial \tau_{xz}}{\partial x} + \frac{1}{\rho}\frac{\partial \tau_{yz}}{\partial y} + \frac{1}{\rho}\frac{\partial \tau_{zz}}{\partial z} + g_z \tag{5}$$

The free water surface is initially assumed as calm water with density of 998.2 kg/m^3 (freshwater). The averaging of Navier-Stokes equations produces a nonlinear Reynolds stress term, which necessitates further modeling, namely a turbulence model, to properly resolve the system. The Reynolds stresses have been computed using Shear Stress Transport (SST) k-ω turbulence model. The velocity profiles very close to the no-slip walls are estimated using a wall function in the turbulence model.
The SST k-ω turbulence model [11] is a two-equation eddy-viscosity model that is used for many aerodynamic applications. The SST k-omega Governing Equations are:
Turbulence Kinetic Energy

$$\frac{\partial k}{\partial t} + U_j \frac{\partial k}{\partial x_j} = P_k - \beta * k\omega + \frac{\partial}{\partial x_j}\left[(v + \sigma_k v_T)\frac{\partial k}{\partial x_j}\right] \tag{6}$$

Specific Dissipation Rate

$$\frac{\partial \omega}{\partial t} + U_j \frac{\partial \omega}{\partial x_j} = \alpha S^2 - \beta\omega^2 + \frac{\partial}{\partial x_j}\left[(v + \sigma_\omega v_T)\frac{\partial \omega}{\partial x_j}\right] + 2(1 - F_1)\sigma_{\omega^2}\frac{1}{\omega}\frac{\partial k}{\partial x_i}\frac{\partial \omega}{\partial x_i} \tag{7}$$

2.1 Trimaran hull configurations

The study investigates the total resistance on model scale trimaran with NPL-4a hull transom stern and symmetrical hull shape

Table 1. Scale model of trimaran

Parameter	Unit	Mainhull	Trimaran Sidehull
LWL	m	0.8000	0.4000
B	m	0.0892	0.0446
T	m	0.0446	0.0223
WSA	m^2	0.0542	0.0136
Δ	kg	1.2850	0.1600

Table 2. Trimaran configuration

S/L	Fr				
	0.2	0.3	0.4	0.5	0.6
0.2	TRI-A1	TRI-A2	TRI-A3	TRI-A4	TRI-A5
0.3	TRI-B1	TRI-B2	TRI-B3	TRI-B4	TRI-B5
0.4	TRI-C1	TRI-C2	TRI-C3	TRI-C4	TRI-C5

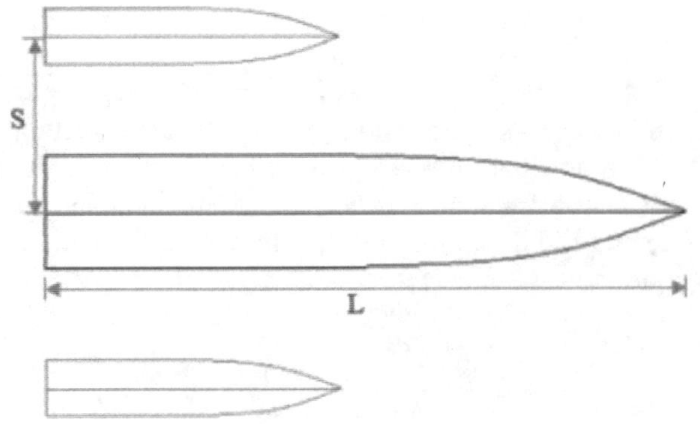

Figure 1. Trimaran hull

IOP Conf. Series: Earth and Environmental Science 1081 (2022) 012056 doi:10.1088/1755-1315/1081/1/012056

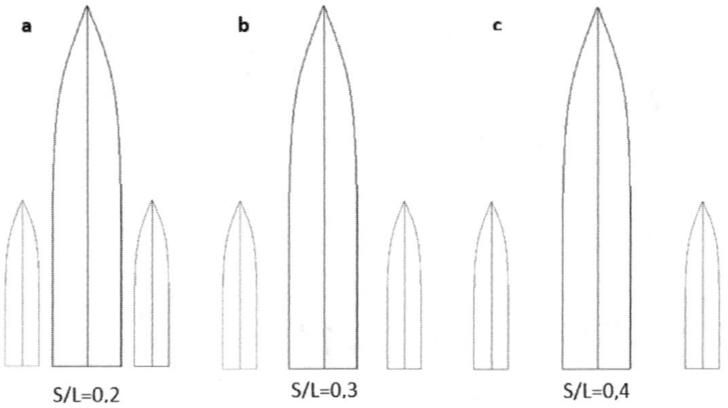

Figure 2. Trimaran configuration

2.2 Simulation cases

The goal is to use the CFD approach to calculate the resistance characteristics of the trimaran hull form (Table 1) for various outrigger positions in longitudinal orientations. In Figure 1, the longitudinal stagger is shown as S/L. As a result, we evaluated and examined nine different outrigger longitudinal position spacing configurations (Table 2). Using three longitudinal spacing variations (Figure 2) and five-speed variations, namely at Fr 0.2, 0.3, 0.4, 0.5 and 0.6). The effect of shifting outrigger location on wave interference and overall resistance could then be studied in depth.

2.3 Numerical Setup

For an incompressible flow with constant viscosity and nobody forces acting on the hull, the averaged continuity and momentum equations in the Cartesian coordinate system are as follows [12]. According to the ITTC (International Towing Tank Conference) CFD recommendations [13], the time domain has been partitioned into time steps as follows:

$$\Delta t = 0.005 - 0{,}01\, {}^{L}\!/_{v} \quad (8)$$

The length between perpendiculars is L, and the ship speed is v. To compute the velocity and pressure of each flow separately, a segregated solver was utilized.

2.4. Boundary Condition

In order to examine the impact of outriggers on the resistance characteristics both experimentally and numerically, a trimaran hull from the NPL-4a series in 1/62.5th scale ratio with a narrow transom-stern mainhull stabilized by two small symmetrical outriggers was used (Figure 3)

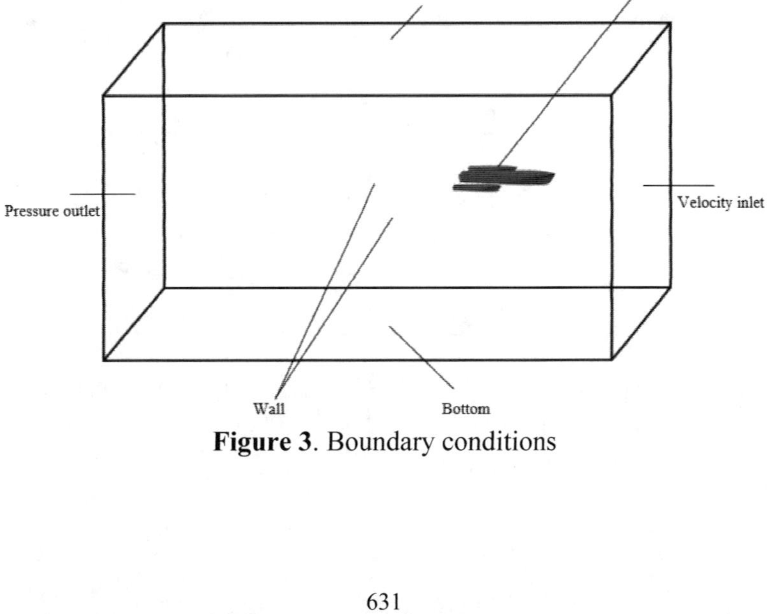

Figure 3. Boundary conditions

2.5 Grid Resolution

For the towing tank simulations of the trimaran hull, a static (rigid) grid approach is used. The grid sizes gradually expand as you get further away from the hull (Figure 4 and Figure 5). Near the hull surfaces, a structured grid system with the boundary layer is used, while an unstructured grid with fully hexahedral elements is used for the rest of the computational domain.

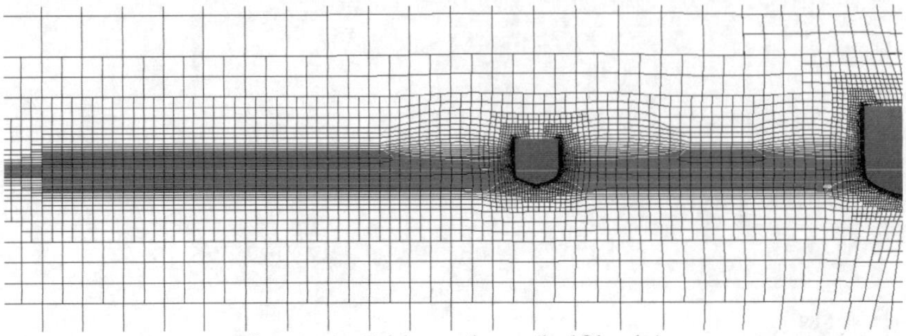

Figure 4. Grid topology (half body)

Figure 5. 3D trimaran geometric form

3. Computational Results for Different Sidehull Positions

3.1 General Arrangement

All cases are examined at the vessel's service speed, previously known as Fr 0.2, 0.3, 0.4, 0.5 and 0.6. Table 3 shows a comparison between CFD and EFD results [4] for each configuration.

Table 3. Coefficient of total resistance on each configuration

Fr	C_T	C_T 10^{-3}		
	Tri-NI	TRI-A	TRI-B	TRI-C
0.2	5.467	6.459	6.628	6.548
0.3	5.756	7.116	7.148	7.070
0.4	7.294	8.317	8.657	8.513
0.5	7.847	8.499	7.976	7.683
0.6	6.753	6.903	6.669	6.589

Figure 6 shows the coefficient of total resistance calculated from CFD results, and figure 7 shows the comparison between CFD results and EFD data adopted from previous research [4]. Both data showed a similar changing trend except for TRI-A3 and TRI-B3. Total resistance increases at speeds of 0.2 to

0.4, but at 0.5 and 0.6 total resistance decreases on trimaran as a result of the lift generated at the higher speed. The total resistance decreases in proportion to the increase in the S/L ratio as the S/L ratio rises. This is due to a change in the center of gravity as the longitudinal spacing ratio increases. The total resistance on high-speed decreases if the sidehull (SH) keeps away from the mainhull (MH) in the longitudinal direction (closest point to the mainhull). In addition, the trimaran's gravity center rises, and trim (by the stern) decrease.

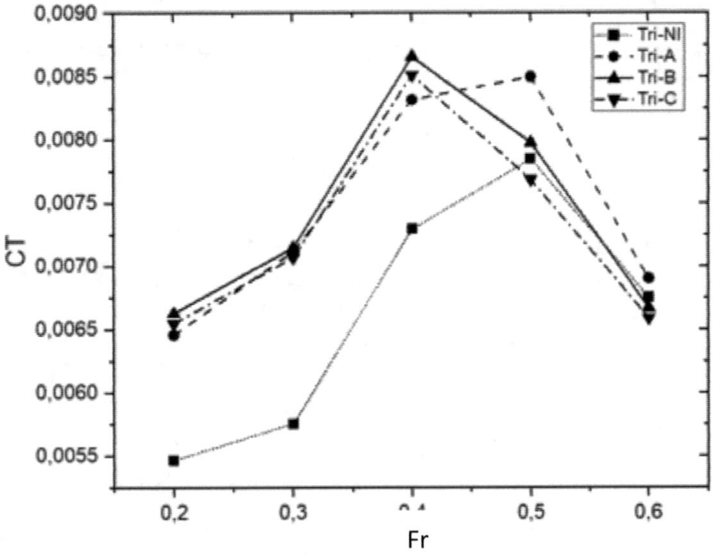

Figure 6. Total resistance coefficient CFD results

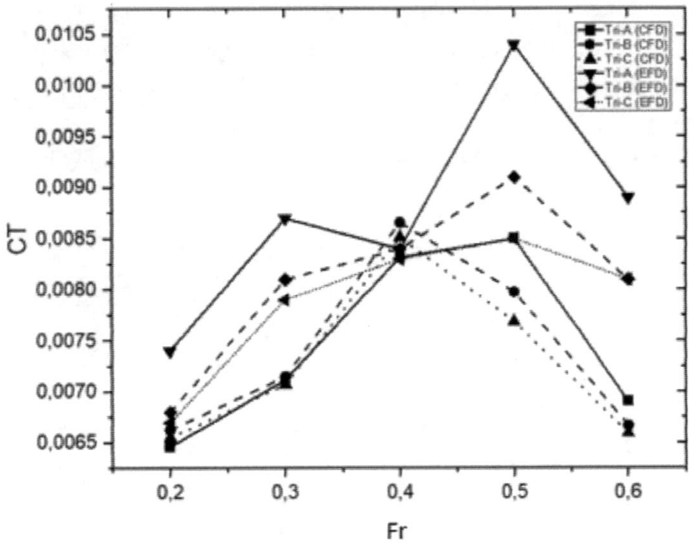

Figure 7. Comparison of C_T between CFD and EFD results

3.2 Interference Effects

When a multihull vessel's resistance performance is studied, the resistance of the multihull concept differs from the total of the resistance of each hull towed separately. Coefficient of total resistance in

IOP Conf. Series: Earth and Environmental Science 1081 (2022) 012056 doi:10.1088/1755-1315/1081/1/012056

non-interference trimaran hull $\mathbf{C_{T_{tri-NI}}}$ and interference $\mathbf{C_{T_{tri-If}}}$ can be calculated by the following Equations 9 and 10,

$$C_{T_{tri-NI}} = C_{T_{tri-MH}} + 2C_{T_{tri-SH}} \quad (9)$$

$$C_{T_{tri-If}} = C_{T_{tri-NI}} + \Delta C_{T_{tri}} \quad (10)$$

Where $C_{T_{tri-MH}}$ is the total resistance coefficient of mainhull, $C_{T_{tri-SH}}$ is the total resistance coefficient of sidehull and $\Delta C_{T_{tri}}$ is the difference between non-interference and interference trimaran resistance coefficient.

The IF interference factor can be estimated as the difference in total drag caught when moving from three independent hulls to one bonded trimaran hull, and it is easy to represent the difference as a ratio of total resistance non-interference

$$\mathrm{IF} = \frac{C_{T_{tri-If}}}{C_{T_{tri-NI}}} - 1 \quad (11)$$

The lowest interference factors indicate effective interference, i.e., the trimaran configuration's resistance is less than the sum of the individual hull resistance, whereas negative interference factors indicate detrimental interference [14]. The interference phenomenon is the differential in pressure between individual and combined multihull vessels.

Table 4 shows the interference factors for each trimaran hull configuration. IF is low on Tri-B and Tri-C when Fr 0.5 and 0.6 indicates that wave interference has a positive effect on reducing trimaran resistance.

Table 4. Interference factor trimaran hull

Fr	IF		
	TRI-A	TRI-B	TRI-C
0.2	0.18149873	0.21235121	0.19776839
0.3	0.23627418	0.24177683	0.22818903
0.4	0.14030808	0.18690660	0.16716158
0.5	0.08308848	0.01644863	0.02078972
0.6	0.02212659	0.01246803	0.02435783

The influence of variances in longitudinal spacing and Fr on trimaran resistance interference may be seen in Figures 8 and 9.

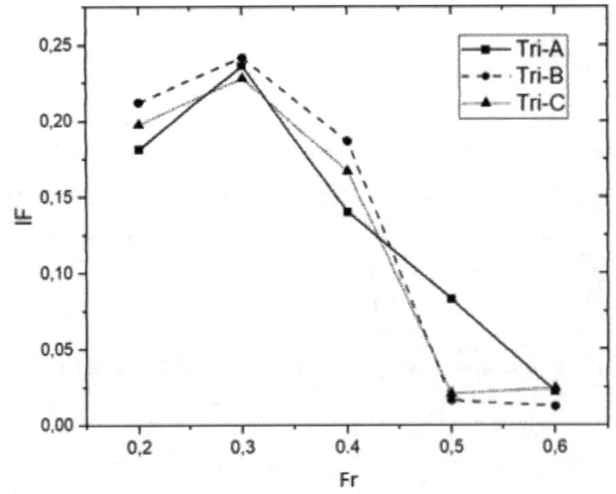

Figure 8. Interference factor ratio

IOP Conf. Series: Earth and Environmental Science 1081 (2022) 012056 doi:10.1088/1755-1315/1081/1/012056

Figure 8 shows the relationship between the interference factor and the increase in Fr speed, which is comparable to the trend established by the change in the total resistance coefficient, in which the rises in proportion to increasing speed, then drops at Fr 0.4 to Fr 0.6.

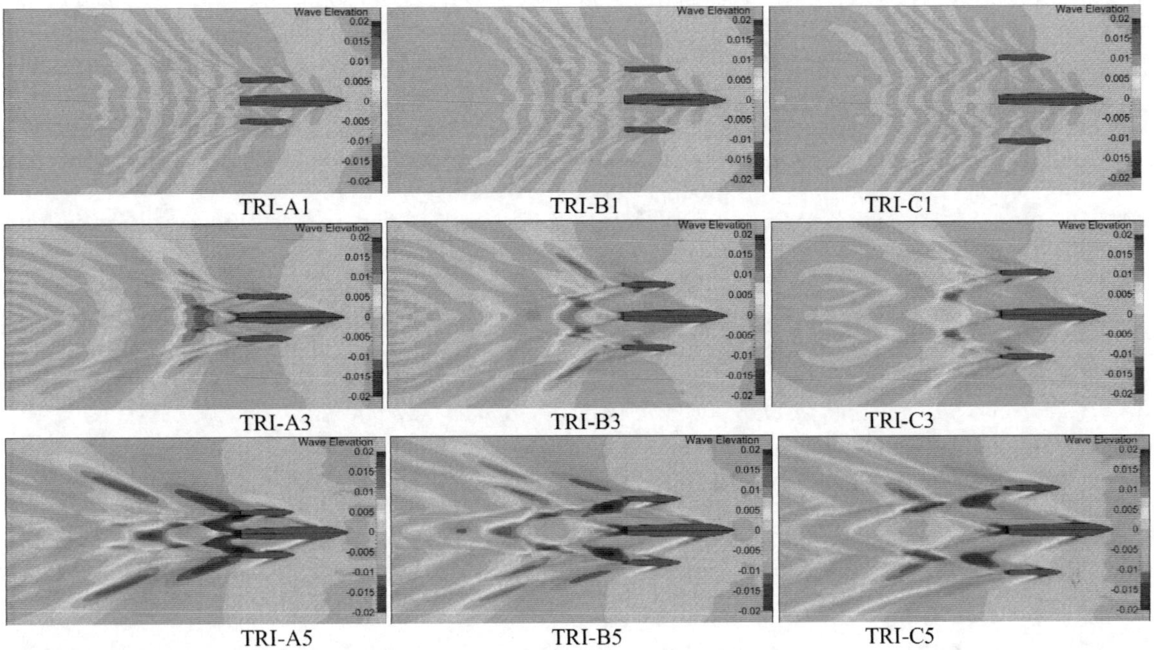

Figure 9. Wave pattern on three different spacing and speed variations

Figure 9 shows the wave pattern caused by wave interference on the trimaran at low (1), medium (3) and high (5) speeds, with small (A), medium (B), and large (C) S/L ratios. It is apparent that the wave pattern formed by the trimaran at low speeds has wave interference in the bow sidehull area, but when the speed is increased at maximum speed, a wave interference pattern forms in the area around the stern even further behind the stern, its causes the wave interference produced by the trimaran to have a positive effect on the trimaran resistance, especially at high speeds.

4. Conclusions

On a model scale trimaran with NPL-4a hull transom stern and symmetrical hull shape, this study shows the findings of CFD simulation with FINE™/MARINE BY NUMECA SOFTWARE that compare EFD data on a model scale trimaran with NPL-4a hull transom stern and symmetrical hull shape. The influence of longitudinal spacing and Fr on total resistance, as measured by the coefficient of total resistance and the interference factor of resistance between the hulls, was investigated.

The longitudinal spacing is primarily responsible for decreasing the total resistance on the trimaran. The trimaran C type generally shows a good overall resistance at maximum speed namely the lowest resistance for the trimaran model is Tri-C5 with C_T of 65.89×10^{-4}, this is due to favorable wave interference with a minimum IF of 0.02. The IF follows the same pattern in each configuration, however at maximum Fr, the minimal IF produced on Tri-A3, Tri-C4, and Tri-C5 is 0.22, 0.12, and 0.02, respectively In a mainhull trimaran's transom, there was a significant interference impact, as seen by lower interference under specific conditions. Further testing at a more detailed position and speed should be investigated in order to acquire more comprehensive data.

5 References

[1] Dubrovsky, Victor A. 2016. "*Multi-hull ship, Specificity and designing of multi-hull ships & boats*", Nova Science Publishers, ISBN 9781634846158, USA, 2016.

[2] Bertorello C, Bruzzone D, Cassella P, Zotti I. 2001. *Trimaran Model Test Results and Comparison with Different High Speed Craft*. Proceedings of the Eighth International Symposium on Practical Design of Ships and Other Floating Structures 16 – 21 September 2001 Shanghai,Chaina. Volume I, 2001, Pages 143-149.

[3] Poundra G.A.P., Utama I.K.A.P, Hardianto, D.,and Suwasono, *Optimizing trimaran hull configuration based on resistance and seakeeping criteria,* in Procedia Engineering,2017,doi:10.1016/j.proeng.2017.08.124.

[4] Luhulima, R.B., Sutiyo, Alia, M.R., Utama, I.K.A.P., 2021. *Experimental Investigation into the Resistance Characteristics of Trimaran and Pentamaran Configurations*. International Journal of Technology. Volume 12(5), pp. 1058-1070

[5] Doctors L, A. Day, *Resistance prediction for transom-stern vessels*, in: Proc. 4th Int. Conference on Fast Sea Transportation (FAST'97), Australia, vol. 2, 1997, pp. 743–750.

[6] Utama, I.K.A.P., Setyawan, D., Jamaluddin, A. dan and Murdijanto (2010), *Experimental and CFD Investigation into the Drag Characteristics of Catamaran Fishing Vessel,* Proc. Regional Conference on Mechanical and Aerospace Technology (RCMeAe), Bali, Indonesia.

[7] Andersson B, Andersson R, Hakansson L, Mortensen M, Sudiyo R, Wachem BV. 2012. *Computational Fluid Dynamics for Engineers.* Cambridge University Press, ISBN 978-1-107-018952. United Kingdom.

[8] Mynard T, P. Sahoo, J. Mikkelsen, D. McGreer, 2008. *Numerical and Experimental Study of Wave Resistance for Trimaran Hull Forms*, Australian Maritime College, Australia, pp. 117–132.

[9] Utama I.K.A.P., Setiawan D. 2014. *Experimental and Numerical Investigation into the Effect of Water Depth on the Resistance Components of Multihull Configuration*. ICSOT Indonesia. 4-5 November 2014; Makasar.

[10] Utama, I.K.A.P., Sutiyo., Suastika I.K., 2021. *Experimental and Numerical Investigation into the Effect of the Axe-Bow on the Drag Reduction of a Trimaran Configuration*. International Journal of Technology 12(3) 527-538 (2021). doi: 10.14716/ijtech.v12i3.4659.

[11] Menter, F. R. 1993, "*Zonal Two Equation k-ω Turbulence Models for Aerodynamic Flows*", AIAA Paper 93-2906.

[12] Ferziger, J.H., Peric, M., 2002. *Computational Methods for Fluid Dynamics, third rev*. Springer, Berlin.

[13] ITTC, 2011. *Practical guidelines for ship CFD applications*. In: Presented at the Proceedings of 26th ITTC (Rio de Janerio, Brazil.).

[14] Hafez K.A., El-kot A.A. (2012). *Comparative Investigation of the Stagger Variation Influence on the Hydrodynamic Interference of High Speed Trimaran*. Alexandria Engineering Journal 51, pp.153-169.

IOP Conf. Series: Earth and Environmental Science 1081 (2022) 012057 doi:10.1088/1755-1315/1081/1/012057

Self propulsion Performance Analysis of Japanese Bulk Carrier with PBCF Based CFD Method

Achdri Fauzi Nugraha Oloan[1], I Made Ariana[1] and Ahmad Baidowi[1]

Institut Teknologi Sepuluh Nopember, Marine Technology, Surabaya
Achdri16@mhs.ne.its.ac.id, ariana@its.ac.id, ahmadbai@gmail.com

Abstract. An energy-saving device (ESD) is used to obtain energy lost and convert additional energy to a propulsive coefficient. This research uses PBCF as an energy-saving device that is attached to an MP687 propeller and a Japanese bulk carrier a hull. Method for solving this research using the open water test and self propulsion method based on computational fluid dynamic (CFD). Added propulsive coefficient obtained from PBCF conventional can reach up to +1.52 %. PBCF with duct make smaller added for propulsive coefficient caused by duct splitting fluid flow from the propeller. PBCF with duct can increase propulsive coefficient up to +0.48 %. PBCF can eliminate a strong hub vortex at propeller MP687 and its ba e good choice for the increased propulsive coefficient of the ship

1. Introduction

Environmentally friendly design in the future development of ship design to achieve the goals of reducing greenhouse gases. The propulsion system contributed 5 ~ 15 % potential to reduce greenhouse gases. An old ship can use a retrofit project that adds Energy Saving Devices (ESD) to achieve this regulation. Additional ESD to existing ships is important to increase the propulsive coefficient. The propulsive coefficient can be obtained from hull efficiency, rotative relative efficiency, and propeller efficiency. This research uses a Japanese Bulk Carrier (JBC) with a single screw propeller. Range value hull efficiency for single screw 1.1 until 1.3, rotative relative efficiency value 1.0 until 1.07, and propeller efficiency typically at 0.55 until 0.70 [6]. To make a good decision can using starting with the propulsive coefficient for this paper using PBCF. PBCF zone of operation at II with function to attempt to recover energy lost from propeller [1]. Research of PBCF can be focused on hub vortex and increasing propeller efficiency. PBCF can diffuse and eliminate with changing flow at propeller boss to make propeller efficiency increase [8] and then make more propeller efficient, and then make more propeller efficient, PBCF can be considered and well match with propeller pitch distribution [9]. Investigation continues for PBCF and gives results it effectively prevents forming hub vortex that bigger contributor to increasing propeller efficiency [10]. Optimization procedure through the numerical procedure on PBCF produces ta total net propulsion system of 2.5 % thrust and 1.39 % torque when compared without PBCF [7]. The optimization of the open water test with PBCF installed on the KP505 propeller achieves increased efficiency at +0.9 %, while the E779A propeller achieves increased efficiency at +4 % [2]. For the latest design of PBCF, the mini duct is added to the design parameter. Conventional model PBCF produces an efficiency of up to +2.2 %, but when it is compared to PBCF with installed duct, it can produce a smaller efficiency with up to +1.8 % [3].

IOP Conf. Series: Earth and Environmental Science 1081 (2022) 012057 doi:10.1088/1755-1315/1081/1/012057

This can happen because the contribution of ducted PBCF gains positive thrust from the flowfield and is not effective for balancing the additional torque from rotation, since pressure field modification by acceleration flow, then negative interaction with hub shape can make decreasing of hub resistance that compared to conventional PBCF and make slightly decreasing at performance than ducted PBCF [3].

2. Method

Method for solving this research using open water test and self propulsion test. Experimental data and design parameters were taken from NMRI EFD towing test model. The model consists of two parts, there are propeller MP687 and Ja Japanese bulk carrier for ships. The simulation was taken to solve this research to compare EFD vs CFD performance and with or without PBCF.

2.1. Open Water Test

Simulation CFD for open water test using 3 models (Propeller MP687 Original, Propeller MP687 Conventional PBCF, and Propeller MP687 with duct). Detail **Figure 1.** shows the basic parameter for running CFD solver with an open water test. Size of the domain can be found starting with 2D (from propeller back to the shaft), starting from propeller face to wal,l and 3D from propeller to the upper boundary and lower boundary when all of the D values take from propeller diameter.

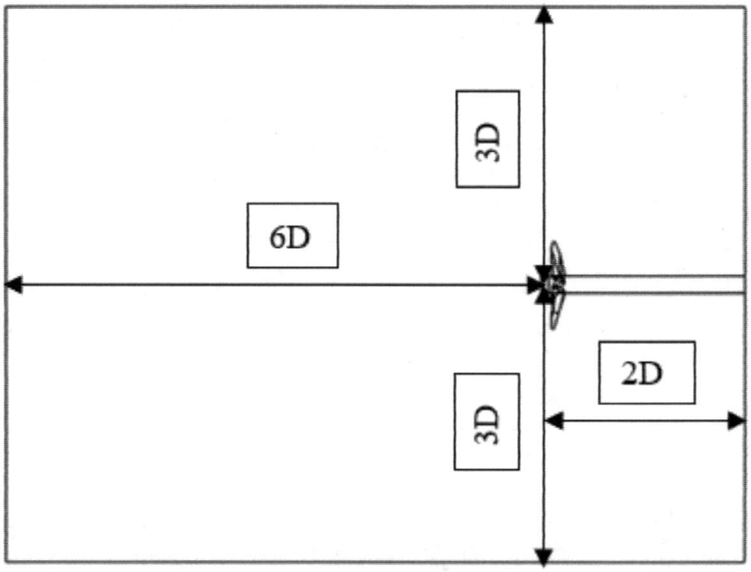

Figure 1. Open water Test Configuration CFD

Table 1. Principal Dimension MP687 Propeller

Description	Item
Type	MP687
Blade	5
Dprop	8.12m
Revolution	7.8rps

The result simulation open water test for the validation test can be seen in **Table 2.** That shows validation of propeller MP687 between experiment fluid dynamic (EFD) vs propeller MP687 computational fluid dynamic (CFD). Thrust result average error for validation about 4 %. Bigger error gaps are located at advance ratio (J) 0thanthen a smaller error gap is located at advance ratio (J) 0.50. Torque average error value at 3 %. Bigger error gaps located at advance ratio (J) 0.50 than a smaller error ga located at advance ratio (J) 0.10.

Table 2. Validation Propeller MP687

Advance Ratio	KT EFD	KT CFD	KT Error	10*KQ EFD	10*KQ CFD	KQ Error
J	Thrust	Thrust	%	Torque	Torque	%
0.10	0.3267	0.3095	5.272	0.3748	0.3802	1.432
0.20	0.2949	0.2809	4.749	0.3500	0.3571	2.032
0.30	0.2598	0.2490	4.144	0.3210	0.3311	3.132
0.40	0.2214	0.2135	3.560	0.2871	0.3008	4.786
0.50	0.1798	0.1741	3.146	0.2479	0.2649	6.857
Average	0.2565	0.2454	4.174	0.3162	0.3268	3.648

Table 3. Propeller MP687 + Conventional PBCF

Advance Ratio	KT CFD	KT PBCF	KT Gain	10*KQ CFD	10*KQ PBCF	KQ Gain
J	Thrust	Thrust	%	Torque	Torque	%
0.10	0.3095	0.3113	+0.584	0.3802	0.3781	-0.540
0.20	0.2809	0.2824	+0.537	0.3571	0.3552	-0.548
0.30	0.2490	0.2503	+0.500	0.3311	0.3291	-0.591
0.40	0.2135	0.2142	+0.324	0.3008	0.2985	-0.762
Average	0.2454	0.2645	+0.486	0.3268	0.3402	-0.610

Table 3. Reflect that shows performance changing between propeller MP687 original vs MP687 with conventional PBCF. Thrust result change about +0.486 % average adding to propeller MP687 performance. Bigger added performance for thrust located at advance ratio (J) 0.10 then a smaller added performance located at advance ratio (J) 0.40. Torque result changes about -0.610 % on average adding to propeller MP687 performance. Bigger reduced performance for torque located at advance ratio (J) 0.40 than a smaller reduced performance located at advance ratio (J) 0.10.

Simulation result PBCF with added duct **Table 4.** Represent that shows performance changing between propeller MP687 original vs MP687 with ducted PBCF. Thrust result change about -0.013 % average decrease to propeller MP687 performance. A bigger decrease performance for thrust located at advance ratio (J) 0.40 than a bigger added performance located at advance ratio (J) 0.10. Torque result change about -0.452 % average decrease to propeller MP687 performance. Bigger reduced performance for torque located at advance ratio (J) 0.10 than a smaller reduced performance located at advance ratio (J) 0.40.

Table 4. Propeller MP687 + Ducted PBCF

Advance Ratio	KT CFD	KT PBCF	KT Gain	10*KQ CFD	10*KQ PBCF	KQ Gain
J	Thrust	Thrust	%	Torque	Torque	%
0.10	0.3095	0.3097	+0.085	0.3802	0.3782	-0.515
0.20	0.2809	0.2811	+0.059	0.3571	0.3556	-0.414
0.30	0.2490	0.2489	-0.038	0.3311	0.3297	-0.404
0.40	0.2135	0.2132	-0.160	0.3008	0.2994	-0.476
Average	0.2454	0.2632	-0.013	0.3268	0.3407	-0.452

After the open water test, we can get results for all propeller MP687 efficiency. A full comparison can be seen in **Table 4.** 2 models compared efficiency performance (PBCF Conventional vs PBCF with duct) which results from the open water test. The validation EFD vs CFD result test can be shown that bigger error efficiency located at advance ratio J = 0.40, smaller error efficiency at J = 0.10, and a total average error of 7.00 %. The final result for comparison of efficiency PBCF conventional vs PBCF with duct can be shown average gap takes 0.65 %. PBCF conventional make good result rather than PBCF with a duct. PBCF with duct make decrease efficiency at propeller MP687.

Table 5. Propeller MP687 ETA Comparation

Advance Ratio	ETA EFD	ETA CFD	ETA Error	ETA PBCF	ETA Duct PBCF	ETA Conv Vs Duct
J	η	η	%	η	η	Gap %
0.10	0.1387	0.1296	6.54	0.1311	0.1304	-0.52
0.20	0.2681	0.2505	6.56	0.2532	0.2517	-0.61
0.30	0.3864	0.3594	7.00	0.3633	0.3607	-0.72
0.40	0.4909	0.4521	7.91	0.4570	0.4535	-0.77
Average	0.3210	0.2979	7.00	0.3012	0.2991	-0.65

2.2. Self Propulsion Test

Self propulsion test made by towing tank or numerical simulation that includes hull and propeller [4]. Self propulsion test has a measuring system, there are model speed, resistance or external tow force, propeller thrust, propeller torque, propeller rate of revolutions, duct/pod thrust, pod unit side force, water temperature, and sinkage force and aft [5].

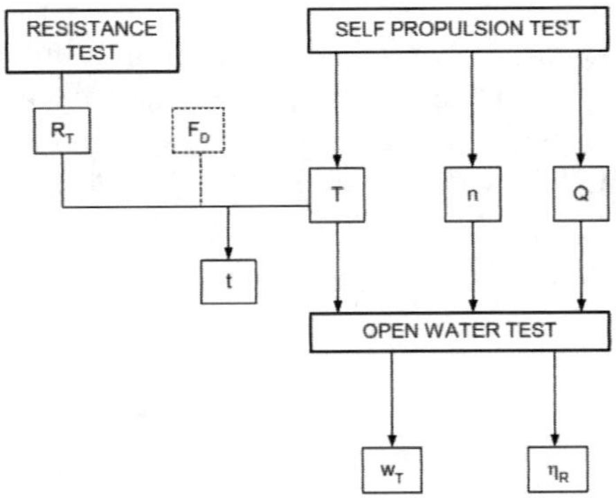

Figure 2. Self Propulsion Test Diagram [5] (ITTC, 2008)

2.2.1. Propulsive Coefficient

The propulsive coefficient has a relation between effective power for driving vessels which contains products from ship resistance, ship speed, and power delivered from engine to propeller [1]. The equation for searching the propulsive coefficient can be seen below:

$$\eta_D = \eta_H \times \eta_O \times \eta_R \quad (1)$$

Table 6. Propeller MP687 Propulsive Coefficient

Item Parameter	PC EFD	PC CFD	PC PBCF	PC PBCF+Duct
No.	ηD	ηD	ηD	ηD
1	0.748	0.762	0.774	0.766

2.2.2. ETA Hull

This efficiency defines the relationship between the thrust deduction coefficient and the wake fraction coefficient. The hull efficiency becomes larger by reducing the power required to propel the ship if the thrust deduction coefficient (t) is smaller than the wake fraction coefficient (w) [6]. The equation shown below is valid for single screw and symmetric crew ships which case twin-screw propeller is the sum of the thrust and torque that mead of rpm at the port and starboard value different [5].

$$\eta_H = \frac{(1-t)}{(1-w_T)} \quad (2)$$

Table 7. Propeller MP687 ETA Hull Comparation

Item Parameter	ETA EFD	ETA CFD	ETA PBCF	ETA PBCF+Duct
No.	ηH	ηH	ηH	ηH
1	1.471	1.583	1.585	1.600

2.2.3. ETA Relative Rotative

The relative rotative efficiency can be found by dividing torque the coefficient obtained from the open water test and the torque coefficient obtained from the self propulsion experiment [5]. Normally ships with a single propeller gain relative rotative efficiency from around 1 until 1.07 but twin screw propellers with conventional hulls gain less than around 0.98 [6].

$$\eta_R = \frac{K_{QT}}{K_Q} \tag{3}$$

Table 8. Propeller MP687 ETA Relative Rotative Comparation

Item Parameter	ETA EFD	ETA CFD	ETA PBCF	ETA PBCF+Duct
No.	ηr	ηR	ηR	ηR
1	1.015	1.125	1.128	1.129

2.2.4. ETA Open Water

Efficiency open water is a relationship between thrust coefficient (KT), torque coefficient (KQ), and advance coefficient (J). The value of propeller efficiency can reach between 0.35 and 0.75 depending on the advanced ratio location [6].

$$\eta_O = \frac{P_T}{P_D} = \frac{T \times V_A}{Q \times 2\pi \times n} = \frac{K_T}{K_Q} \times \frac{J}{2\pi} \tag{4}$$

Table 9. Propeller MP687 ETA Open Water Comparation

Item Parameter	ETA EFD	ETA CFD	ETA PBCF	ETA PBCF+Duct
No.	ηO	ηO	ηO	ηO
1	0.501	0.427	0.433	0.424

2.3. Energy Saving Device

The energy-saving device is a device that is attached based on 3 zones of the hull: zone 1 which works on the growth boundary layer on the sternship with aim of getting a direct advantage or optimizing fluid flow to the propeller, zone 2, and zone 3 work on hull wake and slipstream propeller area to restore energy may be lost [1].

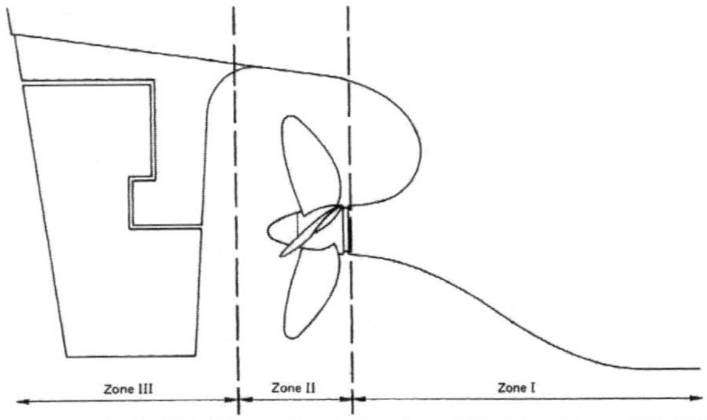

Figure 3. ESD Zone Installation [1] (Carlton, 2007)

For PBCF, can be installed on zone 2 with the propeller. The general principle of PBCF work with small fins formed from flat plates. The number of fins itself depends on the number of propeller blades. The function of the fins reduces the hub vortex and increases the kinetic energy of the rotating flow. At this point, the fins increase the propeller efficiency. PBCF was created by Mr ouchi with the purpose to increase propeller efficiency by reducing the value of energy losses from the vortex generated by the propeller while working [1]. Detailed visualization of different working propeller characteristics before and after PBCF installation can be seen in **Figure 4.**

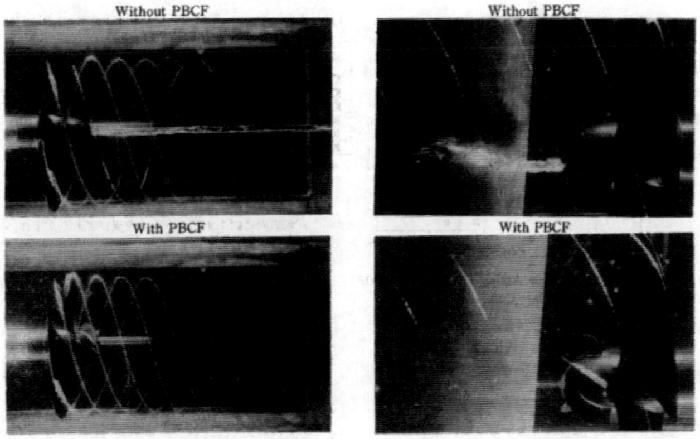

Figure 4. Propeller With and Without PBCF [8].

3. Result and Discussion

The simulation parameter for the self-propulsion test provides a propulsive coefficient value. Solving with computational fluid dynamic to generate KTself, 10*Kqself, wake fraction, and thrust deduction. To get a better value, the impact of PBCF conventional vs PBCF with the ducting need to be compared.

3.1 Model Parameter

Before running the Computational Fluid Dynamic at self propulsion, domain size needs to be adjusted based on AP ships. Domain size needs to be the same domain size for each model and meshing parameter with or without PBCF. For domain, size can see the detailed in **Figure 5.**

IOP Conf. Series: Earth and Environmental Science 1081 (2022) 012057 doi:10.1088/1755-1315/1081/1/012057

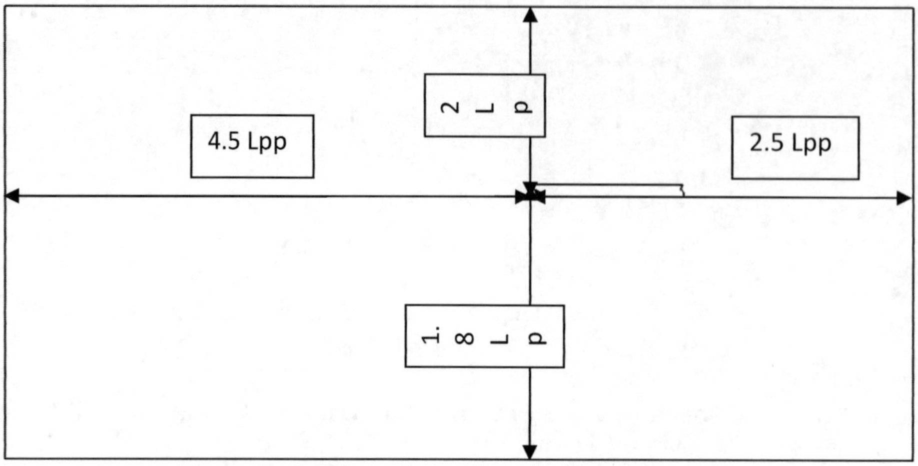

Figure 5. Self Propulsion Test Configuration CFD

Table 10. Propeller Self Propulsion Result

Item Parameter	JBC EFD	JBC CFD	JBC PBCF	JBC PBCF+Duct
KTself	0.217	0.2241	0.2248	0.2246
10*KQself	0.279	0.2754	0.2727	0.2738
ETAself	0.509	0.4812	0.4880	0.4813
1-t	0.812	0.7940	0.7910	0.7920
1-w	0.552	0.5010	0.5010	0.5050

Based on the simulation result, detailed in **Table 10.** Describe thrust coefficient from PBCF conventional get increase up to +0.31 %, torque get decrease up to -0.98 % and ETAself get increase +1.41 %. Then PBCF with ducted get thrust coefficient increase up to +0.19 %, torque get decrease up to -0.56 % and ETAself get increase +0.03 %. It can be shown that PBCF works properly at Japanese bulk carriers. Wake fraction gap value between JBC EFD vs JBC CFD caused by the advance coefficient location. Advanced coefficient at JBC EFD have around 0.4, then for JBC CFD at 0.37 so that can influence wake fraction. Wake fraction from JBC PBCF conventional can make wake fraction decrease up to -0.11 % and for JBC PBCF + duct can make wake fraction increase up to +0.72%. for thrus deduction, JBC PBCF conventional can make decrease thrust deduction by around -0.31 % when compared with JBC CFD. Then JBC PBCF + duct can make a decrease thrust deduction of around -0.19 % when compared with JBC CFD.

3.2 CFD Analysis Result

From this self-propulsion result at Japanese bulk carrier, propeller MP687 with integration PBCF conventional and PBCF with duct can be shown vortex flow among propeller boss cap. This detailed view takes from the isometric view of a Japanese bulk carrier hull. The following is a various fluid flows based on the PBCF model.

IOP Conf. Series: Earth and Environmental Science 1081 (2022) 012057 doi:10.1088/1755-1315/1081/1/012057

Figure 6. PBCF Location Model; (a) JBC Original, (b) JBC with PBCF Conventional and (c) JBC with PBCF + Duct

Based on the result of the flow pattern on the propeller boss. **Figure 6,** explains that the vortex hub gathers on the propeller boss, so it can inform a strong flow pattern in the middle of the propeller boss. **Figure 6a,** represents a strong hub vortex forming into the propeller boss. After installation of PBCF **Figure 6b**, the Strong hub vortex at the centre of the propeller boss can be eliminated which was originally had a small red circular become disappear and change to be green colour. **Figure 6c,** representative working characteristics from PBCF with duct can be seen that the hub vortex is split by duct. It can reduce conversion rate energy and make a small increase in propulsive coefficient when compared with PBCF conventional. Value of increase propulsive coefficient from PBCF conventional have reached about +1.52 %, bigger than PBCF with duct can reach additional propulsive coefficient about +0.48%.

4. Conclusion

PBCF that using energy-saving can make increasing propulsive coefficient at Japanese bulk carrier, especially MP687 propeller performance. From the open water test, can be seen that PBCF can make open water test efficiency increase. PBCF conventional can make increase efficiency by average about +0.30 % and PBCF with duct can make increase efficiency by about +0.29 %. The efficiency gap average is +0.65 % between PBCF conventional vs PBCF with a duct. From the self-propulsion test, PBCF conventional can make increase value at propeller thrust about +0.31 %, torque get decrease at -0.98 % and ETA self makes increase +1.41 %. PBCF with duct can make increase value at propeller thrust about +0.19 %, torque get decrease at -0.65 % and ETA self makes increase +0.03 %. Propulsive coefficient value for PBCF conventional has increment up to +1.52 %, bigger than PBCF with duct gain up to +0.48 % of propulsive coefficient. From this research, we can conclude PBCF can eliminate the hub vortex at propeller MP687 because the strong hub vortex at first has a small red circle at the hub that can be disappeared and change into green colour. Future studies of PBCF can be taken for better design, especially for increased ship propulsive coefficient.

5. References

[1] Carlton, J.S., 2007. *Marine propellers and propulsion,* Second edition. ed. Butterworth-Heinemann, is an imprint of Elsevier, Oxford.

[2] Gaggero, S., 2018. *Design of PBCF energy saving devices using optimization strategies: A step towards a complete viscous design approach*. Ocean Engineering 159, 517–538. https://doi.org/10.1016/j.oceaneng.2018.01.003

[3] Gaggero, S., Martinelli, M., 2021. *Comparison of different propeller boss cap fins design for improved propeller performances*. Applied Ocean Research 116, 102867. https://doi.org/10.1016/j.apor.2021.102867

[4] Gokce, M.K., Kinaci, O.K., Alkan, A.D., 2019. *Self-propulsion estimations for a bulk carrier. Ships and Offshore Structures* 14, 656–663. https://doi.org/10.1080/17445302.2018.1544108

[5] ITTC, 2008. *ITTC-Recommended Procedures and Guidelines Testing and Extrapolation Methods Propultion, Performance Propultion Test*. ITTC.

[6] MAN BW, 2018. *Basic Principles Of Ship Propulsion*. Man Energy Solution, Denmark.

[7] Mizzi, K., Demirel, Y.K., Banks, C., Turan, O., Kaklis, P., Atlar, M., 2017. *Design optimisation of Propeller Boss Cap Fins for enhanced propeller performance*. Applied Ocean Research 62, 210–222. https://doi.org/10.1016/j.apor.2016.12.006

[8] Ouchi, K., Kawasaki, T., Tamashima, M., Koizuka, H., 1989. A Research and Development of PBCF (Propeller Boss Cap Fins) - 2nd Report : *Study on Propeller Slipstream and Actual Ship Performance*.

[9] Ouchi, K., Tamashima, M., Koizuka, H., Arai, K., 1992. *A Study on Correlation between Propeller Pitch Distribution and Improvement of Propeller Efficiency* by PBCF. The Japan Society of Naval Architect and Ocean Engineering.

[10] Seo, J., Lee, S.J., Han, B., Rhee, S.H., 2016. *Influence of Design Parameter Variations for Propeller-Boss-Cap-Fins on Hub Vortex Reduction*. Journal of Ship Research 60, 203–218. https://doi.org/10.5957/JOSR.60.4.160003

IOP Conf. Series: Earth and Environmental Science 1081 (2022) 012058 doi:10.1088/1755-1315/1081/1/012058

Risk Evaluation Concept of Engine Room Module Installation on Modular Construction Shipbuilding

Intan Baroroh[1,a,*], I Made Ariana[1,b] and AAB Dinariyana[1,2,c]

[1]Departement of Marine Engineering, Institut Teknologi Sepuluh Nopember (ITS), Surabaya
[2]Center of Excellence in Maritime Safety and Marine Installation (PUI KEKAL), ITS Surabaya
a. intan.baroroh@hangtuah.ac.id, b. ariana@its.ac.id, c. kojex@its.ac.id
*corresponding author

Abstract. The implementation of the construction of modular construction ships is able to accelerate ship production by up to 50%. However, in practice, delays in the production process can still be found in the engine room module of the modular ship building process. Production delays occur due to inappropriate block dimension accuracy, damaged machine conditions, loading blocks that are not on schedule, many revisions to drawings, delays in equipment and basic ship materials. The production process includes a series of activities carried out including: Hull construction, Painting & Protection, Outfitting Manufacture, Machinery Outfitting, Accommodation Outfitting, Electric, Electrical Outfitting. This research aims to evaluate the performance of engine room module installations in shipyards that apply a modular construction system. At an early stage, the research was conducted by identifying factors and constraints that affect the performance of the engine room installation in shipbuilding which has the potential to cause delays. The conceptual and technical approach are carried out by analyzing the causal relationship of the identification of factors at the production stage. The results of this research is expected to anticipate earlier delays in the engine room module installation with the concept of evaluation using Bayesian Network integration and System Dynamics.

1. Introduction

According to the National shipyard in Surabaya city, there are still many delivery delays. The data was obtained from three national shipyards, for example PT. DM, PT. DS and PT.XYZ (2011-2019). PT. XYZ in 2011 – 2015 experienced delays in 5 vessels, two vessels in the form of a 17,500 Long Ton Deadweight Tanker, two 2400 Horse Power Tug boats and a Strategic Sealift Vessel Philippines Navy ship with a delivery time delay of 0.5 to 6 months. PT. DM Shipyards has experienced delays in 7 vessels from 2015 to 2020, there are 4 ship constructions and a 3500 Deadweight Ton Hull No 111 tanker for 7 months. In 2015 the Fast Patrol Boat 60 meters' series A and B experienced one-month delay in the delivery of orders from the Directorate General of Customs and Excise, Ministry of Finance. In 2018 the 100 TEUs container ship was delayed for one month with an addendum of 4 months, in 2014 the 500 gross tonnage ferry ship was for one month and in 2015 the 2000 gross tonnage pioneer ship was for one month. PT. DS Shipbuilding in the period 2011 to 2015 there was delay in the construction of 7 vessels, including the production of Landing Craft Tanks, Landing Craft Tanks of 100 TEUS, Tanker 6,500 Deadweight Ton, Self-Propelled Barge Cement Carrier.

Delays in ship delivery at national shipyards Surabaya city were due to the handover delay which was influenced by many obstacles such as inappropriate block dimension accuracy, arrival of images not in sequence, damaged engine conditions, loading blocks that were not on schedule, many image revisions and equipment delays. as well as the ship's basic material. Although a modular shipbuilding system has been developed which is proven to be able to shorten the time. Hunt [1] identified cost reductions during production and modernization in modular shipbuilding as the technology of the future by shortening the production process which means it will also reduce costs. Increasing the speed of the ship's production process can reduce the time to about 50% or even up to 75% [2]. Research on ship modularity has been carried out by [3-10], regarding the advantages and flexibility in the design and production of modular ships. However, the national shipyards Surabaya city still have many problems with ship delivery delays. At the beginning of the project losses from delays can actually be reduced or anticipated if the risk management process can be implemented and executed properly in the shipbuilding project. At PT.XYZ, there is already a special section that handles risk management, but why are there still many delays in ship delivery which have the potential to cause considerable losses. This shows that the application of risk management principles and risk assessment as well as how to mitigate these risks have not been carried out properly as part of risk anticipation. Risk assessment research in the maritime sector has been carried out by [11-18], in this study the risk probability value was measured.

Currently PT.XYZ is building a BRS ship (Hospital Assistance Vessel) which keel will be laid on September 16, 2019 and the handover plan is September 30, 2021, however, the handover has been delayed until December 8, 2021. On the BRS ship, there are 6 starting points or zones in the shipbuilding system of 123 blocks, according to Table 1.

Table 1. Name of the zone consisting of the number of blocks, the duration of the process and the weight.

Zone name	Number of blocks	Duration (days)	Weight (tons)
Cargo Hold	23	207	818,5
Engine Room	28	433	1.009,2
Fore Part	14	322	300,2
After Part	21	187	684,2
Wheel House	22	407	662,8
Hangar	15		

The weight of the six zones is calculated based on the weight of the blocks that make up each of these zones, the process of which is carried out according to the erection network and equipment.

The integration of all existing systems in the hull has a center in the engine room, so that the engine room is a critical part of the hull construction process. By looking at the number of blocks that make up the engine room zone, which at most reaches 28 blocks among other zones, the longest time duration reaches 76% of the total ship construction and the heaviest zone reaches 30% of the ship's block weight and the complex integrated system in it, make the engine room the most urgent zone in the object of research in the shipbuilding process. So that during the process of installing the engine room module, it is necessary to observe the obstacles that occur so that the engine room zone which is a critical zone in the process of building the hull does not experience delays in the production duration considering that the longest time duration is 76% of the total ship construction.

The contribution of this research is to create a new method of installing the engine room module in the construction of a modular ship system by observing and evaluating the factors that influence shipbuilding delays.

2. Methodology

The method applied in the risk evaluation uses a conceptual approach and technical problems, namely the Bayesian integration modelling approach and SD based on systemic thinking and using a perspective based on information feedback and delays, to understand the dynamics of the physical behaviour of complex systems that occur in the shipbuilding of modular system on installation engine room module. PT.XYZ is the shipyard chosen to be the research work area. The methodology developed is as follows figure 1.

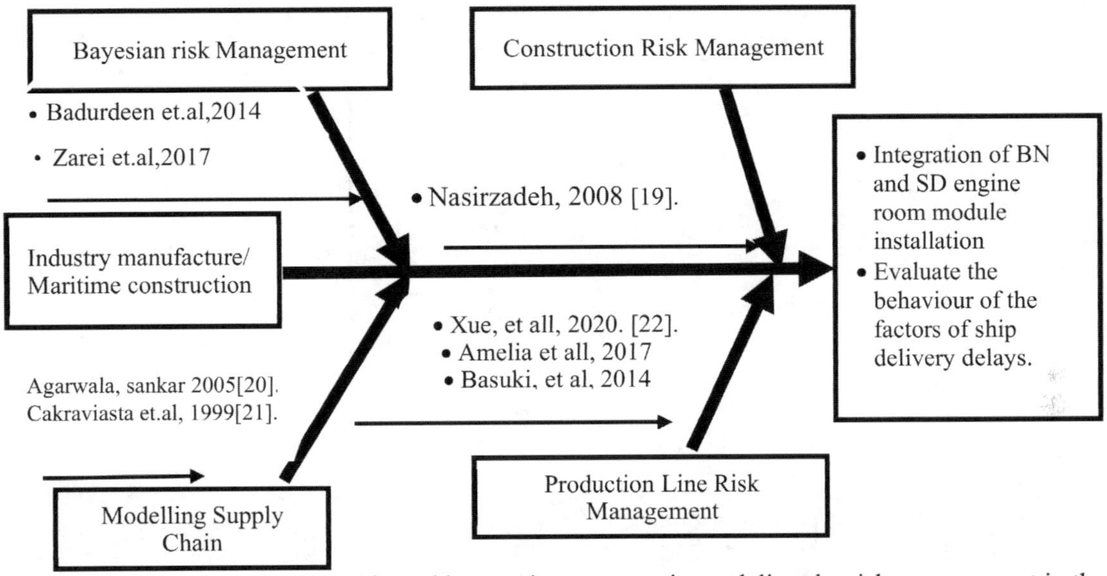

Figure 1. Research methods conducted by previous experts in modeling the risk management in the maritime industry.

In Figure 1, several research methods have been carried out by experts in modelling risk management in the maritime industry. However, no one has developed risk management with the installation of the engine room module, shipbuilding modular systems in shipyards. The novelty in this research is to apply the evaluation of the risk assessment model using BN integration and SD to build a conceptual model and a simulation model with complex constraint factors to understand the behavior of a complex system engine room module in modular construction shipbuilding.

Identification of the risk management evaluation of engine room installation with construction modular. It was carried out through literature study and interviews with stakeholder experts, practitioners and policy makers as shown in Figure 2. The data collected consisted of detailed observations, interviews with respondents from the Supply Chain department, hull construction, machinery outfitting and electrical outfitting. Formulation of the model with simulation SD program Stella 9.1.3 and BN to determine the level of the dominant variable. Bayesian consists of a set of variables (cause and effect) and conditional probabilities represent the strength of the relationship between cause and effect. BN is a technique for depicting probabilistic relationships between model elements, where objective data is lacking and the use of expert opinion is required. However, Bayesian is inadequate to capture dynamic aspects including feedback loops and delays. Meanwhile, SD captures the dynamics of the interaction of the factors that cause uncertainty in the system. Thus, BN models alone may not be enough for risk analysis. Dynamic causal relations can be modeled well using simulation tools such as SD.

Evaluating the system under several scenarios, realistic and far-fetched, can help gain further insight into the behavior of the system and enable companies to prepare for radical or extreme situations [23]. Flowchart of risk evaluation of engine room installations in shipyards can be seen in Figure 2.

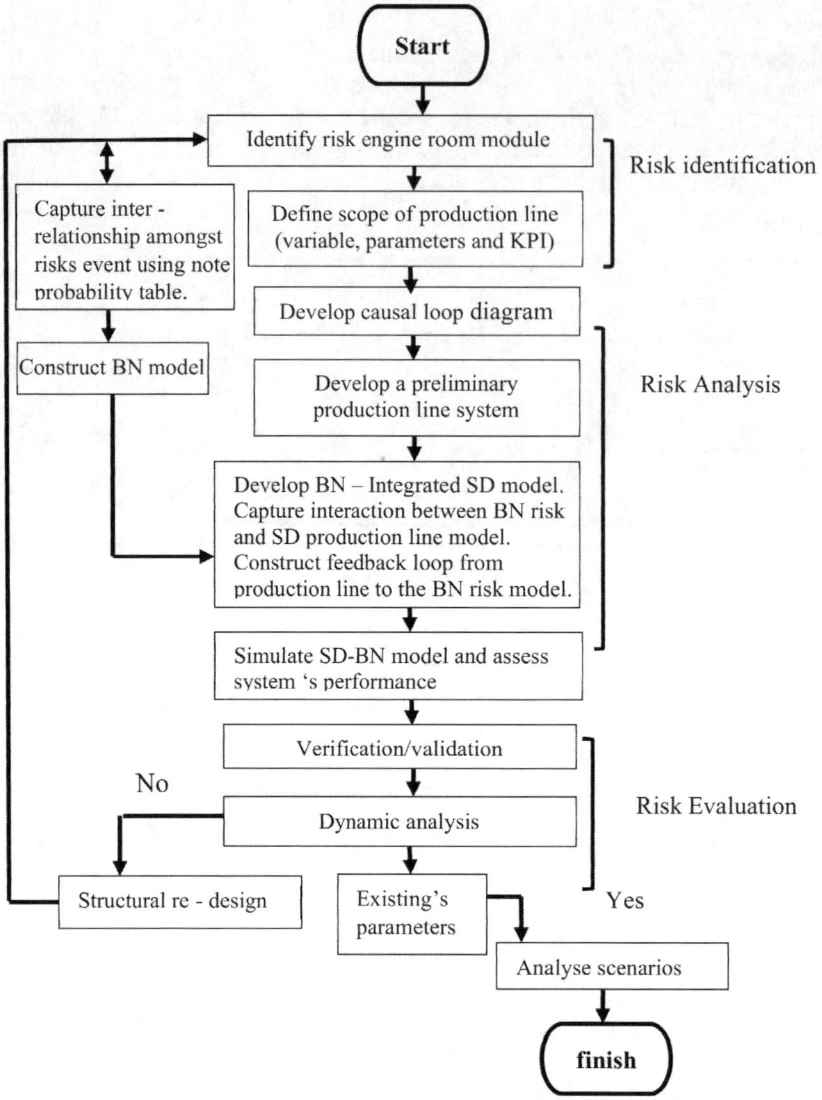

Figure 2. Flow Chart of risk evaluation for engine room installations on shipyards [23].

When there is sufficient information to establish "deterministic" relationships between model factors or parts thereof simulation-based techniques, such as SD are solutions to complex models and analytical solutions are either too time consuming. The SD approach demonstrates a significant ability to model specific decision-making processes and system behaviour. SD's strength is its ability to account for non linearity in dynamics, feedback, and time delays, but is unable to capture the probability values that exist in the system. By combining SD with Bayesian can compensate for the shortcomings of both. SD describes the underlying dynamic causal mechanisms in complex systems and the BN captures the uncertainties involved in its elements. As an application, the SD - BN methodology is integrated with probabilistic risk analysis to enable a complex system risk analysis framework by capturing the dynamics of the interaction of factors causing uncertainty in the system. The model is capable of capturing dynamic nature of variables within the system through SD and BN captures inter-relationships and uncertainty in risk events [24].

Development strategy of the engine room installation for ship building modular systems in a complex manner with various obstacles, a dynamic approach is needed to solve the problems.

IOP Conf. Series: Earth and Environmental Science 1081 (2022) 012058 doi:10.1088/1755-1315/1081/1/012058

3. Discussion

3.1. Risks Identification of engine room module installation.

The ship production process with a modular system emphasizes the parallel development of ship construction with a total of 6 starting points from the modules used. The production process includes a series of activities including: Zone 1 Cargo Hold, Zone 2 Engine room, Zone 3 Fore Part, Zone 4 After Part, Zone 5-wheel house and Zone 6 hangar. In the implementation of Product-oriented Work Breakdown Structure to increase productivity in bulk, it is carried out with a batch production system, namely producing the same blocks in one batch and assigned to different shipyards. There is one carrier shipyard to assemble (hull erection), it means there is one main shipyard that has the function of assembling (hull erection), ship blocks become the ship's hull and made by other supporting shipyards. The quality between blocks has a different quality of work because it is carried out by various shipyards.

The engine room installation activities consist of Hull construction, Painting & Protection, Outfitting Manufacture, Machinery Outfitting, Accommodation Outfitting, Electric, Electrical Outfitting. In the installation of the engine room using a series parallel system means that production is carried out at different start jobs but can be done at the same time with different facilities, so that the time becomes faster than the normal time. Meanwhile, the completeness of the main engine system includes: starting air system, fuel oil system, lubrication oil system, cooling system and exhaust.

The identified hazard is collecting data obtained from the construction process of the Hospital Auxiliary Ship (BRS TNI AL). Especially in zone 2 at PT.XYZ in the form of a hazard in the Engine room installation process with the construction number W000302. The engine room module installation includes the activities of the Machinery Outfitting (MO) & Electrical Outfitting (EO) Department in charge of Machinery Outfitting workshops, Electric workshops, Electrical Outfitting, Piping workshops and Steel Work workshops. The Hull Outfitting & Accommodation Outfitting (HO/AO) department is in charge of Outfitting Manufacture, Accommodation Outfitting and Painting & Protection workshops. The Hull Construction Department oversees the Hull Construction workshop. The three departments are very close which can be shown in the linkage of the workshops for the engine room installation process. The relationship between each workshop from the sample room of the engine room module installation has a potential hazard probability that affects the risk and level of risk production process in the workshops. The identification of internal and external risks in each activity in the workshop can be identified as follows Table 2.

3.2. Integrase Bayesian Network with System Dynamics on engine room module installation.

Risk identification is the most important step for risk assessment. Identifying relevant with potential risks at the production process level forms of continuing strong foundation, risk analysis and risk evaluation stages. One of the most important tasks defined the boundaries of risk will be assessed. It's easy to deviate from production process level risk to industrial level risk. The best to define these boundaries by discussing with the team to assess the risk and consensus on the risk assessment scope.

BN are powerful probabilistic approach and often used for reasoning, diagnosis, prediction, decision making under uncertainty. BN 's ability to exploit quantitative and qualitative data to generate posterior probabilities are very helpful in the area of risk assessment. The conditional probability is the probability that the child node C occurs due to the occurrence of the parent node Pt is true and is denoted as P(Pt |C). The child note C is very dependent on the parent note Pt. The relationship between the nodes in BN is fundamentally based on Bayes' theorem and can be expressed as follows:

$$P(Pt \mid C) = (P(C \mid Pt) * P(Pt))/(P(C)) \tag{1}$$

where, $P(Pt \mid C)$ is the conditional probability of occurrence of the parent node (Pt) given that the child node (C) occurs. Similarly, P(Pt) and P(C) are the probabilities of Pt and C occurring. Alternatively, $P(C \mid Pt)$ is the probability that C given Pt occurs. For risk assessment using BN, each risk event is considered as a node and the complex relationship between these risk events is captured through probability conditionals [19].

IOP Conf. Series: Earth and Environmental Science 1081 (2022) 012058 doi:10.1088/1755-1315/1081/1/012058

Table 2. Activity risk identification of Department Machinery Outfitting/MO & Electrical Outfitting/EO, Department Hull Construction and Department Hull Outfitting & Accommodation Outfitting, HO/AO.

Activities	Constrain Factor
Machinery Outfitting/MO & Electrical Outfitting/EO	
Electric, Electrical Outfitting	Working drawing of late cable cutting Late fabrication and installation drawings Revision of the image due to hitting equipment Late cable material
Steel Work	Late material and production. Revision of installation drawing for foundation or machinery is too late. There are no pictures of installing grating engine room and auxiliary machine room
Piping System	Delay in arrival of pipe material. Part of the crane lifting equipment is broken. Late sketch pipe drawing.
Hull Construction.	
Fabrication	Overhead crane broken Broken Conveyor Short blasting and damaged painting NC plasma cutting is broken.
Assembly	One side welding is broken. Imported material is late or damage. The accuracy of the block dimensions not match
Erection	Loading block is late Loading permission from the old owner Goliath crane broken Broken Conveyor Transporter facilities broken
Hull Outfitting & Accommodation Outfitting, HO/AO	
Painting and Protection	Repair painting due to burning Repair painting due to dust and dirt. Difficulty painting due to complicated position.
Outfitting Manufacturing	Design drawing delay Pipe material delay Complicated work according to sequence takes a long time
Accommodation Package in the Engine Room Area ion	Loading block is late Loading permission from the old owner Goliath crane broken The material does not match the needs to be installed.

The engine room module installation activities can be described by the BN model which was developed and described into several models. Thus model developed from all workshops that carry out activities on the performance of the installation, which can be described as follows:

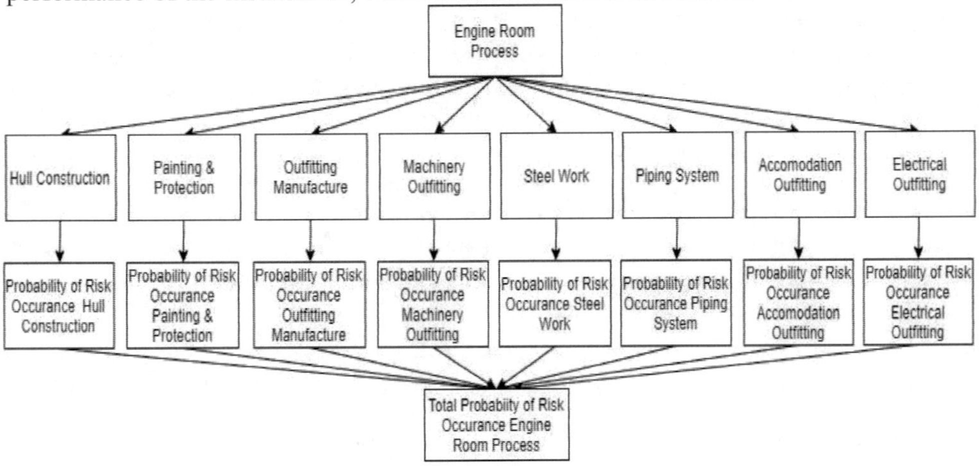

Figure 3. The main network model of BN in new building engine room.

Each activity model and sub-activity model, the probability results obtained are then compared with the standard based on the provisions of The Australian New Zealand Risk Management Standard (AS/NZS 4360: 2004) which includes the criteria for probability (Likelihood) and potential consequences to know the level of risk. The evaluation of the results, the dominant risk in the engine room installation system activity, using SD with several selected scenarios. The input-output diagram described in Figure 4. The input-output diagram is designed to describe the variables that are the inputs and outputs of the real system. In the input-output diagram, the variables that have been defined previously are classified into controllable inputs, uncontrollable inputs, desired outputs, unwanted outputs, and the environment. For the data information in each workshop: weight product, duration process, number of workers, productivity index. The output of utilizing SD is time delay. The time delay is the time delay in the production process that causes a loss penalty is calculated from the day of delay in the completion of the work. The following is an input-output diagram of this research system.

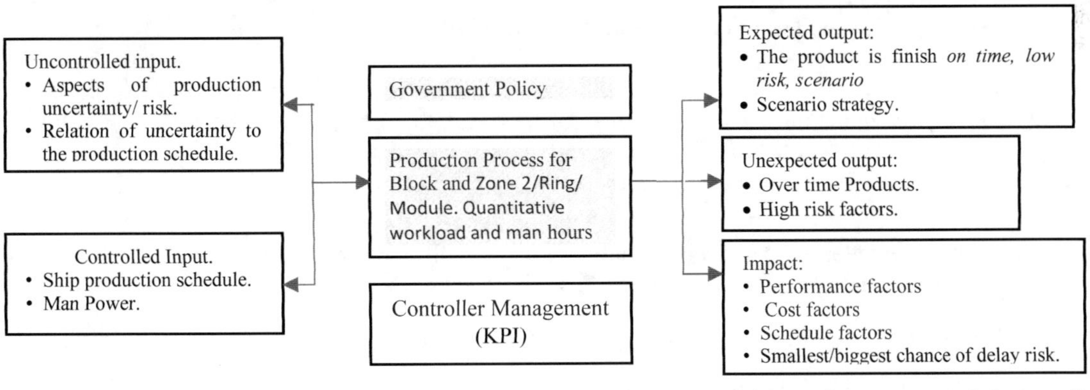

Figure 4. Diagram conceptual of the input, process and output of the engine room module installation.

Stock and flow diagrams are used to explain causal loop diagrams into a more detailed system. This diagram can show the value of each interrelated variable based on the level of influence of the variable and the length of the simulation time. The causal loop is used to determine the relationship between variables in the system. The causal loop is then interpreted into a stock flow diagram simulation. Improvements in the stock flow diagram are carried out in 2 types of scenarios. The first scenario is a parameter scenario. this scenario is done by changing the model parameter values to see the impact on

IOP Conf. Series: Earth and Environmental Science 1081 (2022) 012058 doi:10.1088/1755-1315/1081/1/012058

the model output. The second scenario is the structural scenario. This scenario is done by changing the structure of the model involved in the system and then seeing the impact on the output model. Each variable will be categorized into stock/level, rate/flow and converter as shown in Figure 5. Diagram of causal loop SD engine room module installation. The diagram describes the process of outfitting and equipment stages in each block that composes an engine room consisting of 28 blocks having an erection network and equipment system according to their respective construction functions. The stages of the workshop that carry out the outfitting and equipment process in each block that compose the engine room consists of a piping workshop, machinery outfitting (MO), full outfitting block system & steel work, electrical, electrical outfitting and ends with a function test. Painting & Protection, Outfitting Manufacture Meanwhile, Hull construction, Painting & Protection, Outfitting Manufacture and Accommodation Outfitting activities are carried out in the fabrication, sub-assembly, assembly and erection workshops. Each workshop has a risk aspect - following production activities that produce risk values in the grouping of probabilities and consequences.

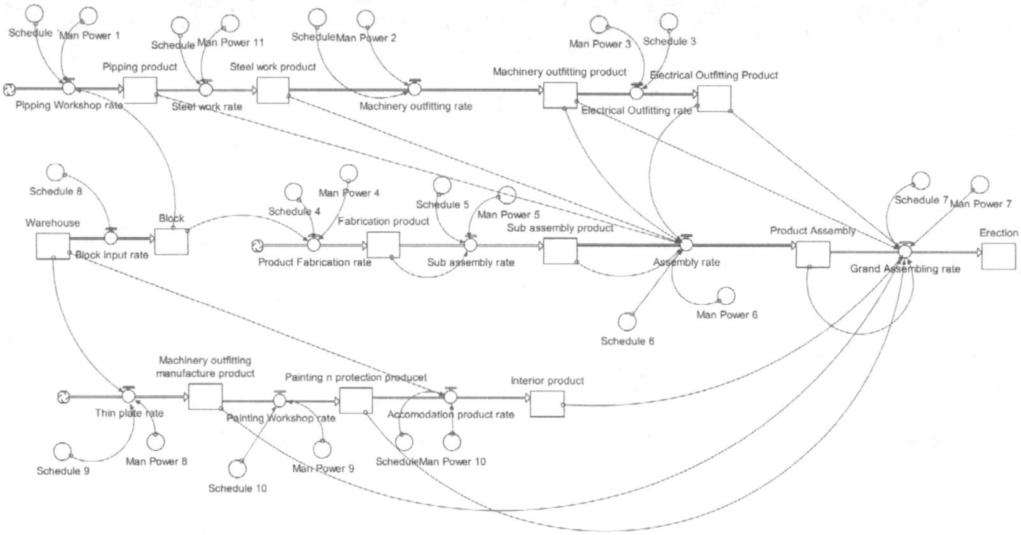

Figure 5. Causal loop diagram of SD for the engine room module installation.

If the formulation is entered into each element of the sub system, the model can be run using scenarios to reduce the risk factors in the model on the shipyard. The simulation equation can be shown as follows:

$$U\,t = Y0 + Y1 + Y2 + \ldots\ldots.. + Yt \quad (2)$$

$$Yt = Lt \times Wt \times Tt - Pt\,(Lt \times Wt \times Tt) \quad (3)$$

The relationship between workshops can be concluded in the following model:

$$Zt = Utx1 + Utx2 + Utx3 + Utx4 + \ldots + Utxn \quad (4)$$

Equation element description:

U1,2,……..t = total product in time interval t
Y1,2……. .t = number of products in time
I1,2,……... t = productivity index
W1,2,…… t = number of workers
T1,2,……..t = duration of work in days or hours
P1,2,……..t = resistance index of each workshop
Zt = total production of all workshops per unit time
x1,2,3,…...n = name of each workshop

The process is carried out in a SD, simulating existing conditions into a dynamic simulation model. The ability of a SD in a production line can provide solutions in the form of selecting the best scenario to minimize the constraint factors that cause delays.

The results of the risk evaluation of the shipyard engine room module installation, it can be providing an overview to increase the shipyard's readiness in developing shipbuilding modular systems. The planned engine room module installation according to available facilities and equipment, as well as improvements and developments more productive, examples:

- The risk identification process in shipbuilding, especially the engine room installation is still not optimal. It is suspected that the project analysis for the implementation of risk management in the engine room has not been implemented.
- Delay in shipbuilding projects will cause loss of profit opportunities and lack of credibility of shipyard capabilities, by anticipating the prevention of inhibiting factors that affect engine room installation system delays in shipbuilding. Then the potential loss due to fines and penalties will be minimized with a system evaluation approach by designing scenarios in SD to minimize the impact of delays.
- The risk assessment and evaluation process in risk management has not been properly defined in new building projects, it is suspected that many new building projects are experiencing delays in the East Java national shipyard. The Bayesian integration method approach and SD provides a solution in minimizing delay of the engine room systems installation.

4. Conclusion

The conclusions are the dominant risk in each production activity that has the potential for delays, lost opportunities for profit and losses due to penalties can be anticipated early with the BN integration approach and SD, which provides an understanding of system behavior and its constraint factors. The ability of SD in the production line of engine room installations provides a solution in minimizing the constraint factors that cause delays in various scenarios. The construction of ships with a modular system with a high level of parallel performance requires a more dynamic production line with engine room installations creating innovative risk management strategies as a dynamic evaluation. The dominant factors for the delay in the installation of the engine room module are as follows:

- Damage to machines due to obsolescence and irregular maintenance resulting in production delays, delivery times and claims risk.
- Consistency and timeliness in procuring materials according to the shipyard's schedule also contributes to the timeliness of production.
- Delays in sending the design results as well as revising the drawings will delay the assembly process.

5. Acknowledgments

Thank you to the Commercial Ship Division of Production Workshop PT. PAL Indonesia, Corporate Strategic Planning, Department of Electrical Outfitting and Machinery Outfitting, Department of Hull outfitting & accommodation outfitting and Department of Hull construction along with Department of Marine Engineering Institut Teknologi Sepuluh Nopember for the realization of the preparation of this paper.

6. References

[1] Hunt, T.E., (2014). *NSWCCD Ship Design Group Hosts Modular Ship Design Pioneer*, Waves, December, pp. 25.

[2] CERF (1996). *Construction Industry Research Prospectuses for the 21st Century*. Report 96-5016.

[3] Bertram, V, (2005), "*Modularization of Ships*", Report within the framework of project 'Intermodule' s/03/G Intermarec, France.

[4] Rubeša, R., N. Fafandjel, and D. Kolić. (2011). "*Procedure for Estimating the Effectiveness of Ship Modular Outfitting*." Eng. Rev. 31(1):55–62.

[5] Doerry, Norbert, (2014), Institutionalizing Modular Adaptable Ship technologies, Journal of Ship Production and Design, Vplume 30, hal 126-141.

[6] Agarwala, N. (2014). Developments in modular construction for shipbuilding. In: Proceedings of the International Seminar on Future Trends and Opportunities in Warship Design and Construction, New Delhi, September 25, pp. 1–11.

[7] Noor Virliantarto, (2017), Measurement of Technology Readiness for Container Ship 100 TEUs Building with Modular System at PT. PAL Indonesia, Jurnal Wave Volume 11 Nomor 1, Juli 2017: Hal: 31-38.

[8] Wee, Tanawan and Marco Aurisicchio. (2017). "Evaluating Modularisation Tools in Construction Evaluating Modularisation Tools in Construction." (June 2018).

[9] Pero, Margherita, (2015) Linking product modularity to supply chain integration in the construction and shipbuilding industries International Journal of Production Economics, 602-615 vol 170.

[10] S.O., Erikstad. (2009). Modularisation in Shipbuilding and Modular Production. Norwegian University of Science and Technology Department of Industrial Economics and Technology Management.

[11] Noorrahman, Dafie Buthlan, Postgraduate Programs, Study Programs, Masters in Systems, and Indonesian Computer University (2011). "Network Security Risk Management Modeling Using Bayesian Networks."

[12] Lee, E, Y. Park, and G. Shin, J. (2009). "Large Engineering Project Risk Management Using a Bayesian Belief Network, Expert Systems with Applocations." International 36(3):5880–87.

[13] Ben, Azher and J.Z. (2008). "*Development Program Risk Assessment Based on Utility Theory*." Risk Management 10:285–99.

[14] Amelia, Putri, Iwan Vanany, International University, Semen Indonesia, College, and Technology, (2017). "Operational Risk Analysis at Warship Division PT. PAL." 2:1–11.

[15] Basuki, M., D. Manfaat, S. Nugroho, and AAB Dinariyana. (2014). "Probabilistic Risk Assasment of The Shipyard Industry Using The Bayesian Method." Technology 5(1):88=97.

[16] Zarei, Esmaeil, Ali Azadeh, Mirzaei Aliabadi, and Iraj Mohammadfam. (2017). "*Dynamic Safety Risk Modeling of Process Systems Using Bayesian Network*." 00(00):1–9.

[17] Badurdeen, F., M. Shuaib, K. Wijekoon, A. Brown, W. Faulkner, Amundson, I. S. J., Jawahir, T. Goldsby, And Iyengar, D., and B. Boden. (2014). "*Quantitative Modeling and Analysis of Supply Chain Risks Using Bayesian Theory*." Manufacturing Technology Management 5:631–54.

[18] Mcdonald, K. S., D. S. Ryder, and M. Tighe. (2015). "Developing Best-Practice Bayesian Belief Networks in Ecological Risk Assessments for Freshwater and Estuarine Ecosystems: A Quantitative Review." Journal of Environmental Management 154:190–200.

[19] Nasirzadeh, Farnad, Abbas Afshar, Mostafa Khanzadi, and Susan Howick. (2008). "Integrating System Dynamics and Fuzzy Logic Modelling for Construction Risk Management." (May 2014).

[20] Agarwal, Ashish and Ravi Shankar, (2005). "*Modeling Supply Chain Performance Variables*." 10(2):47–68.

[21] Cakravastia, Andi and Lucia Diawati, (1999). "*To Diagnose The Logistic Chain Performance Of Shipbuilding Industry In Indonesia*." (January).

[22] Xue, Lei, Guofu Shi, Dashuang Dai, and Yizhuang Xu. (2020). "*Two-Stage Efficiency Structure Analysis Model of Shipbuilding Based on Driving Factor: The Case of Chinese Shipyard*." 182–200.

[23] Punyamurthula, Sudhir. (2018). "*Knowledge Bayesian-Integrated System Dynamics Modelling*."

[24] Mohaghegh, Z., (2010). "*Combining System Dynamics and Bayesian Belief Networks for Socio-Technical Risk Analysis*," IEEE International Conference on Intelligence and Security Informatics (ISI), pp. 196-201, Vancouver, Canada, May23-26, 2010.

IOP Conf. Series: Earth and Environmental Science 1081 (2022) 012059 doi:10.1088/1755-1315/1081/1/012059

Supply Chain Performance of Logistic Business Using SCOR Model in Tg.Priok Port Jakarta

Abdul Rachman[1*], Sindu Rahayu[2], Budi Sambarani[3], Dodi Sugianto[4], Agung Kwartama[5]

[1,2,3,4,5]Sekolah Tinggi Ilmu Pelayaran Jakarta, Marunda, 14150, Indonesia

*Corresponding author's e-mail: abdulrachman_agbstip@yahoo.co.id

Abstract. This study aims to identify and evaluate accurately the measurement process of Supply Chain Management of port operators in Tanjung priok jakarta. The target of this research is several sea transportation companies, especially loading and unloading companies in Jakarta. Loading and unloading companies with experience in these ports will later be able to help direct students in the Independent Campus Self-Learning process in Indonesia through an internship program. The main problem in the delayed loading and unloading process is the loading and unloading workers, especially at the Tanjung Priok Port, Jakarta. This study uses a qualitative descriptive analysis with the SCOR method through five measurement processes, namely Plan, Source, Make, Deliver, and Return. The loading and unloading conditions at three major ports in Indonesia are the benchmarks in this research. The research sample consisted of several company leaders in the transportation and logistics industry as informants from the port authority Regional Tanjung Priok Port Branch Jakarta, the findings of this study were to determine the Dwelling Time related to the desired Container Yard at the port as a temporary warehouse for loading and unloading companies and management of loading and unloading workers at ports that must be considered.

1. Introduction

Supply Chain Management in general is a series of activities that include planning, managing, and scheduling product flows from procurement to distribution to consumers. Through Supply Chain Management, the supply chain is designed in such a way that it can be carried out most effectively and efficiently. The supply chain is a continuous process, starting from the purchase of raw materials to finished products, with several functions such as sales forecasting, purchasing, production, distribution, sales and marketing, and with three main flows, namely materials, information and money. Supply Chain Management has represented a new frontier for gaining competitive advantage. This study provides supply chain decision makers with information on the following questions: How to measure supply chain performance; How to measure supply chain performance using the SCOR method; What are the problems and challenges of the supply chain process in loading and unloading at ports; and How to solve the problem of loading and unloading workers at the port. Based on the background and existing problems, the purpose of this research is to analyse and evaluate the performance value of Supply Chain Management and optimize improvement strategies that can be applied in several ports in Indonesia using the SCOR method. According to [1], the factors that influence differences in the productivity of the loading and unloading process are equipment and Supply Chain Management in general is a series of activities that include planning, regulating, and scheduling product flows from procurement to distribution to consumers. Through Supply Chain Management, the supply chain is designed in such a

IOP Conf. Series: Earth and Environmental Science 1081 (2022) 012059 doi:10.1088/1755-1315/1081/1/012059

way that it can be carried out most effectively and efficiently. mooring pool depth. The use of the SCOR integrated approach is to measure supply chain performance in real-life case studies of companies, especially Third Party Logistics Providers (3PL) [2]–[4]. They discuss how to leverage third-party logistics to improve supply chain performance and conceptually discuss the relationship between 3PL and supply chain performance. Meanwhile, study [5] assumes that the Fourth Party Logistics Provider (4PL) has the necessary skills to manage and provide added value versus internal solutions. Meanwhile [6], also describes the positive impact of 4PL in a complex disaster environment and provides and provides a key driver for enhancing and simplifying collaboration among humanitarian supply chain actors.

2. Literature Review

2.1. Supply Chain Management

Theoretically, according to supply chain [7], [8] is a process from suppliers through factories and distribution to final customers. It is the activity of managing various activities to convert raw materials into finished products and sent through the distribution system to consumers. Supply Chain Management manages the flow of information, products and services across the network, whether customers, companies, or suppliers [9], [10]. In addition, according to Supply Chain Management [11], are individual processes of the supply chain such as: procurement, production, inventory and retail management through the possibility of process integration, digitization and automation, and generating new analytical skills. Supply chain performance measurement often uses SCOR as a guide. According to supply chain [12], [13], many parties are involved, either directly or indirectly, in fulfilling customer demands. The supply chain includes not only manufacturers and suppliers but also carriers, warehouses, retailers, and even the customers themselves.

2.2. Loading-Unloading

Theoretically, loading and unloading activities include; (1) loading and unloading; (2) cargo doring; and (3) accept; and (4) delivery [14],[15]. Loading and unloading at the port needs to pay attention to the safety of the workforce which is significantly mediated by the supervision of safety management [16]. Research result [17] revealed that infrastructure improvements will increase the productivity of port operations. According to [18] the efficiency of the loading and unloading process at the port can minimize the ship's cost budget. Meanwhile, the findings [19] show that simultaneously the weight of the container and the number of aisles have a significant effect on loading and unloading productivity. Furthermore, findings [21] in Australia confirm that the initiation of loading and unloading activities is a challenge for the organization including lack of confidence in the workplace, high individualism, less resources. In the findings [20] there are five variables that have a significant effect on the cause of idle time, and operator activity in the gate-in process has the most significant effect on idle time in port services. Previous findings by [22], show a positive integration of Dwelling Time specifically for loading and unloading of imported containers at Tanjung Priok Port. Finally, the decisive answer is how long Dwelling Time can occur and is influenced by the relationship between Consignee/Freight Forwarder, Customs and Excise, and Terminal [23]. Fourth Party Logistics Provider - Fourth Party Logistics (4PL) has emerged as an ideal solution that enables companies worldwide from various industries to have a single point of accountability in both supply and demand chains. Nearly 90 percent of respondents in the end-user study cited the above reasons as the main drivers for implementing a 4PL strategy. Companies are gradually realizing that it is increasingly important in the global economy not only to focus on core activities but also non-core activities such as remote supply chain management to stay competitive. In addition, they are turning to 4PL to build closer relationships among the participants along the supply chain, to support cost reduction initiatives, to develop flexibility in the face of uncertainty. in supply and demand, and ultimately a positive impact on the bottom-line [35]. 4PL appears to provide competencies related to available knowledge, information technology, and skills in establishing and maintaining successful supply chain relationships [36].

IOP Conf. Series: Earth and Environmental Science 1081 (2022) 012059 doi:10.1088/1755-1315/1081/1/012059

2.3. Information and Communication Technology
Maritime Technology is technology about all the means to achieve goals in order to fulfill the continuity and comfort of human life used in the marine sector, especially related to shipping (navigation) and focuses on economic activities.

2.4. Maritime Logistic Performance Indicators
Maritime Logistics is the process of planning, implementing, and managing the movement of goods and information involved in sea transportation. Logistics is part of the supply chain that handles the flow of goods, information flow, and money flows through the procurement, warehousing, transportation, distribution and delivery services. in accordance with the type, quality, quantity, time and place desired by consumers, safely, effectively and efficiently, starting from the point of origin (point of origin) to the point of destination (point of destination).

Table 1. Operational performance standards of the ship scouting division of Port Tj. Priok

NO	TERMINAL NAME	WT (Hours)	AT (Hours)
1	JICT	0.91	1.9
2	TPK KOJA	0.91	1.7
3	NPCT 1	0.91	2
4	MAL	0.91	1.6
5	IPC TPK	0.91	1.9
6	PTP	0.91	1.9
7	IKT	0.91	2

Table 2. Operational performance standards effective time: berthing time (%)

NO	TERMINAL NAME	ET:BT (%)
1	JICT	85
2	TPK KOJA	85
3	NPCT 1	85
4	MAL	75
5	IPC TPK Area I	
	a. Konventional	65
	b. TPK 009	55
6	IPC TPK Area II	
	a. Konventional	65
	b. TPK	70
7	PTP REGION I	67
8	PTP REGION II	
	a. Domestiv	67
	b. Foreign	68
9	IKT	90

IOP Conf. Series: Earth and Environmental Science 1081 (2022) 012059 doi:10.1088/1755-1315/1081/1/012059

Table 3. Container loading and unloading performance standards and receiving/ delivery

NO	TERMINAL NAME	Countainer Terminal		Counventional Terminal		Receiving (Minutes)	Delivery (Minutes)
		B/C/H	B/S/H	B/C/H	B/S/H		
1	JICT	26	60			85	117
2	TPK KOJA	24	52			60	120
3	NPCT 1	26	60			60	75
4	MAL	23	55			30	60
5	IPC TPK Area I						
	a. Konventional	19	27			57	60
	b. TPK 009	18	27				
6	IPC TPK Area II						
	a. Konventional	18	40			57	60
	b. TPK	22	55				
7	PTP REGION I			15			
8	PTP REGION II			15			

Table 4. Container loading and unloading non container

NO	TERMINAL NAME	GC	BC	UN	UN ALBER	CC	CK	LIFE STOCK
		(T/G/J)	(T/G/J)	(U/S/H)	(U/S/H)	(T/G/J)	(T/J)	(H/G/J)
1	PTP REGION I	67	67	35	4	65	125	-
2	PTP REGION II							
	a. Domestic	75	40	35	-	65	235	-
	b. Foreign	140	57	-	-	135	125	130
3	IKT	-	-	125	4	-	-	-

Table 5. Performance Standards for facility utilization and equipment readiness

NO	TERMINAL NAME	FACILITY UTILIZATION			READY TO OPERATING EQUIPMENT (%)
		BOR (%)	SOR (%)	YOR (%)	
1	JICT	70		65	80
2	TPK KOJA	70		65	80
3	NPCT 1	70		65	80
4	MAL	70		65	80
5	IPC TPK Area I				
	a. Konventional	60		65	70
	b. TPK 009	60		60	
6	IPC TPK Area II				
	a. Konventional	60		65	80
	b. TPK	60		65	
7	PTP REGION I	70	70	65	80
8	PTP REGION II	70	70	65	90
9	IKT	70		65	85

3. Research Methods

One way to measure supply chain performance is to use the SCOR (Supply Chain Operation Reference) method, and the Supply Chain Council (SCC) explains that the SCOR process stretches from suppliers to all customers [37]. SCC added, the reliability of supply chain performance is the company's ability to

IOP Conf. Series: Earth and Environmental Science 1081 (2022) 012059 doi:10.1088/1755-1315/1081/1/012059

do the job as expected [38]. This SCOR model was developed and is a model for supply chain improvement in developing countries. Now the SCOR model includes all supplier, customer, and all market interactions and consists of five core processes; (1) Plan, (2) Source, (3) Create, (4) Deliver and (5) Return which is followed by the process of categories, elements, tasks and activities (Figure 1) [13], [39]. The SCOR model according to [37], is a management tool used to address, improve, and communicate Supply Chain Management decisions in companies and with suppliers and governments as customers. The SCOR model is described by [40] as a model that integrates business concepts of process reengineering, bench marking and measurement into its framework. In its development, five dimensions are used to determine performance measurement, namely; (1) reliability, (2) responsiveness, (3) flexibility, (4) cost, (5) assets. The SCOR model consists of three top levels associated with the three processes at level 1. At level 2, which is the configuration level, concerns the process category, and level 3 is the process element level. According to SCC, all performance attributes are interrelated with metrics [41]. The SCOR model enables capital management leverage, supply chain road map creation, alignment of business functions [37]. The application of the SCOR model within certain limits is quite flexible and can be adapted to increase productivity to meet needs. Measurement uses a structured set of metrics. In general, the SCOR model has been applied through the global market [39], [42], [43]. They add that the SCOR 4.0 model can be used by the public and private sectors to improve supply chain strategies in many countries [44]. Integration is fundamental to the use of all Supply Chain Management tools, resources, strategies and implementations.

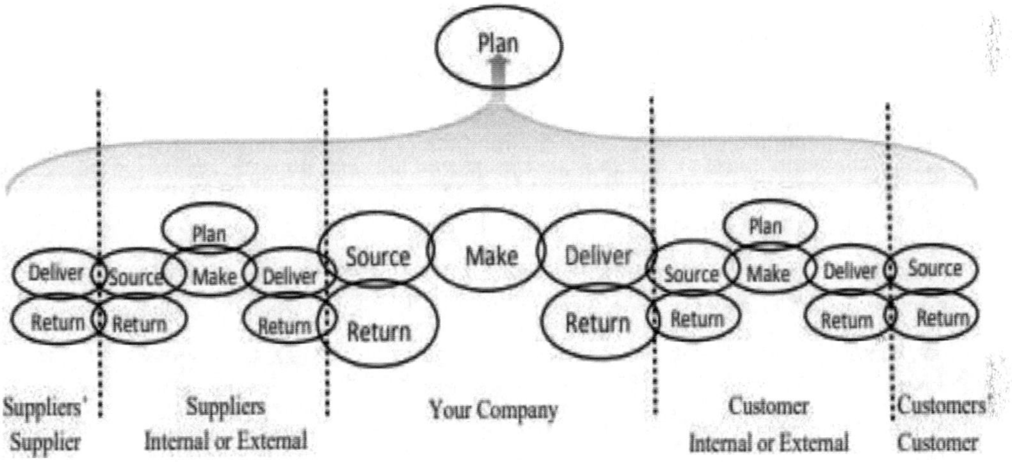

Figure 1. Five Core Processes in SCOR Model

This research was conducted in Jakarta. Data were obtained from several port operators such as the Port authority Jakarta branch, loading and unloading companies. This research was conducted for two weeks at the end of December 2021. This is a qualitative study using the Supply Chain Operating Reference (SCOR) analysis model. The initial step in this research is to identify the supply chain process of the loading and unloading company carried out. This step is carried out through direct observation at the company to identify problems that exist in the research location. Furthermore, from the problems identified, the formulation of the problem and the setting of research objectives were made. The second step is data collection and processing. Data from respondents was collected through interviews with informant. Next, the collected data was analyzed. Data processing steps include; (1) identifying the supply chain by observing the supply chain process carried out by the company, then analyzing it using the conceptual approach of SCOR version 11.0, (2) Determining Performance Indicators based on the results of supply chain identification, then placing them in a questionnaire.

3.1. Port of Tanjung Priok as the busiest and most advanced sea port in Indonesia

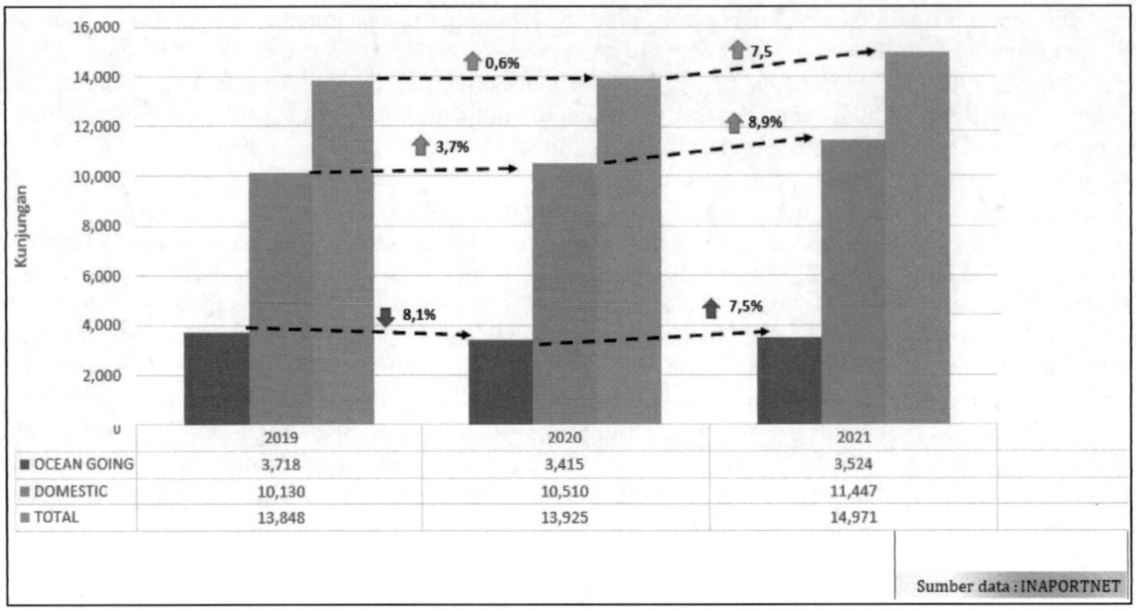

Figure 2. Port Ship Visits Tg. Priok 2019 to 2021

Table 6. Performance Data 2019-2021

Terminal Container	Year	Shipping Service			Goods Service		Facility and Equipment Utilization		
		ET:BT (%)	B/C/H (Box)	B/S/H (Box)	Receiving (menit)	Delivery (menit)	BOR (%)	YOR (%)	Readiness Equipment (%)
JICT	2019	88,48	26,43	68,33	80,08	111,97	48,3	41,98	98,04
	2020	89,18	28,58	69,36	65,11	97,48	40,24	39,31	98,38
	2021	90,42	25,75	57,91	77,13	104,08	49,79	56,24	98,11
KOJA	2019	90,67	26,76	67,58	39,76	98,92	55,16	39,98	94,24
	2020	89	26,16	65,26	43,15	97,46	52	32	96
	2021	90	26	66,19	48,11	104,94	55	46	95
MAL	2019	79,57	23,54	65,47	20,67	52,25	38,63	58,53	89,57
	2020	76,41	24,88	65,38	20,83	43,42	33,1	46,15	88,79
	2021	78,4	23,4	59,7	21,81	44,2	29,2	49	94
NPCT1	2019	80	31,9	96,82	24	34	40	39,67	99
	2020	84,7	30,56	86,33	26	38	40,69	36,75	99
	2021	86,1	28,88	79,27	30	50	49,5	50,4	99
IPC TPK 1	2019	63,15	18,35	26,53	33,9	29,32	30,47	25,47	87,21
	2020	65,94	18,98	26,12	32,42	26,99	41,68	17,56	89,4
	2021	66,78	20,99	32,97	25,72	22,08	46,75	44,87	91,55
IPC TPK 2	2019	67,41	21,83	51,56	33,35	55,71	24,42	27,17	87,2
	2020	68,73	22,11	52,71	32,42	26,99	18,26	36,38	92,84
	2021	63,38	22,14	54,66	31,84	49,78	12,62	25,31	92,93

Terminal Container	Year	Shipping Service	Goods Service		Facility and Equipment Utilization		
		ET:BT (%)	Vehicle (U/S/H)	Truk & Alber (U/S/H)	BOR (%)	YOR (%)	Readiness Equipment (%)
IKT	2019	83,71	139,52	19,86	40,67	54,22	98
	2020	81,96	127,26	8,68	38,84	53,1	98
	2021	79,52	135,29	8,87	40,85	62,77	98

Long Dwelling Time Process

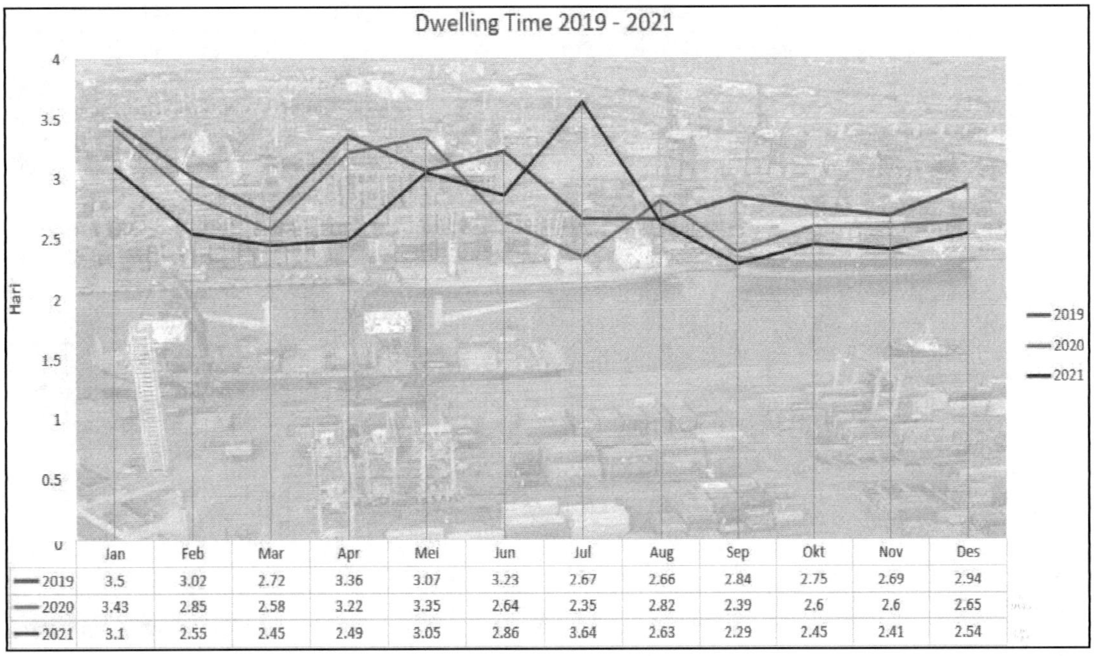

Figure 3. Dwelling Time 2019-2021

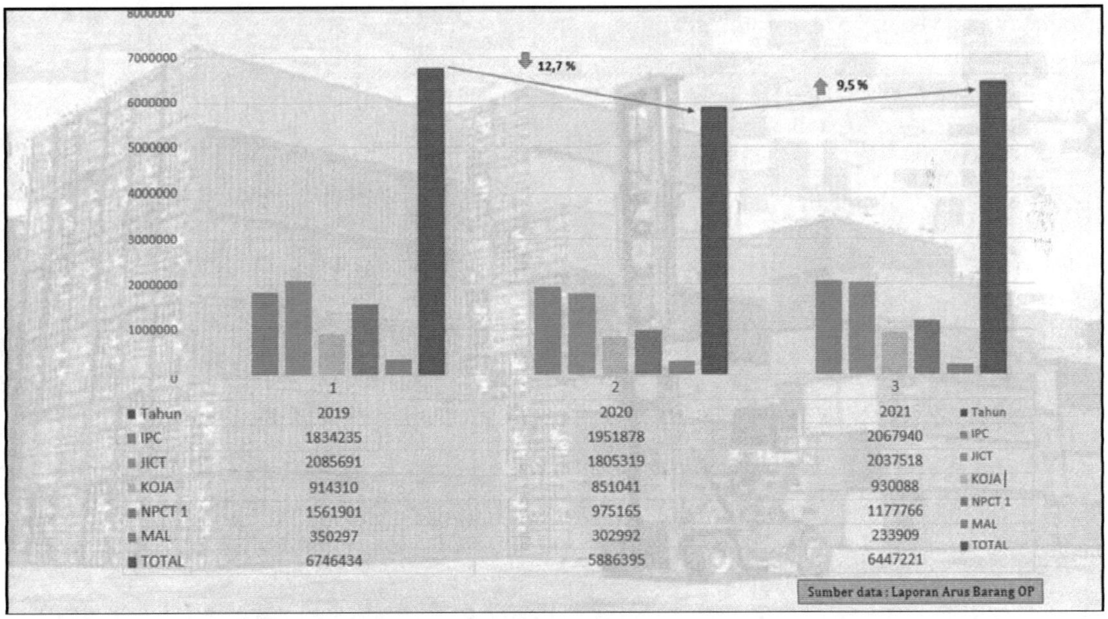

Figure 4. Container Flow Chart

IOP Conf. Series: Earth and Environmental Science 1081 (2022) 012059 doi:10.1088/1755-1315/1081/1/012059

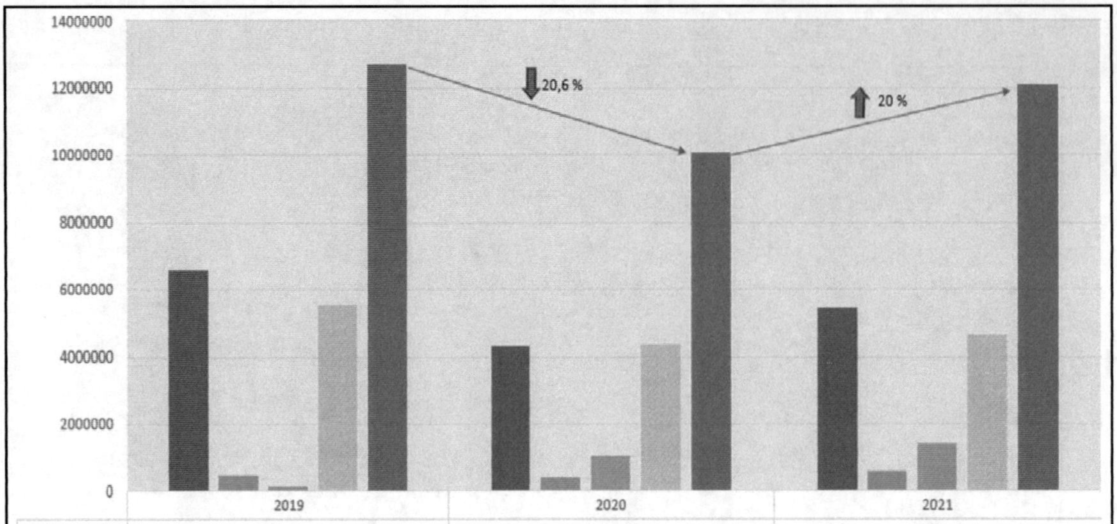

	2019	2020	2021
GENCAR	6593105	4319391	5462341
BAG CARGO	461871	382815	587283
LIQUID CARGO	127132	1010911	1409345
DRY BULK	5515565	4361653	4631888
JUMLAH	12697673	10074770	12090857

Sumber data : Laporan Arus Barang OP

GENCAR BAG CARGO LIQUID CARGO DRY BULK JUMLAH

Figure 5. Non Container Flow Chart

Volume of Domestic Trade Through the Port

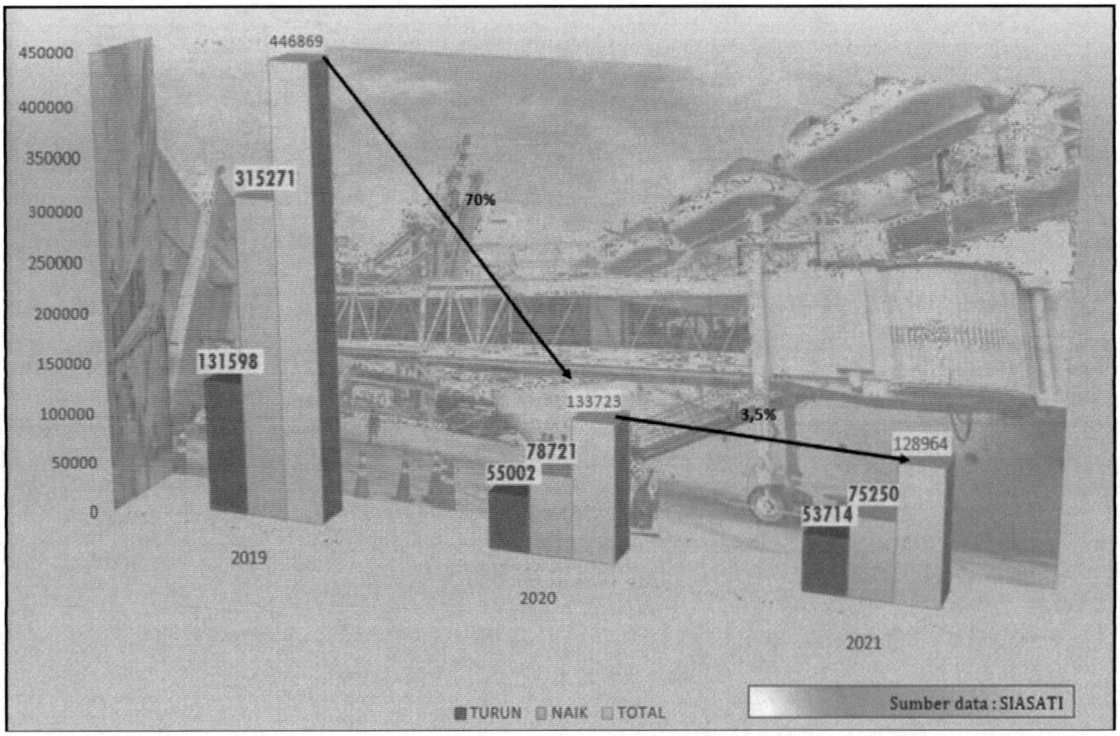

Figure 6. Domestic Passenger Up – Down Chart

IOP Conf. Series: Earth and Environmental Science 1081 (2022) 012059 doi:10.1088/1755-1315/1081/1/012059

Export and Import Volume Through the Port

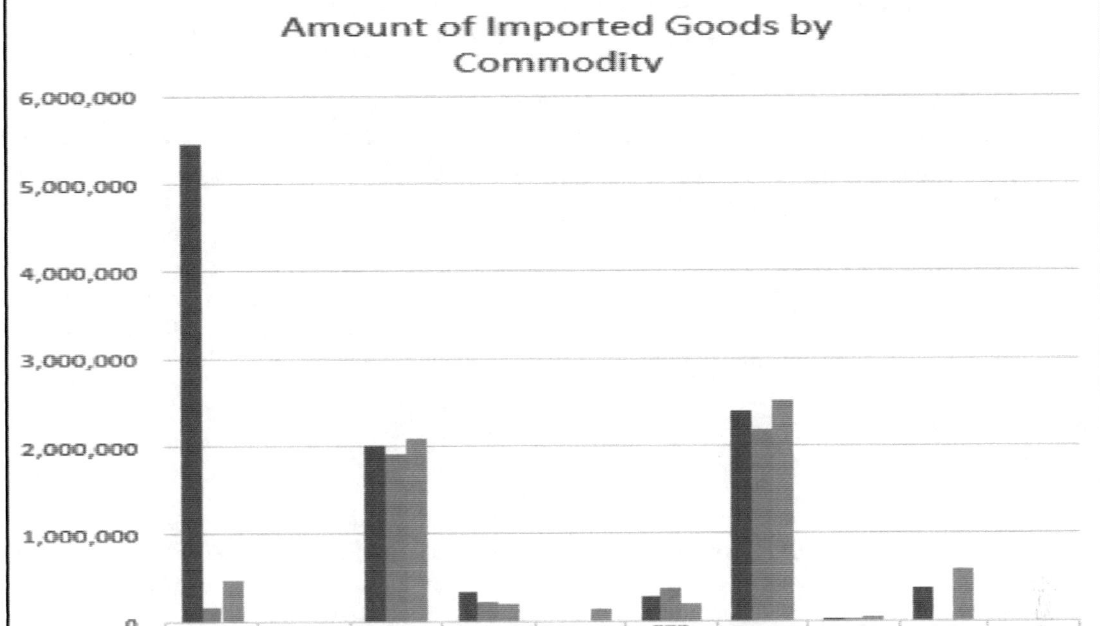

	GC (TON)	C. Cair (TON)	C. Kering (TON)	Hewan (EKOR)	Unitized (TON)	Bag Cargo (TON)	Petikemas (TEUS)	Kendaraan (UNIT)	Alat Berat (TON)	Truk (UNIT)
2019	5,462,9	7,006	2,020,6	336,146	4,320	271,619	2,394,7	37,314	374,790	3,687
2020	161,827	8,001	1,916,0	225,955	4,815	373,163	2,194,3	33,248	2,745	833
2021	476,996		2,091,9	206,948	135,968	207,475	2,525,2	47,006	585,713	3,546

Figure 7. Amount of imported goods by commodity

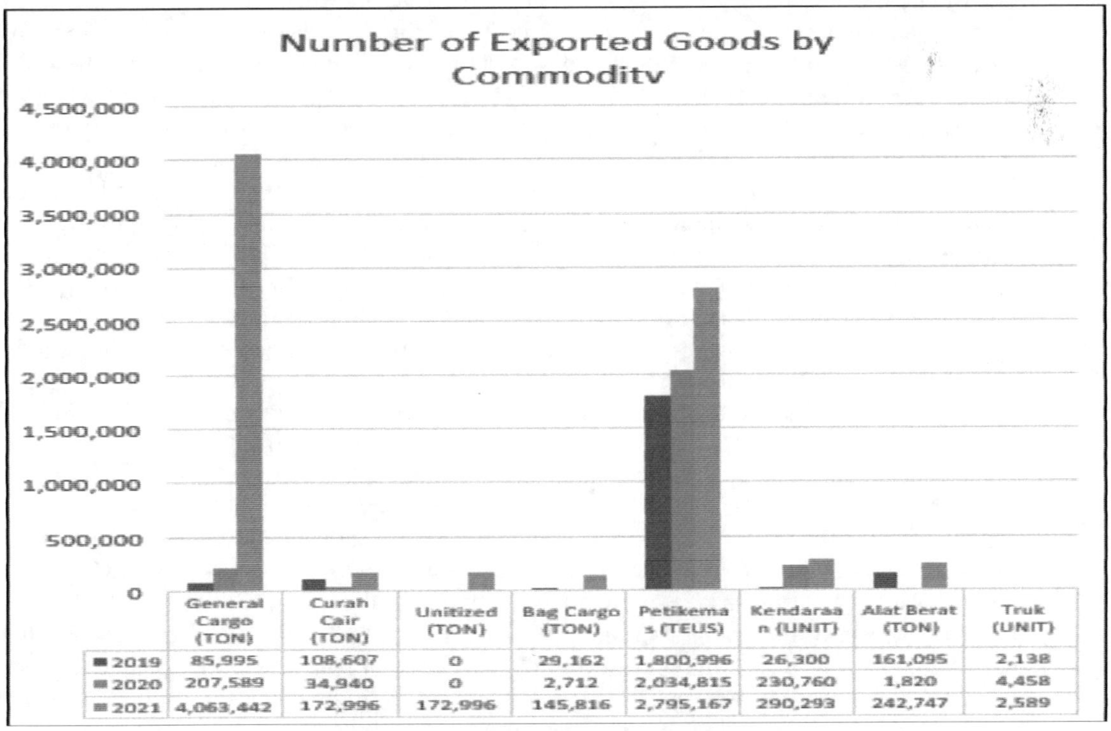

	General Cargo (TON)	Curah Cair (TON)	Unitized (TON)	Bag Cargo (TON)	Petikemas (TEUS)	Kendaraan (UNIT)	Alat Berat (TON)	Truk (UNIT)
2019	85,995	108,607	0	29,162	1,800,996	26,300	161,095	2,138
2020	207,589	34,940	0	2,712	2,034,815	230,760	1,820	4,458
2021	4,063,442	172,996	172,996	145,816	2,795,167	290,293	242,747	2,589

Figure 8. Number of export goods by commodity

Business Entities in the Port. Business Entity In Tanjung Priok Port Business activities at Tanjung Priok Port are carried out by the Port Operator, namely PT. Indonesian Port II. Quoting from its Web Service, the Pelindo Main Business Cluster is divided into four clusters. There are four main business clusters of Pelindo, namely:

- Container clusters;
- Non-container clusters;
- Logistics cluster & coastal area development (hinterland);
- Marine clusters, equipment and port services

The grouping of business clusters is carried out in order to make business development at Pelindo more focused, improve the capabilities and expertise of the HR working in each cluster, so that they are able to work more efficiently and increase customer satisfaction. If this happens, it will open up a wider market share and increase company profits.

4. Result and Discussion

4.1. Analysis
In-depth Interview Results The object of research through this in-depth interview process is the port manager as the operator and the loading and unloading company as the executor at several ports in Jakarta. Port operational activities involve stakeholders and collaborate with academics in port administration. The interview also involved student representatives from the Marketing, innovation and technology a department. The interview took place as a Third Party Logistics operator (3 PL) or Fourth Party Logistics (4 PL).

Participation of Partners in Implementation of Activities through In-depth Interviews

- In-depth Interview with Company Management, Jakarta
  SCOR Method with Five Plan-Source-Make-Deliver-Return Performance Measurements The main entrepreneurs of OP Enterprises are the logistics industry, ship agency, loading and unloading logistics and shipping from ships to factory warehouses. The company's main business is from end to end process to door to door process and many import activities mainly from China. Indeed, there is an imbalance in import and export growth; national growth used to be seven percent, but now it's still five to seven percent. The company's market share comes from domestic spending and now most of it is in the form of medical devices. Broadly speaking, the business processes carried out by 4PL actors are; Customer Permit, Loading and Unloading, Transportation, Heavy Equipment, Warehousing and Factory. Wheat flour factories for imports with wheat raw materials have been prioritized. Monitoring of supply chain movements by OP is through the Marine Traffic application. Other processes are through Port Clearance, Bill of Lading, Stowage Plan, Quarantine for loading and unloading permits and as a 4PL actor or coordinating several loading and unloading companies through a good communication process with the responsibility still with each Loading and Unloading Company. Every Ship Agency must have a Ship Agency Business License. 3PL and 4PL differ in asset ownership, and in the port even though Operator can actually act as 4PL as it also handles dock management. Meanwhile, in increasing the Level of Service (LOS), OP Tanjung Priok in Jakarta as an operator as well as implementer is categorized as 3PL, especially in the Cargo during area. A. The process of supply chain movement in loading and unloading activities at the port is under the supervision of the Port Authority and the Port Authority Office with a work division where the Port Authority Office has to oversee fairness issues at the port while the Port Authority has to check the seaworthiness of ships. ship docked. Especially for the return process. Customer complain. In terms of reliability, the level of customer complaints is still reasonable. Currently, there are many inefficiencies such as loading and unloading outside the rules. For example, loading and unloading workers usually have three shifts but become 6 shifts or 4 gangs become

2 gangs in real operations. That's not the job of the loading and unloading company. Loading and unloading conditions are strongly influenced by truck waiting time. Are there product returns? Product returns still exist, but not always every year with a maximum value of 80 tons of product shipped annually. Product will be returned in case of spill and damage. Identification of product conditions; each product is always identified using IT and has been in accordance with procedures. Defective product return authorization; the authority remains with the company as the sender. Product return scheduling; According to Standard Operating Procedures, product returns are still scheduled. The company remains responsive and committed if there is a delay in product returns.

- In-depth Interview with Tanjung Priok Port Management Jakarta
OP as the operator has provided a detailed explanation of the loading and unloading conditions at the Tanjung Priok Port, Jakarta. Tanjung Priok Port is supported by several equipment: Tugboats, Scouts, Kepil Ships, Warehouses, Stacking Yards and loading and unloading equipment, and provides services that include: Ship Services, Goods Services, Terminal Services, Land, Buildings, Water and Electric Services. Although there have been additions to tg.priok port facilities, the berth of large ships at the Port of Jakarta is still limited. At that time, ships with a maximum draft = 5 m or a deadweight size of ± 3,500 tons could dock at the tg.priok Pier. Meanwhile, ships with a draft > 5 m still have to dock outside the port or offshore which is ± 3 miles from the pier. Hence it is known . Since 1970, the flow of ships and goods through the Port of tg.priok tends to increase every year. According to data from 1970 to 1983, the average annual increase in the flow of goods is more than 10%. In connection with the limitations of port facilities, such as inadequate depth and width of the channel for the entry and exit of ships, the Government decided to expand the port of tg.priok. One of the services provided is ship operation at the port, such as loading and unloading activities. One of the loading and unloading activities carried out is the loading and unloading of oil products at the liquid bulk dock. In the process of loading and unloading of oil products, there may be several obstacles and disturbances that cause delays in the loading and unloading process which reduces the performance of the Port Business Entity. In terms of management, Tanjung Priok Port in Jakarta requires facilities and infrastructure to support loading and unloading of containers. So, the number of equipment needs that must be provided by the port manager will of course be influenced by the number of containers to be unloaded or loaded at the port and the layout of the port. It is necessary to evaluate the capacity of loading and unloading facilities and equipment owned by the container terminal of Tanjung Priok Port, Jakarta, so that it can provide an optimal level of service in the loading and unloading process at the container terminal to anticipate future needs. Constraints in the process of docking ships are the dock, weather, and the availability of facilities and infrastructure provided by the port management unit. These obstacles can be overcome by improving port services, including maximizing the utilization of the wharf by determining based on the existing schedule, providing information to ships regarding daily weather reports. The performance system at Jakarta Container Terminal is not good. This can be seen from the average dwelling time which is still too long. Optimization of dwelling time needs to be done to get a balance in the projected increase in sales and to compensate for the limitations of terminal facilities and infrastructure. One way is to implement container handling procedures using t The volume of cargo shipments using containers through Tanjung Priok Port is increasing from year to year. Shipping cargo using containers allows goods to be put together so that the operation becomes faster, more effective and efficient. In addition, the use of containers is also expected to increase the amount of cargo that can be handled. However, the container service at the Tanjung Priok Port container terminal, Jakarta, has recently faced several obstacles that affect its speed. These obstacles include: (1) not yet maximal use of gantry cranes and rubber tired gantry in serving loading and unloading of containers, (2) untidy arrangement of containers in stacking places, (3) service time of trucks from outside transporting exported goods. the container coincides with the time the chassis truck serves loading to the ship, thus slowing down the process; the same thing happened in the

IOP Conf. Series: Earth and Environmental Science 1081 (2022) 012059 doi:10.1088/1755-1315/1081/1/012059

disassembly process, and (4) the number of chassis he Cost Terminal Operation System application. trucks did not meet the standard operational configuration. Loading and unloading workers are generally an inseparable part of human resources at ports. Phenomena that occur in loading and unloading workers are erratic wages and the risk of occupational diseases and work accidents. Loading and unloading work is classified as heavy work and has a high risk of accidents, especially during manual loading. In 2017 there were seven cases, in 2018 there were 8 cases, and as of September 2019 there have been 6 cases of work accidents. According to the survey, work accidents are caused by unsafe behavior. Currently the port terminal operator services have been carried out well by the Tanjung Priok port, Jakarta, able to complete the terminal operator services used for ship docking and administration in related agencies can be completed properly.

- In-depth Interview with Tanjung Priok Port Management, Jakarta
  Tanjung Priok Port handles more than 30% of Indonesia's non-oil and gas commodities. In addition, 50% of all cargo inflows/outflows to and from Indonesia are shipped through this port. Therefore, Tanjung Priok is a barometer of the Indonesian economy. Complete intermodal facilities at this port can connect Tanjung Priok to all cities in Indonesia. With modern technology and facilities, Tanjung Priok has been able to serve next generation ships that go directly to various international trade centers (direct calls). Port development is directed at anticipating the acceleration of cargo loading and unloading activities through the availability and completeness of special service facilities. The construction of deep roads, widening the channel and ship entrances (to be two-way) and deepening the channel to -14 mLWS are priority programs that must be carried out. Tanjung Priok port is the origin of a city and the development of the city starts from the north side of Jakarta. The main problem according to the Deputy General Manager is that automation will not be applied to trucking activities because there will be something missing. From high logistics costs, it will also have an impact on truck drivers who will lose their jobs. There is now a truck waiting time condition. It's a different situation. In Shanghai China, with few workers and many uses in Information Technology, it is easy to monitor conditions. Especially for truck management, it has now used the Single Truck ID system. There are 3,200 loading and unloading workers consisting of 2,400 old unloading workers and 800 new unloading workers, but they are not under the management of OP. Indeed, it must be clear who is fully responsible for the management of loading and unloading workers. Problems with non-unloading workers must also be considered, such as whether their welfare is under the authority of the Ministry of Cooperatives or the Port Authority; how to manage loading and unloading workers as intermediaries under the 4PL and 3PL, and the fate of truck drivers if Automation is implemented. The truck driver problem is a classic one. As a driver's habit where at closing time only wants to enter the port at the last moment, resulting in waiting times for trucks in a long queue of trucks outside the gate even though outside the port, 4 hectares of vacant land has been provided. There is also a need for legal counseling on the work safety of loading and unloading workers at the Tanjung Priok Port in carrying out loading and unloading activities on ships, because it is a type of high-risk work that requires very high caution from loading and unloading workers and work safety. protection from the employer. Have the loading and unloading workers at the Tanjung Priok Port, Jakarta, carried out their work in accordance with occupational safety and protection standards.

4.2. Discussion

Recently, the problem of Dwelling Time has been reduced from five days to three days. The Port of Tanjug Priok cannot be compared to the Port of Singapore as a Transshipment port, while the Port of Tanjung Priok is only a Gate Way port. Tanjung Priok Port is part of the Sea Toll Road, an end to end business volume management that can reduce costs. The loading and unloading company must start from the ownership of heavy equipment, the market, and profit sharing. To compare and evaluate since 2018, the management of heavy equipment in China with a loading capacity of three to four tons is not the best scheme for Tanjung Priok Port. Today, consignees still want their cargo to stay for three days at Stacking Yard, not be rented out of port. In conclusion, in 2020 the dwelling time for export activities is 3.9 days. This shows a longer time than the average from 2019 to 2020. The average Dwelling Time for export activities at Jakarta International Container Terminal Inc for two years for men is 3.01 days. The median value which is lower than the daily average for two consecutive periods causes the Dwelling Time at Jakarta International Container Terminal Inc. to be still very high, although the average duration is still in accordance with the 2015 ministry of education, which is three days. In the period 2011-2017 it has been studied by [22] that the results of the analysis show the positive integration of Dwelling Time specifically for loading and unloading of imported containers at Tanjung Priok Port and is still categorized as very high, with an average duration of seven years is 3.8 days. This study is in line with previous findings by [45] at Tanjung Priok Port which showed that the role of assistant operator was very meaningful for operators to operate container loading and unloading equipment safely at Tanjung Priok Port, Jakarta. . Moreover, based on field findings [46] regarding the four port performance indicators, it can be concluded that Tanjung Priok Port still has to continue to improve to become a world-class port. Another finding [26] at the Port of Tanjung Priok shows that the main factors affecting the safety of loading and unloading of dangerous goods are port supporting facilities and infrastructure. According to analysis [47] at Tanjung Priok Port, Jakarta, the smooth running of all operations is highly dependent on chassis truck services in the container yard, such as structuring the container stacking pattern or optimizing the use of rubber tyred gantry. The danger of loading and unloading work using cranes at the Tanjung Priok Port in Jakarta has been controlled to the ALARP stage (as low as possible with practical convenience) [48]. In addition, the readiness of loading and unloading personnel is also a determining factor. Another funding by [49], which supports this research, reveals that the loading and unloading process and ship document processing have a significant influence on the performance of the relevant agencies. This research is also in line with the findings [23] that actually Dwelling Time at Tanjung Priok Port does not need to occur if all relevant parties can resolve all problems quickly, the completion of activity documents is carried out by Customs and Excise. and Excise with the consignee/freight forwarder through pre-permit, customs, and post-permit processes. The problem of Dwelling Time, especially at the Container Terminal at Tanjung Priok Port, Jakarta, also supports this research by saying that often the length of the container process is unknown; for example the user does not complete the required documents, the processing of imported containers takes longer than the processing of export containers and the documents must be checked by Quarantine and Customs [50]. Finally, this study also supports the findings [51] at the OP terminal at Tanjung Priok Port, Jakarta, which shows that loading and unloading equipment, unloading workers, and ship services have a significant correlation with loading and unloading performance.

Table 7. Performance Data 2019 to 2021

Terminal Container	Year	Shipping Service			Goods Service		Facility and Equipment Utilization		
		ET:BT (%)	B/C/H (Box)	B/S/H (Box)	Receiving (minute)	Delivery (minute)	BOR (%)	YOR (%)	Readiness Equipment (%)
JICT	2019	88,48	26,43	68,33	80,08	111,97	48,3	41,98	98,04
	2020	89,18	28,58	69,36	65,11	97,48	40,24	39,31	98,38
	2021	90,42	25,75	57,91	77,13	104,08	49,79	56,24	98,11
KOJA	2019	90,67	26,76	67,58	39,76	98,92	55,16	39,98	94,24
	2020	89	26,16	65,26	43,15	97,46	52	32	96
	2021	90	26	66,19	48,11	104,94	55	46	95
MAL	2019	79,57	23,54	65,47	20,67	52,25	38,63	58,53	89,57
	2020	76,41	24,88	65,38	20,83	43,42	33,1	46,15	88,79
	2021	78,4	23,4	59,7	21,81	44,2	29,2	49	94
NPCT1	2019	80	31,9	96,82	24	34	40	39,67	99
	2020	84,7	30,56	86,33	26	38	40,69	36,75	99
	2021	86,1	28,88	79,27	30	50	49,5	50,4	99
	2019	63,15	18,35	26,53	33,9	29,32	30,47	25,47	87,21

IPC TPK 1	2020	65,94	18,98	26,12	32,42	26,99	41,68	17,56	89,4
	2021	66,78	20,99	32,97	25,72	22,08	46,75	44,87	91,55
IPC TPK 2	2019	67,41	21,83	51,56	33,35	55,71	24,42	27,17	87,2
	2020	68,73	22,11	52,71	32,42	26,99	18,26	36,38	92,84
	2021	63,38	22,14	54,66	31,84	49,78	12,62	25,31	92,93

Terminal Container	Year	Shipping Service	Goods Service		Facility and Equipment Utilization		
		ET:BT (%)	Vehicle (U/S/H)	Truk & Alber (U/S/H)	BOR (%)	YOR (%)	Readiness Equipment (%)
IKT	2019	83,71	139,52	19,86	40,67	54,22	98
	2020	81,96	127,26	8,68	38,84	53,1	98
	2021	79,52	135,29	8,87	40,85	62,77	98

Conclusion: it was found that the Container terminal consists of : JICT, KOJA, MAL, NPCT1, IPC TPK 1, IPC TPK 2, IKT TG.PRIOK PORT 1. Facility Utilization Percentage BOR(%) and YOR (%) di 100 % below the target which is not in accordance with the Standard Regulation of the Head of the Tanjung Priok Main Port Authority.

5. Conclusion

The result of this research is to know the determination of Dwelling Time related to the Container Yard which is expected at the port as a temporary warehouse for loading and unloading companies. The management of loading and unloading workers must also be considered because it is not the port operator who is responsible but the main manager must be clear, whether it is the Port Authority. Based on interviews with informants, especially 3PL and 4PL service providers at ports, currently services are still dominated by 3PL as field operators. Only at Tanjung Priok Port, Jakarta, the operator plays the role of implementing the 4PL. Interviews using the SCOR method were systematic and measurable to obtain some useful findings and implications to inform policy makers and operators in government and ports at the national level. Through research activities like this, students are expected to be able to adapt and collaborate quickly through information and communication technology as well as the collaboration process with partners outside the campus. In order to sharpen the Learning Independent Campus programs in the future, especially for strengthening research, students are invited to participate in discussions and Focus Group Discussions with business actors at ports and port operators as well as national papers to assist in the preparation of final reports and draft research journals.

References

[1] K. Agustin, A. Retnaningtyas, I. Chairrudin, and D. A. Baskoro, “The Factor Analysis of The Difference in Productivity of Unloading Process Between West Berth and North Berth at JICT on 2019,” Adv. Transp. Logist. Res., vol. 3, pp. 555–561, 2020.

[2] R. I. Van Hoek, “The contribution of performance measurement to the expansion of third party logistics alliances in the supply chain,” Int. J. Oper. Prod. Manag., vol. 21, no. 1/2, 2001.

[3] D. Jothimani and S. P. Sarmah, “Supply chain performance measurement for third party logistics,” Benchmarking An Int. J., vol. 21, no. 6, 2014.

[4] H. Zhang and S. C. Okoroafo, “Third-party logistics (3PL) and supply chain performance in the Chinese market: a conceptual framework,” Eng. Manag. Res., vol. 4, no. 1, p. 38, 2015.

[5] A. Win, “The value a 4PL provider can contribute to an organisation,” Int.J. Phys. Distrib. Logist. Manag., vol. 38, no. 9, pp. 674–684, 2008.

[6] H. Abidi, S. de Leeuw, and M. Klumpp, “The value of fourth-party logistics services in the humanitarian supply chain,” J. Humanit. Logist. Supply Chain Manag., vol. 5, no. 1, pp. 35-60., 2015.

[7] R. G. Schroeder, Operations management: contemporary concepts and cases. New York: McGraw-Hill, 2007.

[8] B. Render and J. Heizer, Operations Management, 11th Eds. New Jersey: Pearson, 2014.
[9] R. B. Chase, F. R. Jacobs, and N. J. Aquilano, Operations Management for Competitive Advantege with Global Cases. Boston: McGraw-Hill/ Irwin, 2006.
[10] R. S. Russell and B. M. Taylor, Operations and Supply Chain Managemant. New Jersey: John Wiley and Sons Inc, 2016.
[11] H. Fatorachian and H. Kazemi, "Impact of Industry 4.0 on supply chain performance," Prod. Plan. Control, vol. 32, no. 1, pp. 63-81., 2021.
[12] S. Chopra, P. Meindl, and D. V. Kalra, Supply chain management: Strategy, planning, and operation. Boston, MA: Pearson., 2013.
[13] I. N. Pujawan and E. R. Mahendrawathi, Supply Chain Management, 3rd eds. Suirabaya: Guna Widya, 2017.
[14] D. A. Lasse, Port Management. Jakarta: Rajagrafindo Persada, 2014.
[15] S. Sugandi, A. Sahil, and D. A. Lasse, "Efforts to Improve The Performance Loading and Unloading Workforce at The Port of Marunda North Jakarta," Adv. Transp. Logist. Res., vol. 1, pp. 1339-1367., 2018.

AUTHOR INDEX